Environmental Science: In Context

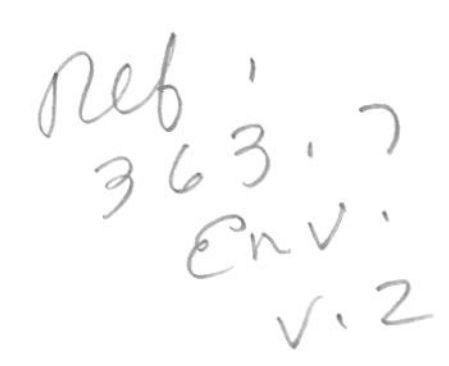

Environmental Science: In Context

VOLUME 2

MAPS AND ATLASES TO WOODS HOLE OCEANOGRAPHIC INSTITUTION

Brenda Wilmoth Lerner & K. Lee Lerner, Editors

Detroit • New York • San Francisco • New Haven, Conn • Waterville, Maine • London

Environmental Science: In Context

Brenda Wilmoth Lerner and K. Lee Lerner

Project Editor: Madeline Harris

Editorial: Kathleen Edgar, Debra Kirby, and Kimberley McGrath

Rights Acquisition and Management: Margaret Abendroth, Tracie Richardson, Timothy Sisler, and Robyn Young

Composition: Evi Abou-El-Seoud

Manufacturing: Wendy Blurton

Imaging: Lezlie Light

Product Design: Jennifer Wahi

Product Management: Julia Furtaw

LIBRARY OF CONGRESS CATALOGING-IN-PUBLICATION DATA

Environmental science : in context / Brenda Wilmoth Lerner & K. Lee Lerner, editors.
v.
Includes bibliographical references and index.
Contents: v. 1. Acid rain to Montreal protocol -- v. 2. National Environmental Policy Act to Woods Hole Oceanographic Institution.
ISBN 978-1-4144-3617-3 (set) -- ISBN 978-1-4144-3618-0 (vol. 1) -- ISBN 978-1-4144-3619-7 (vol. 2) ISBN-13: 978-1-4144-3714-9 (ebook)
1. Environmental sciences--Encyclopedias. I. Lerner, Brenda Wilmoth. II. Lerner, K. Lee.

GE10.E583 2008
628—dc22 2008019961

Gale
27500 Drake Rd.
Farmington Hills, MI, 48331-3535

ISBN-13: 978-1-4144-3617-3 (set)
ISBN-13: 978-1-4144-3618-0 (vol. 1)
ISBN-13: 978-1-4144-3619-7 (vol. 2)

ISBN-10: 1-4144-3617-3 (set)
ISBN-10: 1-4144-3618-1 (vol. 1)
ISBN-10: 1-4144-3619-X (vol. 2)

This title is also available as an e-book.
ISBN-13: 978-1-4144-3714-9 ISBN-10: 1-4144-3714-5
Contact your Gale, a part of Cengage Learning sales representative for ordering information.

Printed in China
1 2 3 4 5 6 7 12 11 10 09 08

Table of Contents

Advisors and Contributors xvii

Introduction xxi

About the *In Context* Series xxiii

About This Book xxv

Using Primary Sources xxxi

Glossary. xxxiii

Chronology liii

VOLUME ONE

Acid Rain 1

Agricultural Practice Impacts 5

Air Pollution 8

Algal Blooms 11

Alternative Fuel Impacts 14

Antarctic Issues and Challenges 21

Antarctic Treaty 25

Aquatic Ecosystems 30

Aquifers. 33

Arctic Darkening and Pack Ice Melting 36

Asbestos Contamination 39

Atmosphere 41

Atmospheric Circulation. 44

Atmospheric Inversions 47

Aviation Emissions 49

Bays and Estuaries 53

Benthic Ecosystems. 55

Biodegradation 57

Biodiversity 60

Biofuels 67

Biogeography. 74

Bioremediation 76

Biostratigraphy 79

Bureau of Land Management 81

Carbon Dioxide (CO_2). 83

Carbon Dioxide (CO_2) *Emissions*. 85

Carbon Sequestration 90

Careers In Environmental Science 95

Chemical Spills 100

Chlorofluorocarbons 102

CITES (Convention on International Trade in Endangered 105
Species of Wild Fauna and Flora) 105

Clean Air Act of 1970 108

Clean Air Mercury Rule of 2005 112

Clean Development Mechanism 115

Clean Energy 118

Clean Water Act 122

Clear-Cutting 125

Climate Change 128

Climate Modeling 139

Closed Ecology Experiments 142

Coal Resource Use 145

Coastal Ecosystems 150

Coastal Zones 154

Commercial Fisheries 160

Comprehensive Environmental Response, Compensation, and 166
Liability Act (CERCLA) 162

Conservation 166

Coral Reefs and Corals 170

Corporate Green Movement 174

Cultural Practices and Environmental Destruction 179

Dams 181

DDT 186

Deforestation 189
Dendrochronology 192
Desertification 194
Deserts 197
Drainage Basins 200
Dredging 202
Drought 205
Dust Storms 205
Earth Day 209
Earth Summit (1992) 212
Earth Summit (2002) 215
Earthquakes 218
Eco-Terrorism 224
Ecodisasters 230
Ecological Competition 236
Ecosystem Diversity 239
Ecosystems 242
El Niño and La Niña 247
Electronics Waste 252
Emissions Standards 254
Endangered Species 259
Environmental Activism 265

Environmental Assessments 269

Environmental Benefits and Liabilities of Petroleum Resource Use 272

Environmental Crime 279

Environmental Protection Agency (EPA) 282

Environmental Protection Agency (EPA) Regulation: Massachusetts v. Environmental Protection Agency (2007) 286

Environmental Protests 289

Estuaries 294

Extinction and Extirpation 298

Factory Farms, Adverse Effects Of 302

Fish Farming 306

Flood Control and Floodplains 310

Floods 314

Forest Resources 317

Forests 321

Fossil Fuel Combustion Impacts 326

Freshwater and Freshwater Ecosystems 329

Gaia Hypothesis 332

Genetic Diversity 335

Geochemistry 338

Geographic Information Systems (GIS) 341

Geological History 345

Geospatial Analysis 351

Geothermal Resources 354

Glacial Retreat 357

Glaciation 362

Global Warming 366

Grasslands 372

Green Movement 375

Greenhouse Effect 383

Greenhouse Gases 390

Groundwater 396

Groundwater Quality 401

Habitat Alteration 404

Habitat Loss 408

Hazardous Waste 413

Herbicides 418

Horticulture 420

Human Impacts 423

Hunting Practices 427

Hurricanes 430

Hurricanes: Katrina Environmental Impacts 437

Hybrid Vehicles 441

Hydrologic Cycle 445

Ice Ages 448

Ice Cores 452

Industrial Pollution 455

Industrial Water Use 459

Industrialization in Emerging Economies 462

Inland Fisheries 467

Insecticide Use 469

Intergovernmental Panel on Climate Change 474

International Environmental Law 477

Invasive Species 481

IPCC 2007 Report 485

Iron Fertilization 490

Irrigation 493

Island Ecosystems 496

Kyoto Protocol 500

Laboratory Methods in Environmental Science 509

Lakes 514

Land Use 518

Landfills 522

Landslides 527

Light Pollution 530

Liquified Natural Gas Resource Use 533

Logging 536

VOLUME TWO

Maps and Atlases 539

Marine Ecosystems 541

Marine Fisheries 546

Marine Water Quality 549

Mathematical Modeling and Simulation 556

559*Media: Environmentally Based News and Entertainment* 559

Medical Waste 562

Migratory Species 568

Mining and Quarrying Species Impacts 570

Montreal Protocol 574

National Environmental Policy Act 577

National Oceanic and Atmospheric Administration (NOAA) 582

National Park Service Organic Act 585

Natural Reserves and Parks 589

Natural Resource Management 598

Nonpoint-Source Pollution 601

Non-Scientist Contributions to Nature and Environment Studies 604

North Atlantic Oscillation 608

Nuclear Power 610

Nuclear Test Ban Treaties 615

Ocean Circulation and Currents 620

Ocean Salinity 624

Ocean Tides 626

Oceanography 628

Oceans and Coastlines 632

Oil Pollution Acts 636

Oil Spills 641

Organic and Locally Grown Foods 645

Organic Farming: Environmental Science and Philosophy 649

Overfishing 652

Overgrazing 655

Ozone Hole 658

Ozone Layer 661

Paper and Wood Pulp 664

Pharmaceutical Development Resources 666

Photography, Environmental 669

Pollinators 672

Precipitation 674

Predator-Prey Relationships 676

Radiative Forcing 680

Radioactive Waste 683

Rain Forest Destruction 686

Real-Time Monitoring and Reporting 688

Recreational Use and Environmental Destruction 690

Recycling 693

Red Tide 699

Reef Ecosystem 702

Reforestation 706

Resource Extraction 712

Rivers and Waterways 714

Runoff 716

Saltwater Intrusion 719

Sea Level Rise 721

Seasonal Migration 730

Shrublands 732

Silent Spring 734

Smog 739

Snow and Ice Cover 741

Soil Chemistry 743

Soil Contamination 745

Soil Resources 748

Solar Power 750

Solid Waste Treatment Technologies 753

Species Reintroduction Program 755

Spill Remediation 757

Streamflow 761

Superfund site 763

Surface Water 765

Surveying 767

Sustainable Development 769

Tellico Dam Project (Snail Darters) and Supreme Court Case (TVA v. Hill, 1977) 776

Temperature Records 779

Teratogens 782

Tidal or Wave Power 785

Tides 788

Toxic Waste 790

Tsunami Impacts 793

Tundra 798

United Nations Conference on the Human Environment (1972) 800

United Nations Convention on the Law of the Sea (UNCLOS) 803

United Nations Framework Convention on Climate Change (UNFCCC) 806

United Nations Policy and Activism 810

United Nations World Commission on Environment and Development (WCED) Our Common Future *Report (1987)* 813

Vegetation Cycles 816

Volcanoes 818

Walden 821

War and Conflict-Related Environmental Destruction 825

Waste Transfer and Dumping 829

Wastewater Treatment Technologies 832

Water Conservation 835

Water Pollution 837

Water Resources 843

Water Supply and Demand 848

Watershed Protection and Flood Prevention Act of 1954 854

Watersheds 856

Weather and Climate 859

Weather Extremes 861

Wetlands 864

Whaling, International Convention for the Regulation of 867

Wilderness Act of 1964 872

Wildfire Control 874

Wildfires 877

Wildlife Population Management 880

Wildlife Protection Policies and Legislation 883

Wildlife Refuge 886

Wind and Wind Power 888

Woods Hole Oceanographic Institution 891

Sources Consulted 895

Index 929

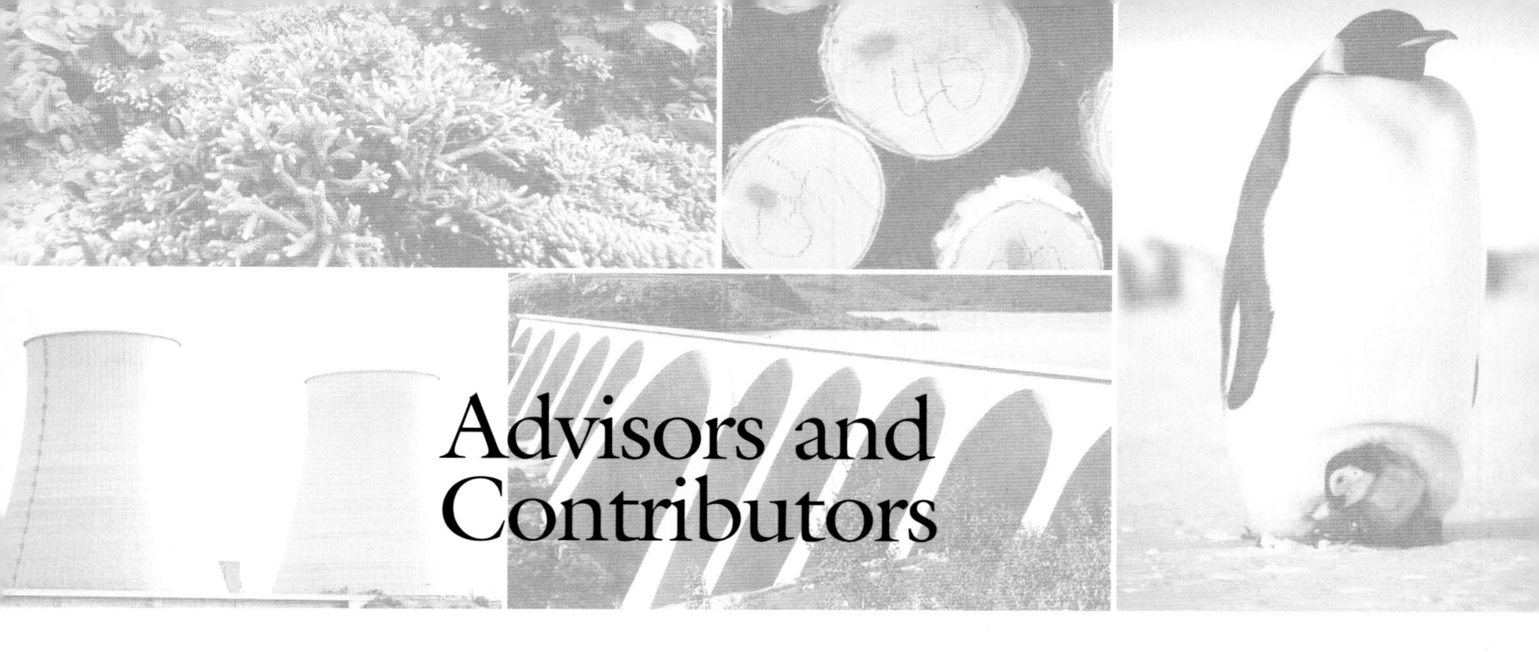

Advisors and Contributors

While compiling this volume, the editors relied upon the expertise and contributions of the following scientists, scholars, and researchers, who served as advisors and/or contributors for *Environmental Science: In Context*:

Susan Aldridge, Ph.D.
Independent scholar and writer
London, United Kingdom

Julie Berwald, Ph.D.
Geologist (Ocean Sciences) and writer
Austin, Texas

Philip Chaney, Ph.D., P.S.
Associate Professor of Geography
Auburn University
Auburn, Alabama

James Anthony Charles Corbett
Journalist
London, United Kingdom

Paul Davies, Ph.D.
Director, Science Research Institute
Adjunct Professor Université Paris
Paris, France

William J. Engle. P.E.
Exxon-Mobil Oil Corporation (Rt.)
New Orleans, Louisiana

Angela Scobey Fedor
Independent scholar
Mobile, Alabama

Larry Gilman, Ph.D.
Independent scholar and journalist
Sharon, Vermont

Amit Gupta, M.S.
Journalist
Ahmedabad, India

Tony Hawas, M.A.
Writer and journalist
Brisbane, Australia

Brian D. Hoyle, Ph.D.
Microbiologist
Nova Scotia, Canada

Joseph Hyder, J.D.
Jacksonville, Florida

Alexandr Ioffe, Ph.D.
Senior Scientist
Geological Institute of the Russian Academy of Sciences
Moscow, Russia

David T. King, Jr. Ph.D.
Professor, Department of Geology
Auburn University
Auburn, Alabama

Kenneth T. LaPensee
Epidemiologist and Medical Policy Specialist
Hampton, New Jersey

Adrienne Wilmoth Lerner, J.D.
Jacksonville, Florida

Miriam C. Nagel
Independent scholar
Avon, Connecticut

Caryn Neumann, Ph.D.
Visiting Assistant Professor
Denison University
Granville, Ohio

Anna Marie Roos, Ph.D.
Research Associate, Wellcome Unit for the History of Medicine
University of Oxford
Oxford, United Kingdom

ACKNOWLEDGMENTS

The editors are grateful to the truly global group of scholars, researchers, and writers who contributed to *Environmental Science: In Context.*

The editors also wish to thank their skilled copyeditors, Sonia Benson, Kate Kretchmann, and Alicia Cafferty Lerner, for excellent eyes and revisions that enhanced the quality and readability of the text.

The editors gratefully acknowledge and extend thanks to Janet Witalec, Julia Furtaw, and Debra Kirby at the Gale, Cengage Learning for their faith in the project and for their sound advice and guidance related to the essential construction and style of *Environmental Science: In Context.* Without the able guidance and efforts of talented technology teams, rights and acquisition management, and imaging at Gale, Cengage Learning, this book would not have been possible. The editors are also indebted to Kimberley McGrath, Kathy Edgar, and others at Gale for their invaluable help in reading and correcting copy. For *Environmental Science: In Context*, Kathy Edgar's skilled reading, corrections, and contributions deserve special praise. The editors also wish to acknowledge the contributions of Marcia Schiff at the Associated Press for her help in securing archival images.

Our most sincere thanks and profound gratitude are extended to project manager Madeline Harris. Whatever support was needed she readily offered, whatever skill was required she possessed. Her good humor, insight, and seemingly boundless "can do" spirit fused with her experience kept us on a steady course.

Introduction

We completed *Environmental Science: In Context* while working on assignment in China and Cambodia in June 2008. In Beijing preparations for the 2008 Olympics were nearing completion. The last bits of construction were finishing in a flurry of activity. The city and its generous people readied themselves to put their best foot forward as host to the world. Flower boxes lined airports and roadways, all overflowing with beauty. The excitement in the air was palpable, but so too was the smog that, at times, tortured the lungs and brought both real and symbolic tears to the eyes.

Walking along a clean swept road one day, we watched as a small piece of paper, no larger than a gum wrapper, flew off the backpack of a young man speeding along on his bicycle to work or school. The young man never saw the litter and soon turned a corner. The small piece of paper, something that would be inconspicuous among the debris routinely encountered along streets in most of the other great capitals of the world, stood in stark contrast to the clean street and generous bike path. Less than a minute later, however, a woman peddling by in the opposite direction spotted the paper, turned, stopped, dismounted from her bicycle, picked up the paper and put in her own backpack before resuming her journey.

The incident contrasted a people meticulous and caring of their environment with inhabitants of a city sometimes choked with smog—and in an instant crystallized our understanding that it is simply not the case that the Chinese do not *care* about pollution in Beijing or other cities. Rather it *is* the case that in the twenty-first century China and other developing economies now face the very same tests of balancing economic growth with environmental concern that many Western governments and cities often failed during the twentieth century.

Surrounded by mountains and hilly terrain on three sides, Beijing suffers to the same or greater degree the fate of most other cities where smog often stagnates. Yet, pollution is pollution, and freshening breezes may seemingly clear the air, but it is a false cleansing that simply transfers and disperses the pollutants elsewhere—and temporary solutions to clear the air for the Olympics are simply cosmetic. Even with clear skies over Beijing, China has a tremendous, and globally threatening, pollution problem.

However, rather than sit in smug judgment regarding pollution in the developing world, it is essential that as global citizens we should be mindful of the tremendous strains on developing economies torn between providing prosperity (which in many places means a hope of adequate food, medicine, and shelter) and protecting the environment.

In addition to continuing to admit our own mistakes and correct our own substantial list of environmental problems, as the West moves to increase efficiency and conservation, it must realize that Earth demands we treat it as our common home. A fresh breeze over Washington, London, or Paris, will do little if the smokestacks of the developing world continue to bellow pollution into our common atmosphere. Only global cooperation, good science, and the willingness to develop the programs and technology needed to facilitate clean development can truly foster global sustainability.

Environmental Science: In Context is a collection of entries on topics covering a diversity of environmental interests in an attempt to stimulate critical thinking about such challenges.

Intended for a wide and diverse audience, every effort has been made to set forth *Environmental Science: In Context* entries in everyday language and to provide accurate and generous explanations of the most important scientific terms. Entries are designed to instruct, challenge, and excite less experienced students, while providing a solid foundation and reference for more advanced students. *Environmental Science: In Context* provides students and readers with essential information and insights that enhance and enable a better understanding of environmental issues.

As editors, we have placed special emphasis on updating entries related to climate change. *Environmental Science: In Context* is one of the first books to be able to draw resources from the entire range of important Intergovernmental Panel on Climate Change (IPCC) reports issued in 2007.

During recent years, advances in technology have allowed us to identify hundreds of planets orbiting suns in other solar systems. Despite the growing population of known planets, Earth still remains uniquely beautiful. So far, Earth is the only planet with blue skies, warm seas, and life.

Humanity travels through the Cosmos, supported and nurtured by Earth. Indeed, our bodies are intrinsically bonded to Earth. We are composed of elements, made in stars, but that now compose Earth. In many important ways we are Earth's children now called to be responsible shepherds of her fate.

As Carl Sagan wrote in *Pale Blue Dot: A Vision of the Human Future in Space*, "The Earth is a very small stage in a vast cosmic arena." For humans to play wisely upon that stage, to secure a future for the children who shall inherit the Earth, we owe it to ourselves to become players of many parts, so that our repertoire of scientific knowledge enables us to use reason and intellect in our civic debates, and to understand the complex harmonies of Earth.

K. Lee Lerner & Brenda Wilmoth Lerner, editors

Bejing, China

June, 2008

With a portfolio of more than three dozen books and films, Lerner & Lerner have managed projects in more than a dozen countries. The Lerner & Lerner and LernerMedia portfolio (viewable at www.lernermedia.co.uk) reflects their continuing focus on science and the relationship of science to global issues.

"The least movement is of importance to all nature. The entire ocean is affected by a pebble." Blaise Pascal.

This book is dedicated to all of our past and present colleagues at Gale. It has been our great pleasure and privilege to work with such a dedicated group of intelligent and skilled people. Thank you for sharing hard work, good thoughts, good laughter, and a bit of your lives with us.

About the *In Context* Series

Written by a global array of experts, yet aimed primarily at high school students and an interested general readership, the *In Context* series serves as an authoritative reference guide to essential concepts of science, the impacts of recent changes in scientific consensus, and the impacts of science on social, political, and legal issues.

Cross-curricular in nature, *In Context* books align with and support national science standards and high school science curriculums across subjects in science and the humanities, and facilitates science understanding important to higher achievement in the No Child Left Behind (NCLB) science testing. Inclusion of original essays written by leading experts and primary source documents serve the requirements of an increasing number of high school and international baccalaureate programs, and are designed to provide additional insights on leading social issues, as well as spur critical thinking about the profound cultural connections of science.

In Context books also give special coverage to the impact of science on daily life, commerce, travel, and the future of industrialized and impoverished nations.

Each book in the series features entries with extensively developed words-to-know sections designed to facilitate understanding and increase both reading retention and the ability of students to understand reading in context without being overwhelmed by scientific terminology.

Entries are further designed to include standardized subheads that are specifically designed to present information related to the main focus of the book. Entries also include a listing of further resources (books, periodicals, Web sites, audio and visual media) and references to related entries.

In addition to maps, charts, tables and graphs, each *In Context* title has approximately 300 topic related-images that visually enrich the content. Each *In Context* title will also contain topic-specific timelines (a chronology of major events), a topic-specific glossary, a bibliography, and an index especially prepared to coordinate with the volume topic.

About This Book

The goal of *Environmental Science: In Context* is to help high-school and early college-age students understand the essential facts and deeper cultural connections of topics and issues related to the scientific study of the environment and its impacts on humanity.

Human biological and cultural origins are, of course, deeply tied to the environment. But just as Earth's environment shaped humanity, human activity (anthropogenic activity) now leaves an unmistakable stamp upon the natural world. *Environmental Science: In Context* places special emphasis on exploring the impacts of human habitation and economic activity on the environment.

This book also reflects the scientific consensus regarding global climate change—that it is real and an urgent global problem—and offers topics devoted to explaining both the science and the social challenges.

In an attempt to enrich the reader's understanding of the mutually impacting relationship between science and culture, as space allows we have included primary sources that enhance the content of *In Context* entries. In keeping with the philosophy that much of the benefit from using primary sources derives from the reader's own process of inquiry, the contextual material introducing each primary source provides an unobtrusive introduction and springboard to critical thought.

General Structure

Environmental Science: In Context is a collection of entries on diverse topics selected to provide insight into increasingly important and urgent topics associated with the study of environmental science.

The articles in the book are meant to be understandable by anyone with a curiosity about topics related to the science of climate change and the first edition of *Environmental Science: In Context* has been designed with ready reference in mind:

- Entries are arranged alphabetically, rather than by chronology or scientific subfield.
- The **chronology** (timeline) includes many of the most significant events in the history of environmental studies and advances of science. Where appropriate, related scientific advances are included to offer additional context.
- An extensive **glossary** section provides readers with a ready reference for content-related terminology. In addition to defining terms within entries, specific Words to Know sidebars are placed within each entry.

- A bibliography section (citations of books, periodicals, Web sites, and audio and visual material) offers additional resources to those resources cited within each entry.
- An **index** guides the reader to topics and persons mentioned in the book.

Entry Structure

In Context entries are designed so that readers may navigate entries with ease. Toward that goal, entries are divided into easy-to-access sections:

- **Introduction**: A opening section designed to clearly identify the topic.
- **Words to Know** sidebar: Essential terms that enhance readability and critical understanding of entry content.
- Established but flexible **rubrics** customize content presentation and identify each section, enabling the reader to navigate entries with ease. Inside *Environmental Science: In Context* entries readers will find a general scheme of organization. All entries contain a brief introduction, words to know, and then a section describing the essential history and scientific foundations of the topic. Sections titled "Impacts and Issues" or "Modern Cultural Connections" then interrelate key scientific, political, or social considerations related to the topic.
- Sidebars added by the editors enhance expert contributions by focusing on key areas, providing material for divergent studies, or providing evidence from key scientific reports (e.g., the 2007 Intergovernmental Panel on Climate Change reports, etc.)
- If an entry contains a related primary source, it is appended to the end of the author's text. Authors are not responsible for the selection or insertion of primary sources.
- **Bibliography:** Citations of books, periodicals, Web sites, and audio and visual material used in preparation of the entry or that provide a stepping stone to further study.
- **"See also" references** clearly identify other content-related entries.

Environmental Science: In Context special style notes

Please note the following with regard to topics and entries included in *Environmental Science: In Context*:

- Primary source selection and the composition of sidebars are not attributed to authors of signed entries to which the sidebars may be associated.
- Equations are, of course, often the most accurate and preferred language of science, and are essential to scientists studying climate change. To better serve the intended audience of *Environmental Science: In Context*, however, the editors attempted to minimize the inclusion of equations in favor of describing the elegance of thought or essential results such equations yield.
- A detailed understanding of physics and chemistry is neither assumed nor required for *Environmental Science: In Context*. Accordingly, students and other readers should not be intimidated or deterred by the sometimes complex names of chemical molecules or biological classification. Where necessary, sufficient information regarding chemical structure or species classification is provided. If desired, more information can easily be obtained from any basic chemistry or biology reference.

Bibliography citation formats (How to cite articles and sources)

In Context titles adopt the following citation format:

BOOKS

Gore, Al. *An Inconvenient Truth: The Planetary Emergency of Global Warming and What We Can Do About It.* Emmaus, PA: Rodale Press, 2006.

Metz, B., et al, eds. *Climate Change 2007: Mitigation of Climate Change: Contribution of Working Group III to the Fourth Assessment Report of the Intergovernmental Panel on Climate Change.* New York: Cambridge University Press, 2007.

Parry, M.L., et al, eds. *Climate Change 2007: Impacts, Adaptation and Vulnerability: Contribution of Working Group II to the Fourth Assessment Report of the Intergovernmental Panel on Climate Change.* New York: Cambridge University Press, 2007.

Solomon, S., et al, eds. *Climate Change 2007: The Physical Science Basis: Contribution of Working Group I to the Fourth Assessment Report of the Intergovernmental Panel on Climate Change.* New York: Cambridge University Press, 2007.

Weart, Spencer. *The Discovery of Global Warming.* Cambridge, MA: Harvard University Press, 2004.

PERIODICALS

Collins, William, et al. "The Physical Science Behind Climate Change." *Scientific American* (August 2007).

Oreskes, Naomi. "The Scientific Consensus on Climate Change." *Science* 306 (2004): 1686.

Thomas, Chris D. "Extinction Risk from Climate Change." *Nature* 427 (2004): 145–148.

WEB SITES

"Climate Change." *U.S. Environmental Protection Agency*, November 19, 2007. <http://epa.gov/climatechange/index.html> (accessed December 9, 2007).

Intergovernmental Panel on Climate Change. <http://www.ipcc.ch> (accessed December 9, 2007).

United Nations Framework Convention on Climate Change. <http://unfccc.int/2860.php> (accessed December 9, 2006).

Alternative citation formats

There are, however, alternative citation formats that may be useful to readers and examples of how to cite articles in alternative formats are shown below.

APA Style

Books:

Reisner, Marc. (1986). *Cadillac Desert.* New York: Viking Penguin. Excerpted in K. Lee Lerner and Brenda Wilmoth Lerner, eds. (2006) *Environmental Issues: Essential Primary Sources*, Farmington Hills, Mich.: Thomson Gale.

Periodicals:

Aldo, Leopold. (October 1925). "Wilderness as a Form of Land Use." *The Journal of Land and Public Utility Economics,* 1 : 398–404. Excerpted in K. Lee Lerner and

Brenda Wilmoth Lerner, eds. (2006) *Environmental Issues: Essential Primary Sources*, Farmington Hills, Mich.: Thomson Gale.

Web Sites:

United States Environmental Protection Agency. "How to Conserve Water and Use It Effectively." Retrieved January 17, 2006 from http://www.epa.gov/ow/you/chap3.html. Excerpted in K. Lee Lerner and Brenda Wilmoth Lerner, eds. (2006) *Environmental Issues: Essential Primary Sources*, Farmington Hills, Mich.: Thomson Gale.

Chicago Style

Reisner, Marc. *Cadillac Desert.* New York: Viking Penguin, 1986. Excerpted in K. Lee Lerner and Brenda Wilmoth Lerner, eds. *Environmental Issues: Essential Primary Sources.* Farmington Hills, Mich.: Thomson Gale, 2006.

Periodicals:

Aldo, Leopold. "Wilderness as a Form of Land Use." *The Journal of Land and Public Utility Economics,* 1 (October 1925): 398–404. Excerpted in K. Lee Lerner and Brenda Wilmoth Lerner, eds. *Environmental Issues: Essential Primary Sources.* Farmington Hills, Mich.: Thomson Gale, 2006.

Web sites:

United States Environmental Protection Agency. "How to Conserve Water and Use It Effectively." <http://www.epa.gov/ow/you/chap3.html>. (accessed January 17, 2006). Excerpted in K. Lee Lerner and Brenda Wilmoth Lerner, eds. *Environmental Issues: Essential Primary Sources.* Farmington Hills, Mich.: Thomson Gale, 2006.

MLA Style

Books:

Reisner, Marc. *Cadillac Desert*, New York: Viking Penguin, 1986. Excerpted in K. Lee Lerner and Brenda Wilmoth Lerner, eds. *Environmental Issues: Essential Primary Sources*, Farmington Hills, Mich.: Thomson Gale, 2006.

Periodicals:

Aldo, Leopold. "Wilderness as a Form of Land Use." *The Journal of Land and Public Utility Economics,* 1 (October 1925): 398–404. Excerpted in K. Lee Lerner and Brenda Wilmoth Lerner, eds. *Environmental Issues: Essential Primary Sources,* Farmington Hills, Mich.: Thomson Gale, 2006.

Web sites:

"How to Conserve Water and Use It Effectively." United States Environmental Protection Agency. 17 January 2006. <http://www.epa.gov/ow/you/chap3.html>. Excerpted in K. Lee Lerner and Brenda Wilmoth Lerner, eds. *Environmental Issues: Essential Primary Sources*, Farmington Hills, Mich.: Thomson Gale, 2006.

Turabian Style (Natural and Social Sciences)

Books:

Reisner, Marc. *Cadillac Desert,* (New York: Viking Penguin, 1986). Excerpted in K. Lee Lerner and Brenda Wilmoth Lerner, eds. *Environmental Issues: Essential Primary Sources*, (Farmington Hills, Mich.: Thomson Gale, 2006).

Periodicals:

Aldo, Leopold. "Wilderness as a Form of Land Use." *The Journal of Land and Public Utility Economics,* 1 (October 1925): 398–404. Excerpted in K. Lee Lerner and Brenda Wilmoth Lerner, eds. *Environmental Issues: Essential Primary Sources*, (Farmington Hills, Mich.: Thomson Gale, 2006).

Web sites:

United States Environmental Protection Agency. "How to Conserve Water and Use It Effectively." available from http://www.epa.gov/ow/you/chap3.html; accessed January 17, 2006. Excerpted in K. Lee Lerner and Brenda Wilmoth Lerner, eds. *Environmental Issues: Essential Primary Sources*, (Farmington Hills, Mich.: Thomson Gale, 2006).

Using Primary Sources

The definition of what constitutes a primary source is often the subject of scholarly debate and interpretation. Although primary sources come from a wide spectrum of resources, they are united by the fact that they individually provide insight into the historical *milieu* (context and environment) during which they were produced. Primary sources include such materials as newspaper articles, press dispatches, autobiographies, essays, letters, diaries, speeches, song lyrics, posters, works of art—and in the twenty-first century, Web logs—that offer direct, first-hand insight or witness to events of their day.

Categories of primary sources include:

- Documents containing firsthand accounts of historic events by witnesses and participants. This category includes diary or journal entries, letters, email, newspaper articles, interviews, memoirs, and testimony in legal proceedings.
- Documents or works representing the official views of both government leaders and leaders of other organizations. These include primary sources such as policy statements, speeches, interviews, press releases, government reports, and legislation.
- Works of art, including (but certainly not limited to) photographs, poems, and songs, including advertisements and reviews of those works that help establish an understanding of the cultural *milieu* (the cultural environment with regard to attitudes and perceptions of events).
- Secondary sources. In some cases, secondary sources or tertiary sources may be treated as primary sources. For example, if an entry written many years after an event, or to summarize an event, includes quotes, recollections, or retrospectives (accounts of the past) written by participants in the earlier event, the source can be considered a primary source.

Analysis of primary sources

The primary material collected in this volume is not intended to provide a comprehensive or balanced overview of a topic or event. Rather, the primary sources are intended to generate interest and lay a foundation for further inquiry and study.

In order to properly analyze a primary source, readers should remain skeptical and develop probing questions about the source. Using historical documents requires that readers analyze them carefully and extract specific information. However, readers must also read "beyond the text" to garner larger clues about the social impact of the primary source.

In addition to providing information about their topics, primary sources may also supply a wealth of insight into their creator's viewpoint. For example, when reading a news article about an outbreak of disease, consider whether the reporter's words also indicate something about his or her origin, bias (an irrational disposition in favor of someone or something), prejudices (an irrational disposition against someone or something), or intended audience.

Students should remember that primary sources often contain information later proven to be false, or contain viewpoints and terms unacceptable to future generations. It is important to view the primary source within the historical and social context existing at its creation. If, for example, a newspaper article is written within hours or days of an event, later developments may reveal some assertions in the original article as false or misleading.

Test new conclusions and ideas

Whatever opinion or working hypothesis the reader forms, it is critical that they then test that hypothesis against other facts and sources related to the incident. For example, it might be wrong to conclude that factual mistakes are deliberate unless evidence can be produced of a pattern and practice of such mistakes with an intent to promote a false idea.

The difference between sound reasoning and preposterous conspiracy theories (or the birth of urban legends) lies in the willingness to test new ideas against other sources, rather than rest on one piece of evidence such as a single primary source that may contain errors. Sound reasoning requires that arguments and assertions guard against argument fallacies that utilize the following:

- false dilemmas (only two choices are given when in fact there are three or more options);
- arguments from ignorance (*argumentum ad ignorantiam*; because something is not known to be true, it is assumed to be false);
- possibilist fallacies (a favorite among conspiracy theorists who attempt to demonstrate that a factual statement is true or false by establishing the possibility of its truth or falsity. An argument where "it could be" is usually followed by an unearned "therefore, it is.");
- slippery slope arguments or fallacies (a series of increasingly dramatic consequences is drawn from an initial fact or idea);
- begging the question (the truth of the conclusion is assumed by the premises);
- straw man arguments (the arguer mischaracterizes an argument or theory and then attacks the merits of their own false representations);
- appeals to pity or force (the argument attempts to persuade people to agree by sympathy or force);
- prejudicial language (values or moral goodness, good and bad, are attached to certain arguments or facts);
- personal attacks (*ad hominem*; an attack on a person's character or circumstances);
- anecdotal or testimonial evidence (stories that are unsupported by impartial observation or data that is not reproducible);
- *post hoc* (after the fact) fallacies (because one thing follows another, it is held to cause the other);
- the fallacy of the appeal to authority (the argument rests upon the credentials of a person, not the evidence).

Despite the fact that some primary sources can contain false information or lead readers to false conclusions based on the "facts" presented, they remain an invaluable resource regarding past events. Primary sources allow readers and researchers to come as close as possible to understanding the perceptions and context of events and thus to more fully appreciate how and why misconceptions occur.

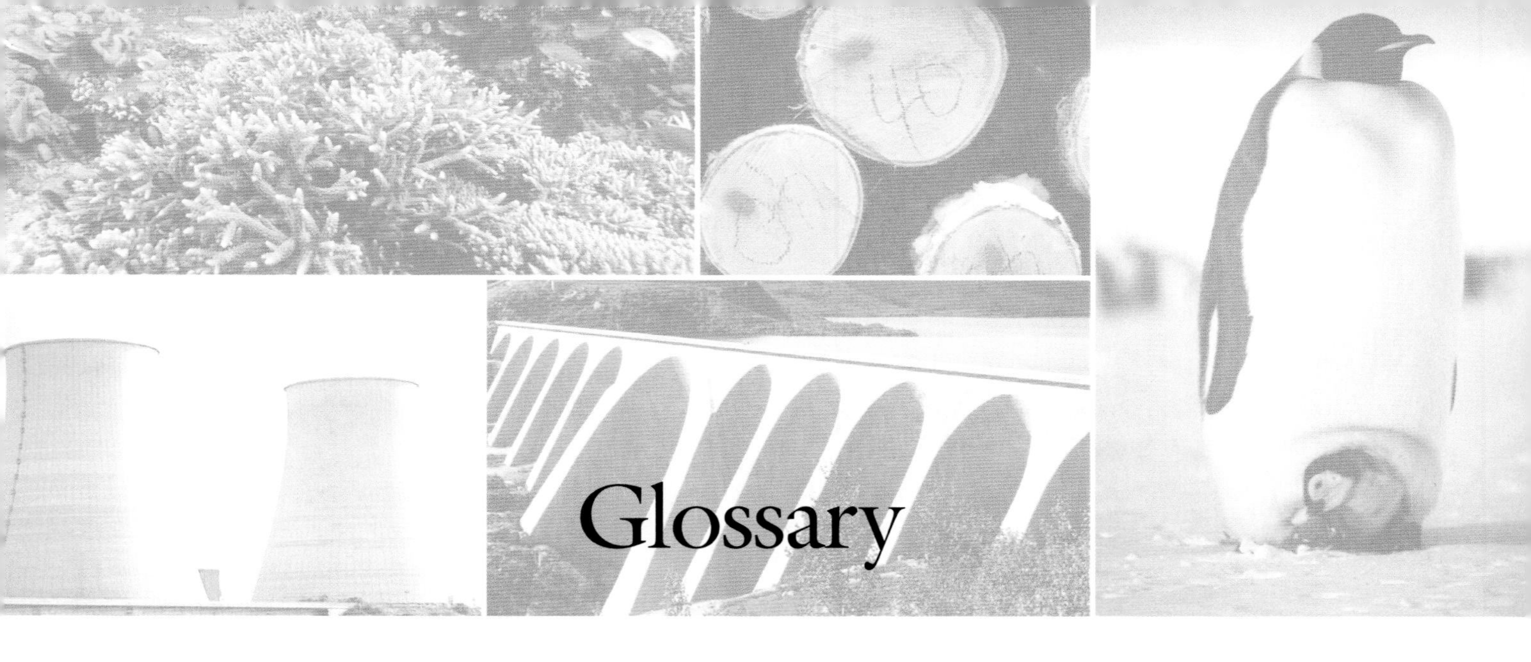

Glossary

1750 VALUE: Refers to pre-industrial greenhouse gas levels. The 2001 Intergovernmental Panel on Climate Change (IPCC) determined that greenhouse gas concentrations prior to 1750 were uninfluenced by human activity.

A

ABIOTIC: A term used to describe the portion of an ecosystem that is not living, such as water or soil.

ABLATION (GLACIAL): The erosive reduction of ice or snow from the surface of a mass of ice.

ACID: A substance that when dissolved in water is capable of reacting with a base to form a salt.

ACID RAIN: A form of precipitation that is significantly more acidic than neutral water, often produced as the result of industrial processes.

ACTIVE FAULT: A fault where movement has been known to occur in recent geologic time.

AEROSOL: Liquid droplets or minute particles suspended in air.

AFTERSHOCK: A subsequent earthquake (usually smaller in magnitude) following a powerful earthquake that originates at or near the same place.

AGRICULTURE: Replacement of a natural ecosystem with animals and plants chosen by people.

AIR POLLUTION: The existence in the air of substances in concentrations that are determined unacceptable. Contaminants in the air we breathe come mainly from manufacturing industries, electric power plants, automobiles, buses, and trucks.

ALBEDO: A numerical expression describing the ability of an object or planet to reflect light.

ALFISOIL: Rich soil formed under deciduous forest.

ALGAE: Single-celled or multicellular plants or plantlike organisms that contain chlorophyll, thus making their own food by photosynthesis.

ALGAL BLOOM: Sudden reproductive explosion of algae (single-celled aquatic green plants) in a large, natural body of water such as a lake or sea. Blooms near coasts are sometimes called red tides.

ALTERNATIVE ENERGY: An energy source that is used as an alternative to fossil fuels. Solar, wind, and geothermal power are examples of alternative energies.

AMBIENT: Existing condition.

AMOSITE: The brown form of asbestos.

ANABOLISM: The process by which energy is used to build up complex molecules.

ANAEROBIC: Pertaining to the absence of oxygen.

ANAEROBIC BACTERIA: Bacteria that grow without oxygen, also called anaerobes.

ANTARCTIC TREATY: A 1959 series of agreements regulating international relations in Antarctica, establishing the area as a demilitarized zone to promote open scientific research.

ANTHROPOGENIC: Made by humans or resulting from human activities.

ANTHROPOGENIC SOURCES: Sources that are due to human activity. An example of anthropogenic air pollution is the burning of wood for fuel.

AQUACULTURE: The farming of fish or shellfish in freshwater or saltwater.

AQUIFER: Rock, soil, or sand that is able to hold and transmit water.

ARIDOSOL: Type of soil found in arid environments.

ARTESIAN SPRING OR WELL: Groundwater that flows from the aquifer to the surface without the need for a pump.

ASBESTOS: An incombustible fibrous mineral once commonly used for fireproofing and electrical insulation.

ASPHYXIATES: Compounds that cause a shortage of oxygen.

ATMOSPHERE: The air surrounding the Earth, described as a series of layers of different characteristics. The atmosphere, composed mainly of nitrogen and oxygen with traces of carbon dioxide, water vapor, and other gases, acts as a buffer between Earth and the sun.

ATOMIC BOMB: A highly destructive weapon that derives its explosive power from the fission of atomic nuclei.

ATOMIC NUCLEUS: The small, dense, positively charged central region of an atom, composed of protons and neutrons.

ATV: Abbreviation for "all-terrain vehicle," a four-wheeled vehicle designed for off-road use that is straddled like a motorcycle and steered with handlebars.

AVIAN INFLUENZA: Also known as bird flu, it is a respiratory disease caused by the H5N1 virus that is thought to have originated in Asian poultry factory farms. The disease, which is lethal when passed from bird-to-bird, is evolving to be capable of person-to-person transmission.

B

BASEL CONVENTION: A 1992 global treaty designed to restrict the movement of hazardous waste between nations, especially from developed to less-developed countries.

BASELINE POWER: The amount of steady, non-intermittent electric power that is constantly being produced by a power source.

BATTERY: Device that can easily convert stored energy in the chemical bonds of an electrolyte into electrical energy.

BEDROCK: Solid layer of rock lying beneath Earth's surface.

BENTHIC: Living on, or associated with, the ocean floor.

BIOAUGMENTATION: The introduction of specific strains of microorganisms to break down pollutants in contaminated soil or water.

BIOCHEMICAL OXYGEN DEMAND: The amount of oxygen required by decomposing microorganisms in a water sample and an important measure of water pollution.

BIODEGRADABLE: Capable of being degraded in the environment by the action of microorganisms.

BIODIESEL: A fuel made from a combination of plant and animal fat. It can be safely mixed with petro diesel.

BIODIVERSITY: Literally, "life diversity." the wide range of plants and animals that exist within any given geographical region.

BIOENGINEERED CROP: Herbicide or insect-resistant plants developed through genetic engineering.

BIOFUEL: A fuel derived directly by human effort from living things, such as plants or bacteria. A biofuel can be burned or oxidized in a fuel cell to release useful energy.

BIOGEOCHEMICAL CYCLE: The chemical interactions that take place among the atmosphere, biosphere, hydrosphere, and geosphere.

BIOLOGICAL MAGNIFICATION: An increase in the concentration of toxins as they pass up the food chain.

BIOMASS: The sum total of living and once-living matter contained within a given geographic area; or, organic matter that can be converted to fuel and is regarded as a potential energy source.

BIOME: A well-defined terrestrial environment (e.g., desert, tundra, or tropical forest) and the complex of living organisms found in that region.

BIOREMEDIATION: The use of living organisms to help repair damage such as that caused by oil spills.

BIOSPHERE: All life forms on Earth and the interactions among those life forms.

BOG: Area of wet, spongy ground consisting of decayed plant matter.

BOILING CRYOGEN: Low-temperature liquid at its bubble point; the temperature and pressure at which evaporation begins.

BOOM: A physical barrier placed around an oil spill to contain it.

BOREAL FORESTS: A forest biome of coniferous trees running across northern North America and Eurasia; its high northern latitudes are often referred to as taiga.

BOTTOM TRAWLING: An industrial fishing practice in which large, heavy nets are dragged across the sea floor.

BREVETOXIN: Any of a class of neurotoxins produced by the algae that cause red tide (coastal algae blooms). Brevetoxins can be concentrated by shellfish and poisonous to humans who eat the shellfish.

BROWNFIELD SITE: A site contaminated with hazardous waste.

BYCATCH: Non-target species killed in the process of fishing.

C

CALVING: Process of iceberg formation in which huge chunks of ice break free from glaciers, ice shelves, or ice sheets due to stress, pressure, or the forces of waves and tides.

CAP-AND-TRADE PROGRAM: An emissions trading program designed to control industrial pollution by providing economic incentives.

CAPTIVE BREEDING: A wildlife conservation method in which rare or endangered species are bred in restricted environments such as zoos or wildlife preserves.

CAPTURE FISHERY: The harvesting of fish stocks occurring naturally in a body of water.

CARBON BANKING: Form of accounting that tracks the carbon emissions, reductions, and offsets of a client in a way that is analogous to the treatment of money in ordinary banking. A helpful adjunct to emissions trading schemes.

CARBON CREDITS: Units of permission or value, similar to monetary units such as dollars or euros, that entitle their owner to emit one metric ton of carbon dioxide (CO_2) into the atmosphere per credit.

CARBON CYCLE: The circulation of carbon (C) atoms through natural processes such as photosynthesis.

CARBON DIOXIDE: An odorless, colorless, non-poisonous gas, with the chemical formula CO_2, that is released by natural processes and by burning fossil fuels. The increased amounts of CO_2 in the atmosphere enhance the greenhouse effect, blocking heat from escaping into space and contributing to the warming of Earth's lower atmosphere.

CARBON FOOTPRINT: The amount of carbon dioxide (or of any other greenhouse gas, counted in terms of the greenhouse-equivalent amount of CO_2) emitted to supply the energy and materials consumed by a person, product, or event.

CARBON SEQUESTERING: Storage or fixation of carbon in such a way that it is isolated from the atmosphere and cannot contribute to climate change.

CARBON SEQUESTRATION: The uptake and storage of carbon (C) from the atmosphere into carbon sinks (such as oceans, forests, or soils).

CARBON SINK: A forest or other ecosystem that absorbs and stores more carbon than it releases.

CARCINOGEN: A cancer-causing agent, such as a chemical or virus.

CARRYING CAPACITY: The population of a species that an ecosystem can support with food, water, and other resources.

CARTOGRAPHY: The science of mapmaking.

CATABOLISM: The process by which large molecules are broken down into smaller ones with the release of energy.

CELLULOSIC FERMENTATION: Digestion of high-cellulose plant materials (e.g., wood chips, grasses) by bacteria that have been bred or genetically engineered for that purpose. The useful product is ethanol, which can be burned as a fuel.

CHANNEL: A water-filled path which connects mudflats and salt marshes to the ocean in an estuary.

CHLOROFLUOROCARBONS (CFCs): A family of chemical compounds consisting of carbon, fluorine, and chlorine that were once used widely as propellants in commercial sprays but regulated in the United States since 1987 because of their harmful environmental effects.

CHRYSOTILE: The white form of asbestos, which is less deadly than the blue or brown form.

CIRRUS CLOUD: A thin cloud of tiny ice crystals forming at 20,000 feet (6 km) or higher; reflects sunlight from Earth and also reflects infrared (heat) radiation back at the ground.

CLAY: The portion of soil comprising the smallest particles, resulting from the weathering and breakdown of rocks and minerals.

CLEAN DEVELOPMENT MECHANISM: One of the three mechanisms set up by the Kyoto Protocol of 2007 to, in theory, allow reductions in greenhouse-gas emissions to be implemented where they are most economical. Under the Clean Development Mechanism, polluters in wealthy countries can obtain carbon credits (greenhouse pollution rights) by funding reductions in greenhouse emissions in developing countries.

CLEAR-CUTTING: A forestry practice involving the harvesting of all trees of economic value at one time.

CLEARCUT: A parcel of forest in which all trees have been removed for harvesting.

CLIMATE MODEL: A quantitative method of simulating the interactions of the atmosphere, oceans, land surface, and ice. Models can range from relatively simple to quite comprehensive.

CLIMATE NEUTRAL: The process of reducing greenhouse emissions so as to create a neutral impact on climate change.

CLOUD: A patch of condensed water or ice droplets.

CO-GENERATION: The simultaneous generation of both heat and electricity at one facility.

COLD WAR: A term describing the ideological, political, economic, and military tensions and struggles between the two dominant superpowers of the era, the United States and the former Union of Soviet Socialist Republics (USSR) between 1945 (the end of the World War II) and the collapse of the Soviet Union in 1991.

COLIFORM: Bacteria present in the environment and in the feces of all warm-blooded animals and humans; useful for measuring water quality.

COLONIZATION: The process by which a species populates a new area.

COMBUSTION: The process of burning a material.

COMMUNITY: All of the populations of species living in a certain environment.

COMPOSTING: Breakdown of organic material by microorganisms.

CONDENSATION: The coalescence of water molecules from the vapor to the liquid or solid phase.

CONSERVATION: The act of using natural resources in a way that ensures that they will be available to future generations.

CONTIGUOUS ZONE: A maritime zone extending 24 nautical miles (44 km) from the outer edge of the territorial sea, in which a coastal state can exert limited control of its laws.

CONTINENTAL DRIFT: A theory that explains the relative positions and shapes of the continents and other geologic phenomena by lateral movement of the continents. This was the precursor to plate tectonic theory.

CONTINENTAL SHELF: A gently sloping, submerged ledge of a continent.

CONTRAIL: A high-altitude cloud formed by the passage of an aircraft.

CONTROLLED BURN: A forest management technique in which small, controlled fires are set to clear brush and prevent larger wildfires in the future.

CORAL: Invertebrate organisms in the phylum Cnidaria that form reefs in tropical ocean waters.

CORAL ATOLL: A low tropical island, often roughly ring-shaped, formed by coral reefs growing on top of a subsiding island. The rocky base of the atoll may be hundreds of feet below present-day sea level.

CORAL BLEACHING: Decoloration or whitening of coral from the loss, temporary or permanent, of symbiotic algae (zooxanthellae) living in the coral.

CORAL POLYP: A living organism that, as part of a colony, builds the rocky calcium carbonate ($CACO_3$) skeleton that forms the physical structure of a coral reef.

CORE: The central region of a star, where thermonuclear fusion reactions take place to produce the energy necessary for the star to support itself against its own gravity.

CORIOLIS EFFECT: A force exemplified by a moving object appearing to travel in a curved path over the surface of a spinning body.

CORIOLIS FORCE: The apparent tendency of a freely moving particle to swing to one side when its motion is referred to a set of axes that is itself rotating in space, such as Earth. Winds are affected by rotation of the Earth so that instead of a wind blowing in the direction it starts, it turns to the right of that direction in the Northern Hemisphere, and left in the Southern Hemisphere.

CRIMINALIZATION: The social or legal process by which a certain behavior comes to be redefined as a crime.

CROCIDOLITE: The blue form of asbestos.

CROP: Plants grown for food, energy, or some other human need.

CRUST: The hard, outer shell of Earth that floats upon the softer, denser mantle.

CULL: The selection, often for destruction, of a part of an animal population.

CYANOBACTERIA: Photosynthetic bacteria, commonly known as blue-green algae.

CYCLONE: A large-scale system of low pressure in which circular winds blow counterclockwise in the Northern Hemisphere and clockwise in the Southern Hemisphere.

D

DDT: One of the earliest insecticides, dichlorodiphenyl-trichloroethane, used until banned by many countries in the 1960s after bird populations were decimated by the substance, and other negative environmental consequences were recognized.

DEAD ZONE: An area of ocean in which nothing can live except bacteria that flourish on fertilizer from agricultural runoff.

DEBRIS: The remains of anything broken down or destroyed.

DECIDUOUS: Plants that shed leaves or other foliage after their growing season.

DEEP SEA MINING: The extraction of valuable mineral deposits from the ocean floor; not yet practiced due to legal, environmental, and monetary concerns.

DEFORESTATION: A reduction in the area of a forest resulting from human activity.

DEGRADATION: The microbial breakdown of a complex compound into simpler compounds with the release of energy.

DEGRADED WATER: Water that has been reduced in quality through industrial use.

DEMANUFACTURING: The disassembly, sorting, and recovering of valuable or toxic materials from electronic products such as televisions and computers.

DENDROARCHAEOLOGY: The analysis of wooden material from archaeological sites using the techniques of dendrochronology (tree-ring dating).

DENDROCLIMATOLOGY: The study of past climates using the techniques of dendrochronology (tree-ring dating).

DEPLETED URANIUM (DU): A byproduct of spent nuclear fuel, DU is a dense metal with a variety of civilian and military uses. It is often used to enhance the armor piercing qualities of munitions, although its deployment is riddled with controversy and it has been linked to increased incidence of cancer rates and birth defects.

DEPTH HOAR: Brittle, loosely arranged crystals at the base of a snowpack.

DESALINATION: Removal of salt from salt water to produce fresh water.

DESERT: A land area so dry that little or no plant or animal life can survive.

DESERTIFICATION: Transformation of arid or semiarid productive land into desert.

DETRITUS: Matter produced by decay or disintegration of living material.

DEVELOPING NATION: A country that is relatively poor, with a low level of industrialization and relatively high rates of illiteracy and poverty.

DEVELOPMENT: The process by which a multicellular organism is produced from a single cell.

DINOFLAGELLATE: Small organisms with both plant-like and animal-like characteristics, usually classified as algae (plants). They take their name from their twirling motion and their whiplike flagella.

DIRECT ACTION TACTICS: Methods of political or social activism involving immediate, confrontative demand for change, such as strikes, sit-ins, and boycotts.

DISTILLATION: The process of purifying a liquid by successive evaporation and condensation.

DISTURBANCE SEVERITY: The amount of vegetation killed by fire or tree cutting activity, and the type of growing space made available for new plants.

DIURNAL: Performed in twenty-four hours, such as the diurnal rotation of Earth; also refers to animals and plants that are active during the day.

DOMOIC ACID: A neurotoxin produced by the algae that cause red tide (coastal algae blooms). Domoic acid can be concentrated by shellfish and poisonous to humans who eat the shellfish.

DREDGING: The excavation of sediment from the bottom of a body of water.

DRIP IRRIGATION: Slow, localized application of water just above the soil surface.

DRY SPELL: A short period of drought, usually lasting fewer than 14 days.

DRYLAND: Land where freshwater supplies are limited.

DUCKS UNLIMITED: An international non-profit organization founded in Canada in the 1940s to preserve and protect wetlands.

E

ECOLOGICAL: Having to do with interactions among organisms.

ECOLOGICAL NICHE: The sum of the environmental requirements necessary for an individual to survive and reproduce.

ECOLOGICAL SERVICES: The benefits to human communities that stem from healthy forest ecosystems, such as clean water, stable soil, and clean air.

ECOLOGY: The branch of science dealing with the interrelationship of organisms and their environments.

ECOSYSTEM: A system of living organisms interacting with each other and their physical environment.

ECOSYSTEM PROCESSES: The dynamic interrelationships among and between living organisms and their particular habitat elements.

ECOSYSTEM SERVICES: Services that a natural or restored ecosystem provides to human communities, including improving water quality, reducing soil erosion, flooding, and landslides and increasing carbon sequestration.

ECO-TERRORISM: Criminal sabotage against persons or property carried out by an environmentally oriented group for symbolic purposes.

ECO-TOURISM: Environmentally responsible travel to natural areas that promotes conservation, has a low visitor impact, and provides for beneficially active socio-economic involvement of local peoples.

ECOZONE: A broad section of Earth's surface that features distinct climate patterns, ocean conditions, types of landscapes, and species of plants and animals.

EFFLUENTS: Waste materials, often as outflow from septic or water treatment systems, that are discharged into the environment.

EL NIÑO/SOUTHERN OSCILLATION: A global climate cycle that arises from interaction of ocean and atmospheric circulations. Every 2-7 years, westward-blowing winds over the Pacific subside, allowing warm water to migrate across the Pacific from west to east. This suppresses normal upwelling of cold, nutrient-rich waters in the eastern Pacific, shrinking fish populations and changing weather patterns around the world.

ELECTRICAL GRID: Network of power lines that carry electricity from the source of generation to where the power can be used.

ELECTROMAGNETIC ENERGY: Energy conveyed by electromagnetic waves, which are paired electric and magnetic fields propagating together through space. X rays, visible light, and radio waves are all electromagnetic waves.

EMBRYOLOGY: The study of early development in living things.

ENDANGERED SPECIES: A species that is vulnerable to extinction.

ENDEMIC SPECIES: A species that is exclusively native to a certain area.

ENERGY RECOVERY: Incineration of solid waste to produce energy.

ENVIRONMENTAL DEGRADATION: The overall deterioration of environmental quality due to a range of issues, such as deforestation, desertification, pollution, and climate change.

ENVIRONMENTAL ESTROGEN: Compounds in toxic waste that mimic estrogen in their effect on humans and other animals.

ENVIRONMENTAL IMPACT STATEMENT: A document outlining the potential environmental impact of any new federal project, required by the U.S. National Environmental Policy Act.

ENVIRONMENTAL INVESTIGATION AGENCY (EIA): An independent, international campaigning organization that investigates and exposes environmental crime.

ENVIRONMENTAL MOVEMENT: A diverse social, political, and scientific movement revolving around the preservation of Earth's environment.

ENVIRONMENTAL PROTECTION AGENCY (EPA): An agency of the United States government since 1970, charged with protecting human health and the environment through research, regulation, and education.

ENZYME: A protein that catalyzes a chemical reaction, usually by lowering the energy at which the reaction can occur, without itself being changed by the reaction.

EPIFAUNA: Animals that live attached to the surface of a substrate such as rocks or pilings.

EQUINOX: Either of the two times during the year when the sun crosses the plane of Earth's equator, making night and day of approximately equal length.

EROSION: The wearing away of soil or rock over time by the action of water, glaciers, or wind.

ESTUARY: Lower end of a river where ocean tides meet the river's current.

ETHANOL: A compound of carbon, hydrogen, and oxygen (CH_3CH_2OH) that is a clear liquid at room temperature; also known as drinking alcohol or ethyl alcohol.

EUPHOTIC ZONE: The uppermost layer of a body of water in which the level of sunlight is sufficient for photosynthesis to occur.

EUSTATIC SEA LEVEL: Change in global average sea level caused by increased volume of the ocean (caused both by thermal expansion of warming water and by the addition of water from melting glaciers). Often contrasted to relative sea level rise, which is a local increase of sea level relative to the shore.

EUTROPHICATION: The process whereby a body of water becomes rich in dissolved nutrients through natural or man-made processes. This often results in a deficiency of dissolved oxygen, producing an environment that favors plant over animal life.

EVAPORITES: Salts deposited by the evaporation of aqueous solutions.

EVAPOTRANSPIRATION: The sum of evaporation and plant transpiration. Potential evapotranspiration is the amount of water that could be evaporated or transpired at a given temperature and humidity, if there was plenty of water available. Actual evapotranspiration cannot be any greater than precipitation, and will usually be less because some water will run off into rivers and flow to the oceans.

EVERGREEN: Bearing green leaves throughout the entire year.

E-WASTE: A term describing electronic equipment at the end of its useful life. E-waste is the fastest-growing type of waste in the world.

EXCLUSION ZONE: A zone established by a sanctioning body to prohibit specific activities in a specific geographic area.

EXCLUSIVE ECONOMIC ZONE (EEZ): A maritime zone extending 200 nautical miles (370 km) from the outer edge of the territorial sea, in which a coastal state has special rights over the exploration and use of marine resources.

EXTINCT: No longer in existence. In geology, it can be used to mean a process or structure that is permanently inactive (e.g., an extinct volcano).

EXTINCTION: The total disappearance of a species or the disappearance of a species from a given area.

***EXXON VALDEZ*:** An oil tanker that spilled millions of gallons of oil in Prince William Sound, Alaska, beginning on March 24, 1989.

F

FACTORY FARMS: Enclosed or open-air facilities that house thousands to tens of thousands of poultry, swine, or cattle.

FAULT: A fracture in the continuity of a rock formation resulting from tectonic movement.

FAUNA: The animal life existing in a defined area.

FEEDSTOCK: Raw material required for an industrial process.

FELONY: A crime that is considered serious and is punishable by a more stringent sentence than what is given for a misdemeanor.

FEMA: The U.S. Federal Emergency Management Agency, founded in 1979 as an agency of the Department of Homeland Security, is responsible for coordinating responses to disasters taking place within the United States.

FERMENTATION: A chemical reaction in which enzymes break down complex organic compounds (for example, carbohydrates and sugars) into simpler ones (for example, ethyl alcohol).

FILTER FEEDERS: Animals that obtain their food from filtering passing water.

FIREBREAK: A strip of cleared or plowed land that acts as a barrier to slow or stop the progress of a wildfire.

FISH FARMING: The commercial production of fish in tanks or enclosures, usually for food; also known as aquaculture.

FISHERY: An industry devoted to the harvesting and selling of fish, shellfish, or other aquatic animals.

FISHING STOCK: A subpopulation of fish occupying a certain area that is in reproductive isolation from the rest of its species.

FISSION: A process in which the nucleus of an atom splits, usually into two daughter nuclei, with the transformation of tremendous levels of nuclear energy into heat and light.

FLOOD IRRIGATION: Irrigation carried out by flooding a crop with water.

FLORA: The plant life of an area.

FLYWAY: A flying route regularly taken by migratory birds.

FODDER: Food for grazing animals.

FOLIAR: Pertaining to the leaf.

FOOD CHAIN: A sequence of organisms, each of which uses the next lower member of the sequence as a food source.

FOOD WEB: An interconnected set of all the food chains in the same ecosystem.

FORAMINIFERA: Single-celled marine organisms that inhabit small shells and float free in ocean surface waters. The shells of dead foraminifera sink to the sea bottom as sediment, forming thick deposits over geological time. Because the numbers and types of foraminifera are climate-sensitive, analysis of these sediments gives data on ancient climate changes.

FORESHOCK: A tremor that precedes a much larger earthquake.

FOREST MONOCULTURE: The development of a forest that is dominated by a single species of tree and which lacks the ecological diversity to withstand disease and parasites over the long term.

FOSSIL: A remnant of a past geological age that was embedded and has been preserved in Earth's crust.

FOSSIL FUELS: Non-renewable fuels formed by biological processes and transformed into solid or fluid minerals over geologic time. Fossil fuels include coal, petroleum, and natural gas.

FOSSIL RECORD: The time-ordered mass of fossils that is found in the sedimentary rocks of Earth. The fossil record is one of the primary sources of knowledge about evolution and is also used to date rock layers.

FRAGMENTATION: Breakup of a continuous habitat into several smaller areas.

FREEDOM-OF-THE-SEAS DOCTRINE: An eighteenth-century agreement establishing the right of neutral shipping in international waters.

FRESHWATER: Water containing less than one gram per liter of dissolved solids.

G

GAIA HYPOTHESIS: The hypothesis that Earth's atmosphere, biosphere, and its living organisms behave as a single system striving to maintain a stability that is conductive to the existence of life.

GAMMA RAYS: Streams of high-energy electromagnetic radiation given off by an atomic nucleus undergoing radioactive decay.

GENE PLUNDER: Exploiting genetic diversity without compensating the country of origin.

GENETIC: Having to do with the genetic material, or DNA, in an organism.

GENETICALLY MODIFIED (GM): Organism containing a gene transferred from another organism.

GENOME: The total content of genetic material in organisms.

GEOCODING: In geographic information systems, the assignment of geological data to particular features on maps or to other data records, such as photographs.

GEOGRAPHIC INFORMATION SYSTEMS (GIS): A set of computer-based tools that collects, analyzes, and maps spatial data.

GEOLOGICAL TIME: The period of time extending from the formation of the Earth to the present.

GEOLOGIST: A person who studies the origin, history, and structure of Earth.

GEO-REFERENCING: The process used for referring information to a geographic region.

GEOSPATIAL INFORMATION: The combination of a huge range of information taken from various sources and referenced by a geographic region.

GIGAWATT: A unit of power that is equal to one billion watts.

GLACIAL: Pertaining to glaciers or ice sheets.

GLACIER: A large mass of ice slowly moving over a land mass, resulting from a multi-year surplus accumulation of snowfall in excess of snowmelt.

GLOBAL POSITIONING SYSTEM (GPS): A system consisting of 25 satellites used to provide highly precise position, velocity, and time information to users anywhere on Earth or in its neighborhood at any time.

GLOBAL SOUTH: Academic term referring to underdeveloped countries.

GLOBAL WARMING: Warming of Earth's atmosphere that results from an increase in the concentration of gases that store heat, such as carbon dioxide (CO_2).

GLOBALIZATION: The integration of national and local systems into a global economy through increased trade, manufacturing, communications, and migration.

GRAFTING: Uniting a shoot or bud with a growing plant.

GRAVEL: The most coarse particles in soil.

GRAVITY: An attractive force that exists between all mass in the universe such as the moon and Earth.

GRAZING CAPACITY: The number of animals that a given area of land can support.

GREEN: Environmentally friendly or safe; non-polluting or not adding to global warming.

GREEN CHEMISTRY: An approach to chemical manufacturing that reduces its negative impact on the environment.

GREEN MOVEMENT: A social ideology focused on environmental and quality-of-life issues, promoting values such as global responsibility, community-based economics, and sustainability.

GREEN PARTIES: Values-oriented political parties based on the environmental and social principles of the green movement.

GREENHOUSE EFFECT: The warming of Earth's atmosphere due to water vapor, carbon dioxide (CO_2), and other gases in the atmosphere that trap heat radiated from Earth's surface.

GREENHOUSE-GAS EMISSIONS: Releases of greenhouses gases into the atmosphere.

GREENHOUSE GASES: Gases that accumulate in the atmosphere absorb infrared radiation, contributing to the greenhouse effect.

GROUNDWATER: Fresh water that is present in an underground location.

GULF STREAM: A warm, swift ocean current that flows along the coast of the Eastern United States and extends northward toward Europe.

GYRE: A zone of spirally circulating oceanic water that tends to retain floating materials, as in the Sargasso Sea of the Atlantic Ocean.

H

HABITABILITY: The degree to which a given environment can be lived in by human beings. Highly habitable environments can support higher population densities, that is, more people per square mile or kilometer.

HABITAT: The location and accompanying conditions where a plant or animal lives.

HALF LIFE: The amount of time it takes for half an initial amount to disintegrate.

HEAT ISLAND: An urban area with significantly higher air and surface temperatures than surrounding areas. Occurs because pavement and buildings absorb solar energy while being little cooled by evaporation compared to vegetation-covered ground.

HERBICIDE: A chemical substance used to destroy or inhibit plant growth.

HERBICIDE SELECTIVITY: The ability to kill weeds while leaving a crop unharmed.

HERBIVORE: A plant-eating organism.

HOLDING POND: A reservoir used to hold polluted or sediment-laden water until it can be treated or recycled.

HOT SPRING: A spring produced by superheated water emerging from Earth's crust.

HOTSPOTS: Locations that are at high risk from natural hazards.

HUNTER-GATHERERS: Human groups that subsist on game and gathered vegetation.

HURRICANE KATRINA: A Category 3 hurricane that caused over 1,800 deaths and catastrophic damage to the Gulf Coast region of the United States in August, 2005.

HYDROCARBONS: Molecules composed solely of hydrogen and carbon atoms.

HYDROELECTRICITY: Electricity generated by causing water to flow downhill through turbines, which are fanlike devices that turn when fluid flows through them. The rotary mechanical motion of each turbine is used to turn an electrical generator.

HYDROELECTRICITY: Electricity generated by the conversion of energy from running water.

HYDROFLUOROCARBONS (HFCs): Gaseous compounds consisting of hydrogen, fluorine, and carbon that have no ozone depletion potential; a suggested replacement for chlorofluorocarbons (CFCs).

HYDROGEN: The simplest and most abundant element in the universe, which is being investigated as a fuel source.

HYDROLOGIC CYCLE: The overall process of evaporation, vertical and horizontal transport of vapor, condensation, precipitation, and the flow of water from continents to oceans.

HYDROLOGY: The study of the distribution, movement, and physical-chemical properties of water in Earth's atmosphere, surface, and near-surface crust.

HYDROPHOBIC: Compounds do not dissolve easily in water, and are usually non-polar. Oils and other long hydrocarbons are hydrophobic.

HYDROSPHERE: The total amount of liquid, solid, and gaseous water present on Earth.

HYDROTHERMAL VENTS: Underground jets of mineral-rich hot water.

HYPOXIA: A condition in which cells of the body are deprived of oxygen.

I

ICE AGE: Period of glacial advance.

ICE CORE: A cylindrical section of ice removed from a glacier or ice sheet in order to study climate patterns of the past.

ICE CRYSTALS: An arrangement of water molecules in which motion among the molecules slows and the structure takes on a rigid shape, as a consequence of temperatures near freezing. Crystals often form around particulate matter (dust, pollutants, etc.).

ICE JAM: An stationary accumulation of river ice that restricts water flow and increases the risk of flooding.

ICE SHEET: Glacial ice that covers at least 19,500 square miles (50,000 square km) of land and that flows in all directions, covering and obscuring the landscape below it.

ICE SHELF: Section of an ice sheet that extends into the sea a considerable distance and that may be partially afloat.

IMMUNOSUPPRESSANT: Something used to reduce the immune system's ability to function, like certain drugs or radiation.

IMPERVIOUS SURFACES: Land surfaces created by human construction that do not allow rainfall to penetrate into the soil and down to aquifers.

***IN SITU* REMEDIATION:** A procedure in which decontamination of a polluted area is performed on-site by using or simulating natural processes.

INCINERATION: The burning of solid waste as a disposal method.

INDIGENOUS PEOPLES: Human populations that migrated to their traditional area of residence sometime in the relatively distant past, e.g., before the period of global colonization that began in the late 1400s.

INDIGENOUS SPECIES: A species that is native to its region, but may occur in other regions as well.

INDUSTRIAL REVOLUTION: The period, beginning about the middle of the eighteenth century, during which humans began to use steam engines as a major source of power.

INERTIA: The tendency of an object to continue in its state of motion.

INFAUNA: Animals living within the material, such as mud or sand, at the bottom of an ocean or sea, or on the beach.

INFRARED RADIATION: Electromagnetic radiation of a wavelength shorter than radio waves but longer than visible light that takes the form of heat.

INNOCENT PASSAGE: The right of all ships to pass through the territorial waters of another state subject to certain restrictions.

INSECTICIDE: A chemical substance used to kill insects.

INSOLATION: Solar radiation received at Earth's surface.

INTEGRATED PEST MANAGEMENT: An ecological pest-control strategy involving minimal application of pesticides.

INTERGLACIAL: Occurring between periods of glacial action.

INTERGLACIAL PERIOD: A geological time period between glacial periods, which are periods when ice masses grow in the polar regions and at high elevations.

INTERGOVERNMENTAL PANEL ON CLIMATE CHANGE (IPCC): The Intergovernmental Panel on Climate Change (IPCC) was established by the World Meteorological Organisation (WMO) and the United Nations Environment Programme (UNEP) in 1988 to assess the science, technology, and socioeconomic information needed to understand the risk of human-induced climate change.

INTERNAL COMBUSTION ENGINE: An engine that relies on the chemical energy released during the combustion of a fuel to create power.

INTERNATIONAL UNION OF GEOLOGICAL SCIENCES (IUGS): Nongovernmental group of geologists founded in 1961 and headquartered in Trondheim, Norway. The group fosters international cooperation on geological research of a transnational or global nature.

INTERNATIONAL WHALING COMMISSION: An international body established to provide for the proper conservation of whale stocks and make possible the orderly development of the whaling industry.

INTERSPECIFIC COMPETITION: The competition between individuals of different species for the same limited resource.

INTERTROPICAL: Pertaining to a narrow belt along the equator where convergence of air masses of the northern and southern hemispheres produces a low-pressure atmospheric condition.

INTRASPECIFIC COMPETITION: The competition among members of the same species for the same limited resource.

INVASIVE SPECIES: A non-native species whose introduction causes, or is likely to cause, economic or environmental harm or harm to human health.

INVERSION: A type of chromosomal defect in which a broken segment of a chromosome attaches to the same chromosome, but in reverse position.

IONIZING RADIATION: Any electromagnetic or particulate radiation capable of direct or indirect ion production in its passage through matter. In general use: Radiation that can cause tissue damage or death.

IRON FERTILIZATION: The process of seeding the ocean with iron to stimulate the growth of microorganisms, in an effort to trap carbon and lessen the release of carbon dioxide to the atmosphere.

ISOBAR: A line on a map connecting points at which the barometric pressure is the same at a specified moment.

ISOTOPE: A form of a chemical element distinguished by the number of neutrons in its nucleus.

J

JET STREAM: Currents of high-speed air in the atmosphere. Jet streams form along the boundaries of global air masses where there is a significant difference in atmospheric temperature.

K

KEELING CURVE: Plot of data showing the steady rise of atmospheric carbon dioxide from 1958 to the present, overlaid with annual sawtooth variations due to the growth of northern hemisphere plants in summer.

KEPONE: A carcinogenic pesticide, banned in the United States in 1975, that caused a 1970s environmental disaster in Hopewell, Virginia.

KEYSTONE SPECIES: A species whose impact on its environment has a disproportionately large effect relative to its abundance.

KRILL: Small marine crustaceans of the order Euphausiacea, which are consumed as food by certain whales.

KYOTO PROTOCOL: Extension in 1997 of the 1992 United Nations Framework Convention on Climate Change (UNFCCC), an international treaty signed by almost all member countries with the goal of mitigating climate change.

L

LA NIÑA: A cycle of cooling in the equatorial Pacific that causes a disruption of global weather patterns; the opposite of El Niño.

LAND DEGRADATION: Gradual land impoverishment caused primarily by human activities such as agriculture.

LANDFILL: A low-lying area in which solid refuse is buried between layers of dirt.

LANDFILL SITE: Solid waste disposal site consisting of a lined, covered hole in the ground.

LANDMINE: A bomb planted on or near the surface of the ground that is triggered by something passing over it.

LEGISLATION: The making or enacting of laws.

LENTIC: The vertically layered nature of a lake.

LEVEE: A raised embankment designed to prevent a river from overflowing.

LIGHT POLLUTION: Also known as photopollution and luminous pollution, refers to the presence of excessive amounts of light in the atmosphere.

LIQUEFACTION: The process of changing the state of something to a liquid.

LITTLE ICE AGE: A cold period lasting from approximately 1550-1850 in Europe, North America, and Asia, marked by rapid expansion of mountain glaciers.

LITTORAL: The region of a lake near the shore.

LOTIC: Flowing water, as in rivers and streams

M

MAGMA: Molten rock formed in the interior of Earth.

MALARIA: A group of parasitic diseases common in tropical and subtropical areas, characterized by attacks of chills, fever, and sweating.

MANTLE: The thick, dense layer of rock that underlies Earth's crust and overlies the core.

MAP SCALE: Portrays the relationship between a distance/size on map and the corresponding distance/size on the land.

MARIANAS TRENCH: A canyon almost 36,000 feet (11,000 m) in depth located in the floor of the Pacific Ocean; it is the deepest part of the world's oceans.

MARINE: Living in, or associated with, the sea.

MASS BALANCE (GLACIAL): The difference between the accumulation and ablation (reduction) of ice mass over a period of time.

MASS EXTINCTION: An extinction event characterized by high levels (or rates) of species extinction in a geologically short period of time.

MESOSPHERE: The third layer of the atmosphere, extending from the stratosphere to about 50 miles (80 km) above Earth.

METHANE: An odorless, colorless, flammable gas, with the formula CH_4, that is the simplest hydrocarbon compound and the major constituent of natural gas.

METHYL ISOCYANATE: A toxic chemical used in the manufacture of pesticides, which caused a disastrous 1984 chemical leak in Bhopal, India.

MICROCOSM: A miniature representation of a system that is used to model the system and study interactions.

MICROFINANCING: An economic development strategy in which small loans (microcredit) and other financial services are provided to very low-income individuals.

MIGRATORY: Traveling from one place to another at regular times of year, often over long distances.

MILANKOVITCH CYCLES: Regularly-repeating variations in Earth's climate caused by shifts in its orbit around the sun and its orientation (i.e., tilt) with respect to the sun.

MISDEMEANOR: A criminal offense that is considered minor and is punishable by a much lesser sentence than a felony.

MODEM: A device that permits information from a computer to be transmitted over a telephone line or cable.

MODIFIED MERCALLI SCALE: A scale used to compare earthquakes based on the effects they cause.

MOLLISOL: Rich soil formed under grasslands.

MONOCULTURE: A single species.

MONSOON: An annual shift in the direction of the prevailing wind that brings on a rainy season and affects large parts of Asia and Africa.

MUDFLAT: An area of low-lying muddy land that is covered at high tide and exposed at low tide.

MUNICIPAL WASTE: Waste that comes from households or is similar to household waste.

N

NATO: The North Atlantic Treaty Organization is a military alliance comprising the United States and 25 other members states, along with 14 major allies. It was involved in the bombing of Serbia in 1999 and, most recently, in Afghanistan.

NATURAL SELECTION: Also known as survival of the fittest; the natural process by which those organisms best adapted to their environment survive and pass their traits to offspring.

NATURALIST: One who studies or is an expert in natural history, especially in zoology or botany.

NEUROTOXIN: A poison that interferes with nerve function, usually by affecting the flow of ions through the cell membrane.

NICHE: A term describing a species' habitat, range of physical and biological conditions, and relationships within its ecosystem.

NITROGEN CYCLE: Biochemical cycling of nitrogen by plants, animals, and soil bacteria.

NITROGEN FIXATION: Conversion of atmospheric nitrogen into nitrate by the roots of leguminous plants.

NITROGEN OXIDES: Compounds of nitrogen (N) and oxygen (O) such as those that collectively form from burning fossil fuels in vehicles.

NON-GOVERNMENTAL ORGANIZATION (NGO): A voluntary organization that is not part of any government; often organized to address a specific issue or perform a humanitarian function.

NONPOINT-SOURCE POLLUTANTS: Pollutants that come from a wide range of sources.

NON-RENEWABLE RESOURCE: A natural resource of finite supply that cannot be regenerated.

NUCLEAR WASTE: Material left over from nuclear processes that is often radioactive or contaminated by radioactive elements.

NUCLEAR WEAPON: A military device whose explosive power is derived from nuclear fission or fusion.

NUCLIDE: A type of atom having a specific number of protons and neutrons in its nucleus.

O

OCEAN HEAT TRANSPORT: Movement by ocean currents of warm water from the tropics toward the poles, effectively transporting heat energy toward the poles where it is more quickly radiated into space.

OIL: Liquid petroleum.

OIL SLICK: A layer of oil floating on the surface of water.

OPPORTUNITY COST: The total cost incurred by choosing one option over another.

ORE: Rock containing a significant amount of minerals.

ORGANIC FARMING: Farming that uses no artificial chemicals or genetically engineered plants or animals.

ORGANIC WATER POLLUTANT: Organic materials from farm or food waste that increase the nutrient content of water.

OVERFISHING: Overharvesting applied to fish.

OVERGRAZING: The grazing of land over extended periods of time or without sufficient recovery periods, to the detriment of vegetation and soil quality.

OVERHARVESTING: Harvesting so much of a resource that its economic value declines and/or its existence is threatened.

OZONE: An almost colorless, gaseous form of oxygen, with an odor similar to weak chlorine, that is produced when an electric spark or ultraviolet light is passed through air or oxygen.

OZONE HOLE: A term invented to describe a region of very low ozone concentration above the Antarctic that appears and disappears with each austral (Southern Hemisphere) summer.

OZONE LAYER: The layer of ozone that begins approximately 9 miles (15 km) above Earth and thins to an almost negligible amount at about 31 miles (50 km), and which shields Earth from harmful ultraviolet radiation from the sun. The highest natural concentration of ozone (approximately 10 parts per million by volume) occurs in the stratosphere at approximately 16 miles (25 km) above Earth. The stratospheric ozone concentration changes throughout the year as stratospheric circulation changes with the seasons. Natural events such as volcano eruptions and solar flares can produce changes in ozone concentration, but man-made changes are of the greatest concern.

P

PACK ICE: Floating sea ice that has been driven together into a single large mass.

PALEOCLIMATE: The climate of a given period of time in the geologic past.

PALEOCLIMATOLOGY: The study of past climates throughout geological history, and the causes of variations among those climates.

PALEONTOLOGY: The study of life in past geologic time.

PARASITE: An organism that lives on or in another organism, and which harms the host.

PASTURE: Low-growing plants suitable for grazing livestock, or land containing such plants.

PEAT: Partially carbonized vegetable matter that can be cut and dried for use as fuel.

PEER REVIEW: The standard process in science for reducing the chances that faulty or fraudulent claims will be published in scientific journals. Before publication of an article, scientists with expertise in the article's subject area review the manuscript, usually anonymously, and make criticisms that may lead to revision or rejection of the article.

PELAGIC: Living in, or associated with, open areas of ocean away from the bottom.

PERMAFROST: Perennially frozen ground that occurs wherever the temperature remains below 32°F (0°C) for several years.

PERSONAL WATERCRAFT: Small boats, steered by handlebars and propelled by a jet of water. Often known under the trade name "Jet Ski."

PESTICIDE: Substances used to reduce the abundance of pests or any living thing that causes injury or disease to crops.

PETROLEUM: A deposit formed from the action of high pressure and temperature on the buried remains of organisms from millions of years ago.

pH: The measure of the amount of dissolved hydrogen ions in solution.

PHASE: In the study of waves, a repeating or periodic wave's phase is the relationship of its pattern of peaks and valleys to a fixed reference (such as time).

PHOCOMELIA: A birth defect in which the upper portion of a limb is absent or poorly developed, so that the hand or foot attaches to the body by a short, flipperlike stump.

PHOTOCHEMICAL SMOG: A type of smog created by the action of sunlight on pollutants.

PHOTOSYNTHESIS: The process by which plants fix carbon dioxide (CO_2) from the atmosphere using the energy of sunlight.

PHYLUM (PLURAL, PHYLA): A biological classification group ranked between kingdom and class.

PHYTOPLANKTON: Microscopic marine organisms (mostly algae and diatoms) that are responsible for most of the photosynthetic activity in the oceans.

PLANKTON: Floating animal and plant life.

PLEISTOCENE EPOCH: The geologic period characterized by ice ages in the Northern Hemisphere, from 1.8 million to 10,000 years ago.

POACHING: Illegal hunting.

POINT SOURCE POLLUTANT: A pollutant that enters the environment from a single point of entry.

POINT SOURCE POLLUTION: Pollution arising from a fixed source, such as a pipe.

POLAR CELLS: Part of a group of atmospheric air circulation patterns occurring at the Earth's poles.

POLLUTION: Physical, chemical, or biological changes that adversely affect the environment.

POLLUTION CREDIT: A credit allowing the holder to legally emit a certain amount of pollutants, and which can be bought and sold as part of a cap-and-trade emissions program.

POPULATION MODEL: A mathematical model that can project the demographic consequences to a species as a result of certain changes to its environment.

PRECESSION: The comparatively slow torquing of the orbital planes of all satellites with respect to Earth's axis, due to the bulge of Earth at the equator which distorts Earth's gravitational field. Precession is manifest by the slow rotation of the line of nodes of the orbit (westward for inclinations less than 90 degrees and eastward for inclinations greater than 90 degrees).

PRECIPITATION: Moisture that falls from clouds as a result of condensation in the atmosphere.

PREDATOR-PREY MODEL: A mathematical model used to analyze interaction between two species in an ecosystem.

PRIMARY ENERGY SUPPLY: The total amount of available energy embodied in natural resources (such as coal) that has not been subjected to any conversion or transformation processes.

PRIMARY POLLUTANT: Any pollutant released directly from a source to the atmosphere.

R

RADIATIVE FORCING: A change in the balance between incoming solar radiation and outgoing infrared radiation, resulting in warming or cooling of Earth's surface.

RADIOACTIVE: Containing an element that decays, emitting radiation.

RADIOACTIVITY: The property possessed by some elements of spontaneously emitting energy in the form of particles or waves by disintegration of their atomic nuclei.

RADIOLARIA: Single-celled animals with silica skeletons.

RADIOMETRIC AGE: The age of an object as determined by the levels of certain radioactive substances present in that object.

RADIOMETRIC DATING: Use of naturally occurring radioactive elements and their decay products to determine the absolute age of the rocks containing those elements.

RADIOSONDES: Set of instruments carried into the atmosphere by weather balloons to measure temperature, humidity, and air pressure at various altitudes.

RECHARGE: Replenishment of an aquifer by the movement of water from the surface into the underground reservoir.

RECLAMATION: The act of restoring to use.

REFORESTATION: The replanting of a forest that had been cleared by fire or harvesting.

REFUSE-DERIVED FUEL: Solid waste from which unburnable materials have been removed.

RELATIVE HUMIDITY: The amount of water vapor in the air compared to the maximum amount it could hold at that temperature.

RELATIVE SEA LEVEL: Sea level compared to land level in a given locality. Relative sea level may change because land is locally sinking or rising, not because the sea itself is sinking or rising (eustatic sea level change).

REMEDIATION: A remedy. In the case of the environment, remediation seeks to restore an area to its unpolluted condition, or at least to remove the contaminants from the soil and/or water.

REMOTE SENSING: The acquisition of data about an object or area without coming into contact with it.

RENEWABLE ENERGY: Energy that can be naturally replenished. In contrast, fossil fuel energy is nonrenewable.

RENEWABLE ENERGY SOURCE: An energy resource that is naturally replenished, such as sunlight, wind, or geothermal heat.

RENEWABLE RESOURCE: Any resource that is renewed or replaced fairly rapidly (on human historical time-scales) by natural or managed processes.

RESERVOIR: The collection of naturally occurring fluids in the porosity of a subsurface rock formation.

RESIDENCE TIME: For a greenhouse gas, the average amount of time a given amount of the gas stays in the atmosphere before being absorbed or chemically altered.

RESOLUTION: Pixels per square inch on a computer-generated display or photograph.

RESPIRATION: The process by which animals use up stored foods (by combustion with oxygen) to produce energy.

RETURN INTERVAL: The average time between occurrences of disturbances in a given stand of trees.

REWILDING: The restoration of a plant or animal to its historic habitat.

RICHTER SCALE: A scale used to compare earthquakes based on the energy released by the earthquake.

RUNOFF: Water that falls as precipitation and then runs over the surface of the land rather than sinking into the ground.

RUN-UP HEIGHT: The vertical distance between the mean-sea-level surface and the maximum point attained on the coast.

S

SABOTAGE: The deliberate act of interference, disruption, or destruction of an opponent's operations as part of a dispute.

SALINITY: Measurement of the amount of sodium chloride (NaCl) in a given volume of water.

SALINIZATION: An increase in salt content. The term is often applied to increased salt content of soils due to irrigation; salts in irrigation water tend to concentrate in surface soils as the water quickly evaporates rather than sinking down into the ground.

SALT MARSH: Wetland which is sometimes flooded with seawater.

SALTATION: The intermittent, leaping movement of small particles due to the force of wind or running water.

SATURATION POINT: The maximum concentration of water vapor that the air can hold at a given temperature.

SAVANNA: A flat, open grassland ecosystem found in tropical or subtropical regions.

SAXITOXIN: A neurotoxin found in a variety of dinoflagellates. If ingested, it may cause respiratory failure and cardiac arrest.

SCORCHED EARTH: A military policy involving widespread destruction of property and resources, especially by burning, so that an advancing enemy cannot use them.

SEA LICE: A type of crustacean that is a parasite of farmed fish including salmon and rainbow trout.

SEA SEDIMENT CORE: A cylindrical, solid sample of a layered deposit of sediment on the ocean floor that can provide information about past climate changes.

SEA SHEPHERD CONSERVATION SOCIETY: An international, non-profit marine wildlife conservation group known for its radical direct-action tactics.

SEDIMENT: Solid unconsolidated rock and mineral fragments that come from the weathering of rocks and are transported by water, air, or ice and form layers on Earth's surface. Sediments can also result from chemical precipitation or secretion by organisms.

SEDIMENT LOADING: Presence of moving mineral particles in rivers or streams. Faster-moving water can carry more sediment. Erosion increases sediment loading, limited by stream capacity for increased sediment loading.

SEDIMENTARY ROCK: Rock formed from compressed and solidified layers of organic or inorganic matter.

SEISMIC WAVE: A wave of energy that travels through Earth as the result of an earthquake or explosion.

SELECTION PRESSURE: Factors that influence the evolution of an organism. An example is the overuse of antibiotics, which provides a selection pressure for the development of antibiotic resistance in bacteria.

SEMIARID: Receiving very little annual rainfall, roughly 10-20 inches (25-50 cm), and characterized by short grasses and shrubs.

SENSOR: A device to detect or measure a parameter as it is occurring.

SESSILE: Any animal that is rooted to one place. Barnacles, for example, have a mobile larval stage of life and a sessile adult stage of life.

SEWAGE: Waste and wastewater discharged from domestic and commercial premises.

SHALLOW-WATER WAVES: Waves that have a greater wavelength and in which the ratio between the depth of water and the wavelength is very small.

SHARPS WOUNDS: Wounds caused by discarded medical instruments including hypodermic needles and, scalpels (knives) and other sharp objects potentially contaminated by human contact.

SHOALING EFFECT: Transformation that takes place when a wave travels from deep water to shallow water resulting in the decrease in wavelength and increase in the height of the wave.

SIERRA CLUB: An environmental organization founded in 1892 by American naturalist John Muir (1838-1914), whose mandate includes the protection and, when needed, restoration, of natural environments.

***SILENT SPRING*:** A seminal 1962 book, written by Rachel Carson (1907–1964), that is credited with inspiring widespread public interest in pollution and the environment.

SILVICULTURE: Management of the development, composition, and long-term health of a forest ecosystem. The objective is often to allow logging of the forest over many years.

SLUDGE: Semisolid material formed as a result of wastewater treatment or industrial processes.

SMOG: A mixture of smoke or other atmospheric pollutants combined with fog.

SNOW PACK: An area of naturally formed, packed winter snow that accumulates in mountain or upland regions.

SOFTWARE: Computer programs or data.

SOIL: Unconsolidated materials above bedrock.

SOOT: A black, powdery, carbonaceous substance produced by the incomplete combustion of coal, oil, wood, or other fuels.

SPECIATION: The evolutionary development of new biological species.

SPECIES: A biological classification group ranked below genus, consisting of related organisms capable of interbreeding.

SPECIES DIVERSITY: The number of different species living in a particular place.

SPECTRAL: Relating to a spectrum, which is an ordered range of possible vibrational frequencies for a type of wave. The spectrum of visible light, for example, orders colors from red (slowest vibrations visible) to violet (fastest vibrations visible) and is itself a small segment of the much larger electromagnetic spectrum.

SPODOSOL: Acidic soil formed under pine forests.

SPRAWL: The unregulated and unplanned spread of urban and suburban development in a metropolitan region.

SPRING: The emergence of an aquifer at the surface, which produces a flow of water.

STAKEHOLDERS: A group of people holding investments, shares, or interest in a commercial enterprise.

STEPPE: Grasslands that occur in places with cold winters and warm summers.

STORM SURGE: Rise of the sea at a coastline due to the effect of storm winds.

STORM WATER: Water that flows over the ground when it rains or when snow melts.

STRATA: A bed or layer of sedimentary rock in which composition is usually the same throughout.

STRATIGRAPHY: The branch of geology that deals with the layers of rocks or the objects embedded within those layers.

STRATOSPHERE: The region of Earth's atmosphere ranging between about 9-30 miles (15-50 km) above Earth's surface.

STRATOSPHERIC OZONE LAYER: Layer of Earth's atmosphere from about 9-22 miles (15-35 km) above the surface in which the compound ozone (O_3) is relatively abundant.

STRICT LIABILITY: Liability that is imposed without a finding of fault, such as negligence or intent.

STRIP MINING: A method for removing coal from seams located near Earth's surface.

SUBMARINE SLIDES: Marine landslides that can transport subsurface rock and sediment down the continental slope.

SUBMERGED SHORELINE: A shoreline formed by the submergence of a landmass, characterized by bays, promontories, and other minor features.

SUBSIDENCE: A sinking or lowering of a part of Earth's surface.

SUBSISTENCE: Having secured enough provisions to cover basic needs.

SUBSISTENCE FARMER: A farmer whose products are intended to provide for his/her own basic needs, with little or no profit.

SUCCESSION: Regrowth of forest following its disturbance by deforestation.

SUPERFUND: Legislation that authorizes funds to clean up abandoned, contaminated sites.

SUPERFUND SITE: A Superfund site is a location contaminated by hazardous waste that has been designated by the U.S. Environmental Protection Agency (EPA) for management and cleanup.

SURFACE WATER: Water collecting on the ground or in a stream, river, lake, wetland, or ocean, as opposed to groundwater.

SUSTAINABILITY: Practices that preserve the balance between human needs and the environment, as well as between current and future human requirements.

SUSTAINABLE: Capable of being sustained or continued for an indefinite period without exhausting necessary resources or otherwise self-destructing: often applied to human activities such as farming, energy generation, or the maintenance of a society as a whole.

SUSTAINABLE AGRICULTURE: Agricultural use that meets the needs and aspirations of the present generation without compromising those of future ones.

SUSTAINABLE DEVELOPMENT: Development (i.e., increased or intensified economic activity; sometimes used as a synonym for industrialization) that meets the cultural and physical needs of the present generation of persons without damaging the ability of future generations to meet their own needs.

SUSTAINABLE RESOURCE: A resource that can be renewed or maintained indefinitely.

SUSTAINABLE WHALING: The process by which a limited number of whales from thriving species, such as the minke, are culled each year.

SYMBIOSIS: A pattern in which two or more organisms of different species live in close connection with one another, often to the benefit of both or all organisms.

SYMBIOTIC: Describes a relationship in which two or more organisms of different species live in close connection with one another, often to the benefit of both or all organisms.

T

TAIGA: The part of the boreal forest occurring in high northern latitudes, consisting of open woodland of coniferous trees growing in a rich floor of lichen.

TECTONIC PLATES: Thick crustal blocks that move relative to one another on the outer surface of Earth, causing continents to shift.

TEMPERATE: Pertaining to a region located between 30 and 60 degrees latitude in both hemispheres in which the climate undergoes seasonal change in temperature and moisture.

TERATOGENIC EFFECT: The combined consequences of the consumption of a harmful substance on a developing fetus.

TERRITORIAL WATER: A maritime zone of coastal waters extending up to 12 nautical miles (22 km) from the baseline of a coastal state, within which a state can exert control of its laws and regulations.

THERMAL PLUME: A discharge of heated waste water into the water supply.

THERMAL POLLUTION: Industrial discharge of heated water into a river or lake, creating a rise in temperature that is injurious to aquatic life.

THERMOHALINE CIRCULATION: Large-scale circulation of the world ocean that exchanges warm, low-density surface waters with cooler, higher-density deep waters; also termed meridional overturning circulation.

THERMOSTERIC EXPANSION: Expansion as a response to a change in temperature, also called thermal expansion.

THREATENED SPECIES: A species that is likely to become an endangered species over all or much of its range.

THROUGHFLOW: The horizontal flow of water through soil.

TIDE: The changing surface level of a large body of water such as the ocean or a lake that results from changes in the gravitational pull on the water by the moon and the sun.

TIDE CYCLE: A period that includes a complete set of tide conditions or characteristics, such as a tidal day or a lunar month.

TIDE GAUGE: A device, usually stationed along a coast, that measures sea level continuously. Measurements from tide gauges were the main source of sea-level data prior to the beginning of satellite measurements in the 1970s.

TOPOGRAPHY: The surface features of an area, such as hills or valleys.

TOXIN: A poisonous substance produced by living cells or organisms.

TRADE WINDS: Surface air from the horse latitudes that moves back toward the equator and is deflected by the Coriolis force, causing the winds to blow from the northeast in the Northern Hemisphere and from the southeast in the Southern Hemisphere.

TRANSCENDENTALISM: A literary and philosophical movement asserting the existence of an ideal spiritual reality that transcends the empirical and scientific and is knowable through intuition.

TRANSIT PASSAGE: The right of free navigation and overflight for the purpose of continuous transit through a strait between one part of the high seas and another.

TRANSPIRATION: Loss of water taken in by roots from leaves through evaporation.

TREE RINGS: Marks left in the trunks of woody plants by the annual growth of a new coat or sheath of material. Tree rings provide a straightforward way of dating organic material stored in a tree trunk.

TROPICAL: The area between 23.5 degrees north and south of the equator. This region has small daily and seasonal changes in temperature, but great seasonal changes in precipitation.

TROPICAL DEPRESSION: A rotating system of thunderstorms with low atmospheric pressure in the center and maximum windspeed less than 39 mph (63 km/h).

TROPICAL STORM: A low-pressure storm system that forms over warm tropical waters, with winds ranging from 30 to 75 mph (48 to 121 km/h).

TROPOSPHERE: The lowest layer of Earth's atmosphere, ranging to an altitude of about 9 miles (15 km) above Earth's surface.

TSUNAMI: A series of ocean waves resulting from an undersea disturbance such as an earthquake.

TUNDRA: A type of ecosystem dominated by lichens, mosses, grasses, and woody plants. It is found at high latitudes (arctic tundra) and high altitudes (alpine tundra). Arctic tundra is underlain by permafrost and is usually very wet.

TURBIDITY: A measure of the degree to which water loses its clarity due to the presence of suspended particles.

TURBINE: An engine that moves in a circular motion when force, such as moving water, is applied to its series of baffles (thin plates or screens) radiating from a central shaft.

U

UPWELLING: The vertical motion of water in the ocean by which subsurface water of lower temperature and greater density moves toward the surface of the ocean.

URBAN HEAT ISLAND EFFECT: Warming of atmosphere in and immediately around a built-up area. Occurs because pavement and buildings absorb solar energy while being little cooled by evaporation compared to vegetation-covered ground.

V

VAPORIZATION: The conversion of a solid or liquid into a gas.

VARVE: An annual layer or series of layers of sediment deposited in a body of still water.

VELD: An area of elevated open grassland, characteristic of parts of southern Africa.

VORTEX: A rotating column of a fluid such as air or water.

W

WASTE TRANSFER STATION: A structure in which wastes are temporarily held and sorted before being transferred to larger facilities.

WASTEWATER: Water that carries away the waste products of personal, municipal, and industrial operations.

WATER CONSUMPTION: Use of water in such a way that it cannot be re-used.

WATER FOOTPRINT: The total volume of water used to produce goods and services consumed by an individual, business or nation.

WATER INTENSITY: The amount of water used to produce specific goods or services.

WATER STRESS: Inability to provide enough water to meet basic needs.

WATER TABLE: The level of the aquifer under the ground.

WATER VAPOR: The most abundant greenhouse gas, it is the water present in the atmosphere in gaseous form. Water vapor is an important part of the natural greenhouse effect. While humans are not significantly increasing its concentration, it contributes to the enhanced greenhouse effect because the warming influence of greenhouse gases leads to a positive water vapor feedback. In addition to its role as a natural greenhouse gas, water vapor plays an important role in regulating the temperature of the planet because clouds form when excess water vapor in the atmosphere condenses to form ice and water droplets and precipitation.

WATER WITHDRAWAL: Removal of water from a water supply, some of which will later be returned.

WATERSHED: The expanse of terrain from which water flows into a wetland, water body, or stream.

WEATHERING: The natural processes by which the actions of atmospheric and other environmental agents, such as wind, rain, and temperature changes, result in the physical disintegration and chemical decomposition of rocks and earth materials in place, with little or no transport of the loosened or altered material.

WETLAND: A shallow ecosystem where the land is submerged for at least part of the year.

WETLANDS: Areas that are wet or covered with water for at least part of the year.

WHALE OIL: Oil rendered from the blubber of a whale.

WILDLAND FIRE USE: A fire-management technique that uses naturally ignited fires to benefit natural resources.

WIND: A natural motion of the air, especially a noticeable current of air moving in the atmosphere parallel to Earth's surface, caused by unequal heating and cooling of Earth and its atmosphere.

WIND CELLS: Vertical structures of moving air formed by warm (less-dense) air welling up in the center and cooler (more-dense) air sinking around the perimeter; also called convective or convection cells.

WIND SHEAR: Variation in wind speed within a current of wind: that is, when wind shear is present, part of a wind current is moving at one speed while another, nearby part of the current, moving parallel to the first, is moving at a different speed.

WOOD PULP: A suspension of wood tissue in water that is often used to make paper.

WORLD BANK: International bank formed in 1944 to aid in reconstruction of Europe after World War II (1939–1945), now officially devoted to the eradication of world poverty through funding of development projects.

X

XENOBIOTICS: Synthetic organic compounds that are hard to break down in the environment.

Z

ZOONOSIS (ZOONOTIC DISEASE): Any disease that can be transmitted from animals to humans.

ZOOPLANKTON: Small, herbivorous animal plankton that float or drift near the surface of aquatic systems and that feed on plant plankton (phytoplankton and nanoplankton).

ZOOXANTHELLAE: Algae that live in the tissues of coral polyps and, through photosynthesis, supply them with most of their food.

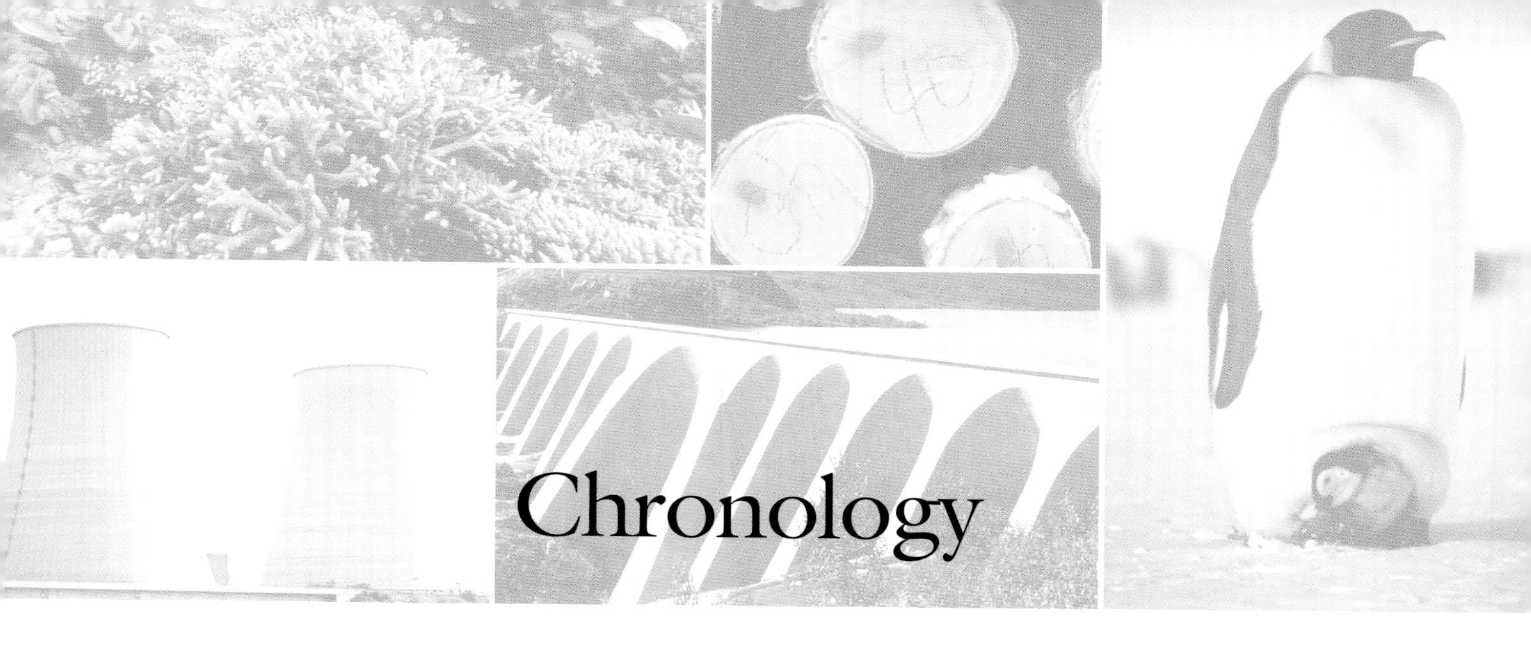

Chronology

A chronology of events related to environmental science studies in context with other scientific, environmental, and news milestones.

4.6 billion years BC to 100 BC

c. 4.6 billion years BC

Origin of Earth

c. 79000 BC

Animal fat is used to fuel lamps.

c. 50000 BC

Wall painting depictions are used for instruction in hunting and daily activities.

c. 18000 BC

Nile valley sees rise of agricultural techniques; cereal grains are specifically cultivated.

c. 10000 BC

Neolithic Revolution: transition from a hunting and gathering mode of food production to farming and animal husbandry, that is, the domestication of plants and animals.

c. 8000 BC

With the beginning of settled agriculture comes the first simple digging and harvesting tools.

c. 6000 BC

The rise of the great ancient civilizations, beginning in Mesopotamia, begats institutions and persons devoted to the collection and teaching of agricultural science and technology to support infrastructure.

c. 4000 BC

The Egyptian calendar based on 365 days is established. Because the Egyptian economy depends on the annual flooding of the Nile River, it is essential to know the onset of flooding season. When the priests see Sirius, the brightest star in the sky, rising in line with the sun in the morning, they know that this date usually coincides with the yearly flood.

c. 3000 BC

Assyrians develop sewer network to carry wastewater.

c. 3000 BC

The construction of the Egyptian pyramids at Giza begins. The fact that they are oriented to the cardinal points of the compass, and that their interiors have inclined passages which point to the pole star of the period, are strong evidence that the Egyptians possessed precise sighting or astronomical instruments.

c. 3000 BC

The Sumerians in the lower Euphrates valleys build dams, canals, statues, and pyramids that demonstrate a working knowledge of the lever and other basic tools.

c. 2900 BC

Stonehenge is begun on the Salisbury Plain in England. This early version consists of an earthen bank and a ditch along 56 holes.

c. 2600 BC

The Chinese make a primitive sundial and estimate the time of day by using a rod placed in the ground to project the sun's shadow.

c. 2500 BC
Apiculture, or beekeeping, begins in Egypt around this time. One of the oldest forms of animal husbandry, it involves the care and manipulation of colonies of honeybees (of the "Apis" species) so that they will produce and store a quantity of honey above their own requirements.

c. 1500 BC
Rig-Veda in the Indian Vedas, a collection of religious, philosophical, and educational writings, states that Earth is a globe.

c. 1370 BC
Large-scale manufacture of glass begins in Egypt.

c. 450 BC
Empedocles (c.492–c.432 BC), Greek philosopher, first offers his concept of the composition of matter, being made of four elements—earth, air, fire, and water. This notion is later adapted by Aristotle and becomes the basis of chemical theory for over two thousand years.

c. 450 BC
Herodotus (484-426 BC), Greek historian, states that the fertile soil of Egypt was taken from countries upstream and deposited there by a great river.

c. 450 BC
Leucippus (fl. 5th century BC), Greek philosopher, first states the rule of causality (that every event has a natural cause). He is also the teacher of Democritus (c.470–c.380 BC) and the supposed inventor of that philosopher's theory of atomism.

c. 450 BC
Anaxagoras of Athens (500–428 BC) teaches that the moon shines with the light of the sun.

c. 425 BC
Democritus (c.470–c.380 BC), Greek philosopher, states his atomic theory that all matter consists of infinitesimally tiny particles that are indivisible. These atoms are eternal and unchangeable, although they can differ in their properties. They can also recombine to form new patterns. His intuitive ideas contain much that is found in modern theories of the structure of matter.

c. 400 BC
Aristotle (384–322 BC), Greek philosopher, classifies animals into species and classes.

c. 350 BC
Aristotle offers observational proofs that Earth is not flat.

c. 350 BC
Aristotle states that Earth changes ceaselessly although gradually, and that erosion and silting cause major changes in its physical geography.

c. 100 BC
The wheel-of-pots driven by the river current appears in Egypt as a method of drawing water.

44 to 1600

c. 44 Pomponius Mela (fl. c.44), Roman geographer, divides Earth into five zones (North Frigid, North Temperate, Torrid, South Temperate, and South Frigid). This system is used even in modern times.

c. 600 Oil wells dug by hand are known to reach 600 feet (183 m) deep in Japan.

c. 900 Rhazes (c.860–c.930), Persian physician and alchemist, is the first to divide all substances into the grand classification of animal, vegetable, and mineral. He also subclassifies minerals into metals, volatile liquids (spirits), stones, salts, and others.

914 Al Masudi (c.871–957), Arab historian and geographer, records accurate principles on evaporation and the causes of ocean salinity. He is the first in the East to combine history and scientific geography in a major work. He is known as the "Herodotus of the Arabs."

c. 1100 In western North America, the Hopi tribe of Native Americans uses coal for both heating and cooking purposes.

c. 1175 Chu Hsi (c.1130–c.1200) of China writes his *Chu Tsi Shu Chieh Yao* in which he states that fossils were once living organisms.

1307 The burning of coal in London is so widespread that lime burners are forbidden from burning it so as not to add to the pollution.

1335 Richard Wallingford (c.1292-1336), English scholar, builds a mechanical clock that indicates low and high tide.

c. 1350 Giacomo Dondi (1298–1359), Italian physician and astronomer, writes his *Tractatus de Cause Salsedinis* in which he discusses the salinity of the oceans as well as the phenomenon of the tides. He suggests that the combined effects of the sun, moon, and planets have an effect on Earth's tides. He also is the first to recommend the extraction of salts from mineral waters for medicinal purposes.

c. 1370 Konrad von Megenburg, German scholar, writes his *Buch der Natur* in which he says that earthquakes are the result of an accumulation of winds in the Earth's interior forcing their way towards the surface.

1473 *Philobiblon* by medieval writer Richard de Bury (1287–1345) is first published. This book describes the science or law of the earth as opposed to "theology" which is the science of the divine.

1513 Juan Ponce de León (c.1460–1521), Spanish explorer, offers the first written description of the Gulf Stream in the North Atlantic.

1519 Leonardo da Vinci (1452–1519), Italian artist and scientist, writes his (unpublished) treatise on water in which he says that seawater rose to the mountain tops, formed rivers, and then returned to the oceans.

1564 Gabriel Fallopius (1523–1562), Italian anatomist, writes in his *De Medicatis Aquis atque de Fossilibus* that water from the oceans descends into the earth, is evaporated by fire, and then rises to the surface where it is condensed as spring water since its salt has been deposited below.

1584 Giordano Bruno (1548–1600), Italian philosopher, writes that there never was a great flood but only local ones. He also says that there have been many frequent alterations in Earth's topography, and that the oceans are deeper than the mountains are high.

1584 In Moscow, the Kamennyj Prizak is founded as an organization for identifying and exploiting natural resources.

1592 Galileo Galilei (1564–1642), Italian astronomer and physicist, argues that the physical laws of Cosmos are the same as those on Earth.

1600–1699

1600 William Gilbert (1544–1603), English physician and physicist, publishes his *De Magnete* in which he concludes that Earth's interior is largely iron and that Earth itself is one great magnet. In this major scientific work which is one of the earliest based on actual experimentation, he describes his experiments with his magnetized model of Earth called a "terrella."

1619 Johannes Kepler (1571–1630), German astronomer, publishes *Harmonices Mundi*, a mystical work of astronomical speculation in which he also says that there is an ongoing process in which earth decomposes matter to form other materials.

1625 Nathanael Carpenter (1589–1635), English mathematician, publishes his *Geography* in which he says mountains are continually being built up even as the weather wears others down.

1627 Francis Bacon (1561–1626), English philosopher, considers oceanographic issues such as tides, waves, currents, depth, and salinity in his posthumously-published *Sylva Sylvarum.*

1661 John Evelyn (1620–1706) of England publishes his attack on air pollution, *The Smoke of London.*

1669 Jan Swammerdam (1637–1680), Dutch naturalist, begins his pioneering work on the metamorphosis of insects and the anatomy of the mayfly (1675).

1674 Pierre Perrault (c.1611–1680), French hydrologist, publishes *De l'Origine des Fontaines* anonymously. He shows conclusively that precipitation is more than adequate to sustain the flow of rivers. His field work on the Seine River is instrumental in establishing the science of hydrology on a quantitative basis.

1680 The dodo bird becomes extinct because of excessive hunting by sailors in the Indian Ocean.

1691 Edmund Halley (1656–1742), English astronomer, writes an article in which he describes Earth's water cycle. The sun evaporates sea water which rises to mountain heights and is condensed into rain which falls and penetrates into Earth to emerge in springs and rivers.

1694 Joseph Pitton de Tournefort (1656–1708), French botanist, publishes his classic three-volume work *Elemens de Botanique* in Paris. It makes a major contribution to the emerging concept of genus in botany (being groups of all related species).

1700–1799

1735 Carolus Linnaeus (1707–1778), Swedish botanist, first publishes his *Systema Naturae*, a methodical and hierarchical classification of all living things. He develops the binomial nomenclature that will be used in classifying plants and animals. In this system, each type of living thing is given first a generic name (for the group to which it belongs), and then a specific name for itself.

1754 Joseph Black (1728–1799), Scottish chemist, proves by quantitative experiments the existence of carbon dioxide which he calls "fixed air." His proof is contained in his thesis done for his medical degree, and two years later he openly publishes the results.

1754 Pierre-Louis Moreau de Maupertuis (1698–1759), French mathematician, suggests that species transform over time.

1759 Bernard de Jussieu (1699–1777), French physician, is invited by French King Louis XV (1710–1774) to develop a botanical garden at the Petit Trianon garden at Versailles. Jussieu creates a living illustration of his natural classification system.

1765 Karl Wilhelm Scheele (1742–1786), Swedish chemist, discovers prussic acid. Known also as hydrogen cyanide, this highly volatile, colorless liquid is extremely poisonous and is eventually used in industrial chemical processing.

1768 Captain James Cook (1728–1779), English navigator, makes the first of his three famous voyages into the Pacific. The first of the really scientific navigators, he sails under the auspices of the Royal Society.

1772 Daniel Rutherford (1749–1819), Scottish chemist, discovers nitrogen. It is also independently discovered shortly thereafter by Priestley, Scheele, and Cavendish.

1774 Joseph Priestley (1733–1804), English chemist, discovers oxygen or what he calls "dephlogisticated air." It had been isolated sometime between 1770 and 1773 by the Swedish chemist Karl Scheele (1742–1786) who calls it "fire air," but Priestley publishes his results first. It is named by Antoine L. Lavoisier on September 5, 1779.

1776 Alessandro G.A.A. Volta (1745–1827), Italian physicist, discovers marsh gas (methane) which is inflammable in the reed-beds of Lake Maggiore.

1779 Horace-Bénédict de Saussure (1740–1799), Swiss physicist, publishes his *Voyages dans les Alpes*. Saussure is a pioneer in climbing mountains in order to investigate them scientifically, he writes four volumes of accurate observations on mountains and glaciers.

1779 Jan Ingenhousz (1730–1799), Dutch physician and plant physiologist, publishes his *Experiments upon Vegetables* in London and shows that light is necessary for plants to produce oxygen and that sunlight is the energy source for photosynthesis. He also says that carbon dioxide is taken in by plants in the daytime and given off at night.

1779 Antoine-Laurent Lavoisier (1743–1794), French chemist, first proposes the name "oxygen" for that part of the air that is breathable and which is responsible for combustion.

1780 Antoine Laurent Lavoisier (1743–1794), French chemist, and Pierre Simon Laplace (1749–1827), French astronomer and mathematician, collaborate to demonstrate that respiration is a form of combustion.

Breathing, like combustion, liberates heat, carbon dioxide, and water.

c. 1780 Last wolf on the British Isles is killed.

1785 James Hutton (1726–1797), Scottish geologist, reads his first paper, "Concerning the System of the Earth," to the Royal Society of Edinburgh. It is out of this brief paper that geology has its beginnings as a real science or as an organized field of study. Hutton offers a theory called "uniformitarianism" that contains one of the most fundamental principles of geology—that natural processes have been constantly at work shaping Earth over enormously long periods of time. The prime force behind this shaping is the internal heat of Earth. These ideas meet considerable resistance, but Hutton eventually prevails and comes to be known as the "father of geology."

1794 Erasmus Darwin (1731–1802), English physician and grandfather of Charles Darwin, publishes his *Zoonomia* in which he elaborates on Buffon's evolutionary ideas and stresses the role of the environment in changes in organisms.

1796 Jan Ingenhousz (1730–1799), Dutch physician and plant physiologist, concludes that plants utilize carbon dioxide. He finds that the activity of plants and animals complement each other, with plants taking in people's carbon dioxide and giving them oxygen in return.

1798 James Hall (1761–1832), Scottish geologist and chemist, begins a series of papers between this year and 1812 in which he provides experimental evidence for Hutton's geological theories. Because of his ability to back Hutton's views by facts obtained experimentally, he is considered the founder of experimental geology and of geochemistry.

1798 Benjamin Henry Latrobe (1764–1820), American architect and engineer, writes in his journal that rocks weathered into soil and that river erosion could cut out valleys.

1799 Alexander von Humboldt (1769– 1859), German naturalist, begins his scientific explorations in South America (1799–1804).

1799 John Dalton (1766–1844), English chemist, writes that rain and dew compensate for water that is carried away by rivers and evaporation. His first love is meteorology, and Dalton keeps careful daily records for fifty-seven years.

1800–1849

1802 John Dalton introduces modern atomic theory into the science of chemistry.

1815 William Smith (1769–1839) publishes first geological map of UK.

1830 Charles Lyell (1797–1875) publishes *Principles of Geology* and argues Earth is a least millions of years old.

1830 World population is approximately one billion.

1831 Charles Robert Darwin (1809–1882) begins his historic voyage on the H.M.S. *Beagle* (1831–1836). His observations during the voyage lead to his theory of evolution by means of natural selection.

1840 Louis Agassiz (1807–1873) publishes his *Etudes sur les glaciers.* Also discovers glacial feature in Scotland away from an ice covered area and advances the theory of glaciation.

1842 Charles Robert Darwin wrote out an abstract of his theory of evolution, but he did not plan to have this theory published until after his death.

1849 U.S. Department of the Interior established.

1850–1899

1850 Rudolph Julius Emanuel Clausius (1822–1888), German physicist, publishes a paper which contains what becomes known as the second law of thermodynamics, stating that, "heat cannot, of itself, pass from a colder to a hotter body." He later refines the concept.

1851 Armand Hippolyte Fizeau (1819–1896), French physicist, measures the speed of light as it flows with a stream of water and as it goes against the stream. He finds that the velocity of light is higher in the former.

1851 Jean Bernard Léon Foucault (1819–1868), French physicist, conducts his spectacular series of experiments associated with the pendulum. He swings a heavy iron ball from a wire more than 200 feet (60 m) long and demonstrates that the swinging pendulum maintains its plane while Earth slowly twists under it. The crowd of spectators who witnesses this demonstration come to realize that they are watching the Earth rotate under the pendulum—experimental proof of a moving Earth.

1851 William Thomson (1824–1907), later known as Lord Kelvin, Scottish mathematician and physicist, publishes *On the Dynamical Theory of Heat* in which he explores Carnot's work and deduces that all energy tends to rundown and dissipate itself as heat. This is another form of the second law of thermodynamics and is advanced further by Clausius at about the same time. Kelvin's work is considered the first nineteenth-century treatise on thermodynamics.

1852 Abraham Gesner (1797–1864), Canadian geologist, prepares the first kerosene from petroleum. He obtains the liquid kerosene by the dry distillation of asphalt rock, treats it further, and calls the product kerosene after the Greek word *keros*, meaning oil.

1852 Alexander William Williamson (1824–1904), English chemist, publishes his study which shows for the first time that catalytic action clearly involves and is explained by the formation of an intermediate compound.

1853 Hans Peter Jorgen Julius Thomsen (1826–1901), Danish chemist, works out a method of manufacturing sodium carbonate from the mineral cryolite. This mineral will soon become important to the production of aluminum.

1853 William John Macquorn Rankine (1820–1872), Scottish engineer, introduces into physics the concept of potential energy, also called the energy of position.

1854 George Airy (1801–1892) estimates the mass of Earth from measurements of gravity from within deep mine shafts.

1854 Gregor Mendel (1822–1884) began studying 34 different strains of peas. He selected 22 kinds for further experiments. From 1856 to 1863, Mendel grows and tests over 28,000 plants and analyzes seven pairs of traits.

1854 Henry David Thoreau (1817–1862) publishes *Walden*.

1855 Charles-Adolphe Wurtz (1817–1884), French chemist, develops a method of synthesizing long-chain hydrocarbons by reactions between alkyl halides and metallic sodium. This method is called the Wurtz reaction.

1856 Neanderthal fossil identified.

1857 Louis Pasteur (1822–1895) demonstrated that lactic acid fermentation is caused by a living organism. Between 1857 and 1880, he performed a series of experiments that refuted the doctrine of spontaneous generation. He also introduced vaccines for fowl cholera, anthrax, and rabies, based on attenuated strains of viruses and bacteria.

1858 Charles Darwin and Alfred Russel Wallace (1823–1913) agree to a joint presentation of their theory of evolution by natural selection.

1860 Cesium is the first element discovered using the newly developed spectroscope. Robert Wilhelm Bunsen (1811–1899), German chemist, and Gustav Robert Kirchhoff (1824–1887), German physicist, name their new element cesium after its "sky blue" color in the spectrum.

1860 Stanislao Cannizzaro (1826–1910), Italian chemist, publishes the forgotten ideas of Italian physicist Amedeo Avogadro (1776–1856)—about the distinction between molecules and atoms—in an attempt to bring some order and agreement on determining atomic weights.

1861 Alexander Mikhailovich Butlerov (1828–1886), Russian chemist, introduces the term "chemical structure" to mean that the chemical nature of a molecule is determined not only by the number and type of a atoms but also by their arrangement.

1861 Friedrich August Kekulé von Stadonitz (1829–1896), German chemist, publishes the first volume of *Lehrbuch der organischen Chemie*, in which he is the first to define organic chemistry as the study of carbon compounds.

1861 William Thomson (1824–1907), later known as Lord Kelvin, Scottish mathematician and physicist, publishes his *Physical Considerations Regarding the Possible Age of the Sun's Heat* which contains the theme of the heat death of the universe. This is offered in light of the principle of dissipation of energy stated in 1851.

1862 Anders Angstrom (1814–1874) observes the spectrum of light from the sun (spectrographic analysis) and discovers the frequencies and wavelengths of light associated with the element Hydrogen.

1863 Ferdinand Reich (1799–1882), German mineralogist, and his assistant Hieronymus Theodor Richter (1824–1898), examine zinc ore spectroscopically and discover the new, indigo-colored element iridium. It is used in the next century in the making of transistors.

1863 William Huggins (1824–1910) discovers stellar spectra indicate that stars are made of same elements found on Earth.

1864 George Perkins Marsh (1801–1882) publishes *Man and Nature.*

1864 Yosemite in California becomes the first state park in the United States.

1865 Alexander Parkes (1813–1890), English chemist, produces celluloid, the first synthetic plastic material. After working since the 1850s with nitrocellulose, alcohol, camphor, and castor oil, he obtains a material that can be molded under pressure while still warm. Parkes is unsuccessful at marketing his product, however, and it is left to the American inventor, John Wesley Hyatt (1837–1920) to make it a success.

1866 August Adolph Eduard Eberhard Kundt (1839–1894), German physicist, invents a method by which he can make accurate measurements of the speed of sound in the air. He uses a Kundt's tube whose inside is dusted with fine powder which is then disturbed by traveling sound waves.

1869 Dimitri Ivanovich Mendeleev (1834–1907), Russian chemist, and Julius Lothar Meyer (1830–1895), German chemist, independently put forth the Periodic Table of Elements, which arranges the elements in order of atomic weights. However, Meyer does not publish until 1870, nor does he predict the existence of undiscovered elements as does Mendeleev.

1869 Ernst Haeckel coins the term ecology to describe "the body of knowledge concerning the economy of nature."

1871 Charles Robert Darwin (1809–1882) published *The Descent of Man, and Selection in Relation to Sex.* This work introduces the concept of sexual selection and expands his theory of evolution to include humans.

1872 Ferdinand Julius Cohn (1828–1898) published the first of four papers entitled "Research on Bacteria," which established the foundation of bacteriology as a distinct field. He systematically divided bacteria into genera and species.

1872 Yellowstone in Wyoming becomes the first U.S. national park.

1873 James Clerk Maxwell (1831–1879), Scottish mathematician and physicist, publishes *Treatise on Electricity and Magnetism* in which he identifies light as an electromagnetic phenomenon. He determines this when he finds his mathematical calculations for the transmission speed of both electromagnetic and electrostatic waves are the same as the known speed of light. This landmark work brings together the three main fields of physics—electricity, magnetism, and light.

1873 Johannes Van der Waals (1837–1923), Dutch physicist, offers an equation for the gas laws which contains terms relating to the volumes of the molecules themselves and the attractive forces between them. It becomes known as the Van der Waals equation.

1873 Walther Flemming (1843–1905) discovers chromosomes, observes mitosis, and suggests the modern interpretation of nuclear division.

1875 American Forestry Association founded to encourage wise forest management.

1876 Henry Augustus Rowland (1848–1901), American physicist, establishes for the first time that a moving electric charge or current is accompanied by electrically charged matter in motion and produces a magnetic field.

1879 U.S. Geological Survey established.

1880 Carl Oswald Viktor Engler (1842–1925), German chemist, begins his studies on petroleum. He is the first to state that it is organic in origin.

1882 Robert Koch (1843–1910), German bacteriologist, discovers the tubercle bacillus and enunciates "Koch's postulates," which define the classic method of preserving, documenting, and studying bacteria.

1883 Ernst Mach (1838–1916), Austrian physicist, publishes his *Die Mechanik in ihrer Entwickelung historisch-kritisch dargestellt* in which he offers a radical philosophy of science that calls into question the reality of such Newtonian ideas as space, time, and motion. His work influences Einstein and prepares the way for relativity.

1883 Frank Wigglesworth Clarke (1847–1931), American chemist and geophysicist, is appointed chief chemist to the U.S. Geological Survey. In this position, he begins an extensive program of rock analysis and is one of the founders of geochemistry.

1883 Johann Gustav Kjeldahl (1849–1900), Danish chemist, devises a method for the analysis of the nitrogen content of organic material. His method uses concentrated sulfuric acid and is simple and fast.

1886 Paul-Louis-Toussint Héroult (1863–1914), French metallurgist, and Charles Martin Hall (1863–1914), American chemist, independently invent an electrochemical process for extracting aluminum from its ore. This process makes aluminum cheaper and forms the basis of the huge aluminum industry. Hall makes the discovery in February of this year, and Héroult achieves his in April.

1887 Herman Frasch (1851–1914), German-American chemist, patents a method for removing sulfur compounds from oil. Once the foul sulfur smell is removed through the use of metallic compounds, petroleum becomes a marketable product.

1890 Yosemite becomes a national park.

1892 Adirondack Park established by New York State Constitution, which mandates that the region remain forever wild.

1892 Henry S. Salt (1851–1939) publishes *Animal Rights Considered in Relation to Social Progress*, a landmark work on animal rights and welfare.

1892 John Muir (1838–1914) founds the Sierra Club to preserve the Sierra Nevada mountain chain.

1893 Augusto Righi (1850–1920), Italian physicist, demonstrates that Hertz (radio) waves differ from light only in wavelength and not because of any essential difference in their nature. This helps to establish the existence of the electromagnetic spectrum.

1893 Ferdinand-Frédéric-Henri Moissan (1852–1907), French chemist, produces artificial diamonds in his electric furnace.

1894 Guglielmo Marconi (1874–1937), Italian electrical engineer, uses Hertz's method of producing radio waves and builds a receiver to detect them. He succeeds in sending his first radio waves 30 feet (9 m) to ring a bell. The next year, his improved system can send a signal 1.5 miles (2.4 km).

1894 John William Strutt Rayleigh (1842–1919), English physicist, and William Ramsay (1852–1916), Scottish chemist, succeed in isolating a new gas in the atmosphere that is denser than nitrogen and combines with no other element. They name it "argon," which is Greek for inert. It is the first of a series of rare gases with unusual properties whose existence had not been predicted.

1895 Pierre Curie (1859–1906), French chemist, studies the effect of heat on magnetism and shows that there is a critical temperature point above which magnetic properties will disappear. This comes to be called the Curie point.

1895 Wilhem Conrad Röntgen (1845–1923), German physicist, submits his first paper documenting his discovery of x rays. He tells how this unknown ray, or radiation, can affect photographic plates, and that wood, paper, and aluminum are transparent to it. It also can ionize gases and does not respond to electric or magnetic fields nor exhibit any properties of light. This discovery leads to such a stream of groundbreaking discoveries in physics that it has been called the beginning of the second scientific revolution.

1895 William Ramsay (1852–1916), Scottish chemist, discovers helium in a mineral named cleveite. It had been speculated earlier that helium existed only in the sun, but Ramsay proves it also exists on Earth. It is discovered independently this year by Swedish chemist and geologist Per Theodore Cleve (1840–1905). Helium is an odorless, colorless, tasteless gas that is also insoluble and incombustible.

1896 Antoine Henri Becquerel (1852–1908), French physicist, studies fluorescent materials to see if they emit the newly-discovered x rays and discovers instead that uranium produces natural radiation that is eventually called radioactivity in 1898 by the Polish-French chemist, Marie Curie (1867–1934).

1897 Joseph John Thomson (1856–1940), English physicist, discovers the electron. He conducts cathode ray experiments and concludes that the rays consist of negatively charged "electrons" that are smaller in mass than atoms.

1898 Marie Curie (1867–1934), Polish-French chemist, discovers thorium, which she proves is radioactive.

1898 Rivers and Harbors Act established in an effort to control pollution of navigable waters.

1899 Ernest Rutherford (1871–1937), British physicist, discovers that radioactive substances give off different kinds of rays. He names the positively charged ones alpha rays and the negative ones beta rays.

1900–1949

1900 Carl Correns (1864–1933), Hugo de Vries (1848–1935), and Erich von Tschermak (1871–1962) independently rediscover Mendel's laws of inheritance. Their publications mark the beginning of modern genetics. Using several plant species, de Vries and Correns perform breeding experiments that parallel Mendel's earlier studies and independently arrive at similar interpretations of their results. Therefore, upon reading Mendel's publication, they immediately recognize its significance. William Bateson (1861–1926) describes the importance of Mendel's contribution in an address to the Royal Society of London.

1900 Friedrich Ernst Dorn (1848–1916), German physicist, analyzes the gas given off by (radioactive) radium and discovers the inert gas he names radon. This is the first clear demonstration that the process of giving off radiation transmutes one element into another during the radioactive decay process.

1900 Lacey Act regulating interstate shipment of wild animals in the United States is passed.

1900 Paul Ulrich Villard (1860–1934), French physicist, discovers what are later called gamma rays. While studying the recently discovered radiation from uranium, he finds that in addition to the alpha rays and beta rays, there are other rays, unaffected by magnets, that are similar to x rays, but shorter and more penetrating.

1901 Antoine Henri Becquerel (1852–1908), French physicist, studies the rays emitted by the natural substance uranium and concludes that the only place they could be coming from is within the atoms of uranium. This marks the first clear understanding of the atom as something more than a featureless sphere.

1901 Guglielmo Marconi (1874–1937), Italian electrical engineer, successfully sends radio signal from England to Newfoundland.

1902 Discovery of Tyrannosaurus Rex fossil.

1902 Oliver Heaviside (1850–1925), English physicist and electrical engineer, and Arthur Edwin Kennelly (1861–1939), British-American electrical engineer, independently, and almost simultaneously, make the first prediction of the existence of the ionosphere, an electrically conductive layer in the upper atmosphere that reflects radio waves. They theorize correctly that wireless telegraphy works over long distances because a conducting layer of atmosphere exists that allows radio waves to follow Earth's curvature instead of traveling off into space.

1902 U.S. Bureau of Reclamation established.

1903 Antoine Henri Becquerel (1852–1908), French physicist, shares the Nobel Prize in physics with the husband-and-wife team of Marie Curie (1867–1934), Polish-French chemist, and Pierre Curie (1859–1906), French chemist. Becquerel wins for his discovery of natural or spontaneous radioactivity, and the Curies win for their later research on this new phenomenon.

1903 Ernest Rutherford (1871–1937), British physicist, and Frederick Soddy (1877–1956), English chemist, explain radioactivity by their theory of atomic disintegration. They discover that uranium breaks down and forms a new series of substances as it gives off radiation.

1903 William Ramsay (1852–1916), Scottish chemist, and Frederick Soddy (1877–1956), English chemist, discover that helium is continually produced by naturally radioactive substances.

1904 Ernest Rutherford postulates age of Earth by radioactvity dating.

1904 William Ramsay (1852–1916), Scottish chemist, receives the Nobel Prize in chemistry for the discovery of the inert gaseous elements in air, and for his determination of their place in the periodic system.

1905 Albert Einstein (1879–1955), German-Swiss physicist, publishes Special theory of relativity.

1905 Albert Einstein uses Planck's theory to develop a quantum theory of light which explains the photoelectric effect. He suggests that light has a dual, wave-particle quality.

1905 National Audubon Society formed.

1907 Georges Urbain (1872–1938), French chemist, discovers that last of the stable rare earth elements, and names it lutetium after the Latin name of Paris.

1907 Pierre Weiss (1865–1940), French physicist, offers his theory explaining the phenomenon of ferromagnetism. He states that iron and other ferromagnetic materials form small domains of a certain polarity pointing in various directions. When some external magnetic field forces them to be aligned, they become a single, strong magnetic force. This explanation is still accurate.

1908 Alfred Wegener (1800–1930) proposes the theory of continental drift.

1908 Chlorination is used extensively in U.S. water treatment plants for the first time.

1908 Ernest Rutherford (1871–1937), British physicist, and Hans Wilhelm Geiger (1882–1945), German physicist, develop an electrical alpha-particle counter. Over the next few years, Geiger continues to improve this device which becomes known as the Geiger counter.

1908 Ernest Rutherford (1871–1937), English physicist, is awarded the Nobel Prize in chemistry for his investigations into disintegration of the elements and the chemistry of radioactive substances.

1908 Tunguska event occurs when a comet or asteroid enters the atmosphere, and causes major damage to a forested region in Siberia.

1911 Arthur Holmes (1892–1965) publishes the first geological time scale with dates based on radioactive measurements.

1911 Ernst Rutherford (1871–1937), British physicist, discovers that atoms are made up of a positive nucleus surrounded by electrons. This modern concept of the atom replaces the notion of featureless, indivisible spheres that dominated atomistic thinking for 23 centuries—since Democritus (c. 470–c.380).

1911 Victor Hess (1883–1964) identifies high altitude radiation from space.

1912 Friedrich Karl Rudolf Bergius (1884–1949), German chemist, discovers how to treat coal and oil with hydrogen to produce gasoline.

1913 Charles Fabry (1867–1945), French physicist, first demonstrates the presence of ozone in the upper atmosphere. It is found later that ozone functions as a screen, preventing much of the sun's ultraviolet radiation from reaching Earth's surface. Seventy-five years after the discovery of ozone, in 1985, a hole in the ozone layer over Antarctica is discovered via satellite.

1913 Construction of Hetch-Hetchy Valley Dam approved to provide water to San

Francisco; however, the dam also floods areas of Yosemite National Park.

1913 Niels Henrik David Bohr (1885–1962), Danish physicist, proposes the first dynamic model of the atom. It is seen as a very dense nucleus surrounded by electrons rotating in orbitals (defined energy levels).

1914 Ernest Rutherford (1871–1937), British physicist, discovers a positively charged particle he calls a proton.

1914 Martha, the last passenger pigeon, dies in the Cincinnati Zoo.

1915 Albert Einstein (1879–1955), German-Swiss physicist, completes four years of work on his theory of gravitation, or what becomes known as the general theory of relativity.

1915 Richard Martin Willstätter (1872–1942), German chemist, is awarded the Nobel Prize in chemistry for his researchers on plant pigments, especially chlorophyll.

1916 U.S. National Park Service is established.

1918 Save-the-Redwoods League founded.

1918 The United States and Canada sign treaty restricting the hunting of migratory birds.

1919 Arthur Eddington (1882–1944) records data on the sun's gravitational deflection of starlight during a solar eclipse, confirming Einstein's general theory of relativity.

1920 Ernest Rutherford (1871–1937), English physicist, names the positively charged part of the atom's nucleus a "proton."

1920 Mineral Leasing Act enacted to regulate mining on federal land.

1922 Izaak Walton League founded.

1924 Gila National Forest in New Mexico is designated the first wilderness area.

1927 Hermann Joseph Muller (1890–1967) induces artificial mutations in fruit flies by exposing them to x rays. His work proves that mutations result from some type of physical-chemical change. Muller writes extensively about the danger of excessive x rays and the burden of deleterious mutations in human populations.

1928 George Gamow (1904–1968), Russian-American physicist, develops the quantum theory of radioactivity which is the first theory to successfully explain the behavior of radioactive elements, some of which decay in seconds and others after thousands of years.

1929 Walther Wilhelm Georg Franz Bothe (1891–1957), German physicist, invents coincidence counting by using two Geiger counters to detect the vertical direction of cosmic rays. This allows the measurement of extremely short time intervals, and he uses this technique to demonstrate that the laws of conservation and momentum are also valid for subatomic particles.

1933 Dust Bowl conditions due to extended drought in United States exacerbate Depression-era economic and environmental woes.

1932 James Chadwick (1891–1974), English physicist, proves the existence of the neutral particle of the atom's nucleus, called the neutron. It proves to be by far the most useful particle for initiating nuclear reactions.

1932 Karl Jansky (1905–1950) makes first attempts at radio astronomy.

1932 Lev Davidovich Landau (1908–1968) proposes the existence of neutron stars.

1932 Ruska builds first electron microscope.

1932 Thomas H. Morgan (1866–1945) received the Nobel Prize in medicine or physiology for his development of the theory of the gene. He was the first geneticist to receive a Nobel Prize.

1933 Tennessee Valley Authority is created to assess impact of hydropower on the environment.

1934 Arnold O. Beckman (1900–2004), American chemist and inventor, invents the pH meter, which uses electricity to accurately measure a solution's acidity or alkalinity.

1934 Frédéric Joliot-Curie (1900–1958) and Irène Joliot-Curie (1897–1956), husband-and-wife team of French physicists, discover what they call artificial radioactivity. They bombard aluminum to produce a radioactive form of phosphorus. They soon learn that radioactivity is not confined only to heavy elements like uranium, but that any element can become radioactive if the proper isotope is prepared. For producing the first artificial radioactive element they win the Nobel Prize in chemistry the next year.

1934 Taylor Grazing Act enacted to regulate grazing on federal land.

1935 U.S. Soil Conservation Service established to study and curb soil erosion.

1935 Wilderness Society founded by Aldo Leopold (1887–1948).

1936 Carl David Anderson (1905–1991), American physicist, discovers the muon. While studying cosmic radiation, he observes the track of a particle that is more massive than an electron but only a quarter as massive as a proton. He initially calls this new particle, which has a lifetime of only a few millionths of a second, a mesotron, but it later becomes known as a muon to distinguish it from Yukawa's meson.

1936 National Wildlife Federation established.

1937 Emilio Segrè (1905–1989), Italian-American physicist, and Carlo Perrier (1886–1948) bombard molybdenum with deuterons and neutrons to produce element 43, technetium. This is the first element to be prepared artificially that does not exist in nature.

1938 Bethe, Critchfield, von Weizsacker, argue that stars are powered by thee CNO-cycle of nuclear fusion.

1939 Leo Szilard (1898–1964), Hungarian-American physicist, and Canadian-American physicist, Walter Henry Zinn (b. 1906), confirm that fission reactions (nuclear chain reactions) can be self-sustaining using uranium.

1939 Linus Carl Pauling (1901–1994), American chemist, publishes *The Nature of the Chemical Bond*, a classic work that becomes one of the most influential chemical texts of the twentieth century.

1939 Lise Meitner (1878–1968), Austrian-Swedish physicist, and Otto Robert Frisch (1904–1979), Austrian-British physicist, suggest the theory that uranium breaks into smaller atoms when bombarded. Meitner offers the term "fission" for this process.

1939 Niels Hendrik David Bohr (1885–1962), Danish physicist, proposes a liquid-drop model of the atomic nucleus and offers his theory of the mechanism of fission. His prediction that it is the uranium-235 isotope that undergoes fission is proved correct when work on an atomic bomb begins in the United States.

1941 R. Sherr, Kenneth Thompson Bainbridge (1904–1996), American physicist, and H.H. Anderson produce artificial gold from mercury.

1942 Astronomer Grote Reber (1911–2002) constructs radio map of the sky.

1942 Enrico Fermi (1901–1954), Italian-American physicist, heads a Manhattan Project team at the University of Chicago that produces the first controlled chain reaction in an atomic pile of uranium and graphite. With this first self-sustaining chain reaction, the atomic age begins.

1943 Alaska Highway is completed, linking lower United States and Alaska.

1943 First operational nuclear reactor is activated at the Oak Ridge National Laboratory in Oak Ridge, Tennessee.

1943 J. Robert Oppenheimer (1904–1967), American physicist, is placed in charge of United States atomic bomb production at Los Alamos, New Mexico. He supervises the work of 4,500 scientists and oversees the successful design construction and explosion of the bomb.

1944 Norman Borlaug (1914–) begins his work on high-yielding crop varieties.

1944 Otto Hahn (1879–1968), German physical chemist, receives the Nobel Prize in chemistry for his discovery of nuclear fission.

1945 First atomic bomb is detonated by the United States near Almagordo, New Mexico. The experimental bomb generates an explosive power equivalent to 15–20 thousand tons of TNT.

1945 Joshua Lederberg (1925–2008) and Edward L. Tatum (1909–1975) demonstrate genetic recombination in bacteria.

1945 United States destroys the Japanese city of Hiroshima with a nuclear fission bomb based on uranium-235 on August 6. Three days later, a plutonium-based bomb destroys the city of Nagasaki. Japan surrenders on August 14 and World War II ends.

This is the first use of nuclear power as a weapon.

1946 Atomic Energy Commission established to study the applications of nuclear power. It was later dissolved in 1975, and its responsibilities were transferred to the Nuclear Regulatory Commission and Energy Research and Development Administration.

1946 George Gamow (1904–1968) proposes the big bang hypothesis.

1946 U.S. Bureau of Land Management created.

1947 A U.S. aircraft travels faster than the speed of sound.

1947 Defenders of Wildlife founded, superseding Defenders of Furbearers and the Anti-Steel-Trap League, to protect wild animals and their habitat.

1947 First carbon-14 dating.

1948 Gamow and others assert theory of nucleosynthesis is consistent with hot big bang.

1949 Aldo Leopold (1887–1948) publishes *A Sand County Almanac*, in which he sets guidelines for the conservation movement and introduces the concept of a land ethic.

1950–1999

1950 Astronomer Jan Oort (1900–1992) offers explanation of origin of comets.

1952 First thermonuclear device is exploded successfully by the United States at Eniwetok Atoll in the South Pacific. This hydrogen-fusion bomb (H bomb) is the first such bomb to work by nuclear fusion and is considerably more powerful than the atomic bomb exploded over Hiroshima on August 6, 1945.

1952 First use of isotopes in medicine.

1952 Oregon becomes first state to adopt a significant program to control air pollution.

1953 James D. Watson (1928–) and Francis H.C. Crick (1916–2004) publish two landmark papers in the journal *Nature*: "Molecular structure of nucleic acids: a structure for deoxyribose nucleic acid" and "Genetical implications of the structure of deoxyribonucleic acid." Watson and Crick propose a double helical model for DNA and call attention to the genetic implications of their model. Their model is based, in part, on the x-ray crystallographic work of Rosalind Franklin (1920–1958) and the biochemical work of Erwin Chargaff (1905–2002). Their model explains how the genetic material is transmitted.

1953 Stanley Miller (1930–2007) produces amino acids from inorganic compounds similar to those in primitive atmosphere with electrical sparks that simulate lightning.

1954 Humane Society is founded in United States.

1954 Linus Carl Pauling (1901–1994), American chemist, receives the Nobel Prize in chemistry for his research into the nature of the chemical bond and its applications to the elucidation of the structure of complex substances.

1955 First synthetic diamonds are produces in the General Electric Laboratories.

1956 Construction of Echo Park Dam on the Colorado River is aborted, due in large part to the efforts of environmentalists.

1957 Francis Crick proposes that during protein formation each amino acid is carried to the template by an adapter molecule containing nucleotides and that the adapter is the part that actually fits on the RNA template. Later research demonstrated the existence of transfer RNA.

1957 Soviet Union launches Earth's first artificial satellite, Sputnik, into Earth orbit.

1958 Astronomer Martin Ryle (1918–1984) argues evidence for evolution of distant cosmological radio sources.

1958 George W. Beadle (1903–1989), Edward L. Tatum (1909–1975), and Joshua Lederberg (1925–2008) are awarded the Nobel Prize in medicine or physiology. Beadle and Tatum were honored for the work in *Neurospora* that led to the one gene-one enzyme theory. Lederberg was honored for discoveries concerning genetic recombination and the organization of the genetic material of bacteria.

1958 National Aeronautics and Space Administration (NASA) is established.

1959 Soviet Space program sends space probe to impact the moon.

1959 St. Lawrence Seaway is completed, linking the Atlantic Ocean to the Great Lakes.

1961 Agent Orange is sprayed in Southeast Asia, exposing nearly 3 million American servicemen to dioxin, a probable carcinogen.

1961 Edward Lorenz (1917–2008) advances chaos theory and offers possible implications on atmospheric dynamics and weather.

1961 Murray Gell-Mann (1929–), American physicist, and Israeli physicist Yuval Ne'eman (1925–), independently introduce a new way to classify heavy subatomic particles. Gell-Mann names it the eight-fold way, and this system accomplishes for elementary particles what the periodic table did for the elements.

1961 Soviet Union launches first cosmonaut, Yuri Gagarin (1934–1968), into Earth orbit.

1962 James D. Watson (1928–), Francis Crick (1916–2004), and Maurice Wilkins (1916–2004) are awarded the Nobel Prize in medicine or physiology for their work in elucidating the structure of DNA.

1962 *Silent Spring* published by Rachel Carson (1907–1964) to document the effects of pesticides on the environment.

1963 First Clean Air Act passed in the United States.

1963 Fred Vine (1939–) and Drummond Matthews (1931–1997) offer important proof of plate tectonics by discovering that oceanic crust rock layers show equidistant bands of magnetic orientation centered on a site of sea floor spreading.

1963 Nuclear Test Ban Treaty signed by the United States and the Soviet Union to stop atmospheric testing of nuclear weapons.

1964 Wilderness Act passed, which protects wild areas in the United States.

1965 Arno Allan Penzias (1933–) and Robert Woodrow Wilson (1936–) detect cosmic background radiation.

1965 Water Quality Act passed, establishing federal water quality standards.

1966 Eighty people die in New York City due to pollution-related causes.

1966 Marshall Nirenberg (1927–) and Har Gobind Khorana (1922–) lead teams that decipher the genetic code. All of the 64 possible triplet combinations of the four bases (the codons) and their associated amino acids are determined and described.

1967 American Cetacean Society founded to protect whales, dolphins, porpoises, and other cetaceans. Considered the oldest whale conservation group in the world.

1967 Environmental Defense Fund established to save the osprey from DDT.

1967 Supertanker Torrey Canyon spills oil off the coast of England.

1968 Wild and Scenic Rivers Act and National Trails System Act passed to protect scenic areas from development.

1969 *Apollo 11* mission to the moon. U.S. astronauts Neil Armstrong (1930–) and Buzz Aldrin (1930–) become first humans to walk on another planet.

1969 Greenpeace founded.

1969 Max Delbrück (1906–1981), Alfred D. Hershey (1908–1997), and Salvador E. Luria (1912–1991) are awarded the Nobel Prize in medicine or physiology for their discoveries concerning the replication mechanism and the genetic structure of viruses.

1970 Environmental Protection Agency (EPA) created.

1970 First Earth Day celebrated on April 22.

1970 National Environmental Policy Act is passed, requiring environmental impact statements for projects funded or regulated by federal government.

1971 Consultative Group on International Agricultural Research (CGIAR) is founded to improve food production in developing countries.

1972 Clean Water Act passed.

1972 Coastal Zone Management Act and Marine Protection, Research, and Sanctuaries Act passed.

1972 Discovery of 2 million-year-old humanlike fossil, *Homo habilis*, in Africa.

1972 *Limits to Growth* published by the Club of Rome, calling for population control.

1972 Oregon becomes first state to enact bottle-recycling law.

1972 Paul Berg (1926–) and Herbert Boyer (1936–) produce the first recombinant DNA molecules. Recombinant technology emerges as one of the most powerful techniques of molecular biology. Scientists are able to splice together pieces of DNA to form recombinant genes. As the potential uses, therapeutic and industrial, became increasingly clear, scientists and venture capitalists establish biotechnology companies.

1972 U.N. Conference on the Human Environment is held in Stockholm, Sweden, to address environmental issues on a global level.

1972 Use of DDT is phased out in the United States.

1973 Arab members of the Organization of Petroleum Exporting Countries (OPEC) institute an embargo preventing shipments of oil to the United States.

1973 Convention on International Trade in Endangered Species of Wild Fauna and Flora (CITES) is signed to prevent the international trade of endangered or threatened animals and plants.

1973 Cousteau Society founded by Jacques-Yves Cousteau (1910–1997) and his son to educate the public and conduct research on marine-related issues.

1973 E. F. Schumacher (1911–1977) publishes *Small Is Beautiful*, which advocates simplicity, self-reliance, and living in harmony with nature.

1973 Endangered Species Act is passed.

1974 Safe Drinking Water Act is passed, requiring EPA to set quality standards for the nation's drinking water.

1975 Atlantic salmon is found in the Connecticut River after a 100-year absence.

1975 Scientists at an international meeting in Asilomar, California, call for the adoption of guidelines regulating recombinant DNA experimentation.

1975 *The Monkey Wrench Gang* is published by Edward Abbey (1927–1989), who advocates radical and controversial methods for protecting the environment, including "ecotage."

1976 Land Institute is founded by Wes and Dana Jackson to encourage more natural and organic agricultural practices.

1976 Poisonous gas containing 2,4,5-TCP and dioxin is released from a factory in Seveso, Italy, causing massive animal and plant death. Although no human life is lost, a sharp increase in deformed births is reported.

1976 Resource Conservation and Recovery Act is passed, giving EPA authority to regulate municipal solid and hazardous waste.

1976 U.S. Viking spacecraft lands and conducts experiments on Mars.

1977 Robotic submarine "Alvin" explores mid-oceanic ridge and discoveres chemosynthetic life.

1977 Voyager spacecraft is launched; contains golden record recording of Earth sounds.

1978 Oil tanker *Amoco Cadiz* runs aground, spilling 220,000 tons of oil.

1978 Residents of Love Canal, New York, are evacuated after Lois Gibbs discovers that the community was once the site of a chemical waste dump.

1979 Three Mile Island Nuclear Reactor almost undergoes nuclear melt-down when the cooling water systems fail. Since this accident no new nuclear power plants have been built in the United States.

1980 Alaska National Interest Lands Conservation Act enacted, setting aside millions of acres of land as wilderness.

1980 Comprehensive Environmental Response, Compensation, and Liability Act (Superfund) enacted to clean up abandoned toxic waste sites.

1980 Earth First! is founded by Dave Foreman (1947–), with the slogan "No compromise in the defense of Mother Earth."

1980 *Global 2000 Report* is published, documenting trends in population growth, natural resource depletion, and the environment.

1980 Mount St. Helens explodes with a force comparable to 500 Hiroshima-sized bombs.

1980 Thomas Lovejoy proposes the idea of debt-for-nature swap that helps developing countries alleviate national debt by implementing policies to protect the environment.

1984 Emission of poisonous methyl isocyanate vapor, a chemical by-product of agricultural insecticide production, from the Union

Carbide plant kills more than 2,800 people in Bhopal, India.

1984 Ozone hole over Antarctica discovered.

1985 Rainforest Action Network founded.

1986 Chernobyl Nuclear Power Station undergoes nuclear core melt-down, spreading radioactive material over vast parts of the Soviet Union and northern Europe.

1986 Evacuation of Times Beach, Missouri, due to high levels of dioxin.

1987 *Ecodefense: A Field Guide to Monkeywrenching* published by Dave Foreman, in which he describes spiking trees and other "environmental sabotage" techniques.

1987 Montreal Protocol on Substances that Deplete the Ozone Layer signed by 24 nations, declaring their promise to decrease production of chlorofluorocarbons (CFCs).

1987 *Our Common Future* (The Brundtland Report) is published.

1987 World population is five billion.

1987 Yucca Mountain designated the first permanent repository for radioactive waste by the U.S. Department of Energy.

1988 Global ReLeaf program inaugurated with the motto "Plant a tree, cool the globe" to address the problem of global warming.

1988 Ocean Dumping Ban Act established.

1988 The Human Genome Organization (HUGO) is established by scientists in order to coordinate international efforts to sequence the human genome. The Human Genome Project officially adopts the goal of determining the entire sequence of DNA comprising the human chromosomes.

1989 Oil tanker *Exxon Valdez* runs aground in Prince William Sound, Alaska, spilling 11 million gallons of oil.

1990 Clean Air Act amended to control emissions of sulfur dioxide and nitrogen oxides.

1990 Hubble Space Telescope launched.

1990 Oil Pollution Act signed, setting liability and penalty system for oil spills as well as a trust fund for clean up efforts.

1991 Andrew A. Griffith, American chemist, uses an atomic force microscope to obtain extraordinarily detailed images of the electrochemical reactions involved in corrosion.

1991 K-T event impact crater identified near the Yucatan Peninsula.

1991 Mount Pinatubo in Philippines erupts, shooting sulfur dioxide 25 miles (40 km) into the atmosphere.

1991 Over 4,000 people die from cholera in Latin American epidemic.

1991 Persian Gulf War begins. During the war Saddam Hussein's (1937–2006) army burns oil fields.

1991 Train containing the pesticide *meta sodium* falls off the tracks near Dunsmuir, California, releasing chemicals into the Sacramento River. Plant and aquatic life for 43 miles (70 km) downriver die as a result.

1992 Captive-bred California condors and black-footed ferrets reintroduced into the wild.

1992 Mexico City, Mexico, suffers general shut down as a result of incapacitating air pollution.

1992 U.N. Earth Summit held in Rio de Janeiro, Brazil.

1992 United Nations calls for an end to global drift net fishing by the end of 1992.

1993 Oil tanker runs aground in the Shetland Islands, Scotland, spilling its oil into the sea.

1993 Eight people from Biosphere 2 emerge after living two years in a self-sustaining, glass dome.

1993 Forest Summit convened in Portland, Oregon, by President Bill Clinton (1946–), who meets with loggers and environmentalists concerned with the survival of the northern spotted owl.

1993 George Washington University researchers clone human embryos and nurture them in a petri dish for several days. The project provokes protests from ethicists, politicians and critics of genetic engineering.

1993 Norway resumes hunting of minke whales in defiance of a ban on commercial whaling instituted by the International Whaling Commission.

1994 Astronomers observe comet Shoemaker-Levy 9 (S-L 9) colliding with Jupiter.

1994 Researchers at Fermilab discover the top quark. Scientists believe that this discovery may provide clues about the genesis of matter.

1995 Ken-Sara Wiwa (1941–1995) is executed in Nigeria for protesting and speaking out about oil industry practices in the country.

1995 Mayor and Queloz identify first extra-solar planet. A Jupiterlike planet orbiting an ordinary star.

1995 Paul Crutzen (1933–), Dutch meteorologist, Mario Molina (1943–), Mexican American chemist, and R. Sherwood Rowland (1927–), American atmospheric chemist, receive the Nobel Prize in chemistry for their work in atmospheric chemistry, particularly concerning the formation and decomposition of ozone.

1997 Forest fires worldwide burn a total of five million hectares of forest.

1997 Julia Butterfield Hill climbs a 180-foot (55 meter) redwood tree in California to protest the logging of the surrounding forest as well as to protect the tree. Hill removes herself from the tree in 1999 after she negotiates a deal to save the tree and an additional three acres of the forest.

1997 Kyoto Protocol mandates a reduction of reported 1990 emissions levels by 6-8% by 2008.

1997 Microscopic analysis of Murchison meteorite lead some scientists to argue evidence of ancient life on Mars.

1997 Monserrat volcano erupts.

1998 Ian Wilmut (1944–) announces the birth of Polly, a transgenic lamb containing human genes.

1998 Two research teams succeed in growing embryonic stem cells.

1999 Scientists announce the complete sequencing of the DNA making up human chromosome 22. The first complete human chromosome sequence is published in December 1999.

1999 World population reaches six billion.

1999 World Trade Organization (WTO) conference in Seattle, Washington, is marked by heavy protests, highlighting WTO's weak environmental policies.

2000–2008

2000 During his presidency, Bill Clinton (1946–) appropriates a total of 58 million acres of wilderness as conservation land—the largest amount of land to be set aside for conservation by any other president to date.

2000 Russian nuclear submarine the Kursk sinks off the coast of Minsk, Russia. Despite concerns, no nuclear waste escapes.

2000 West Nile virus is discovered in the eastern United States.

2000 On June 26, 2000, leaders of the public genome project and Celera announce the completion of a working draft of the entire human genome sequence. Ari Patrinos of the DOE helps mediate disputes between the two groups so that a fairly amicable joint announcement could be presented at the White House in Washington, DC.

2000 The National Cancer Institute (NCI) estimates that 3,000 lung cancer deaths, and as many as 40,000 cardiac deaths per year among adult nonsmokers in the United States can be attributed to passive smoke or environmental tobacco smoke (ETS).

2001 In February 2001, the complete draft sequence of the human genome is published. The public sequence data is published in the British journal *Nature* and the sequence obtained by Celera is published in the American journal *Science*.

2001 The United States fails to ratify the Kyoto Protocol.

2001 The World Trade Center towers in New York collapse after being struck by two commercial airplanes commandeered by terrorists. A third airplane is crashed into the Pentagon building just outside Washington, D.C., causing loss of life and major damage to the building.

2002 EPA Administer Christine Todd Whitman (1946–) resigns.

2002 EPA adopts California emissions standards for off-road recreation vehicles to be implemented by 2004.

2002 EPA announces its Strategic Plan for Homeland Security to support the National Strategy for Homeland Security enacted after the September 11, 2001, terrorist attacks in the United States.

2002 President George W. Bush (1946–) introduces the Clear Sky Initiative.

2002 President George W. Bush signs bill approving the using of Yucca Mountain as a nuclear waste storage site.

2002 Satellites capture images of icebergs more than ten times the size of Manhattan Island breaking off Antarctic ice shelf.

2002 In June 2002 traces of biological and chemical weapon agents are found in Uzbekistan on a military base used by U.S. troops fighting in Afghanistan. Early analysis dates and attributes the source of the contamination to former Soviet Union biological and chemical weapons programs that utilized the base.

2002 Severe acute respiratory syndrome (SARS) virus is found in patients in China, Hong Kong, and other Asian countries. The newly discovered corona virus is not identified until early 2003. The spread of the virus reaches epidemic proportions in Asia and expands to the rest of the world.

2002 U.N. Earth Summit is held in Johannesburg, South Africa.

2003 Three Gorges Dam in China begins filling.

2003 United States invades Iraq and finds chemical, biological, and nuclear weapons programs but no actual weapons.

2003 Energy bill introduced in Congress includes ethanol use mandates.

2003 Electric power failure causes blackout from New York to Ontario.

2003 EPA rejects petition to regulate emissions from vehicles, EPA claims lack of authority under the Clean Air Act.

2004 Russia ratifies Kyoto treating, putting it into effect worldwide even without United States ratification.

2004 Kenyan environmentalist and human rights activist Wangari Maathai (1940–) wins the Nobel Peace Prize.

2004 On December 26, the most powerful earthquake in more than 40 years occurs underwater off the Indonesian island of Sumatra. The tsunami produces a disaster of unprecedented proportion in the modern era. Less than two weeks after the tsunami impact, the International Red Cross puts the death toll at over 150,000 lives and most experts expect that number to continue to climb. Many experts claim this will be the costliest, longest, and most difficult recovery period ever endured as a result of a natural disaster.

2005 Kyoto Protocol officially goes into force February 16, 2005.

2005 H5N1 virus, responsible for avian flu, moves from Asia to Europe. The World Health Organization attempts to coordinate multinational disaster and containment plans. Some nations begin to stockpile antiviral drugs.

2005 Hurricane Katrina slams into the U.S. Gulf Coast, causing levee breaks and massive flooding to New Orleans. Damage is extensive across the coasts of Louisiana, Mississippi and Alabama. Federal Emergency Management Agency (FEMA) is widely criticized for a lack of coordination in relief efforts. Three other major hurricanes make landfall in the United States within a two-year period stressing relief and medical supply efforts. Long term health studies begin of populations in devastated areas.

2005 A massive 7.6-magnitude earthquake leaves more than 3 million homeless and without food and basic medical supplies in the Kashmir mountains between India and Pakistan. 80,000 people die.

2005 Dorothy Stang (1931–2005), an American nun, is murdered in Brazil by contract killers after spending decades fighting efforts by loggers and ranchers to clear large areas of the Amazon rainforest. Less than a week later, Brazil's government awards a disputed patch of Amazon rainforest to a sustainable development project championed by Stang.

2005 The United Nations reports that the hole in the ozone layer above Antarctica has grown to near record size, suggesting 20 years of attempted pollution controls have so far had little effect.

2005 The European Project for Ice Coring in Antarctica reports that carbon dioxide in the current atmosphere is greater than at any time during the last 650,000 years.

2006 Norway announces plans to build a "doomsday vault" in a mountain close to the North Pole that will house a two-million-crop seed bank in the event of catastrophic climate change, nuclear war, or rising sea levels.

2006 China orders all restrictions on small cars lifted in order to encourage energy conservation.

2006 NASA research shows that the Antarctic ice sheet melting has raised global sea level by 1.2 millimeters since 2002.

2006 NASA launches two satellites designed to provide the first 3-D views of Earth's clouds and help predict how cloud cover contributes to global warming.

2006 China's official news agency reports that glaciers in the Qinghai-Tibet plateau, also known as the "roof of the world," are melting at a rate of seven percent annually due to global warming.

2006 U.S. scientists attribute a reported four-fold increase in the number of fires in western United States to climate change.

2006 Researchers report that carbon dioxide from industrial emissions is raising the acidity of the world's oceans, threatening plankton and other organisms that form the base of the entire marine food web.

2006 NASA satellite data shows Greenland's ice sheet is melting at a rate that exceeds scientists' estimates.

2006 Researchers report that Earth's temperature has been warming by 0.3°F per year for the last three decades, and is at a current 12,000-year high.

2007 Scientists warn that glaciers could all but disappear from the European Alps by 2050, and that most would be gone by 2037.

2007 Fourth IPCC report is issued (the first segment in February, and the last in November). The Intergovernmental Panel on Climate Change (IPCC) Scientists, composed of scientists from 113 countries, issues a consensus report stating that global warming is caused by man, and predicting that warmer temperatures and rises in sea level will continue for centuries, no matter how much humans control their pollution.

2007 In London, airline tycoon Richard Branson (1950–) announces a $25 million prize for the first person to devise a way of removing greenhouse gases from the atmosphere.

2007 As of January 2007, the warmest year in recorded weather history in terms of global average temperature was 1998, which was 0.94°F (0.52°C) above the average for 1961–1990. The warmest year on record so far for the United States, prior to 2007, was actually 2006, which was slightly warmer than the prior record set in 1998.

2007 In August 2007, a month before annual Arctic sea-ice melting normally peaks, a new record is already established for shrinkage of the north polar ice cap. No summer since 1979, when satellites first allowed accurate tracking of the annual melting, recorded such a large loss of ice. William Chapman (1964–), an expert at the University of Illinois, Urbana-Champaign, says that the melting rate in June and July of 2007 is "simply incredible." Arctic sea-ice melting is predicted to have many impacts, including increased Arctic regional temperatures, increased Arctic coastal erosion, declines in polar bear populations, damage to traditional hunting practices of indigenous Arctic peoples, and disruption of large-scale ocean circulations.

2007 Increased extreme weather is predicted consequence of global climate change. In August 2007, scientists at the World Meteorological Organization, an agency of the United Nations, announce that during the first half of 2007, Earth shows significant increases above long term global averages in both high temperatures and frequency of extreme weather events (including heavy rainfalls, cyclones, and wind storms). Global average land temperatures for January and April of 2007 are the warmest recorded for those two months since records began in 1880.

2007 The World Wildlife Fund (WWF) conservation group states that climate change, pollution, over extraction of water, and encroaching development are killing some of the world's major rivers including China's Yangtze, India's Ganges and Africa's Nile.

2007 In Algeria, an international conference on desertification closes with a call (dubbed the Algiers Appeal) to all African countries to ratify the Kyoto Protocol in an effort to help slow the rapid expansion of deserts on the continent.

2007 The United Nations reports that air pollution contributes to the premature death of over 400,000 Chinese people each year, and that 16 of the most polluted cities in the world are in China.

2007 The area of ocean covered by ice around the North Pole varies seasonally, reaching a minimum every September. In August and September 2007, the north polar sea-ice cap shrinks to the smallest size ever recorded. As of September 16, U.S. government scientists announce, the ice is 1.59 million square miles (4.14 million km^2) in size, about a fifth smaller than the previous record, set in September 2005. The Northwest Passage, which is the sea route from the Atlantic to the Pacific along the northern edge of North America, is ice-free for the first time in recorded history. Although surprised and concerned by the extent of the 2007 melting, climate scientists predict increased melting of the ice due to global climate change. Thus far, Arctic temperatures have warmed twice as fast as the rest of the world.

2007 The journal *Science* reports that the crucial "carbon sink" that holds fifteen percent of the world's excess carbon dioxide, Antarctica's Southern Ocean, is nearing saturation and could soon be unable to absorb more.

2007 Former United States vice president Al Gore (1948–) and the IPCC share the 2007 Nobel Peace prize for raising public awareness about global warming and cliamte change and for establishing the foundations to begin to solve the problem. Gore's film about climate change and global warming, "An Inconvenient Truth" also wins major film awards.

2007 World leaders at a G8 summit agree to "seriously consider" proposals to cut the emissions of greenhouse gases by 50 percent by 2050.

2007 The largest driver of global climate change is carbon dioxide (CO_2) in the atmosphere, which is directly increased by the burning fossil of fuels. In October 2007 scientists announce that CO_2 levels had increased since 2000 faster than even the most pessimistic predicted forecasts of the late 1990s. Growth in atmospheric CO_2 was only 1.1% per year for 1990–1999, but accelerated sharply to more than 3% per year for 2000–2004. Climate scientists attribute most of the increased emissions to increases in human population and industrial activity.

2007 Researchers from the University of North Carolina in Chapel Hill report that coral reefs in the Indo-Pacific, an area stretching from the island of Sumatra island to French Polynesia, dropped 20 percent since 1985 due largely to climate change and coastal development.

2008 Computer models of global climate change predict that East Antarctica will gain ice as snowfall increases while West Antarctica will lose ice as melting accelerates. Both effects have been observed. If Antarctica loses more ice than it gains, it contributes to sea-level rise: if the reverse, it slows sea-level rise. Not until 2008 did climatologists know whether gain or loss was dominant.

2008 In January, an article in *Nature Geoscience* reports that Antarctica is definitely losing ice overall, and losing it an accelerating pace. Ice loss is near zero in East Antarctica, but in the West Antarctic Peninsula, ice loss increased from 1996 to 2006 by 140%, to about 60 billion metric tons per year, and in the rest of West Antarctica it increased by 59%, to about 132 billion tons per year. Most of the increased loss is happening in channels along which glaciers flow to the sea: as ocean currents are changed by global warming, they shrink the parts of the glaciers protruding into the sea, in effect uncorking the glaciers. According to the new figures, Antarctica is now losing ice almost as fast as Greenland.

2008 The UN's Intergovernmental Panel on Climate Change announces in January 2008 that its next report will study the possibility of accelerated melting of both the Greenland and Antarctic ice shelves, which could cause several meters of sea-level rise by the end of this century.

2008 At the Royal Astronomical Society's National Astronomy Meeting, astronomers from St. Andrews University announce the discovery of a planetary system about five thousand light-years away that is similar to regions of Earth's solar system. Using micro-lensing techniques they discover that the system that orbits a star about half the size of Earth's sun. The discovery fuels speculation among astronomers that the discovery of systems that could contain Earth-like terrestrial planets is immanent. As of early 2008 more than 300 extrasolar planets (exoplanets) are identified, but none yet approximating Earth conditions.

2008 Ongoing research confirms that global sea level—the average height of the ocean's surface apart from the daily changes of the tides—is rising. Sea level rise threatens fresh water supplies, coastal land use, and vital economic activity. For some islands, significant sea level rise could be devastating.

2008 In tropical cities, crowded conditions with little sanitation infrastructure can lead to an outbreak of dengue fever during the rainy season. Dengue fever, a mosquito-borne viral disease reaches epidemic proportions in areas of Brazil in April 2008.

2008 China overtakes the United States as the world's largest emitter of greenhouse gases.

2008 Oil and food prices rise sharply on a global scale, increasing dangers of famine and poverty. Critics contend increased prices for petroleum lead to the diversion of food crops to biofuel production.

2008 Global agricultural experts issue warning that UG99, a plant rust fungus that kills to up to 80% of current wheat strains, threatens crops (most immediately those in developing countries). The spread of UG99 raises the specter of widespread wheat crop destruction on a global scale. Such destruction would, of course, result in massive poverty for farmers, widespread economic damage, increased global wheat prices (at a time when wheat prices have already experienced sharp increases), and possible famine.

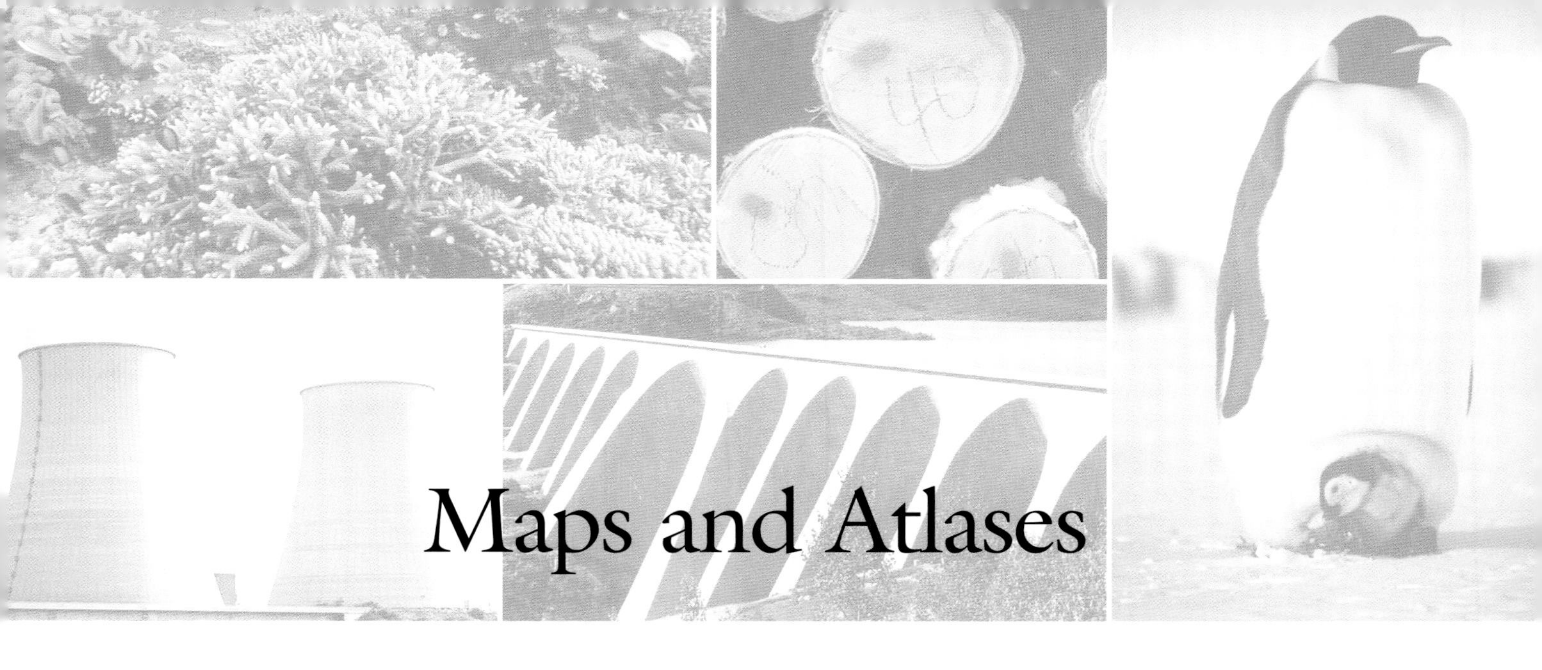

Maps and Atlases

Introduction

A map is a visual representation or scale model of spatial concepts such as geographical regions, locations, and their attributes. Maps may focus on a few selected characteristics of the region or give an overview of the region under consideration. They may project physical, biological, or cultural features of the region or depict correlation between several features of the region. Maps are made up of several components—symbols, title, legend, direction, map scale, source, and insets.

An atlas is a compilation of maps presented in the form of a print publication or in a multimedia format. The purpose of an atlas is to help the user by providing additional information and analyses of maps. Atlases often contain social, religious, economic, and geopolitical information for a specific region.

Maps and atlases are the primary tools for spatial analysis. From flying airplanes and military planning to forecasting impending natural hazards, these tools have a wide application.

Historical Background and Scientific Foundations

References to maps can be found in documents pertaining to the Chinese civilization of seventh century BC. However, the earliest specimen of a map can be traced back to Babylon in 1000 BC. *Geographike hyphygesis* of Claudius Ptolemy (c.AD 90–c.168) is a significant work on world geography. In fact, the earliest known printed atlases were Ptolemy's editions of this text. The Renaissance period saw the invention of printed maps. During the fifteenth and sixteenth centuries, navigation charts, compass lines, and other navigation aids greatly influenced cartography. Later, in the early sixteenth century, world maps were published. Early examples are Waldseemüller's world map of 1507 and Rosselli's world map of 1508.

The earliest historical evidence of atlases being compiled is found in China in the third century. In Europe, however, it happened during the Renaissance period. In 1570, Abraham Ortelius (1527–1598) published the first modern atlas, called *Theatrum Orbis Terrarum*. The term "atlas" appeared after the posthumous publication of Gerard Mercator's *Atlas sive Cosmographicae meditationes de Fabrica Mundi et fabricati figura* in 1595. In this publication, Titan Atlas—a mythical king—is depicted holding up Earth on the cover page. The first atlases published in English were *Epitome* and the *Theatrum* in 1601.

In the nineteenth century, Europeans applied the metric system for map scale and the Greenwich Prime Meridian was also established. In the twentieth century, computer technology, electronic distance-measuring instruments, inertial navigation systems, remote sensing, and other technological developments led to advanced mapping methods. With the advent of Geographic Information Systems (GIS), the use of cartography saw a decline. In the twenty-first century, the output of GIS mapping can be accessed through the Internet as well.

With GIS technology, scientists can study the spatial aspects of issues like climate change, poverty, natural calamities, and shortage of food. GIS maps can be used to identify vulnerable hotspots in a specific region. The vulnerability of a region refers to the extent to which it is at risk from a natural crisis.

Impacts and Issues

A map is merely a visual representation of a given set of information. Hence, the reliability of a map depends entirely on the accuracy of the input data. If the data used are incorrect, incomplete, or misleading, the result-

WORDS TO KNOW

GEOGRAPHIC INFORMATION SYSTEM (GIS): A set of computer-based tools that collects, analyzes, and maps spatial data.

HOTSPOTS: Locations that are at high risk from natural hazards.

MAP SCALE: Portrays the relationship between a distance/size on map and the corresponding distance/size on the land. It is given as a figure having two parts, for example, 1:50,000, meaning 1 unit of measure on the map is equal to 50,000 units of measure on the land.

ing map will be inaccurate. Even with GIS technology, a map is susceptible to errors like human oversight, inconsistent scale, processing errors, or technical errors.

Another drawback with mapping is its inability to represent everything. A map cannot be relied upon as a comprehensive source of information on the world. First, field measurements are error-prone. Even advanced technologies like satellite images are restricted to showing only a portion of the light spectrum. In general, it is impossible to create a map that can represent all of the features (physical, biological, and cultural) of a specific region. Moreover, all maps are assessments, generalizations, and analysis of geographic conditions. They are produced using basic assumptions. There is also a possibility of changes in the surveyed area due to natural factors or human activities.

SEE ALSO *Geographic Information Systems (GIS); Geospatial Analysis*

BIBLIOGRAPHY

Books

Harley, John Brian, and David Woodward. *The History of Cartography*. Chicago: University of Chicago Press, 1987.

Robinson, Arthur H., et al. *Elements of Cartography*, 6th ed. Hoboken, NJ: Wiley, 1995.

Web Sites

Emporia State University. "Brief History of Maps and Cartography," 2004. http://academic.emporia.edu/aberjame/map/h_map/h_map.htm (accessed April 20, 2008).

MapForum Ltd. "The Earliest Atlases: From Ptolemy to Ortelius." http://www.mapforum.com/01/atlas.htm (accessed April 20, 2008).

Rice University. "What Are Maps?" http://math.rice.edu/~lanius/pres/map/mapdef.html (accessed April 20, 2008).

The flight director for LightHawk stands in front of a picture of an airplane flying over a forest where areas have been clear-cut, in 1995. LightHawk, a conservation group, does aerial photographing and mapping of environmentally-damaged areas. *AP Images*

Marine Ecosystems

Introduction

The seas and oceans are rich in animal life, from whales and fish to starfish and sponges. The nature of marine communities depends upon depth, with the surface and the ocean floor being particularly rich in biodiversity. Similar communities are often found at similar depth, even though they may be widely separated geographically. There are also complex marine habitats, such as mangrove swamps, estuaries, and coral reefs, occurring near the shore.

Marine ecosystems depend largely upon phytoplankton, which are photosynthetic algae living near the surface of the water where the sun penetrates. Tiny herbivores feed on the phytoplankton and these, in turn, are eaten by increasingly larger animals, ending with larger fish and sharks at the top of the marine food web. The seas and oceans are important for humans as a resource for fish and other marine products, but its ecosystems are threatened by exploitation, such as overfishing and pollution.

Historical Background and Scientific Foundations

Until the middle of the nineteenth century, it was assumed that few, if any, animals and plants lived in the seas and oceans because the waters were dark and cold. The first clues to the existence of rich and complex marine ecosystems came from broken underwater telegraph wires. When these were retrieved, various unusual and previously unknown creatures were found clinging onto them. The *HMS Challenger*, a British naval vessel, carried out the first-ever oceanographic survey between 1872 and 1876, exploring as deep as 18,700 ft (5,700 m) in the Pacific Ocean. The expedition returned with thousands of specimens, many of which were previously unknown to science.

The marine environment is where life evolved in the first place. Seawater contains sodium chloride (NaCl) and other mineral salts, and has remained at roughly the same salt concentration for millions of years. Salinity, as the salt concentration is called, is around one ounce per liter, of which 90% is composed of sodium chloride. This happens to be the same sodium chloride concentration as living cells, making it a natural environment for organisms to deal with.

Water covers around 70% of Earth's surface. Its average depth is just over 2 mi (3.2 km)—ranging from a few inches close to the shore to around 7 mi (11.2 km) at its greatest depth. The oceans alone provide more than 170 times more living space than land, air, and freshwater put together. The weight of water does exert pressure upon those organisms living there. This pressure will increase by one atmosphere for every 33 ft (10 m) of depth. But, like organisms living with atmospheric pressure on land, deep-sea animals such as fish and snails have the same pressure inside and outside their bodies. Most sea creatures are composed mainly of water and, since liquids are incompressible, they do not experience adverse effects on moving from one depth to another.

The seas are constantly in motion due to surface currents and deeper circulation currents, which means that cold salty water, which is heavier, sinks and is replaced by warmer, less salty water. This mixes the water, making its chemistry uniform as well as carrying oxygen (O) from the surface to deeper layers, making life down there possible. The temperature of surface water varies from 104°F (40°C) in tropical waters to about 35°F (1.9°C) for seawater in the Arctic and Antarctic. The depths of the oceans are always very cold, even in tropical regions, at 32 to 37°F (0 to 3°C).

Compared to the land, the marine environment is not so rich in different species. Only one tenth of the

WORDS TO KNOW

BENTHIC: Living on, or associated with, the ocean floor.

MARINE: Living in, or associated with, the sea.

PELAGIC: Living in, or associated with, open areas of ocean away from the bottom.

PHYLUM (PLURAL, PHYLA): A biological classification group lying between kingdom and class.

nearly 2 million animal species known are found in the sea, and only around 4,000 plants, compared to a quarter of a million on land. But marine ecosystems tend to be more diverse, with 28 different phyla existing in the oceans, compared to only 11 on land.

Temperature, salinity, and the availability of oxygen, light, and nutrients shape the marine ecosystems. Thousands of different invertebrates make their home in the ocean, often providing food for larger species. For instance, there are about 5,000 species of sponges, of a wide variety of colors and shapes. These animals have the least complex body structure of all multi-celled creatures, consisting of an outer layer of tissue and an inner layer of either silica (SiO_2) or calcium carbonate ($CaCO_3$). The echinoderms, which include starfish, sea urchins, and sea cucumbers, are a group of exclusively marine bottom-dwelling invertebrates, characterized by hard, spiny skin. Around 6,000 species dwell in the world's salt waters, often in rocky pools around beaches.

The cnidarians are another important marine invertebrate group and include the jellyfish, corals, and sea anemones. These organisms are characterized by their soft, watery bodies. The molluscs, which are found on land and sea, include the gastropods (limpets, slugs, and snails), cephalopods (octopus, squid, and cuttlefish), and bivalves (clams, mussels, cockles, oysters, and scallops). Most crustaceans, of which there are about 39,000 species, are marine and include crabs, lobsters, shrimps, barnacles, and woodlice. The copepods, which are tiny crustaceans ranging from 0.02 to 0.7 in (0.05 to 1.8 cm) in length are particularly abundant. There could be hundreds of thousands of copepods in a cubic meter of surface water and they are an important source of food for predators like marine worms and the smaller jellyfish.

Marine worms are another diverse group, many of which are completely different from the well-known earthworm. For instance, the arrow worm is a major predator of the copepods. Only 1-in (2.5-cm) long, it is also known as the chaetognath worm, which means "bristle jaw." The worm darts and grabs its prey in its jaws. The nematodes, or thread worms, are among the tiniest of creatures in the marine environment, being just a fraction of an inch long. They live in sediments and feed off bacteria. As a group, the nematode worms have not been studied much, and researchers believe there may be thousands of species remaining to be discovered. At the other extreme are the tube worms found on the ocean floor, which may be up to 6 ft (1.8 m) in length. These creatures are so unusual that they have been placed in a phylum of their own. They have no mouths, but their bodies are filled with chemosynthetic bacteria, which extract energy from minerals rather than sunlight, and these provide much of their food supply.

Birds, mammals, and fish all live in or around the marine environment, but fish are the only vertebrates that are purely aquatic and are found in both freshwater and saltwater around the world. Around half of the 25,000 known species of fish live in the marine environment, mainly in shallower, warmer waters. Around 1,000 fish species occupy the open ocean. Fish that live in the deep ocean are generally black, brown, or gray, without the silvery camouflage that characterizes those living nearer the surface. Some are buoyant and swim up and down the depths of the water, searching for food, while others, including sharks and rays, are heavier and sink if they cease swimming. They tend to stay in place, catching food as it passes, or making just short hunting excursions.

Whales are marine mammals, breathing with lungs, and the largest animals in the oceans. There are two suborders: the baleen whales, which do not have teeth and filter feed on massive amounts of plankton; and the toothed whales, which feed on fish and squid. Toothed whales have the remarkable ability to navigate through the ocean by echolocation. The whales, sharks, and giant squid represent the top end of the marine ecosystem. Their sheer size has made them famous in marine culture and folklore.

In the open ocean, the ecosystems are vertically stratified. That is, different plants and animals are found at different depths. Sunlight penetrates to a depth of only 3 ft (0.9 m) or so in the cloudy waters of an estuary, compared to up to 300 ft (90 m) in the clearer waters of the open ocean. In this relatively light region, known as the euphotic zone, there is a net primary production of food by photosynthesis carried out by 4,000 or so species of phytoplankton.

The column of water extending down from the surface to a depth of about 2.5 mi (4 km) is called the pelagic zone and is composed of the epipelagic (top), mesopelagic (middle or twilight), and bathypelagic (bottom) zones. Below this, extending to about 4 mi (6.4 km), are the abyssal and hadal zones. Hadal comes from the Greek word for unseen, abyssal from the Greek word for bottomless. There is no light in these zones close to the ocean floor other than what is emitted from the organisms themselves. Thousands of species, including bacteria, squid, and fish, emit flashes of light by a process

Jean-Michel Cousteau views 82 tons of marine debris collected by NOAA's Debris team at Pearl and Hermes Atoll in a brief two-and-a-half week period in July 2003. *AP Images*

IN CONTEXT: THE GULF OF MEXICO

The Gulf of Mexico is a unique, semi-enclosed sea located between the Yucatán and Florida peninsulas, at the southeast shores of the United States. The Gulf of Mexico borders 5 of the 50 United States (Alabama, Florida, Louisiana, Mississippi, and Texas), and also Cuba and the eastern part of Mexico. Sometimes it is also called America's Sea. The Straits of Florida divides the Gulf from the Atlantic Ocean, while the Yucatán Channel separates it from the Caribbean Sea. The Gulf of Mexico covers more than 600,000 square mi (almost 1.5 million square km), and in some areas its depth reaches 12,000 ft (3,658 m; called Sigsbee Deep, or the "Grand Canyon under the sea"). About two thirds of the contiguous United States (31 states between the Rocky Mountains and the Appalachian Mountains) belongs to the watershed area of the Gulf of Mexico, while it receives freshwater from 33 major river systems, and many small rivers, creeks, and streams. This watershed area covers a little less than two million square mi (almost 5 million square km).

The Gulf of Mexico has several environmental quality problems originating either from natural processes, or from anthropogenic (human-caused) pollution, or their combination. The problems range from erosion and topsoil washing from the land into the Gulf, to oil spills and hazardous material spills, or trash washing ashore. These problems not only affect the estuaries, wetlands, and water quality in the Gulf, but have led to problems such as hypoxia (a zone of oxygen-depleted water), declining fish catch, contaminated fish, fish kills, endangered species, and air and water quality problems.

known as bioluminesence to distract predators and locate prey.

The oceans and seas contain some specific ecological niches. For instance parts of the ocean floor, or benthos, contain hydrothermal vents, where jets of water containing sulfur (S) compounds gush up, heated by the magma beneath the ocean floor. The ecological communities of microbes, worms, and mussels living around hydrothermal vents were known only since 1977. They can withstand temperatures of up to 660°F (350°C), and the microbes are capable of extracting biochemical energy from the sulfur compounds.

Shallow areas of the seas and oceans have rich ecosystems too. The sea floor slopes gradually from the shore out to the deeper ocean. The area near the shore is called the littoral zone, and many fish and shellfish are found here. Filter feeders such as mussels and barnacles live in rocky-bottomed shores, while soft-bottomed beaches are home to scavengers like shrimps and polychaete worms. The number of species increases nearer to water. On sand that is never completely submerged, it is necessary to dig in to find worms and sand crabs, while sea anemones and sea cucumbers are found near rocky shores. Estuaries, where river meets the sea, mixes freshwater and mud sediments with seawater and are often rich in fish because the nutrient levels in the water are high.

Mangrove swamps or forests are another important area near the tropical shores. Mangroves are salt-resistant trees that support fish or shrimp. Coral reefs such as the Great Barrier Reef, which stretches for around 1,200 mi (1,930 km) along the eastern coast of Australia, are probably one of the best-known marine ecosystems. They consist of animals called stony corals, each of which is a polyp with tentacles that can trap organisms. The polyps live in symbiosis with photosynthetic algae. The algae provide them with food, and the polyps provide the algae with protection. Each polyp lies in a shell of calcium carbonate and they are joined together by a sheet of tissue to form a giant organism and ecosystem in its own right. Coral reefs tend to occur in warm, shallow, clear waters where they provide a home for a diverse community of fish, worms, and crustaceans, protecting the small fish from larger predatory fish.

The marine food web is based upon the photosynthetic activity of phytoplankton, which is often greatest near the shore, where the water tends to be richer in nutrients like nitrogen (N) and phosphorus (P) because of runoff from land. Many tiny herbivores feed directly on phytoplankton, as well as the juvenile stages of certain squid and fish. These miniature predators will, in turn, be meals for slightly bigger creatures. The pattern goes on, up to the larger fish and sharks that are the equivalent of wolves and lions on the land. Generally, animals do not feed on those that are more than a tenth their size. However, there are exceptions to this. Some sharks will feed on animals as big or bigger than themselves, taking bite-sized lumps out of them. Fish with wide mouths and big teeth in the deeper layers can swallow bigger animals. The blue whale, the largest animal in the ocean, dines solely on tiny shrimp called krill. Meanwhile, dead phytoplankton and other organisms sink to the bottom of the ocean, forming a "marine snow" that benthic organisms, such as crabs and fish, depend upon as a food source.

Impacts and Issues

Human activity impacts upon marine ecosystems just as it does on land, with the areas nearest the coast being most affected. For instance, the coral reefs are among the world's most threatened biomes. They are affected by a range of factors, including destructive fishing practices, pollution, and sewage. Meanwhile, global warming is already causing bleaching of coral reefs. Increased temperatures destroy the symbiotic relationship between the coral and the algae that live there. According to research from the United Nations, one third of coral reefs around the world are destroyed already, 60% are damaged and will likely be dead by 2030. Mangrove swamps are similarly at risk.

Estuaries and shore areas are also at risk from pollution, which can cause eutrophication by raising nutrient levels in the water. Eutrophication encourages the growth of decomposers that consume available oxygen in the water. As oxygen levels fall, fish and other organisms begin to die off. Eutrophication also causes an overgrowth of algae in the water, often visible as a red, yellow, or green scum on the surface and a visible sign of a threatened ecosystem. These dead zones are found in many areas around the world, such as the Mediterranean and the East Coast of the United States.

The deep ocean has also been used as a dump for low-level radioactive waste, although this was banned in 1993. There have also been discussions on burying medium and high-level radioactive waste from nuclear power stations. If land alternatives prove too risky, these options may be put into practice, but no one knows what the long-term impact might be upon marine ecosystems, however secure the waste was made.

Humans do not just put things into the seas and oceans. They take things out as well. Fishing is a traditional activity, with fish being an important protein source in the human diet. However, global fish harvest-

Despite the death of much of the hard coral during the El Niño weather pattern, the ecosystem near Mahe Island, Seychelles, is making a comeback. *AP Images*

ing has increased 5 times during the last 50 years or so, partly because fishing technology has become more efficient and partly because of the increase in human population, which has increased demand for food. The oceans can probably support a fish harvest of about 100 million tons (90 million metric tons) of fish caught per year. As the fishing industry has expanded, these limits are being reached. Fishing vessels now have to travel farther and farther to get catches. Not only does this hurt the economy, there could also be as-yet-unknown disturbances to the marine ecosystem by driving fish stocks down in this way.

People are also seeking to exploit the ocean for oil and gas. Drilling offshore began in 1947 in the Gulf of Mexico, and now there are thousands of such developments. It may be that efforts to extract oil and gas will go deeper still, despite the difficulties of the technology, with unknown effects on the marine environment. There has also been discussion about whether it might be possible to exploit the sea bed as a source of minerals. As the world's population grows and industrial development spreads, the pressure to use the ocean as a resource can only increase, with unknown impacts on marine ecosystems.

SEE ALSO *Algal Blooms; Aquatic Ecosystems; Bays and Estuaries; Benthic Ecosystems; Coastal Ecosystems; Coastal Zones; Commercial Fisheries; Estuaries; Marine Fisheries; Ocean Salinity; Ocean Tides; Oceans and Coastlines; Overfishing; Reef Ecosystems*

BIBLIOGRAPHY

Books

Cunningham, W.P., and A. Cunningham. *Environmental Science: A Global Concern*. New York: McGraw-Hill International Edition, 2008.

Kaufmann, R., and C. Cleveland. *Environmental Science*. New York: McGraw-Hill International Edition, 2008.

Rice, Tony. *Deep Ocean*. London: Natural History Museum, 2000.

Susan Aldridge

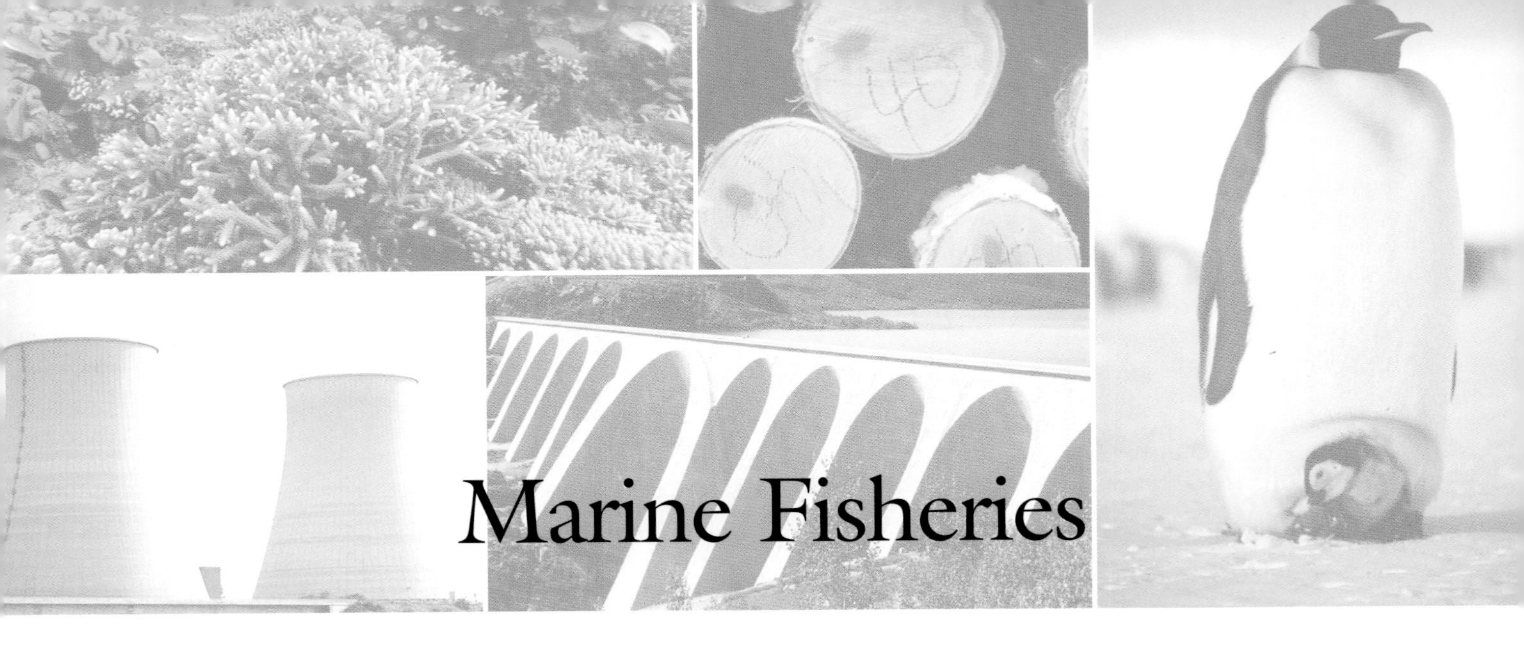

Marine Fisheries

Introduction

Around half of the world's fish species are found in the marine environment, and those that are edible have long provided an important food resource for humans and other animals. Fishing is now a huge commercial operation and marine fisheries have grown in size and sophistication in the last 50 years. Fish are caught in nets or on lines, sorted on board a fishing vessel, and then brought back to shore.

Although the oceans are vast, their fish stocks are not limitless. Many marine fisheries are now taking fish faster than stocks can renew themselves. This means that some fishes, such as the Atlantic cod, are now regarded as endangered species and several marine ecosystems are under threat. Therefore, many nations have now put limits on fishing, although it is not known to what extent fish stocks can recover. Consumers can help by purchasing from sustainable marine fisheries, which catch only those species whose populations are not threatened.

Historical Background and Scientific Foundations

There is a long tradition of fishing, both for pleasure and for food, around the world. Fish are highly valued as a source of animal protein, comparing favorably with meat in terms of cost and health benefits. Many health authorities actively promote regular consumption of fish, such as salmon and sardines, that are rich in heart-healthy omega-3 fatty acids. The average consumption of fish worldwide is about 30 pounds a week. Fishing covers harvesting fish themselves, shellfish, and other aquatic animals, such as squid and octopus, but not whales. Some marine fish are farmed in tanks, rather than in the open water. Besides food, fish are used for their oils, enzymes, and other manufactured products, as well as having a decorative and educational function in aquaria.

Of the 30,000 or so of identified fish species, about 500 are known to be edible. Around half of all known fish species live in a marine environment; the rest are found in freshwater. Tuna is found in the open ocean, whitefish and salmon around temperate coastlines, shrimp in coastal seas and oysters, clams, and mussels in estuaries. There are also areas called upwellings, where nutrient-rich deep water comes to the surface and can support high fish productivity. These are found off the western coasts of Africa, North and South America, across the Western Pacific, and around the Southern Ocean and are a popular site for marine fisheries. Countries with major fish industries include China, the United States, Chile, Iceland, and Norway.

Because the seas and oceans are so large, covering 70% of Earth's surface, it was once assumed that the supply of marine organisms to meet human needs is limitless. The fishing industry therefore expanded dramatically from the 1960s, with more vessels being put to sea and advances in fishing technology bringing in higher yields of fish. The way fish are caught varies with the size of the operation and the type of fish involved. Most fishing involves pots, lines, and nets, but the size of the vessel may vary from a single fisher with a few lines to massive ocean-going trawlers casting thousands of square miles of strong plastic netting each day. Fish harvesting has increased 5 fold in the last 50 years. From 1970 to 1990, it went up from 60 million to 80 million tons per year and then leveled off. Fish now account for about 16% of animal protein consumption by humans, and the industry employs about 10 million people directly, with many millions more being dependent on it for sustenance.

Removal of large numbers of fish species cause shifts in the marine food web. There are other effects on the marine ecosystem. Fishers divide their catch into two groups. Undersized and diseased fish that will not sell are thrown back to avoid taking up valuable space on the fishing vessel. The keepers, as they are known, are taken back to shore. The discards provide a food source for large sea birds, like the great skua in the North Sea. However, fewer of these discards have been available in recent years, because fish hauls have been restricted. Great skua populations have declined and those birds remaining have had to turn to alternative food sources, such as other, smaller, sea birds.

WORDS TO KNOW

FOOD WEB: An interconnected set of all the food chains in the same ecosystem.

OVERFISHING: Overharvesting applied to fish.

OVERHARVESTING: Harvesting so much of a resource that its economic value declines and/or its existence is threatened.

SUSTAINABLE: Capable of being sustained or continued for an indefinite period without exhausting necessary resources or otherwise self-destructing: often applied to human activities such as farming, energy generation, or the maintenance of a society as a whole.

■ Impacts and Issues

The unsustainable approach of marine fisheries over the last 50 years or so has now hit fish stocks hard. The marine environment can support the taking of around 100 million tons of fish a year. The global fishing fleet is now about two and a half times larger than what the oceans can sustainably support. Motors have replaced sails, boats are bigger, and onboard refrigerators allow fishers to stay at sea longer. Fisheries are sometimes subsidized unfairly and may be allowed to encroach upon the waters of developing nations, who have a more urgent need for the fish. Around 52% of the global marine fish stocks are classed as being fully exploited and 24% are overexploited, depleted, or recovering from depletion, the United Nations Food and Agriculture Organization (FAO) reports. If the current situation continues, there will be a major collapse in most species fished for food by 2048.

The Canadian bluefin fishery is managed with conservation, enforcement, and monitoring measures, including the type of gear used, fish size, and quota of fish to be caught. Federal observers are placed on various boats and at dockside to ensure compliance. *AP Images*

Over 1,000 species of fish are now endangered, either directly because of overfishing or because of the impact that marine fisheries have upon ecosystems. One of the most important, economically, is the Atlantic cod. Large open ocean fish, like tuna, swordfish, and marlin, are also in trouble, along with large groundfish such as halibut, skate, and flounder. Stocks of all these are down to around 10% according to the FAO.

Meanwhile, fish can even pose a health hazard. Fish species exposed to contaminated waters may concentrate toxins in their flesh, including pesticides, mercury, and polychlorinated biphenyls. These substances may enter the water at low concentrations, but as they are absorbed through the marine food chain, they accumulate to levels that may be harmful. Farmed fish may also be contaminated with the antibiotics used to ensure high yields.

Marine fisheries can give rise to other problems. Sometimes they harm or destroy coastal habitats, mangrove swamps, and coral reefs, the latter being especially endangered by bottom trawling. The method by which fish are harvested is also not always sustainable. For instance, nets used to catch shrimp, tuna, and scallops may harm other species. Unusual fish species from coral reefs may be taken to stock aquaria. Meanwhile, fish farms may deplete wild fish stocks when they take supplies for ponds. There are, however, a few fish species left that can be taken sustainably from the marine environ-

The livelihoods of many Bangladesh coastal communities are directly or indirectly dependent on fisheries resources in the local coastal and marine waters. Due to overfishing, the fisheries resources are being depleted and biodiversity is degraded.
AP Images

ment. These include farmed catfish or tilapia, Pacific pollack, squid, crayfish, crabs, and wild salmon.

The way forward for marine fisheries is to become more sustainable and consumers more aware. The Environmental Defense Fund recommends avoiding top predator species including swordfish, marlin, shark, bluefin, and albacore tuna when choosing a seafood meal. These species grow slowly and take several years to reproduce. Groundfish and deepwater fish that have been affected by overfishing include orange roughy; grouper; flounder or sole; cod; haddock; pollack, which is used in various fish products; and monkfish.

Many countries where marine fishing is an important industry now have legislation in place to protect stocks. In the United States, this is the Magnuson-Stevens Fishery Conservation and Management Reauthorization Act of 2006, which sets catch limits and requires sustainable management to protect against overfishing.

SEE ALSO *Commercial Fisheries; Inland Fisheries; Marine Ecosystems; Overfishing*

BIBLIOGRAPHY

Books

Cunningham, W.P., and A. Cunningham. *Environmental Science: A Global Concern*. New York: McGraw-Hill International Edition, 2008.

Kaufmann, R., and C. Cleveland. *Environmental Science*. New York: McGraw-Hill International Edition, 2008.

Rice, Tony. *Deep Ocean*. London: Natural History Museum, 2000.

Web Sites

Environmental Defense Fund. "Eco-Friendly Seafood Selector." 2008. http://www.edf.org/page.cfm?tagID=1521 (accessed March 31, 2008).

WWF. "Our Oceans Are Being Plundered." February 29, 2008. http://www.panda.org/about_wwf/what_we_do/marine/problems/problems_fishing/index.cfm (accessed March 31, 2008).

Susan Aldridge

Marine Water Quality

Introduction

Marine water quality refers to the presence or absence of any number of pollutants in ocean waters. Some of the more important pollutants include oil, sedimentation, sewage, nutrients, heavy metals, and thermal pollution. Water quality monitoring relies on taking a suite of measurements of ocean water. These include temperature, salinity, density, light transmission, the concentration of dissolved oxygen, and the concentration of chlorophyll (an indicator of the amount of phytoplankton in the water). Water samples are analyzed in the laboratory for the presence and concentration of various forms of bacteria, chlorophyll, phaeopigment (an indicator of the amount of dead phytoplankton in the water), nitrate, nitrite, ammonium, orthophosphate, and silicate. Significant legislation has been enacted to improve marine water quality such as the Clean Water Act, the Water Quality Act, and the Ocean Dumping Act. Several non-profit organizations also monitor ocean water quality and educate the public on marine water quality issues.

Historical Background and Scientific Foundations

Human activities influence water quality in numerous ways and as a result the concept of water quality has emerged, both as a scientific term and as a term that can be used for legislation of environmental policy. Water quality in the marine environment has a different emphasis than water quality in the freshwater environment where one of the concerns has to do with the quality of drinking water. In the marine environment, water quality is most often tied to the presence or absence of pollutants. Pollutants are substances or energy that affect any component of the ecosystem, be they chemical, biological, or physical substances. Some of the more influential pollutants in marine waters are oil, sediments, sewage, nutrients, heavy metals, and thermal pollution.

Oil seeps naturally into the ocean, however the amount of oil that reaches the ocean has well exceeded natural concentrations. As early as 1970, oceanographic ecologists noted that they could rarely pull a net through the surface of the ocean without collecting some form of tar or oil. About one-quarter of a percent of the oil produced each year, or about 6 million metric tons of oil, ends up in the ocean. A significant amount of oil input results from shipping oil from one place to another. Another important input is runoff from urban streets and sewers. The presence of oil in marine waters severely degrades water quality by clogging an animal's feeding-structures, killing larvae, and blocking available sunlight for photosynthesis.

Sediments are small particles of soil or sand that enter the ocean because of erosion. Erosion occurs because of natural weathering of rocks, and human activities often exacerbate erosion. Deforestation and overgrazing of range land can expose soils that get washed into the ocean. Construction generates sediments that also contribute to sedimentation. When sediments enter the water they absorb light, reducing the amount of photosynthesis that plants can perform. Because plants are the base of the food web in most marine systems, sedimentation has an effect on the overall ecosystem. Sediments can also cover the delicate tentacles of sessile species, disrupting or destroying their ability to filter food out of the water. This is of particular significance for coral reefs, which are often near shore in places where significant construction for recreation occurs. Sediments can also act as vehicles for other forms of marine pollution: toxic compounds and disease-causing agents.

Sewage is human waste along with water, food, detergents, and other material that people wash down drains. Sewage treatment facilities usually separate the liquid portion of sewage from the solid portion and then

WORDS TO KNOW

EUTROPHICATION: The process whereby a body of water becomes rich in dissolved nutrients through natural or man-made processes. This often results in a deficiency of dissolved oxygen, producing an environment that favors plant over animal life.

FOOD WEB: An interconnected set of all the food chains in the same ecosystem.

SESSILE: Any animal that is rooted to one place. Barnacles, for example, have a mobile larval stage of life and a sessile adult stage of life.

THERMAL POLLUTION: Industrial discharge of heated water into a river or lake, creating a rise in temperature that is injurious to aquatic life.

treat the components to remove harmful organisms. However, when heavy rains overwhelm sewer systems, raw sewage can enter the ocean. Disease-causing organisms can then be released into ocean water. In these cases, beachgoers and swimmers run the risk of becoming infected with bacteria or viruses that have contaminated the water. Sewage also releases heavy metals as well as organic and inorganic nutrients into ocean waters.

The addition of nutrients containing nitrogen and phosphorus to ocean waters can result in a condition known as eutrophication. Nutrients most often enter the ocean as runoff from agricultural land and through sewage. In eutrophic waters, the additional nutrients act as fertilizer stimulating the growth of phytoplankton (microscopic ocean plants). This often allows a single species, and in some cases a harmful species, to overgrow the rest of the phytoplankton community. In the wake of the outburst of growth, bacteria on the decaying phytoplankton coupled with cellular respiration of phytoplankton at night deplete the water of dissolved oxygen. When the oxygen concentrations in water decrease, aquatic animals, which rely on dissolved oxygen for respiration, can become stressed or die.

Heavy metals often enter the ocean through industrial pollution and in residues from burning fossil fuels. The most dangerous to the ecosystem are mercury and lead, which lead to neurological disorders in both humans and other animals. Human activity contributes about 5 times as much mercury and about 17 times as much lead to ocean waters, as do natural sources. In addition, copper, which is extremely toxic to marine invertebrates, has had negative impacts on numerous benthic communities.

Thermal pollution occurs when heated water that is produced by industry is released into the ocean. Many electrical power plants use seawater to produce steam and to cool moving parts. The water returned to the ocean is significantly warmer than ambient water. When this occurs, the organisms that usually live in an environment are unable to grow and reproduce optimally. Organisms from more tropical ecosystems take over available habitats. This changes the ecosystem dynamics in the region.

A sign warns swimmers about high levels of ocean water bacteria at Doheny State Beach in Dana Point, California, 2004. *AP Images*

The network of 52 small islands in the center of the Pacific Ocean is being purchased by the Nature Conservancy to guarantee the preservation of its pristine marine wilderness. *AP Images*

In order to measure water quality, ocean scientists use sampling equipment that measures some basic parameters of the water. The equipment may consist of a moored instrument that takes water quality measurements continuously. These instruments can also be lowered from the surface to the bottom of the ocean, giving a sense of how the water quality changes spatially. Some of the typical measurements made by these types of instruments include temperature, salinity, density, light transmission, the concentration of dissolved oxygen, and the concentration of chlorophyll (a proxy for the amount of phytoplankton in the water). In other cases, water samples are taken and analyzed in a laboratory. Often laboratory samples are measured for the presence and concentration of various forms of bacteria, chlorophyll, phaeopigment (a proxy for the amount of dead phytoplankton in the water), nitrate, nitrite, ammonium, orthophosphate, and silicate. Another measurement known as Secchi disk depth is an assessment of the particulate material such as sediments, phytoplankton, and bacteria in the water.

Temperature, salinity, and density are basic oceanographic measurements that determine what types of marine animals and plants can easily grow in a particular location. Dissolved oxygen helps scientists determine if the water has become eutrophic. Light transmission tells scientists how far light can penetrate into the water and how deep in the water phytoplankton are able to perform photosynthesis. Chlorophyll and phaeopigment measure the amount of living and dead plant material suspended in the water, respectively. The concentrations of different types of bacteria are often measured in water samples and may include total coliform, fecal coliform, fecal strep, *Enterococcus*, and *Clostridia*. All of these bacteria are found in the gastrointestinal tract of humans. These bacteria are not necessarily harmful in and of themselves, but may indicate the presence of disease-causing organisms such as the microorganisms that cause hepatitis, dysentery, and cholera.

■ Impacts and Issues

Public awareness that water quality is affected by dumping refuse into the water has influenced policy for more than a century. In 1899, the United States passed the Refuse Act, which was intended to reduce dumping pollution into public waterways. Since then numerous laws have been passed that attempt to impact water quality. The Clean Water Act of 1977 was initially passed as the Water Pollution Control Act in 1972 and was amended in 1981 and 1987. This act regulates the maximum quantity of pollutants that can be discharged into the ocean from point sources such as factories and sewage treatment plants. The Clean Water Act has had reasonable success at improving water quality. The Water Quality Act of 1987 has provisions to limit pollution from non-point sources, such as agricultural and urban runoff, however these forms of pollution are much more difficult to regulate. In 1991, the Ocean Dumping Ban Act was passed. This act forbids dumping sewage sludge and industrial waste into the ocean.

On a state level, many regions have additional regulations that mandate both the quantity of pollutants released into ocean water and the type of monitoring required to protect visitors to beaches. California has regulations that require local officials to monitor ocean water quality at highly trafficked beaches. These beaches are tested for total coliform bacteria, fecal coliform bacteria, and *Enterococcus* every week between April and October. A 1996 epidemiology study at the University of Southern California showed that when concentrations of these bacterial indicators are high, the probability that people entering the water will contract stomach flu, upper respiratory infections, and major skin rashes increases greatly. In addition, local law mandates that warning signs must be placed on beaches to indicate when water quality standards are not met.

In response to degraded water quality on popular beaches, several non-profit organizations have developed programs that perform independent water quality monitoring and provide public education and advocacy. In Southern California, the environmental group "Heal the Bay" posts weekly and monthly report cards of water quality at local beaches. The Surfrider Foundation has raised awareness of ocean water quality issues throughout the United States. They have lobbied state and local governments to improve and increase the frequency of water quality monitoring and have established water quality monitoring programs in many communities.

■ Primary Source Connection

Marine debris is a recurring problem and poses a grave threat to the pristine marine environment. It is a considerable hazard to marine life and is a constant danger for marine navigation.

Disasters like shipwrecks, oil tanker accidents, and offshore oil rig fires contribute to this problem, as does the disposal of urban garbage into oceans, dumping of waste oil by passing ships, commercial coastal and offshore entertainment, and deep-sea industrial and nuclear waste dumps.

Expanding oil and gas exploration activities into deeper seas are a serious threat to the marine environment. Over a prolonged period, accumulated debris fouls the seas and kills marine life. Marine debris, which is also a growing problem for fishing communities and shipping, causes a loss in fishing harvest, a vital source of sustenance for many fishing communities. Vessels damaged by marine debris collisions can cost a shipping company a fortune to repair.

To mitigate this danger and damage to the environment, Congress passed the Marine Debris Research Prevention and Reduction Act that was signed into law on December 22, 2006.

MARINE DEBRIS RESEARCH PREVENTION AND REDUCTION ACT

SEC. 2. FINDINGS AND PURPOSES.

(a) FINDINGS. The Congress makes the following findings:

1. The oceans, which comprise nearly three quarters of Earth's surface, are an important source of food and provide a wealth of other natural products that are important to the economy of the United States and the world.
2. Ocean and coastal areas are regions of remarkably high biological productivity, are of considerable importance for a variety of recreational and commercial activities, and provide a vital means of transportation.
3. Ocean and coastal resources are limited and susceptible to change as a direct and indirect result of human activities, and such changes can impact the ability of the ocean to provide the benefits upon which the Nation depends.
4. Marine debris, including plastics, derelict fishing gear, and a wide variety of other objects, has a harmful and persistent effect on marine flora and fauna and can have adverse impacts on human health.
5. Marine debris is also a hazard to navigation, putting mariners and rescuers, their vessels, and consequently the marine environment at risk, and can cause economic loss due to entanglement of vessel systems.
6. Modern plastic materials persist for decades in the marine environment and therefore pose the greatest potential for long-term damage to the marine environment.
7. Insufficient knowledge and data on the source, movement, and effects of plastics and other marine debris in marine ecosystems has hampered efforts to develop effective approaches for addressing marine debris.
8. Lack of resources, inadequate attention to this issue, and poor coordination at the Federal level has undermined the development and implementation of a Federal program to address marine debris, both domestically and internationally.

(b) PURPOSES. The purposes of this Act are:

1. to establish programs within the National Oceanic and Atmospheric Administration and the United States Coast Guard to help identify, determine sources of, assess, reduce, and prevent marine debris and its adverse impacts on the marine environment and navigation safety, in coordination with other Federal and non-Federal entities;

2. to re-establish the Inter-agency Marine Debris Coordinating Committee to ensure a coordinated government response across Federal agencies;
3. to develop a Federal information clearinghouse to enable researchers to study the sources, scale and impact of marine debris more efficiently; and
4. to take appropriate action in the international community to prevent marine debris and reduce concentrations of existing debris on a global scale.

SEC. 3. NOAA MARINE DEBRIS PREVENTION AND REMOVAL PROGRAM.

(a) ESTABLISHMENT OF PROGRAM—There is established, within the National Oceanic and Atmospheric Administration, a Marine Debris Prevention and Removal Program to reduce and prevent the occurrence and adverse impacts of marine debris on the marine environment and navigation safety.

(b) PROGRAM COMPONENTS—Through the Marine Debris Prevention and Removal Program, the Under Secretary for Oceans and Atmosphere (Under Secretary) shall carry out the following activities:

(1) MAPPING, IDENTIFICATION, IMPACT ASSESSMENT, REMOVAL, AND PREVENTION—The Under Secretary shall, in consultation with relevant Federal agencies, undertake marine debris mapping, identification, impact assessment, prevention, and removal efforts, with a focus on marine debris posing a threat to living marine resources (particularly endangered or protected species) and navigation safety, including—

(A) the establishment of a process, building on existing information sources maintained by Federal agencies such as the Environmental Protection Agency and the Coast Guard, for cataloguing and maintaining an inventory of marine debris and its impacts found in the United States navigable waters and the United States exclusive economic zone, including location, material, size, age, and origin, and impacts on habitat, living marine resources, human health, and navigation safety;

(B) measures to identify the origin, location, and projected movement of marine debris within the United States navigable waters, the United States exclusive economic zone, and the high seas, including the use of oceanographic, atmospheric, satellite, and remote sensing data; and

(C) development and implementation of strategies, methods, priorities, and a plan for preventing and removing marine debris from United States navigable waters and within the United States exclusive economic zone, including development of local or regional protocols for removal of derelict fishing gear.

(2) REDUCING AND PREVENTING LOSS OF GEAR—The Under Secretary shall improve efforts and actively seek to prevent and reduce fishing gear losses, as well as to reduce adverse impacts of such gear on living marine resources and navigation safety, including—

(A) research and development of alternatives to gear posing threats to the marine environment, and methods for marking gear used in specific fisheries to enhance the tracking, recovery, and identification of lost and discarded gear; and

(B) development of voluntary or mandatory measures to reduce the loss and discard of fishing gear, and to aid its recovery, such as incentive programs, reporting loss and recovery of gear, observer programs, toll-free reporting hotlines, computer-based notification forms, and providing adequate and free disposal receptacles at ports.

(3) OUTREACH—The Under Secretary shall undertake outreach and education of the public and other stakeholders, such as the fishing industry, fishing gear manufacturers, and other marine-dependent industries, on sources of marine debris, threats associated with marine debris and approaches to identify, determine sources of, assess, reduce, and prevent marine debris and its adverse impacts on the marine environment and navigational safety. Including outreach and education activities through public-private initiatives. The Under Secretary shall coordinate outreach and education activities under this paragraph with any outreach programs conducted under section 2204 of the Marine Plastic Pollution Research and Control Act of 1987 (33 U.S.C. 1915).

(c) Grants—

(1) IN GENERAL—The Under Secretary shall provide financial assistance, in the form of grants, through the Marine Debris Prevention and Removal Program for projects to accomplish the purposes of this Act.

(2) 50 percent matching requirement—

(A) IN GENERAL—Except as provided in subparagraph (B), Federal funds for any project under this section may not exceed 50 percent of the total cost of such project. For purposes of this subparagraph, the non-Federal share of project costs may be provided by in-kind contributions and other noncash support.

(B) WAIVER—The Under Secretary may waive all or part of the matching requirement under subparagraph (A) if the Under Secretary determines that no reasonable means are available through which applicants can meet the matching requirement and the probable benefit of such project outweighs the public interest in such matching requirement.

(3) Amounts paid and services rendered under consent—

(A) CONSENT DECREES AND ORDERS—The non-Federal share of the cost of a project carried out under this Act may include money paid pursuant to, or the value of any in-kind service performed under, an administrative order on consent or judicial consent decree that will remove or prevent marine debris.

(B) OTHER DECREES AND ORDERS—The non-Federal share of the cost of a project carried out under this Act may not include any money paid pursuant to, or the value of any in-kind service performed under, any other administrative order or court order.

(4) ELIGIBILITY—Any natural resource management authority of a State, Federal or other government authority whose activities directly or indirectly affect research or regulation of marine debris, and any educational or nongovernmental institutions with demonstrated expertise in a field related to marine debris, are eligible to submit to the Under Secretary a marine debris proposal under the grant program.

(5) GRANT CRITERIA AND GUIDELINES—Within 180 days after the date of enactment of this Act, the Under Secretary shall promulgate necessary guidelines for implementation of the grant program, including development of criteria and priorities for grants. Such priorities may include proposals that would reduce new sources of marine debris and provide additional benefits to the public, such as recycling of marine debris or use of biodegradable materials. In developing those guidelines, the Under Secretary shall consult with:—

(A) the Interagency Marine Debris Committee;

(B) regional fishery management councils established under the Magnuson-Stevens Fishery Conservation and Management Act (16 U.S.C. 1801 et seq.);

(C) State, regional, and local governmental entities with marine debris experience;

(D) marine-dependent industries; and

(E) non-governmental organizations involved in marine debris research, prevention, or removal activities.

(6) PROJECT REVIEW AND APPROVAL—The Under Secretary shall review each marine debris project proposal to determine if it meets the grant criteria and supports the goals of the Act. Not later than 120 days after receiving a project proposal under this section, the Under Secretary shall:—

(A) provide for external merit-based peer review of the proposal;

(B) after considering any written comments and recommendations based on the review, approve or disapprove the proposal; and

(C) provide written notification of that approval or disapproval to the person who submitted the proposal.

(7) PROJECT REPORTING—Each grantee under this section shall provide periodic reports as required by the Under Secretary. Each report shall include all information required by the Under Secretary for evaluating the progress and success in meeting its stated goals, and impact on the marine debris problem.

SEC. 4. COAST GUARD PROGRAM.

The Commandant of the Coast Guard shall, in cooperation with the Under Secretary, undertake measures to reduce violations of MARPOL Annex V and the Act to Prevent Pollution from Ships (33 U.S.C. 1901 et seq.) with respect to the discard of plastics and other garbage from vessels. The measures shall include—

1. the development of a strategy to improve monitoring and enforcement of current laws, as well as recommendations for statutory or regulatory changes to improve compliance and for the development of any appropriate amendments to MARPOL;
2. regulations to address implementation gaps with respect to the requirement of MARPOL Annex V and section 6 of the Act to Prevent Pollution from Ships (33 U.S.C. 1905) that all United States ports and terminals maintain receptacles for disposing of plastics and other garbage, which may include measures to ensure that a sufficient quantity of such facilities exist at all such ports and terminals, requirements for logging the waste received, and for Coast Guard comparison of vessel and port log books to determine compliance;
3. regulations to close record keeping gaps, which may include requiring fishing vessels under 400 gross tons entering United States ports to maintain records subject to Coast Guard inspection on the disposal of plastics and other garbage, that, at a minimum, include the time, date, type of garbage, quantity, and location of discharge by latitude and longitude or, if discharged on land, the name of the port where such material is offloaded for disposal;
4. regulations to improve ship-board waste management, which may include expanding to smaller vessels existing requirements to maintain ship-board

receptacles and maintain a ship-board waste management plan, taking into account potential economic impacts and technical feasibility;

5. the development, through outreach to commercial vessel operators and recreational boaters, of a voluntary reporting program, along with the establishment of a central reporting location, for incidents of damage to vessels caused by marine debris, as well as observed violations of existing laws and regulations relating to disposal of plastics and other marine debris; and
6. a voluntary program encouraging United States flag vessels to inform the Coast Guard of any ports in other countries that lack adequate port reception facilities for garbage.

U.S. Congress

U.S. CONGRESS. "MARINE DEBRIS RESEARCH PREVENTION AND REDUCTION ACT," FEBRUARY 10, 2005.

SEE ALSO *Algal Blooms; Aquatic Ecosystems; Bays and Estuaries; Benthic Ecosystems; Clean Water Act; Coastal Ecosystems; Coastal Zones; Coral Reefs and Corals; Hazardous Waste; Industrial Water Use; Marine Ecosystems; Nonpoint-Source Pollution; Oceans and Coastlines; Oil Pollution Acts; Real-Time Monitoring and Reporting; Runoff; Toxic Waste; Wastewater Treatment Technologies; Water Pollution*

BIBLIOGRAPHY

Books

Garrison, Tom. *Oceanography: An Invitation to Marine Science,* 5th ed. Stamford, CT: Thompson/Brooks Cole, 2004.

Raven, Peter H., Linda R. Berg, and George B. Johnson. *Environment.* Hoboken, NJ: Wiley, 2002.

Web Sites

Environmental Science and Technology Online. "Marine Waters." January 19, 2005. http://pubs.acs.org/subscribe/journals/esthag-w/2005/jan/science/jp_nutrient.html (accessed February 28, 2008).

Heal the Bay. "Beach Report Card." January 16, 2007. http://www.healthebay.org/brc/ (accessed February 28, 2008).

Los Angeles County Department of Public Health. "State Ocean Water Quality Standards." http://www.lapublichealth.org/eh/progs/envirp/rechlth/ehrecocstand.htm (accessed February 28, 2008).

Surfrider Foundation. "Water Quality." http://www.surfrider.org/waterquality.asp (accessed February 28, 2008).

U.S. Congress. "Marine Debris Research Prevention and Reduction Act," February 10, 2005. http://www.govtrack.us/congress/billtext.xpd?bill=s109-362 (accessed April 10, 2008).

Washington State Department of Ecology. "Marine Waters." http://www.ecy.wa.gov/programs/eap/mar_wat/mwm_intr.html (accessed February 28, 2008).

Juli Berwald

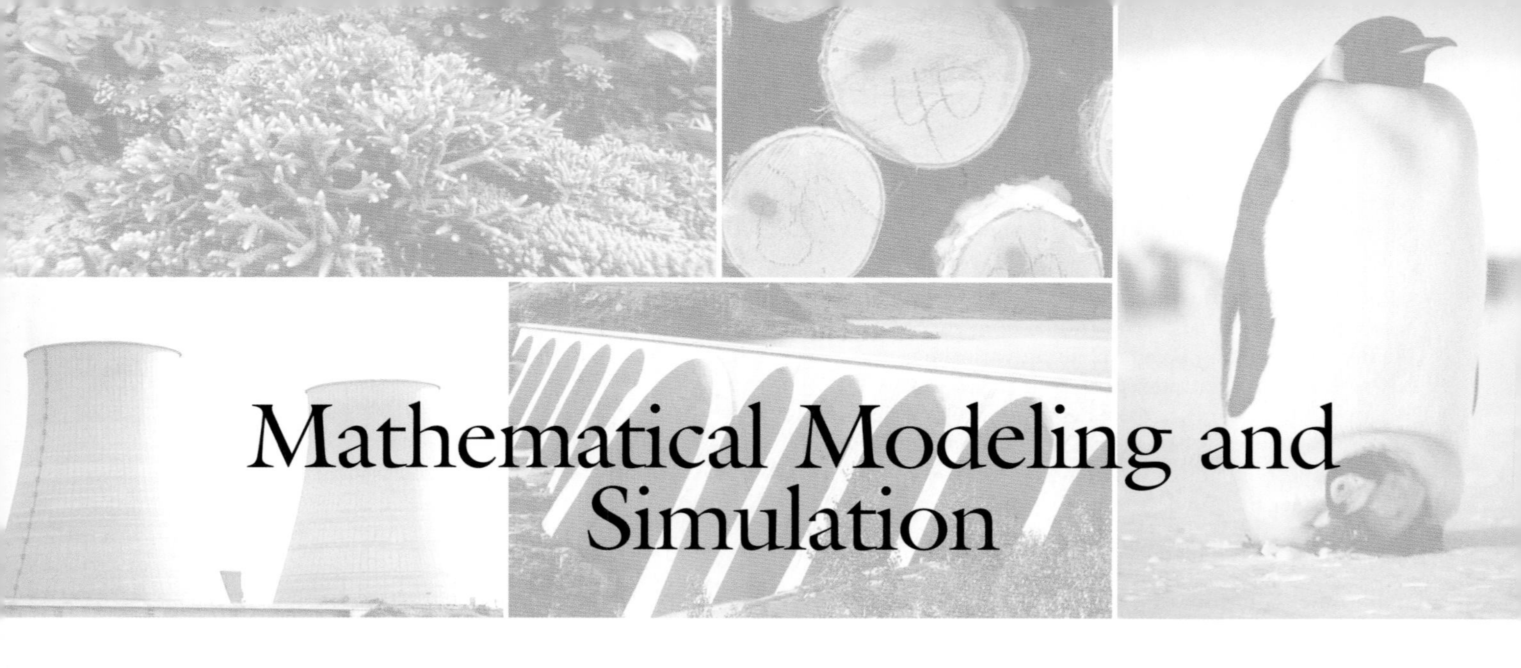

Mathematical Modeling and Simulation

Introduction

Mathematical modeling and simulation are important research and monitoring tools used to understand biological communities and their relationships to the environment. Mathematical models are collections of variables, equations, and starting values that form a cohesive representation of a process or behavior. Because interactions among the members of biological communities and components of the abiotic environment are extremely complex, mathematical models are useful for understanding how ecosystems function and for making predictions about managing ecosystems.

There are numerous types of mathematical models used by ecologists and environmental scientists. Some models isolate the key factors that drive elements of a system. Other mathematical models are comprehensive simulations that include as many components and interactions as possible. Mathematical models also cover different spatial and temporal scales: from the smallest tidepool ecosystem to the entire planet; from a single day to millions of years.

Historical Background and Scientific Foundations

Mathematical models and simulations are used scientifically as a tool for improving the understanding of the ecology of a region and managerially as a tool for making decisions regarding resource and environmental issues. Depending on the scope and sensitivity of the system, developing mathematical models of an environment can be an extremely interdisciplinary undertaking. It might require input from ecologists, biologists, chemists, physicists, engineers, and computer scientists, as well as social scientists, economists, and politicians. Results from mathematical models might be shared among scientists, policy makers, and various government officials, often from different localities and jurisdictions.

Mathematical models usually have three basic parts. These are the variables and their definitions, the equations into which the variables are incorporated, and starting values for the variables. When mathematical models are applied to ecological situations, more information is required. For example, an ecological model requires the user to assign a meaning to the variables, to know the units of the variables, and to bind the ranges of values over which the variables are realistic. It might also be useful to know the way that the values of the variables are measured, the ecological context of the variables, the reproducibility of the values of the variables, among other information.

There are two basic approaches to building a mathematical model of an ecosystem. The first is called the compartmental system approach or a conceptual model. This type of model is often depicted using box and flow illustrations. The boxes represent a pool or a compartment of something found in an ecosystem. For example, a box might represent the total weight of carbon in the phytoplankton in a certain region of the ocean. The box is then impacted by several flows, visually represented by arrows into and out of the box. An arrow into the phytoplankton carbon box might represent the rate of conversion of atmospheric carbon dioxide into carbohydrate by photosynthesis. An arrow out of the box might represent the loss of carbon from the community of phytoplankton because of predation.

Conceptual models most often put emphasis on the gross dynamics of a whole ecosystem. They tend to be rather general models that can be applied to many different systems. For example, the phytoplankton model discussed earlier may be altered for use in either the Pacific Ocean or the Atlantic Ocean. However, conceptual models tend to be imprecise. The rates of photosynthesis in the Atlantic and Pacific Oceans are likely very different. This is because the climate conditions, the nutrients that drive photosynthesis, and the species of phytoplankton all differ from place to place. These factors are largely omitted from conceptual or compartmental system models.

In contrast, the qualitative model, which is also referred to as the experimental components approach, control model, or stressor model, incorporates as many interactions and components of a system as possible. Instead of using a single compartment to represent the phytoplankton carbon, a qualitative model would represent the various populations of phytoplankton individually. It would take experimental results of photosynthetic rates for the different species and use this information to come up with a rate of carbon dioxide assimilation for the entire community. Qualitative models tend to be more precise than compartmental models, but they are also very specific. It would be difficult to use a qualitative model of phytoplankton growth designed to represent Pacific Ocean phytoplankton when modeling the phytoplankton in the Atlantic Ocean. Qualitative models can be extremely useful for understanding how ecosystems will respond to environmental stresses.

Depending on the needs of the users and the ecological questions addressed, several types of models may be developed to represent a single ecosystem. Ecological models can also be embedded within one another. One part of the system may be represented by a qualitative model, while another part is represented by a conceptual model. Many ecological models have a geographical component. Geographical information systems (GIS) and geospatial analyses are often used in conjunction with ecological simulations.

Mathematical models and simulations are used to better understand a multitude of ecological issues. For example, simulations are commonly used to model biogeochemical cycling within aquatic systems. They are used to understand the way that toxic materials move through ecosystems. These toxic materials include pesticides, heavy metals, and radionuclides. Ecological models of ground water flow and plume diffusion in air are often incorporated into these ecotoxicology models. In terrestrial systems, scientists have developed agricultural models and forestry models. One of the most common types of mathematical models involves population dynamics. Population models are used to predict and understand the age structure of fish, the age structure of trees, and rates of primary production and photosynthesis. Predator-prey models (the Lotka-Volterra equations) are commonly used to predict the population dynamics of predator and prey when external environmental impacts are minimal.

WORDS TO KNOW

ABIOTIC: A term used to describe the portion of an ecosystem that is not living, such as water or soil.

CONCEPTUAL MODEL: the collection of all populations of organisms in an area.

ECOSYSTEM: all of the physical properties of a location.

POPULATION MODEL: A mathematical model that can project the demographic consequences to a species as a result of certain changes to its environment.

PREDATOR-PREY MODEL: A mathematical model used to analyze interaction between two species in an ecosystem.

Impacts and Issues

Using mathematical models appropriately involves some challenges. Users need to be aware of the assumptions in any model of an ecological system. For example, boundary conditions, time steps, valid ranges of variables, appropriateness of equations chosen to represent actual behavior or measurements, and the stability of the underlying mathematics can all impact whether or not the model actually answers the questions it sets out to answer. Mathematical modelers spend a significant amount of time testing simulations to assess how well the model corresponds to expected patterns from the real system and how the model changes when perturbations to the system occur.

Perhaps the most well publicized and most controversial types of ecological simulations are climate models. Because the global climate is such a complex system, it is very difficult to incorporate all of the factors that influence it. As a result, climate simulations tend to be conceptual models. Conceptual models are fundamentally imprecise, and depending on starting values and assumptions, a wide range of predictions can be made. This led to controversy over the validity of the concept of climate change resulting from anthropogenic factors. Through years of research and study, climate change models have became more refined and the predictions made by the models more cohesive. The Intergovernmental Panel on Climate Change (IPCC) has followed and assessed much of the climate change modeling over the last decade. The models that IPCC scientists have studied are broad-scale and identify the drivers and stressors that affect global climate change.

SEE ALSO *Careers in Environmental Science; Climate Modeling; Ecosystems; Environmental Assessments; Geographic Information Systems (GIS); Geospatial Analysis; Global Warming; Intergovernmental*

A wave is generated in the tsunami wave basin at the Hinsdale Wave Research Laboratory at Oregon State, December 2005. The lab is used by researchers so they can simulate tsunami waves as they approach a coastline and assess the level of damage they can cause. *AP Images*

Panel on Climate Change; IPCC 2007 Report; Laboratory Methods in Environmental Science; Predator-Prey Relationships; Radiative Forcing; Real-Time Monitoring and Reporting

BIBLIOGRAPHY

Books

Grant, William, and Todd Swannack. *Ecological Modeling*. Malden, MA: Blackwell Publishing, 2007.

Hadlock, Charles R. *Mathematical Modeling in the Environment*. Boca Raton, FL: CRC Press, 2007.

Odum, Eugene, and Gary W. Barrett. *Fundamentals of Ecology*, 5th ed. St. Paul, MN: Brooks Cole, 2004.

Pastorok, Robert A., Steven M. Bartell, Scott Ferson, and Lev R. Ginzberg, eds. *Ecological Modeling in Risk Assessment: Chemical Effects on Populations, Ecosytems, and Landscapes*. Boca Raton, FL: CRC Press, 2002.

Web Sites

Intergovernmental Panel on Climate Change. "Home Page." http://www.ipcc.ch/ (accessed March 4, 2008).

International Society for Ecological Modeling. "Home Page." October 25, 2007. http://www.isemna.org/ (accessed March 2, 2008).

National Park Service. "Developing Conceptual Models of Relevant Ecosystem Components." February 20, 2008. http://science.nature.nps.gov/im/monitor/ConceptualModels.cfm (accessed March 2, 2008).

Smithsonian Environmental Research Center. "Ecological ModelingLab." http://www.serc.si.edu/labs/ecological_modeling/index.jsp (accessed March 2, 2008).

Juli Berwald

Media: Environmentally Based News and Entertainment

Introduction

As awareness and concern of environmental issues have grown, the reporting of environmental events and issues by news organizations has also expanded. Also, the use of an environmental issue as the premise for entertainment has become attractive. A well-known recent example of the use of environmental issues as entertainment is the 2004 movie titled *The Day After Tomorrow*, a fictional tale of the tribulations caused by a sudden global climate change. The consensus among climate scientists was negative, and the accuracy of the movie's science was criticized.

The accuracy of environmental reporting has also become an issue in an industry where reporters are often from a journalist background, which does not necessarily include training in science. Furthermore, the arresting images of environmental damage and extreme weather can be sensationalized in the few minutes allotted on a mainstream national news broadcast to drive up viewer or readership numbers rather than to convey the complexities of the issues.

However, there has been recognition that environmental reporting often requires a more in-depth exploration. Many newspapers now have reporters specifically assigned to cover environmental topics and many Web sites act as a clearinghouse of environmental information.

Historical Background and Scientific Foundations

In 1962 the book *Silent Spring* was published. In the work, American author Rachel Carson (1907–1964) warned of the consequences of the widespread and extensive use of the pesticide DDT on the natural environment and humans. Given that relatively little was known of the interactions of DDT with living creatures, the book proved to be prophetic; research over the next decade linked DDT exposure to adverse effects in humans, animals, and birds, both in the short and long term.

The book was influential in bringing the state of the environment to the attention of many people and was one reason for the surge in environmentalism. News organizations responded by reporting more on the adverse aspects of environmental issues than had been done prior to the 1960s. Partially as a result of the publication of *Silent Spring* and the public concern that followed, the U.S. Environmental Protection Agency (EPA) was established in 1970.

Part of the motivation for environmental reporting by the mainstream newspapers and television networks was, and has remained, readership and viewers. Descriptions and, particularly, images of environmental disasters and extreme events such as tornados, hurricanes, and the aftermath of a tsunami are attention getting. In the few articles devoted to such an issue in a newspaper or the brief time allotted on a network evening news broadcast, the riveting information often has proven to be more attractive than an exploration of the environmental issue.

A main reason for this is the complexity of many environmental events and topics. Reporting on the changes to an ecosystem, for example, is complicated because many living and nonliving components are part of the ecosystem, and it is their many interactions that determine how the ecosystem functions. Likewise, climate change is a very complex topic that remains not fully understood by many people.

Environmental reporting involves an understanding of science. Many news reporters are trained as journalists rather than as scientists, although this is changing. Indeed, several universities in the United States and elsewhere offer degrees in science journalism. Still, the task of translating environmental science information into a

WORDS TO KNOW

ANTHROPOGENIC: Made by humans or resulting from human activities.

DDT (DICHLORO-DIPHENYL-TRICHLOROETHANE): One of the earliest insecticides, dichloro-diphenyl-trichloroethane, used until banned by many countries in the 1960s after bird populations were decimated by the substance, and other negative environmental consequences occurred. Lately, selective DDT use has returned in targeted areas in Africa in order to eliminate high concentrations of the mosquitoes that carry the parasite that causes malaria.

form that is relevant and meaningful to a reader or viewer who may have little connection with science is challenging, and it is sometimes done poorly.

Environmentally based entertainment generally focuses on the sensational aspects of environment events. *The Day After Tomorrow* is a recent example of how scientific facts can be manipulated to produce a film with broad popular appeal.

Despite the more sensational presentation of environmentally based news, many newspapers have reporters specifically assigned to cover environmental happenings. Additionally, environmental reporting is suited to the more extensive treatment of an issue that is possible in weekend editions of a newspaper and as a book.

The internet is proving to be an ideal format for environmental reporting. A variety of Web sites dedicated to environmental news exist. Often, information from a wide variety of sources is assembled and links are provided to the sites. This allows the user to acquire information at a pace and to a depth that s/he chooses.

Impacts and Issues

A hazard of environmental news reporting is the journalist tradition of presenting all sides of an issue. While presenting a single view of an issue could be biased, environmental issues such as global warming are now more consensual, that is the majority of scientists come to accept a view that is supported by the bulk of the available evidence and which can be used to make logical predictions. Thus, environmental reporting about global warming that devotes as much space to the view that global warming is not influenced by human activities as to the consensual view that human activities are the main driver of the atmospheric warming, while being journalistically correct, is inaccurate and misleading.

The release of the 2007 report by the Intergovernmental Panel on Climate Change (IPCC) was extensively covered in the mainstream news media. Interestingly, the panel's conclusion that the accelerating atmospheric temperature increase evident since the 1950s "is very likely due to the observed increase in anthropogenic greenhouse gas concentrations" was widely quoted. The term anthropogenic became an often-quoted word in public conversation, and the growing public awareness and understanding of the term attests to the growing environmental savvy of even those who are nonscientists. Studies conducted in the United States and the United Kingdom have shown that the news coverage of climate change has risen markedly since the year 2000. For example, in 2003 the United Kingdom's three largest newspapers by subscribers published an average of about 75 climate change related arti-

Australian Steve Irwin and his wife Terri holding a 9-foot (2.7-m) female alligator in 1999 at their "Australia Zoo" in Beerwah, Queensland, Australia. Irwin, the Australian television personality and environmentalist known as the Crocodile Hunter, was killed in 2006 by a stingray barb during a diving expedition. *AP Images*

cles every month. By 2006 the average monthly total was about 250 articles.

Environmental reporting has spawned a new type of journalist—one who is trained in both reporting and science. Indeed, environmental reporters can come from a science background, having subsequently acquired the journalistic training and skills. The ability to convey environmental science information in a way that is meaningful and relevant to the general public is an increasingly important skill.

SEE ALSO *Corporate Green Movement; Environmental Protests; IPCC 2007 Report; Non-Scientist Contributions to Nature and Environment Studies; Photography, Environmental*

BIBLIOGRAPHY

Books

Jarrell, Melissa. *Environmental Crime and the Media: News Coverage of Petroleum Refining Industry Violations.* New York: LFB Scholarly Publishing LLC, 2007.

Parker, Lee. *Environmental Communication: Messages, Media, and Methods.* Dubuque: Kendall/Hunt Publishing, 2005.

Periodicals

Berg, Rebecca. "Environmental Health and the Media, Part 3: Make Noise, Make the News." *Journal of Environmental Health* 68: 73–78 (2006).

Brian D. Hoyle

Medical Waste

Introduction

Medical waste, also termed health-care waste, is garbage and effluent (liquid waste) produced by hospitals, clinics, research facilities, laboratories, and veterinary facilities. Most health-care waste (up to 85%) is similar to ordinary non-hazardous household trash, but about 10% consists of potentially infectious biological materials and 5% of hazardous wastes (chemical, radioactive, and pharmaceutical substances). Sometimes, the term medical waste is restricted to the infectious/hazardous component of health-care waste. In the United States, medical waste is governed by the Medical Waste Tracking Act of 1988, which required the U.S. Environmental Protection Agency (EPA) to establish a medical-waste tracking program.

Historical Background and Scientific Foundations

Medical waste first became a major political issue in the United States in 1987 and 1988, when large amounts of medical waste and other trash washed up on New Jersey beaches. The waste included disposable syringes (injection needles), so the event came to be known as the Syringe Tide. Hearing about the repulsive and possibly dangerous pollution, thousands of tourists stayed away from the Jersey shore, usually a prime tourist destination, costing the state almost one billion dollars in lost business revenue. New Jersey blamed New York City for the waste, and indeed, it was eventually traced to New York's Fresh Kills Landfill on Staten Island. In late 1987, a federal judge approved an agreement in which New York City accepted partial responsibility for the medical and other trash washing up on New Jersey beaches, and agreed to pay one New Jersey town, Woodbridge, $1 million in compensation and to pay for cleanup of the town's shoreline. The suit actually long predated the Syringe Tide, being brought by the town in 1979. Regular collection of floating trash by New York City and the U.S. Army Corps of Engineers has kept closings of beaches due to washed-up trash much lower since 1989.

Publicity over the medical waste washing up on New Jersey beaches led to hundreds of pieces of legislation (according to the EPA) originating from town and city governments up to Congress. The most far-reaching law pertaining to medical waste to come from the Syringe Tide incident was the Medical Waste Tracking Act of 1988. This laid out a federal definition of what constitutes medical waste and mandated a cradle-to-grave tracking system for June 1989 to June 1991 under which hospitals would have to keep track of their medical waste so that its proper disposal, usually by incineration, would be verifiable. Penalties for mismanagement of medical waste were established. The tracking system was applied only in New York, New Jersey, Connecticut, Rhode Island, and Puerto Rico and expired in 1999.

What Is Medical Waste?

The hazardous fraction of medical waste can be broken down into a number of sub-categories. Different expert sources use somewhat different categories, but the World Health Organization of the United Nations recognizes the following:

- Infectious waste: includes used bandages and dressings; swabs, sponges, tubes, and other disposable products used in surgery; infectious cultures from laboratories; all wastes from infectious-disease wards.
- Pathological waste: tissue samples, body parts removed during surgery or autopsy, blood and blood products, dead infected laboratory animals, body fluids, etc.

- Sharps: all items that can cut or puncture the skin: most commonly needles, used surgical tools, and broken glass.
- Pharmaceutical waste: unused drugs of all kinds.
- Genotoxic waste: waste containing substances that interfere with the functioning of DNA, whether by causing cancer, interfering with fetal development, or causing mutations. The main substances in this category are cytotoxic (cell-toxic) drugs, widely used in chemotherapy for cancer to interfere with the replication of cancer cells.
- Chemical waste: waste containing chemical substances such as disinfectants, solvents, cleansers, preservatives (e.g., formaldehyde), photographic chemicals, oils, pesticides, acids, or the like.
- Heavy metals: wastes containing a high heavy-metal content. The most common toxic heavy metals are lead and mercury, which are found in batteries, old thermometers, and wall-mounted sphygmomanometers (blood-pressure gauges), and other wastes.
- Pressurized containers: all containers containing gas under pressure, such as handheld aerosol cans and oxygen tanks.
- Radioactive waste: all radioactive materials, including sealed sources of radiation from radiotherapy machines and substances used in nuclear medicine.

WORDS TO KNOW

RADIOACTIVITY: The property possessed by some elements of spontaneously emitting energy in the form of particles or waves by disintegration of their atomic nuclei.

SHARPS WOUNDS: Wounds caused by discarded medical instruments including hypodermic needles and, scalpels (knives) and other sharp objects potentially contaminated by human contact.

TOXIC: Something that is poisonous and that can cause illness or death.

An incinerator disposes of medical waste by burning it at extremely high temperatures to destroy harmful microorganisms and toxic chemicals in the waste. The emission of smoke and combustion gases is tightly regulated. *Robert Brook/Photo Researchers, Inc.*

Impacts and Issues

The type of hazard posed by medical waste depends on what type, or mixture of types, of waste it contains. It may contain infectious agents (viruses or bacteria that can cause disease), genotoxic agents, toxic chemicals, radioactivity, or sharps that can cause wounds and transmit infections (e.g., of HIV, the virus that causes AIDS). Persons working in health-care facilities are most exposed to these hazards; the next most exposed category includes people who handle waste directly from such facilities, such as garbage collectors.

Injury by sharps is the most common form of harmful encounter with medical waste. For example, from 46,000 to 70,000 nurses are injured by used sharps every year in the United States; from 500 to 7,300 waste handlers are injured. Cases of disease transmission by hospital waste have been documented: most of these involve hospital workers handling sharps. For example, by June 1996, the U.S. Centers for Disease Control and Prevention (CDC) had identified 51 cases of HIV infection due to occupational exposure. Most of the cases were from sharps wounds; four involved contact with infected blood. Data from Japan show that puncture by a used hypodermic needle in medical waste conveys a 0.3% chance of contracting HIV, a 3% chance of contracting viral hepatitis B, and a 3 to 5% chance of contracting viral hepatitis C.

The EPA, after the two years when it tracked medical waste in part of the United States, concluded that the potential of medical waste to cause disease is greatest when it is newly generated. At that point it contains the most living disease organisms and its chemicals are most concentrated. Risk from exposure to medical waste decreases as the waste ages.

Few data are available on the risks to health workers and the public of chemical, genotoxic, radioactive, and other sub-categories of medical waste. However, these are among the most potentially harmful types of medical waste, likely to cause much more disease over time than infectious agents. Of particular concern are heavy metals. According to the EPA, over 90% of potentially infectious medical waste is disposed of by incineration. Incineration is not appropriate for heavy metals, since they are dispersed more widely by burning and not destroyed. Medical waste incinerators (most of which are located at hospitals) were once a major source of mercury pollution: In 1990, they were responsible for 25% of U.S. mercury emissions. The EPA states that its regulations governing emissions from medical-waste incinerators, promulgated in 1997, have reduced mercury emissions from medical-waste incineration by 94%.

One well-documented case illustrates the importance of containing radioactive medical wastes. In the Brazilian town of Goiano, a radiotherapy institute was closed in 1985. (Radiotherapy is the selective use of radiation to kill cancer cells.) A machine containing a large quantity of the radioactive nuclide ^{137}Cesium, whose radiation was once used to treat cancer, was abandoned in the facility. Local people, scavenging gear from the site, obtained the lead-encased source and broke it open, revealing its blue contents. Having no idea what it was, they gave the sticky radioactive powder to children to play with, who painted their skin with it, and otherwise spread it widely. Four people died from radiation exposure and hundreds more were exposed, causing long-term health consequences that have not been tracked by the Brazilian government.

Medical waste is also generated by home use of pharmaceuticals, home dialysis, and other products. In early 2008, a five-month investigation by the Associated Press revealed that a wide variety of prescription drug metabolites (drugs requiring a doctor's order) are found in most U.S. drinking water: these drugs include mood stabilizers, anti-epileptics, sex hormones, and antibiotics. The likely source of these drugs was not only health-care facilities, but also home consumers of the drugs, who pass the unmetabolized drug out of their bodies in urine and dump unused drugs into the toilet. Flushed drugs enter the municipal wastewater streams and spread out into the environment. Whether any of the drugs exist in amounts large enough to be harmful is debated by epidemiologists (scientists who study the causes of disease in populations). Since February, 2007, the U.S. government has advised persons who are disposing of unused drugs not to flush them down the toilet, but mix them with some repellent material such as coffee grounds (to insure that they will not be retrieved from the garbage and eaten by children) and to put them out with the regular trash for eventual landfill burial.

A worker takes a break while collecting medical waste at a World Health Organization (WHO) warehouse in Banda Aceh, Indonesia, 2005. One of the most pressing environmental problems left by the tsunami is the challenge of disposing the millions of cubic meters of waste, some of it laced with oil, asbestos, and other human waste. *AP Images*

Primary Source Connection

In response to the pollution of beaches by medical waste, Congress enacted the Medical Waste Tracking Act of 1988 (MWTA), an excerpt of which follows, as an amendment to the Solid Waste Disposal Act. The MWTA established a two-year program that studied the disposal of medical wastes in several states on the East Coast as well as states adjacent to the Great Lakes. The program had four major goals. The first was to develop clear definitions of medical waste. The second was to develop a tracking system for medical waste from the point of origin through the final disposal point. The program also developed standardized measures for packing, labeling, and storing medical waste. Finally it established penalties for not adhering to the medical waste tracking system.

The formal outcome of the Medical Waste Tracking Act of 1988 was a set of recommendations called Standards for the Tracking and Management of Medical Waste. Although the Standards were only in place for two years, the MWTA resulted in stricter laws and regulations for the disposal of medical wastes in state and local governments. A multi-billion dollar industry developed in response to these requirements. Facilities that are properly equipped to autoclave, incinerate, microwave, chemically disinfect, or thermally inactivate medical wastes have become an important part of the medical waste disposal stream. Businesses that contract with hospitals, doctors and dentists offices, veterinarians, and research laboratories are required to obtain permits and to adhere to strict performance standards.

MEDICAL WASTE TRACKING ACT OF 1988

TITLE 42, CHAPTER 82, SUBCHAPTER X

Sec. 6992. Scope of demonstration program for medical waste

(a) Covered States

The States within the demonstration program established under this subchapter for tracking medical wastes shall be New York, New Jersey, Connecticut, the States contiguous to the Great Lakes and any State included in the program through the petition procedure described in subsection (c) of this section, except for any of such States in which the Governor notifies the Administrator under subsection (b) of this section that such State shall not be covered by the program.

(b) Opt out

1. If the Governor of any State covered under subsection (a) of this section which is not contiguous to the Atlantic Ocean notifies the Administrator that such State elects not to participate in the demonstration program, the Administrator shall remove such State from the program.
2. If the Governor of any other State covered under subsection (a) of this section notifies the Administrator that such State has implemented a medical waste tracking program that is no less stringent than the demonstration program under this subchapter and that such State elects not to participate in the demonstration program, the Administrator shall, if the Administrator determines that such State program is no less stringent than the demonstration program under this subchapter, remove such State from the demonstration program.
3. Notifications under paragraphs (1) or (2) shall be submitted to the Administrator no later than 30 days after the promulgation of regulations implementing the demonstration program under this subchapter.

(c) Petition in

The Governor of any State may petition the Administrator to be included in the demonstration program and the Administrator may, in his discretion, include any such State. Such petition may not be made later than 30 days after promulgation of regulations establishing the demonstration program under this subchapter, and the Administrator shall determine whether to include the State within 30 days after receipt of the State's petition.

(d) Expiration of demonstration program

The demonstration program shall expire on the date 24 months after the effective date of the regulations under this subchapter.

Sec. 6992a. Listing of medical wastes

(a) List

Not later than 6 months after November 1, 1988, the Administrator shall promulgate regulations listing the types of medical waste to be tracked under the demonstration program. Except as provided in subsection (b) of this section, such list shall include, but need not be limited to, each of the following types of solid waste:

1. Cultures and stocks of infectious agents and associated biologicals, including cultures from medical and pathological laboratories, cultures and stocks of infectious agents from research and industrial laboratories, wastes from the production of biologicals, discarded live and attenuated vaccines, and culture dishes and devices used to transfer, inoculate, and mix cultures.
2. Pathological wastes, including tissues, organs, and body parts that are removed during surgery or autopsy.

3. Waste human blood and products of blood, including serum, plasma, and other blood components.
4. Sharps that have been used in patient care or in medical, research, or industrial laboratories, including hypodermic needles, syringes, pasteur pipettes, broken glass, and scalpel blades.
5. Contaminated animal carcasses, body parts, and bedding of animals that were exposed to infectious agents during research, production of biologicals, or testing of pharmaceuticals.
6. Wastes from surgery or autopsy that were in contact with infectious agents, including soiled dressings, sponges, drapes, lavage tubes, drainage sets, underpads, and surgical gloves.
7. Laboratory wastes from medical, pathological, pharmaceutical, or other research, commercial, or industrial laboratories that were in contact with infectious agents, including slides and cover slips, disposable gloves, laboratory coats, and aprons.
8. Dialysis wastes that were in contact with the blood of patients undergoing hemodialysis, including contaminated disposable equipment and supplies such as tubing, filters, disposable sheets, towels, gloves, aprons, and laboratory coats.
9. Discarded medical equipment and parts that were in contact with infectious agents.
10. Biological waste and discarded materials contaminated with blood, excretion, excudates [1] or secretion from human beings or animals who are isolated to protect others from communicable diseases.
11. Such other waste material that results from the administration of medical care to a patient by a health care provider and is found by the Administrator to pose a threat to human health or the environment.

(b) Exclusions from list

The Administrator may exclude from the list under this section any categories or items described in paragraphs (6) through (10) of subsection (a) of this section which he determines do not pose a substantial present or potential hazard to human health or the environment when improperly treated, stored, transported, disposed of, or otherwise managed

Sec. 6992b. Tracking of medical waste

(a) Demonstration program

Not later than 6 months after November 1, 1988, the Administrator shall promulgate regulations establishing a program for the tracking of the medical waste listed in section 6992a of this title which is generated in a State subject to the demonstration program. The program shall:

(1) provide for tracking of the transportation of the waste from the generator to the disposal facility, except that waste that is incinerated need not be tracked after incineration,

(2) include a system for providing the generator of the waste with assurance that the waste is received by the disposal facility,

(3) use a uniform form for tracking in each of the demonstration States, and

(4) include the following requirements:

(A) A requirement for segregation of the waste at the point of generation where practicable.

(B) A requirement for placement of the waste in containers that will protect waste handlers and the public from exposure.

(C) A requirement for appropriate labeling of containers of the waste.

(b) Small quantities

In the program under subsection (a) of this section, the Administrator may establish an exemption for generators of small quantities of medical waste listed under section 6992a of this title, except that the Administrator may not exempt from the program any person who, or facility that, generates 50 pounds or more of such waste in any calendar month.

(c) On-site incinerators

Concurrently with the promulgation of regulations under subsection (a) of this section, the Administrator shall promulgate a recordkeeping and reporting requirement for any generator in a demonstration State of medical waste listed in section 6992a of this title that:

1. (1) incinerates medical waste listed in section 6992a of this title on site and
2. (2) does not track such waste under the regulations promulgated under subsection (a) of this section. Such requirement shall require the generator to report to the Administrator on the volume and types of medical waste listed in section 6992a of this title that the generator incinerated on site during the 6 months following the effective date of the requirements of this subsection.

(d) Type of medical waste and types of generators

For each of the requirements of this section, the regulations may vary for different types of medical waste and for different types of medical waste generators

U.S. Congress

U.S. CODE. "MEDICAL WASTE TRACKING ACT OF 1988. AN AMENDMENT TO THE SOLID WASTE DISPOSAL ACT." TITLE 42, CHAPTER 82, SUBCHAPTER X, SECTIONS 6992, 6992A AND 6992B.

SEE ALSO *Toxic Waste*

BIBLIOGRAPHY

Books

Dutta, Subijoy. *Environmental Treatment Technologies for Hazardous and Medical Wastes: Remedial Scope and Efficacy*. New York: McGraw-Hill, 2007.

Periodicals

Narvaez, Alfonso. "New York City to Pay Jersey Town $1 Million Over Shore Pollution." *New York Times* (December 8, 1987).

Web Sites

CNN.com "Prescription Drugs Found in Drinking Water Across U.S." March 10, 2008. http://www.cnn.com/2008/HEALTH/03/10/pharma.water1.ap/ (accessed May 10, 2008).

U.S. Environmental Protection Agency. "Medical Waste." http://www.epa.gov/epaoswer/other/medical/ (accessed May 10, 2008).

U.S. Environmental Protection Agency. "Medical Waste Tracking Act (1988)." http://www.epa.gov/epaoswer/other/medical/mwpdfs/mwta.pdf (accessed May 10, 2008).

World Health Organization. "Medical Waste." http://www.who.int/topics/medical_waste/en/ (accessed May 10, 2008).

World Health Organization. "Safe Management of Wastes from Health-Care Activities." http://www.healthcarewaste.org/en/documents.html?id=1 (accessed May 10, 2008).

Larry Gilman

Migratory Species

Introduction

Migratory species are species that move from one habitat to another during different times of the year, as they cannot live in the same environment all year round due to seasonal limitations in factors such as food, sunlight, and temperature. The movement between habitats, which can exceed thousands of miles/kilometers in length for some migratory birds and mammals such as whales, is referred to as migration.

A migration route can involve resting and nourishment stops along the way, and also requires the availability of habitats before and after each migration. Thus, migrating species are at particular risk of changes in environment or land use.

Efforts to protect migratory species include government, private, and individual initiatives within the concerned countries, organizations such as the Sierra Club and Ducks Unlimited, and international agreements such as the Convention on Migratory Species (the Bonn Convention). As of January 2008, 104 nations have signed the Bonn Convention.

Historical Background and Scientific Foundations

Migrating birds were recorded about 3,000 years ago by the ancient philosophers Homer and Aristotle. Also, the Old Testament of the Bible has been interpreted as describing the migration of several bird species.

Migratory species can be capable of very long journeys. For example, the Arctic tern route between the Arctic and Antarctic is over 13,670 mi (22,000 km) long. The Bar-tailed Godwit migratory route, for example, is over 6,800 mi (11,000 km) in length, between Alaska and New Zealand, a journey that is made non-stop.

Other species that migrate include bats, whales, seals, turtles, and insects. An example of the latter is the Monarch butterfly, which migrates from southern Canada to winter in central Mexico.

Migratory species exploit the resources that are present in different areas of the globe at different times of the year. Because they cannot live in the same area all year—due to seasonal limitations in temperature, food availability, weather, or another factors—they have evolved to be capable of movement from one region to another.

Stopping en route to rest and feed is common among migratory species. The fact that they migrate and the need for space and resources such as food and water en route can make migratory species at risk from changing environments. For example, the drying up of a wetland or its development can take away a rest stop that is vital for the success of the migration. Similarly, any threat to the environments occupied at either end of the migratory route threatens species survival.

An example is the Monarch butterfly, whose returning populations from Mexico have been dwindling since the mid-1990s. Conservation scientists who have studied their decline feel that land use changes in central Mexico, specifically logging, have adversely altered the over-wintering grounds of the butterfly.

Climate change can affect migratory species. For example, the increased warming of more northern latitudes has shifted the territory of some migrating bird species farther north, adding length to their migratory journey. Migration itself has been affected for species whose route takes them over deserts. Regions such as the Sahel region of Africa have become drier, decreasing the opportunity to stop and refuel with food and water.

Impacts and Issues

Migratory species can be an indicator of environmental change, particularly of changes due to human activities, since their lives involve movement between different environments. These changes can be direct such as logging, or can be indirect. The warming of the atmosphere due to the production of greenhouse gases by various human-related activities is an important indirect change. As of 2008, only the Ramsar Convention on Wetlands—which was signed in 1971 and which, as of 2008, has 158 parties and includes 161 million hectares of wetland—considers the influence of climate change on migratory species.

Climate changes that could threaten migratory species include a sea level rise, which could bury Caribbean and Mediterranean beaches used as breeding, nesting, or rearing sites by some turtle and seal species. Also the warming of shallow coastal waters would adversely affect whales, dolphins, and manatees, which require cold and shallow water as breeding sites.

An environmental scientist holds a Little Wood Satyr butterfly after tagging and examining the specimen. *AP Images*

WORDS TO KNOW

DUCKS UNLIMITED: An international non-profit organization founded in Canada in the 1940s to preserve and protect wetlands.

HABITAT: The natural location of an organism or a population.

LIGHT POLLUTION: Also known as photopollution and luminous pollution, refers to the presence of excessive amounts of light in the atmosphere.

SIERRA CLUB: Environmental organization founded in 1892 by American naturalist John Muir (1838–1914), whose mandate includes the protection and, when needed, restoration, of natural environments.

Climate change could also affect migratory species by altering the geographical distribution of other species that are food sources. Because migration can be over national boundaries, international cooperation is necessary to help protect migratory species. The Convention on Migratory Species, which was struck in Bonn, Germany, in 1979 under the United Nations Environment Programme (UNEP), and which is concerned with the global conservation of migratory wildlife and habitats, involves 104 nations. Although North America and coastal waters are part of the migratory routes for a variety of bird and mammal migratory species, the United States, Canada, and Mexico have not signed the convention.

SEE ALSO *Biodiversity; Ecological Competition; Sustainable Development; Wetlands; Wildlife Population Management*

BIBLIOGRAPHY

Books

Chiras, Daniel D., John P. Reganold, and Oliver S. Owen. *Natural Resource Conservation: Management for a Sustainable Future.* New York: Prentice-Hall, 2004.

Freyfogle, Eric T. *Why Conservation Is Failing and How It Can Regain Ground.* New Haven: Yale University Press, 2006.

Maher, Neil M. *Nature's New Deal: The Civilian Conservation Corps and the Roots of the American Environmental Movement.* New York: Oxford University Press, 2007.

Web Sites

Monarch Watch. http://www.monarchwatch.org/ (accessed April 21, 2008).

Mining and Quarrying Impacts

Introduction

Mining and quarrying extract a wide range of useful materials from the ground such as coal, metals, and stone. These substances are used widely in building and manufacturing industry, while precious stones have long been used for adornment and decoration. Mining and quarrying involve investigating potential sites of extraction, then getting the required material out of the ground, and finally processing with heat or chemicals to get out the metal or other substance of interest. All these operations may use large amounts of water.

Mining and quarrying can be very destructive to the environment. They have a direct impact on the countryside by leaving pits and heaps of waste material. The extraction processes can also contaminate air and water with sulfur dioxide and other pollutants, putting wildlife and local populations at risk. More careful use of natural resources, including recycling, and also restoration efforts after mining and quarrying can help limit these environmental impacts.

Historical Background and Scientific Foundations

People have always extracted useful materials from the ground. The Stone, Bronze, and Iron Ages were underpinned by knowledge of how to obtain these materials. Archaeological studies have shown evidence for copper mining in Africa around 6,000 years ago and, a little later, in ancient Egypt and North America. Meanwhile, the Romans developed many mining techniques to make the process more efficient. Today practically every manufactured item contains material that has been mined or quarried.

Mining involves taking an economically useful material from the ground. Substances that are mined include ores, coal, evaporites, and precious stones and metals. Quarrying is the cutting or digging of stone, and related materials, from an excavation site or pit and it usually leaves behind a large hole in the ground. An ore is a deposit containing an economically viable amount of a mineral, which itself is a crystalline inorganic compound, usually containing a metal. It is the metal that is of value. The main groups of minerals that are mined are oxides, sulfides, and silicates. Economically, the most significant metals are aluminum, manganese, copper, chromium, and nickel.

The evaporites are materials that are deposited in the ground from evaporation of chemical solutions. They include rock salt, used for culinary purposes and in water softening, and gypsum, which is used to make plasterboard. Substances like diamond, precious metals, and stones are always in great demand for decorative purposes, including jewelry. Meanwhile, gravel, clay, sand, and limestone are quarried in vast quantities for use in building materials like concrete, cement, and glass. Crushed stone from quarries is used in large amounts to build roads. A mile of a motorway could require nearly a quarter of a million tons of crushed stone. Sulfur deposits are mined mainly to make sulfuric acid, which is a mainstay of the chemical industry.

Mining and quarrying involve three distinct stages. First there is exploration and assessment to see whether a resource is worth exploiting. This might involve a certain amount of drilling into the ground. Then the substance is extracted by whatever technique is most appropriate to its location. This is often dictated by the depth of the resource under the surface. Open pit and shallow strip mining are commonly used to extract resources up to 600 ft (180 m) below ground. The mining process removes the source and the rock and soil, known as overburden, on top of it. The overburden is stacked up into a so-called spoil heap close by. Deeper resources will be extracted by underground mining that can go to about 8,000 ft (2,440 m). Beyond this, temperatures increase to a level that makes mining impracticable. Rock removed to create tunnels for mining is generally added to the spoil heap. Finally, the ore or other resource must be processed to extract the metal or material of interest. This usually involves some kind of heat or chemical treatment. For example, smelting is a common form of processing and involves roasting an ore to release the metal it contains.

Impacts and Issues

Mining and quarrying have often been criticized for their social and environmental impact. Far fewer lives are claimed by the industry in modern times, thanks to improved technology and safety measures. However, mining was a difficult and dangerous job. Valuable materials like gold and diamonds have often helped finance corrupt regimes, crime, and terrorism while inhuman labor conditions have often been employed in their extraction.

The environmental impacts of mining and quarrying are several. While the extractions are underway, the landscape is visibly disfigured and habitat loss can be extensive. The mining operations themselves and the accompanying spoil heaps cause a drastic change in the location with direct destruction of habitat and blocking or burying nearby bodies of water. Mining can often affect local hydrology, causing changes in the water flow as well as quality. The pits left behind by large mining operations often fill with groundwater, which then becomes polluted. Mining companies now acknowledge that they need to invest in restoring land they have exploited. This involves leveling spoil heaps, filling in holes and re-grassing the area. However, it can take many years for vegetation to become re-established at a former mine site. There are also many abandoned mines where environmental impact is ongoing.

Emissions from mining and quarrying can contaminate both air and water. The U.S. Environmental Protection Agency (EPA) lists as many as 100 different air pollutants issuing from the nation's mining industry, including dust particles and sulfur dioxide, which can create acid precipitation. Meanwhile, the Mineral Policy Center in Washington, D.C., says that 12,000 mi

WORDS TO KNOW

ACID RAIN: A form of precipitation that is significantly more acidic than neutral water, often produced as the result of industrial processes.

EVAPORITES: Salts deposited by the evaporation of aqueous solutions.

ORE: Rock containing a significant amount of minerals.

STRIP MINING: A method for removing coal from seams located near Earth's surface.

Dead Monarch butterflies litter a snowy field west of Mexico City (January 1996). Snowfall and a cold snap killed millions of monarch butterflies at their wintering grounds in mountainous western Mexico. A preliminary survey of the butterfly sanctuaries indicated at least 30 percent of the 50–60 million monarchs that migrated from the United States and Canada perished. *AP Images*

The effects of gold mining on Venezuela's Amazon rain forest are shown in 1997, at the site of the Las Cristinas gold mine. Environmentalist groups in Venezuela are fighting the recent legalization of further gold mining in Imataca reserve, an area of the Amazon rain forest. *AP Images*

(19,312 km) of rivers and streams in the United States are polluted by abandoned and current mining operations. The problem is that a great deal of water is used in extraction of ores, especially those containing only low concentrations of metals, and this leaches heavy metals and sulfur from the rocks so that it enters the water supply. Ore processing can sometimes be more polluting than the extraction itself. For example, smelting often releases sulfur dioxide into the atmosphere to create acid precipitation, including acid rain.

The National Science Foundation selected the Henderson Mine (Colorado) as a potential site for a high-tech laboratory where scientists will conduct research in high-energy physics, astrophysics, and earth sciences in caverns below ground. *AP Images*

Fires occurring in underground mines are another environmental impact. These can be difficult to extinguish and may actually burn for many years. There are hundreds of such mine fires around the world, in China, the United States, Russia, and India, for example. They emit a substantial amount of methane and carbon dioxide into the atmosphere, thereby adding to the greenhouse effect.

The polluting nature of mining and quarrying is underlined by work carried out by the Blacksmith Institute, which is focused on solving pollution problems in the developing world. Each year they publish a list of the world's most polluted sites. In 2007, six of the top ten most polluted sites were mines or smelter facilities.

Mining and quarrying are, by their very nature, destructive to the environment. As the global population grows and many countries improve their standards of living, demand for industrial materials is sure to grow. This creates increasing pressure on existing mineral resources, which are finite. Prospectors will go further afield in search of new supplies. There have even been discussions about trying to exploit the pristine environment of Antarctica. However, there is a growing awareness that mineral resources are indeed finite and that they should be conserved. Efforts to recycle metals and other materials could help prevent the depletion of resources.

SEE ALSO *Acid rain; Industrial Pollution; Resource Extraction*

BIBLIOGRAPHY

Books

Cunningham, W.P., and A. Cunningham. *Environmental Science: A Global Concern*. New York: McGraw-Hill International Edition, 2008.

Web Sites

Bioethics Education Project. "Pollution: Mining and Quarrying." http://www.beep.ac.uk/content/231.0.html (accessed April 30, 2008).

Blacksmith Institute. "The World's Worst Polluted Places 2007." http://www.blacksmithinstitute.org/ten.php (accessed April 30, 2008).

Susan Aldridge

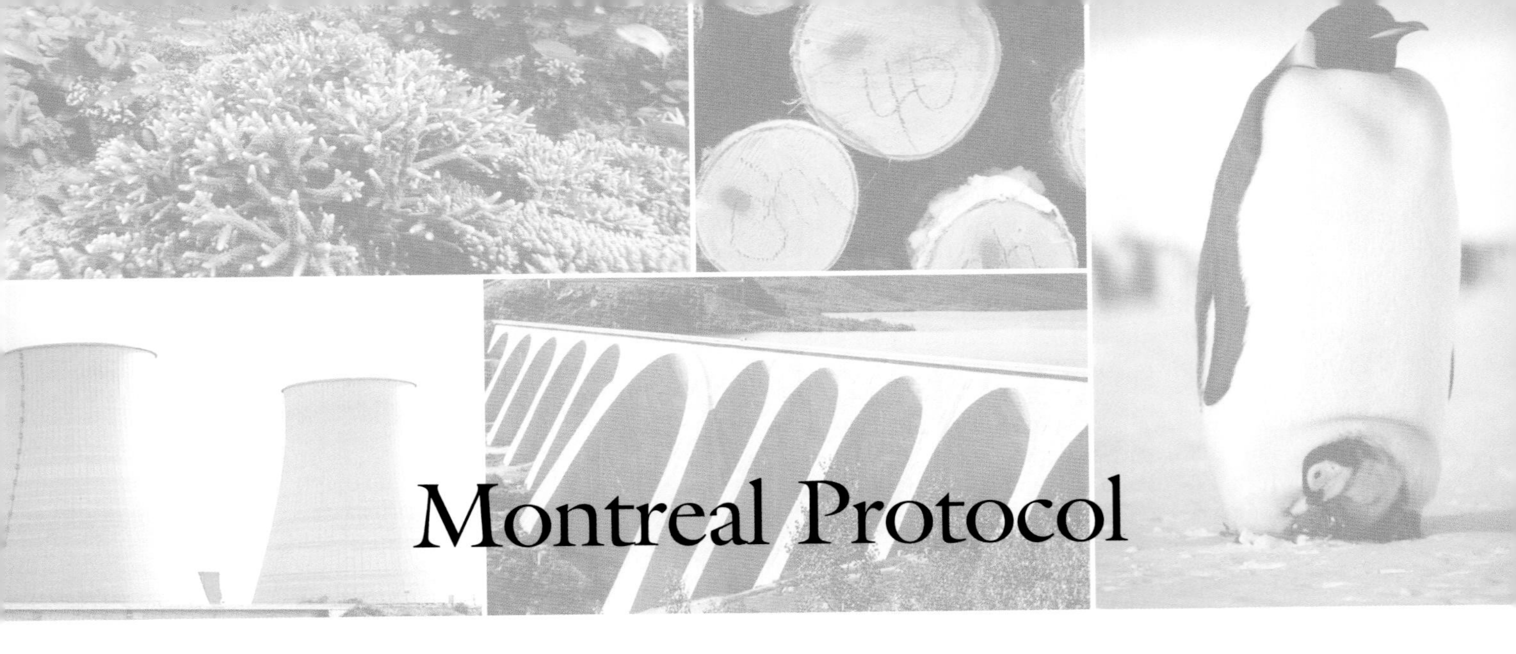

Montreal Protocol

Introduction

The Montreal Protocol on Substances that Deplete the Ozone Layer is an international environmental treaty aimed at reducing and phasing out the production of ozone-depleting substances (ODS), including chlorofluorocarbons (CFCs). Adopted in 1987, the Montreal Protocol was the first binding international multilateral environmental agreement. The discovery of a "hole" in Earth's protective stratospheric ozone layer in the 1980s prompted the international community to act quickly to reduce ODS. The widespread international support and the effectiveness of the Montreal Protocol made it one of the most successful international environmental treaties.

Historical Background and Scientific Foundations

The ozone layer is a relatively concentrated area of ozone (O_3) in the lower portion of the stratosphere that absorbs most of the harmful ultraviolet (UV) radiation from the sun. Ultraviolet radiation, particularly UV-B, causes skin cancer and harms oxygen-producing plants and phytoplankton. Ozone-depleting substances (ODS) are chemical compounds that destroy ozone molecules. All ODS are haloalkanes, a group of chemical compounds comprised of alkenes (a type of hydrocarbon) combined with one or more halogens, typically chlorine or fluorine. Haloalkanes were widely used in industry as aerosol propellants, refrigerants, solvents, and fire extinguishing agents.

Chlorofluorocarbons (CFCs) are the most widely known group of haloalkanes that destroy ozone. CFCs are a group of chemical compounds consisting of two halogens, chlorine and fluorine, and carbon. CFCs, which do not contain hydrogen, remain in the atmosphere for 50 to 100 years before degrading. A single CFC molecule typically degrades 10,000 ozone molecules but can degrade millions of ozone molecules.

Unlike CFCs, hydrochlorofluorocarbons (HCFCs) do contain hydrogen. HCFCs are less destructive to the ozone layer than CFCs. Therefore, HCFCs are commonly used to replace CFCs in industrial uses. Hydrofluorocarbons (HFCs) are a group of haloalkanes comprised of hydrogen, carbon, and fluorine. HFCs do not contain chlorine. HFCs do not destroy ozone and, therefore, are not regulated under the Montreal Protocol.

In addition to depleting the ozone layer, CFCs and HCFCs are powerful greenhouse gases that contribute to global climate change. One metric ton of CFCs is the equivalent of 5,000–8,000 metric tons of carbon dioxide. The longevity of haloalkanes in the atmosphere contributes to their potency as greenhouse gases. Although HFCs are not an ozone depleting substance, they are an even more powerful greenhouse gas.

In the early 1970s, research hypothesized that CFCs destroyed stratospheric ozone molecules. Scientists stated that CFC molecules break down in the atmosphere after 50 to 100 years. The chlorine atoms released during this process then degrade tens of thousands to millions of ozone molecules. In 1976, the National Academy of Sciences issued a report asserting the academy's acceptance of the role of CFCs in ozone destruction.

In 1985, a study by the British Antarctic Survey bolstered the ozone depletion hypothesis. The study showed that a hole had developed in the ozone layer over Antarctica. Scientists attributed the depletion of the ozone layer over Antarctica to the widespread use of CFCs. The international community responded quickly with over 20 nations signing the Vienna Convention for the Protection of the Ozone Layer in 1985. Although the Vienna Convention did not set CFC-reduction goals, the convention serves as a framework for international agreements on reducing ozone-depleting substances.

With the framework provided by the Vienna Convention, the international community began working toward a binding international treaty with specific ODS-reduction goals. On September 16, 1987, the Montreal Protocol on Substances that Deplete the Ozone Layer was opened for signature. The Montreal Protocol went into effect on January 1, 1989 after ratification by 29 countries and the European Economic Community. Today, 191 countries are parties to the treaty.

The Montreal Protocol and its amendments provide a goal-oriented mechanism for reducing and phasing out the production of ozone-destroying substances. The Montreal Protocol established a two-tiered system with different phase-out dates for industrialized and developing countries. The Montreal Protocol set a final CFC phase-out date of 1996 for industrialized countries and 2010 for developing countries. Industrialized nations must gradually phase out HCFCs by 2030, while developing countries have until 2040. HFCs are not regulated by the Montreal Protocol because they do not degrade ozone. The Kyoto Protocol to the United Nations Framework Convention on Climate Change (UNFCCC), however, regulates HFCs as greenhouse gases.

In 1991, the Second Meeting of the Parties to the Montreal Protocol established the Multilateral Fund for the Implementation of the Montreal Protocol. The Multilateral Fund provides financial and technical assistance to developing nations to assist in phasing out ODS. Between 1991 and 2005, the Multilateral Fund collected over $2.2 billion. The Multilateral Fund has been instrumental in assisting developing nations in meeting their ODS-reduction goals under the Montreal Protocol, and it has contributed assistance to over 5,500 projects in 144 nations, resulting in an annual reduction of over 150,000 metric tons of ozone-depleting substances.

WORDS TO KNOW

AEROSOL: Liquid droplets or minute particles suspended in air.

GREENHOUSE GAS: A gas whose accumulation in the atmosphere increases heat retention.

KYOTO PROTOCOL: Extension in 1997 of the 1992 United Nations Framework Convention on Climate Change (UNFCCC), an international treaty signed by almost all member countries with the goal of mitigating climate change.

OZONE: An almost colorless, gaseous form of oxygen, with an odor similar to weak chlorine, that is produced when an electric spark or ultraviolet light is passed through air or oxygen.

PROPELLANT: Gas or liquid ejected by a rocket or other vehicle to make the vehicle move in the opposite direction.

ULTRAVIOLET RADIATION: The energy range just beyond the violet end of the visible spectrum. Although ultraviolet radiation constitutes only about 5 percent of the total energy emitted from the sun, it is the major energy source for the stratosphere and mesosphere, playing a dominant role in both energy balance and chemical composition.

Impacts and Issues

The Montreal Protocol has been a successful international treaty in both compliance and achievement. Former United Nations Secretary General Kofi Annan called the Montreal Protocol "perhaps the most successful international agreement to date." The governments of 191 countries signed the Montreal Protocol in its first 20 years. Andorra, Iraq, San Marino, Timor Leste, and the Vatican City are the only countries that have not ratified the Montreal Protocol.

By the end of 2006, the parties to the Montreal Protocol had phased out 95% of all ozone-depleting substances. Production of ODS fell from more than 1.8 million metric tons in 1987 to 83,000 metric tons in 2005. Due to the long half-life of CFCs, however, the recovery of the atmospheric ozone levels has been slow. Atmospheric chlorine levels did not begin to drop until 1997. Although the thinning in the ozone layer over Antarctica has improved to some degree, the largest hole in the ozone ever recorded occurred in October 2006. The hole measured 11.4 million square mi (29.5 million square km), which is an area larger than North America. By late 2007, the hole in the ozone layer had returned to its ten-year average. Scientists predict that Antarctic ozone levels could recover completely by the year 2075.

SEE ALSO *Chlorofluorocarbons; Ozone Hole; Ozone Layer*

BIBLIOGRAPHY

Books

Anderson, Stephen O., and K. Madhava Sarma. *Protecting the Ozone Layer: The United Nations History.* London: Earthscan Publications, 2005.

Parson, Edward A. *Protecting the Ozone Layer: Science and Strategy.* Oxford: Oxford University Press, 2003.

Web sites

NASA/Goddard Space Flight Center. "NASA Keeps Eye on Ozone Layer Amid Montreal Protocol's Success." September 18, 2007. http://www

In 2005, Mexican Environment Secretary Tamargo (right) announced that Mexico had stopped producing ozone-depleting chemicals four years before a deadline set by the 1987 Montreal Protocol. *AP Images*

.sciencedaily.com /releases/2007/09/070913181730.htm (accessed January 30, 2008).

United Nations Department of Public Information. "Montreal Protocol on Ozone-Depleting Substances Effective, But Work Still Unfinished, Says Secretary-General in Message for International Day." September 7, 2006. http://www.un.org/News/Press/docs/2006/sgsm10620.doc.htm (accessed May 6, 2008).

United Nations Development Programme. "20 Years of Success: Montreal Protocol on Substances that Deplete the Ozone Layer." 2007. http://www.energyandenvironment.undp.org/undp/indexAction.cfm?module=Library&action=GetFile&DocumentAttachmentID=2304 (accessed May 6, 2008).

United Nations Environment Programme. "The Montreal Protocol on Substances that Deplete the Ozone Layer." 2000. http://www.unep.org/ozone/pdfs/Montreal-Protocol2000.pdf (accessed January 28, 2008).

United Nations Environment Programme. "The Ozone Secretariat." 2007. http://ozone.unep.org (accessed January 28, 2008).

United Nations Environment Programme. "The Vienna Convention for the Protection of the Ozone Layer." 2001. http://www.unep.org/Ozone/pdfs/viennaconvention2002.pdf (accessed May 6, 2008).

Joseph P. Hyder

National Environmental Policy Act

Introduction

The National Environmental Policy Act (NEPA) is a piece of U.S. federal legislation, signed into law in 1970, that governs all major government actions that may alter the environment. It requires that before any project is undertaken by any agency of the federal government that may affect the environment, such as building a highway, the agency must complete and make public a report called an Environmental Impact Statement (EIS). The law also requires that reviews of proposed projects be open to input from members of the public. According to a 2007 statement by the Council on Environmental Quality of the Executive Office of the President, NEPA was the first major U.S. environmental law.

Historical Background and Scientific Foundations

NEPA was passed by Congress in 1969 and signed into law by President Richard M. Nixon (1913–1994) on January 1, 1970, at a time of high public concern for pollution and the environment in the United States. The U.S. Environmental Protection Agency (EPA) was also established in 1970. As stated by the NEPA itself, it is intended to "declare a national policy which will encourage productive and enjoyable harmony between man and his environment; to promote efforts which will prevent or eliminate damage to the environment and biosphere and stimulate the health and welfare of man; to enrich the understanding of the ecological systems and natural resources important to the Nation; and to establish a Council on Environmental Quality." NEPA mandates that any federal agency prepare a science-based Environmental Impact Statement before performing any action that may significantly affect the quality of the environment.

NEPA's writers were careful to not declare any specific relationship between the government of the United States and the "environment." In particular, NEPA does not forbid environmentally harmful projects. It requires, rather, that whatever decisions are made must be informed by detailed studies of environmental impact. These studies are to be reported in documents termed Environmental Impact Statements. In principle, an agency's decision, once informed by an Environmental Impact Statement, may be to go ahead without any type or amount of environmental destruction. In practice, public admission that a given project would be highly destructive would probably, for political reasons, not be attempted. Therefore, agencies considering potentially destructive projects sometimes seek exemption from NEPA's requirement to produce an Environmental Impact Statement.

Three bodies oversee implementation of NEPA. First, the Council for Environmental Quality, a group established by NEPA, is part of the Executive Office of the President and has primary responsibility for overseeing implementation of NEPA. In 1978, the council set forth rules requiring each federal agency to set its own procedures for implementing the terms of NEPA, resulting in somewhat varying procedures among agencies. Second, once Environmental Impact Statements are produced, they are reviewed by the Environmental Protection Agency, which comments publicly on them. Third, disputes among citizens, government agencies, businesses, and other groups about NEPA issues may be resolved with the help of the U.S. Institute for Environmental Conflict Resolution in Tucson, Arizona.

Impacts and Issues

The Council for Environmental Quality is part of the Executive Office of the President. This means that administration of NEPA is not independent of politics,

WORDS TO KNOW

SONAR: Sound Navigation and Ranging (SONAR) is a remote sensing system with important military, scientific and commercial applications. Active SONAR transmits acoustic (i.e., sound) waves. Passive SONAR is a listening mode to detect noise generated from targets. SONAR allows the determination of important properties and attributes of the target (i.e., shape, size, speed, distance, etc.).

because the president appoints the members of the council, and so may pack the membership of the council with persons friendly to her or his own view of NEPA. For example, an environmentalist president could appoint former members of Greenpeace, the Sierra Club, or other environmental groups to the council, seeking to make enforcement of NEPA stricter. In reality, it has more often been the case that presidents install pro-business figures on the council in order to weaken barriers against projects that damage the environment. For example, Philip Cooney (1959–) served as a lobbyist for the American Petroleum Institute, which is affiliated with the U.S. oil industry, before being made head of the Council for Environmental Quality by President George W. Bush in 2001. In 2005, he resigned his post a few days after the media revealed that Cooney, who had no scientific training, had altered scientific reports to exclude language about the dangers of global climate change. A few days later, Cooney was hired by the oil company ExxonMobil.

NEPA has been at the center of at least two security-related controversies in recent years. In 2005, environmental groups sued the U.S. Navy to prevent use of high-powered sonar off the coast of California. Sonar detects objects under water by sending out pulses of sound that bounce back from obstacles; some naval radars are so powerful that their pulses can injure or kill whales and other marine mammals. According to the Navy's internal Environmental Assessment report, the exercises would have permanently injured the hearing of over 450 whales. However, the Navy alleged that it did not have to produce a more complete, scientifically rigorous, NEPA-style Environmental Impact Statement. A federal court ruled in favor of the environmentalists and forbade use of high-power sonar within 12 mi (19 km) of the California coast, but the president, stating that the training exercises constituted an "emergency" under the terms of NEPA, issued a waiver (special permission to the Navy to ignore the federal court's ruling). In January 2008, a federal court ruled that the waiver had no legal standing and that the Navy was indeed banned from using high-power sonar in the exclusion zone. In particular, the judge stated that the Navy is not "exempted from compliance with the National Environmental Policy Act."

In 2007, Congress ordered the Department of Homeland Security to build a 670 mi (1,080 km) fence along the border between the United States and Mexico. Such a fence could have many impacts on the environment (e.g., blocking migration of land animals). A draft Environmental Impact Statement for the fence project was released in 2007, with an invitation for public comment and a promise of more detail in a final statement. However, on April 2, 2008, Homeland Security Secretary Michael Chertoff announced that that his department would be ignoring NEPA and about 30 other environmental laws (including, according to the *New York Times* for April 8, 2008, laws protecting Indian graves, endangered species, and religious freedom) to build the fence. Chertoff argued that his department had been authorized by Congress to void any or all U.S. laws that interfered with the construction of the fence. As of April 2008, Congress's action had been appealed to the U.S. Supreme Court as unconstitutional. The plaintiffs argued that Congress has no right give its own powers away to any other branch of government, including its power to make law. However, in mid-2008 it is not yet known whether the Supreme Court will hear the case or how it would decide.

Primary Source Connection

Public law 91-190, also known as the National Environmental Policy Act of 1969, or NEPA, is a comparatively short piece of legislation that has had far-reaching effects during the decades since it took effect on January 1, 1970. The four major purposes of the act were to establish a national policy promoting harmony between humans and their environment; to decrease or eliminate damage to the environment; to promote a better understanding of the ecological systems and resources deemed important to the country; and to create the Council on Environmental Policy within the executive branch of government.

NEPA mandated that an integrated and interdisciplinary approach be used when making decisions or plans that might adversely affect the environment, that qualitative environmental values be considered along with economic and technical factors when making decisions, and that proposed federal projects include a document that evaluates the environmental impact of the project. The requirement for an environmental evaluation of all federally funded projects is probably the most widely known provision of the act.

NATIONAL ENVIRONMENTAL POLICY ACT OF 1969

National Environmental Policy Act of 1969, as amended

A revised plan for managing the elusive and threatened Canadian lynx still includes provisions that environmentalists fear would jeopardize the cat's habitat. The draft environmental impact statement released in November 2006 to comply with the National Environmental Policy Act, adds the White River National Forest habitat to the rest of forests in Colorado. *AP Images*

Sec. 2.

The purposes of this Act are: To declare a national policy which will encourage productive and enjoyable harmony between man and his environment; to promote efforts which will prevent or eliminate damage to the environment and biosphere and stimulate the health and welfare of man; to enrich the understanding of the ecological systems and natural resources important to the Nation; and to establish a Council on Environmental Quality.

TITLE I

CONGRESSIONAL DECLARATION OF NATIONAL ENVIRONMENTAL POLICY

Sec. 101.

(a) The Congress, recognizing the profound impact of man's activity on the interrelations of all components of the natural environment, particularly the profound influences of population growth, high-density urbanization, industrial expansion, resource exploitation, and new and expanding technological advances and recognizing further the critical importance of restoring and maintaining environmental quality to the overall welfare and development of man, declares that it is the continuing policy of the Federal Government, in cooperation with State and local governments, and other concerned public and private organizations, to use all practicable means and measures, including financial and technical assistance, in a manner calculated to foster and promote the general welfare, to create and maintain conditions under which man and nature can exist in productive harmony, and fulfill the social, economic, and other requirements of present and future generations of Americans.

(b) In order to carry out the policy set forth in this Act, it is the continuing responsibility of the Federal Government to use all practicable means, consistent with other essential considerations of national policy, to improve and coordinate Federal plans, functions, programs, and resources to the end that the Nation may:

1. fulfill the responsibilities of each generation as trustee of the environment for succeeding generations;
2. assure for all Americans safe, healthful, productive, and aesthetically and culturally pleasing surroundings;
3. attain the widest range of beneficial uses of the environment without degradation, risk to health or safety, or other undesirable and unintended consequences;
4. preserve important historic, cultural, and natural aspects of our national heritage, and maintain, wherever possible, an environment which supports diversity, and variety of individual choice;
5. achieve a balance between population and resource use which will permit high standards of living and a wide sharing of life's amenities; and
6. enhance the quality of renewable resources and approach the maximum attainable recycling of depletable resources.

(c) The Congress recognizes that each person should enjoy a healthful environment and that each person has a responsibility to contribute to the preservation and enhancement of the environment.

Sec. 102.

The Congress authorizes and directs that, to the fullest extent possible: (1) the policies, regulations, and public laws of the United States shall be interpreted and administered in accordance with the policies set forth in this Act, and (2) all agencies of the Federal Government shall

(A) utilize a systematic, interdisciplinary approach which will insure the integrated use of the natural and social sciences and the environmental design arts in planning and in decisionmaking which may have an impact on man's environment;

(B) identify and develop methods and procedures, in consultation with the Council on Environmental Quality established by title II of this Act, which will insure that presently unquantified environmental amenities and values may be given appropriate consideration in decisionmaking along with economic and technical considerations;

(C) include in every recommendation or report on proposals for legislation and other major Federal actions significantly affecting the quality of the human environment, a detailed statement by the responsible official on

(i) the environmental impact of the proposed action,

(ii) any adverse environmental effects which cannot be avoided should the proposal be implemented,

(iii) alternatives to the proposed action,

(iv) the relationship between local short-term uses of man's environment and the maintenance and enhancement of long-term productivity, and

(v) any irreversible and irretrievable commitments of resources which would be involved in the proposed action should it be implemented.

Prior to making any detailed statement, the responsible Federal official shall consult with and obtain the comments of any Federal agency which has jurisdiction by law or special expertise with respect to any environmental impact involved. Copies of such statement and the comments and views of the appropriate Federal, State, and local agencies, which are authorized to develop and enforce environmental standards, shall be made available to the President, the Council on Environmental Quality and to the public as provided by section 552 of title 5, United States Code, and shall accompany the proposal through the existing agency review processes;

(D) Any detailed statement required under subparagraph (C) after January 1, 1970, for any major Federal action funded under a program of grants to States shall not be deemed to be legally insufficient solely by reason of having been prepared by a State agency or official, if:

(i) the State agency or official has statewide jurisdiction and has the responsibility for such action,

(ii) the responsible Federal official furnishes guidance and participates in such preparation,

(iii) the responsible Federal official independently evaluates such statement prior to its approval and adoption, and

(iv) after January 1, 1976, the responsible Federal official provides early notification to, and solicits the views of, any other State or any Federal land management entity of any action or any alternative thereto which may have significant impacts upon such State or affected Federal land management entity and, if there is any disagreement on such impacts, prepares a written assessment of such impacts and views for incorporation into such detailed statement.

The procedures in this subparagraph shall not relieve the Federal official of his responsibilities for the scope, objectivity, and content of the entire statement or of any other responsibility under this Act; and further, this subparagraph does not affect the legal sufficiency of statements prepared by State agencies with less than statewide jurisdiction.

(E) study, develop, and describe appropriate alternatives to recommended courses of action in any proposal which involves unresolved conflicts concerning alternative uses of available resources;

(F) recognize the worldwide and long-range character of environmental problems and, where consistent with the foreign policy of the United States, lend appropriate support to initiatives, resolutions, and programs designed to maximize international cooperation in anticipating and preventing a decline in the quality of mankind's world environment;

(G) make available to States, counties, municipalities, institutions, and individuals, advice and information useful in restoring, maintaining, and enhancing the quality of the environment;

(H) initiate and utilize ecological information in the planning and development of resource-oriented projects; and

(I) assist the Council on Environmental Quality established by title II of this Act.

Sec. 103.

All agencies of the Federal Government shall review their present statutory authority, administrative regulations, and current policies and procedures for the purpose of determining whether there are any deficiencies or inconsistencies therein which prohibit full compliance with the purposes and provisions of this Act and shall propose to the President not later than July 1, 1971, such measures as may be necessary to bring their authority and policies into conformity with the intent, purposes, and procedures set forth in this Act.

Sec. 104.

Nothing in section 102 or 103 shall in any way affect the specific statutory obligations of any Federal agency (1) to comply with criteria or standards of environmental quality, (2) to coordinate or consult with any other Federal or State agency, or (3) to act, or refrain from acting contingent upon the recommendations or certification of any other Federal or State agency.

Sec. 105.

The policies and goals set forth in this Act are supplementary to those set forth in existing authorizations of Federal agencies.

TITLE II

COUNCIL ON ENVIRONMENTAL QUALITY

Sec. 201.

The President shall transmit to the Congress annually beginning July 1, 1970, an Environmental Quality Report (hereinafter referred to as the "report") which shall set forth (1) the status and condition of the major natural, manmade, or altered environmental classes of the Nation, including, but not limited to, the air, the aquatic, including marine, estuarine, and fresh water, and the terrestrial environment, including, but not limited to, the forest, dryland, wetland, range, urban, suburban an rural environment; (2) current and foreseeable trends in the quality, management and utilization of such environments and the effects of those trends on the social, economic, and other requirements of the Nation; (3) the adequacy of available natural resources for fulfilling human and economic requirements of the Nation in the light of expected population pressures; (4) a review of the programs and activities (including regulatory activities) of the Federal Government, the State and local governments, and nongovernmental entities or individuals with particular reference to their effect on the environment and on the conservation, development and utilization of natural resources; and (5) a program for remedying the deficiencies of existing programs and activities, together with recommendations for legislation.

Sec. 202.

There is created in the Executive Office of the President a Council on Environmental Quality (hereinafter referred to as the "Council"). The Council shall be composed of three members who shall be appointed by the President to serve at his pleasure, by and with the advice and consent of the Senate. The President shall designate one of the members of the Council to serve as Chairman. Each member shall be a person who, as a result of his training, experience, and attainments, is exceptionally well qualified to analyze and interpret environmental trends and information of all kinds; to appraise programs and activities of the Federal Government in the light of the policy set forth in title I of this Act; to be conscious of and responsive to the scientific, economic, social, aesthetic, and cultural needs and interests of the Nation; and to formulate and recommend national policies to promote the improvement of the quality of the environment.

U.S. Congress

U.S. CONGRESS. "NATIONAL ENVIRONMENTAL POLICY ACT OF 1969." WASHINGTON, DC: U.S. CONGRESS, 1969. HTTP://CEQ.EH.DOE.GOV/NEPA/REGS/NEPA/NEPAEQIA.HTM (ACCESSED APRIL 10, 2008).

SEE ALSO *Environmental Assessments; Environmental Protection Agency (EPA)*

BIBLIOGRAPHY

Books

Lindstrom, Matthew, and Zachary A. Smith. *The National Environmental Policy Act: Judicial Misconstruction, Legislative Indifference, & Executive Neglect.* College Station, TX: Texas A&M University Press, 2002.

Periodicals

Liptak, Adam. "Power to Build Border Fence Is Above U.S. Law." *New York Times* (April 8, 2008).

Web Sites

Council on Environmental Quality, Executive Office of the President. "A Citizen's Guide to the NEPA." http://ceq.eh.doe.gov/nepa/Citizens_Guide_Dec07.pdf (accessed April 16, 2008).

MSNBC. "Final Border Fence Impact Report Bypassed with Waiver." http://www.msnbc.msn.com/id/23927996/ (accessed April 21, 2008).

Natural Resources Defense Council. "Bush Attempts Illegal Override of Court Order Protecting Whales from Sonar." http://www.nrdc.org/legislation/fnepa.asp (accessed April 17, 2008).

Natural Resources Defense Council. "Defending NEPA from Assault." http://www.nrdc.org/legislation/fnepa.asp (accessed April 17, 2008).

U.S. Government Printing Office. "National Environmental Policy Act of 1969." http://frwebgate.access.gpo.gov/cgi-bin/getdoc.cgi?dbname=browse_usc&docid=Cite:+42USC4321 (accessed April 16, 2008).

National Oceanic and Atmospheric Administration (NOAA)

Introduction

The National Oceanic and Atmospheric Administration (NOAA) is an agency that is part of the U.S. Department of Commerce. NOAA's mandate is concerned with the oceans and the atmosphere. Specifically, the agency focuses on the safety of both environments, such as by research into the development of severe storms and making navigational charts that help avoid danger on marine waterways. In addition, the agency assists in protecting coastal regions of the ocean under U.S. jurisdiction. This includes working to increase public participation in protection and conservation efforts.

According to the organizations's vision statement, NOAA seeks "an informed society that uses a comprehensive understanding of the role of the oceans, coasts, and atmosphere in the global ecosystem to make the best social and economic decisions."

Part of this goal is concerned with the sustainable use of the ocean and atmosphere for economic gain while preserving as much as possible the natural environments. This balance between commercial interests and preservation is especially challenging near coasts, where the influence of land use such as agriculture can adversely alter water quality.

Additionally, research into the causes of global warming and related aspects of climate change is being done; one aim is to help prepare the United States for a world in which climate change will call for adaptation. The agency also focuses on the present, and is active in the forecasting of storms, floods, droughts, and other forms of severe weather, and in providing information that makes daily life safer and more productive.

Historical Background and Scientific Foundations

NOAA was formed on December 3, 1970 by President Richard M. Nixon (1913–1994). Earlier the same year, the president had formed the Environmental Protection Agency (EPA). As with the EPA, NOAA was formed by bringing together the functions of several different agencies. In both cases, the intent was to make the activities of the organizations more efficient and accountable by eliminating the duplication of efforts between different parts of the government.

The agencies that were the basis of the NOAA were some of the oldest in the United States: the U.S. Coast and Geodetic Survey (formed in 1897), the Weather Bureau (formed in 1870), and the Bureau of Commercial Fisheries (formed in 1871).

In the years following its establishment, NOAA became structured to be capable of performing its marine and atmospheric monitoring, safety, and stewardship roles. There are now six major divisions: National Weather Service, National Ocean Service, National Marine Fisheries Service, National Environmental Satellite, Data and Information Service, the Research Division, and Program Planning and Integration.

NOAA's activities involve both civilians and a uniform branch of the federal civil service known as the NOAA Commissioned Corps. The commissioned personnel operate the agencies's ships and airplanes, as well as having some scientific and administrative roles.

The National Weather Service is comprised of several national centers, others in various regions of the country, and over 120 local offices. The service issues daily weather updates, forecasts, and notifications of adverse weather that have various scales of importance depending on the potential severity and danger of the weather event. Their collective output is huge; nearly 735,000 weather forecasts and over 45,000 severe weather warnings are issued every year. The data are publicly available; the decades of nationwide data that have been complied are proving to be invaluable in tracking, as two examples, the progression of global warming in terms of storm severity and frequency, and the ozone hole in Earth's atmosphere.

The National Ocean Service is responsible for the protection of the dozen marine sanctuaries—areas set aside from commercial development and large-scale fishing, with the intent of preserving unique and/or threatened marine species—that have been established in waters under U.S. jurisdiction. This branch of NOAA is also responsible for the production and updating of navigation charts (paper and electronically available versions, utilizing geographical positioning system technology) for marine transportation routes, through a division called the National Geodetic Survey (the previous Coast and Geodetic Survey was the oldest scientific agency in the United States).

The National Marine Fisheries Service is concerned with science-based research on fisheries that aims to protect and manage current fisheries to sustain the industry, and to restore fisheries that have been compromised by environmental degradation or overfishing. As well, the service has an enforcement role.

The National Environmental Satellite, Data, and Information Service is charged with operating NOAA's satellite program, which is used to gather information on the atmosphere and some surface data, and in helping manage the satellite programs of other federal agencies including the Army, Air Force, and Federal Aviation Administration (FAA). These operational and management functions are carried out at national centers dedicated to either climate, geophysics, oceanography, snow and ice, and coastal waters.

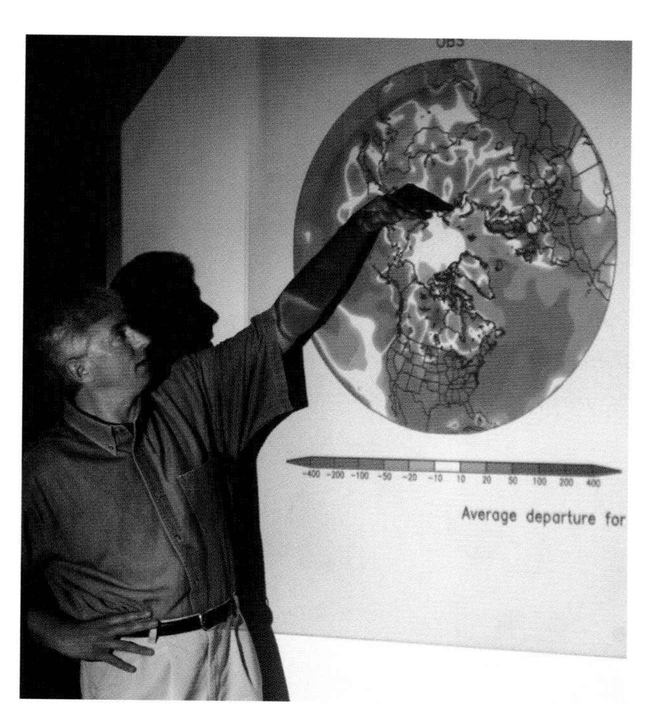

NOAA meteorologist talks about precipitation and how it relates to El Niño during a 2002 news conference at NOAA headquarters in Boulder, Colorado. *AP Images*

WORDS TO KNOW

GREENHOUSE GAS: A gas whose accumulation in the atmosphere increases heat retention.

OZONE HOLE: A term invented to describe a region of very low ozone concentration above the Antarctic that appears and disappears with each austral (Southern Hemisphere) summer.

SALINITY: Measurement of the amount of sodium chloride in a given volume of water.

SUSTAINABILITY: Practices that preserve the balance between human needs and the environment, as well as between current and future human requirements.

NOAA's science research is carried out through the Office of Oceanic and Atmospheric Research. Rather than being basic in scope, the research is geared toward protecting property and people from ocean- and atmospheric-related hazards, and in helping government efforts to increase the country's economic prosperity. Examples of research include studies to better understand and predict tornadoes and hurricanes; modeling the formation and wind-aided dispersion of air pollution; the sustainability of fisheries; ocean current patterns and environmental influences on currents; and threats to coastal ecosystems such as runoff of pesticides and sewage.

This research takes place at NOAA-funded and operated labs, and by funding a variety of research programs at U.S. universities and at oceanographic and atmospheric centers around the world.

Impacts and Issues

From its formation, NOAA has been a science-based agency. This science foundation supports NOAA's mandate of predicting changes in the ocean and atmosphere that will affect the safety and economic prosperity of Americans.

These long-term efforts are becoming more important in an era when global changes due to atmospheric warming are becoming increasingly evident. For example, the increasing severity and frequency of storms in the Gulf Coast of the United States has been docu-

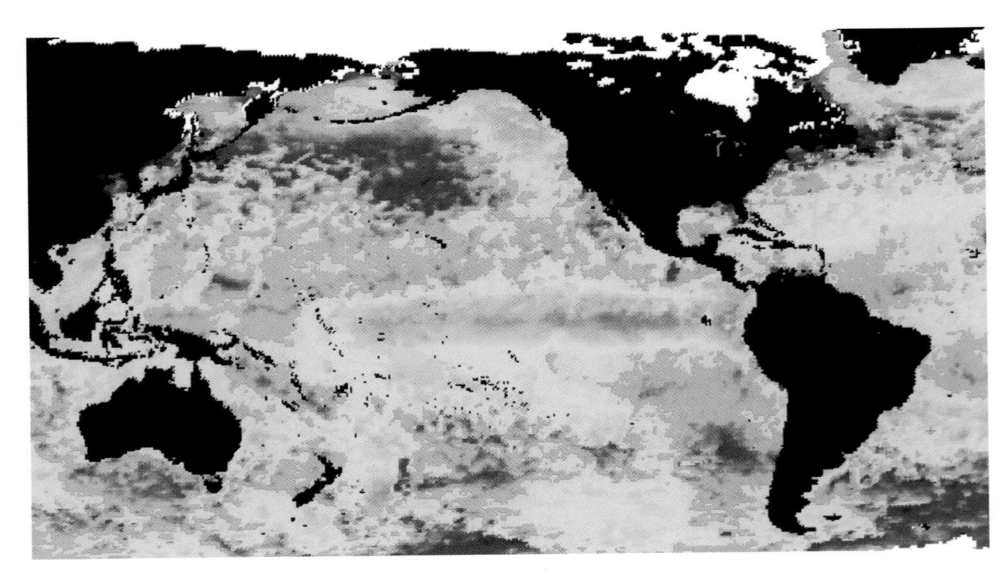

Satellite imaging is used by NOAA meteorologists to aid in predicting weather conditions, as shown by this image of the Pacific Ocean taken in 2002. *AP Images*

mented since the 1990s. In particular, the 2005 hurricane season, which spawned Hurricane Katrina, was the worst on record in terms of hurricane strength and damage-related costs.

Another aspect of NOAA that is becoming more important is fishery research. As the collapse of the cod fishery on the Grand Banks off of the East Coast of North America exemplifies, the loss of a fishery can be economically disastrous for a region, and a burden for government employment support programs. More reliable estimates of present and future fish stocks help ensure the use of fisheries at a level that achieves the maximum profit in both the present and the future.

The importance of the agencies research and other activities were reflected in the latest U.S. budget. In fiscal year 2009, NOAA's funding of $4.1 billion is a nearly 8% increase over the previous year, and the most money ever allocated to the agency. This comes at a time when budgets at other agencies such as National Institutes of Health have not changed or are decreasing.

The latest funding includes increases for the launch of geostationary satellites, which remain in orbit over a designated portion of Earth's surface, to provide better prediction of climate and monitoring of weather, and for ocean protection and restoration programs.

SEE ALSO *Climate Change; Environmental Protection Agency (EPA); Greenhouse Effect; Greenhouse Gases; Ozone Hole; Ozone Layer; Weather and Climate; Weather Extremes*

BIBLIOGRAPHY

Books

Committee on the Future of Rainfall Measuring Missions. *NOAA's Role in Space-Based Global Precipitation Estimation and Application.* Washington, DC: National Academies Press, 2007.

Goodwin, Lei. *Discover Your World with NOAA.* Washington: NOAA, 2007.

Sverdrup, Keith A. *An Introduction to the World's Oceans.* New York: McGraw-Hill, 2008.

National Park Service Organic Act

Introduction

The National Park Service Act (NPSA), often referred to as the "Organic Act," is the U.S. federal law that established the National Park Service in 1916. (The term "organic" here refers to the creation of the service as an organization, not to "organic" as in foods raised without chemicals.) According to the act, the National Park Service is to serve "the fundamental purpose" of the nation's parks, monuments, and reservations, namely "to conserve the scenery and the natural and historic objects and the wild life therein and to provide for the enjoyment of the same in such manner and by such means as will leave them unimpaired for the enjoyment of future generations."

The NPSA established the first national park system in the world. Today, the U.S. National Park Service manages over 380 monuments and preserves, including 58 national parks, and has an annual budget of a little over $2 billion and about 15,000 full-time employees. Over 270 million people visit the National Park System every year.

Historical Background and Scientific Foundations

Conservation of particularly striking landscapes as part of the national park system began in the United States in 1872, when an area surrounding Yellowstone Lake was set apart by an act of Congress. However, there was not yet any federal bureau dedicated to managing Yellowstone Park or the other national parks that were soon created.

Enthusiasm for founding a dedicated federal bureau to oversee the park system came originally from outside the federal government. Urban planner J. Horace McFarland, president of the American Civic Association (a private group encouraging planned development to beautify cities) from 1904 to 1925, led his organization in lobbying for a national park law. The group even participated in drafting the language of the act, especially the statement of purpose quoted above, and in 1911 proposed the National Park Service as the name of the new bureau. The service's statement of purpose, McFarland said in 1917 after the NPSA's passage, was "the essential thing ... the reason we feel that it 'the NPSA' is worthwhile."

McFarland and other supporters of the park system were not primarily interested in preserving untouched wilderness, but in developing tourism and recreation as a boon to the national economy. Unimpaired, in the language of the NPSA, meant primarily unimpaired for purposes of tourism, not unimpaired as a natural ecosystem. The national forest system (today managed by the Department of Agriculture) should, McFarland thought, be "the nation's woodlot," while the national park system—managed for recreation rather than resource extraction—would be "the nation's playground."

Congressional hearings on the proposed act were held in 1915 and 1916. Despite opposition from ranching and mining interests, the act was passed and was signed into law by President Woodrow Wilson on August 25, 1916. The functions of the new National Park Service were expanded in 1933, when an Executive Order by President Franklin D. Roosevelt transferred 63 federally owned monuments and military sites to the Park Service from other federal departments. The Park Service thus became charged with historic preservation as well as with the management of scenic natural areas. Today, only an act of Congress can create a new national park, and such actions have become infrequent. As of 2008, the U.S. National Park Service included over 83 million acres in every state but Delaware and in a number of American territories (e.g., the Virgin Islands).

WORDS TO KNOW

CONSERVATION: The act of using natural resources in a way that ensures that they will be available to future generations.

ECOSYSTEM: The community of individuals and the physical components of the environment in a certain area.

Impacts and Issues

The U.S. National Park Service has preserved many spectacular landscapes from development. It has also been controversial, and since 1916 its ecological policies have changed with shifting attitudes toward nature. For example, after the passage of the NPSA, one of the Park Service's earliest acts was to begin the extermination of large predators such as wolves, coyotes, and mountain lions in Yellowstone National Park. The last wolf pack in Yellowstone was destroyed in 1926. This was typical of manipulations of ecosystems carried on by the Park Service in the pursuit of a more recreation-oriented environment.

However, in 1995, the Park Service reintroduced wolves to Yellowstone, despite opposition from ranchers in adjacent areas. Top predators such as wolves—animals which prey on others, but which are not hunted themselves except by human beings—are crucially important in determining the structure of any ecosystem, as scientists realized clearly in the latter decades of the twentieth century. Thirty-one wolves were introduced in 1995 and 1996; by 2001, there were 220 wolves running in 21 packs. By 2003, there was evidence that the wolves were improving certain aspects of the Yellowstone ecosystem, slightly reducing deer, elk, and moose herds, and changing their grazing behaviors so as to allow riverside tree species to recover that had been damaged by overgrazing.

Wolves also leave unfinished carcasses for scavengers such as coyotes, magpies, and golden eagles, benefiting those species. Seasonal hunts by human beings supply large quantities of dead animal material over a short period, but wolf hunting continues year-round and so is of greater ecosystem benefit. The wolf reintroduction program in Yellowstone is typical of the more science-based decisions of the modern National Park Service.

A small group of people who oppose "big government" in any form would prefer that the federal government owned little or no land, and have sometimes urged selling off national park lands to private owners. For example, Congressman Richard Pombo (R-CA) drafted a bill in 2005 that would have sold off 15 national parks. The bill was circulated among lawmakers (but never voted on) partly as a pressure tactic in debates over whether to open up the Arctic National Wildlife Refuge for oil drilling. However, any political move to privatize the national park service would probably fail, as the national park service has been extremely popular since even before the 1916 creation of the National Park Service Act.

Primary Source Connection

The National Park Service Organic Act of 1916 remains significant because it formally established an organizational structure specifically for a growing network of parks and monuments that included the world's first national park. Before passage of the act, the National Park System was operated by the Department of the Interior using soldiers rather than professional park rangers to administer its lands. Since passage of the act, the National Park System within the United States has grown nearly tenfold, from 40 to nearly 400 parks, monuments, and other units. Although the language of the Organic Act has been amended and supplemented by subsequent acts of Congress, the core idea of a nationwide system of parks with unique scenic, natural, or historical value has remained intact for nearly a century.

THE NATIONAL PARK SERVICE ORGANIC ACT

An act to establish a National Park Service, and for other purposes.

Be it enacted by the Senate and House of Representatives of the United States of America in Congress assembled, That there is hereby created in the Department of the Interior a service to be called the National Park Service, which shall be under the charge of a director, who shall be appointed by the Secretary and who shall receive a salary of $4,500 per annum. There shall also be appointed by the Secretary the following assistants and other employees at the salaries designated: One assistant director, at $2,500 per annum, one chief clerk, at $2,000 per annum; one draftsman, at $1,800 per annum; one messenger, at $600 per annum; and, in addition thereto, such other employees as the Secretary of the Interior shall deem necessary: Provided, That not more than $8,100 annually shall be expended for salaries of experts, assistants, and employees within the District of Columbia not herein specifically enumerated unless previously authorized by law. The service thus established shall promote and regulate the use of the Federal areas known as national parks, monuments, and reservations hereinafter specified by such means and measures as conform to the fundamental purposes of the said parks, monuments, and reservations, which purpose is to conserve the scenery and the natural and historic objects and the wildlife therein and to provide for the enjoyment of

A late autumn sunset drapes El Capitan in Yosemite National Park. *AP Images*

the same in such manner and by such means as will leave them unimpaired for the enjoyment of future generations.

SEC. 2. That the director shall, under the direction of the Secretary of the Interior, have the supervision, management, and control of the several national parks and national monuments which are now under the jurisdiction of the Department of the Interior, and of the Hot Springs Reservation in the State of Arkansas, and of such other national parks and reservations of like character as may be hereafter created by Congress: Provided, That in the supervision, management, and control of national monuments contiguous to national forests the Secretary of Agriculture may cooperate with said National Park Service to such extent as may be requested by the Secretary of the Interior.

SEC. 3. That the Secretary of the Interior shall make and publish such rules and regulations as he may deem necessary or proper for the use and management of the parks, monuments, and reservations under the jurisdiction of the National Park Service, and any violations of any of the rules and regulations authorized by this Act shall be punished as provided for in section fifty of the Act entitled "An Act to codify and amend the penal laws of the United States," approved March fourth, nineteen hundred and nine, as amended by section six of the Act of June twenty-fifth, nineteen hundred and ten (Thirty-sixth United States Statutes at Large, page eight hundred and fifty-seven). He may also, upon terms and conditions to be fixed by him, sell or dispose of timber in those cases where in his judgment the cutting of such timber is required in order to control the attacks of insects or diseases or otherwise conserve the scenery or the natural or historic objects in any such park, monument, or reservation. He may also provide in his discretion for the destruction of such animals and of such plant life as may be detrimental to the use of any of said parks, monuments, or reservations. He may also grant privileges, leases, and permits for the use of land for the accommodation of visitors in the various parks, monuments, or other reservations herein provided for, but for periods not exceeding thirty years; and no natural curiosities, wonders, or objects of interest shall be leased, rented, or granted to anyone on such terms as to interfere with free access to them by the public: Provided, however, That the Secretary of the Interior may, under such rules and regulations and on such terms as he may prescribe, grant the privilege to graze live stock within any national park, monument, or reservation herein referred to when in his judgment such use is not detrimental to the primary purpose for which such park, monument, or reservation was created, except that this provision shall not apply to the Yellowstone National Park: And provided further, That the Secretary of the Interior may grant said privileges, leases, and permits and enter into contracts relating to the same with responsible persons, firms, or corporations without advertising and without securing competitive bids: And provided further, That no contract, lease, permit, or privilege granted shall be assigned or transferred by such grantees, permittees, or licensees, without the approval of the Secretary of the Interior first obtained in writing: And provided further, That the Secretary may, in his discretion, authorize such grantees, permittees, or licensees to execute mortgages and issue bonds, shares of stock, and other evidences of interest in or indebtedness upon their rights, properties, and franchises, for the purposes of installing, enlarging or improving plant and equipment and extending facilities for the accommodation of the public within such national parks and monuments.

Sac. 4. That nothing in this Act contained shall affect or modify the provisions of the Act approved February fifteenth, nineteen hundred and one, entitled "An Act relating to rights of way through certain parks, reservations, and other public lands."

U.S. Congress

U.S. CONGRESS. "NATIONAL PARK SERVICE ORGANIC ACT." WASHINGTON, D.C.: U.S. CONGRESS, 1916.

SEE ALSO *Conservation; Natural Reserves and Parks; Wilderness Act of 1964*

BIBLIOGRAPHY

Books

Wright, Gerald, ed. *National Parks and Protected Areas: Their Role in Environmental Protection.* Cambridge, MA: Blackwell Science, 1996.

Periodicals

Biello, David. "No Arctic Oil Drilling? How About Selling Parks?" *San Francisco Chronicle* (September 24, 2005).

Web Sites

National Geographic. "Wolves' Leftovers Are Yellowstone's Gain, Study Says." December 4, 2003. http://news.nationalgeographic.com/news/2003/12/1204_031204_yellowstonewolves.html (accessed April 12, 2008).

National Park Service. "Codifying Tradition: The National Park Service Act of 1916." http://www.nps.gov/history/history/online_books/sellars/chap2.htm (accessed April 12, 2008).

National Park Service. "National Park Service Organic Act." http://www.nps.gov/legacy/organic-act.htm (accessed April 12, 2008).

National Park Service. "National Park Service Science in the 21st Century." http://www.nature.nps.gov/scienceresearch/pdf/SCIENCEREPORTMar2004.pdf (accessed April 12, 2008).

Larry Gilman

Natural Reserves and Parks

Introduction

Natural reserves and parks are areas of land or water set aside by governments or private groups to preserve them from uncontrolled development and exploitation. In international discussion, the term protected areas is used to denote all types of land and ocean areas where access and development are restricted in ordered to protect, at least partially, the natural character of that area. Not all protected areas are pure wilderness; in some cases, human settlements remain within the boundaries of the area.

Historical Background and Scientific Foundations

The Wilderness Ideal and the World's First Parks

The setting aside of parks and other protected areas for the sake of preserving their original or natural character is historically recent. Until the nineteenth century, a landscape set aside as a park was generally a private or royal land on which deer and other wild animals were kept for recreational hunting by the ruling classes. These landscapes were not intended for public use, or to preserve wild landscapes as such. For centuries, the penalty for illegal hunting in a European royal park was typically death.

The colonization of North America brought Europeans into sustained contact with a less-tamed land than could be found anywhere in Europe, including vast forests, lakes, mountain ranges, and rivers. Although Native American practices had modified many of these ecosystems, their influence was minor by modern standards or even by those of the early European settlers, who brought with them plow-based agriculture, livestock herds, and logging. With the gun, steel ax, and steel plow—tools not yet available to Native Americans—the settlers shaved off whole forests, caused widespread soil erosion, drove indigenous species such as the wolf and deer out of New England, and made other drastic changes. The earliest Puritan settlers did not look kindly on wilderness, as they saw the relatively unmodified landscapes inhabited by Native Americans, and did not wish to preserve it. On the contrary, they viewed the wilderness as raw, meant for conquering, cultivating, and transforming into a new Eden.

By the mid-nineteenth century, even after two centuries of taming and exploitation, much wilderness still remained in North America, and the early Puritan aversion to wilderness was being replaced by admiration. The literary and artistic movement known as Romanticism, which began in Europe in the late 1700s, called attention to the moods and beauties of nature and was an influence on American naturalists. In 1832, the artist George Catlin (1796–1872) suggested that some of the Great Plains region of the American West be set aside as a "nation's park" in order to preserve Indian culture and the buffalo (an animal unique to the New World). In 1862, in his essay "Huckleberries," Massachusetts native Henry Thoreau (1817–1862) became the first person to express the modern ideal of the wilderness park when he wrote, "Each town should have a park, or rather a primitive forest, of five hundred or a thousand acres, where a stick should never be cut for fuel, a common possession forever, for instruction and recreation." At that time, such an idea was radical; publicly owned land set aside and kept wild for the sake of its wildness did not yet exist.

Wilderness tourism began to flourish in the mid-nineteenth century, as ordinary citizens and families from cities and towns sought out wild areas for amusement and relaxation. Admirers of nature, joined by entrepreneurs who hoped to exploit the business poten-

WORDS TO KNOW

BUSHMEAT: Meat from wild animals killed for food. Poaching (illegal hunting) for bushmeat has contributed to population stress for many species of animals in Africa, including the great apes, and has also resulted in disease outbreaks in man from consuming the meat from wild animals.

ECO-TOURISM: Environmentally responsible travel to natural areas that promotes conservation, has a low visitor impact, and provides for beneficially active socio-economic involvement of local peoples.

INDIGENOUS SPECIES: A species that is native to its region, but may occur in other regions as well.

SUSTAINABILITY: Practices that preserve the balance between human needs and the environment, as well as between current and future human requirements.

tial of what would now be called eco-tourism, urged that the Yosemite Valley of California be set aside as public land. In 1864, President Abraham Lincoln (1809–1865) granted the valley to the State of California "for public use, resort and recreation." There was as yet no thought of preserving the wild character of the region for its own sake. Business leaders also moved Congress to set aside the Yellowstone region of what is now Wyoming in 1872, creating the world's first national park. Scottish immigrant John Muir (1838–1914), naturalist and lover of wilderness, agitated successfully for the creation of a larger, nationally owned park around the Yosemite valley, which Congress did in 1890. These were the beginnings of the world's first national park system.

Meanwhile, the national forests and other categories of protected federal and state land were also being set aside. The first national forest, reserved not primarily for recreation but to secure a sustainable supply of timber, was the Yellowstone Timberland Reserve, created by an executive order in 1891. The Adirondack and Catskill State Parks in New York were established by the government of that state in 1892, making them the oldest and, in the case of the Adirondacks, still the largest non-federal parks in the United States. The 2.7 million acres (11,000 square km) of the Adirondack Park that are state-owned comprise a larger area than most national parks; the whole park is about the size of Vermont. In 1903, the first National Wildlife Refuge was created on Pelican Island, Florida, by order of President Theodore Roosevelt (1858–1919).

In 1916, Congress passed legislation creating the National Park Service in order to oversee the growing system of national parks. In 1933, scores of federally owned monuments and historical sites were transferred to the care of the National Park Service. To this day, the Park Service remains in charge of many human-made sites such as the Lincoln Memorial in Washington, D.C.

U.S. Protected Lands Today

The more than 380 monuments and preserves managed by the National Park Service cover 83 million acres (336,000 square km), with holdings in every state but Delaware. The National Forests cover about 190 million acres (769,000 square km), over twice as much area as the national parks. The Bureau of Land Management oversees an additional 264 million acres (1 million square km) of land. The degrees of protection accorded to land managed by the federal government range from zero, as in the case of lands leased for strip-mining by the Bureau of Land Management, to complete prohibition on human entry, as in especially sensitive park locations such as nesting areas.

City and state governments maintain tens of thousands of parks or public recreational spaces. These range in size and degree of wildness from playgrounds to large wilderness preserves such as New York's Adirondacks Park or Alaska's Wood-Tikchick State Park, which is as large as the state of Delaware.

The Global Picture

During the twentieth century, the concept of the national park became global. By the early 2000s, over 90 countries had set aside areas of wild land as national parks. The creation and scientific management of protected areas is encouraged by the International Union for Conservation of Nature (IUCN), an international group whose members include over 1,000 governments and private organizations. The IUCN, which is headquartered in Geneva, Switzerland, and works closely with the United Nations Environment Programme (UNEP), states that its mission is "to influence, encourage, and assist societies throughout the world to conserve the integrity and diversity of nature and to ensure that any use of natural resources is equitable and ecologically sustainable."

According to the *2003 United Nations List of Protected Areas,* as of 2003 there were at least 102,102 protected areas in the world covering 7.3 million square mi (18.8 million square km). The United States contained about 8% of this protected area. Excluding marine protected areas, about 6.6 million square mi (17.1 million square km) of land were protected, some 11.5% of the world's land surface—an area almost as large as South America. The IUCN defines a protected area as "an area of land and/or sea especially dedicated to the protection and maintenance of biological diversity, and of natural and associated cultural resources, and managed through legal or other effective means" (2003 UN List). Biological diversity (also called biodiversity) is defined by the Convention on Biological Diversity, a

1992 treaty ratified by 189 countries (not including the United States) as "the variability among living organisms from all sources ... 'including' diversity within species, between species and of ecosystems." Slowing the global loss of biodiversity is one of the main priorities of most groups managing protected areas today.

As already mentioned, reserves and parks span a range of wildness, from monuments to pristine wildernesses. The IUCN recognizes six main categories of protected area:

1. Stricture Nature Reserve or Wilderness Area, managed mainly for science or wilderness protection;
2. National Park, managed mainly for ecosystem protection and recreation;
3. Natural Monument, managed mainly for conservation of specific natural features;
4. Habitat/Species Management Area, managed mainly for conservation through management intervention (rather than, say, by direct exclusion of certain activities, such as logging or fishing);
5. Protected Landscape/Seascape, where "the interaction of people and nature over time has produced an area of distinct character with significant aesthetic, ecological and/or cultural value" (2003 UN List), managed mainly for landscape or seascape conservation and recreation; and
6. Managed Resource Protected Area, an area managed mostly for sustainable exploitation of natural ecosystems (for example, a national forest).

Antarctica

The world's largest single protected area and largest wilderness, the continent of Antarctica, is centered on the South Pole and covers almost 5.4 million square mi (14 million square km). It is not counted in the United Nations's 2003 reckoning of global protected areas, although the UN does state that the continent is part of "the world's protected areas estate." Antarctica is not under the jurisdiction of any single government, but is governed by the 1959 Antarctic Treaty, which applies to all points south of 60° south latitude. The Environmental Protection Protocol to the Antarctic Treaty, in force since 1998, declares that Antarctica is a "natural reserve devoted to peace and science."

Private Preservation

Due to bureaucratic delays and political opposition to the acquisition of land by governments, not all sensitive natural areas can be preserved by government action. In the United States, in 1946, a small group of ecologists (scientists who study the interactions of organisms with each other and the environment), frustrated by the lack of government action to preserve key natural areas, formed a private group called the Ecologists Union. In 1950, this group changed its name to the Nature

Environmentalists worried about development of the pristine Georgia barrier island, Cumberland Island, are looking to Congress for money to expand the National Park Service's holdings by 1,000 acres (405 ha). *AP Images*

The whooping crane is one of the first species that appears to have rebound from extinction thanks to legislation and public awareness. *AP Images*

Conservancy. In 1961, the Nature Conservancy purchased a 60-acre (0.25 square km) parcel of land in New York State. By 2008, the conservancy was working in over 30 countries, employed more than 700 full-time staff scientists, had protected more than 183,000 square mi (473,000 square km) of land worldwide, and had cooperated with the U.S. Congress, the U.S. Department of Defense, corporations, state governments, foreign governments, and other parties in the preservation of landscapes.

For example, the conservancy cut a $50 million deal with cash-strapped paper companies in Maine in 2002 that resulted in the protection of 377 square mi (975 square km) of forest. The deal guaranteed public access, traditional recreational uses, and sustainable forestry practices. Similar deals followed. By 2008, the Maine chapter of the Nature Conservancy had protected over 1 million acres, about 4.6% of the state. In 2004, after more than a decade of negotiations, the Nature Conservancy purchased 130 square mi (340 square km) of land in Colorado and transferred it to the federal government. The land was declared the Great Sand Dunes National Park by an act of Congress six months later. The primary goal of the Conservancy is the preservation of biodiversity, with recreational access a secondary priority.

Also prominent in international private preservation is the World Land Trust, a group founded in the United Kingdom in 1989. The trust focuses on the direct preservation of rain forests, the world's most biodiverse and rapidly disappearing type of terrestrial ecosystem.

Scores of smaller, locally oriented land trusts and conservancy organizations in the United States and elsewhere are also protecting significant areas of land. One of the primary tools used by such groups is the conservation easement. A conservation easement is a contract between a landowner and a government or conservation group that limits what the owner can do with his/her land. For example, an easement might specify that the owner can log the land, but cannot erect buildings or roads on it. An easement might specify that the public be allowed access to the protected area, or might allow the landowner to exclude the public. An easement may be donated by a landowner or sold and goes with the land, that is, continues to govern what can be done with the land even if it is passed to a new owner. The benefit of selling a conservation easement is that the owner retains certain rights to the land while gaining cash income. The disadvantage of an easement, from the owner's point of view, is that it may reduce the market value of the land or forbid profitable development, such as subdivision and the erection of houses. By 2003, conservation easements had been used by the Nature Conservancy and local land trusts to protect 8,000 square mi (20,600 square km) of land in the United States.

Impacts and Issues

The area devoted to parks and other protected areas around the world has grown steadily since the 1960s. In 1982, there were 3.4 million square mi (8.8 million square km) of protected areas; in 1992, 4.8 million square mi (12.3 million square km), a 42% increase from 1982; and in 2003, 7.3 million square mi (18.8 million square km), a 53% increase from 1992. However, development, rain-forest destruction, and other forms of habitat destruction have also been accelerating, leading to a global wave of species extinctions. As of 2003, about 12% of the world's biome area was protected, with coverage varying greatly depending on the type of biome. Protection ranged from 1.54% of lake systems to 23.32% of tropical humid forests.

Tourism

Tourism can both benefit and damage protected areas. It can be of benefit when the income from the tourist industry motivates governments to set aside and protect lands that preserve biodiversity; it can cause damage when too many tourists enter a sensitive area, trampling or polluting it or disturbing the life cycles of native plants and animals, or when sensitive areas are developed for hotels and other tourist facilities. Coral reefs, for example, may be easily damaged by poorly planned tourism, with sewage from hotels causing algae to smother living corals, careless tourists standing on or kicking corals or damaging them with boats, skis, anchors, or souvenir-collecting.

The Galápagos islands of Ecuador, a biodiversity hotspot and protected area, are suffering from excessive tourism. In 2007, after 53 sea lions were slaughtered with clubs in the Galápagos, President Rafael Correa of Ecuador declared that the island group was at risk from tourism and immigration. A spokesperson for Acción Ecológica, a group of Ecuadorian ecologists, said in a 2008 *National Geographic News* interview that "The biggest problem we have in the Galápagos is excessive tourism.... The most fundamental change we need is to limit the number of tourist visitors to the area. We are putting too much pressure on the zone, and there should be stricter regulations"

Antarctica, too, is threatened by the friendly attentions of thousands of tourists. In 1990–1991, only about 4,700 tourists visited Antarctica; in 2003–2004, over 24,000 visited, and the number continued to grow rapidly. Although most tour operations in Antarctica are governed by the rules of the International Association of Antarctica Tour Operators, both membership in the group and obedience to its rules are voluntary. Large cruise ships may spew sewage or chemicals, dump garbage, and disturb wildlife by landing large parties of tourists. Incidents have been reported where tourists, who are supposed to stay at least 15 ft (5 m) away from all sea-birds, have chased penguins with video cameras in hand. Tour helicopters have also flown near colonies, possibly disrupting nesting and mating behaviors.

Conflict with Neighbors

Some farmers, herders, and ranchers who live on lands bounding parks view the animals that survive in the parks as threats to their livelihood. For example, in the 1990s and 2000s, ranchers in the American West strongly opposed the re-introduction of wolves into various federal lands. Bison that wander out of Yellowstone National Park in the winter in search of food are routinely shot to accommodate the cattle industry, which fears that the bison will spread a disease (brucellosis) to domestic cattle and seeks to prevent the bison from competing with cattle for grass. Since 1990, thousands of the bison have been killed by hunters or sent by the federal government to slaughterhouses. In 2007 and 2008, in Kenya's Masai Mara National Reserve and Uganda's Queen Elizabeth National Park, native herders used a widely available pesticide (carbofuran) to poison lions and hyenas, which occasionally preyed on their flocks. Animal poaching for bushmeat also continues to be a problem in national parks within developing countries.

Governments can sometimes ease such conflicts by offering compensation and incentives to herders and others in conflict with wildlife. For example, in Montana and Idaho, the U.S. government has granted special permits to ranchers allowing them to kill wolves that attacked livestock, and it has operated compensation programs to pay cash to ranchers whose livestock are killed by wolves.

Primary Source Connection

Our National Parks is a collection of essays by American conservationist and founder of the Sierra Club, John Muir (1838–1914), originally written for the *Atlantic Monthly*. Muir wrote about the beauty and grandeur of the nation's forests and mountain ranges, hoping to encourage people to visit these areas and to realize the importance of maintaining a portion of the country in its original, natural state.

Muir's encouragement and vivid descriptions kept the park in the public eye, increasing the number of visitors, and also drew the attention of the government, eventually leading to the enactment of the National Park Service in 1916, which protected not only Yellowstone, but the other parks that had been designated, unifying them as a single entity designed to benefit the citizens of the United States. It also created the post of a director to oversee the properties and their management, thereby ensuring that large tracts of land would remain untouched by industrial progress, as well as unspoiled in the years to come.

OUR NATIONAL PARKS

Of the four national parks of the West, the Yellowstone is far the largest. It is a big, wholesome wilderness on the broad summit of the Rocky Mountains, favored with abundance of rain and snow,—a place of fountains where the greatest of the American rivers take their rise. The central portion is a densely forested and comparatively level volcanic plateau with an average elevation of about eight thousand feet above the sea, surrounded by an imposing host of mountains belonging to the subordinate Gallatin, Wind River, Teton, Absaroka, and snowy ranges. Unnumbered lakes shine in it, united by a famous band of streams that rush up out of hot lava beds, or fall from the frosty peaks in channels rocky and bare, mossy and bosky, to the main rivers, singing cheerily on through every difficulty, cunningly dividing and finding their way east and went to the two far-off seas.

Glacier meadows and beaver meadows are outspread with charming effect along the banks of the streams, parklike expanses in the woods, and innumerable small gardens in rocky recesses of the mountains, some of them containing more petals than leaves, while the whole wilderness is enlivened with happy animals.

Beside the treasures common to most mountain regions that are wild and blessed with a kind climate, the park is full of exciting wonders. The wildest geysers in the world, in bright, triumphant bands, are dancing and singing in it amid thousands of boiling springs, beautiful and awful, their basins arrayed in gorgeous colors like gigantic flowers; and hot paint-pots, mud springs, mud volcanoes, mush and broth caldrons whose contents are of every color and consistency, plash and heave and roar in bewildering abundance. In the adjacent mountains, beneath the living trees the edges of petrified forests are exposed to view, like specimens on the shelves of a museum, standing on ledges tier above tier where they grew, solemnly silent in rigid crystalline beauty after swaying in the winds thousands of centuries ago, opening marvelous views back into the years and climates and life of the past. Here, too, are hills of sparkling crystals, hills of sulphur, hills of glass, hills of cinders and ashes, mountains of every style of architecture, icy or forested, mountains covered with honey-bloom sweet as Hymettus, mountains boiled soft like potatoes and colored like a sunset sky. A' that and a' that, and twice as muckle's a' that, Nature has on show in the Yellowstone Park. Therefore it is called Wonderland, and thousands of tourists and travelers stream into it every summer, and wander about in it enchanted.

Fortunately, almost as soon as it was discovered it was dedicated and set apart for the benefit of the people, a piece of legislation that shines benignly amid the common dust-and-ashes history of the public domain, for which the world must thank Professor Hayden above all others; for he led the first scientific exploring party into it, described it, and with admirable enthusiasm urged Congress to preserve it. As delineated in the year 1872, the park, contained about 3344 square miles. On March 30, 1891 it was to all intents and purposes enlarged by the Yellowstone National Park Timber Reserve, and in December, 1897, by the Teton Forest Reserve; thus nearly doubling its original area, and extending the southern boundary far enough to take in the sublime Teton range and the famous pasture-lands of the big Rocky Mountain game animals. The withdrawal of this large tract from the public domain did not harm to any one; for its height, 6000 to over 13,000 feet above the sea, and its thick mantle of volcanic rocks, prevent its ever being available for agriculture or mining, while on the other hand its geographical position, reviving climate, and wonderful scenery combine to make it a grand health, pleasure, and study resort,—a gathering-place for travelers from all the world.

The national parks are not only withdrawn from sale and entry like the forest reservations, but are efficiently managed and guarded by small troops of United States cavalry, directed by the Secretary of the Interior. Under this care the forests are flourishing, protected from both axe and fire; and so, of course, are the shaggy beds of underbrush and the herbaceous vegetation. The so-called curiosities, also, are preserved, and the furred and feathered tribes, many of which, in danger of extinction a short time ago, are now increasing in numbers,—a refreshing thing to see amid the blind, ruthless destruction that is going on in the adjacent regions. In pleasing contrast to the noisy, ever changing management, or mismanagement, of blundering, plundering, money-making vote-sellers who receive their places from boss politicians as purchased goods, the soldiers do their duty so quietly that the traveler is scarce aware of their presence.

This is the coolest and highest of the parks. Frosts occur every month of the year. Nevertheless, the tenderest tourist finds it warm enough in summer. The air is electric and full of ozone, healing, reviving, exhilarating, kept pure by frost and fire, while the scenery is wild enough to awaken the dead. It is a glorious place to grow in and rest in; camping on the shores of the lakes, in the warm openings of the woods golden with sunflowers, on the banks of the streams, by the snowy waterfalls, beside the exciting wonders or away from them in the scallops of the mountain walls sheltered from every wind, on smooth silky lawns enameled with gentians, up in the fountain hollows of the ancient glaciers between the peaks, where cool pools and brooks and gardens of precious plants charmingly embowered are never wanting, and good rough rocks with every variety of cliff and scaur are invitingly near for outlooks and exercise.

From these lovely dens you may make excursions whenever you like into the middle of the park, where the geysers and hot springs are reeking and spouting in their beautiful basins, displaying an exuberance of color and strange motion and energy admirably calculated to surprise and frighten, charm and shake up the least sensitive out of apathy into newness of life.

However orderly your excursions or aimless, again and again amid the calmest, stillest scenery you will be brought to a standstill hushed and awe-stricken before phenomena wholly new to you. Boiling springs and huge deep pools of purest green and azure water, thousands of them, are plashing and heaving in these high, cool mountains as if a fierce furnace fire were burning beneath each one of them; and a hundred geysers, white torrents of boiling water and steam, like inverted waterfalls, are ever and anon rushing up out of the hot, black underworld. Some of these ponderous geyser columns are as large as sequoias,—five to sixty feel in diameter, one hundred and fifty to three hundred feet high,—and are sustained at this great height with tremendous energy for a few minutes, or perhaps nearly an hour, standing rigid and erect, hissing, throbbing, booming, as if thunderstorms were raging beneath their roots, their sides roughened or fluted like the furrowed boles of trees, their tops dissolving in feathery branches, while the irised spray, like misty bloom is at times blown aside, revealing the massive shafts shining against a background of pine-covered hills. Some of them lean more or less, as if storm-bent, and instead of being round are flat or fan-shaped, issuing from irregular slits in silex pavements with radiate structure, the sunbeams sifting through them in ravishing splendor. Some are broad and round-headed like oaks; others are low and bunchy, branching near the ground like bushes; and a few are hollow in the centre like big daisies or water-lilies. No frost cools them, snow never covers them nor lodges in their branches; winter and summer they welcome alike; all of them, of whatever form or size, faithfully rising and sinking in fairy rhythmic dance night and day, in all sorts of weather, at varying periods of minutes, hours, or weeks, growing up rapidly, uncontrollable as fate, tossing their pearly branches in the wind, bursting into bloom and vanishing like the frailest flowers,—plants of which Nature raises hundreds or thousands of crops a year with no apparent exhaustion of the fiery soil.

The so-called geyser basins, in which this rare sort of vegetation is growing, are mostly open valleys on the central plateau that were eroded by glaciers after the greater volcanic fires had ceased to burn. Looking down over the forests as you approach them from the surrounding heights, you see a multitude of white columns, broad, reeking masses, and irregular jets and puffs of misty vapor ascending from the bottom of the valley, or entangled like smoke among the neighboring trees, suggesting the factories of some busy town or the camp-fires of an army. These mark the position of each mush-pot, paint-pot, hot spring, and geyser, or gusher, as the Icelandic words mean. And when you saunter into the midst of them over the bright sinter pavements, and see how pure and white and pearly gray they are in the shade of the mountains, and how radiant in the sunshine, you are fairly enchanted. So numerous they are and varied, Nature seems to have gathered them from all the world as specimens of her rarest fountains, to show in one place what she can do. Over four thousand hot springs have been counted in the park, and a hundred geysers; how many more there are nobody knows.

These valleys at the heads of the great rivers may be regarded as laboratories and kitchens, in which, amid a thousand retorts and pots, we may see Nature at work as chemist or cook, cunningly compounding an infinite variety of mineral messes; cooking whole mountains; boiling and steaming flinty rocks to smooth paste and mush,—yellow, brown, red, pink, lavender, gray, and creamy white,—making the most beautiful mud in the world; and distilling the most ethereal essences. Many of these pots and caldrons have been boiling thousands of years. Pots of sulphurous mush, stringy and lumpy, and pots of broth as black as ink, are tossed and stirred with constant care, and thin transparent essences, too pure and fine to be called water, are kept simmering gently in beautiful sinter cups and bowls that grow ever more beautiful the longer they are used. In some of the spring basins, the waters, though still warm, are perfectly calm, and shine blandly in a sod of overleaning grass and flowers, as if they were thoroughly cooked at last, and set aside to settle and cool. Others are wildly boiling over as if running to waste, thousands of tons of the precious liquids being thrown into the air to fall in scalding floods on the clean coral floor of the establishment, keeping onlookers at a distance. Instead of holding limpid pale green or azure water, other pots and craters are filled with scalding mud, which is tossed up from three or four feet to thirty feet, in sticky, rank-smelling masses, with gasping, belching, thudding sounds, plastering the branches of neighboring trees; every flask, retort, hot spring, and geyser has something special in it, no two being the same in temperature, color, or composition.

In these natural laboratories one needs stout faith to feel at ease. The ground sounds hollow underfoot, and the awful subterranean thunder shakes one's mind as the ground is shaken, especially at night in the pale moonlight, or when the sky is overcast with storm-clouds. In the solemn gloom, the geysers, dimly visible, look like monstrous dancing ghosts, and their wild songs and the earthquake thunder replying to the storms overhead seem doubly terrible, as if divine government were at an end. But the frembling hills keep their places. The sky clears, the rosy dawn is reassuring, and up comes the sun like a god, pouring his faithful beams across the mountains and forest, lighting each peak and tree and ghastly geyser alike, and shining into the eyes of the reeking springs, clothing them with rainbow light, and dissolving the seeming chaos of darkness into varied forms of harmony. The ordinary work of the world goes on. Gladly we see the flies dancing in the sun-beams, birds feeding their young, squirrels gathering nuts, and hear the blessed ouzel singing confidingly in the shallows of the river, —most faithful evangel, calming every fear, reducing everything to love.

John Muir

JOHN MUIR. *OUR NATIONAL PARKS.* BOSTON: HOUGHTON-MIFFLIN, 1901.

■ Primary Source Connection

The Congressional Act of 1872, making Yellowstone a National Park under the control of the U.S. Secretary of the Interior, is divided into two sections. The first, relying on the work of the expeditionary parties, defines the boundaries of the park and declares it federal property. The second section outlines what it means for a tract of land to be so designated and lays down a set of environmental guidelines. It put the newly created park under the "exclusive control" of the Secretary of the Interior and specified his responsibilities, first and foremost the creation of a body of rules, inside the congressional guidelines, to govern the use and determine the development of the park.

By creating Yellowstone National Park, the U.S. Congress recognized the wonders contained in that tract of land. It also recognized and accepted two responsibilities. The first was to preserve the land in its natural condition and protect its resources from alteration or despoliation. The second was to develop it in accord within the limits of the act. These two responsibilities were not always compatible.

YELLOWSTONE NATIONAL PARK ACT

SECTION 1. *Be it enacted by the Senate and House of Representatives of the United States of America in Congress assembled*, That the tract of land in the Territories of Montana and Wyoming, lying near the head-waters of the Yellowstone River, and described as follows, to wit, commencing at the junction of Gardiner's river with the Yellowstone river, and running east to the meridian passing ten miles to the eastward of the most eastern point of Yellowstone lake; thence south along said meridian to the parallel of latitude passing ten miles south of the most southern point of Yellowstone lake; thence west along said parallel to the meridian passing fifteen miles west of the most western point of Madison lake; thence north along said meridian to the latitude of the junction of the Yellowstone and Gardiner's rivers; thence east to the place of beginning, is hereby reserved and withdrawn from settlement, occupancy, or sale under the laws of the United States, and dedicated and set apart as a public park or pleasuring-ground for the benefit and enjoyment of the people; and all persons who shall locate or settle upon or occupy the same, or any part thereof, except as hereinafter provided, shall be considered trespassers and removed therefrom.

SECTION 2. That said public park shall be under the exclusive control of the Secretary of the Interior, whose duty it shall be, as soon as practicable, to make and publish such rules and regulations as he may deem necessary or proper for the care and management of the same. Such regulations shall provide for the preservation, from injury or spoliation, of all timber, mineral deposits, natural curiosities, or wonders within said park, and their retention in their natural conditions. The secretary may in his discretion, grant leases for building purposes for terms not exceeding ten years, of small parcels or ground; at such places in said park as shall require the erection of buildings for the accommodation of visitors; all of the proceeds of said leases, and all other revenues that may be derived from any source connected with said park, to be expended under his direction in the management of the same, and the construction of roads and bridle-paths therein. He shall provide against the wanton destruction of the fish and game found within said park, and against their capture or destruction for the purposes of merchandise or profit. He shall also cause all persons trespassing upon the same after the passage of this act to be removed therefrom, and generally shall be authorized to take all such measures as shall be necessary or proper to fully carry out the objects and purposes of this act.

U.S. Congress

U.S. CONGRESS. 17 STAT. 32. "YELLOWSTONE NATIONAL PARK ACT," MARCH 1, 1872.

SEE ALSO *Antarctic Treaty; Bureau of Land Management; Extinction and Extirpation; National Park Service Organic Act; Recreational Use and Environmental Destruction; Wilderness Act of 1964*

BIBLIOGRAPHY

Books

Conard, Rebecca, and Wayne Franklin. *Places of Quiet Beauty: Parks, Preserves, and Environmentalism.* Iowa City, IA: University of Iowa Press, 1997.

Lockwood, Michael, et al., eds. *Managing Protected Areas: A Global Guide.* Sterling, VA: Earthscan Publications, 2006.

Mitchell, Nora, et al, eds. *The Protected Landscape Approach: Linking Nature, Culture and Community.* Gland, Switzerland: World Conservation Union, 2004.

Web Sites

National Geographic News. "Dozens of Sea Lions Found Massacred in Galápagos." February 4, 2008. http://news.nationalgeographic.com/news/2008/02/080204-sea-lions.html (accessed May 4, 2008).

National Park Service. http://awww.nps.gov/ (accessed May 3, 2008).

United Nations Environment Programme. "Tourism and Biodiversity: Mapping Tourism's Global Footprint." http://www.unep.fr/pc/tourism/library/mapping_tourism.htm (accessed May 3, 2008).

United Nations Environment Programme. "United Nations List of Protected Areas" http://www.unep-wcmc.org/protected_areas/UN_list/ (accessed May 3, 2008).

Natural Resource Management

Introduction

Natural resource management refers to strategies intended to sustain both renewable and non-renewable resources for present and future use. By considering how best to use the particular resource, its productivity is prolonged, and its relationship with the environment is protected.

One example of natural resource management could include the efforts of a number of organizations to preserve water availability and quality around the world. Management of water is becoming especially important as regions of the globe become warmer and drier under the influence of global warming. Natural resources that are especially vulnerable to exploitation include forests, fisheries, and the agricultural capability of land.

Historical Background and Scientific Foundations

In sectors such as agriculture, forestry, and fishing, resource management focuses on preventing the overexploitation of the resource. For example, as farms have become larger and the practice of farming has shifted from manual labor to mechanization, and as the use of pesticides has increased, agricultural practices have become more damaging to the environment. Clearing the land can increase the erosion of the soil, and runoff of pesticides can contaminate surface and groundwater.

A mature, natural forest is a complex, biodiverse ecosystem. Many different types of tree species and other vegetation are present, which support many insect, bird, animal, and fish species. In contrast, forests in which lumber has been harvested for a long time tend to be less diverse, as only one or only a few species of tree remain. Such a monoculture type of forest, which is often the result of management of the resource by timber companies, can still thrive and support life, as the rate of tree cutting is controlled and locations for harvest are rotated. These practices help to make the forest a sustainable source of lumber for centuries.

Many forested regions in North America and Europe have been managed by clear-cutting. Although clear-cutting is a less expensive means of lumbering, critics of the practice decry this form of resource management as being destructive to an ecosystem because erosion of the destabilized soil can occur and the former biodiversity is obliterated. Elsewhere, the Brazilian rain forest, which comprises about 30% of all global rain forest, is being logged at a rate of over 5 million acres each year. If allowed to continue, the rain forests of Brazil, which have an important influence on moderating the global climate, could be eliminated by the year 2050.

Tides are a natural resource that can be managed to produce hydroelectric power. Depending on the geography of the coast, tides can cycle tremendous amounts of water each day. The Bay of Fundy, located between the Canadian maritime provinces of Nova Scotia and New Brunswick, which progressively narrows and become shallower, has differences between low and high tides that exceed 50 ft (15 m). Harnessing this energy on a large scale, which is currently feasible, could be a source of renewable energy that could power hundreds of thousands of homes, while emitting no greenhouse gases in the process. The largest tidal power system in the world today is located at St. Malo in France. The dam-based system generates almost 240 megawatts of electricity, enough to power about 100,000 homes for a year. Ironically, the production of hydroelectric power by the damming of watercourses can sometimes be destructive to inland ecosystems, and displaces wildlife and people.

Impacts and Issues

Natural resource management requires forethought and planning, and some efforts are more successful than others. An example of a resource strategy gone awry has occurred in Brazil, where selective logging of the rain forest was encouraged by construction of public roads into the rain forest. The management strategy, which was intended to limit the environmental effects of clear-cutting, has had the opposite effect. A survey of the region carried out in 2006 revealed that selective logging has led to more extensive clear-cutting, because the roads have provided access to previously pristine regions. Selective logging is producing a loss of rain forest that is twice as fast as was previously estimated.

The importance of natural resource management is exemplified by the loss of the Amazon rain forest. Even though rain forests occupy only about 2% of Earth's surface, they contain 60 to 70% of all life on the globe. The loss of this biodiversity could remove thousands of insect species and types of microorganisms that have not yet been discovered, and could deprive humans of as yet undiscovered beneficial compounds such as active ingredients for medicines.

WORDS TO KNOW

CLEAR-CUTTING: A forestry practice involving the harvesting of all trees of economic value at one time.

EROSION: The wearing away of the soil or rock over time.

GREENHOUSE GAS: A gas whose accumulation in the atmosphere increases heat retention.

NON-RENEWABLE RESOURCE: A natural resource of finite supply that cannot be regenerated.

RENEWABLE RESOURCE: Any resource that is renewed or replaced fairly rapidly (on human historical time-scales) by natural or managed processes.

RUNOFF: Water that falls as precipitation and then runs over the surface of the land rather than sinking into the ground.

SUSTAINABILITY: Practices that preserve the balance between human needs and the environment, as well as between current and future human requirements.

WATERSHED: The expanse of terrain from which water flows into a wetland, water body, or stream.

China's vast northern region's dry climate, coupled with a booming economy and huge population size, means water resource conservation and management is becoming a critical issue. *AP Images*

Advocates for a more sustainable approach to logging argue for an ecosystem-based management of the forests. In this approach, logging is viewed as being an important contributor to the overall stability of the ecosystem. Some areas may be logged intensively, but with attention paid to minimizing environmental degradation, while other regions are logged more selectively.

The maritime provinces of Canada once had a robust Atlantic cod fishery with a seemingly infinite number of cod living on the Grand Banks off the east coast of Newfoundland. However, a number of factors, including centuries of overfishing, decimated the cod stocks to the point of near extinction. The fishery was closed in the 1980s, and the resultant economic loss and social upheaval have been considerable. In 1994, two years after banning cod fishing in the North Atlantic, scientists again observed schools of spawning cod off the coast of Newfoundland. With a continued ban on fishing, the species is recovering slowly, and through future study and regulation, scientists and authorities aim to manage the still-renewable resource of codfish in a manner that is sustainable for future generations.

SEE ALSO *Biofuels; Deforestation; Fish Farming; Mining and Quarrying Impacts; Overfishing; Overgrazing; Sustainable Development*

Part of a tropical rain forest outside Totuguero National Park, Costa Rica. Ninety-five percent of the world's tropical forests remain unprotected, despite a significant increase in their environmentally friendly use over the past two decades. *AP Images*

BIBLIOGRAPHY

Books

Chiras, Daniel D., John P. Reganold, and Oliver S. Owen. *Natural Resource Conservation: Management for a Sustainable Future*. New York: Prentice-Hall, 2004.

Diamond, Jared. *Collapse: How Societies Choose to Fail or Succeed*. New York: Viking, 2004.

Freyfogle, Eric T. *Why Conservation Is Failing and How It Can Regain Ground*. New Haven: Yale University Press, 2006.

Web Sites

U.S. Geological Survey. "Recent Highlights: Natural Resources." January 29, 2003. http://www.usgs.gov/themes/FS-187-97/ (accessed April 20, 2008).

Nonpoint-Source Pollution

Introduction

Nonpoint-source pollution is pollution that enters a waterway from diverse sources. Runoff from precipitation and atmospheric deposition are two of the most common forms of nonpoint-source pollution. Some of the more important sources of nonpoint-source pollution are fertilizers, herbicides, insecticides, oil, grease, toxic chemicals, sediment, salt, and infectious agents. Nonpoint-source pollution is extremely difficult to manage and monitor because it is unpredictable and the source is often hard to trace. Several forms of legislation, such as the Clean Water Act and Coastal Zone Act Reauthorization Amendments have, however, attempted to decrease nonpoint-source pollution in waterways in the United States.

Historical Background and Scientific Foundations

Pollution is a substance or form of energy that affects the natural functioning of the environment, be it chemical, biological, or physical. Major types of pollutants that affect water bodies include oil, sediments, sewage, nutrients, heavy metals, salts, and thermal pollution. Many pollutants enter water from a single point of entry, such as a pipe from a factory, sewage treatment plant, or agricultural entity with a specific point of discharge. These forms of pollutants are called point source pollutants because the location of discharge is known and can easily be monitored. In the United States, cities, industries, storm water runoff from cities with over 100,000 people, and animal feedlots are known point source polluters and are required to have a permit to release pollutants called a National Pollutant Discharge Elimination System Permit (NPDES). Pollution that enters the environment through other pathways that are not governed by NPDES is considered nonpoint-source pollution.

Nonpoint-source pollution comes from many diffuse sources. One of the major forms of nonpoint pollution is runoff from precipitation. When rain and other forms of precipitation fall to the ground, they may not soak directly into the soil, especially if the soil has been damaged by erosion. The water gathers and runs along the surface collecting contaminants as it flows along. Eventually this runoff empties into streams, rivers, lakes, or oceans, bringing with it the pollutants it has collected. Atmospheric deposition of contaminants is also a significant nonpoint-source pollutant. Airborne contaminants, such as pesticides and acids, can precipitate directly into water bodies or onto snow or other surfaces, and then enter water bodies through runoff.

The pollutants transported into waterways via runoff have a diverse range of sources. Oil, grease, heavy metals, and other toxic chemicals from agricultural lands, construction sites, roads, and parking lots are incorporated into runoff. Sediment from construction sites, croplands, and logged areas can be suspended into runoff. In northern latitudes, salt used to melt snow and ice in the winter is a major pollutant in runoff. Salt can also contribute to runoff pollution in areas where abandoned mines are common. Pet wastes, livestock wastes, and overloaded sewage treatment systems can contribute to infectious bacteria and nutrients in runoff.

In contrast to point source pollution, the timing and predictability of nonpoint-source pollution is often difficult to assess. The first heavy rainfall of the year can wash heavy loads of pollutants that have collected on impermeable surfaces like roads and parking lots into waterways. Rainstorms following this initial flush, may not be as damaging. In the spring, melting snow can wash high concentrations of atmospheric deposition containing contaminants into lakes and streams. The unpredictability of nonpoint-source runoff makes it difficult to monitor and manage.

WORDS TO KNOW

EUTROPHICATION: The process whereby a body of water becomes rich in dissolved nutrients through natural or man-made processes. This often results in a deficiency of dissolved oxygen, producing an environment that favors plant over animal life.

POINT SOURCE POLLUTANT: A pollutant that enters the environment from a single point of entry.

SEDIMENT: Solid unconsolidated rock and mineral fragments that come from the weathering of rocks and are transported by water, air, or ice and form layers on Earth's surface. Sediments can also result from chemical precipitation or secretion by organisms.

Impacts and Issues

The U.S. Environmental Protection Agency (EPA) reports that approximately 40% of all natural freshwater sources that have been mapped have been affected by nonpoint-source pollution. The damage caused by nonpoint-source pollution has affected drinking supplies, recreation, fisheries, and wildlife.

One of the major effects of nonpoint-source pollution is the addition of nutrients, generally nitrogen and phosphorus compounds found in fertilizers, wastes, and garbage that enter waterways through runoff. These nutrients are a source of food for microorganisms that live in the water and can cause giant blooms of phytoplankton. As these phytoplankton die, decomposers break them down and in the process consume large amounts of dissolved oxygen. This lowers the oxygen concentration in the water and stresses, or even kills, fish and other animals that require oxygen. This process of overfertilization is known as eutrophication. In the 1970s, the once-flourishing ecosystem in the Chesapeake Bay deteriorated so significantly from eutrophication that fisheries collapsed and recreation came to a halt. Pollution prevention programs initially established by grassroots efforts eventually decreased the input of phosphorus by at least 40%, significantly improving the health of the bay.

The Clean Water Act of 1972 is an important control on pollution in waterways. Part of the act established the NPDES permit system. This act led to significant improvements in water quality from point sources. In 1987, the act was amended to include section 319, the Nonpoint Source Management Program. Under this section, states and territories receive money to support a range of activities that help monitor and prevent nonpoint-source projects.

In 1990, Congress passed the Coastal Zone Act Reauthorization Amendments. This legislation was aimed at controlling nonpoint-source pollution in coastal waters. The regulations provide funding to manage and monitor development on land that borders

Fertilizers used in agriculture can enter waterways, increasing the amounts of nitrogen and phosphorus, and upsetting the balance of plant and animal life. *AP Images*

Beneath the placid surface is an ongoing life-or-death conflict over chemical makeup of Everglades water, virtually ruined by decades of fertilizer runoff and other non-point pollution sources. *AP Images*

waterways that empty into the ocean. The goal of this amendment is to create a situation in which the water that enters the ocean is as free of pollution as possible.

One of the most important ways of decreasing nonpoint-source pollution is for citizens to play an active role in preventing harmful substances from becoming incorporated into runoff. Properly disposing of wastes such as motor oil, antifreeze, and paints, and using minimal fertilizers and pesticides is one way to reduce contaminants. Planting ground cover in gardens and lawns to minimize areas that are prone to erosion decreases sediment runoff. Using detergents that are low in phosphorus decreases the nutrient concentrations that flow into waterways and prevents eutrophication.

SEE ALSO *Acid Rain; Algal Blooms; Aquatic Ecosystems; Bays and Estuaries; Clean Water Act; Coastal Ecosystems; Environmental Protection Agency (EPA); Groundwater; Groundwater Quality; Industrial Pollution; Industrial Water Use; Insecticide Use; Lakes; Logging; Marine Ecosystems; Marine Water Quality; Oceans and Coastlines; Precipitation; Runoff; Water Pollution*

BIBLIOGRAPHY

Books

Cunningham, W.P., and A. Cunningham. *Environmental Science: A Global Concern*. New York: McGraw-Hill International Edition, 2008.

Garrison, Tom. *Oceanography: An Invitation to Marine Science*, 5th ed. Stamford, CT: Thompson/Brooks Cole, 2004.

Web Sites

U.S. Environmental Protection Agency (EPA). "Polluted Runoff (Nonpoint Source Pollution)." February 25th, 2008. http://www.epa.gov/owow/nps/ (accessed March 5, 2008).

U.S. Environmental Protection Agency, Hawaii State Department of Health, and the City and County of Honolulu Department of Environmental Services. "Nonpoint Source Pollution." 2002. http://protectingwater.com/index.html (accessed March 5, 2008).

Juli Berwald

Non-Scientist Contributions to Nature and Environment Studies

Introduction

Studies concerning the natural environment and environmental change are organized and run by scientists whose expertise is in the area of study. Their knowledge is necessary to design experiments and analyze the obtained data in ways that will produce meaningful interpretations of the results.

However, non-scientists can contribute in important ways in these studies. People who are not trained in the particular subject area may have technical skills that are useful, such as in the maintenance and operation of equipment used in transport or in data collection.

Additionally, non-scientists can be trained to participate in sample collection. A well-known example is the Christmas bird population count organized by the National Audubon Society. The data collected from many thousands of volunteers is much more than could be collected by a single research team, and so is of great value. Volunteers in organizations devoted to preserving water quality have also been recruited to collect water samples and even to perform tests.

Historical Background and Scientific Foundations

The participation of volunteers in annual census of bird populations dates back to 1900, when ornithologist Frank Chapman proposed a Christmas Day census of birds. The first count involved 27 people spread across North America. Since then, the annual count organized by the Audubon Society has expanded in time, being conducted from December 14 to January 5, and has grown in participation, with an estimated 50,000 observers taking part each year.

The data collected for over a century have proven valuable in assessing changes in the population and distribution of particular species of birds, and in revealing wholesale changes in the multiple species that can be an indicator of habitat change due to, as examples, surface or groundwater contamination, presence of pesticides, or climate change. The huge numbers of people involved obtain quantities of data that would be impossible for a single laboratory or group of scientists to collect.

Another study that relied on the participation of non-scientists as observers conducted in the Rocky Mountain region of the United States since 1973 has documented that the earlier blooming of plants in spring has occurred coincidently with climate-related temperature increase. The altered blooming timing of the flowers may be disrupting pollinator activity, threatening the survival of both the plant species and the pollinators. The individual observations (in this case, observing when blooming occurs) are easy to do and volunteers were trained for the task. Moreover, the task was part of a volunteer organization dedicated to natural preservation, and so became part of the organization's annual roster of activities. This kind of dedicated sampling can be easier to maintain than in a lab faced with changes in research priorities and funds.

Non-scientists can be as dedicated to environmental preservation as a professional scientist. This zeal can be beneficial in keeping track of local environments and in reporting environmental incidents such as accidental spills and unauthorized disposal of noxious compounds. As this sort of information will be entered into databases, it becomes information that can be extracted and analyzed. Without the vigilance of the observer, the information would never have been obtained.

Non-scientists are also useful in transporting scientific staff to the locations of study, transporting collected samples back to the lab for analysis, and in maintaining equipment that is required for the study. Such support studies are vital to the success of any study conducted outside of the laboratory.

Organizations such as the Sierra Club, Ducks Unlimited, and the World Wildlife Fund, which are made up of many nonscientists, have and continue to be influential in protecting the environment directly through remediation activities, and also through their lobbying power and educational programs. As one example, the World Wildlife Fund is active in 100 countries and has a global membership of almost 5 million people.

Individuals can have a profound effect on nature and environmental awareness. A well-known recent example is former Vice President of the United States Al Gore, whose longtime environmental interests led him to star in the documentary film *An Inconvenient Truth*. The film helped make the issue of global warming and the role of human activities in the warming of Earth's atmosphere widely known and accepted. Together with the Intergovernmental Panel on Climate Change (IPCC), Gore received the 2007 Nobel Peace Prize.

Phoebe Snetsinger has sighted more than 7,000 birds and was the first woman and the first American to reach the 7,000 mark, 1992. *AP Images*

WORDS TO KNOW

HABITAT: The natural location of an organism or a population.

GROUNDWATER: Fresh water that is present in an underground location.

Impacts and Issues

Non-scientists contribute fundamentally to nature and environment studies through their expertise in maintaining the equipment and facilities required to carry out the studies and, perhaps more importantly, in the case of volunteers who help conduct a census or who obtain samples, through their sheer numbers.

A census involving tens of thousands of dedicated non-scientists can amass far more data than could a small team of scientists. Moreover, the care taken by the volunteers produces data that are as scientifically robust as that collected by the trained scientist.

The development of water tests that can be done on-site with minimal sample manipulation has made it possible for volunteers to collect water samples and actually perform the test. This participation from local grassroots organizations dedicated to water quality preservation has been enlisted by some government agencies. The results can be the collection of data at less expense than would otherwise be incurred by the particular agency, and the fostering of a more collegial relationship between the agency and those it serves.

Primary Source Connection

In his study, his painting, and his biographies of the birds of the United States, American naturalist and ornithologist John James Audubon (1785–1851) created more than a decorative descriptive catalog of birds. He rendered a map of the natural environment of America, of its forests, marshes, and swamps; of its dark, nearly unapproachable regions; of its insects and its trees; and of its birds. His work served as a bridge for the human species to cross in order to enter a new and unexplored territory, one that was more than a landmass, but a living environment that people were endeavoring to appropriate, to become part of, and to make theirs. Audubon showed the contours and the nature of this world. As much as the great territorial explorers who revealed both the vastness and the wonders of the North American continent, Audubon, through his exploration and his art, revealed the diversity and the particularities of the continent's natural environment. He also simply made a record of

The annual Great Backyard Bird Count encourages birdwatchers of all levels to count birds and report their results; the count's results are analyzed and summaries are posted highlighting the year's trends and findings. *AP Images*

the continent's environmental wealth, not only regarding its birds, but their habitats as well.

Audubon's inventory and the wonders it presented inspired a preservationist consciousness in Americans, one that was expressed concretely through the formation of the Audubon Society for the Protection of Birds in 1886 by American naturalist George Bird Grinnell (1849–1938). Grinnell's group, which he soon disbanded because its membership became too large for him to handle, was the beginning of the National Audubon Society, founded in 1905. From its beginning as a society of bird lovers dedicated to safeguarding birds, the Audubon Society has grown into a citizen's organization devoted to environmental protection and conservation as well as bird-watching and education.

BIRDS OF AMERICA

...I wish, kind reader, it were in my power to present to your mind's eye the favourite resort of the Ivory-billed Woodpecker. Would that I could describe the extent of those deep morasses, overshadowed by millions of gigantic dark cypresses, spreading their sturdy moss-covered branches, as if to admonish intruding man to pause and reflect on the many difficulties which he must encounter, should he persist in venturing farther into their almost inaccessible recesses, extending for miles before him, where he should be interrupted by huge projecting branches, here and there the massy trunk of a fallen and decaying tree, and thousands of creeping and twining plants of numberless species! Would that I could represent to you the dangerous nature of the ground, its oozing, spongy, and miry disposition, although covered with a beautiful but treacherous carpeting, composed of the richest mosses, flags, and water-lilies, no sooner receiving the pressure of the foot than it yields and endangers the very life of the adventurer, whilst here and there, as he approaches an opening, that proves merely a lake of black muddy water, his ear is assailed by the dismal croaking of innumerable frogs, the hissing of serpents, or the bellowing of alligators! Would that I could give you an idea of the sultry pestiferous atmosphere that nearly suffocates the intruder during the meridian heat of our dogdays, in those gloomy and horrible swamps! ...

The flight of this bird is graceful in the extreme, although seldom prolonged to more than a few hundred yards at a time, unless when it has to cross a large river, which it does in deep undulations, opening its wings at first to their full extent, and nearly closing them to renew the propelling impulse. The transit from one tree to another, even should the distance be as much as a hundred yards, is performed by a single sweep, and the bird appears as if merely swinging itself from the top of the one tree to that of the other, forming an elegantly curved line. At this moment all the beauty of the plumage is exhibited, and strikes the beholder with pleasure. It never utters any sound whilst on wing, unless during the love-season; but at all other times, no sooner has this bird alighted than its remarkable voice is heard, at almost every leap which it makes, whilst ascending against the upper parts of the trunk of a tree, or its highest branches. Its notes are clear, loud, and yet rather plaintive. They are heard at a considerable distance, perhaps half a mile, and resemble the false high note of a clarionet. They are usually repeated three times in succession, and may be represented by the monosyllable pait, pait, pait. These are heard so frequently as to induce me to say that the bird spends few minutes of the day without uttering them, and this circumstance leads to its destruction, which is aimed at, not because (as is supposed by some) this species is a destroyer of trees, but more because it is a beautiful bird, and its rich scalp attached to the upper mandible forms an ornament for the war-dress of most of our Indians, or for the shot-pouch of our squatters and hunters ...

The Ivory-billed Woodpecker nestles earlier in spring than any other species of its tribe. I have observed it boring a hole for that purpose in the beginning of March. The hole is, I believe, always made in the trunk of a live tree, generally an ash or a hagberry, and is at a great height. The birds pay great regard to the particular situation of the tree, and the inclination of its trunk; first, because they prefer retirement, and again, because they are anxious to secure the aperture against the access of water during beating rains ...

Both birds [male and female] work most assiduously at this excavation, one waiting outside to encourage the other, whilst it is engaged in digging, and when the latter is fatigued, taking its place. I have approached trees whilst these Woodpeckers were thus busily employed in forming their nest, and by resting my head against the bark, could easily distinguish every blow given by the bird. I observed that in two instances, when the Woodpeckers saw me thus at the foot of the tree in which they were digging their nest, they abandoned it for ever. For the first brood there are generally six eggs. They are deposited on a few chips at the bottom of the hole, and are of a pure white colour. The young are seen creeping out of the hole about a fortnight before they venture to fly to any other tree ...

The food of this species consists principally of beetles, larvae, and large grubs. No sooner, however, are the grapes of our forests ripe than they are eaten by the Ivory-billed Woodpecker with great avidity. I have seen this bird hang by its claws to the vines, in the position so often assumed by a Titmouse, and, reaching downwards, help itself to a bunch of grapes with much apparent pleasure ...

The Ivory-bill is never seen attacking the corn, or the fruit of the orchards, although it is sometimes observed working upon and chipping off the bark from the belted trees of the newly-cleared plantations. It seldom comes near the ground, but prefers at all times the tops of the tallest trees. Should it, however, discover the half-standing broken shaft of a large dead and rotten tree, it attacks it in such a manner as nearly to demolish it in the course of a few days...The strength of this Woodpecker is such, that I have seen it detach pieces of bark seven or eight inches in length at a single blow of its powerful bill, and by beginning at the top branch of a dead tree, tear off the bark, to an extent of twenty or thirty feet, in the course of a few hours, leaping downwards with its body in an upward position, tossing its head to the right and left, or leaning it against the bark to ascertain the precise spot where the grubs were concealed, and immediately after renewing its blows with fresh vigour, all the while sounding its loud notes, as if highly delighted ...

When wounded and brought to the ground, the Ivory-bill immediately makes for the nearest tree, and ascends it with great rapidity and perseverance, until it reaches the top branches, when it squats and hides, generally with great effect. Whilst ascending, it moves spirally round the tree, utters its loud pait, pait, pait, at almost every hop, but becomes silent the moment it reaches a place where it conceives itself secure. They sometimes cling to the bark with their claws so firmly, as to remain cramped to the spot for several hours after death. When taken by the hand, which is rather a hazardous undertaking, they strike with great violence, and inflict very severe wounds with their bill as well as claws, which are extremely sharp and strong. On such occasions, this bird utters a mournful and very piteous cry.

John James Audubon

AUDUBON, JOHN JAMES. *BIRDS OF AMERICA*. PHILADELPHIA: J. B. CHEVALIER, 1842.

SEE ALSO *Conservation; Earth Day; Environmental Activism; Environmental Protests; Human Impacts; Hunting Practices; Organic and Locally Grown Foods; Sustainable Development*

BIBLIOGRAPHY

Books

Cunningham, W.P., and A. Cunningham. *Environmental Science: A Global Concern*. New York: McGraw-Hill International Edition, 2008.

Enger, Eldon, and Bradley Smith. *Environmental Science: A Study of Interrelationships*. New York: McGraw-Hill, 2006.

Miller, G. Tyler. *Environmental Science: Working with the Earth*. New York: Brooks Cole, 2005.

Periodicals

Biesmeijer, J. C., et al. "Parallel Declines in Pollinators and Insect-Pollinated Plants in Britain and the Netherlands." *Science* 313 (2006): 354–357.

Web Sites

We Campaign. "We Can Solve the Climate Crisis." http://www.wecansolveit.org/ (accessed May 2, 2008).

Brian D. Hoyle

North Atlantic Oscillation

Introduction

The North Atlantic Oscillation (NAO) is a natural shift in weather patterns that occurs in the North Atlantic Ocean and influences the weather of surrounding areas from the eastern United States to Siberia. The phenomenon has been observed for decades, but its mechanism is still poorly understood and the shifts of the NAO remain difficult to predict.

Historical Background and Scientific Foundations

NAO is a shift in the pattern of atmospheric pressures over the Atlantic. Over the Azores Islands in the subtropical Atlantic (at about the same latitude as Virginia), there is a permanent high-pressure center; in the vicinity of Iceland, there is a low-pressure center. NAO has two phases, called positive and negative. During NAO's positive phase, the pressure of the high-pressure Icelandic center is higher and that of the low-pressure Azores center is lower. Also, the Icelandic center shifts to the east, and the Azores center to the west. All of these effects are reversed during negative NAO. There is generally about one NAO oscillation per year.

By themselves, NAO pressure shifts would not be significant. However, they are part of a system of changes that alters the exchange of heat and moisture between the Atlantic and the landmasses surrounding it. Storms moving eastward across the Atlantic, in the direction of the prevailing westerly winds, are routed differently depending on the state of NAO. During positive NAO, these westerly winds are strengthened, which has particularly strong effects on the weather, making it cooler and drier over the northwestern Atlantic and Mediterranean countries and warmer and wetter in northern Europe, the eastern United States, and much of Scandinavia. During negative NAO, these effects are reversed.

Weather is the most important direct effect of NAO from the human point of view, since it influences agriculture and water management. However, NAO affects much more than the weather. Because NAO affects surface winds over the Atlantic, it alters the exchange of heat and freshwater at the surface. This affects the strength of the large-scale ocean circulation in the Atlantic that exchanges surface water for deep water, the thermohaline circulation or Atlantic conveyor belt. Slowdown of this conveyor belt may be caused by global climate warming, which accelerates melting of Greenland's ice cap and so dumps larger amounts of freshwater into the far northern Atlantic. A strengthening trend in NAO might, for a while, mask any such slowdown of the conveyor belt. NAO also affects ecosystems on land and at sea. Changes include movements of populations of fish, shellfish, and zooplankton, and later or earlier blooming of plants and reproductive behaviors of animals on land.

Interest in NAO was low for decades because it was considered highly unpredictable. However, since the late 1990s, increasingly sophisticated computer climate models have made it possible for scientists to probe possible cause-and-effect mechanisms controlling the timing and strength of NAO. Reliable forecasts of NAO would be useful for agriculture in Europe, but NAO remains difficult to forecast.

Impacts and Issues

A strong trend has been observed in NAO over almost the last 30 years. The trend is toward a more strongly positive NAO. This trend accounts for a number of climate shifts across the Northern Hemisphere, including changes in Arctic ocean water composition, shifts in regional precipitation patterns (causing some European glaciers to advance and others to retreat), more severe winters in eastern Canada, changes in zooplankton production in the Atlantic, and shifted fish distributions. Although this shift coincides with global climate change caused by anthropogenic emissions of gases such as carbon dioxide, scientists are uncertain about whether NAO's trend toward more a positive phase has been caused by global climate change or by the variability of the natural process itself.

Some scientists have pointed out that much of the observed climate warming over Europe and Asia in recent decades can be explained by the NAO trend, and suggested that this has contributed to overestimation of human-caused global warming. However, most climate scientists agree that NAO's own changes may well be driven by climate change, making it a link in the global warming process. By 2001, some scientists were arguing that the recent trend in NAO is driven by warming of tropical sea-surface temperatures, especially in the tropical Indian Ocean and Pacific, which in turn is attributed to increased levels of greenhouse gases in the atmosphere. In short, global warming is probably causing the positive trend in NAO, rather than the other way around. However, as of 2008, NAO remained one of the more puzzling large-scale features of Earth's climate.

SEE ALSO *El Niño and La Niña; Climate Change*

BIBLIOGRAPHY

Periodicals

Hurrell, James W. "The North Atlantic Oscillation." *Science* 291 (2001): 605–605.

Visbeck, Martin H., et al. "The North Atlantic Oscillation: Past, Present, and Future." *Proceedings of the National Academy of Sciences* 98 (2001): 12876–12877.

Web Sites

Climate Research Unit, School of Environmental Sciences, University of East Anglia, UK. "North Atlantic Oscillation." http://www.cru.uea.ac.uk/cru/info/nao/ (accessed April 12, 2008).

WORDS TO KNOW

ANTHROPOGENIC: Made by humans or resulting from human activities.

THERMOHALINE CIRCULATION: Large-scale circulation of the world ocean that exchanges warm, low-density surface waters with cooler, higher-density deep waters; also termed meridional overturning circulation.

ZOOPLANKTON: Small, herbivorous animal plankton that float or drift near the surface of aquatic systems and that feed on plant plankton (phytoplankton and nanoplankton).

The Houses of Parliament are seen through rain puddles in London as heavy rain falls on the capital in December 2007. A positive North Atlantic Oscillation index can mean warmer, wetter winters in Europe and colder, dryer winters in Canada and Greenland. *AP Images*

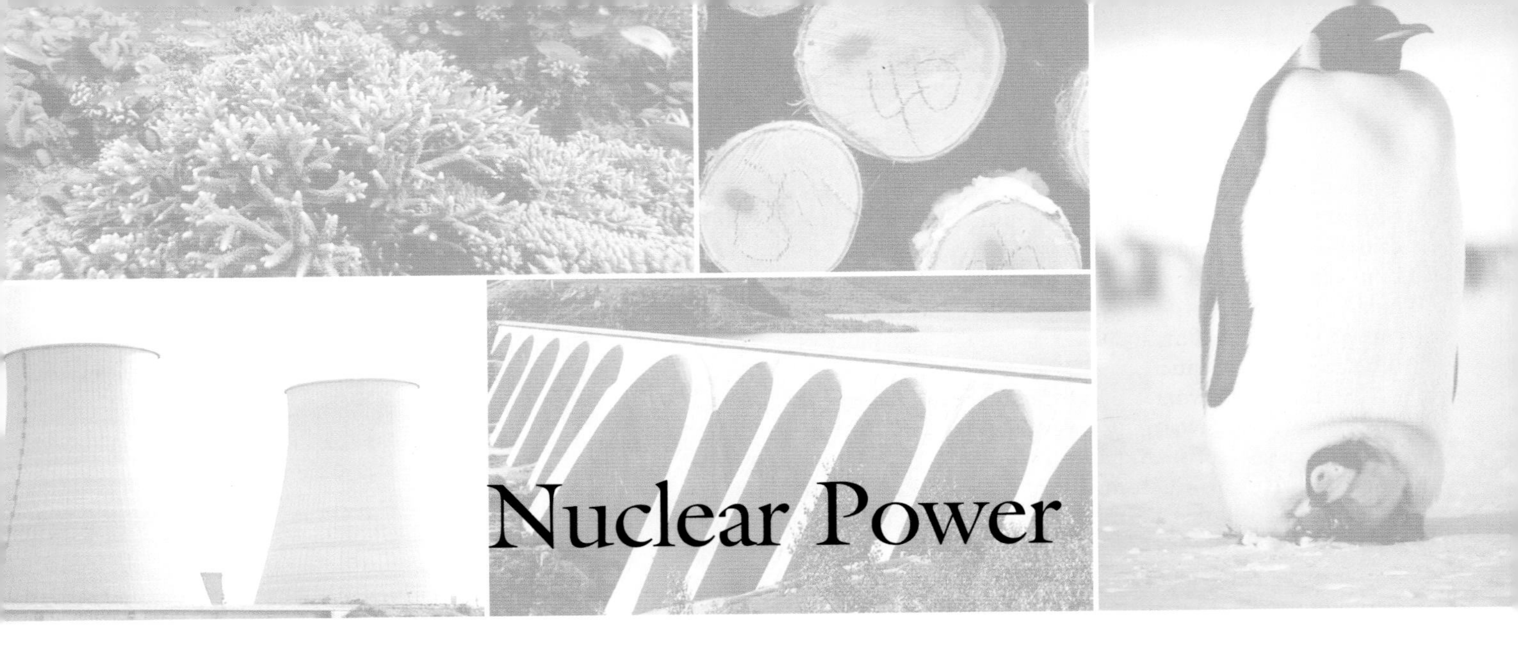

Nuclear Power

Introduction

Nuclear power is energy from controlled reactions involving the nuclei (dense, tiny centers) of atoms. Nuclear power plants are large industrial facilities fueled by radioactive metals such as uranium 235 and plutonium. Uranium 235 is obtained by refining natural ores, while plutonium is an artificial element created by exposing a common but otherwise mostly useless form of uranium (uranium 238) to radiation, usually in nuclear power plants. Most nuclear power plants are fueled by uranium 235.

In a nuclear power plant, nuclear fuel is concentrated in a structure called a nuclear reactor. In the reactor, radiation from disintegrating fuel atoms is made to impinge on other fuel atoms. This triggers further disintegrations, leading to a sustained reaction that releases a flow of energy. This energy is harvested as heat, which in a typical plant is used to boil water to make steam to spin turbines that turn electric generators.

As of 2008, about 16% of the world electricity supply was produced by nuclear power plants, including 19% of U.S. supply. As most energy is consumed in forms other than electricity, especially heat and vehicular power, the nuclear contribution to world energy supply was smaller than its contribution to world electricity supply (less than 7%). There is ongoing global debate about whether building a new wave of nuclear power plants is necessary to mitigate global climate change while meeting global energy needs.

Historical Background and Scientific Foundations

The first application of nuclear fission was not to power production but was used in war—the U.S. destruction of the Japanese cities of Hiroshima and Nagasaki in August 1945 during World War II (1939–1945), which caused about 274,000 civilian deaths. Electricity from a nuclear reactor was first fed to a power grid (system of wires distributing power) in the Soviet Union in 1954.

When nuclear power first became practical, it was presented to the public as a miracle of science, promising future electricity "too cheap to meter" (in the words of the chairman of the U.S. Atomic Energy Commission, 1954). It was widely greeted by the American public on these terms. However, by the 1970s nuclear power was widely argued in the United States and Europe to entail an unacceptable threat of long-lived radioactive pollution released by disastrous accidents at nuclear power plants or waste-storage facilities. These fears were strengthened by a near-disaster at the Three Mile Island nuclear plant in Pennsylvania in 1979 and a full-fledged disaster at the Chernobyl power plant in the Ukraine (then a member state of the Soviet Union) in 1986, which spread radioactivity over much of Europe and killed thousands.

However, construction of nuclear power plants had already stalled in the United States (the world's largest generator of nuclear electricity) before these accidents happened. As of early 2008, no new nuclear power plants had been licensed in the United States since 1978. Nuclear construction halted not because of public opposition to nuclear power but because the cost of nuclear power plants had turned out to be too high for power companies to afford.

All energy sources except the tides are ultimately nuclear in origin. The sun's radiant energy derives from nuclear reactions in its core, and sunlight drives biological growth, winds, and precipitation. Nuclear reactions inside Earth supply the heat of geothermal energy. However, in reactors the nuclear reactions are concentrated on Earth's surface. This creates both opportunity and danger for human-controlled nuclear reactions: The high power density of a nuclear reactor makes it possible to harvest a massive, continuous flow of energy in a nuclear power plant, but this flow is dangerous to the planet itself. Also, nuclear materials and reactions similar to those used in power generation are used to make nuclear weapons. In 2008, there was international concern over whether nuclear material had been diverted from the Iranian nuclear power program to a secret nuclear-weapons program, although the International Atomic Energy Agency of the United Nations had not found any proof of such diversion. Concern over Iran's nuclear behavior highlights the close link between nuclear power-generating technology and bomb-building technology.

The burning of a conventional fuel such as gasoline is a chemical, not a nuclear reaction, and involves only the electrons surrounding the nucleus. A nuclear reaction changes the nucleus itself, making it into a new element or a different isotope of the same element. Thus, after fuel metal has been exposed to radiation in a nuclear reactor for some months, many of its atoms have been changed into other elements. Most of these atoms are, like the original fuel, radioactive, that is, tend to break apart spontaneously and emit fast-moving subatomic particles and high-energy electromagnetic waves (gamma rays). These particles and rays are dangerous to life because they disrupt chemical reactions in cells. At high doses, radiation kills cells directly: At low doses, it can cause cancer in multicellular organisms (such as people) by altering DNA.

As the reaction products (altered atoms) in nuclear fuel accumulate, the fuel does not work as well in the reactor. Eventually this spent fuel must be removed. Spent or used-up fuel from a reactor is a form of high-level radioactive waste, that is, material that emits intense radioactivity. Some of the radioactive substances in this waste, such as plutonium, remain dangerously radioactive for hundreds of thousands of years.

WORDS TO KNOW

ATOMIC NUCLEUS: The small, dense, positively charged central region of an atom, composed of protons and neutrons.

FISSION: A process in which the nucleus of an atom splits, usually into two daughter nuclei, with the transformation of tremendous levels of nuclear energy into heat and light.

GREENHOUSE GAS: A gas whose accumulation in the atmosphere increases heat retention.

ISOTOPE: A form of a chemical element distinguished by the number of neutrons in its nucleus.

NUCLEAR WASTE: Material left over from nuclear processes that is often radioactive or contaminated by radioactive elements.

OPPORTUNITY COST: The total cost incurred by choosing one option over another.

RADIOACTIVE: Containing an element that decays, emitting radiation.

■ Impacts and Issues

Both supporters and opponents of nuclear power agree that high-level radioactive waste must be isolated from the environment lest it injure human beings and other creatures. They disagree, however, on whether this goal can be dependably met both for the short term (tens of years, while nuclear waste is stored inside nuclear power plants or in above-ground storage facilities) or the long term (tens to hundreds of thousands of years, the lifetime of high-level radioactive waste). Supporters of nuclear power argue that nuclear power plants will only rarely release their radioactive materials to the environment, as happened in the Chernobyl accident of 1986, and that deep burial in rock layers will isolate nuclear waste for the long term. Opponents of nuclear power argue that nuclear power plants and storage facilities can and have failed, quoting pre-Three Mile Island reassurances from pro-nuclear experts that nothing could possibly go wrong, and note that nuclear facilities may be deliberately targeted by terrorists or military enemies. They also argue that there is no way to be sufficiently confident that any waste-storage scheme, even deep burial, will be effective for hundreds of thousands of years.

Starting in the 1990s, advocates of nuclear power have turned the environmentalists' anti-nuclear arguments on their head, urging that nuclear power, far from being an environmental threat, is the only way to generate enough electricity for modern society without worsening global warming. This argument is based on the fact that the nuclear reactions that release energy in a nuclear power plant, unlike the chemical reactions that release energy in a coal- or gas-fired power plant, do not release carbon dioxide or other greenhouse gases. An operating nuclear power plant is therefore, considered in isolation, greenhouse-neutral.

In 2005, U.S. President George W. Bush advocated massive government grants for a new round of nuclear power plants on the grounds that nuclear plants are "environmentally friendly." In 2005, the World Nuclear Association, a nuclear industry advocacy group, recommended that between then and the end of the century, the global number of nuclear power plants be increased over tenfold, from 441 to 5,000, in order to combat global warming while meeting increased electricity demands in developing countries. The editors of the *New York Times* opined in 2006, "Not so many years ago, nuclear energy was a hobgoblin to environmentalists, who feared the potential for catastrophic accidents and long-term radiation contamination. But this is a new era.... Suddenly nuclear power is looking better."

Critics of nuclear power—and a very small percentage of environmentalists, despite some media claims, have switched to supporting nuclear power—respond that (1) the nuclear power cycle as a whole, including mining, refining of fuel, building plants, disposing of waste, and taking old plants apart, is in fact a significant emitter of greenhouse gases; (2) even if nuclear power were greenhouse-neutral, it is so expensive compared to other available ways of reducing greenhouse-gas emissions that money spent on nuclear power actually increases greenhouse emissions by hogging limited funds; (3) and nuclear power plants take too long to build (at least 5 years) to be part of a rapid response to climate change.

For example, prominent antinuclear energy expert Amory Lovins wrote in 2005 that the higher cost of nuclear power "per unit of CO_2 displaced," compared to efficiency, wind power, cogeneration (making electricity and useful heat together instead of discarding heat from electric plants as waste), and other renewables, "means that every dollar invested in nuclear expansion will *worsen* climate change by buying less solution per dollar.... In this sense of 'opportunity cost' ... nuclear power is far *more* carbon-intensive than a coal plant." Lovins also argued that nuclear power plants have been rejected by market forces, so that all nuclear power plants proposed today would depend on government grants. Nuclear power advocates respond that non-nuclear alternative energy sources simply cannot produce enough energy to power modern industrial society. This debate continues. In 2005, the U.S. government allocated $6 billion in benefits for new nuclear reactors, but as of 2008, construction had not yet commenced on any new U.S. reactor.

■ Primary Source Connection

There are vehement arguments for and against nuclear power. The main advantage of nuclear power plants is that they do not cause atmospheric pollution. No smokestacks are needed because nothing is being burned. Nuclear power plants do not contribute to potential global warming and help countries lower their carbon footprint. For example, France initiated a large-scale nuclear program after the Arab oil embargo in 1973 and has been able to reduce its acid rain and carbon dioxide emissions by more than 40%. Shipments of fuel are also minimal and so the hazards of coal transportation and oil spills are avoided.

As with other forms of producing electricity, nuclear power generation can, however, have serious and unintended environmental impacts. The main objections to nuclear power plants are the fear of possible accidents, the unresolved problem of nuclear waste storage, and the possibility of plutonium diversion for weapons production by a terrorist group. The issue of waste storage becomes particularly emotional because leakage from a waste depository could contaminate groundwater.

An engineer inspects the damage of the machine room at reactor two of the atomic power plant in Chernobyl, Russia, 1986 after a fire destroyed the roof construction. This was the second accident in one year. Environmentalists demanded immediate closure of the reactor. *AP Images*

Despite increased safety measures, as the following article shows, the fear of accidents remains a potent deterrent toward widespread development of nuclear power facilities.

ACCIDENTS DIM HOPES FOR GREEN NUCLEAR OPTION

The recent earthquake in Japan and accidents at two German power plants raise questions on the safety of nuclear energy as a cleaner alternative.

As concern about global warming has swelled in recent years, so has renewed interest in nuclear energy. The main reason: Nuclear plants produce no carbon dioxide or other greenhouse gases tied to climate change, at least not directly.

New reactor designs make plants safer than those operating in the days of the accidents at Chernobyl and Three Mile Island decades ago, advocates say. And there's no group of OPEC countries in unstable parts of the world controlling the main raw material—uranium.

But that was before an earthquake in Japan this week rattled the Kashiwazaki nuclear power plant. The plant's operator "said it had found more than 50 problems at the plant caused by Monday's earthquake," The New York Times reported, adding:

> While most of the problems were minor, the largest included 100 drums of radioactive waste that had fallen over, causing the lids on some of the drums to open, the company said. The company said that the earthquake also caused a small fire at the plant, the world's largest by amount of electricity produced, and the leakage of 317 gallons of water containing trace levels of radioactive materials into the nearby Sea of Japan.

Meanwhile, accidents at two German nuclear reactors last month prompted German Environment Minister Sigmar Gabriel to call for the early shutdown of all older reactors there, reports Bloomberg News.

Concern about the safety of Germany's 17 reactors has grown after a fire at Vattenfall's Kruemmel site June 28 and a network fault at its Brunsbuettel plant on the same day. Der Spiegel online adds:

> It took the fire department hours to extinguish the blaze. Even worse, the plant operator's claim that a fire in the transformer had no effect on the reactor itself proved to be a lie.

In short, the incident made clear that nuclear energy is by no means the modern, well-organized, high-tech sector portrayed until recently by politicians and industry advocates. Indeed, the frequency of problems occurring at Germany's aging reactors is on the rise. Just as old cars succumb to rust, nuclear power plants built in the 1970s and '80s are undergoing a natural aging process.

IN CONTEXT: NUCLEAR POWER INSPIRES PASSIONATE DEBATE

Environmental activists are not always united, especially on matters such as the benefits vs. costs of nuclear energy, which alternative energy sources provide the greatest potential relief from the pollution caused by the use of fossil fuels (e.g., petroleum, coal, natural gas), how to best mitigate the environmental impacts of large hydroelectric dams, etc.

Nuclear power, for example, remains in a gray zone—in some ways cleaner for the environment, yet with the potential to poison Earth for generations beyond any benefit it offers. Regardless, for many years now it has been a significant part of electricity production in some countries. France, for example, obtains 78% of its electricity from nuclear power. The use of nuclear power, however, remains a lightning rod for protest and extent of its use, or indeed whether it should be used at all, often leads to fractious debate and passionate protest.

On Wednesday, the chief executive of Vattenfall Europe AG stepped down. Klaus Rauscher was the second manager to depart this week amid mounting criticism for the utility's handling of a fire at a nuclear plant in northern Germany, reports the AP.

> 'When it comes to security at nuclear power plants, I can only say, that when it comes to the information policy, this really has not been acceptable and therefore my sympathy for the industry is limited,' Chancellor Angela Merkel said.
>
> Merkel, a physicist by training, normally favors nuclear power, but the June 28 fire at the Kruemmel plant, near Hamburg, has put the industry in a bad light.

Still, nuclear power has won some powerful allies in the environmental community, writes E Magazine editor Jim Motavalli on the web site AlterNet.

He quotes Fred Krupp, president of Environmental Defense, as saying, "We should all keep an open mind about nuclear power." Jared Diamond, best-selling author of "Collapse," adds, "To deal with our energy problems we need everything available to us, including nuclear power," which, he says, should be "done carefully, like they do in France, where there have been no accidents."

Stewart Brand, who founded The Whole Earth Catalog and Whole Earth Review, concludes, "The only technology ready to fill the gap and stop the carbon dioxide loading of the atmosphere is nuclear power."

A worker walks past the nuclear reactor at Sizewell B Nuclear Power Station, in Sizewell, eastern England in January 2008. The British government announced support for the construction of new nuclear power plants, backing atomic energy as a clean source of power to fight climate change. *AP Images*

Environmentalists continue to push for more benign sources—wind, biomass, geothermal, and solar, as well as greater conservation—to power the world.

But The Washington Post reports that "Most of the technologies that could reduce greenhouse gases are not only expensive but would need to be embraced on a global scale." The article continues:

> Many projections for 2030 include as many as 1 million wind turbines worldwide; enough solar panels to cover half of New Jersey, massive reforestation; a major retooling of the global auto industry; as many as 400 power plants fitted with pricey equipment to capture carbon dioxide and store it underground; and, most controversial, perhaps 350 new nuclear plants around the world.

That kind of nuclear expansion in the US seems unlikely. The country hasn't licensed a new plant in more than 30 years, and the devilish political and scientific subject of radioactive waste disposal has yet to be fully addressed.

But that hasn't stopped other countries from pushing ahead. Russia "hopes to export as many as 60 nuclear power plants in the next two decades," The Christian Science Monitor reported this week, including what would be "the first-ever floating atomic power station" at sea.

But noting an estimate from the Massachusetts Institute of Technology, The Salt Lake Tribune in Utah reports that "at least 1,000 new nuclear plants would be needed worldwide in the next 50 years to make a dent in global warming."

Brad Knickerbocker

Kickerbocker, Brad. "Accidents Dim Hopes for Green Nuclear Option." *Christian Science Monitor* (July 19, 2007).

SEE ALSO *Carbon Dioxide (CO_2) Emissions; Climate Change; Global Warming; Nuclear Test Ban Treaties; Radioactive Waste*

BIBLIOGRAPHY

Periodicals

"The Greening of Nuclear Power." *New York Times* (May 13, 2006).

Hoffert, M., et al. "Advanced Technology Paths to Global Climate Stability: Energy for a Greenhouse Planet." *Science.* 298 (2002): 981–987.

Marshall, Eliot. "Is the Friendly Atom Poised for a Comeback?" *Science* 309 (2005): 1168–1169.

Web Sites

Rocky Mountain Institute. "Nuclear Power: Economics and Climate-Change Potential." https://www.rmi.org/images/PDFs/Energy/E05-08_NukePwrEcon.pdf (accessed April 15, 2008).

Larry Gilman

Nuclear Test Ban Treaties

Introduction

Nuclear test-ban treaties are international agreements to forbid test explosions of nuclear weapons. There have been three nuclear test-ban treaties, two of which have entered into force (become legally binding on their signatories) and one of which, the Comprehensive Test Ban Treaty (CTBT), remains in legal limbo. The first test-ban treaty was the Partial Test Ban Treaty (PTBT), which entered into force in 1963. It forbade nuclear explosions in the atmosphere, in outer space, and under water, but did not forbid underground explosions. The second was the Threshold Test Ban Treaty (TTBT), which forbade underground test explosions above a certain size. The CTBT would forbid all nuclear test explosions in all environments. Some countries, including India, North Korea, Pakistan, the United States, and China, have refused to ratify the CTBT, arguing that it is unenforceable or against their national interests.

Historical Background and Scientific Foundations

The first nuclear weapon test was also the first nuclear explosion in history, the blast code-named "Trinity" set off by the United States on July 16, 1945, near Alamogordo, New Mexico. The Soviet Union was the second nation to conduct a nuclear test, in August 29, 1949. Other nations have followed: the United Kingdom (1952), France (1960), China (1964), India (1974), Pakistan (1998), and North Korea (2006). There is some evidence that Israel and South Africa may have conducted a test over the Indian Ocean in 1979.

From 1945 to 2008, well over 2,000 nuclear test explosions have been conducted worldwide, most by the United States and the Soviet Union. The United States has conducted the most tests, exploding at least 1,150 bombs; the Soviet Union is second with over 960 bombs; France is third with at least 210 bombs; and China fourth with 45 bombs.

The first tests by any nation are conducted to verify that their nuclear bomb design can be made to explode. Later tests are conducted in order to verify new bomb designs (especially designs for smaller, lighter-weight bombs that can be mounted on missiles or for other special purposes) and to verify that aging bombs in an arsenal will still explode on demand.

There are four basic categories of nuclear tests, according to location: atmospheric tests, which involve exploding a weapon anywhere from the surface of the ground on up to the stratosphere; exo-atmospheric tests, above the atmosphere in outer space; underwater tests; and underground tests. In the 1940s and 1950s, most tests were atmospheric, with bombs exploded at the ground surface or not far above it. This practice caused radioactive dust to spread downwind from the test explosions. This material, called fallout, was a hazard to health, containing radioactive isotopes such as americium-241, cesium-137, iodine-131, and strontium-90. Several of these, especially iodine-131 and strontium-90, can be concentrated up the food chain and in the human body. In large quantities, fallout can cause almost immediate death; in smaller quantities, it can cause cancer. Iodine-131 concentrates in the thyroid gland, increasing the chances of thyroid cancer; strontium-90 is chemically similar to calcium, and so is concentrated by children in their bones and bone marrow. Other animals concentrate fallout too, but the amount of fallout from surface testing was not large enough to cause widespread environmental harm.

In 1954, the United States tested a large hydrogen bomb, the Castle Bravo test, at Bikini Atoll in the Pacific Ocean. The bomb turned out to be about three times more powerful than expected and produced more fallout than expected also. This was carried by winds over the

WORDS TO KNOW

COLD WAR: At term describing the ideological, political, economic, and military tensions and struggles between the two dominant superpowers of the era, The United States and the former Union of Soviet Socialist Republics (USSR) between 1945 (the end of the WWII) and the collapse of the Soviet Union in 1991.

RADIOACTIVITY: The property possessed by some elements of spontaneously emitting energy in the form of particles or waves by disintegration of their atomic nuclei.

Pacific, raining radioactivity on a Japanese fishing boat and on several inhabited Pacific islands. One member of the crew of the Japanese boat died from radiation burns. In the continental United States, fears grew over radioactive contamination from U.S. surface tests. However, the full extent of such contamination was not even known by the public; in 1997, a report by the National Cancer Institute revealed that American children had been exposed to 15 to 70 times as much radiation in 1951–1958 from the Nevada Test Site (where the United States carried out most of its atmospheric tests) as had been previously reported. Nevertheless, enough was known in the 1950s to trigger a nationwide scare over fallout and cancer.

Public fears put pressure on ongoing test-ban negotiations between the United States, the Soviet Union, Canada, France, and the United Kingdom, which had begun in May 1955. The Soviet Union proposed linking a test ban to a general disarmament treaty that would reduce both nuclear and non-nuclear weapons. The United States and its allies opposed such a linkage. The Soviets were also opposed to on-site inspections to verify a test ban, because they feared that the United States would use inspections as an opportunity to spy. The United States feared that without on-site inspections, there would be no way to detect small nuclear tests by the Soviets in violation of a comprehensive test-ban treaty. Public concern over fallout moved the Soviet Union and United States to stop nuclear testing in November 1958. The suspension held until September 1961, when the Soviets resumed testing. The United States resumed testing on April 25, 1962.

Despite resumed testing, agreement on a partial test-ban treaty was near. After the Cuban Missile Crisis of October 1962, during which the United States and the Soviet Union came perilously close to an all-out nuclear war, U.S. President John F. Kennedy (1917–1963) and Soviet Premier Nikita Khrushchev (1894–1971) resumed negotiations toward a test-ban treaty in July 1963. Khrushchev gave up the Soviet demand for a ban on underground testing as well as atmospheric, exo-atmospheric, and underwater testing. After just 12 days of negotiations, the PTBT was agreed upon, and the two countries signed the treaty on August 5, 1963. The U.S. Congress ratified the treaty on September 23, 1963. By 2008, 113 other countries had also signed the treaty. China and France have never signed the treaty and Pakistan has signed but not ratified; all three of these countries do, however, now restrict nuclear tests to underground explosions. France continued atmospheric testing until 1975.

The second test-ban treaty was the TTBT, signed by the United States and the Soviet Union in 1974. These two countries remain this treaty's only signatories. It forbids the testing of nuclear devices with yields greater than 150 kilotons. In nuclear-weapons jargon, a kiloton is a unit of explosive force equal to that of 1,000 tons of the conventional chemical explosive trinitrotoluene (TNT). Compliance with the terms of the treaty could be monitored independently by each country by seismic monitoring, that is, recording of vibrations in the ground.

The third test-ban treaty was the CTBT, which bans nuclear explosions in all environments. It was opened for signature in September 1996, after long and contentious negotiations. As of 2008, it had been signed by 178 governments and ratified by 144, not including the United States, India, Pakistan, and North Korea, all nuclear powers.

Impacts and Issues

Soon after the signing of the PTBT in 1963, levels of radioactivity from fallout began to steadily decline worldwide as the radioactive isotopes began to disappear. (Every such isotope has an innate half-life, or time period after which half of any given amount of the substance has emitted radioactivity and transformed into some other isotope; eventually, all radioactive substances become non-radioactive, some quickly, some slowly.) Most scientists agree that the PTBT has been at least a partial success, diminishing harm from fallout.

The TTBT was less-warmly received. In the mid-1970s, many scientists who had called for treaties limiting or abolishing nuclear weapons and weapons testing opposed the TTBT on the ground that it was inadequate. The Arms Control Association opposed the treaty as worse than not having any treat at all; the Pugwash group of nuclear scientists characterized it as a "mockery"; and the Federation of American Scientists stated that it was a "counter-productive sham." Two of the objections of these groups were that (1) the treaty's explosive threshold, 150 kilotons (about 10 times larger than the bomb with which the United States destroyed the Japanese city of Hiroshima on August 6, 1945), was far higher than it needed to be to permit verification, and would allow the superpowers unrestrained development of new nuclear weapons in the range of explosive power they were most interested in anyway; and (2) the treaty allowed for nuclear explosions of any size for peaceful purposes, even though these are not physically different from weapons tests.

In the United States, political opposition to the CTBT centered around two claims, namely that (1) a complete ban on nuclear explosions would be technologically impossible to verify, allowing cheating, and (2) continued testing was necessary to assure the reliability of the U.S. nuclear arsenal. The U.S. Senate rejected the treaty on October 13, 1999, in a partisan split. Lack of participation by the United States, one of the world's largest nuclear powers, has crippled the treaty. A 2002 study by a panel convened by the National Academy of Sciences concluded that none of the technical objections raised against the CTBT were serious problems, should the United States choose to ratify and implement the treaty. As of 2008, it was still possible that the United States could choose to ratify the CTBT.

Despite the CTBT's initial rejection by the United States, the signatory nations have proceeded to construct a global system of vibration sensors to allow the treaty to be verified. By 2005, the system consisted of 85 sensor stations listening to vibrations in the air, ground, and water, feeding their data in real time to the International Data Center in Vienna, Austria. For global coverage and sensitivity, it outclasses all other seismic sensor networks and has become a powerful tool for geologists, seismologists, and other earth scientists.

■ Primary Source Connection

At the height of the Cold War in 1962, the Cuban Missile Crisis accelerated the debate of a nuclear arms embargo, as U.S. naval ships blocked a convoy of Soviet ships delivering intermediate-range nuclear ballistic missiles to Cuba. The events that followed nearly brought the United States and the Soviet Union to the brink of war.

In response to the general feeling of insecurity and the fear of total nuclear destruction of the world in a future war, the Treaty Banning Nuclear Weapon Tests in the Atmosphere, in Outer Space, and Under Water was formed on August 5, 1963, and took effect on October 10, 1963. The agreement, also known as the Nuclear Test Ban Treaty (NTBT), the Limited Test Ban Treaty (LTBT), or the Partial Test Ban Treaty (PTBT), was a result of substantial arms control efforts taken by both the United States as well as the Soviet Union.

The treaty also banned tests that could cause radioactive fallout to settle beyond the territorial limits of the country conducting the tests. Therefore, it set certain territorial limits for radioactive tests and banned countries from conducting those tests that affected regions beyond the specified territory limits. The treaty proclaimed its principal aim was the speedy end to the

A Greenpeace demonstration against China's nuclear tests in front of the Chinese Embassy in Tokyo, August 1995. *AP Images*

arms race and to stamp out incentives for nuclear weapons production and testing. The treaty also declared itself to be of unlimited duration.

As environmental consciousness grew, the treaty was also seen as vital in protecting the environment from radioactive fallout and other negative impacts of nuclear testing.

TREATY BANNING NUCLEAR WEAPON TESTS IN THE ATMOSPHERE, IN OUTER SPACE, AND UNDER WATER (LIMITED TEST BAN TREATY)

Signed at Moscow: August 5, 1963 Ratification advised by U.S. Senate: September 24, 1963 Ratified by U.S. President: October 7, 1963 U.S. ratification deposited at Washington, London, and Moscow: October 10, 1963 Proclaimed by U.S. President: October 10, 1963 Entered into force: October 10, 1963

The Governments of the United States of America, the United Kingdom of Great Britain and Northern Ireland, and the Union of Soviet Socialist Republics, hereinafter referred to as the "Original Parties,"

Proclaiming as their principal aim the speediest possible achievement of an agreement on general and complete disarmament under strict international control in accordance with the objectives of the United Nations which would put an end to the armaments race and eliminate the incentive to the production and testing of all kinds of weapons, including nuclear weapons,

Seeking to achieve the discontinuance of all test explosions of nuclear weapons for all time, determined to continue negotiations to this end, and desiring to put an end to the contamination of man's environment by radioactive substances,

Have agreed as follows:

Article I

1. Each of the Parties to this Treaty undertakes to prohibit, to prevent, and not to carry out any nuclear weapon test explosion, or any other nuclear explosion, at any place under its jurisdiction or control:

(a) in the atmosphere; beyond its limits, including outer space; or under water, including territorial waters or high seas; or

(b) in any other environment if such explosion causes radioactive debris to be present outside the territorial limits of the State under whose jurisdiction or control such explosion is conducted. It is understood in this connection that the provisions of this subparagraph are without prejudice to the conclusion of a Treaty resulting in the permanent banning of all nuclear test explosions, including all such explosions underground, the conclusion of which, as the Parties have stated in the Preamble to this Treaty, they seek to achieve.

Environmentalists covering their faces with masks protest in front of the French Embassy in Seoul, South Korea (December 1995) as they demand an end to French nuclear tests in the South Pacific. *AP Images*

2. Each of the Parties to this Treaty undertakes furthermore to refrain from causing, encouraging, or in any way participating in, the carrying out of any nuclear weapon test explosion, or any other nuclear explosion, anywhere which would take place in any of the environments described, or have the effect referred to, in paragraph 1 of this Article.

Article II

1. Any Party may propose amendments to this Treaty. The text of any proposed amendment shall be submitted to the Depositary Governments which shall circulate it to all Parties to this Treaty. Thereafter, if requested to do so by one-third or more of the Parties, the Depositary Governments shall convene a conference, to which they shall invite all the Parties, to consider such amendment.

2. Any amendment to this Treaty must be approved by a majority of the votes of all the Parties to this Treaty, including the votes of all of the Original Parties. The amendment shall enter into force for all Parties upon the deposit of instruments of ratification by a majority of all the Parties, including the instruments of ratification of all of the Original Parties.

Article III

1. This Treaty shall be open to all States for signature. Any State which does not sign this Treaty before its entry into force in accordance with paragraph 3 of this Article may accede to it at any time.

2. This Treaty shall be subject to ratification by signatory States. Instruments of ratification and instruments of accession shall be deposited with the Governments of the Original Parties—the United States of America, the United Kingdom of Great Britain and Northern Ireland, and the Union of Soviet Socialist Republics—which are hereby designated the Depositary Governments.

3. This Treaty shall enter into force after its ratification by all the Original Parties and the deposit of their instruments of ratification.

4. For States whose instruments of ratification or accession are deposited subsequent to the entry into force of this Treaty, it shall enter into force on the date of the deposit of their instruments of ratification or accession.

5. The Depositary Governments shall promptly inform all signatory and acceding States of the date of each signature, the date of deposit of each instrument of ratification of and accession to this Treaty, the date of its entry into force, and the date of receipt of any requests for conferences or other notices.

6. This Treaty shall be registered by the Depositary Governments pursuant to Article 102 of the Charter of the United Nations.

Article IV

This Treaty shall be of unlimited duration.

Each Party shall in exercising its national sovereignty have the right to withdraw from the Treaty if it decides that extraordinary events, related to the subject matter of this Treaty, have jeopardized the supreme interests of its country. It shall give notice of such withdrawal to all other Parties to the Treaty three months in advance.

Article V

This Treaty, of which the English and Russian texts are equally authentic, shall be deposited in the archives of the Depositary Governments. Duly certified copies of this Treaty shall be transmitted by the Depositary Governments to the Governments of the signatory and acceding States.

IN WITNESS WHEREOF the undersigned, duly authorized, have signed this Treaty.

DONE in triplicate at the city of Moscow the fifth day of August, one thousand nine hundred and sixty-three.

For the Government of the United States of America

DEAN RUSK

For the Government of the United Kingdom of Great Britain and Northern Ireland

SIR DOUGLAS HOME

For the Government of the Union of Soviet Socialist Republics

A. GROMYKO

John F. Kennedy

KENNEDY, JOHN F. "TREATY BANNING NUCLEAR WEAPON TESTS IN THE ATMOSPHERE, IN OUTER SPACE, AND UNDER WATER (LIMITED TEST BAN TREATY)." AUGUST 5, 1963. HTTP://WWW.UCSUSA.ORG/ASSETS/DOCUMENTS/GLOBAL_SECURITY/LIMITED_TEST_BAN_TREATY.PDF (ACCESSED APRIL 12, 2008).

SEE ALSO *Radioactive Waste*

BIBLIOGRAPHY

Books

Hansen, Keith A. *The Comprehensive Nuclear Test Ban Treaty: An Insider's Perspective.* Stanford, CA: Stanford University Press, 2006.

Periodicals

Glanz, James. "Panel Finds No Major Flaws in Nuclear Treaty." *New York Times* (August 1, 2002).

Hansen, Keith. "CTBT: Forecasting the Future." *Bulletin of the Atomic Scientists* (March/April 2005).

Panofsky, Wolfgang K. H. "Nuclear Insecurity." *Foreign Affairs* (September/October 2007).

Web Sites

Center for Nonproliferation Studies. "Summary of the Comprehensive Nuclear Test-Ban Treaty." http://cns.miis.edu/pubs/inven/pdfs/aptctbt.pdf (accessed May 11, 2008).

Department of State. "Treaty Banning Nuclear Weapon Tests in the Atmosphere, in Outer Space and Under Water." http://www.state.gov/t/ac/trt/4797.htm (accessed May 11, 2008).

Office for Disarmament Affairs. "Comprehensive Nuclear-Test-Ban Treaty." http://disarmament.un.org/TreatyStatus.nsf (accessed May 11, 2008).

Larry Gilman

Ocean Circulation and Currents

Introduction

Ocean circulation is the globally connected system of water movements in the oceans. It involves both surface and deep currents. Surface currents are generated by winds and an uneven distribution of heat between equatorial regions and regions farther north and south of the tropics. The deeper currents are produced by the sinking of colder, saltier water and the upwelling of warmer water. The global ocean circulation system transfers heat from low to higher latitudes, making the oceans responsible for about 40% of global heat transport, a vital determinant of the global climate and a crucial influence on both terrestrial and marine ecosystems worldwide.

Disruptions in the global ocean currents, which could occur as polar ice melts and the freshwater is added to the ocean, could markedly change Earth's climate. As of the early 2000s, scientists considered major disruptions unlikely but possible.

Historical Background and Scientific Foundations

Ocean currents make up about 10% of the volume of ocean waters, although all ocean waters are eventually entrained, over decades or centuries, in currents. Currents are analogous to rivers of water flowing within the surrounding ocean. These movements of water form a continuing, interconnected pattern that is global in scale.

The Gulf Stream is one example of a warm ocean current. This current, which flows past the eastern coast of North America before arcing over to flow in a southward direction past the United Kingdom and Europe, originates as a current in the Pacific Ocean. Britain experiences a much more temperate climate than regions at a similar northern latitude, such as parts of the United States and Canada, because of these warm waters. Ultimately, this current flows back to the Pacific Ocean, and the cyclical global route of the current is repeated.

The Gulf Stream is part of the global system of ocean currents that is known as the ocean conveyor belt (or, more technically, the meridional overturning circulation). The flow of water in the conveyor mixes colder waters nearer to the polar regions with the warmer waters in the mid-latitudes and equatorial regions. The warmer water from equatorial regions rises to the surface, while the colder polar water sinks to the bottom. The colder water then moves to more tropical latitudes, where it warms and rises upward. It is this continuous cycling of water that drives the formation of currents. Changes in salinity also influence the large-scale movement of water—saltier water is more dense than less-salty water, at a given temperature—so the large-scale circulation of the world's ocean waters is also referred to, on occasion, as the thermohaline (heat-and-salt) circulation.

Currents also develop closer to the surface of the ocean because of prevailing winds—winds that blow predictably. The friction between the moving air and the water molecules pushes the water in the direction of the wind, piling the water up similar to the piling up of water as waves form. Gravity acts to pull the piled water downward. If Earth did not rotate, the water would flow in a straight line, but as Earth rotates on its axis, a force called the Coriolis force rotates the water to the right (clockwise) in the Northern Hemisphere and to the left (counterclockwise) in the Southern Hemisphere. The results are gyres—large mounds of water with a flow of water around them. Gyres produce currents in the Northern and Southern Hemispheres of the Atlantic and Pacific Oceans. The Gulf Stream is part of the North Atlantic Gyre.

The clockwise rotating gyre in the northern part of the Atlantic Ocean consists of the warm North Equatorial Current flowing near the equator, joining the western boundary of the warm Gulf Stream, which turns toward the east and becomes the warm North Atlantic Drift, while the cold Labrador Current, West Greenland Drift, and East Greenland Drift return cold water from the North. The eastern boundary, the southward Canary Current close to West Africa, also brings cold water toward the equator, closing the North Atlantic Ocean gyre. In the southern part of the Atlantic Ocean, in the counterclockwise gyre, the warm South Equatorial Current joins the warm Brazil Current going south toward Antarctica at the east coast of South America, meeting the cold Falkland Current, and the cold West Wind Drift heads east around Antarctica. The gyre is closed by the cold Benguela Current going north at the southwestern coast of Africa, while the eastward flowing warm Equatorial Countercurrent connects the gyres from the North and South Atlantic Oceans.

The system of currents in the Pacific Ocean is somewhat similar to that in the Atlantic. The clockwise rotating gyre in the northern part of the Pacific Ocean consists of the warm North Equatorial Current flowing westward north of the equator, joining the western boundary warm Kuroshio Current, which turns toward the East and becomes the warm North Pacific Drift, while the cold Oyashio Current returns cold water from the North. At the coast of Alaska loops the warm Alaska Current. The eastern boundary California Current, close to the coast of California, also brings cold water toward the equator, closing the North Atlantic Ocean gyre. This is where upwelling, the rising of cold water replacing sur-

WORDS TO KNOW

CORIOLIS FORCE: The apparent tendency of a freely moving particle to swing to one side when its motion is referred to a set of axes that is itself rotating in space, such as Earth. Winds are affected by rotation of Earth so that instead of a wind blowing in the direction it starts, it turns to the right of that direction in the Northern Hemisphere, and left in the Southern Hemisphere.

GULF STREAM: A warm, swift ocean current that flows along the coast of the Eastern United States and extends northward toward Europe.

GYRE: A zone of spirally circulating oceanic water that tends to retain floating materials, as in the Sargasso Sea of the Atlantic Ocean.

HYDROSPHERE: The total amount of liquid, solid, and gaseous water present on Earth.

OCEAN HEAT TRANSPORT: Movement by ocean currents of warm water from the tropics toward the poles, effectively transporting heat energy toward the poles where it is more quickly radiated into space.

SALINITY: Measurement of the amount of sodium chloride in a given volume of water.

UPWELLING: The vertical motion of water in the ocean by which subsurface water of lower temperature and greater density moves toward the surface of the ocean.

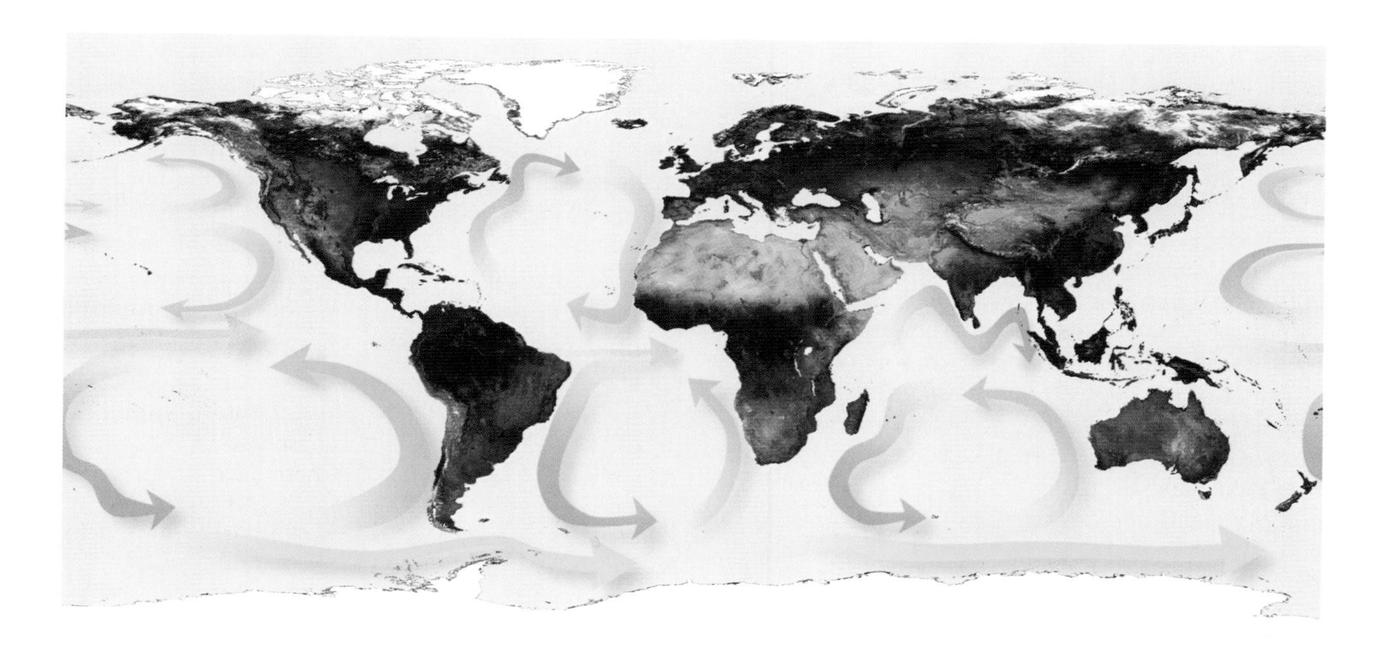

Computer artwork showing surface ocean currents: warm water is red and cold water is blue. The Gulf Stream is a warm current that arises in the Caribbean Sea and travels up the U.S. East coast before crossing the Atlantic. *Karsten Schneider/Photo Researchers, Inc.*

The Brazilian Air Force said it would help some 50 stray penguins dragged by ocean currents to the warm waters off Rio de Janeiro get back to their icy natural habitat. Brazil began operating penguin airlifts in 2000, returning as many as 100 a year to their natural habitat. *AP Images*

face water that drifts away due to the winds, occurs. Although it brings cold water, low clouds, and sometimes fog in summer, the nutrient-rich, cold water helps the fishing industry. South of this gyre, but still in the Northern Hemisphere, the eastward flowing warm North Equatorial Countercurrent can be found. In the southern part of the Pacific Ocean, the warm, westward South Equatorial Current flows opposite of the eastward warm South Equatorial Countercurrent, and continues toward the South at the east coast of Australia. While the West Wind Drift is heading east around Antarctica, the cold Peru or Humboldt Current going northward at the southwestern coast of South America closes this gyre.

The system of surface ocean currents is simpler in the Indian Ocean, because instead of two figure eight-shaped gyres, only one is present. The gyre is influenced by seasonally changing winds. It consists of the warm North Equatorial Current, the South Equatorial Countercurrent, the South Equatorial Current, and the eastward, circumpolar current around Antarctica, the cold West Wind Drift.

Impacts and Issues

The global movement of ocean water is crucial to establishing and maintaining Earth's current climate pattern. Put another way, disruption in the global pattern of ocean circulation could cause marked climate changes.

A number of computer models that simulate the ocean and climate have predicted that the addition of large amounts of freshwater to the North Atlantic Ocean have repeatedly disrupted the Atlantic overturning circulation (conveyor belt), greatly reducing the northward flow of warm tropical water to the North Atlantic. Melting of the Greenland ice cap, which has greatly accelerated from the late 1990s to the early 2000s, is adding freshwater to this region of the ocean. According to scientists studying ocean circulations, there is a possibility that this freshwater might cause a significant slowdown of the meridional overturning circulation, with possible drastic cooling in the vicinity of northern Europe and Scandinavia. However, the complete shutdown of the Atlantic circulation, which would have the most drastic environmental effects, is thought to be unlikely.

SEE ALSO *Climate Change; Global Warming; Greenhouse Effect; Sea Level Rise*

BIBLIOGRAPHY

Books

Bigg, Grant R. *The Oceans and Climate.* Cambridge: Cambridge University Press, 2004.

Desonie, Dana. *Oceans: How We Use the Seas.* New York: Chelsea House Publications, 2007.

Fujita, Rodney. *Heal the Ocean: Solutions for Saving Our Seas.* Gabriola Island, British Columbia, Canada: New Society Publishers, 1998.

Roberts, Callum. *The Unnatural History of the Sea.* Washington, DC: Island Press, 2007.

Web Sites

National Aeronautics and Space Administration (NASA). "A Chilling Possibility." http://science.nasa.gov/headlines/y2004/05mar_arctic.htm (accessed April 2, 2008).

Agnes Galambosi

Ocean Salinity

Introduction

Ocean salinity, or the saltiness of the seas, ranges from 32 to 37 parts per thousand of dissolved salts, averaging about 35 parts per thousand. Most of the saltiness of the seas is in the form of dissolved sodium chloride, which is the same as table salt. There are, however, many other dissolved salts in the seas as well. Dissolved materials in sea water are derived from weathering of minerals and rocks on land. The dissolved elements are transported to the sea by rivers. Some salts are also introduced into sea water by underwater eruptions of sea-floor volcanoes and volcanic vents. Evaporation concentrates salts in sea water, up to the point where the water becomes saturated with those salts. At the point of saturation, salt crystals precipitate from sea water and are thus removed from sea water. In this way, the saltiness of the sea is maintained over long time periods.

Historical Background and Scientific Foundations

The sea was not always salty. It is thought that the original oceans on Earth were freshwater bodies that rapidly became salty as chemical weathering of minerals and rocks on dry land contributed to salts in the sea. Over time, the seas became quite salty, and marine sediments are quite full of salts and other chemical compounds precipitated from sea water over time. In addition, the evolutionary advent of hard parts and shells of marine organisms caused those organisms to extract many chemical compounds from sea water, including calcium carbonate, a common mineral in shells and other hard parts of marine animals and plants. Over time, the chemistry of the seas has varied and this is recorded in the chemical nature of marine sediments over geological time.

There are over 70 chemical elements in sea water. The most common constituents of sea water are (in order): chlorine; sodium; sulfate; magnesium; calcium; potassium, bicarbonate; and bromide. Together these account for 99.999% of all chemical components dissolved in sea water. When evaporated, either in nature or in the laboratory, sea water yields a predictable sequence of chemical precipitates, which are (in order): calcium carbonate (calcite); hydrated calcium sulfate (gypsum); sodium chloride (halite); and then other salts including magnesium chloride and sulfate. Studies of layers of sedimentary rock formed during evaporation of sea water indicate this sequence holds in nature. There are sequences of sedimentary rock showing layers of limestone (a rock containing calcite), rock gypsum, and rock salt—in that order—in some sedimentary basins.

Impacts and Issues

Salinity of the oceans varies from place to place, but over a limited range, as noted earlier. The most saline of open ocean waters are in the mid-latitudes of the Atlantic Ocean basin. The least saline of open ocean waters are in the Northern Ocean adjacent to landmasses and in the oceans adjacent to India and southeastern Asia. In waters that are not open, for example in bays, lagoons, estuaries, and other water bodies that are connected to the sea, but partially surrounded by land, salinity can vary greatly. In such areas where there is much freshwater from rivers, ocean salinity can be substantially lowered, a condition called hyposalinity. The opposite, hypersalinity, or salinity elevated above the high end of the normal range, can occur in bays, lagoons, estuaries, or other restricted bodies of water where freshwater input is limited and/or the connection to the open sea is somehow limited. This causes evaporation to concentrate salts and other dissolved chemical compounds in the sea water.

Changes in salinity greatly affect the ocean ecosystem and many shoreline ecosystems as well. Some species are well adapted to changes in salinity, whereas others cannot survive when salinity changes. Thus, both natural and human-made salinity changes can profoundly affect ocean ecosystems or other ecosystems that are strongly connected to sea water. Natural salinity changes occur when there are long-term changes in climate and freshwater run off. Freshwater run off may be climate related or may be related to mountain building on the continents. Human-made salinity changes may be connected to human-induced climate change, but more locally are related to human-made changes in local drainage systems and the discharge of water-borne wastes. Human activities may restrict oceanic connections with bays, lagoons, estuaries, and other branches of the sea, thus causing salinity elevation. Natural processes, such as sand bar growth or sand spit growth may restrict oceanic connections and thus accomplish the same thing in terms of salinity elevation.

SEE ALSO *Marine water quality; Runoff*

WORDS TO KNOW

ECOSYSTEM: The community of individuals and the physical components of the environment in a certain area.

ESTUARY: Lower end of a river where ocean tides meet the river's current.

RUNOFF: Water that falls as precipitation and then runs over the surface of the land rather than sinking into the ground.

BIBLIOGRAPHY

Books

Kurlansky, Mark. *Salt: A World History.* New York, Walker and Company, 2002.

Web Sites

U.S. Geological Survey. "Why Is the Ocean Salty?" (accessed 26 March 2008).

David T. King Jr.

Scientists are studying salinity, temperature, minerals, light transmission, plant life, bacteria and larger animals looking for possible indicators of climate change. *AP Images*

Ocean Tides

Introduction

Tides are periodic changes in the height of the ocean surface at a certain location. Ocean tides are caused by a complicated combination of forces generated by the gravitational pull of the moon and sun, the rotation of Earth, and the geography of a particular location. Tides influence the ecology of many coastal ecosystems. They have also been used to generate power in places where tides are extreme.

Historical Background and Scientific Foundations

Ocean tides are brief, episodic changes in the height of the ocean surface at a certain place caused by the gravitational pull of the moon and the sun as well as the spinning of Earth. The wavelength of tides is half Earth's circumference. The average global height of ocean tides is about 7 ft (2 m), although they can vary from undetectable to 50 ft (15 m) depending on a collection of complex factors that influence tides in different locations.

The relationship between the position of the moon and the height of tides was first observed by Greek navigator Pytheas around 300 BC. Ancient Hawaiians and Chinese also recognized the tides and theorized about their source. Isaac Newton (1642–1727), English mathematician and physicist, put forward the first mechanistic description of tides. He showed that the pull of gravity between any two objects depends on their mass and the distance between the objects. Using this central finding, Newton developed the equilibrium theory of tides, which explains the major characteristics of tides. Newton's ideas were later refined by French mathematician Pierre-Simon Laplace (1749–1827), who developed what has become known as the dynamic theory of tides.

To understand the tides, imagine Earth as a ball covered in water. The gravitational pull of Earth holds the water on its surface. The moon also exerts a gravitational pull on the water, creating a bulge in the water in the location that is the closest to the moon. The moon and Earth are fixed in a rotating system. Inertia, which explains why an object continues moving in the same direction if no force acts on it, produces an outward-flinging force in the direction opposite the moon. This results in a second bulge of water on the opposite side of Earth.

Earth also spins on its axis once a day. The bulges remain fixed in position relative to the moon. Imagine Earth spinning beneath the two bulges of water. A particular location on Earth passes below the bulges during the course of a day. The bulges create high tides and the positions between the bulges create low tides.

Although the moon is the major factor that drives tides, numerous other factors also impact their size and timing. In a similar manner to the moon, the sun exerts a gravitational force on the water on the planet. However, because it is much farther away, its impact is only about 40% as great as that of the moon. In addition, both the moon and the sun move north and south of the equator depending on the time of year and time of month. This movement offsets the positions of the bulges and influences both the size of the tides and their frequency. The depth of the ocean, the geography of coastlines, and the actual energy consumed by water friction due to tides all affect the timing and height of tides in any specific location. All in all, there are approximately 140 tide-generating forces and factors that are required to completely predict the tides in any location.

Impacts and Issues

Coastal organisms are necessarily affected by tides. The tides influence the frequency and duration of time that organisms are covered in water or exposed to air. Some organisms require submergence for extended periods of time. These animals live in a zone below the low tide position. Other organisms can withstand only infrequent periods when they are submerged. They aggregate above the high tide mark. The effects of the tides create thin zones within the coastal region that are host to very different types of ecosystems.

Some organisms have developed adaptations that depend on changes in the tides. Certain species of diatoms, small plantlike organisms, rise to the surface of sandy beaches at low tides in order to perform photosynthesis. When the tide rushes in, they burrow down into the sand, protected from wave surge and predation. A species of fish called the grunion swim onto sandy beaches along the Pacific coast in large numbers during the highest tides. They lay their eggs in the sand, free from predation by aquatic animals. The eggs hatch nine days later at the return of the first tide that reaches the same point on the beach.

Bahraini fisherman walks through the mud at low tide to shore September 2007, in Sitra, Bahrain, carrying bags of fish after clearing out his traps. *AP Images*

WORDS TO KNOW

ESTUARY: Lower end of a river where ocean tides meet the river's current.

GRAVITY: An attractive force that exists between all mass in the universe such as the moon and Earth.

INERTIA: The tendency of an object to continue in its state of motion.

Ocean tides are used as a source of power in some places where tidal variation is extreme. On the estuary of the River Rance in France and along the Annapolis River in Nova Scotia, Canada, power plants have been built that harness the energy from tides. Energy is harvested as the water rushes inland as the tide becomes high and again as the water rushes seaward as the tide becomes low. Proposals for tidal plants in Scotland, New York City, San Francisco Bay, and in several inlets in Australia have also been proposed. In some cases tidal power plants can negatively affect the ecosystem by changing the turbidity and salinity of the enclosed bay. In addition, fish may swim through turbines and become injured or killed. As a result, there are often significant environmental objections to building tidal-powered electrical facilities.

SEE ALSO *Aquatic Ecosystems; Bays and Estuaries; Benthic Ecosystems; Coastal Ecosystems; Coastal Zones; Dams; Estuaries; Marine Ecosystems; Marine Fisheries; Marine Water Quality; Ocean Circulation and Currents; Oceans and Coastlines; Tidal or Wave power; Tides*

BIBLIOGRAPHY

Books

Garrison, Tom. *Oceanography: An Invitation to Marine Science*, 5th ed. Stamford, CT: Thompson/Brooks Cole, 2004.

Web Sites

Renewable Energy and AEoogle. "Tidal Power." http://www.alternative-energy-news.info/technology/hydro/tidal-power/ (accessed March 7, 2008).

Tulane University. "Coastal Zones." April 9, 2007. http://www.tulane.edu/~sanelson/geol204/coastalzones.htm (accessed February 8, 2008).

Juli Berwald

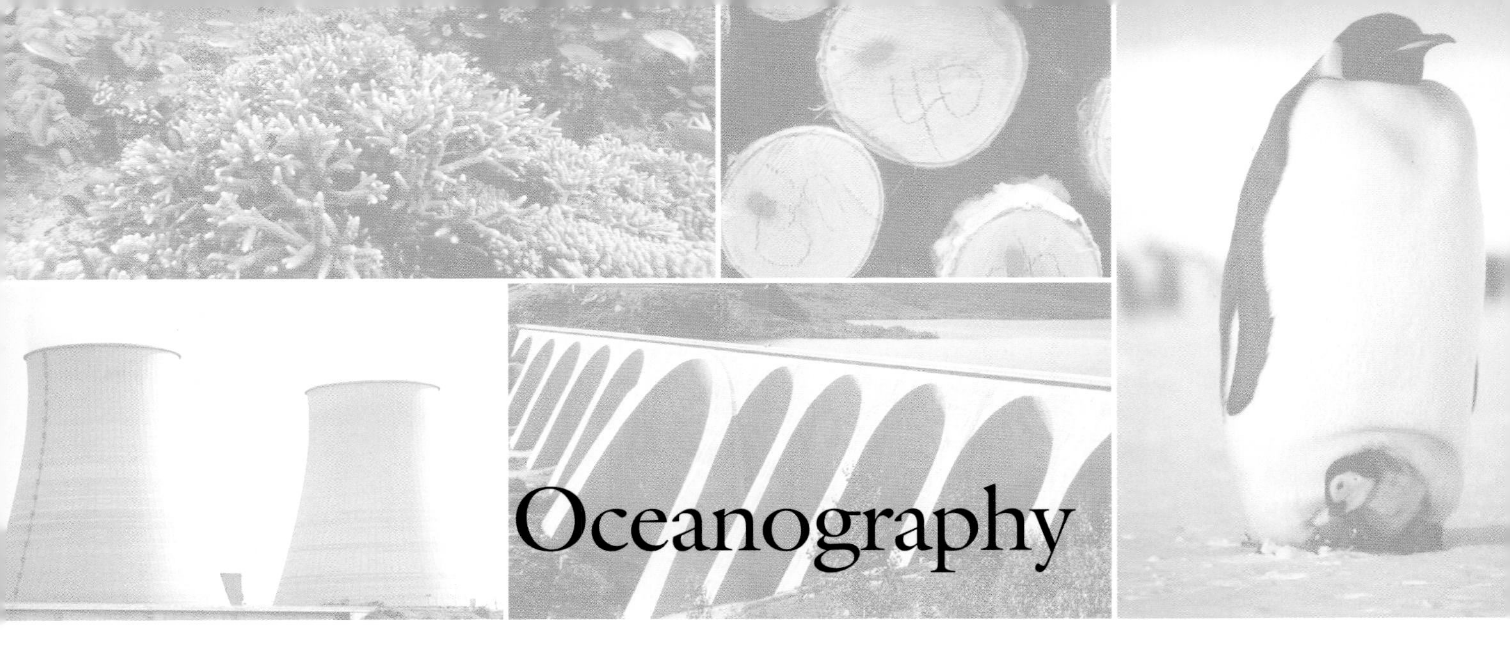

Oceanography

Introduction

Oceanography, alternately known as oceanology or marine science, is a branch of science that studies the open oceans, coastal regions, and seas. The science of oceanography covers a wide variety of topics and scientific disciplines. Some examples are the movement of water as currents, influences of environmental factors on the direction and strength of currents, marine life including its diversity and threats to that diversity, the generation and dynamics of waves, and the interaction of the ocean and the atmosphere.

Given this diversity, oceanography is a varied science and includes physicists, microbiologists, biologists, ecologists, computer scientists, mathematicians, modelers, and chemists. Research often involves teams of scientists at facilities such as Woods Hole Oceanographic Institution, a large research and educational facility in the eastern United States.

Oceanography has taken on more importance since the 1970s, with the recognition that human activities are altering the ocean. Only a few decades ago, the notion that the chemistry and biology of the global ocean could be altered was dismissed, as the ocean was assumed to be so vast and its volume so great that any influence would be accommodated. However, changes in parameters, including water temperature and pH, have been verified. Some of these changes could be affecting the atmosphere.

Historical Background and Scientific Foundations

The roots of oceanography date back to the fourth century BC, when the Greek philosopher Aristotle made observations on tides. In the eighteenth century, James Cook included observations of the South Pacific Ocean in the chronicles of his voyages. By then, it had been recognized that knowledge of ocean patterns of wind and storms was essential to a safe voyage and exploitation of ocean currents as trade routes.

In addition to his observations that provided the underpinnings for his treatise on evolution, another aspect of Charles Darwin's voyages on the HMS *Beagle* in the 1830s was his work on the formation of reefs. Such information proved valuable to seafarers, since collisions with reefs have claimed countless ships and many lives throughout recorded history.

Darwin and his contemporaries were true oceanographers, who were conducting research on the oceans. In Darwin's time, more was being learned about ocean currents and the variation in ocean depth. The steep drop-off in depth beyond continental shelves was discovered in 1849; the upwelling of colder water from the deeper ocean at the edge of continental shelves brings nutrients closer to the surface, providing food for huge numbers of fish. Knowing where continentals shelves ended was key in increasing the productivity and reliability of the global fishery.

The first oceanography textbook was published in 1855, and oceanography emerged as a scientific discipline in 1871, with the first dedicated oceanography expedition funded by the British government.

By the beginning of the twentieth century, oceanographic research was flourishing around the world. In the United States, the Scripps Institution of Oceanography was established in San Diego in 1903, followed in 1930 by Woods Hole Oceanographic Institute in Woods Hole, Massachusetts.

A few examples of dedicated oceanographic facilities around the globe are the Bedford Institute of Oceanography (Canada), Ocean University of China, Institute of Oceanography and Fisheries (Croatia), National Institute of Oceanography (India), Institute of

Marine Research (Norway), P. P. Shirov Institute of Oceanography (Russia), and the National Oceanography Centre (United Kingdom).

By the time Woods Hole was formed, explorations of the ocean were becoming more sophisticated. For example, the technique of echo sounding—in which pulses of sound energy are directed down into the water and the time and pattern of the returning reflected sound waves are used to measure depth and the presence of objects on the seabed—had been used to measure the Atlantic Ocean to depths of 70,000 ft (21,300 m).

By the 1970s, the use of unmanned robots to explore the ocean was accompanied by vehicles capable of carrying researchers. A famous example occurred in 1977, when the manned submersible *Alvin* was piloted by two Woods Hole researchers to the hydrothermal vents located at the Galápagos Rift, a volcanically active region in the Pacific. Subsequently, vents have been discovered at many locations around the globe, and the unique forms of life in the vicinity of the vents have been well-characterized.

Alvin and the other manned vehicles are termed "human-occupied vehicles" (HOVs) to distinguish them from unmanned vehicles that are tethered to a ship, which are termed "autonomous undersea vehicles" (AUVs), and vehicles that can move independently, which are called "remotely operative vehicles" (ROVs).

Jacques Cousteau was an underwater adventurer, filmmaker, author, environmentalist, and scuba pioneer who opened the mysterious world beneath the seas to millions of landlocked readers. *AP Images*

WORDS TO KNOW

HYDROSPHERE: The total amount of liquid, solid, and gaseous water present on Earth.

IRON FERTILIZATION: A proposal to seed the ocean with iron to stimulate the growth of microorganisms, in order to trap carbon and so lessen the release of carbon dioxide to the atmosphere.

MARIANAS TRENCH: A canyon almost 36,000 ft (11,000 m) in depth located in the floor of the Pacific Ocean; it is the deepest point in the ocean.

SALINITY: Measurement of the amount of sodium chloride in a given volume of water.

Echo sounding is just one of many scientific techniques that contribute to oceanography. Classical techniques of physics, chemistry, and biology have been joined by modern computational methods including modeling. Some of the data are collected from instruments that are deployed from ships. There are many designs of unmanned probes that are tethered to a ship by a cable that can be thousands of feet long. The cable also joins the AUV to the command computer aboard ship, allowing the probe to be guided remotely.

The supporting cable once hindered the depth to which a probe could be sent. The weight of a cable tens of thousands of feet in length would be more than could be supported by a ship's winch and by the ship itself without capsizing. However, in 2006, researchers at Woods Hole completed development of fiber-optic cables that housed the electrical connection to the remote shipboard computer. The fiber-optic cable is a fraction of the weight of the old version, making it feasible to manufacture a tethering cable capable of sending unmanned probes to the Marianas Trench, which at 7 mi (11 km) is the deepest part of the ocean.

ROVs operate independently, "flying" underwater while being remotely controlled onboard ship. Samples of water can be collected during flight of these vehicles for real-time analysis or can be stored for collection upon return of the probe to the ship.

Since 1964, Woods Hole has operated the National Deep Submergence Facility—a fleet of HOVs, AUVs, and ROVs for the benefit of the entire U.S. oceanographic research community.

Still other probes have been designed that are stationary. Whether deployed as buoys on the surface of the

IN CONTEXT: FUTURE CHALLENGES FOR OCEANOGRAPHERS

According to the Intergovernmental Panel on Climate Change (IPCC), climate change already impacts the oceans and seas. The research and monitoring of Earth's marine systems provide important challenges for current and future oceanographers.

"At continental, regional and ocean basin scales, numerous long-term changes in climate have been observed. These include changes in arctic temperatures and ice, widespread changes in precipitation amounts, ocean salinity, wind patterns and aspects of extreme weather including droughts, heavy precipitation, heat waves and the intensity of tropical cyclones."

SOURCE: *Solomon, S., et al, eds. "IPCC, 2007: Summary for Policymakers." In: Climate Change 2007: The Physical Science Basis. Contribution of Working Group I to the Fourth Assessment Report of the Intergovernmental Panel on Climate Change. New York: Cambridge University Press, 2007.*

ocean or on the seabed, these probes can carry an array of instruments that permit a great deal of data to be gathered simultaneously. As well, buoys can be deployed to drift with a current, collecting data over time to assess the current conditions at different locations.

In 2005, the University of Victoria in Canada deployed a series of seabed instruments in the Pacific Ocean off Canada's west coast. The project, dubbed VENUS (Victoria Experimental Network Under the Sea) linked the probes via fiber-optic cable to create an undersea observatory that is connected to the Internet. Chemical data including temperature, salinity, turbidity, dissolved gas concentration, current speed and direction, sounds, and both video and still images are collected and sent in real time to labs, classrooms, science centers, and homes around the world. As of 2008, another array is being readied for deployment in the region.

Molecular biology techniques have also been harnessed for oceanographic research and can be used to detect target sequences of bacteria to allow one type of bacterium to be distinguished from another. This gives researchers the ability to analyze water samples to determine the diversity of microbial life in the ocean. In one example, a 2004 cruise to the Sargasso Sea by molecular biologist Craig Venter (1946–) revealed evidence of over one million new genes and thousands of previously unknown species of bacteria. In a second example, a team of Woods Hole researchers published a 2007 study documenting the discovery of a previously unknown bacterium that is able to use gases, such as butane (C_4H_{10}) and ethane (C_2H_6), as food in the absence of oxygen (O). Such research is establishing that a huge diversity of microbial life exists in the ocean.

Impacts and Issues

Oceanography has revealed the huge diversity of life that exists in the ocean, with much still left to discover. New studies are reinforcing the view that the ocean, which covers about 70% of Earth's surface, is under threat from human activity.

Climate change that is being driven by human activity is altering the world's oceans by increasing the temperature and the acidity of seawater, and altering the global circulation patterns of water. Studies conducted from virtually all regions of the globe, including both poles, are showing that marine life is being stressed by the changing oceans.

There are few areas of the ocean that have not been affected, according to a 2008 study published in the journal *Science*. Over 80% of the ocean has been affected by human activities, including overfishing, runoff of agricultural pesticides, pharmaceuticals, and sewage from coastal regions, with more than 40% of the ocean being in need of remediation. Regions of most concern include the North Sea, South and East China Seas, Caribbean Sea, Mediterranean Sea, Red Sea, Persian Gulf, Bering Sea, the western Pacific Ocean, and the east coast of North America.

As well, studies conducted since the 1990s have shown that the ability of the ocean to trap carbon (C) is diminishing, perhaps due to the increasing acidity of the water. As a result, more carbon has been escaping to the atmosphere, where it has contributed to atmospheric warming.

Among the proposals to deal with the diminishing carbon-trapping capacity of the ocean is the seeding of regions of the ocean with iron (Fe), which would stimulate the explosive growth of microorganisms such as algae. These algal blooms could sequester carbon. This approach, which has been termed "iron fertilization," is controversial, with some critics arguing of the potential harm that could result from the deliberate manipulation of the ocean environment.

In February 2008, the private company Planktos indefinitely shelved its plans for iron fertilization. The initiative had been intended to generate carbon credits that would have been sold to countries and industries seeking to offset their greenhouse-gas emissions without as rigorous an emissions control program as would otherwise have been necessary.

SEE ALSO *Coral Reefs and Corals; Marine Ecosystems; Marine Water Quality; Ocean Circulation and Currents; Sea Level Rise; Woods Hole Oceanographic Institution*

BIBLIOGRAPHY

Books

Garrison, Tom S. *Oceanography: An Invitation to Marine Science*. New York: Brooks Cole, 2007.

Parsons, Tim. *The Sea's Enthrall: Memoirs of an Oceanographer*. Victoria, British Columbia, Canada: Trafford Publishing, 2007.

Sverdrup, Keith A. *An Introduction to the World's Oceans*. New York: McGraw-Hill, 2008.

Periodicals

Halpern, B. S., et al. "A Global Map of Human Impact on Marine Ecosystems." *Science* 319 (2008): 948–952.

Kniemeyer, O., et al. "Anaerobic Oxidation of Short-chain Hydrocarbons by Marine Sulphate-reducing Bacteria." *Nature* 449 (2007): 898–901.

Venter, J. C., et al. "Environmental Genome Shotgun Sequencing of the Sargasso Sea." *Science* 304 (2004): 66–174.

Web Sites

Victoria Experimental Network Under the Sea (VENUS). April 8, 2008. http://www.venus.uvic.ca (accessed April 21, 2008).

Brian D. Hoyle

Oceans and Coastlines

Introduction

Water covers around two thirds of Earth's surface. Its depth ranges from seven mi (11 km) at the greatest depth of the oceans to just a few inches (centimeters) near the line of the coast. Oceans and coastlines vary in their temperature, oxygen content, and access to sunlight. These factors create a wide range of different ecosystems containing thousands of invertebrate species, as well as fish, birds, and whales.

Oceans and coastlines are important zones for human activity for economic reasons, such as fishing, and for recreation and sport. However, both oceans and coastlines face severe threats. Oceans are at risk from overfishing because fish supplies are finite if not managed as a renewable resource, and the fishing industry has expanded greatly in recent years. Coastlines are close to land, and so are threatened by development and pollution from sewage, agriculture, and dumping rubbish. Some important ecosystems like mangrove swamps and coral reefs now face destruction because of the impact of human activities.

Historical Background and Scientific Foundations

Oceans and seas differ only in their size, with the oceans comprising the world's five great bodies of water, namely the Pacific, Antarctic, Indian, Atlantic, and Arctic. Seas are smaller bodies of water, like the North Sea or Barents Sea. Both are composed of saltwater of varying temperatures from 104°F (40°C) in the tropics to frozen water near the Poles. Surface waters may be warmed by the sun, but deeper waters are cold. Near the shore, which is the part of an ocean closest to land, the water is naturally shallow and cloudy so the sun can only penetrate a few feet. The waters of the open ocean are clearer and sunlight may penetrate to a depth of up to 300 ft (91 m).

The type of ecosystem found in the ocean depends largely upon the availability of sunlight, oxygen and nutrients, and the temperature. The surface layer where sunlight penetrates is called the euphotic zone, and is where photosynthesis is carried out by phytoplankton. Phytoplankton act as food for larger species, forming the basis for marine food webs. The bottom layers are nutrient rich, because dead organisms sink and provide food for scavengers. There is, however, a certain amount of mixing of the upper and lower layers of the oceans driven by surface currents and deeper circulation patterns. This provides oxygen to lower layers and nutrients to upper layers. These currents lead to some particularly productive marine biomes called upwellings where the mixing is particularly pronounced. One such upwelling is found along the equator in the Pacific Ocean, another along the Antarctic coast, which provides food supplies in the form of krill for both baleen whales, which are filter feeders, and large seabird populations.

The open ocean is not just a featureless mass of water. It exists in distinct layers. The column of water extending down from the surface to a depth of about 2.5 mi (4 km) is called the pelagic zone and is composed of the epipelagic (top), mesopelagic (middle or twilight), and bathypelagic (bottom) zones. Below this, extending to about 4 mi (6 km) are the abyssal and hadal zones. Hadal comes from the Greek word for unseen, abyssal from the Greek word for bottomless. Marine ecosystems are stratified too, with respect to the vertical layer they can adapt to. For example, there is no light in the layers close to the ocean floor save for that emitted from the organisms themselves. Thousands of species, including bacteria, squid, and fish, exhibit bioluminescence, which is the ability to emit flashes of light to distract predators and locate prey.

The oceans have the same salt composition as living cells so, in that sense, it is a viable environment and indeed life probably evolved in the oceans. Compared to the land, however, the marine environment is not so biodiverse. But thousands of different invertebrates do exist in the oceans, some of which live on the shore. These include sponges and echinoderms, such as starfish and cnidarians, including jellyfish and corals. Among vertebrates, whales are the most important mammals living in the ocean, but there are many birds and fish there, too. Around 1,000 fish species occupy the open ocean, inhabiting all of its different vertical zones.

The coast is the margin of land next to an ocean, or sea. It gives way to the beach, the very edge of the water, which may regularly be covered with water by the tides. The ocean or sea floor slopes gradually down from the coast to the deeper water and may be soft or rocky. The area nearest to the sea is called the littoral zone and many fish and shellfish are found here. For instance, molluscs, which are found on both land and sea, include the bivalves such as clams, mussels, cockles, oysters, and scallops, which are economically important. So too are the crustaceans, which include crabs, lobsters, and shrimps.

A rocky-bottomed beach encourages epifauna, which are animals that are attached to the solid surface and tend to filter feed, such as mussels. A soft bottom is generally populated by infauna, which live within the mud. Some are scavengers, like small shrimp and worms, while others, like tube worms, are filter feeders. The sand on a beach is created by erosion, which shapes a coastline over time. Not all coastlines have a beach and some coastlines occur inland. Barrier islands and sea cliffs are important features of some coastlines. The nature of a

WORDS TO KNOW

BIOME: A well-defined terrestrial environment (e.g., desert, tundra, or tropical forest) and the complex of living organisms found in that region.

EPIFAUNA: Animals that live attached to rocks.

EUTROPHICATION: The process whereby a body of water becomes rich in dissolved nutrients through natural or man-made processes. This often results in a deficiency of dissolved oxygen, producing an environment that favors plant over animal life.

FILTER FEEDERS: Animals that obtain their food from filtering passing water.

INFAUNA: Animals living within the material, such as mud or sand, at the bottom of an ocean or sea, or on the beach.

PHYTOPLANKTON: Microscopic marine organisms (mostly algae and diatoms) that are responsible for most of the photosynthetic activity in the oceans.

UPWELLING: The vertical motion of water in the ocean by which subsurface water of lower temperature and greater density moves toward the surface of the ocean.

Earth Day demonstrators trying to dramatize environmental pollution conclude their rally at the Interior Department in Washington, April 1970, leaving spilled oil in their wake. The oil was used to protest pollution by off-shore oil drilling. *AP Images*

Wooden fences stabilize and encourage the formation of sand dunes along the coastline at St. Joseph Peninsula State Park on Cape San Blas, Florida. *AP Images*

coastline depends on geological processes originating on land and in the ocean or sea.

Mangrove swamps and coral reefs are among the specialized biomes found in coastal areas. Mangroves are salt-resistant trees that support fish or shrimp. Coral reefs, like Australia's Great Barrier Reef that stretches for around 1,200 mi (1,930 km) along the eastern coast of Australia, consist of animals called stony corals, each of which is a polyp with tentacles that can trap organisms. Coral reefs tend to occur in warm, shallow, clear waters where they provide a home for a diverse community of fish, worms, and crustaceans, protecting the small fish from larger predatory fish.

■ Impacts and Issues

Although the oceans are vast, they are not inexhaustible. Overfishing is perhaps the greatest threat they now face. Global fish harvesting has increased 5 times during the last 50 years or so. The oceans can support a fish harvest of about 100 million tons of fish per year and this limit is close to being reached. Not only will this hit the fishing industry, but it also impacts on marine ecosystems.

Nearer to the coast, there are a number of ecosystem threats caused by human activities. For instance, the coral reefs are among the world's most threatened biomes. They are affected by a range of factors, including destructive fishing practices, pollution, and sewage, while global warming is beginning to destroy them through bleaching. Mangrove swamps are similarly at risk.

Shore areas are also at risk from pollution, which can cause eutrophication by raising nutrient levels in the water because of sewage outlet and agricultural runoff. As oxygen levels fall, fish and other organisms begin to die off. Eutrophication also causes an overgrowth of algae in the water, often visible as a red, yellow, or green scum on the surface. They are a visible sign of pollution, as are found in many areas around the world, such as the Mediterranean and the east coast of the United States. Meanwhile, many beaches are rendered unsightly by rubbish such as plastic bags, which can strangle wildlife.

SEE ALSO *Coastal Zones; Marine Ecosystems; Ocean Tides; Reef Ecosystems*

BIBLIOGRAPHY

Books

Cunningham, W.P., and A. Cunningham. *Environmental Science: A Global Concern.* New York: McGraw-Hill International Edition, 2008.

Kaufmann, R., and C. Cleveland. *Environmental Science*. New York: McGraw-Hill International Edition, 2008.

Web Sites

Office of Naval Research. "Habitats: Beaches." http://www.onr.navy.mil/focus/ocean/habitats/beaches1.htm (accessed April 3, 2008).

Susan Aldridge

Oil Pollution Acts

Introduction

Oil is a resource for energy and the chemical industry, but it needs to be handled with care. It is usually transported from its source in large tanker ships, and occasionally these tankers spill some oil into the sea. The resulting pollution can have severe and long-lasting effects on sea birds and fish. The widespread devastation caused by major oil spills, most notably the disastrous *Exxon Valdez* spill off the coast of Alaska in 1989, led to legislation on oil pollution.

The United States Oil Pollution Act of 1990 provides a legal framework for the U.S. Environmental Protection Agency (EPA) to do its job of protecting national waters from oil pollution. It covers facilities that handle oil and requires them to take measures to prevent oil from escaping. They also need to have an emergency plan to deal with any spills that do occur.

Historical Background and Scientific Foundations

Every year, 1.5 billion tons (1.3 billion metric tons) of oil are shipped in large tankers, of which around 1% escapes as a spill. The EPA says that there are around 14,000 spills of oil each year in the United States occurring on inland or coastal waters, some of which extend to beaches. There are many ways in which an oil spill may occur. The ship may develop a mechanical fault or crack in its hull that leads to leaking, it may run aground, or oil may be discharged through washing procedures. Sometimes oil is dumped illegally at sea, and oil spills can also be used as an act of war or terrorism. An oil spill can also result from extreme adverse weather conditions such as a hurricane.

One of the most notorious oil spills came from the tanker *Amoco Cadiz*, which was carrying a cargo from the Persian Gulf to Rotterdam in the Netherlands. At the time it was the largest-ever oil spill. The ship encountered a severe storm in the English Channel and broke up near the coast of Brittany, France, on March 16, 1978. It lost its entire cargo to the water, consisting of 68.7 million gallons (260 million liters) of oil. Around 200 mi (320 km) of the French coastline was contaminated with oil, with fishing and tourism being affected to the cost of around $1 billion.

On March 24, 1989, the *Exxon Valdez* ran aground on a reef in Prince William Sound in Alaska, creating an 8-mi (13-km) oil slick. Shortly after, the U.S. government passed the Oil Pollution Act of 1990 in an attempt to protect the nation from oil spills. Any facility that stores, processes, transfers, refines, or otherwise handles oil must do everything it can to prevent an oil release from happening. Should a spill occur, despite these precautions, then there must be a response plan to contain and clean up the oil and reduce its impact on the waterway. This is known as an Oil Spill Prevention, Control, and Countermeasure plan. Moreover, owners of a boat that creates an oil spill will have to pay compensation for every ton that escapes into the water. One obvious way of preventing oil spills is to make tankers have a double hull, so that escapes are contained between two layers. All boats in the United States will be required to have a double hull by 2015.

Over recent years, technologies have been developed to help clean up oil pollution and limit the environmental damage it causes. A boom, or physical barrier, is placed around the spill to stop it spreading farther on the water or reaching the beach. Then skimmers, which are special boats or vacuum machines, physically remove the oil into a container. They work best on still waters. Oil dispersants may be sprayed onto the spill from a boat or from a plane. These are chemicals that can break down the oil into tiny droplets, which may then be degraded by marine microorganisms. Environmental legislation will control which chemicals can be put into the ocean as a dispersant. Sometimes the oil is burned off, but this is controversial because it creates air pollution, particularly if there is wind. The method employed will depend on the location of the spill and prevailing tidal and weather conditions.

Oil spilled on a beach is dealt with differently. This type of spill is treated by high-pressure hosing and materials called sorbents that soak up the oil. The affected sand will often be shoveled away. Hand cleaning and ongoing care is necessary for animals, like birds, that have been affected by spilled oil. It is extremely difficult to remove oil from plants and often they have to be destroyed. Bioremediation, which is the use of microorganisms that are known to break down the compounds in oil, may be used for cleaning up the oil over a period of time. All of these remedial actions have to be carried out in accordance to the provisions of the Oil Pollution Act of 1990 or equivalent legislation elsewhere.

Impacts and Issues

The impact of an oil spill can be severe and long-lasting, which is why environmental legislation is needed for prevention and remedy. Clearly, oil pollution on a waterway or beach is unsightly and will hit recreation and tourism, but its effect on marine ecosystems is more serious. Oil is a complex mixture of hydrocarbons. The lighter fractions may evaporate from the surface of the water over a period of time. The heavier fractions that remain are rich in toxic materials called polyaromatic hydrocarbons. Media images of the aftermath of an oil spill often show sea birds covered in oil. Unable to fly because the oil weighs it down, the bird will naturally try to clean itself. Once ingested, the oil may enter the lungs or liver and poison the bird. Oil may also blind a bird so that it cannot see predators. Specialized hand cleaning of the bird in a sanctuary involves removal of the oil from the exterior and interior of its body. Unless they receive this kind of attention, most sea birds will be killed by an oil spill.

Fish covered in oil may be eaten by whales. The oil then is transferred to the whale's blowhole and causes it to suffocate. Sea otters and seals are also severely affected by oil spills. The oil blocks the air insulation in their coats and makes them prone to hypothermia. Meanwhile, oil can also poison the plankton and marine invertebrates, having an adverse impact upon the entire marine food web.

Oil spills have a large visual impact upon the general public, both directly or indirectly. Seeing the damage that oil can do to vulnerable wildlife can help people question their use of oil and look to whether they can find ways of reducing their consumption. The less oil is transported, the less the overall environmental impact of spills.

WORDS TO KNOW

BOOM: A physical barrier placed around an oil spill to contain it.

HYDROCARBONS: Molecules composed solely of hydrogen and carbon atoms.

OIL: Liquid petroleum.

PETROLEUM: A deposit formed from the action of high pressure and temperature on the buried remains of organisms from millions of years ago.

Primary Source Connection

The *Exxon Valdez* spill is considered to be the largest oil spill in American shipping history. The incident acted as a driving force to amend and strengthen the provisions of the original Oil Pollution Act of 1924. These provisions are now incorporated in the Oil Pollution Control Act (OPA) 1990. OPA 1990 is a federal act that defines *Valdez* and the surrounding areas as an exclusive economic zone. It prohibits and thereby imposes severe liability on discharge of oil into navigable waters in this area. It also establishes liabilities for resulting injuries, and also for loss of natural resources.

The Oil Spill Recovery Institute (OSRI) was established by the U.S. Congress in response to the 1989 *Exxon Valdez* oil spill. The act identifies the Prince William Sound Science and Technology Institute (known as the PWS Science Center) situated in Cordova, Alaska, as administrator and headquarters for OSRI. The purpose of the institute was to understand the impact of oil spills on the environment, especially the marine ecosystem, to enhance the response ability of rescue teams, to gather information and bring general awareness, and to collaborate with other organizations and take benefit of their research and analytical knowledge.

OIL POLLUTION ACT OF 1990

Section 2731. Oil Spill Recovery Institute

(a) Establishment of Institute

The Secretary of Commerce shall provide for the establishment of a Prince William Sound Oil Spill Recovery Institute (hereinafter in this section referred to as the "Institute") through the Prince William Sound Science and Technology Institute located in Cordova, Alaska.

(b) Functions

The Institute shall conduct research and carry out educational and demonstration projects designed to:

1. identify and develop the best available techniques, equipment, and materials for dealing with oil spills in the arctic and subarctic marine environment; and
2. complement Federal and State damage assessment efforts and determine, document, assess, and understand the long-range effects of Arctic or Subarctic oil spills on the natural resources of Prince William Sound and its adjacent waters (as generally depicted on the map entitled "EXXON VALDEZ oil spill dated March 1990"), and the environment, the economy, and the lifestyle and well-being of the people who are dependent on them, except that the Institute shall not conduct studies or make recommendations on any matter which is not directly related to Arctic or Subarctic oil spills or the effects thereof.

Section 2732. Terminal and tanker oversight and monitoring

(a) Short title and findings

(1) Short title. This section may be cited as the "Oil Terminal and Oil Tanker Environmental Oversight and Monitoring Act of 1990."

(2) Findings. The Congress finds that:

(A) the March 24, 1989, grounding and rupture of the fully loaded oil tanker, the EXXON VALDEZ, spilled 11 million gallons of crude oil in Prince William Sound, an environmentally sensitive area;

(B) many people believe that complacency on the part of the industry and government personnel responsible for monitoring the operation of the Valdez terminal and vessel traffic in Prince William Sound was one of the contributing factors to the EXXON VALDEZ oil spill;

(C) one way to combat this complacency is to involve local citizens in the process of preparing, adopting, and revising oil spill contingency plans;

(D) a mechanism should be established which fosters the long-term partnership of industry, government, and local communities in overseeing compliance with environmental concerns in the operation of crude oil terminals;

(E) such a mechanism presently exists at the Sullom Voe terminal in the Shetland Islands and this terminal should serve as a model for others;

(F) because of the effective partnership that has developed at Sullom Voe, Sullom Voe is considered the safest terminal in Europe;

(G) the present system of regulation and oversight of crude oil terminals in the United States has degenerated into a process of continual mistrust and confrontation;

(H) only when local citizens are involved in the process will the trust develop that is necessary to change the present system from confrontation to consensus;

(I) a pilot program patterned after Sullom Voe should be established in Alaska to further refine the concepts and relationships involved; and

(J) similar programs should eventually be established in other major crude oil terminals in the United States because the recent oil spills in Texas, Delaware, and Rhode Island indicate that the safe transportation of crude oil is a national problem.

(b) Demonstration programs

(1) Establishment. There are established 2 Oil Terminal and Oil Tanker Environmental Oversight and Monitoring Demonstration Programs (hereinafter referred to as "Programs") to be carried out in the State of Alaska.

(2) Advisory function. The function of these Programs shall be advisory only.

(3) Purpose. The Prince William Sound Program shall be responsible for environmental monitoring of the terminal facilities in Prince William Sound and the crude oil tankers operating in Prince William Sound. The Cook Inlet Program shall be responsible for environmental monitoring of the terminal facilities and crude oiltankers operating in Cook Inlet located South of the latitude at Point Possession and North of the latitude at Amatuli Island, including offshore facilities in Cook Inlet.

With respect to the Cook Inlet Program, the terminal facilities, offshore facilities, or crude oil tanker owners and operators enter into a contract with a voluntary advisory organization to fund that organization on an annual basis and the President annually certifies that the organization fosters the general goals and purposes of this section and is broadly representative of the communities and interests in the vicinity of the terminal facilities and Cook Inlet.

Section 2734. Vessel traffic service system

The Secretary of Transportation shall within one year after August 18, 1990:

1. acquire, install, and operate such additional equipment (which may consist of radar, closed circuit television, satellite tracking systems, or other shipboard dependent surveillance), train and locate such personnel, and issue such final regulations as are necessary to increase the range of the existing

VTS system in the Port of Valdez, Alaska, sufficiently to track the locations and movements of tank vessels carrying oil from the Trans-Alaska Pipeline when such vessels are transiting Prince William Sound, Alaska, and to sound an audible alarm when such tankers depart from designated navigation routes; and

2. submit to the Committee on Commerce, Science, and Transportation of the Senate and the Committee on Merchant Marine and Fisheries of the House of Representatives a report on the feasibility and desirability of instituting positive control of tank vessel movements in Prince William Sound by Coast Guard personnel using the Port of Valdez, Alaska, VTS system, as modified pursuant to paragraph (1).

Section 2735. Equipment and personnel requirements under tank vessel and facility response plans

(a) In general

In addition to the requirements for response plans for vessels established by section 1321(j) of this title, a response plan for a tanker loading cargo at a facility permitted under the Trans-Alaska Pipeline Authorization Act (43 U.S.C. 1651 et seq.), and a response plan for such a facility, shall provide for:

1. prepositioned oil spill containment and removal equipment in communities and other strategic locations within the geographic boundaries of Prince William Sound, including escort vessels with skimming capability; barges to receive recovered oil; heavy duty sea boom, pumping, transferring, and lightering equipment; and other appropriate removal equipment for the protection of the environment, including fish hatcheries;
2. the establishment of an oil spill removal organization at appropriate locations in Prince William Sound, consisting of trained personnel in sufficient numbers to immediately remove, to the maximum extent practicable, a worst case discharge or a discharge of 200,000 barrels of oil, whichever is greater;
3. training in oil removal techniques for local residents and individuals engaged in the cultivation or production of fish or fish products in Prince William Sound;
4. practice exercises not less than 2 times per year which test the capacity of the equipment and personnel required under this paragraph; and
5. periodic testing and certification of equipment required under this paragraph, as required by the Secretary.

Section 2737. Limitation

Notwithstanding any other law, tank vessels that have spilled more than 1,000,000 gallons of oil into the marine environment after March 22, 1989, are prohibited from operating on the navigable waters of Prince William Sound, Alaska.

Section 2738. North Pacific Marine Research Institute

(a) Institute established

The Secretary of Commerce shall establish a North Pacific Marine Research Institute (hereafter in this section referred to as the "Institute") to be administered at the Alaska SeaLife Center by the North Pacific Research Board.

(b) Functions

The Institute shall:

1. conduct research and carry out education and demonstration projects on or relating to the North Pacific marine ecosystem with particular emphasis on marine mammal, sea bird, fish, and shellfish populations in the Bering Sea and Gulf of Alaska including populations located in or near Kenai Fjords National Park and the Alaska Maritime National Wildlife Refuge; and
2. lease, maintain, operate, and upgrade the necessary research equipment and related facilities necessary to conduct such research at the Alaska SeaLife Center.

(f) Availability of research

The Institute shall publish and make available to any person on request the results of all research, educational, and demonstration projects conducted by the Institute. The Institute shall provide a copy of all research, educational, and demonstration projects conducted by the Institute to the National Park Service, the United States Fish and Wildlife Service, and the National Oceanic and Atmospheric Administration.

U.S. Congress

U.S. CONGRESS. "OIL POLLUTION ACT." 33 U.S.C.A. 40. II, SEC. 2731–8. WASHINGTON, D.C.: U.S. CONGRESS, 1990.

SEE ALSO *Marine Ecosystems*

BIBLIOGRAPHY

Books

Cunningham, W.P., and A. Cunningham. *Environmental Science: A Global Concern.* New York: McGraw-Hill International Edition, 2008.

Web Sites

U.S. Environmental Protection Agency (EPA). "Oil Program." http://www.epa.gov/region09/waste/sfund/oilpp (accessed April 21, 2008).

U.S. Environmental Protection Agency (EPA). "Threats from Oil Spills." http://www.epa.gov/emergencies/content/learning/effects.htm (accessed April 21, 2008).

Susan Aldridge

Oil Spills

Introduction

Oil spills refer to the release of oil (liquid petroleum hydrocarbon) onto land or into water. The release can be accidental and due to human activity; the most well-known example is the loss of crude oil, a processed oil product such as gasoline or diesel fuel from a tanker following an accident near a seacoast or coast of a large freshwater river or lake.

More commonly, oil spills are the result of the leakage of oil or oily liquid from ships during their normal operation, and from runoff of oily wastes from the land into the water. Oil spills can also occur naturally and are the result of the seepage of oil from a seafloor deposit.

The environmental consequences of an oil spill can be devastating. Animals and birds are killed if they ingest the oil, and the oily coating on the fur or feathers can prove to be lethal. Also, the oily fouling of coastal beaches and vegetation can take months to remove, and can disrupt normal ecological functions for years.

Clean-up efforts have been aided by the use of microorganisms that have naturally adapted to use the hydrocarbons in an oil spill as a food source. As well, microbes can be genetically manipulated to be capable of digesting the pollutant, although the use of these genetically modified organisms in a natural environment is not usually done. However, if the oil spill can be collected and the fouled material transported to a secure site, use of the genetically modified bacteria can be useful.

Historical Background and Scientific Foundations

Oil spills caused by human error or accident have occurred ever since oil has been recovered from the ground and from the sea. Oil recovery is a complex process that involves drilling down into the ground or below the seabed to deposits, and—unless the oil formation is under great pressure that can push the oil to the surface—pumping the oil to the surface.

A century ago, oil production was more of a domestic issue, with seagoing transport of oil confined to small quantities carried aboard sailing ships. By the 1970s, however, with the demand for oil growing in countries including the United States, crude and processed oil was being loaded onto large tankers for transport all over the globe. The capacity of the tankers grew. In 2008, "supertankers" that stretch up to 1,770 ft (540 m) in length—almost the distance of six football fields—can hold up to two million barrels of oil.

From the time shipping began, it has been a hazardous practice. Storms, mechanical failure, and other calamities have caused countless shipping accidents. When this happens to an oil transport ship, the result can be an environmental disaster.

There have been hundreds of large ship-related oil spills since the beginning of the 1970s. The following are just a few noteworthy examples. In March 1972, the tanker *Torrey Canyon* ran aground off the coast of Cornwall in England and spilled nearly 120,000 tonnes of oil (one tonne is 1,000 kilograms or over 2,200 pounds). In 1976, the *Argo Merchant* ran aground off Nantucket in the United States, and spilled 25,000 tonnes of oil. The resulting film of oil on the surface of the water (an oil slick) was 100 mi (160 km) long and more than 50 mi (80 km) wide. Two years later, over 136,000 tonnes of crude oil were spilled onto more than 100 mi (160 km) of coastline near Portsall, France, when the tanker *Amoco Cadiz* ran aground.

In February 1983, a rupture at the Nowruz oil field in the Persian Gulf caused the release of 260,000 tonnes of oil. In August of the same year, the tanker *Castillo de Bellver* caught fire off the coast of South Africa, spilling

WORDS TO KNOW

BIOREMEDIATION: The use of living organisms to help repair damage such as that caused by oil spills.

ECOSYSTEM: The community of individuals and the physical components of the environment in a certain area.

GROUNDWATER: Fresh water that is present in an underground location.

POLLUTION: Physical, chemical or biological changes that adversely affect the environment.

252,000 tonnes of oil. In 1990, the tanker *American Trader* gashed its hull and the resulting oil spill off the coast of southern California created a 15-mi (24-km) long slick that contaminated the Boca Chica nature reserve. In the 1991 Persian Gulf War, a massive amount of oil was spilled from refinery terminals, anchored tankers, and oil wells during the Iraqi invasion of Kuwait. Estimates of the amount of oil that was spilled range up to 780,000 tonnes. In 1999, the *Erika* broke up in stormy seas off the Atlantic coast of France; the resulting slick polluted over 100 mi (160 km) of coastline. Finally in this brief list, a 2004 spill from the Terra Nova, an oil drilling platform in the Atlantic Ocean off the Canadian province of Newfoundland, killed up to 100,000 seabirds.

Oil spills can involve what is termed crude oil. The planet's crude oil was made millions of years ago by the decay of plants and animals that settled onto seabeds. This decay formed deposits of crude oil, which today is recovered as a black liquid (oil that is contained in sandy formations in northern Alberta, Canada, is near molasseslike in consistency, and has to be recovered using very sophisticated and energy-intensive techniques). Crude oil is also known as petroleum. Tankers are one of the transportation routes used to transfer the crude oil to a facility called an oil refinery, where components including gasoline are separated in the process of refining and collected.

Both crude and refined oil are composed of molecules called hydrocarbons. These are sticklike molecules built of linked carbon atoms containing attached hydrogen molecules. This structure makes hydrocarbon virtually unable to dissolve in water; such compounds are termed "hydrophobic" ("water hating"). Because of its hydrophobic property, spilled oil will tend float on the surface of fresh and marine water, creating the oil slick.

As the slick is washed ashore by waves, the thick oil coats rocks, vegetation, sand, and living creatures that it contacts. The effects on wildlife can be disastrous. When oil coats a bird's feathers, for example, it destroys the feather's normal ability to repel water and retain air between the layer of feathers and the skin. As a result, the bird losses its ability to regulate its body temperature and dies of hypothermia, or drowns from the loss of buoyancy and the weight of the oil. Death can also result from poisoning, as the bird ingests oil in its attempt to clean itself. Birds that are poisoned but survive may have damage to organs including the liver, kidneys, and intestines, and may have difficulty in breeding. Females can become less capable of laying eggs.

Thus, an oil spill may not only affect bird populations immediately, but can disrupt population numbers for years afterward. Plants and animals can also be harmed or killed.

A frequent sight after an oil spill has fouled a coastline is the army of people attempting to clean up the spill. They may spray steam onto the spill, which can heat the oily coat and make it easier to wash off. As well, chemicals termed dispersants can be applied that cluster

Crude oil from the tanker Exxon Valdez swirls on the surface of Alaska's Prince William Sound near Naked Island, April 9, 1989, 16 days after the tanker ran aground, spilling millions of gallons of oil and causing widespread environmental damage. *AP Images*

IN CONTEXT: THE *EXXON VALDEZ* OIL SPILL, AN INFAMOUS AND LASTING LEGACY

Just after midnight on March 24, 1989, the fully loaded oil tanker *Exxon Valdez*, sailing from the port of Valdez, Alaska, and carrying just over 53 million gallons (200.6 million liters) of oil, hit the Bligh Reef in Prince William Sound, Alaska, at top speed. Nearly 11 million gallons (41.6 million liters) of oil were spilled into the Sound's waters and fragile ecosystem.

Driven by winds and currents, the spill ultimately stretched 460 mi (740 km) and affected 1,300 mi (2,100 km) of shoreline. It is estimated that 250,000 seabirds, 2,800 sea otters, and 300 harbor seals were killed by the oil spill. In addition, 250 bald eagles and at least 22 Orca whales were killed. Billions of salmon and herring eggs were also destroyed.

Following the accident, the U.S. National Transportation Safety Board conducted an investigation into the causes. They found five events contributed to the grounding of the ship. First, the third mate, Gregory T. Cousins, did not maneuver the ship properly, probably because he was tired. Second, the ship's captain, Joseph J. Hazelwood, did not provide an appropriate navigation watch. In his criminal trial, Hazelwood was found not guilty of operating a vessel while intoxicated, but guilty of negligent discharge of oil. Third, the Exxon Shipping Company was negligent in its supervision of Hazelwood and his crew. Fourth, the U.S. Coast Guard failed to adequately monitor the shipping traffic. Finally, the ship should have had a more effective pilot and escort service.

The event is considered one of the worst environmental disasters in U.S. history. It happened in an ecologically rich habitat, with rugged and inaccessible coastlines. Results from studies of the area affected by the oil spill showed that for at least twelve years, oil attributed to the spill could easily be found in the area. Surveys of beaches, for example, found significant amounts of oil below the surface in areas affected by the oil spills.

around the oil and form droplets that enclose the oil molecules. The dispersed oil needs to be collected before it washes back into the water. Other means of cleaning up spills include skimming the slick off the water's surface and confining a slick inside a plastic ring known as a boom.

Clean up of an oil spill can be difficult, time consuming, and expensive. As an example, the 1989 spill from the *Exxon Valdez* tanker released enough oil to fill almost 125 Olympic-sized swimming pools. The slick, which fouled over 1,000 mi (1,600 km) of Alaskan coastline, took four summers to clean up. The clean-up involved 10,000 workers and 1,000 boats. Exxon's bill for the clean-up was approximately $2.1 billion.

Impacts and Issues

The environmental consequences of oil spills cannot be understated. Environments and the wildlife in those environments can be affected for decades. For example, the *Exxon Valdez* spill fouled Alaska's Prince William Sound, a productive fishery and area teeming with birds and wildlife. Studies conducted nearly 15 years later still demonstrated environmental damage. As of 2008, the region has essentially returned to the pre-spill state, but the damage lasted almost 20 years.

In the United States, the consequences of major oil spills that occurred in Pennsylvania in 1988 and the *Exxon Valdez* disaster spurred Congress to amend the Clean Water Act by passage of the Oil Pollution Act of 1990. This legislation improved the ability of the Environmental Protection Agency (EPA) and Coast Guard to response to spills and to prevent spills.

The response to the *Exxon Valdez* spill involved the use of bacteria that could use the oil hydrocarbons as food. This strategy had not been tried before, due to environmental concerns over the use of bacteria in a natural setting. The results, especially the application of a fertilizer spray to accelerate the bacterial breakdown of the hydrocarbons, were promising enough to spur the technology to become incorporated as a response option by the EPA, and spawned the new industry of bioremediation.

Remediation of oil spills has become an unfortunate legacy of the world's dependence on fossil fuel. According to the EPA, even with more stringent shipping regulations and improved ship hull design, nearly 14,000 oil spills occur in freshwater and marine coastal waters every year.

SEE ALSO *Chemical Spills; Oil Pollution Acts; Soil Contamination; Water Pollution*

BIBLIOGRAPHY

Books

Owens, Peter. *Oil and Chemical Spills.* New York: Lucent Books, 2003.

Wang, Zhendi, and Scott Stout. *Oil Spill Environmental Forensics: Fingerprinting and Source Identification.* New York: Academic, 2006.

Alaska Department of Environmental Conservation worker shows oil covering his gloves after he dug beneath the surface of a rocky beach on Perry Island, Alaska (February 1990). Almost a year after the worst oil spill in U.S. history when the tanker Exxon Valdez grounded on a reef, much of the oil is now hidden, but not gone. *AP Images*

Web Sites

U.S. Environmental Protection Agency (EPA). "Oil Spills." 2008. http://www.epa.gov/oilspill/ (accessed March 24, 2008).

Brian D. Hoyle

Organic and Locally Grown Foods

Introduction

Organic foods are produced without using antibiotics in animal feed, genetically engineered organisms, chemical preservatives, radiation, or artificial pesticides, herbicides, or fertilizers. Organic farming must also preserve the soil and treat livestock humanely. Local food is less strictly defined. It may or may not be organic, but is always purchased relatively near its place of production (usually, within 250 mi [400 km]). Although once seen as the concern of the eco-fringe, organic and local foods have been changing the U.S., European, and global food markets rapidly over the last two decades. Since the early 1990s, the market for organic food has grown at about 20% per year, a rapid rate in any business. U.S. sales of organics were about $12 billion in 2005, as compared to about $500 billion for the U.S. food industry as a whole. Worldwide, sales of organics were at about $40 billion in 2006. In the United States, about 1% of land was farmed organically as of 2008; in Europe, it was about 4% (counting only cropped fields, not grazing land).

Organic and local foods are generally more expensive than similar foods grown by conventional means. Consumers, therefore, are motivated to buy them because of certain values or convictions, such as the concept that one's food should be produced by sustainable methods, or the assumption that organic foods are healthier.

Critics of organic food argue that there is no scientific evidence that most organics are healthier to eat. However, advocates of organic and local foods hope for several benefits. First, there is the wish to be part of an agricultural system that cares for the long-term well-being of the natural world rather than exploiting it for short-term profit. Conventional large-scale agriculture is one of the most polluting industries and is causing rapid loss of soil in many parts of the world; organic agriculture causes less pollution, slows soil loss, and increases local biodiversity. Second, organic and local-food advocates hope to reinforce local communities and economies, funneling consumer dollars to small farmers—many of whom struggle to survive financially. Third, many shoppers seeking fresher foods choose locally grown foods with less transit time to market. Moreover, there is some scientific evidence that eating organic foods may, in fact, be healthier because such foods contain fewer hazardous chemicals (such as pesticide residues) and more nutrients.

The point of view underlying the organic and local-foods movement might be summed up in a phrase from Kentucky farmer and writer Wendell Berry: "Eating is an agricultural act."

Historical Background and Scientific Foundations

The organic-food movement began in the early twentieth century as a conscious reaction to the trend, then just beginning, toward larger farms and dependence on machinery, artificial fertilizer, and pest-killing chemicals. In chemistry, the term "organic" refers to compounds containing carbon, but in agriculture it refers to the quality of resembling an organism—that is, a living system in which a number of parts cooperate to the benefit of all. The organic movement remained obscure until the 1960s, when environmental concerns began to spread in the popular culture. A 1962 best-selling book by Rachel Carson, *Silent Spring*, criticized the use of DDT and other pesticides as being hazardous to both human beings and wild plants and animals, and helped turn public opinion against pesticides. Throughout the 1970s, individuals wrote books and founded organizations around the world to promote organic farming as a non-destructive, healthier alternative to industrialized agriculture.

WORDS TO KNOW

BIODIVERSITY: Literally, "life diversity": the wide range of plants and animals that exist within any given geographical region.

HERBICIDE: A chemical substance used to destroy or inhibit plant growth.

IONIZING RADIATION: Any electromagnetic or particulate radiation capable of direct or indirect ion production in its passage through matter. In general use: Radiation that can cause tissue damage or death.

ORGANIC FARMING: Farming that uses no artificial chemicals or genetically engineered plants or animals.

PESTICIDE: Substances used to reduce the abundance of pests or any living thing that causes injury or disease to crops.

SUSTAINABILITY: Practices that preserve the balance between human needs and the environment, as well as between current and future human requirements.

Despite receiving almost no government funding and little scientific attention, organic farming became steadily more popular. In 1979, California passed the California Organic Food Act, establishing a legal definition of what could be sold to consumers as "organic" and what could not. In 1990, the U.S. Congress passed the Organic Foods Production Act as part of that year's Farm Bill. The 1990 act required the federal government to commence a National Organic Program that would define and eventually enforce standards for the creation and handling of "organic" products. In 1995, the U.S. National Organic Standards Board (controlling body of the National Organic Program) made official the following definition of organic farming: "Organic agriculture is an ecological production management system that promotes and enhances biodiversity, biological cycles and soil biological activity. It is based on minimal use of off-farm inputs and on management practices that restore, maintain and enhance ecological harmony." U.S. federal rules governing organic products took effect in February 2001, after a decade of often controversial development. The European Union and many other countries have also established legal definitions of the term organic.

Interest in local foods first became widespread and organized in the 1990s. Persons interested in buying local are usually in favor of organic farming as well, but not uncritically so: They may, for example, choose conventionally raised but local raspberries over organic strawberries flown by jet from the far side of the planet. The local-food or "localvore" movement is inspired by an interest in the social and economic results of shopping choices as well as the environmental impact of the industrial food system. In the United States, the average vegetable is transported about 1,500 mi (2,400 km) to the store where it is sold: This entails the burning of large amounts of fuel, which contribute to air pollution and global warming.

Impacts and Issues

Since the early 1990s, there has been vigorous debate about the detailed definition of "organic," especially in U.S. federal law. Industry groups such as the Organic Trade Association, which represents large growers, seek a broader definition that will allow cheaper production and the use of certain artificial ingredients. Consumer groups such as the Organic Consumers Association have resisted any dilution or weakening, as they see it, of the organic concept. For example, in the 1990s, during the development of U.S. federal standards for organics, there was an attempt to allow irradiation of food (sterilization using high levels of ionizing radiation) into the definition of "organic."

In the early 2000s, debate swirled (sometimes in court) around when milk from a cow is organic and when it is not; the National Organic Standards Board was trying to get the U.S. Department of Agriculture to agree that "organic" milk must be produced by cows that are pastured for at least 120 days a year, excluding milk from cows kept in crowded corrals and fed mostly grain. Almost all conventionally produced milk comes from cows that are housed in confined quarters and fed grain-rich diets laced with antibiotics to maximize their milk output; the debate over cow milk thus turned to the question of whether feeding cows organically grown grain and not giving them antibiotics is enough to make their milk "organic." Such debates are ultimately about values, not about facts that can be scientifically examined, but that does not make them any less real.

Debates have simmered not only over the values but the science of organic food. In particular, the questions of whether organic farming really benefits the environment, and whether organic foods are actually healthier to eat, have been increasingly examined. In the early 2000s, evidence began to accumulate that organic farming does indeed benefit the environment and that organic foods may be at least marginally healthier than conventional foods. A study published in *Science* in 2002 reported that organic farms, compared to conventional farms, enhance soil fertility and biodiversity while using less energy to produce each unit of food. A 2006 study found that switching children to organic food reduced levels of organosphosphorus pesticides in their urine to undetectable levels, and that returning them to conventional foods restored urine pesticide levels. A 2008 survey of scientific papers since 2003 found that in the majority of studies, organic foods contained higher levels of 11 nutrients than identical quantities of conventionally grown foods. However, the main thrust of the organic and local-foods movements, independent of the question of enhanced personal health, is stewardship of Earth.

■ Primary Source Connection

The following news article identifies the growing trend of restaurants to become environmentally savvy through such actions as using local meats and vegetables from small local farmers, filtering local water, recycling, composting, and using renewable energy. The Green Restaurant Association is one organization helping to promote the green movement in restaurants by certifying restaurants who take the green initiative in their philosophies of food, materials, and energy.

MORE RESTAURANTS ARE GOING GREEN BY GOING LOCAL

One course at a time, David Siegel consumes five gourmet dishes remarkable for their flavor and also for where the ingredients came from: sardines and sand dabs from Monterey Bay, Calif., squab and veal from the state's central coast, and strawberries from Oxnard, Calif.

"Ordinarily, I would be gun-shy and run the other way when I hear the word 'sardine,'" says Mr. Siegel. But "because they didn't have to preserve it in salt, this had a freshness and nonfishy taste I've never experienced. It was delightful."

The comment is music to the ears of Neal Fraser, chef of the well-known Grace Restaurant here, who designed a "Close to Home" menu where 90 percent of the ingredients are sourced within 400 miles. Advancing a so-called "socially and environmentally responsible" agenda throughout his restaurant—which includes serving filtered local tap water rather than bottled water from afar and fueling his own car with leftover vegetable oil—Mr. Fraser is part of a growing nationwide restaurant movement to go "green." The ideas are not new, say experts, but they are gaining fresh currency because of the burgeoning global environmental movement and new generations of youth with budding enthusiasm for long-established notions of sustainability, ecological health, and food safety.

As exemplified by Grace Restaurant, one key idea is to leave less of a carbon footprint wherever possible—choosing local meats, vegetables, fish, and fruit over those shipped from thousands of miles away. Another push is to support smaller local ranchers and farmers who avoid the kinds of animal diets and pesticides that are typically used for produce and meat and are often served in the nation's 1 million restaurants.

There is a laundry list of other strategies to reduce global warming: from recycling and composting waste to conserving water and lights, using nontoxic cleaners, tapping wind or other "green" power, and designing minimal-impact buildings. Like Grace, many restaurants are moving away from bottled water because of environ-

Buying local has become more important as customers look for food that was not shipped from far parts of the world. *AP Images*

mental concerns about bottle waste, refrigeration needed, transportation costs, and shipping containers.

"I just began to think about the future of the planet that my daughters would be inheriting and their children and so forth," says Fraser, who decided it was time to change after seeing the movie "An Inconvenient Truth," starring Al Gore.

Supporters report more interest by owners and diners than at any time since such notions began coalescing in the late 1960s. "The movement today is really huge and the debate is getting a far broader audience now," says Wynnie Stein, co-owner of Moosewood Restaurant in Ithaca, N.Y., considered one of the national pioneers of locally sourced organic farming. "It's everybody from restaurants to colleges to food-service directors in schools, hospitals. People are very concerned about the environment for themselves and future generations and there is a new urgency to dramatically expand on ideas that have been around for years."

One measure of the interest is the growth of the Green Restaurant Association, which certifies restaurants coast to coast and encourages them to take four new steps to help the environment each year. Founded in 1990 in San Diego, the association has seen the number of its certified restaurants skyrocket from 60 to 300 in the past two years. The group is also negotiating with major restaurant chains, which could rapidly boost membership to 5,000.

"In the last year we have gotten more interest than in the previous 16 years combined," says founder and director Michael Oshman. "It's beginning to build exponentially from interest of previous decades."

Such restaurants are also drawing attention to the plight of smaller farms, ranches, and suppliers whose practices fit the model but are in danger of being lost.

"Government policies are making it very hard for the smaller, independent, family-run businesses, which operate with higher environmental standards," says Mike Antoci, who runs Superior Anhausner Foods, a Los Angeles distributor. "Restaurants like Fraser's are starting to raise public consciousness about what is at stake."

The new spotlight is creating a domino effect, say observers, in which restaurant customers begin to ask more questions about the local-food movement.

"We have seen a dramatic increase in the number of people who value the availability of food produced right in their own community," says Linda Halley, who runs Fairview Gardens, a nonprofit organic farm and education center in Goleta, Calif.

Some observers question some of the claims of the local-food movement. They say it's entirely possible that food grown locally could have a considerably larger carbon footprint than food flown halfway around the world because transportation represents only a tiny fraction—some experts say as little as 2 percent—of the energy required to grow, store, process, and package the food.

Supporters say that the movement is raising questions that society needs to ask. Mr. Oshman notes that Coffee Bean & Tea Leaf, with 200 stores nationwide, was the first large chain to be certified green last year and this year has announced it will run its stores on windpower exclusively. "If larger chains like this can do it, it shows this is not a fringe thing anymore."

Daniel B. Wood

WOOD, DANIEL B. "MORE RESTAURANTS ARE GOING GREEN BY GOING LOCAL." *CHRISTIAN SCIENCE MONITOR* (JULY 30, 2007).

SEE ALSO *Agricultural Practice Impacts; Herbicides; Silent Spring*

BIBLIOGRAPHY

Books

Berry, Wendell. *The Unsettling of America.* San Francisco: Sierra Club Books, 1977.

Periodicals

Lu, Chensheng, et al. "Organic Diets Significantly Lower Children's Dietary Exposure to Organophosphorus Pesticides." *Environmental Health Perspectives* 114 (2006): 260–263.

Mäder, Paul, et al. "Soil Fertility and Biodiversity in Organic Farming." *Science* 296 (2002): 694–1697

Stokstad, Erik. "Organic Farms Reap Many Benefits." *Science.* 296 (2002): 1589.

Warner, Melanie. "What Is Organic? Powerful Players Want a Say." *New York Times* (November 1, 2005).

Web Sites

The Organic Center. "New Evidence Confirms the Nutritional Superiority of Plant-Based Organic Foods." http://www.organic-center.org/science.latest.php?action=view&report_id=126 (accessed March 29, 2008).

Larry Gilman

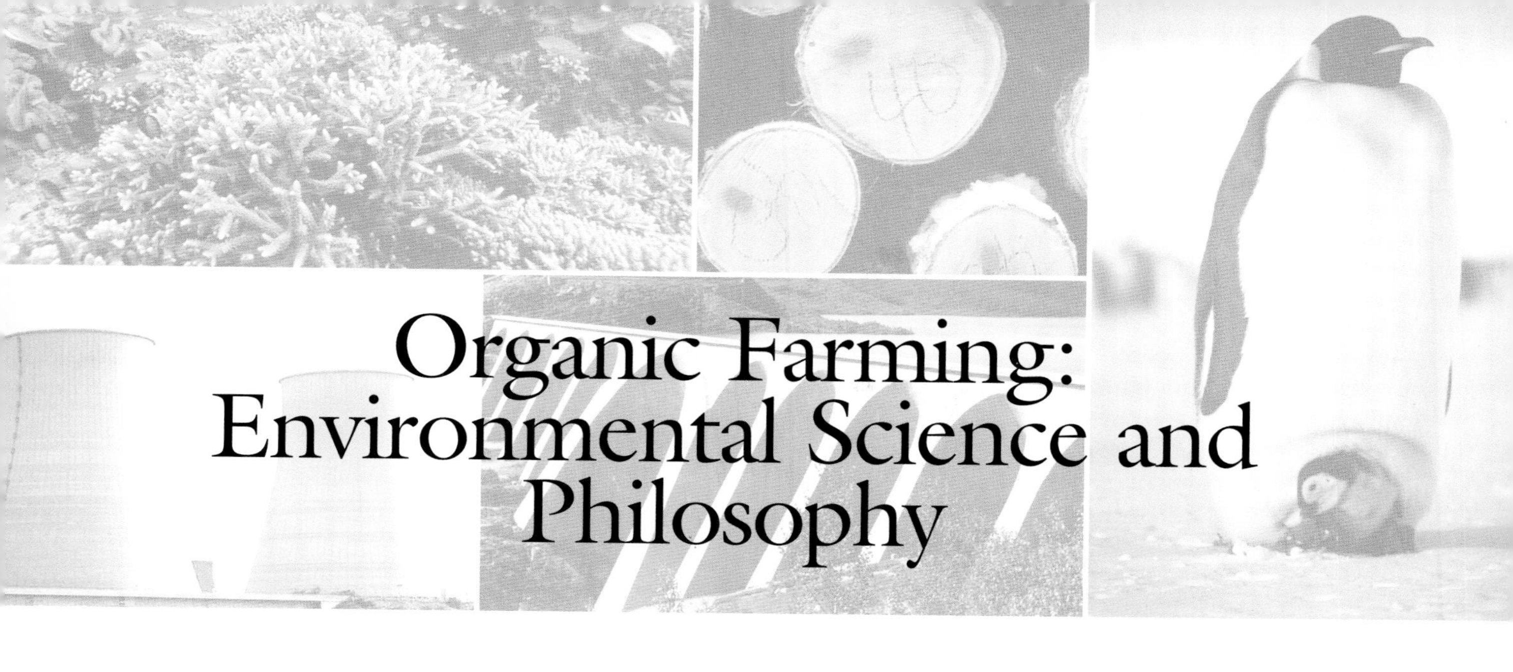

Organic Farming: Environmental Science and Philosophy

Introduction

Organic farming refers to a style of farming that avoids the use of synthetic fertilizers, pesticides, and genetically modified organisms. Beyond this rule, organic farming comprises a wide range of techniques, many of which are as old as farming itself. These include fertilizing with animal waste, composting, hand weeding, mulching, and companion planting. Newer techniques promoted by organic farming include crop-rotation, conservation tillage (absence of or minimal use of mechanical tilling of soil), and new biologically derived pesticides.

The organic movement applies a cohesive philosophy to these techniques and a focus on developing a system of agriculture as an explicit alternative to conventional agriculture. The organic philosophy can be understood as follows: conventional agriculture treats the soil as an inert medium that will reliably transform chemical inputs into agricultural outputs, like a component in a machine. Problems such as nutrient deficiency or insect infestation are treated as endemic pathologies in constant need of treatment by way of various chemicals, and the farm is not seen as a cohesive interconnected system. By contrast, the organic method describes a radically different understanding of soil not as a machine, but as a living entity whose innate fertility can be enhanced through the proper techniques. The farm is understood as a responsive, living system and an effort is made to understand it as a whole rather than in isolated parts.

Historical Background and Scientific Foundations

Organic farming is not new. Although perceived to be a recent trend, the methods of agriculture that would be termed organic have, to a large extent, been in use since humans developed farming thousands of years ago. What is relatively new are the farming methods that have arisen in western society since the beginning of the Industrial Revolution and ballooned in scale after World War II (1939–1945). These include mechanization, large-scale cultivation of a single crop, and the use of synthetic fertilizers, pesticides, and genetically modified organisms. This set of practices has come to be termed conventional agriculture.

The organic movement began from an ideological premise rather than a scientific one. Early thinkers of the organic movement—among them the botanist Albert Howard (1873–1947) and farmer Eve Balfour (1899–1990), both in England, and the Austrian philosopher Rudolph Steiner (1861-1925)—based their ideas on observation, yet they also relied on tradition and anecdotal evidence to support their claims. Howard traveled extensively in India and Asia documenting the efficacy of different agricultural methods throughout the world and used this to argue for the importance of manure and composting in organic farming. Balfour sought to remedy the recognized lack of scientific evidence for organic methods by founding an ongoing agricultural research project called the Haughley Experiment. Steiner, however, whose methods became known as biodynamic agriculture, promoted techniques that bordered on mysticism and could claim no basis in science.

Rigorous scientific experimentation on organic techniques began to develop in the 1940s with the emergence of the Rodale Institute. Founded by the American publisher J. I. Rodale (1898–1971) in Pennsylvania, the institute applied scientific rigor to identifying best practices in organic farming. The institute remains a leader in organic farming research. In addition, programs devoted to researching sustainable agriculture began to develop at universities in the 1990s. Despite institutional participation, exponential growth in the market for organic goods, and increased acres farmed organically, research into organic methods at land-grant universities in the United States totals less than 0.2% of all research acreage

WORDS TO KNOW

COMPOSTING: Breakdown of organic material by microorganisms.

RENEWABLE RESOURCE: Any resource that is renewed or replaced fairly rapidly (on human historical time-scales) by natural or managed processes.

RUNOFF: Water that falls as precipitation and then runs over the surface of the land rather than sinking into the ground.

SUSTAINABILITY: Practices that preserve the balance between human needs and the environment, as we

each year. Thus, institutional research lags well behind the public desire for information.

The organic movement was a reaction, not to a scientific discovery, but to the results of conventional agriculture, observed by its founders in the 1930s. By this time, the use of ammonia-based fertilizers and tractors had become widespread. Albert Howard whose book, *An Agricultural Testament* (1940) is widely considered foundational to the organic movement, argues that soil fertility is the foundation of sustainable agriculture and that conventional farming practices had been disastrous. He notes an increase in plant disease, extensive soil erosion, and a general loss of soil fertility. Lady Eve Balfour also makes reference to the consequences of modern farming in her book, *The Living Soil* (1942). She likewise notes erosion as evidenced by the dramatic Dustbowl phenomena of the 1930s, and she points out the disturbing process of desertification underway in Africa, Asia, and the Americas, whereby semi-arid land becomes desert largely through land mismanagement by humans. These alarming phenomena appeared to be the direct result of poor agricultural practices, making the need for alternatives obvious.

Since those early years, the scale and impact of conventional agriculture have grown. The Green Revolution, an effort begun in the 1940s and continuing today, sought to increase global capacity for food production through the expanded use of agrochemicals, mechanization, irrigation, and new plant varieties.

Impacts and Issues

Mounting concerns about climate change, environmental degradation, and food safety have fueled increased interest in organic products and farming methods. The scientific community is also responding to organic farming as a possible solution to environmental problems with increased research into its purported benefits.

The claims of environmental disaster propounded by Howard and Balfour have now been confirmed and added to by experts citing evidence that conventional agriculture is responsible for soil run-off, waterway contamination (including a phenomenon known as the Gulf Dead Zone in which fertilizer run-off leads to water with so little oxygen that fish cannot survive), soil erosion, increased soil salinity, loss of genetic diversity, and an unsustainable reliance on fossil fuel inputs. These claims

The executive director of the Washington Dairy Federation believes more feed crops need to be planted to meet local demand among organic farms. The fields behind the cattle are planted with corn, sunflower, canola, and safflower to feed the dairy cows. *AP Images*

Police clash with demonstrators as thousands of environmental activists, wildlife conservationists and students took to the streets of Genoa, Italy (May 2000) to march toward the site of an international biotechnology conference to protest against genetically modified foods. *AP Images*

are countered by claims that conventional agriculture is the only method capable of successfully feeding the burgeoning human population. The debate is ongoing.

Organic practitioners suggest that simply eschewing inorganic chemicals in agriculture is not enough to correct the depletion of soil fertility and other problems caused by conventional methods. Soil health must be maintained through practices that promote the production of humus—the organic material that provides structure, nutrients, and a medium for the population of living organisms within soil. Humus is a mixture of partially decomposed plant and animal matter in addition to the organisms that break it down into nutrients usable to plants. Humus acts to bind soil particles into aggregates to improve airflow and drainage through soil. At the same time, it acts as a sponge, retaining water needed for plants. Without it, soil structure degrades and is more vulnerable to erosion, drought conditions, and increased run-off of pollutants, and it can no longer support life. Conventional synthetic fertilizers provide nutrients to plants in the form of salts that dissolve into the soil; however they contribute nothing to soil structure. Organic fertilizing techniques—including composting, manuring, and mulching—are designed specifically to add nutrients to the soil in a way that also builds humus. Emerging science will shed light on whether organic agriculture's focus on soil health and avoidance of synthetic chemicals can indeed solve current environmental problems and prove to be a sustainable, viable alternative to conventional agriculture.

Since 1980, organic farming has come under the umbrella of the larger concept of sustainable agriculture. Sustainability is understood as a broader concept than organic, referring to practices that use resources in a way that can be sustained in the future; for example, through the use of a renewable resource rather than nonrenewable fossil fuels. Sustainable agriculture does not prohibit specific practices as organic agriculture does; however, it maintains the understanding of the farm as an interconnected system and in turn, the farm as part of the larger global ecology.

SEE ALSO *Corporate Green Movement; Organic and Locally Grown Foods*

BIBLIOGRAPHY

Books

Balfour, Eve. *The Living Soil.* London: Faber and Faber, 1948.

Periodicals

Trewaves, Anthony. "Urban Myths of Organic Farming." *Nature* 410: 409–410.

Web Sites

University of California Berkley: College of Natural Resources. "Can Organic Farming Feed the World?" http://www.cnr.berkeley.edu/˜christos/articles/cv_organic_farming.html (accessed April 1, 2008).

Angela Fedor

Overfishing

Introduction

Overfishing refers to a situation in which the fishing industry removes more fish than can be replaced by reproduction of the fish left in the ecosystem. Due to improvements in technology, fishing yields have grown dramatically over the last 50 years. The quantity of fish removed from the ocean by fishing probably approaches the maximum sustainable yield of the ocean. There are several examples of fisheries that have been decimated by overfishing, such as the cod fishery in the North Atlantic and the New Zealand orange roughy industry.

One of the major problems associated with the fishing industry is the high percentage of bycatch that is removed from the ocean during fishing practices. Advances in fishing technology and pressure by nonprofit organizations are helping to improve fisheries management to prevent overfishing.

Historical Background and Scientific Foundations

Fish represent about 16% of the protein in the human diet and this percentage is much greater in developing countries. According to the United Nation's Food and Agricultural Organization (FAO), in Asia, 30% of the animal protein in people's diet comes from fish; in Africa about 20% comes from fish; and in Latin America fish contribute about 8% of the animal protein to people's diets. More than 100 million people depend on the fishing industry for employment in developing countries alone.

Approximately 90% of all fisheries' yield are marine fish. The most common types of commercial fish include small pelagic species like anchovies, anchovetta, sardines, and herring. Larger open ocean predators like the jacks, tunas, billfish, bonito, redfish, mackerels, and basses are also important to commercial fisheries. Other fished species include swimming invertebrates such as krill, shrimp, crab, lobster, oyster, scallop, octopus, and squid. Marine algae including *Ulva, Poryphyra* and *Laminaria* account for roughly 3% of the marine fisheries.

The quantity of fish caught by fishermen has increased rapidly over the last five decades. Since 1950 the annual fish yield increased from 21 million tons to around 90 million tons. The Peruvian anchoveta is the most caught marine species, with a yield of around 10 million tons. China catches the most fish worldwide, with an annual yield approaching 17 million tons.

The great growth in the fishing yield is a result of several factors. Most fish are caught by commercial fishermen from large fleets of ships. The technology employed by these fishing businesses is vast, including satellite sensors, sonar tracking, scouting ships, and spotting airplanes. Enormous factory ships act as entire processing facilities with spaces for cleaning, canning, and freezing all on board. These great fleets of ships have been able to exploit fishing grounds that were largely inaccessible in the past, particularly in the Southern Hemisphere.

Even with the immense fishing effort, the yield of fish taken from the ocean has leveled off, and even declined since the 1980s. Fisheries experts estimate that the maximum sustainable yield that can be removed from the oceans is somewhere near 110 to 150 million tons. The current yield is extremely close to the lower end of this estimate. When so many fish are taken from the ocean that the remaining fish cannot replenish those removed by breeding, the condition is known as overfishing.

In 2006, the FAO's report on the state of the world's fisheries showed that almost 80% of the fisheries were threatened by overfishing. More than 25% of the world's fish stocks were overexploited or depleted. More than half of the total fish stocks were exploited. The report estimates that at least 90% of the world's largest predatory fish, like the oldest tuna and swordfish, are no longer in existence.

One of the most dramatic examples of overfishing occurred in Newfoundland, Canada. The cod fishery had thrived in the region for hundreds of years. However, years of overfishing driven by poor management practices depleted the standing stocks so greatly that the entire fishery collapsed in one year. In 1992, more than 35,000 people lost their jobs when the cod did not return at the beginning of the fishing season. The fishery, and the region's economy, are still in a state of disrepair.

Other examples of similar fisheries crashes as a result of overfishing exist. Orange roughy from waters near New Zealand became extremely popular in the 1980's for its mild, delicate flavor. The fish grows extremely slowly and can live as old as 100 years. It is not able to reproduce until it is 25 to 30 years old. As fishing pressure removed the older, larger fish, the remaining orange roughy were too young to reproduce. After only 13 years of fishing for orange roughy, the species was nearly extinct.

Impacts and Issues

Management of the fisheries industries has the goal of preventing overfishing. The objectives are to maximize the yield of fish, while at the same time maximizing the economic yield. This has led to poor decisions that deplete standing stocks of fish and disrupt ecosystems that sustain commercial fish populations. When fish catch drops, management practices often increase the number of fishing licenses and the number of boats allowed in the fishing fleet. Some organizations estimate that there are currently 2 to 3 times the number of vessels fishing the oceans than are required to catch the maximum yield. In order to support these fishing efforts the governments of the United States, Japan, the European Union, and Russia all subsidize fishing fleets, which significantly exacerbates overfishing.

One of the most notable problems contributing to the depletion of fisheries stocks in the ocean a known as bycatch or bykill. Bycatch refers to the collection of non-target species by fishermen. Most often bycatch is removed from the catch and thrown overboard back into the ocean, where, presumably, it dies. Bycatch has enormous ramifications. Approximately 20 million tons of fish food are lost to bycatch per year. It is estimated that the tuna fisheries kill one million sharks a year. In the shrimp industry, approximately 80% of the catch is bycatch. Bycatch is not restricted to fish; marine mammals, birds, and turtles are all killed as bycatch.

Advances in fishing technology can significantly reduce bycatch. Circle hooks are specially designed to catch the species of fish that fishermen can sell in markets. Specially shaped nets exclude species that will otherwise become bycatch. Harpooning large species with spot planes, rather than using long lines or drift nets, also significantly decreases bycatch. Finally, producing fish in aquaculture farms brings fish to market, without creating bycatch.

Government oversight of fishing is often inconsistent and poorly regulated. There are few international regulations that control fishing in the open ocean. Exclusive Economic Zones (EEZs) are waters within 200 mi (321 km) of a country's coastline. Within these strips of ocean, governments can regulate fishing. However 64% of the ocean falls outside these regions and international laws are few and poorly enforced.

Since the fishing industry is market driven, some non-profit organizations have begun publishing seafood guides to urge consumers to make careful choices about the types of fish that they eat. By discouraging consumption of fish that are caught with significant bycatch or in poorly managed fisheries, these organizations hope to influence the industry and its managers to make better decisions about sustaining fisheries. In particular, Monterey Bay Aquarium has developed a Seafood Watch program that helps businesses and consumers make choices for healthy oceans. One of the major successes has been the publication and distribution of pocket guides recommending consumers buy sustainable species of fish when they are shopping.

Other groups have put pressure on fish sellers to identify the source of the seafood they sell as well as the methods used to harvest it. For example, the grocery chain Whole Foods Market is committed to only selling fish from well-managed fisheries. The Marine Stewardship Council (MSC) is a British organization that identifies and certifies sustainable fisheries.

SEE ALSO *Aquatic Ecosystems; Commercial Fisheries; Endangered Species; Fish Farming; Hunting Practices; Inland Fisheries; Marine Ecosystems; Marine Fisheries; Reef Ecosystems*

WORDS TO KNOW

AQUACULTURE: The farming of fish or shellfish in freshwater or saltwater.

BYCATCH: Non-target species killed in the process of fishing.

SUSTAINABILITY: Practices that preserve the balance between human needs and the environment, as well as between current and future human requirements.

Two workmen cover up three dead dolphins which had been left outside the Department of the Environment Food and Rural Affairs in London, in March 2005. Greenpeace deposited the three dolphins to protest pair trawling for sea bass in the English Channel, which uses large nets and is thought to be responsible for killing thousands of dolphins every year. *AP Images*

BIBLIOGRAPHY

Books

Clover, Charles. *The End of the Line: How Overfishing Is Changing the World and What We Eat.* Berkeley: University of California Press, 2008.

Ellis, Richard. *The Empty Ocean.* Washington DC: Island Press, 2004.

Garrison, Tom. *Oceanography: An Invitation to Marine Science*, 5th ed. Stamford, CT: Thompson/Brooks Cole, 2004.

Web Sites

Food and Agriculture Organization of the United Nations. "Fisheries and Aquaculture Department." 2008. http://www.fao.org/fishery/ (accessed March 24, 2008).

Food and Agriculture Organization of the United Nations. "The State of World Fisheries and Aquaculture." 2004. http://www.fao.org/docrep/007/y5600e/y5600e00.htm (accessed March 24, 2008).

MarionBio.org. "Sustainable Fisheries." February 17, 2008. http://marinebio.org/ (accessed March 24, 2008).

Overfishing.org. "Overfishing—A Global Disaster." 2007. http://overfishing.org/ (accessed March 24, 2008).

United Nations. "Overfishing: A Threat to Marine Biodiversity." http://www.un.org/events/tenstories_2006/story.asp?storyID=800 (accessed March 24, 2008).

World Wildlife Fund. "Problems: Poorly Managed Fishing." February 29, 2008. http://www.panda.org/about_wwf/what_we_do/marine/problems/problems_fishing/index.cfm (accessed March 24, 2008).

Juli Berwald

Overgrazing

Introduction

Overgrazing is the practice of allowing more animals to feed on a given piece of land for a longer time than it is able to support. Overgrazing can subtly damage the environment by altering the kinds of plants able to grow in a certain area, or more conspicuously by destroying most plants and leaving bare ground. Many kinds of animals are responsible for overgrazing in different parts of the world. In the United States, cattle cause much of the damage by overgrazing. Goats, sheep, horses, and even yaks are responsible for the majority of damage in some other countries. Although overgrazing is often associated with domesticated animals, uncontrolled populations of wild animals, including deer, can cause similar problems.

Ranchers and herders can minimize or eliminate the problem of overgrazing by careful management of their herds. Moving animals to new pasture before the vegetation is cropped too short is one important strategy. Allowing sufficient recovery time before returning animals to an already grazed pasture is also vital. Even one animal, if kept in a single pasture for long periods of time, will overgraze the plants by eating the newest growth first and selecting only the plants that it likes best.

Historical Background and Scientific Foundations

Herding animals, particularly sheep, is a very ancient occupation in Europe, Asia minor, and Africa. Sheep provided meat, milk, and skins for early farmers; this versatility made them a valuable supplement to agricultural subsistence. The value placed on sheep in the ancient world can be seen in Homer's *Odyssey*, in which the giant Cyclops tenderly cares for his flocks. With the advent of spinning, sheep became the major source of fiber for clothing. Later, other animals such as cattle and goats were also domesticated. When relatively small numbers of animals were able to roam on large areas of land, overgrazing was not a problem. However, as more people and animals started to compete for finite grazing resources, additional stress was placed on the ecosystem, leading to environmental damage.

Overgrazing results when too many animals are kept on too little land for the plants to recover sufficiently before being grazed again. Because most grazing animals are selective about the plants they eat, the desirable, nutritious plants are usually consumed first, leaving less-appetizing plants to grow to maturity and disperse their seeds. If allowed to continue, this eventually results in the less-desirable plants displacing the desirable ones, leaving the land unsuitable for grazing. This problem is especially serious if the unwanted plants are thorny or poisonous.

Furthermore, if animals are allowed to graze on most grasses for too long before being moved, they can damage the root growth of the plants. Just as when a lawn is mowed too short, the plants cannot generate enough energy through photosynthesis to grow. If its roots are too small during periods of drought, the plant is likely to die because it cannot access enough water. In this way, overgrazing can lead to the death of all the plants in a particular area, leaving the ground too dry and sterile to support new growth.

Impacts and Issues

Overgrazing is a serious problem because it harms both the environment and the animals it is intended to benefit. Plants that have been overgrazed are less healthy because their carbohydrate reserves have been depleted. Animals who are fed on this less-nutritious fodder are not as healthy as their well-fed counterparts, are more

WORDS TO KNOW

FODDER: Food for grazing animals.

GRAZING CAPACITY: The number of animals that a given area of land can support.

PASTURE: Low-growing plants suitable for grazing livestock or land containing such plants. Some pastureland is fenced, while some is part of the open range.

vulnerable to severe winter weather, produce weaker offspring, and calve less often. All of these factors increase costs and decrease income for herders and ranchers, many of whom lack the financial reserves to weather hardship.

Among the environmental costs of overgrazing is the destructive way in which it alters the ecosystem of grasslands. By eating certain plants and not others, the balanced composition of the turf is altered. Some grazing animals will pull up plants by the roots if forced to feed in overgrazed pastures. Plant roots are important soil stabilizers and removing them can lead to erosion of the most fertile topsoil. Erosion can also occur after overgrazed plants die off during drought. In extreme cases, overgrazing can lead to desertification, the process by which an ecosystem is transformed into a desert. In China, geological studies have determined that overgrazing on the steppes directly led to the burial of fertile soils under desert sands about 300 years ago, turning formerly productive land into unusable desert. It is also possible to directly observe the difference in climate between well-managed and overgrazed pasture land. For example, the average ground temperature in the overgrazed Mexican Sonora desert is higher than the ground temperature in adjacent areas of the United States by about 7°F (4°C), leading to increased air temperatures.

Access to pasture land and disagreement over land use have been the subject of heated litigation in the United States. In many western states, ranchers pay for access to federal land on which they graze their herds. Although some federal land is closed to grazing to protect the ecosystem, some ranchers assert that they have the right to use the land despite the government's prohibitions. Furthermore, environmental groups have sued the federal government, alleging environmental damage from overgrazing. In some instances, they have succeeded in halting grazing on federal land, further antagonizing the relationship between ranchers, regulatory bodies, and environmentalists. Although all these groups agree that overgrazing is a dangerous problem, they differ on the definition of overgrazing. Ranchers become concerned that overly strict definitions of overgrazing will damage their livelihoods, while environmentalists worry that irreparable damage is being done by lax management of overgrazing.

Often the problems associated with overgrazing are particularly difficult to solve because they are the product of poverty and economic necessity rather than of

Deforestation, overgrazing and harmful irrigation practices are transforming vast areas of the world's poorest continent into virtual wasteland, such as this fruit farm in the Ivory Coast. *AP Images*

IN CONTEXT: DEADLY CONSEQUENCES OF OVERGRAZING

A 7.6-magnitude earthquake in Kashmir, Pakistan, on October 8, 2005 killed at least 79,000 people. Another 65,000 people were reported to have been injured and more than 32,000 buildings collapsed. Although substantial, the Kashmir earthquake produced waves only 1/100 as large and released energy only 1/1,000 as great as the 1964 Alaskan earthquake (magnitude 9.2). Each increment of magnitude, for example from 7 to 8, corresponds to a ten-fold increase in seismic wave height and a thirty-two-fold increase in the amount of energy released. According to statistics compiled by the U.S. Geological Survey, there are on average fourteen earthquakes of magnitude 7.0 to 7.9 and one of magnitude 8.0 to 8.9 somewhere in the world each year. Thus, the Kashmir earthquake was a large but not unprecedented event.

Most of the deaths and injuries associated with the 2005 Kashmir earthquake occurred when buildings collapsed and buried their inhabitants. There were many reports of people swept away or buried by landslides during the days immediately after the earthquake. In other cases, landslides closed roads and hindered rescue efforts in the remote and mountainous region.

Analysts assigned much of the blame for the post-earthquake landslides to deforestation and to overgrazing on steep hills and mountainsides. Both deforestation and overgrazing are chronic problems throughout the Himalayan region.

simple mismanagement. Worldwide, the demand for meat has increased dramatically in recent years and is forecast to double by 2050. Herders, many of whom are quite poor, are raising more animals to maximize their benefit in the booming market. In fact, by overgrazing the land, they are raising weaker animals; the loss from overgrazed lands amounts to billions of dollars annually. In many places, particularly in Asia and Africa, herders are exceeding the grazing capacity of the land by 50% or more. Mongolia and several western Chinese provinces have seen severe overgrazing in an attempt to feed China's economic boom and subsequent demand for meat. The effects of overgrazing can already be seen in the more than 3,950 square mi (2,000 square km) of Chinese land that becomes desert every year.

In the United States, ranchers and agricultural academics have made significant progress toward minimizing overgrazing. Research into the best grazing practices for different types of animals and climates is continuing, and some simple guidelines for pasture rotation have been developed. Most overgrazing in the United States is typically confined to small areas, but overgrazing on Native American reservations is a particular problem.

Around the world, the extent of overgrazing is much greater, with more people relying on the scarce resources provided by their stressed land and herds. Good management practices—such as supplementing animal fodder with agricultural waste like corn stalks—and an understanding of appropriate grazing rotation will help minimize the damage of overgrazing. However, as long as animal numbers remain so high above the land's capacity for support, the problems associated with overgrazing will continue.

BIBLIOGRAPHY

Periodicals

Rayburn, Ed. "Overgrazing Can Hurt Environment, Your Pocketbook." *West Virginia Farm Bureau News* (November, 2000).

Web Sites

BEEF Magazine. "What Is Overgrazing?" May 1, 1999. http://beefmagazine.com/mag/beef_overgrazing (accessed April 21, 2008).

Environment News Service. "U.S. Sues Property Rights Ranchers over Grazing on Federal Lands." August 30, 2007. http://www.ens-newswire.com/ens/aug2007/2007-08-30-092.asp (accessed April 21, 2008).

Inter Press Service News. "Researchers Highlight Overgrazing." http://ipsnews.net/fao_magazine/environment.shtml (accessed April 21, 2008).

University of Arizona. "Anthropogenic Desertification vs. 'Natural' Climate Trends." August 10, 1997. http://ag.arizona.edu/˜lmilich/desclim.html (accessed April 21, 2008).

Kenneth T. LaPensee

Ozone Hole

Introduction

The ozone hole refers to the thinning (but not the complete absence) of the ozone layer of Earth's atmosphere that occurs over the continent of Antarctica. The thinning occurs mainly during the winter.

Ozone absorbs the incoming ultraviolet (UV) portion of sunlight, reducing the amount of this light that reaches Earth's surface. This natural buffering of sunlight is beneficial, as the high energy UV light waves are able to penetrate into the uppermost layers of skin. The result can be the cell damage that is evident as sunburn and, more ominously, the destruction of portions of the genetic material (deoxyribonucleic acid or DNA) that resides inside every cell. Some DNA damage in skin cells is associated with the development of certain cancers.

The winter human population of the Antarctic consists only of several hundred researchers. The warm winter clothing protects them from the increased UV light, and so the concerns over the health effects of the ozone hole are minimal. However, the existence of the hole, and its increase in size through the 1990s and up until 2006, when the hole was the largest ever recorded, has become a concern for those studying the influence of human activities on the atmosphere, as the basis of ozone destruction is the release of human-generated compounds into the atmosphere. In addition, a similar thinning elsewhere on the globe could have much more serious consequences on climate and human health.

As of early 2008, the size of the ozone hole was smaller than in 2006, as was the measured loss of ozone, according to data from the U.S. National Oceanic and Atmospheric Administration (NOAA). Whether this is the beginning of a longer-term trend that reflects the reduction of the release of ozone-destroying compounds—agreed to by the 191 nations that signed the Montreal Protocol on Substances That Deplete the Ozone Layer—is not yet clear.

Historical Background and Scientific Foundations

Beginning in 1957–1958, which was designated the International Geophysical Year, scientists began to annually record atmospheric measurements, including the concentration of ozone. At first these measurement were taken from instruments placed in weather balloons. With the satellite era, more accurate measurements could be obtained from instruments while in an orbit around Earth.

Through the 1960s and until the late 1970s, ozone levels in the South Pole region were consistently higher in the late spring than in the wintertime. But, in 1978 and 1979, the ozone levels were less at the end of winter than ever before. For the next several years, the late winter ozone level continued to decline.

British researchers reported these findings in 1984. The following year, the U.S. satellite Nimbus-7 confirmed these findings and produced an image of the thinned Antarctic ozone layer. The term ozone hole was coined to describe the phenomenon.

Ozone is a gas that is composed of three oxygen atoms; its chemical formula is O_3. The form of oxygen that we breathe consists of a pair of atoms (O_2). Although the O_2 form of oxygen makes up about 21% of the volume of the atmosphere, ozone occupies only 0.000004% of the atmosphere's volume, which, if it was present as a discrete layer, would be only about one-eighth of an inch thick.

Ozone is not dispersed all through the atmosphere. Rather, it is confined to the uppermost layer of the atmosphere called the stratosphere, which begins 6 to 12 mi (10 to 19 km) above sea level and extends to nearly 30 mi (48 km) high. The stratosphere is not where clouds normally form and where there is a lot of air movement (these occur in a lower layer called the troposphere). This is advantageous, as the ozone that is concentrated in the lower portion of the stratosphere remains relatively undisturbed. The stratosphere's thickness is not uniform; rather, it tends to be thinnest above the equator and thickest at both poles.

Ozone forms when the UV portion of sunlight splits an O_2 molecule into two oxygen (O) atoms; one of these subsequently forms a chemical bond with another O_2 molecule to generate O_3. Even though UV light is required to form ozone, once the molecule has formed, it is capable of absorbing incoming UV light. This reduces the amount of UV radiation that passes through the atmosphere to the surface.

The basis for ozone destruction are chemicals called chlorofluorocarbons (CFCs), hydrochlororfluorocarbons (HCFCs, which are very similar in structure to CFCs), bromine-containing hydrocarbons, and nitrogen oxide. CFCs were once widely used as a coolant in air conditioners and refrigerators, and to propel gas out of aerosol cans. Ironically, a main reason for their use was the belief that they were chemically non-reactive, and so safe to use around people. Nitrogen oxide is a by-product of the burning of fuel and is released to the atmosphere in aircraft exhaust and other sources.

As the use of CFCs became more popular, their escape into the atmosphere accelerated. Subsequently, it was discovered that CFCs can persist for up to 100 years in the stratosphere, and that during that time incoming UV light can break the CFC molecule apart. This releases the chlorine component of CFC, which then can destroy ozone.

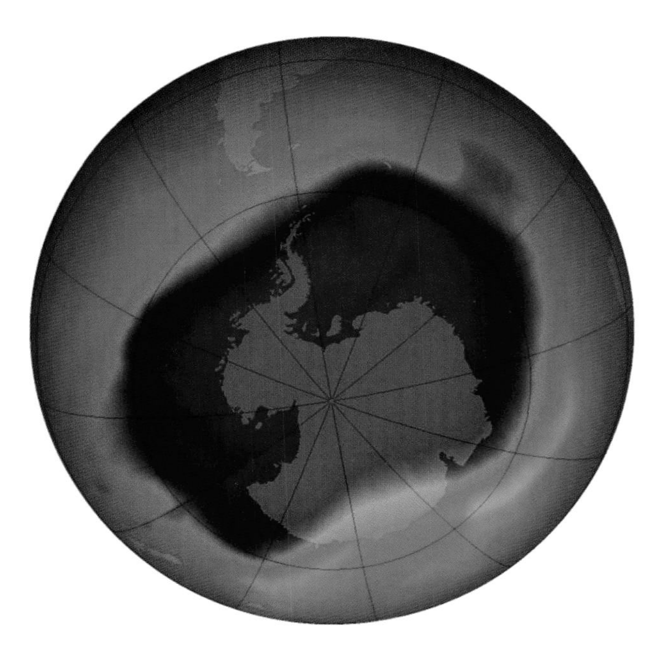

In this image, from September 24, 2006, the Antarctic ozone hole was equal to the record single-day largest area of 11.4 million square miles, reached in September 2000. The blue and purple colors are where there is the least ozone, and the greens, yellows, and reds are where there is more ozone. *AP Images*

WORDS TO KNOW

ATMOSPHERE: The air surrounding Earth, described as a series of layers of different characteristics. The atmosphere, composed mainly of nitrogen and oxygen with traces of carbon dioxide, water vapor, and other gases, acts as a buffer between Earth and the sun.

CLIMATE MODEL: A quantitative method of simulating the interactions of the atmosphere, oceans, land surface, and ice. Models can range from relatively simple to quite comprehensive.

GREENHOUSE GAS: A gas whose accumulation in the atmosphere increases heat retention.

PRIMARY POLLUTANT: Any pollutant released directly from a source to the atmosphere.

The ozone hole would not form if ozone had not been depleted from the stratosphere. The hole forms over Antarctica in the winter months. During this time of the year, strong winds blow around the continent, in effect cutting off the air of Antarctica from the air elsewhere. The ozone-destroying CFCs that are dispersed all through the stratosphere become more concentrated in this trapped region of the atmosphere.

As well, clouds known as Polar Stratospheric Clouds form. Clouds do not usually form in this layer of the atmosphere and, when they do, they concentrate the ozone-destroying compounds still further. The result is an accelerated breakdown of ozone, which produces the ozone thinning over the continent.

As spring returns to the Antarctic and the wind pattern shifts, the atmospheric conditions that triggered the ozone hole disappear, as does the ozone hole. Essentially, the redistribution of ozone restores the ozone to a level that is similar to other parts of the stratosphere.

The reduced amount of ozone in the ozone hole means that less UV light is absorbed. This form of sunlight has enough energy to be able to penetrate into the top-most layers of the skin, and is capable of slicing apart the two strands of DNA that make up the double helix of genetic material inside cells as varied as those of humans, other animals, and microorganisms. When DNA is broken, cell repair mechanisms may be able to restore the structure with minimal effect on the cell. But, damage can be too major to repair; if this damage occurs in regions of the DNA that are critical to the regulation of cell growth and division, the result can be the uncontrolled cell growth that is the hallmark of cancer.

Increased exposure to UV light has been linked to increased rates of skin and other cancers and eye damage in humans. As well, DNA damage has reduced the numbers of microorganisms called phytoplankton in the ocean. Since other marine species that feed on phytoplankton are eaten by additional species that, in turn, are eaten by still other species, a change in the base of this food chain is a serious disruption.

Impacts and Issues

The first report of the thinned ozone layer over Antarctica in 1984 prompted yearly satellite re-examinations of the area as well as the remainder of the atmosphere. While a similar thinning has not so far been detected anywhere else, especially over the other pole, the general depletion of ozone from the stratosphere has been confirmed. Given the sparse number of people in the Antarctic, the ozone hole has not been a health concern. However, the same would not be true if a similar thinning of the ozone occurred over the Arctic. An Arctic ozone hole of the same size as the Antarctic version would expose over 700 million people, plants, and wildlife in the upper Northern Hemisphere to UV levels that have been linked to the development of cancer and eye damage.

From 1985 through 2006, the ozone hole varied in size. Measurements conducted in 2006 revealed that the hole was as big as had been recorded, at 10.6 million square mi (27.5 million square km). This prompted much concern. However, the latest measurements taken in 2007 found that the hole had shrunk by about 30% over the previous year.

Whether this shrinkage will continue is unclear. The atmospheric temperature is an important factor in determining the extent of ozone thinning. A warmer atmosphere tends to diminish the thinning. So, in this sense, global warming—the increasing warming of the atmosphere that is related to human activities—may reduce the severity of the ozone hole.

In addition, as the release of CFCs and other ozone-depleting compounds into the atmosphere has been curbed following the implementation of the Montreal Protocol by the 191 participating nations, ozone depletion could eventually ease. However, given the long lifetime of atmospheric CFCs, the depletion may continue through the 21st century.

SEE ALSO *Air Pollution; Carbon Dioxide (CO_2) Emissions; Greenhouse Effect*

BIBLIOGRAPHY

Books

DiMento, Joseph F. C., and Pamela M. Doughman. *Climate Change: What It Means for Us, Our Children, and Our Grandchildren.* Boston: MIT Press, 2007.

Gore, Al. *An Inconvenient Truth: The Planetary Emergency of Global Warming and What We Can Do About It.* New York: Rodale Books, 2006.

Seinfeld, John H., and Spyros N. Pandis. *Atmospheric Chemistry and Physics: From Air Pollution to Climate Change.* New York: Wiley Interscience, 2006.

Web Sites

NASA. "Ozone Hole Watch." December 4, 2007. http://ozonewatch.gsfc.nasa.gov/ (accessed April 7, 2008).

Brian D. Hoyle

Ozone Layer

Introduction

The ozone layer refers to ozone—a gas composed of three oxygen atoms—that resides in the stratosphere, which is the layer of Earth's atmosphere between about 6 and 30 mi (10 and 48 km) above the surface of Earth. Over 90% of the total atmospheric ozone is found in the stratosphere.

The atmospheric ozone layer differs from the layer of ozone that can accumulate near the ground in areas with heavy air pollution arising from vehicle emissions. Ground-level ozone is undesirable because it can be irritating to the lungs when breathed. People with chronic respiratory ailments are particularly prone to discomfort and illness from ground-level ozone.

In contrast, the atmospheric ozone layer is beneficial, as it provides a vital shield for protecting Earth's surface from harmful levels of ultraviolet (UV) radiation emitted from the sun.

Human activities have been responsible for the emission of a number of gases into the atmosphere that participate in the destruction of ozone. As a result, the amount of ultraviolet radiation reaching Earth's surface has gradually increased throughout the twentieth century and into the twenty-first century. With recognition of the ozone depletion—graphically evident as a periodic and seasonal thinning of ozone over the Antarctic, which has been dubbed the ozone hole—efforts to diminish ozone depletion began in the 1980s.

Historical Background and Scientific Foundations

Ozone is a gas composed of three oxygen atoms (O_3). The form of oxygen that we breathe consists of a pair of atoms (O_2). O_2 comprises about 21% of the volume of the atmosphere, while ozone occupies much less volume (0.000004%). If all the atmospheric ozone was gathered together, the layer that encircled Earth would be only about one-eighth of an inch thick.

Most of the ozone in the atmosphere is not dispersed all through the atmosphere. Rather, it is confined to the uppermost layer of the atmosphere called the stratosphere, which begins 6 to 12 mi (10 to 19 km) above sea level and extends to nearly 30 mi (48 km) high, and tends to be thinnest above the equator and thickest at both poles. The stratosphere is not where clouds normally form and where there is a lot of air movement (these occur in a lower layer called the troposphere). This is an advantage, as the ozone remains relatively undisturbed.

Stratospheric ozone is formed and consumed naturally by photochemical reactions involving ultraviolet (UV) radiation. At any time, the formation and consumption of ozone proceeds simultaneously. The concentration of ozone in the stratosphere naturally varies with latitude and with time. Rates of ozone formation are largest over the equatorial regions of Earth because solar radiation is most intense over those latitudes. However, stratospheric winds carry tropical ozone to polar latitudes, where it tends to accumulate.

Ozone forms when the UV portion of sunlight splits an oxygen (O_2) molecule into two oxygen (O) atoms; one of these subsequently forms a chemical bond with another O_2 molecule to generate O_3. Even though UV light is required to form ozone, once the molecule has formed, it is capable of absorbing incoming UV light. This reduces the amount of UV radiation that passes through the atmosphere to Earth's surface.

Reduction in UV ration at Earth's surface is beneficial, as the high-energy UV waves are able to penetrate into the upper layers of skin. When they contact the genetic material of skin cells as well as other cells, the energy is sufficient to break one or both of the double strands of the cell's deoxyribonucleic acid (DNA).

WORDS TO KNOW

ATMOSPHERE: The air surrounding Earth, described as a series of layers of different characteristics. The atmosphere, composed mainly of nitrogen and oxygen with traces of carbon dioxide, water vapor, and other gases, acts as a buffer between Earth and the sun.

CLIMATE MODELS: Mathematical representations of climate processes. Climate models are computer programs that describe the structure of Earth's land, ocean, atmospheric, and biological systems and the laws of nature that govern the behavior of those systems. Detail and accuracy of models are limited by scientific understanding of the climate system and by computer power. Climate models are essential to understanding paleoclimate, present-day climate, and future climate.

GREENHOUSE GAS: A gas whose accumulation in the atmosphere increases heat retention.

PRIMARY POLLUTANT: Any pollutant released directly from a source to the atmosphere.

Sometimes the damage can be repaired by the cell. But sometimes the damage cannot be repaired, causing cell death, or is enough to cause the cell to function differently than before. The uncontrolled cell growth and division that is the hallmark of cancer can be a result of radiation-induced genetic change.

The depletion of the ozone layer was first observed in the mid-1980s. The basis for ozone destruction includes carbon, chlorine, and fluorine-containing compounds called chlorofluorocarbons (CFCs), hydrochlororfluorocarbons (which are very similar in structure to CFCs), bromine-containing hydrocarbons, and nitrogen oxide. CFCs were once widely used as a coolant in air conditioners and refrigerators, and to propel gas out of aerosol cans. Ironically, a main reason for their use was the assumption that they were chemically non-reactive, and so safe to use around people. Nitrogen oxide is a by-product of the burning of fuel and is released to the atmosphere in aircraft exhaust and other sources.

As the use of CFCs became more popular, their escape into the atmosphere accelerated. Subsequently, it was discovered that CFCs can persist in the stratosphere for up to 100 years, and that during that time incoming UV light can break the CFC molecule apart. This releases the chlorine component of CFC, which then can

Housed in Barrow, Alaska, the Dobson Spectrophotometer is used to measure the ozone. From this remote scientific outpost on the continent's frigid northern edge, information is gathered about the planet's climate. *AP Images*

destroy ozone. It has been estimated that one chlorine atom could destroy up to 100,000 ozone molecules.

Impacts and Issues

The reason that ozone depletion is of concern is because of ozone's ability to absorb the genetically destructive wavelengths of ultraviolet radiation. As such, stratospheric ozone helps to protect humans and other organisms on Earth's surface from some of the harmful effects of exposure to high-energy electromagnetic radiation from the sun. In fact, without the protective action of the stratospheric ozone layer, it is likely that life would not be possible on Earth's surface, and that life in the ocean would be restricted to the depths where sunlight does not penetrate.

The most common effect of ultraviolet overexposure is a sunburn. Although this is seldom serious, other threats are. Basal cell carcinomas account for about 75% of human skin cancers, and squamous cell carcinomas about 20%. These can usually be successfully treated if detected early enough. However, malignant melanoma, which accounts for about 5% of total skin cancers, is often ultimately fatal.

Other human-health effects of ultraviolet exposure include increased risks of developing cataracts and other damage to the cornea of the eye, along with damage to the retina, suppression of the immune system, skin allergies, and accelerated aging of the skin.

As a result of widespread awareness and concerns about the role of CFCs in the depletion of stratospheric ozone, the uses and emissions of these chemicals were curtailed. Their use as propellants in aerosol spray cans was banned in the 1980s. In 1987, the United Nations Environment Programme (UNEP) conducted a conference in Montreal, Quebec, Canada, which addressed the issue of CFCs and ozone depletion. The international agreement that was the culmination of the conference (the Montreal Protocol), pledged to alleviate ozone depletion by reducing the atmospheric release of CFCs. A 1990 revision of the protocol established more stringent timetables calling for the complete phase-out of global CFC use by the year 2000. The United States, Canada, Australia, and other developed countries have completely phased out production of CFCs. However, as of 2008, a complete global phase-out has not been achieved; developing countries now have until 2010 to meet this target.

IN CONTEXT: LOWER ATMOSPHERIC OZONE AND CLIMATE CHANGE

In July 2007 researchers based at the University of Exeter in the United Kingdom published data in the science journal *Nature* that suggested that ozone (O_3) played a greater potential role in climate change than scientists previously assumed. In addition to ozone's role as a greenhouse gas in the upper atmosphere, data suggested that increased levels of ozone in the lower atmosphere damaged plant life by impairing photosynthesis and thereby decreased the ability of plants to act as a carbon sink that removes carbon dioxide (CO_2) from Earth's atmosphere.

SEE ALSO *Air Pollution; Carbon Dioxide (CO_2) Emissions; Greenhouse Effect; Ozone Hole*

BIBLIOGRAPHY

Books

DiMento, Joseph F. C., and Pamela M. Doughman. *Climate Change: What It Means for Us, Our Children, and Our Grandchildren.* Boston: MIT Press, 2007.

Gore, Al. *An Inconvenient Truth: The Planetary Emergency of Global Warming and What We Can Do About It.* New York: Rodale Books, 2006.

Seinfeld, John H., and Spyros N. Pandis. *Atmospheric Chemistry and Physics: From Air Pollution to Climate Change.* New York: Wiley Interscience, 2006.

Web Sites

NASA. "Ozone Hole Watch." December 4, 2007. http://ozonewatch.gsfc.nasa.gov/ (accessed May 10, 2008).

Brian D. Hoyle

Paper and Wood Pulp

Introduction

Paper is one of the most versatile products of everyday life, with hundreds of different uses. It is vital for communication and education, as well as in sanitary and household applications and packaging. Paper is basically a mat of fibers derived from plant material, and it is mostly made of fibers derived from wood. Making paper from wood has long been an important industry in countries such as Canada and Finland.

Paper is biodegradable, recyclable, and a source of energy if it is burned. If it is produced in a sustainable manner, then paper is an environmentally friendly product. The introduction of personal computers and electronic mail could possibly have led to a paper-free office, but the reverse has happened with demand for paper growing in recent years. This need not be a problem if paper is recycled and reused wherever possible.

Historical Background and Scientific Foundations

To the naked eye, paper looks smooth, but if examined under a microscope it reveals itself as a network of plant fibers laid down as a sheet. It is made by draining most of the water from a suspension of these fibers known as pulp. Today, most paper is made from wood pulp, but it can also be made from other plant sources such as hemp, cotton, and, of course, recycled paper.

The ancient Egyptians made a type of paper that they called papyrus by pressing together the strips of the grasslike sedge *Cyperus papyrus*. Then, in the second century AD, the Chinese began to make paper that was more like the material in use today. They used a pulp of mulberry fibers and lifted it up in a silk sieve to drain the water, leaving a sheet of paper, which was dried in the sun. This paper was high quality and long lasting. Indeed, samples still survive in the British Museum in London. It was not until the nineteenth century that wood pulp was used as the main source of paper, and papermaking was accomplished by machine instead of by hand. Previously papermakers used cotton and hemp as a raw material.

Wood consists of cellulose fibers held together by a substance called lignin. In wood pulp production, timber is sawed from the tree and its bark removed. Then the fibers are separated by either mechanical or chemical treatment to create pulp, which may then be bleached, depending on application. Trees suitable for paper production include pine, spruce, birch, and eucalyptus. The property of the paper is tailored to its intended end use. Therefore, durability is important for bank notes, which are made of cotton and flax fibers. Tissues, toilet paper, and sanitary towels need good absorbency. Cardboard is a heavy-duty type of paper, sometimes layered, which is widely used in packaging. Different types of pulp are used for these various applications.

Impacts and Issues

Around four billion trees are cut down each year to make paper. This is about one third of all trees harvested for commercial purposes. World consumption of paper has increased four-fold in the last 40 years, with around 300 million tons being used each year. Around one third of this now comes from recycled paper. Because of the varying grades of paper and the different applications, it is difficult to estimate how much paper one tree yields. However, according to the Wisconsin Paper Council, a single tree could give rise to 250 copies of a newspaper or 90,000 sheets of writing paper.

The paper industry is often portrayed as being wasteful and harmful to the environment, but it should be remembered that paper pulp is not harvested from

Law enforcement members use a ladder from a fire truck to reach an Earth First protester (and his banner) hanging from a rope on a crane at the Willamette Chip Mill in North Carolina (May 1998). *AP Images*

WORDS TO KNOW

BIODEGRADABLE: capable of being degraded in the environment by the action of microorganisms

CARBON SINK: A location like a forest where there is net storage of carbon as sequestration exceeds release.

RENEWABLE RESOURCE: Any resource that is renewed or replaced fairly rapidly (on human historical time-scales) by natural or managed processes.

SUSTAINABLE: Capable of being sustained or continued for an indefinite period without exhausting necessary resources or otherwise self-destructing: often applied to human activities such as farming, energy generation, or the maintenance of a society as a whole.

WOOD PULP: a suspension of wood tissue in water which is the source of most paper

tropical rain forests. Moreover, paper can have a beneficial effect on the carbon cycle. If wood is burned, carbon dioxide is released immediately into the atmosphere, contributing to global warming. If it is made into paper, the carbon is trapped and will be released over a longer period of time.

Unlike fossil fuels, wood is a renewable resource and planting more forests for wood production creates a carbon sink, as trees absorb carbon dioxide from the atmosphere. However, the paper industry does use various harmful chemicals, such as chlorine dioxide, in production, and has been responsible for various pollution incidents. Sustainable forest management, emission control, and recycling of paper products wherever possible are key to an environmentally friendly paper industry.

SEE ALSO *Forest Resources; Logging*

BIBLIOGRAPHY

Web Sites

Confederation of European Paper Industries. "Paperonline." http://www.paperonline.org/ (accessed March 19, 2008).

Ecology.com. "Paper Chase." http://www.ecology.com/feature-stories/paper-chase/index.html (accessed March 20, 2008).

Royal Botanic Gardens Kew. "Paper Information Sheet." http://www.kew.org/ksheets/paper.html (accessed March 19, 2008).

Wisconsin Paper Council. "Paper in Wisconsin." http://www.wipapercouncil.org/fun3.htm (accessed March 20, 2008).

Pharmaceutical Development Resources

Introduction

The genetic diversity in the world's plants, animals, and microbes can be a rich resource for potential medicines. Examples of drugs whose active ingredients were first isolated from environmental sources include penicillin, aspirin, and the anti-cancer drug taxol. There are probably many more potential drugs available in the biodiversity of rain forests, deep oceans, and other biomes. However, many countries where pharmaceutical development resources originate are now demanding that companies work out agreements that share profits equitably if a successful drug is developed from a nation's genetic resources.

The pharmaceutical industry also consumes other natural resources, particularly water, and generates waste. To reduce the impact of pharmaceutical development on the environment, many in the industry are now adopting green chemistry. This involves introducing new chemical reactions that do not require organic solvents and produce fewer toxic by-products. There is also a new emphasis on reducing water and energy consumption, and waste generation in pharmaceutical development.

Historical Background and Scientific Foundations

Traditional systems of medicine rely on the use of extracts from plants and animals for therapeutic purposes. Developments in chemistry during the eighteenth century led to the more scientific application of these compounds, with an early example being the use of digitalis, extracted from the foxglove plant, for the treatment of heart failure. This was followed at the start of the twentieth century with the extraction of salicylic acid from willow bark. Salicylic acid and related compounds have an analgesic and anti-inflammatory effect. Salicylic acid was then chemically modified by chemists at the German company Bayer and the result, acetylsalicylic acid, which is better known as aspirin, went on to become one of the best known drugs of all time. Aspirin is still used to relieve pain and has found new applications in the prevention of heart disease, stroke, and maybe even cancer.

Another important drug derived from natural resources is penicillin, the first major antibiotic, which was isolated from the airborne mold *Penicillium notatum* by Scottish bacteriologist Alexander Fleming (1881–1955) in 1928. Antibiotics are compounds that kill or slow the growth of bacteria and fungi. They treat a wide range of infections including septicemia, pneumonia, and tuberculosis. Penicillin was the first effective antibiotic, and the advent of World War II (1939–1945), with the prospect of many lives lost to wound infections, drove its large-scale production. *Penicillium notatum* proved not to be the best species for this purpose. It was replaced by a strain of *Penicillium chryosgenum* isolated from a moldy melon found in a market stall in Peoria, Illinois.

The search for a mold that could produce a good yield of penicillin in the factories of a pharmaceutical company typifies the way natural resources can be turned into modern medicines. Pharmaceutical development resources are found all over the planet, including rain forest plants, soil microbes, and invertebrates and bacteria from the deep ocean. Around half of today's top-selling drugs originally come from natural sources. For example, the immunosuppressant drug cyclosporin is extracted from a fungus and it has dramatically transformed the outlook for people having an organ transplant. Meanwhile, although many bacteria have developed resistance to penicillin, limiting its use, there are thousands of bacteria and fungi living in the soil that secrete other antibiotic compounds, like streptomycin and cephalosporin.

Plants are an important source of anti-cancer drugs. One of these is vincristine, used for treating leukemia, which is extracted from the rosy periwinkle found in the Madagascar rain forest. Another is taxol, which was originally obtained from the leaves of the Pacific yew tree. Meanwhile, both the deep ocean floor and the layers just beneath the surface of the water are a source of bacteria and invertebrates, such as sponges, which are being screened for their anti-bacterial and anti-cancer potential. New types of antimicrobial peptides are also being isolated from the skin of certain frogs and toads.

Pharmaceuticals are derived either from natural resources, as described earlier, or from synthetic chemistry. Some medicines are semi-synthetic, using a natural resource that is then modified by chemical reactions. Aspirin was, originally, one example of a semi-synthetic drug, although it is a simple enough molecule for total chemical synthesis and this is how it is now manufactured.

Besides its chemical raw material, whether it is natural or synthetic, pharmaceutical development also requires the use of water, energy, and other components in order to produce drugs. Industrial use of water accounts for 20% of global usage, varying from country to country from less than 5% to around 7%, depending on how industrialized the country is. The pharmaceutical industry uses water both in manufacturing processes and in cleaning equipment. The latter accounts for 60 to 80% of water usage in a pharmaceutical manufacturing plant.

Because pharmaceutical products are intended for human consumption, the regulatory authorities are very particular about the grade of water that is used. There are waters at various grades of purification that are required for specific operations in the plant. Water for injection (WFI) is the purest water used in manufacturing and should be fit for the purpose that the name implies. It is virtually sterile and has to be distilled as part of the purification process. The demand for WFI is increasing, not least because of the more stringent demands of the regulatory authorities for purity in the final pharmaceutical product. Since WFI is so costly to produce, manufacturers are looking increasingly at ways of optimizing its use in manufacture.

Impacts and Issues

Any species, anywhere on the planet, may contain genes that make it suitable as a pharmaceutical development resource. Traditionally, companies have regarded the sampling of soils and plants around the world for such genetic knowledge as being acceptable. However, a debate over who owns these genetic resources has arisen. There has been particular concern that the activities of the pharmaceutical companies amount to gene plunder, particularly where resources like rain forest plants are concerned. In other words, the companies exploit the genetic resources originating in a poor country and turn them into profits for rich countries.

WORDS TO KNOW

BIODIVERSITY: Literally, "life diversity": the wide range of plants and animals that exist within any given geographical region.

BIOME: A well-defined terrestrial environment (e.g., desert, tundra, or tropical forest) and the complex of living organisms found in that region.

CARBON FOOTPRINT: The amount of carbon dioxide (or of any other greenhouse gas, counted in terms of the greenhouse-equivalent amount of CO_2) emitted to supply the energy and materials consumed by a person, product, or event.

GENE PLUNDER: Exploiting genetic diversity without compensating the country of origin.

GREEN CHEMISTRY: An approach to chemical manufacturing that reduces its negative impact on the environment.

IMMUNOSUPPRESSANT: Something used to reduce the immune system's ability to function, like certain drugs or radiation.

The United Nations Convention on Biodiversity attempts to stop gene plunder. Pharmaceutical companies are now expected to enter into research agreements with governments and scientists in places where they wish to exploit genetic resources. If a useful drug is developed, then the country of origin can expect to share in the profits. In one example, the pharmaceutical giant Merck Inc. has paid $1.35 million to Costa Rica for the right to genetic information in a local rain forest. Costa Rica has to dedicate this payment to preserving the rain forest habitat.

If biodiversity agreements can help preserve rain forests, then the pharmaceutical industry is benefiting the environment. However, there are other ways in which the industry can have a negative impact. Pharmaceutical manufacture often involves the use of organic solvents, which may be toxic and comprise a hazardous waste. The industry also generates large amounts of wastewater that require treatment and may have an adverse effect on local water supplies.

The pharmaceutical industry is working to reduce its negative impacts on the environment with a green chemistry approach, developed by scientists Paul Anastas of the U.S. Environmental Protection Agency (EPA) and John Warner of the University of Massachusetts. Green chemistry tries to devise new chemical syntheses that use less, or no, organic solvent, and that eliminate the formation of toxic by-products. It also involves the use of online real time analysis of processes to control the formation of hazardous substances. Reduction of energy use and waste are other goals of green chemistry. All the major pharmaceutical companies now have a green chemistry program and ambitious goals for reducing their carbon footprint.

SEE ALSO *Hazardous Waste; Industrial Pollution; Industrial Water Use; Water Supply and Demand*

BIBLIOGRAPHY

Books

Chivian, Eric, and Aaron Bernstein, eds. *Sustaining Life: How Human Health Depends on Biodiversity.* Cambridge, MA: Oxford, 2008.

Cunningham, W.P., and A. Cunningham. *Environmental Science: A Global Concern.* New York: McGraw-Hill International Edition, 2008.

Web Sites

American Institute of Biological Sciences. "Searching for Nature's Medicines." http://www.actionbioscience.org/biodiversity/plotkin.html (accessed May 2, 2008).

Susan Aldridge

Photography, Environmental

Introduction

Environmental photography refers to photographs of the natural environment for artistic, research, or monitoring purposes.

Photography of natural settings can inspire feelings of wonder. Alternately, photographs of oil-soaked birds in the aftermath of a seagoing spill, a toxic dumpsite, or a flaming oil reservoir in Kuwait following a 1991 Iraq incursion can inspire entirely different feelings. Environmental photography has inspired the creation of Yellowstone National Park and, when used to record images of the same site over time, has revealed the effects of climate change.

In the era of environmental photographers such as the American photographer Ansel Adams (1902–1984), photography recorded images on an emulsion bonded to photographic film. Now, in the era of digital photography, images can be sent electronically from one location to another, and can be digitally manipulated. Environmental photography is even possible from Earth-orbiting satellites at a resolution that allows incidents of environmental degradation and natural events such as algal blooms at sea to be detected and tracked over time.

Historical Background and Scientific Foundations

Environmental photography in the United States dates back to the early 1870s. Then, William Henry Jackson (1843–1942) was part of a survey of the Yellowstone River and Rocky Mountains. His photographs of the area were the first taken of the region. Despite trying conditions and cumbersome equipment, Jackson was able to photograph the Grand Tetons, the Old Faithful geyser, and the Yellowstone region. These images were instrumental in swaying the U.S. Congress to declare the Yellowstone region the country's first National Park in 1872. Jackson continued his photographic pursuits, capturing images of the long-abandoned, Native American cliff dwellings in Colorado's Mesa Verde, which later became a national park as well.

Beginning in the 1920s, Ansel Adams began to market his photographs of regions including Yellowstone and Yosemite National Parks. His sharply focused, starkly contrasted, black and white photos proved to be very popular and revealed the national beauty of the United States to many people. Adam's photos brought environmental photography into the mainstream. His images have maintained their allure; at a Sotheby's auction of his prints held in New York in 2006, one print sold for nearly $610,000.

During the Great Depression in the 1930s, when drought and poor farming practices combined to turn the Great Plains of the United States into a "Dust Bowl," thousands migrated from the area in search of work and food. Photographer Margaret Bourke-White (1904–1971) documented this exodus, focusing on the environmental impacts of this condition on the people. Bourke-White, who later rose to prominence for her work during World War II (1939–1945), also photographed scenes of the industrialization of Russia.

Television, which gained popularity in the 1950s, provided a new outlet for environmental photography. The use of television and movie cameras to record environmental images, especially as the modern environmental movement took off in the 1960s, began to reach a wide audience of viewers. Mass distribution of environmental photography intensified with the advent of the Internet and remains popular today. For example, in 2008, Web sites such as *National Geographic* offered videos (without charge) via its site that pertained to a variety of environmental issues. Among the many topics covered were: a look at the nuclear disaster site at Chernobyl in the Ukraine 20 years after the accident; a

WORDS TO KNOW

DEAD ZONE: An area of ocean in which nothing can live except bacteria that flourish on fertilizer from agricultural runoff.

DEFORESTATION: A reduction in the area of a forest resulting from human activity.

RESOLUTION: Pixels per square inch on a computer-generated display or photograph; a higher resolution results in a clearer picture.

new tree replanting project in Haiti designed to reforest one of the world's most barren areas; renewed efforts to save Lake Erie from deterioration; images of the devastation of the 2004 tsunami in Southeast Asia; scenes of drought and dust storms in Australia; and the destruction of wildfires in the western United States.

Satellite Imagery

As satellites were launched into orbit around Earth beginning in the 1950s, satellites containing sophisticated cameras allowed large swaths of the surface to be photographed. Images can now be taken to detect different wavelengths emitted from the surface, and the WorldView satellite launched in 2007 is able to distinguish two objects on the surface that are separated by only 20 in (50 cm). The satellite is capable of photographing 290,000 square mi (750,000 square km) each day.

Satellite images have been used to do such things as determine the rate of deforestation in the Amazon rain forest, to detect algal blooms and hypoxic areas (dead zones) in the ocean, and to document the disruption of the sea bed caused by a method of fishing called bottom trawling

Recording Climate Change

Another use of environmental photography has been to document the effects of climate change on natural features. A well-known example is the series of pictures taken of various glaciers in the United States, Canada, and elsewhere. The pictures document the retreat of the glaciers as a consequence of a warming environment, and are evocative evidence of the reality of global warming.

The Extreme Ice Survey, created by photojournalist James Balog, was begun in 2006. It uses time-lapse photography to record changes to glaciers in places such as Alaska, the Rocky Mountains, Greenland, Iceland, Bolivia, and the Alps. The project uses 26 cameras to record images hourly during daylight hours. Ultimately, the project will release a documentary film illuminating its findings in spring 2010.

■ Impacts and Issues

Environmental photography has helped increase environmental awareness by visually displaying both the beauty of nature and the consequences to that beauty of accidental and deliberate environmental degradation. Orbiting satellites are able to photograph nearly all of Earth's surface, leaving little surface environmental damage undetected.

Environmental photography can be combined with measurements of temperature and other parameters taken at the site of the photograph to provide a detailed survey of the site at that moment in time. When similar information is gathered over time, trends can become evident. In this way, photographs have been valuable in demonstrating the changes taking place in the natural world, which can be correlated with increasing atmospheric temperature.

IN CONTEXT: ANSEL ADAMS

American photographer Ansel Adams (1902–1984) is best known for his dramatic black and white photographs of the American West, particularly the Sierra Nevada and Yosemite Valley. A native of San Francisco, Adams lived through the great earthquake of 1906, but was permanently scarred by a broken nose suffered when he fell during an aftershock. The teenaged Adams fared poorly in traditional schools and completed his education with the help of private tutors.

He began a lifelong association with the Yosemite Valley during a family trip in 1916, and three years later became an active member of the Sierra Club. Adams trained as a concert pianist but, motivated in large part by mountain trips, developed a passion for photography, and in 1930 decided to pursue it as a career. One of his early books, the limited edition volume *Sierra Nevada: The John Muir Trail*, was assembled during the 1930s and is considered to have been instrumental in the creation of Kings Canyon National Park in California. Adams also worked as a commercial photographer and spent time in New York City. He wrote three seminal books on photographic technique that remain highly respected among modern photographers: *The Camera*, *The Negative*, and *The Print*. Adams was elected a fellow of the American Academy of Arts and Sciences in 1968 and received the Presidential Medal of Freedom in 1980.

SEE ALSO *Geospatial Analysis; Glacial Retreat; Global Warming; Maps and Atlases; National Park Service Organic Act*

BIBLIOGRAPHY

Books

Chiras, Daniel D., John P. Reganold, and Oliver S. Owen. *Natural Resource Conservation: Management for a Sustainable Future*. New York: Prentice-Hall, 2004.

Morgan-Griffiths, Lauris. *Ansel Adams: Landscapes of the American West*. London: Quercus, 2008.

Parker, Lee. *Environmental Communication: Messages, Media, and Methods*. Dubuque: Kendall/Hunt Publishing, 2005.

Web sites

Extreme Ice Survey. http://www.extremeicesurvey.org/ (accessed June 3, 2008).

National Geographic. http://video.nationalgeographic.com/video/player/news/ (accessed June 3, 2008).

National Snow and Ice Data Center. "Repeat Photography of Glaciers." http://nsidc.org/data/glacier_photo/repeat_photography.html (accessed May 2, 2008).

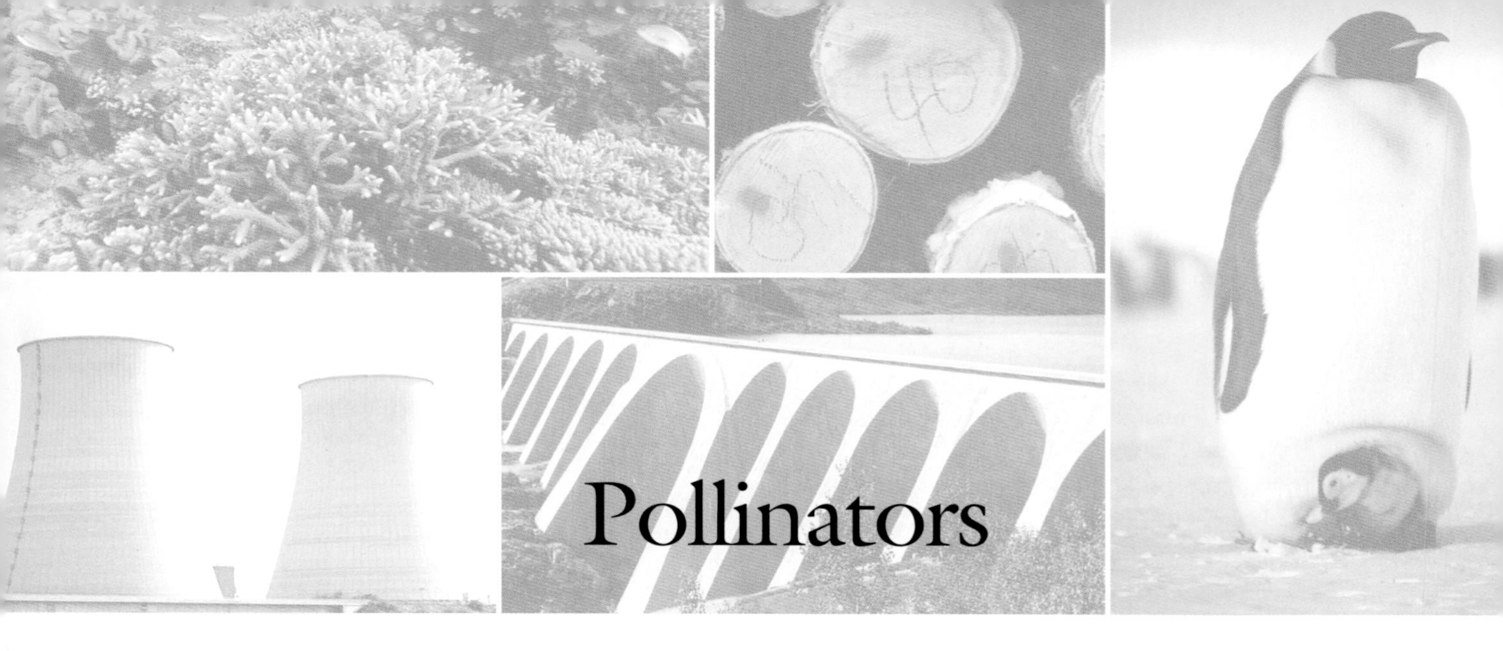

Pollinators

Introduction

Pollinators are creatures that can move pollen from the pollen-producing portion of a flower to a portion of another flower where fertilization (pollination) will occur.

In the natural world, many creatures are pollinators; estimates range up to 200,000 species, most of which are insects. However, other cold- and warm-blooded creatures, including humans, are pollinators. Examples of pollinators include bees, wasps, butterflies, bats, primates, reptiles, and birds (including hummingbirds). Often they are attracted to the bright color of the plant, which has adapted to exploit the reproductive advantage gained by attracting pollinating insects and other species.

Some pollinators have also been efficiently designed for the task during their evolution. A good example is the honeybee, which can pick up pollen on specialized regions of its limbs during its foraging for nectar from plants. Upon visiting another plant the pollen can be deposited. Other creatures including humans are usually accidental pollinators, transferring pollen by brushing against plants. However, beekeepers and others who are interested in raising plants for commercial purposes or as a preservation strategy can be pollinators by choice.

Pollinators are essential for a healthy ecosystem, since only about 10% of flowering plants are able to transfer pollen without the assistance of pollinators. Although pollination can be achieved by wind, with the pollen alighting on a flower after drifting in the breeze, pollinators greatly increase the likelihood of pollination since their feeding deliberately takes them from flower to flower.

In another environmental role, pollinators that are migratory species can be an early indicator of environmental deterioration. For example, honeybee populations can decline markedly in number suddenly. One reason for this so-called colony collapse syndrome may be the infection of the population with a virus. As another example, the decline in numbers of Monarch butterflies following their winter migration in Mexico has been linked to climate change and loss of their habitat due to logging.

Historical Background and Scientific Foundations

Pollen is a powdery appearing material that is composed of grains that are termed microgametophytes. These are the sperm cells produced by the male portion of a plant called the anther. Pollinators function by transferring pollen from the anther to the carpel of another plant. The carpel contains the ovule, which is the part of the plant where the reproductive cells are located. In some varieties of plants, the target area of the carpel is known as the stigma, while in other types of plants it is called the micropyle.

The best-known pollinator is the bee. Images of a bee with a ball of adhering pollen on its rear limbs are a feature of many biology textbooks. The fuzzy texture of the limbs and body, and electrostatic charge, allow pollen to easily stick to the bee. As well, bees have specialized pollen carrying containers ("pollen basket") on their rear limbs.

The transfer of pollen occurs when the creature lands or brushes against a flower. Often, this can be to acquire some of the flowers' sweet nectar. When the same thing occurs on another plant, some of the pollen can be dislodged.

The deliberate human version of pollination is typically done using a brush or cotton swab to transfer the pollen. Just shaking a self-pollinating flower like the tomato can be sufficient.

Impacts and Issues

Pollinators are a vital part of the global ecosystem. Pollination is necessary for the production of seeds and fruit by about 80% of all the plants that produce a flower. This includes over 60% of the food plants in developed and underdeveloped regions of the world.

As well, pollinators are vital in maintaining the genetic diversity of plants, since their transfer of pollen is random. Genetic diversity is desirable especially for crop plants. For example, the deliberate creation of genetic uniform plant species (as is increasingly accomplished with genetically engineered crop plants) increases the risk that a disease that can infect the plant will decimate the population. With diversity, however, the population is less affected by disease, since some plants will survive and thrive.

Aside from their value to nature, pollinators can be sentinels of environmental change. For example, the decline in the Monarch butterfly population has been traced to its increased difficulty in overwintering in areas of Mexico. These regions have become colder in recent decades, perhaps due to the effects of global climate change. As well, the butterfly's habitat has diminished due to logging of forests.

Because pollinators rely on the presence of flowers for their activity, climate changes that affect flowers will affect pollinators. A study that has been conducted in the Rocky Mountain region of the United States since 1973 has documented that the earlier blooming of plants in spring has occurred coincident with climate-related temperature increase. The altered blooming timing of the flowers may be disrupting pollinator activity, which, in the longer term, could threaten the survival of both the plants' species and the pollinators.

Similar declines in plant and pollinator species have been documented in Britain and the Netherlands in a 2006 study published in the journal *Science*.

WORDS TO KNOW

BIOMASS: The sum total of living and once-living matter contained within a given geographic area; or, organic matter that can be converted to fuel and is regarded as a potential energy source.

ECOLOGY: The branch of science dealing with the interrelationship of organisms and their environments.

ECOSYSTEM: The community of individuals and the physical components of the environment in a certain area.

HABITAT: The natural location of an organism or a population.

A persistent bee parasite, a tiny mite, is reducing honey bee populations, leaving fruit and nut growers scrambling to secure a part of the dwindling supply bees for pollination for their crops. *AP Images*

SEE ALSO *Biodiversity; Habitat Loss*

BIBLIOGRAPHY

Books

Belk, Colleen, and Virginia Borden. *Biology: Science for Life with Physiology.* New York: Benjamin Cummings, 2006.

Horn, Tammy. *Bees in America: How the Honey Bee Shaped a Nation.* Lexington: University of Kentucky, 2006.

Wasser, Nickolas, and Jeff Ollerton. *Plant-Pollinator Interactions: From Specialization to Generalization.* Chicago: University of Chicago Press, 2006.

Periodicals

Biesmeijer, J. C., et al. "Parallel Declines in Pollinators and Insect-Pollinated Plants in Britain and the Netherlands." *Science* 313: 354–357 (2006).

Precipitation

Introduction

Falling rain or snow, known as precipitation, is an essential part of the hydrological cycle. It replenishes water supplies for plants, agriculture, and other human uses. Precipitation levels vary widely from place to place and help to shape local weather and overall climate. There are some very dry places on Earth, such as Antarctica, and also locations such as rain forests where there may be 100 in (250 cm) or more of rain every year. Drought, which is a reduction in the usual level of precipitation, can spell disaster by causing crops to fail.

Precipitation is pure compared to seawater, which contains mineral salts. However, it may react with certain gases (such as sulfur dioxide, or SO_2, which is emitted from power stations). These reactions can form acid precipitation or acid rain, which can fall far from the source of pollution. This kind of acidic pollution can damage buildings, lakes and rivers, and trees.

Historical Background and Scientific Foundations

In the hydrological cycle, also known as the water cycle, liquid water evaporates to form water vapor in the atmosphere. The liquid water comes from bodies of water such as oceans and lakes, as well as from respiration and plant transpiration. Evaporation is the escape of the more energetic water molecules from the surface of water, leaving salts and other materials behind. Therefore, water vapor is purer than the water that is its source.

The amount of water vapor held in the air is known as humidity and it increases with temperature. A useful measure is relative humidity, which is humidity compared to the maximum amount of water vapor the air could hold at that temperature. Saturation point occurs when the air is holding as much water vapor as possible at a given temperature. Beyond this it will undergo condensation, where vapor turns into liquid or solid droplets of water. Condensation occurs around tiny, often invisible, particles of dust, ash, spores, or smoke called nuclei. Then the droplets accumulate, forming clouds. Air currents can keep cloud particles suspended in the atmosphere when they are small. However, they tend to increase in size and at some point gravity forces them to fall to Earth as precipitation. Many clouds do not form precipitation because their droplets are too small and continue to be supported in the atmosphere by air currents.

There are various kinds of precipitation. Rain refers to liquid droplets of a diameter between 0.02 and 0.2 in (0.05 and 0.5 cm). Smaller droplets are known as drizzle. Sleet is transparent pellets of frozen water that start off as rain but freeze as they fall through colder air. Snow forms when water vapor condenses as hexagonal solid particles, while hail is ice pellets more than 0.2 in (0.5 cm) in diameter.

Levels of precipitation vary widely around the world. Global air circulation patterns are very influential in shaping the amount of precipitation occurring in a region. Generally, where there is high pressure, there is low precipitation. This means that there tends to be low precipitation between latitude 20° and 40° North and South, but high precipitation between latitude 40° and 60° North and South, as well as near the equator. There is also more precipitation near the oceans and on the windward side of mountains, where air pressure is lower. Conversely, there tends to be less precipitation on the other, leeward, side of a mountain where air pressure rises.

Impacts and Issues

Precipitation is essential for replenishing water resources on land. But it can be damaging. Excessive rainfall can cause flooding that destroys crops and buildings in addition to putting lives at risk. Persistent rainfall can cause soil erosion and hail can devastate crops.

Sulfur (S) emissions from power stations running on coal can react with water vapor to create a dilute form of sulfuric acid known as acid precipitation or acid rain. This form of pollution was discovered by the Scottish chemist Robert Angus Smith in the mid-nineteenth century. Acid rain caused rapid forest decline at high elevations in many parts of Europe and North America in the 1980s. It also caused many lakes in Sweden to become so acidic that many fish species could no longer live there. Buildings and statues in cities around the world, from India's Taj Mahal to Michaelangelo's statue of David in Florence, are also under threat from acid rain. However, legislation against air pollution in recent years has started to limit the damage caused by this form of precipitation.

SEE ALSO *Drought; Fossil Fuel Combustion Impacts; Water Resources*

WORDS TO KNOW

CLOUD: A patch of condensed water or ice droplets.

CONDENSATION: The coalescence of water molecules from the vapor to the liquid or solid phase.

RELATIVE HUMIDITY: The amount of water vapor in the air compared to the maximum amount it could hold at that temperature.

SATURATION POINT: The maximum concentration of water vapor that the air can hold at a given temperature.

BIBLIOGRAPHY

Books

Cunningham, W.P., and A. Cunningham. *Environmental Science: A Global Concern.* New York: McGraw-Hill International Edition, 2008.

Kaufmann Robert, and Cutler Cleveland. *Environmental Science.* New York: McGraw-Hill, 2007.

Web Sites

PhysicalGeography.net. "Precipitation and Fog." http://www.physicalgeography.net/fundamentals/8f.html (accessed April 16, 2008).

Water recovery tanks and troughs at the University of Texas Health Science Center at Houston's new school of nursing building will collect rainwater for re-use. The new building is one of the largest and most sophisticated "green," or environmentally friendly academic buildings in the Southwest. *AP Images*

Predator-Prey Relationships

Introduction

Predator-prey relations refer to the interactions between two species where one species is the hunted food source for the other. The organism that feeds is called the predator and the organism that is fed upon is the prey.

There are literally hundreds of examples of predator-prey relations. A few of them are the lion-zebra, bear-salmon, and fox-rabbit. A plant can also be prey. Bears, for example, feed on berries, a rabbit feeds on lettuce, and a grasshopper feeds on leaves.

Predators and prey exist among even the simplest life forms on Earth, single-celled organisms called bacteria. The bacteria *Bdellovibrio* feed on other bacteria that are bioluminescent (they produce internal light due to a chemical reaction). Indeed, the study of *Bdellovibrio* predation has revealed a great deal of the mechanics of predation and how the predator and prey populations fluctuate in number over time in a related fashion.

Predator and prey populations respond dynamically to one another. When the numbers of a prey such as rabbits explode, the abundance at this level of the food chain supports higher numbers of predator populations such as foxes. If the rabbit population is over-exploited or drops due to disease or some other calamity, the predator population will soon decline. Over time, the two populations cycle up and down in number.

In many higher organisms, the prey can be killed by the predator prior to feeding. For example, a cheetah will stalk, run down, and kill its prey (examples include the gazelle, wildebeest, springbok, impala, and zebra). In contrast, fish and seals that are the prey of some species of shark are examples of prey that is fed on while still alive.

The key aspect of a predator-prey relationship is the direct effect that the predation has on numbers of their prey.

Historical Background and Scientific Foundations

Predators and prey have evolved together, and their relationship is ancient. For example, fossils dating back nearly 400 million years have revealed evidence that extinct animals known as Hederellids were the prey of an as yet unknown creature that killed them by drilling holes through their tubular shells.

As species developed and flourished, other species exploited them as their food. A species that has become a successful predator and has survived has developed a few or a number of strategies to acquire the prey. The predator may use speed; stealth (the ability to approach unnoticed by being quiet and deliberate in its movements, or by approaching from upwind); camouflage; a highly developed sense of smell, sight, or hearing; tolerance to poison produced by the prey; production of its own prey-killing poison; or an anatomy that permits the prey to be eaten or digested. Likewise, the prey has strategies to help it avoid being killed by a predator. A prey species can also use the aforementioned attributes listed for the predator to avoid being caught and killed.

The fitness of the prey population—the number of individuals in the population, chance of being able to reproduce, and chance of survival—is controlled by the predator population.

The ways in which predators stalk, kill, and feed on their prey can be used in a classification scheme. A so-called true predator kills the prey and then feeds on it. True predation usually does not involve harm to the prey prior to death. For example, prior to being chased down and killed by a cheetah, a gazelle is healthy. Cattle that graze on grass are not considered a predator-prey relationship, as only a portion of the grass is eaten, with the intact roots permitting re-growth of the grassy stalk to occur.

A predator and its prey can both be microscopic, as is the case with the bacterium *Bdellovibrio* and other Gram-negative bacteria. But, the size difference between predator and its prey can be immense. An example is the Bowhead whale, which reaches up to 65 ft (20 m) in length, but whose survival is based on straining through its baleen (bony structures in the whale's jaw) millions of microscopic zooplankton that reach only several centimeters in length.

Predator-prey relationships can be more complex than a simple one-to-one relationship, because a species that is the predator or the prey in one circumstance can be the opposite in a relationship with different species. For example, birds such as the blue jay that prey on insects can become the prey for snakes, and the predatory snakes can be the prey of birds such as hawks. This pattern is known as a hierarchy or a food chain. The hierarchy does not go on indefinitely, and ends at what is described as the top of the food chain. For example, in some ocean ecosystems, sharks are at the pinnacle of the food chain. Other than humans, such so-called apex predators are not prey to any other species. This relationship applies only to the particular ecosystem that the apex predator is in. If transferred to a different ecosystem, an apex predator could become prey. For example, the wolf, which is at the top of the food chain in northern forests and tundra environments, could become the prey of lions and crocodiles if it were present in an African ecosystem.

Predator-prey relationships involve detection of the prey, pursuit and capture of the prey, and feeding. Adaptations such as camouflage can make a prey species better able to avoid detection. By blending into the background foliage or landscape and remaining motionless, an insect or animal offers no visual cue to a predator since it mimics its surroundings. There are many examples of mimicry in predator-prey relationships. Some moths have markings on their outer wings that resemble the eyes of an owl or that make the creature look larger in size. Insects popularly known as walking sticks appear similar to the twigs of the plants they inhabit. Another insect species called the praying mantis appears leaflike. As a final example, the stripes on a zebra are a different form of camouflage that exploits animals' tendency to herd together. The vertical stripes cause individual zebras in a herd to blend together when viewed for a distance. To a predator like a lion, the huge shape is not recognized as a potential source of food.

Camouflage can also be a strategy used by a predator to avoid detection by prey. An example is the polar bear, whose white color blends in with snow, reducing the likelihood that the bear will be detected as it approaches its prey. In this case, the same strategy and color can be utilized by young seals, since their color allows them to be invisible as they lie on the snowy surface.

WORDS TO KNOW

ECOSYSTEM: The community of individuals and the physical components of the environment in a certain area.

FOOD CHAIN: A sequence of organisms, each of which uses the next lower member of the sequence as a food source.

FOOD WEB: An interconnected set of all the food chains in the same ecosystem.

HABITAT: The natural location of an organism or a population.

SELECTION PRESSURE: Factors that influence the evolution of an organism. An example is the overuse of antibiotics, which provides a selection pressure for the development of antibiotic resistance in bacteria.

The opposite of camouflage can occur. A prey can be vividly colored or have a pattern that is similar to another species that is poisonous or otherwise undesirable to the predator. This sort of strategy, which is known as aposematism, is meant to repel a potential predator based on the predator's previous undesirable experience with the genuine noxious species.

A successful predator must judge when pursuit of a prey is worth continuing and when to abandon the chase. This is because the pursuit requires energy. A predator that continually pursues prey without a successful kill will soon become exhausted and will be in danger of starvation. Predatory species such as lions are typically inactive during the hot daytime hours, when prey is often also resting, but become active and hunt at night when conditions are less energy taxing and prey is more available. Similarly, bats emerge at night to engage in their sonar-assisted location of insects that have also emerged into the air.

When supplied with food in a setting such as a zoo, predators will adopt a sedentary lifestyle. Predation is an energy-consuming activity that is typically done only when the creature is hungry or to supply food for offspring. In settings such as an aquarium, predators and prey will even co-exist.

Being a prey does not imply that the creature is completely helpless. The prey may escape from the predator by strategies such as mimicry, or can simply outrun or hide from the predator. Some species act coordinately to repel a predator. For example, a flock of birds may collectively turn on a predator such as a larger bird or an animal such as a cat or dog to drive off the predator.

This mobbing type of repulsion can be highly orchestrated. For example, when attacked by an animal such as a dog, mockingbirds have been observed to coordinate their attack, with some birds flying close to the animal's face with others pestering it from the rear when it lunges in response. As well, some bird species use different calls, which are thought to be a specific signal to other birds in the vicinity to join the attack. Even birds of a different species may respond to such a call.

The fluctuation in the numbers of a predator species and its prey that occurs over time represents a phenomenon that is known as population dynamics. The dynamics can be modeled mathematically. The results show that a sharp increase in the numbers of a prey species (an example could be a rabbit) is followed soon thereafter by a smaller increase in numbers of the relevant predator (in this case the example could be the fox). As the prey population decreases due to predator killing, the food available for the predators is less, and so their numbers subsequently decline. With the predator pressure reduced, the numbers of the prey can increase once again and the cycle goes on. The result is a cyclical rising and falling of the numbers of the prey population, with a slightly later cyclical pattern of the predator.

A famous predator-prey model is the Lotka-Volterra version. The two equations were formulated in the mid-1920s by Italian mathematician Vito Volterra (1860–1940) to explain the decline in a fish population observed in the Adriatic Sea during World War I (1914–1918). At the same time, American mathematician Alfred Lotka (1880–1949) was using the equations to explain the behavior of some chemical reactions. Their efforts were recognized as the Lotka-Volterra model, which represents one of the first examples of ecological modeling.

Other examples include the Kermack-McKendrick model and the Jacob-Monod model (used to model predation of one bacterial species on another).

■ Impacts and Issues

Predator-prey relations are an important driving force to improve the fitness of both predator and prey. In terms of evolution, the predator-prey relationship continues to be beneficial in forcing both species to adapt to ensure that they feed without becoming a meal for another predator. This selection pressure has encouraged the development and retention of characteristics that make the individual species more environmentally hardy, and thus collectively strengthens the community of creatures that is part of various ecosystems.

For example, lions that are the fastest will be most successful in catching their prey. Over time, as they survive and reproduce, the number of fast lions in the population will increase. Similarly, the superior attributes that enable prey species to survive will be passed on to succeeding generations. Over time, the fitness of the prey population will also increase. Left to operate naturally, the predator-prey relation will be advantageous for the fitness of both species in relation to how they compete against other species in the same ecosystem. However, since each species improves, their relationship with each other remains unchanged, and the challenge remains to kill or escape from being killed.

The fossil record of Hederellids, which date back almost 400 million years, indicate that the survival race between predator and prey has been a driver of evolution perhaps since evolution began. If so, the predator-prey relationship is fundamentally important to life on Earth.

Predator-prey relationships are also vital in maintaining and even increasing the biological diversity of the particular ecosystem, and in helping to keep the ecosystem stable. This is because a single species is kept under control by the species that uses it for food. Without this population check, a species such as a rabbit could explode in numbers, which can destroy the ability of the ecosystem to support the population. A well-known example is the introduction of rabbits to Australia. An initial population of 24 rabbits was introduced in 1788 to permit hunting. In the absence of natural predators, the population rose unchecked, and by 1859 the numbers exceeded tens of millions. The ecological pressure of this immense population has decimated vegetation, leading to erosion, and the over-competition for food has caused the extinction of plants and nearly 10% of the country's natural mammalian species.

The predator-prey balance of an ecosystem can be disrupted by other changes to the ecosystem including climate related changes such as drought, or human activities that include urban development, foresting, and overuse of resources.

For example, a 2007 study from the Scripps Institution of Oceanography chronicled how overfishing of sharks by people has disrupted the food chain in Caribbean waters. Depriving the food chain of its apex predator causes carnivorous fish that are their usual prey to increase in number, and they in turn decimate the populations of other fish including parrotfish that feed on the algae that grows on the coral in the region. The explosive algal growth can smother the coral.

Modeling of predator-prey population dynamics can be useful in indicating whether the population of a species could tax the capacity of a particular ecosystem to support their numbers. For example, allocating licenses to hunt deer and elk is based on a census of the populations, and modeling. It can be that the reduction of the deer and elk population during annual fall hunting seasons enables the survivors to better make use of the available resources. As well, the information is useful in avoiding the issuing of too many licenses, which could result in a dramatic and harmful reduction in the animal population. Put another way, information on population dynamics is valuable in conservation strategies.

Knowledge of predator-prey relations can be exploited in controlling the numbers of a pest or diseases. For example, a strategy that is being explored in Africa to control the spread of malaria is the release of female mosquitoes that are incapable of breeding. In this case, the mosquito, which can transfer the bacterium responsible for malaria between animals and people or person-to-person when it takes a blood meal, represents the predator and the source of the blood is the prey. By circumventing the production of a new generation of mosquito, the population plummets, leaving insufficient mosquitoes to widely disseminate the disease.

SEE ALSO *Commercial Fisheries; Ecosystem Diversity; Endangered Species; Extinction and Extirpation; Habitat Loss; Habitat Alteration; Human Impacts; Silent Spring; Species Reintroduction Programs*

BIBLIOGRAPHY

Books

Bolen, Eric, and William Robinson. *Wildlife Ecology and Management.* New York: Benjamin Cummings, 2008.

Chiras, Daniel D., John P. Reganold, and Oliver S. Owen. *Natural Resource Conservation: Management for a Sustainable Future.* New York: Prentice-Hall, 2004.

Molyneaux, Paul. *Swimming in Circles: Aquaculture and the End of Wild Oceans.* New York: Thunder's Mouth Press, 2006.

Brian D. Hoyle

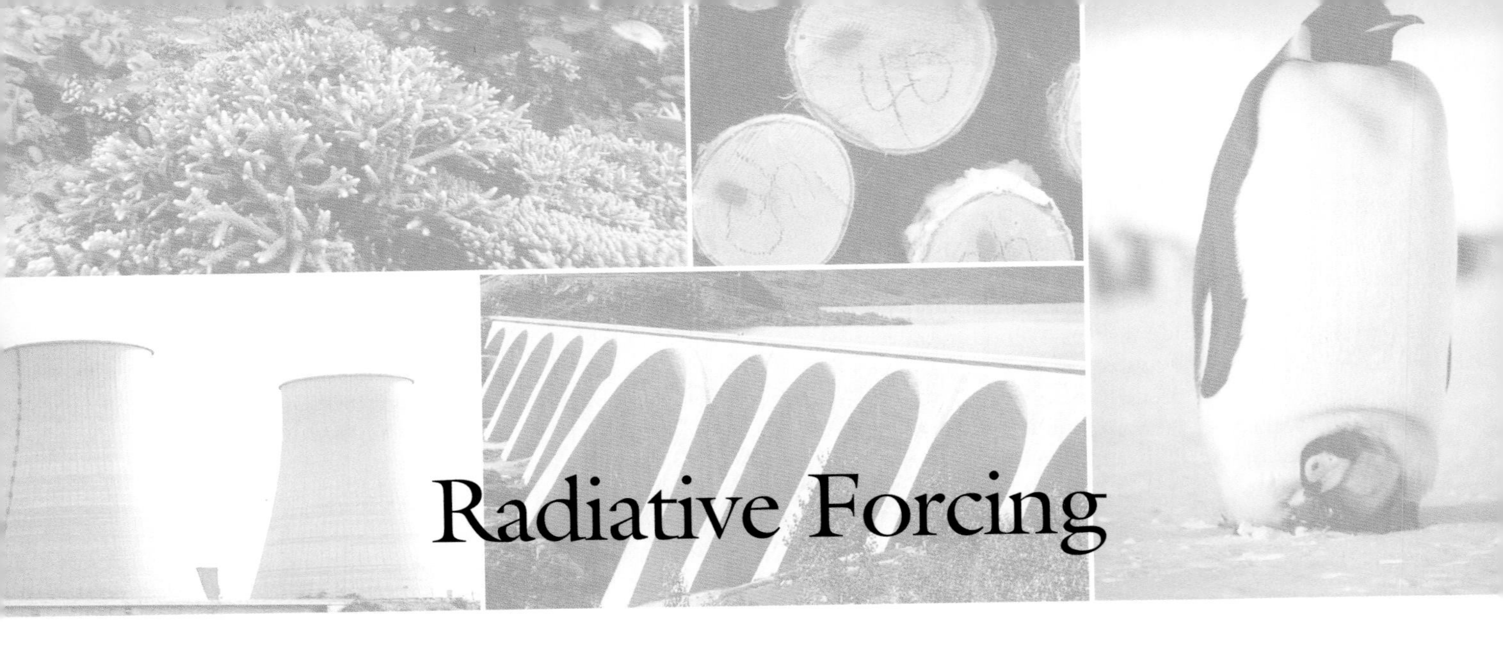

Radiative Forcing

Introduction

The climate-science term radiative forcing (RF) refers to an imbalance in the energy gained and lost by Earth. The capture of more or less solar energy by Earth can be caused by greenhouse gases in the atmosphere, aerosols (small solid particles suspended in the air), clouds, ice and snow, and vegetation. Radiative forcing (often called climate forcing or forcing) is positive if it adds energy to the climate system: Greenhouse gases, dark aerosol particles, and vegetation produce positive radiative forcing. Radiative forcing is negative if it subtracts energy from the climate system; ice, snow, and light-colored aerosol particles produce negative radiative forcing. Scientific understanding of radiative forcing is essential to understanding the history and future of global climate change.

Historical Background and Scientific Foundations

Almost all of the heat energy present in Earth's soil, air, and water comes originally from the sun in the form of light. A small fraction (about 1/7500) comes from the interior of the planet, which is heated by the decay of radioactive elements; an even smaller fraction comes from the rest of the universe.

Light is a form of radiation, that is, energy traveling freely through space. All light, if absorbed by substance, transfers its energy to that substance, heating it or changing it chemically. However, some materials are transparent to some kinds of light: Glass and air, for example, are transparent to visible light, the kind of light the eyes can see. Also, all objects reflect at least some of the light that falls on them. Snow, for example, reflects most visible light, which is why it is white.

When visible light from the sun arrives at Earth, it passes easily through the cloudless parts of the atmosphere and strikes the ground or ocean. Most of it (65–70%, a global average) is absorbed, warming the surface; the rest (30–35%) is reflected back up into space. As all warm objects radiate or emit a type of invisible light called infrared light, Earth's surface, warmed by visible light, glows invisibly in the infrared. This infrared glow shines up toward space. Some of it reaches space and its energy leaves Earth forever, but the rest is intercepted by gases in the atmosphere that are opaque to infrared light, that is, which absorb it. The more concentrated these gases are in the atmosphere, the less infrared light escapes into space, the more is absorbed, and the warmer the atmosphere becomes. This entrapment of energy is termed the greenhouse effect, and the gases that cause the greenhouse effect are called greenhouse gases. Greenhouse gases in the atmosphere act like an invisible blanket, keeping Earth warmer than it would be otherwise.

Most of the greenhouse effect is natural. Without the greenhouse effect, Earth's surface temperature would be well below freezing, even at the equator. However, human activities have been increasing the greenhouse effect for at least 200 years. The burning of fossil fuels and deforestation have been adding greenhouse gases to the atmosphere, warming the globe. At the same time, other human activities, especially the emission of small, light-colored particles of air pollution, have tended to reflect more light from Earth and so cool the globe. Each of these changes in the natural energy balance of Earth is a radiative forcing. Radiative forcings can also be natural, as when a volcano spews millions of tons of light-colored particles into the atmosphere (a temporary negative radiative forcing).

Radiative forcing can be calculated in several different forms. The simplest form, termed instantaneous RF, is the amount of energy gained or lost by Earth per unit of time per square meter of an imaginary surface surrounding the planet at the top of the atmosphere. Energy flow is expressed in watts and area in square meters, so the units of RF are watts per square meter, W/m^2.

The largest radiative forcing today is from greenhouse gases present in the atmosphere in amounts above natural levels. These gases include carbon dioxide (CO_2), methane (CH_4), nitrous oxide (N_2O), low-altitude ozone, and halocarbons. As of 2005, according to the United Nations' Intergovernmental Panel on Climate Change (IPCC), anthropogenic greenhouse gases were causing an RF of 2.63 W/m^2. Carbon dioxide alone was responsible for 1.66 W/m^2, 63% of the total RF from anthropogenic gases and up by 20% from its 1995 value. Low-altitude ozone pollution added another 0.4 W/m^2 or so of positive RF. A smaller positive RF, about 0.1 or 0.2 W/m^2, was caused by darkening of snow and ice by black carbon particles (soot) released by burning fossil fuel and biofuels such as wood.

Negative or cooling RFs were caused by land-use changes that lightened the land (for example, turning forests into grasslands), adding light-colored aerosol particles to the air, and extra cloud formation due to aerosols. The total positive anthropogenic RF was about 2.9 W/m^2, the total negative RF a little more than –1 W/m^2, for a total anthropogenic RF of about 1.75 W/m^2.

Climate forcings are distinguished from climate feedbacks. A climate feedback is any aspect of the Earth system that is changed by radiative forcing, and then changes the forcing itself, making it stronger or weaker. For example, the melting of ice and snow is a positive climate feedback, because positive forcing tends to increase melting, which exposes dark ground or water where there used to be a reflective surface, which causes Earth to absorb more energy, which increases the radiative forcing that caused the melting initially. Water vapor in the atmosphere, which is the most abundant greenhouse gas, is a feedback, not a forcing, because warmer climate increases evaporation (the shift of water from liquid to vapor), which increases radiative forcing by trapping more solar energy. Also, water vapor does not remain in the atmosphere very long—on the order of weeks, as opposed to decades for other greenhouse gases).

Impacts and Issues

Over Earth's total area, a positive RF of about 1.75 W/m^2 amounts to a large amount of energy—on the order of a thousand trillion watts. This energy is continuously entering Earth's climate system over and above the level which was entering it before human beings began to unintentionally modify the climate. It accounts for the rise in average global surface air temperatures that has been observed, about 1.33°F (0.74°C) from 1905 to 2005.

There is a high level of scientific confidence in the RF calculated for long-lived greenhouse gases: that is, the chances are small that the real value is very different from the calculated value. These gases are mixed throughout Earth's whole atmosphere and capture heat in a relatively simple way. However, there are much greater uncertainties for smaller sources of RF, both positive and negative. In particular, the negative RF from aerosols is highly uncertain. In 2005, the National Academy of Sciences went so far as to say that new studies on non-gaseous sources of RF, especially aerosols and their effects on clouds, had "raised doubts as to the continued viability of the radiative forcing concept" and had "raised the question of whether the radiative forcing concept has outlived its usefulness and, if so, what new climate change metrics should be used."

However, the National Academy did not conclude that the radiative forcing concept was obsolete, only that it needs continued study. The concept was used by the Intergovernmental Panel on Climate Change in its 2007 report and continues to be a mainstay of computerized climate modeling. Uncertainties in negative RF terms do not call into question the reality of positive anthropogenic climate forcing because the terms scientists are most sure of (RFs from gases) are large and positive, while the terms they are least sure of (aerosol effects) are small and negative. Large positive certainties outweigh small negative uncertainties: Human beings are indeed warming the planet. Without anthropogenic radiative forcing, computer models of climate cannot be made to reflect the amount of warming that has actually been observed.

WORDS TO KNOW

AEROSOL: Liquid droplets or minute particles suspended in air.

ANTHROPOGENIC: Made by humans or resulting from human activities.

EVAPORATION: Change from a liquid (more dense) to a vapor or gas (less dense). When water is heated it becomes a vapor that increases humidity. Evaporation is the opposite of condensation.

INFRARED LIGHT: Portion of the electromagnetic spectrum with wavelengths slightly longer than optical light that takes the form of heat.

SEE ALSO *Arctic Darkening and Pack-Ice Melting; Carbon Dioxide (CO_2); Carbon Dioxide (CO_2) Emissions; Climate Change; Climate Modeling; Greenhouse Effect; Greenhouse Gases*

BIBLIOGRAPHY

Books

Solomon. S., et al, eds. *Climate Change 2007: The Physical Science Basis. Contribution of Working*

Group I to the Fourth Assessment Report of the Intergovernmental Panel on Climate Change. New York: Cambridge University Press, 2007.

Periodicals

Anderson, Theodore L., et al. "Climate Forcing by Aerosols—A Hazy Picture." *Science* 300 (2003): 1103–1104.

Web Sites

National Research Council of the National Academy of Sciences. "Radiative Forcing of Climate Change: Expanding the Concept and Addressing Uncertainties." http://www.nap.edu/openbook.php?isbn=0309095069 (accessed April 11, 2008).

Larry Gilman

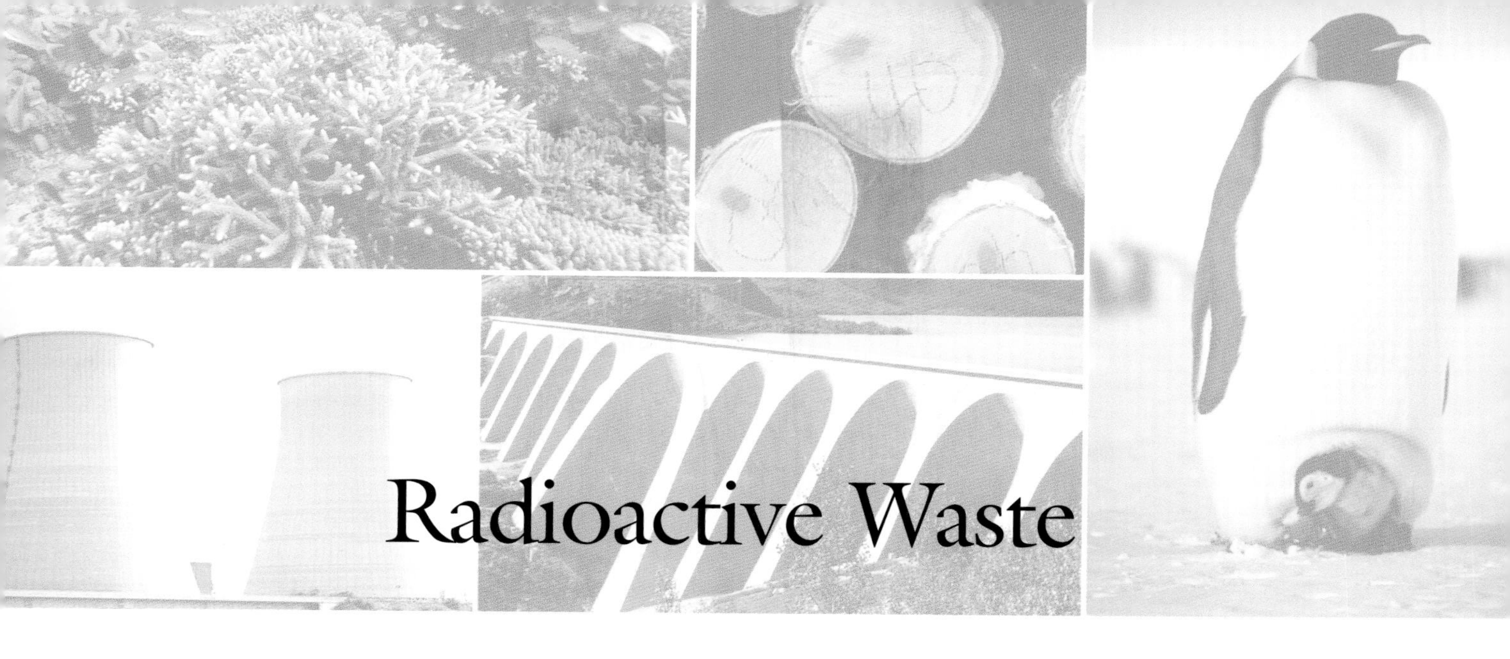

Radioactive Waste

Introduction

Radioactive waste, also called nuclear waste, is any unwanted material that contains a significant number of radioactive atoms. Such material is produced in large quantities by the facilities where nuclear power and nuclear weapons are made. Critics of nuclear power contend that nuclear waste cannot be reliably isolated from the environment for the many tens of thousands of years that must pass until its radioactivity declines to safe levels. They also argue that nuclear waste is dangerously attractive to terrorists and military opponents, who may seek to breach waste facilities and spread the material in inhabited areas using dirty bombs (chemical bombs packed with nuclear waste), hijacked aircraft, or other methods. Because of the long-lived nature of many radioactive materials, in large enough quantities they can render landscapes uninhabitable for decades or longer, causing long-term economic loss in addition to lingering effects such as higher mutation and cancer rates in plants and animals (including people).

Supporters of nuclear power say that radioactive waste can be processed to make it less dangerous and that it can be safely isolated from the environment by burying it in deep, dry rock formations.

Historical Background and Scientific Foundations

Radioactive atoms are atoms that, due to the structure of their nuclei, can break apart at any moment. When they do, they eject fast-moving subatomic particles and high-energy electromagnetic waves called gamma rays, closely akin to light rays. These particles and rays can harm living tissue. If they are intense, they can kill directly, while if they are weak, they can damage DNA and so increase the risk of cancer. Although the breakdown of any single atom of a radioactive substance cannot be predicted, each type (nuclide) of radioactive atom has an average lifespan. When large numbers of atoms are present, as is the case in radioactive waste, the behavior of that mass of atoms is predictable. For example, after a certain time, almost exactly half of the atoms will have changed into other nuclides. This time is termed the half-life of the nuclide. The half-lives of various nuclides vary from a fraction of a second to billions of years. The half-life of plutonium, a substance found in waste from nuclear reactors, is about 24,000 years. After about 10 half-lives, only one one-thousandth of any quantity of a radioactive nuclide will remain, so after 240,000 years any given amount of plutonium will only be about one one-thousandth as dangerous as it was originally. This period is often cited as the amount of time that nuclear waste containing plutonium must be isolated from the environment.

Nuclear waste did not begin to be generated in large quantities until after World War II (1939–1945), when several nations began building thousands of nuclear weapons along with scores of nuclear power plants to generate electricity. Because national survival was thought to depend on building atomic bombs, radioactive waste was treated as unimportant. As a result, poorly contained wastes were created by weapons programs. In 1954, a chemical explosion at a Soviet (now Russian) waste-storage site at Kyshtym in the southern Ural Mountains contaminated about 5,800 square mi (15,000 square km) and forced the evacuation of over 10,000 people. The Soviet Union denied that the event had occurred until 1989, and reports of hundreds of deaths from the accident are still denied by the Russian government.

In the United States, from the 1940s to the 1970s, much of its military nuclear waste was created, stored in underground tanks, or dumped into the environment at the Hanford Nuclear Reservation along the banks of the

WORDS TO KNOW

GAMMA RAYS: Streams of high-energy electromagnetic radiation given off by an atomic nucleus undergoing radioactive decay.

HALF LIFE: The amount of time it takes for half an initial amount to disintegrate.

NUCLIDE: A type of atom having a specific number of protons and neutrons in its nucleus.

RADIOACTIVITY: The property possessed by some elements of spontaneously emitting energy in the form of particles or waves by disintegration of their atomic nuclei.

Columbia River in Washington. Scores of millions of gallons of highly radioactive waste were pumped into leaky tanks or poured into open ditches: An amount of radioactivity was released each day into the Columbia River that would now be considered a major nuclear accident. Radioactive nuclides are still moving through groundwater under the Hanford site, gradually entering the river. A number of these nuclides can concentrate in human tissues and cause cancer. The government did not release this information until the 1980s, when information about the waste leaks was obtained by journalists and environmentalists under the Freedom of Information Act.

Nuclear waste from power plants has been better contained. The primary form of waste from nuclear reactors is spent fuel, that is, metal rods containing pellets of radioactive metal whose mixture of nuclides is no longer suitable for generating power. These spent fuel rods, after removal from the reactor core, are first immersed in boric acid in steel-lined concrete pools. Later, when they are cooler (because some of its shorter-lived nuclides have decayed into other nuclides), they are moved into heavy steel canisters and stored near the nuclear power plant. Such facilities are not intended to contain the waste permanently. The nuclear industry and government agencies envision two possible futures for such waste: (1) It may be reprocessed to extract its plutonium, which can then be burned as a nuclear fuel. This plan must overcome the obstacle that reprocessing produces larger volumes of high-level radioactive waste than it begins with, though they contain less plutonium. (2) It may be buried in deep, dry-rock formations, where it will hopefully remain until its radioactivity has declined to safe levels.

Impacts and Issues

As of 2008, radioactive waste was not yet being stored in any deep-rock permanent facilities anywhere in the world, although it seemed likely that a Finnish facility would begin accepting waste in 2020. In the United States, plans for deep disposal of nuclear waste from

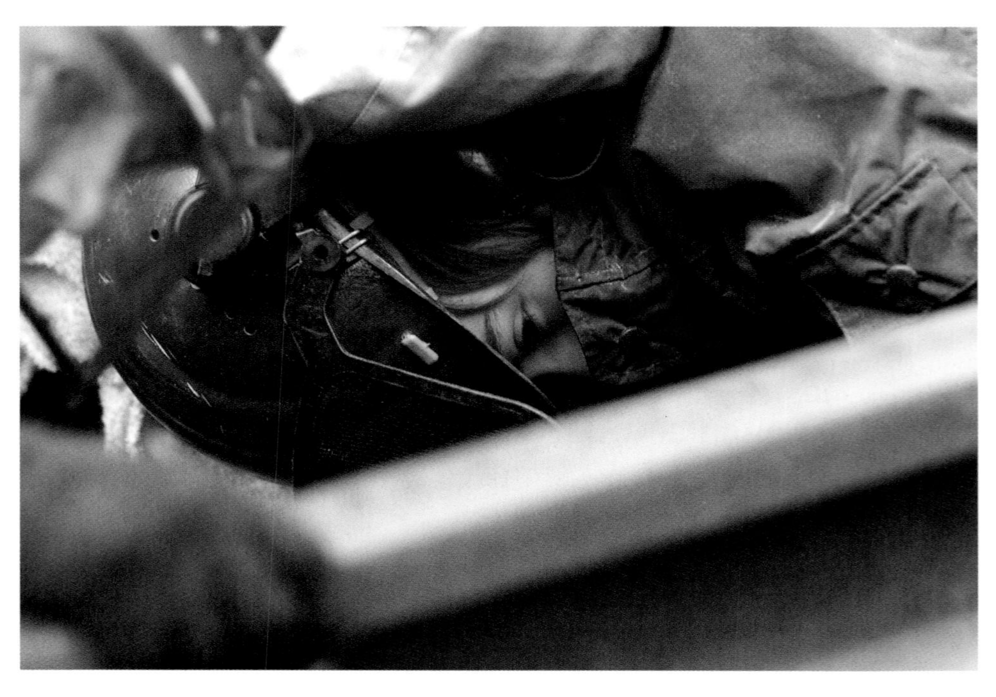

Two environmental activists are attached to train tracks in Germany (March 2001), blocking the way of a train carrying nuclear waste. *AP Images*

IN CONTEXT: COUSTEAU'S CRUSADE AGAINST RADIOACTIVE WASTE

Jacques-Yves Cousteau (1910–1997) was one of the most famous ocean explorers of the twentieth century. He authored more than fifty books and encyclopedias about the oceans, produced numerous films and television shows featuring his adventures at sea, and founded a society for the protection of the oceans. He was a member of the National Academy of Sciences in the United States and was awarded the United Nations International Environmental Prize in 1977.

Cousteau's voyages on his ship *Calypso* enchanted the public and exposed people to the beauty and fragility of many different, and often inaccessible, environments. He was able to demonstrate the threats that pollution and overexploitation posed through his dramatic images and descriptive prose. His writing, in which scientific concepts are explained in a compelling manner similar to storytelling, is a major part of his legacy, as it increased the public's awareness of the great beauty and diversity found in environments throughout the world.

Cousteau's success in film and writing made him a major spokesperson in the environmental movement of the late twentieth century.

Cousteau chose several environmental problems on which he focused his attention and the attention of the public. This often resulted in important changes to government policy. In 1960, Cousteau became concerned about the environmental effects of dumping radioactive waste in the Mediterranean Sea by the European Atomic Energy Community. He launched a publicity campaign pressuring governments to stop the practice. Eventually, dumping radioactive waste into the Mediterranean was banned.

more than 100 reactors around the country remain stalled. The Yucca Mountain facility was originally meant to begin receiving waste in 1998, but as of 2008, government officials said it would be at least 2017, probably later, before this could occur. The state government of Nevada strenuously opposes storage of radioactive waste at the site, and scientific concerns about the site's geological integrity have been raised. Meanwhile, by early 2008, owners of U.S. nuclear power plants had filed 60 lawsuits against the federal government to recover costs due to delays in opening Yucca Mountain (temporary waste-storage sites are expensive for plant owners.) Government payments to industry are forecast to run from $7 billion to $35 billion or more.

SEE ALSO *Nuclear Power*

BIBLIOGRAPHY

Books

McFarlane, Allison, and Rodney C. Ewing, eds. *Uncertainty Underground: Yucca Mountain and the Nation's High-Level Nuclear Waste.* Cambridge, MA: MIT Press, 2006.

Periodicals

Kanter, James. "Radioactive Nimby: No One Wants Nuclear Waste." *New York Times* (November 7, 2007).

Wald, Matthew L. "Agency Is Seen as Unfazed on Atom Waste." *New York Times* (June 12, 2004).

Wald, Matthew L. "As Nuclear Waste Languishes, Expense to U.S. Rises." *New York Times* (February 17, 2008).

Web Sites

National Academy of Sciences. "Health Effects of Exposure to Low-Level Ionizing Radiation: BEIR V." http://www.nap.edu/openbook.php?isbn=0309039959 (accessed April 16, 2008).

Sierra Club. "Nuclear Waste." http://www.sierraclub.org/nuclearwaste/nucw.asp (accessed April 16, 2008).

U.S. Department of Energy. "Yucca Mountain Repository." http://www.ocrwm.doe.gov/ym_repository/index.shtml#skiptop (accessed April 16, 2008).

Larry Gilman

Rain Forest Destruction

Introduction

Rain forest destruction refers to the loss of tropical and temperate rain forests due to logging and burning, and due to the toxic by-products of activities such as mining. In temperate regions on the West Coast of the United States and Canada, the destruction of rain forest is due to logging of centuries-old trees for their lumber. The same is true in tropical rain forests, but in addition, the trees are removed to clear land for conversion to agriculture and raising of livestock.

Sizable portions of the rain forests in Borneo, the Amazon watershed, Vancouver Island in British Columbia, Sri Lanka, Malaysia, and Central America, as some examples, have already been destroyed.

The Brazilian rain forest, which makes up approximately about 30% of total rain forest area of the globe, is being destroyed at a rate of over 5 million acres each year. Continuing this pace will completely decimate this rain forest by 2050. Similarly, if action is not taken to curb the rate of destruction, estimates are that by that time over 80% of the world's remaining rain forests will have been lost.

Clear-cutting for agricultural purposes is only a short-term benefit, since the removal of the forest stops the cycling of nutrients into the soil. In the Amazon, for example, once fertile soils fail to support crop growth within several years, without the addition of fertilizers.

Historical Background and Scientific Foundations

Rain forest destruction benefits only the landowner, perhaps only for a short time. Removal of trees near watercourses eliminates a zone that can restrict the movement of pollutants and sediment from washing into the stream or river. In the absence of the trees, the watercourse can become polluted. As well, run off of agriculture fertilizer can increase the levels of nitrogen and phosphorus in the water, which can stimulate the explosive increase in the numbers of microorganisms such as algae. The growth of the microbes can deplete the oxygen content of the water, which makes the water uninhabitable for fish and vegetation. In addition, fires can occur more easily since regions of the forest are open, increasing the movement of air.

In a tropical rain forest, the loss of the forest canopy makes the ground sunnier and drier, and more prone to erosion of nutrients in the soil during rainfall. Within several years, the formerly rich soil becomes claylike and infertile. This limits the productive lifetime of a cattle farm or cropland that has been created by the rain forest destruction.

Loss of temperate forests also increases soil erosion, but for a different reason. The temperate soil is no longer as stabilized by tree roots, and heavy precipitation will increasingly remove the soil.

In the Amazon, an initiative that was intended to decrease the pace of rain forest destruction led to road construction to selected sited designated for clear-cutting. Although in theory this allows greater control over forest logging, the reality has been very different. Since stringent monitoring of such a large land area as the Amazon watershed is virtually impossible, roads have provided access to the rain forest for illegal logging activities. In a study published in 2005 in the journal *Science*, researchers reported on the analysis of the pace of rain forest destruction in Brazil. Satellite photos demonstrated that the selective logging campaign has doubled the pace of rain forest destruction.

Impacts and Issues

Even though the rain forest destruction yields only a short-term benefit, much of the destruction is accomplished by sustenance farmers who survive on the income they can derive by clearing fields for crops.

Strategies to effectively stop rain forest destruction will have to address the need for a substitute local economic benefit. Ironically, in tropical regions, economic prosperity could be more attainable by using the natural bounty of the rain forest. Calculations have indicated that rain forest land cleared for agriculture yields about $60 per acre, while land used for the raising of cattle nets an owner about $600 per acre for the few years that the land is productive. However, the renewable and sustainable resource of the rain forest, including fruits and nuts, can yield up to $2,400 per acre.

The solution is not as simple as re-planting, since it takes a long time for the nutrients present in the plants to be transferred to the soil in amounts that make the soil fertile. Fertilization, which is intended to supplement a balanced and productive soil, not to supply everything needed for growth, is also not a solution

A vital nutrient needed for forests and cropland is water. Typically, this comes in the form of rain. In tropical regions, the massive loss of trees due to rain forest destruction reduces the water vapor released from leaves into the atmosphere. This reduces cloud formation. In tropical regions, rain forest destruction may create deserts in what are now among the wettest regions on Earth.

Environmental and human rights activists marked the one-year anniversary (2006) of the killing of American nun Dorothy Stang, organizing several protests in the jungle region where she lived and worked to empower peasants and protect the rain forest. *AP Images*

WORDS TO KNOW

EROSION: The wearing away of the soil or rock over time.

GLOBAL WARMING: Warming of Earth's atmosphere that results from an increase in the concentration of gases that store heat, such as carbon dioxide (CO_2).

SILVICULTURE: Management of the development, composition, and long-term health of a forest ecosystem. The objective is often to allow logging of the forest over many years.

SUSTAINABLE RESOURCE: A resource that can be renewed or maintained indefinitely.

Rain forest destruction affects biological diversity; rain forests harbor 60–70% of all biological species even though they occupy about 2% of Earth's surface. Much of the life in rain forests remains undiscovered.

The warming of the atmosphere has been accelerating since the mid-twentieth century. As acknowledged in the 2007 report by the Intergovernmental Panel on Climate Change (IPCC), this warming is most likely due to human activities such as the accelerated release of carbon dioxide and other so-called greenhouse gases. Rain forests are a natural carbon sink, or region that retains carbon. Their loss is increasing the release of carbon dioxide, thereby driving more global warming.

SEE ALSO *Agricultural Practice Impacts; Cultural Practices and Environmental Destruction; Human Impacts; Landslides; Reforestation; Runoff*

BIBLIOGRAPHY

Books

Diamond, Jared. *Collapse: How Societies Choose to Fail or Succeed.* New York: Viking, 2004.

Starr, Christopher. *Woodland Management.* Ramsbury, UK: Crowood Press, 2005.

Wild, Anthony. *Coffee: A Dark History.* New York: Norton, 2005.

Periodicals

Asner, Gregory, et al. "Selective Logging in the Brazilian Amazon." *Science* 310 (2005): 480–482.

Web Sites

Save the Rainforest. http://www.savetherainforest.org/index.htm (accessed April 7, 2008).

Real-Time Monitoring and Reporting

Introduction

Real-time is a phrase that refers to the ability to respond to something so quickly that the response takes place almost as the event is occurring. In environmental science, computer programs and devices can sample environments and produce results in real-time.

This environmental monitoring can be qualitative in nature, such as detecting the presence or absence of a specific substance—or can be quantitative in nature, as in measuring the amount of a specific substance that is present. In real-time monitoring, the information gained can be relayed to its final destination in near real-time by telephone, a signal beamed to a satellite, or over the Internet.

This technology allows developing and potentially dangerous weather systems to be detected and monitored over time, and also allows scientists to monitor environments that are difficult or too dangerous for humans to reach (such as the air above a volcano or its eruption cloud). Moreover, the devices can operate independently and with no maintenance for weeks, months, and, in some cases, decades.

Real-time monitoring and reporting can be done by manually positioning probes in the environment being sampled or by sensors mounted on orbiting satellites. As well, robotic probes and instruments can be positioned in the environment to automatically sample and relay information for analysis. This has allowed, for example, the positioning of instruments on the sea floor to monitor various aspects of the ocean environment over time.

Historical Background and Scientific Foundations

Real-time reporting of environmental conditions dates back to the invention of the telegraph in the 1830s. Until then, weather reports and forecasts could only be transmitted manually, usually involving a train journey from one location to another. Use of the telegraph enabled weather reports and warnings to be sent almost instantaneously to the same locations.

Many real-time measurements are made using sensors. The first such sensor that was developed was the pH meter, a device that measures the quality of hydrogen ions in the solution being tested. It was constructed in 1934.

The principle of a pH meter is similar to that of sensors that measure other chemicals. A pH meter measures the electrical potential between a chamber that is separated from the external environment by a membrane (originally the membrane was glass). This measurement requires comparison with a reference chamber; pH meters now can incorporate the measurement and reference electrodes in to the same probe. As well, pH meters can now obtain measurements from very small volumes of liquid; indeed, measurement of pH from damp leather or concrete is possible.

The difference between internal and external environments are utilized by other sensors to detect and quantify parameters that include dissolved oxygen, temperature, and ions such as ammonium, bromide, calcium, chloride, and fluoride. Modern probes used for environmental monitoring are rugged and hard to break, and do not require frequent checking to make sure the measurement accuracy is acceptable. This allows the probes to be positioned in environments that are not easily accessible where they can be left for a long time.

As an example of the potential of real-time monitoring and reporting, in 2005 the University of Victoria in Canada positioned instrumentation packages on the seafloor in the Pacific Ocean. The VENUS (Victoria Experimental Network Under the Sea) initiative linked the instrument via fiber optic cable to create an undersea grid of monitoring stations. Each station is connected to the Internet and the data that are regularly updated on the web site are publicly accessible. Data on temperature, salinity, turbidity, dissolved gas concentration, current speed and direction; sounds; and both still and video images are collected and sent in real-time to labs, classrooms, science centers, and homes around the world. Another array will be deployed in 2008 or 2009.

Impacts and Issues

Real-time monitoring of the environment allows for measurements over weeks and months of river flow, air and water temperature, ocean chemistry and biology, wind speed and direction, and many other parameters. These individual measurements can be combined in a single probe, and can be simultaneously displayed as different windows in various software programs that have been developed. This allows for a more precise monitoring of environmental conditions and the linking of different environmental aspects (wind speed and temperature, for example) than used to be possible.

One result is that trends in weather are more evident. For example, meteorologists can track a weather system to see if it has the potential to develop into something more extreme and hazardous such as a tornado. If so, warnings can be issued.

The ability to send and receive real-time information using mobile devices such as cell phones and personal digital assistants has expanded the ability to monitor environmental conditions. Similarly, refinements in fiber optic technology now allow multiple measurement devices to be packaged in a fiber optic cable that can be deployed with unmanned robotic probes that are tethered to ocean-going research vessels. Such probes are being readied to explore the deepest regions of the ocean.

WORDS TO KNOW

MODEM: A device that permits information form a computer to be transmitted over a telephone line or cable.

pH: The measure of the amount of dissolved hydrogen ions in solution.

SENSOR: A device to detect or measure a parameter as it is occurring.

SOFTWARE: Computer programs or data.

SEE ALSO *Geospatial Analysis; Mathematical Modeling and Simulation; Surveying; Temperature Records*

BIBLIOGRAPHY

Books

Wang, Zhendi, and Scott Stout. *Oil Spill Environmental Forensics: Fingerprinting and Source Identification.* New York: Academic, 2006.

Web Sites

University of Victoria. "VENUS: Victoria Experimental Network Under the Sea." April 8, 2008. http://www.venus.uvic.ca/ (accessed April 15, 2008).

Recreational Use and Environmental Destruction

Introduction

In the United States, national parks were established with the aim of conserving remarkable natural areas, their environments, and the animals and plants that inhabit them. For much of their history, people have enjoyed national parks through low-impact activities such as hiking, canoeing, and camping. In recent years, motorized off-road vehicles such as snowmobiles, personal watercraft, dirt bikes, and all-terrain vehicles (ATVs) have become increasingly popular forms of recreation in national parks and other natural areas.

Environmentalists, park administrators, and outdoor enthusiasts have become concerned that the use of off-road vehicles is having a negative effect on the environment. Many older snowmobiles, ATVs, and personal watercraft use inefficient engines that produce a great deal of pollution for their size. Many people find the noise levels uncomfortable or are repelled by the smell of the exhaust from these vehicles. Users of these vehicles and advocacy groups for their manufacturers dispute that they are causing damage, and they assert their right to enjoy the parks as other users do.

Historical Background and Scientific Foundations

Snowmobiles were developed in the beginning of the twentieth century, at first by modifying existing automobiles to function better in the snow. Joseph-Armand Bombardier, a Canadian auto mechanic, developed large vehicles with tracks for propulsion and skis for steering. These practical vehicles served as school buses, ambulances, and mail trucks during harsh Canadian winters. When a law was passed in 1948 that required snow to be cleared from all roads, Bombardier began developing the snowmobile as we know it today. Designed for use by mining and other industries located in remote parts of Canada, it also became an important mode of transportation for isolated villages, many of them populated by Native Americans.

ATVs took their current form in the 1970s, when motorcycle companies were seeking to expand their markets. The early 1980s saw technological improvements making the vehicles faster, as well as the introduction of automatic gearboxes and four-wheel drive. These improvements led to a great increase in popularity, with as many as 2.5 million ATVs in use by 1982. ATV users took to the countryside to enjoy their new hobby, many unaware of the environmental damage they could cause. Around this time, the popularity of snowmobiles for recreation also began to increase. Seeking out open, natural areas and trails, many users began to drive their vehicles in national parks, national forests, or other state and federal wildlife areas.

Many enthusiasts of traditional outdoor activities like hiking and camping have questioned whether the use of off-road vehicles is compatible with the purpose of U.S. national parks. In their mission statement, the National Park Service declares that it seeks to preserve natural resources for the enjoyment of current and future generations. Because of off-road vehicles' ability to cause environmental damage, many conservationists argue that they should not be used in natural areas. Conversely, off-road vehicle users insist that, consistent with the national parks' mission statement, they also have a right to enjoy the parks as they see fit. Forestry officials have sought to balance the needs of the two groups while simultaneously preserving the environment. In this spirit, the National Forest Service policy does not completely ban the use of ATVs, snowmobiles, and other off-road vehicles, but it does seek to limit use to properly established trails. Use of off-road vehicles or personal watercraft is banned in some areas deemed particularly sensitive.

Issues and Impacts

The use of snowmobiles in Yellowstone National Park has been a subject of contention for several years. Through the 1990s, about 800 snowmobiles entered the park daily during the riding season until the Clinton administration prohibited their use. The Bush administration then reversed the ban, saying that new technology reduced the snowmobiles' pollution. This left many critics skeptical, citing the need for park rangers to wear respirators at check-in points and a Bush administration study that concluded the park would be best preserved by banning snowmobiles.

Environmentalists also raised concern that animals such as bison and elk are weakened by frequently fleeing from the snowmobile noise. The use of groomed trails was also alleged to have encouraged bison to leave the park for Montana, where they were killed to prevent the spread of disease to domestic cattle. A study conducted to measure the effect of snowmobiles in Yellowstone National Park found that the noise from 250 snowmobiles in the park was audible to visitors to the Old Faithful geyser 50% of the time.

The use of ATVs on public and private land has been equally controversial. Although many users are responsible, riders who do not follow accepted rules of conduct cause serious damage. Riding in wetlands is particularly serious, as it compacts soil and decreases the ability of the ecosystem to filter and clean water. Some riders like to drive along streams, where the ground is relatively unobstructed. This creates accelerated erosion into the stream, fouling the water with sediment. The use of illegal or improvised trails can cause erosion, kill vegetation, and create access for larger trucks to travel formerly wild areas. Wildlife is also affected by ATVs; their noise can disrupt the nesting rituals of birds, leading to abandoned nests. Larger animals fear the vehicles and flee their noise, and ATV use may actually be counterproductive for hunters. Some irresponsible ATV users enjoy their hobby on others' private property, leaving damaged land and litter behind. Disputes between these riders and offended landowners have led to vandalism.

ATV enthusiasts and advocacy groups representing vehicle manufacturers have always contended that the vast majority of riders are responsible. They further assert that the lack of public trails encourages riders to use areas that are forbidden. Those who object to the use of off-road vehicles in natural areas frequently cite noise as a major complaint. They argue that the tranquility of nature is spoiled by the noise of engines, whether it is from snowmobiles, ATVs, or personal watercraft. In a major policy decision of 2006, the National Park Service adopted a policy that would favor conservation over recreation if recreational uses endanger the environment. Overall, Bush administration policy has favored commercial and recreational uses of national parks and wildlife areas, though it is a policy that a future administration may change.

WORDS TO KNOW

ATV: Abbreviation for "all-terrain vehicle," a four-wheeled vehicle designed for off-road use that is straddled like a motorcycle and steered with handlebars.

PERSONAL WATERCRAFT: Small boats, steered by handlebars and propelled by a jet of water. Often known under the trade name "Jet Ski."

Disagreement between off-road riders and environmental conservationists is all but certain to continue. With the adoption of quieter, more efficient engines in some new snowmobiles and ATVs, objections over air quality and noise may soon decrease. Quieter engines may also decrease disturbance to wildlife, though many animals may still remain sensitive to human presence. Problems such as erosion and soil compaction are unlikely to be improved by new technology and are worsened by the creation of illegal trails and use of vehicles in prohibited areas. It is likely that a small group of irresponsible off-road vehicle users will continue to cut new trails and use protected land illegally. Poor enforcement of environmental laws contributes to the attitude of impunity that some users take. However, with the creation of venues and trails where off-road vehicles can be used safely, more riders might be encouraged to practice responsible riding.

State wildlife authorities and off-road vehicle users' groups have started to teach more responsible riding in their safety courses and literature. With improvements in technology and education, and well-defined policy from wildlife management, riders hope to be able to continue to utilize most national parks and other natural areas. Allowing responsible vehicle use can also increase the public constituency for the conservation of wild, open land. Thus it behooves recreational vehicle user groups, government, and environmental groups to work toward such reduced-impact technology, better vehicle user education, and well-defined wildlife management policy.

SEE ALSO *National Park Service Organic Act; Natural Reserves and Parks*

BIBLIOGRAPHY

Periodicals

Barringer, Felicity. "Park Service to Emphasize Conservation in New Rules." *New York Times* (August 31, 2006).

The removal of a dam (which produced hydropower from 1915–1945) from this New York river marks the first time in state history a dam was removed for purely environmental reasons. There are recreational benefits, too—the river was renowned as the birthplace of American fly-fishing. *AP Images*

"Parks to Study Snowmobiles' Effect on Bison." *New York Times* (September 28, 2007).

Seelye, Katharine Q. "Approval of Snowmobiles Contradicts Park Service Study." *New York Times* (January 31, 2003).

Smith, Jessica. "Park Destruction Blamed on ATVs." *East Brunswick [NJ] Sentinel* (January 25, 2007).

Web Sites

City of West Springfield, Massachusetts. "ATV Use on Public Lands: Who's Having the Fun? And Who's Paying the Price?" http://www.west-springfield.ma.us/Public_Documents/WSpringfieldMA_DPW/ENVIRONMENTAL/ATV_damage.pdf (accessed April 16, 2008).

U.S. National Forest Service. "Success Stories: ATVs and the Forest Service." May 18, 2004. http://www.fs.fed.us/plan/par/2003/success/stories/atv_fs.shtml (accessed April 16, 2008).

Kenneth T. LaPensee

Recycling

Introduction

Recycling is the process of collecting used materials, separating them into types, and using them to manufacture new products. All industrialized countries encourage recycling today, for several reasons. First, landfills—open pits into which garbage is dumped, then covered with dirt—have become expensive to operate. Recycling, by reducing the amount of trash that has to be buried, saves cities money. Second, recycling reduces the extraction and manufacture of fresh materials by mining, drilling, and logging, which reduces harm to the environment. Third, making products from recycled materials uses less energy than making the same products entirely from fresh materials. Since making most types of energy causes greenhouse gases such as carbon dioxide to be released to the atmosphere, recycling combats global warming.

Although recycling of basic metals such as steel and copper has been a profit-making business for over a century, recycling of municipal waste—ordinary garbage—is sometimes criticized by skeptics as inefficient, a useless practice motivated by environmental guilt-feelings rather than economic rationality. Few governments have, however, been persuaded by such arguments, and several scientific studies have shown that most recycling programs do benefit the environment.

Historical Background and Scientific Foundations

Since metalworking was first discovered about 4,500 years ago, metals have been recycled because it is easier to re-use metal than to extract it from ore (metal-bearing rock). Since the Industrial Revolution began in the late 1700s, societies have became more and more dependent on metal, making bridges, the frames of tall buildings, and steam engines out of iron or steel (an alloy of iron and carbon) and, later, electrical machinery out of copper. In the mid-twentieth century, aluminum also began to be used in large quantities. Businesses devoted to recovering used metal sprang up in the 1800s along with increased demand for metal. These business turned a profit by collecting scrap, sorting it, and selling it to metal manufacturers.

National security has also motivated recycling. During World War II (1939–1945), the United States and other combatant countries experienced shortages of rubber and metal because international trade was disrupted by the war. They therefore ran campaigns marketing recycling as a patriotic act. For example, one slogan promoted in the United States in the early 1940s urged, "A rubber tire saves seven men / When melted down and used again" (to help make an inflatable life-raft).

The modern environmental ethic of recycling, which seeks to reduce usage of primary resources and lessen environmental harm, is a more recent invention. The very word "recycling" did not begin to be used for the re-use of materials from consumer trash until 1970, when the modern environmental movement was just coming into being. In the space of a few years from the late 1960s through the early 1970s, millions of people became aware that their industrial society was producing vast amounts of pollution and garbage. The idea that industrial materials might be used in an endless cycle, like the water, air, and minerals of natural ecosystems, was attractive because it promised to re-shape industrial society along sustainable, natural lines.

A few curbside pickup recycling programs, where trucks traverse the streets of a town to pick up separated trash materials in front of homes, were begun by U.S. cities and towns in the early 1970s. However, the recycling movement grew slowly because of an economic chicken-and-egg problem: There were already well-established markets for steel and aluminum, but without markets for recycled paper, glass, and plastic, there were

WORDS TO KNOW

COMPOSTING: Breakdown of organic material by microorganisms.

E-WASTE: A term describing electronic equipment at the end of its useful life. E-waste is the fastest-growing type of waste in the world.

SUSTAINABILITY: Practices that preserve the balance between human needs and the environment, as well as between current and future human requirements.

too few places to send such materials once they had been collected. At the same time, a market could not develop until sufficient amounts of recyclables were available. Supply and demand had to grow together, which took time.

In the meantime, bottle bills (also called deposit laws) began to be passed. Bottle bills are laws that require a surcharge, usually a nickel or a dime, to be charged on certain beverages (beer, soda, sometimes bottled water). This money is collected by sellers and bottlers. When consumers return the beverage containers, they are given back the deposit in cash. The first bottle bill in North America was passed in the Canadian province of British Columbia in 1970; in 1971, Oregon became the first U.S. state to pass a deposit law. By 1985, 10 states had passed deposit laws. Hawaii became the eleventh state with such a law in 2005. At that time, about 30% of the U.S. population lived in a deposit-law state.

Meanwhile, similar concerns about the environment and recycling were being felt in Europe. On July 15, 1975, the European Union adopted its Council Directive on Waste, which read, in part, "Member States shall take appropriate steps to encourage the prevention, recycling and processing of waste, the extraction of raw materials and possibly of energy there from and any other process for the re-use of waste." In response, European countries began to develop recycling programs of their own. In 1991, Germany adopted its Green Dot (Grüne Punkt) program, an industry-funded system for recycling packaging. By 2008, 23 European countries were participating in the program, which requires companies selling packaged products to contribute to the cost of recovering and recycling their own packaging. The more packaging a company puts on its product, the higher the fee it must pay to the Green Dot program. This encourages companies to use less packaging, as well as to fund recycling.

In the United States, interest in recycling surged in the late 1980s, when national discussion of landfill costs and other trash-disposal options was triggered by the strange journey of the garbage scow *Mobro*, which was loaded with over 3,100 tons of municipal trash from Long Island in New York State. The *Mobro* wandered the Caribbean and the East Coast of the United States looking in vain for a place to unload. Mexico, Belize, and the Bahamas all refused the garbage, as did Alabama, Florida, Louisiana, Mississippi, North Carolina, and Texas. The trash was eventually incinerated in the Long Island town where it came from, but not before triggering national discussion of landfilling, incineration, and recycling. At that time, 80% of U.S. trash was put in landfills, 10% was incinerated (burned), and only 10% was recycled.

Weeks of headlines about the *Mobro*, combined with rising prices for landfilling, moved communities to institute recycling programs. By 2008, the United States was recycling 32.5% of its solid municipal waste, including household trash. Fifty-two percent of paper, 31% of plastic carbonated-beverage bottles, 45% of aluminum beverage cans, 63% of steel cans, and 67% of refrigerators, stoves, and other large appliances were being recycled. This was similar to recycling rates in other industrialized countries, though some ranked as low as 18% (United Kingdom) and others as high as 80% (Japan). Yet progress had not been inexorable. In 1992, the U.S. beverage container recycling rate was 53%; by 2005, it was down to 33%, and Americans were failing to recycle 144 billion beverage containers per year, including 54 billion aluminum cans, 52 billion plastic containers, 30 billion glass containers, and 10 billion miscellaneous cartons and pouches.

How Recycling Works: One Successful City Program

In 2008, San Francisco, with a population of about 760,000, had a recycling rate of 70%, the highest of any large U.S. city. Using a fleet of specially designed garbage trucks that run on biodiesel fuel and have a separate compartment for recyclables, the city collects mixed loads of paper, cans, and bottles.

Except for organic wastes such as food scraps and lawn clippings, all recyclables from curbside pickup are trucked to a central facility where 155 employees processed 750 tons of glass, metals, plastic, and paper every day (as of 2008). When a truck arrives at the centers, its recyclables are dumped onto conveyor belts that take them to a station for sorting by hand. Workers pull out plastic bags, which can jam machinery further down the line, and any objects of unusual shape or size. Spinning disks then separate paper from dense objects like bottles and cans.

On the paper line, brown (corrugated) cardboard is separated by hand from mixed paper (magazines, envelopes, office paper, paperboard). Plastics are separated by hand into three streams, namely HDPE (high-density polyethylene, type 2), PET (polyethylene terephthalate, type 1), and mixed (all other plastics). Magnets pull out steel cans from the stream of glass and metal containers. Aluminum is not normally affected by magnets, but a method called eddy-current separation is used to push aluminum out. In eddy-current separation, rapidly rotating magnets below the trash stream cause electrical currents to flow in aluminum objects. The aluminum objects are temporarily magnetized by these eddy currents and are pushed away from the rotating magnets.

After plastic, steel, and aluminum have been removed from the container stream, only glass is left. This is separated into clear, brown, amber, and green streams, all re-used separately.

The materials thus recovered are then sold to processors to be made into new materials. San Francisco's recycled paper, for example, is shipped to China, where it is used to manufacture paper packaging for manufactured goods, many of which are shipped back to the United States. The city can charge a higher price for its waste paper because its special garbage trucks do not crush glass bottles when they are collecting recyclables, meaning that the city's paper has an unusually low rate of contamination by glass fragments.

In 1996, researchers hired by the city confirmed that 19% of the waste going into its landfills consisted of food scraps and garden waste. (Nationwide, about 12% of the landfill waste stream consists of food, lawn clippings, and the like.) In 1997, the city began a program to compost food scraps from restaurants; by 2008 it was composting 300 tons of food scraps per day. Composting entails grinding up the scraps of food, plant trimmings, soiled paper, and other organic waste and allowing them to be partly digested by bacteria for about 60 days. The resulting soil-like material has little odor and makes a high-quality fertilizer that is sold to California vineyards as organic compost.

In 2008, the mayor of the city, Gavin Newsom, was planning to send the city's Board of Supervisors a proposal to make recycling of household trash, including food scraps and yard waste, mandatory rather than voluntary. Residents who did not recycle would face suspension of garbage pickup. The goal was to raise the city's recycling rate to 75%, a figure mandated for 2010 by the city council.

Paper and cardboard

In 2007, 56% of paper consumed in the United States—54.3 million tons—was recycled, about 360 pounds (163 kg). This was over twice the 25 million tons recycled in 1990.

Paper and cardboard are recycled by shredding them and adding their fibers to the pulp stage of paper manufacture. The strength of paper or cardboard depends on its structure of interlocking fibers: If the fibers are too short, the material tears easily. Each time a paper fiber is recycled, it tends to get broken, so recycled paper has

Members of an environmentalist group wear clusters of wasted cans as they attend a rally observing the 26th Earth Day in downtown Seoul, South Korea, 1996. *AP Images*

Environmentalists dump bags containing about 120,000 plastic bottles outside the Environmental Protection Administration in Taipei (December 1996) to demand enactment of laws to require beverage makers to retrieve used bottles. *AP Images*

shorter fibers. Paper fibers can be recycled up to six times before they become too short to use.

Recycling paper saves energy and has other environmental benefits. Making one ton of paper from recycled fibers rather than virgin fibers from trees saves 7,000 gallons (27,000 liters) of water, 4,000 kWh of electricity, and 60 pounds (27 kg) of air pollutants.

Beverage Containers

Packaging, including containers, is the largest single category of municipal waste (32%). About 15% of all packaging waste consists of beverage containers. Two hundred billion beverage containers are sold each year in the United States (over 500 million per day), and two-thirds are sent to landfills, thrown on the ground as litter, or incinerated. Americans wasted more than two times as many aluminum cans in 2001 as they did in 1981. The wasting of aluminum cans is particularly unfortunate, not because the cans are a form of toxic pollution—like glass and steel, aluminum is ecologically harmless—but because they are a wasted opportunity. Large amounts of electricity are needed to refine aluminum from raw ore, whereas the aluminum in a beverage container is already pure. Making an aluminum can from recycled aluminum therefore takes only 5% as much energy as making it out of virgin ore. Recycling a single aluminum can saves enough energy to run a 100-watt lightbulb for about 4 hours. In 2001, America's wasted aluminum, with an average scrap value of $0.58 per pound, was worth almost $800 million—revenue that has literally been poured into holes in the ground. There are a number of reasons for this market failure, one being the erosion by inflation of the deposit on beverage containers in 11 U.S. states. In 1971, when the first U.S. bottle bill was enacted, a 5-cent deposit had as much buying power as 25 cents had in 2007. Yet deposits in most states have remained at five cents. Bottle deposits are thus worth a fifth as much today as they were when they were first enacted.

Glass recycling does not save as much energy as aluminum recycling. However, the U.S. National Institutes of Health, an arm of the federal government, states that recycling a single glass bottle saves enough energy to run a 100-watt lightbulb for four hours. Less energy is saved per pound by recycling glass than by recycling aluminum, but a single glass beverage container weighs much more than an aluminum beverage can, so the energy saving per container ends up being about the same.

Electronics

One of the fastest-growing types of municipal solid waste is e-waste, that is, discarded electronics such as computers, televisions, photocopiers, printers, scanners, and MP3 players. In 2000, over 2,100,000 tons of e-waste were generated in the United States; about 55% of it consisting of televisions and 10% of personal computers. In 2004 alone, 315 million personal computers were discarded in the United States (many were stored in closets, basements, and garages, rather than going immediately to landfills or recycling). Each year, about 130 million cell phones are retired, about 98% of them going into landfills. E-waste contains lead, mercury, cadmium, brominated fire retardants, and other toxic substances.

In 2007, the U.S. national rate of recycling for e-waste was only 10%. California's rate was up to 17%, thanks to its 2004 Cell Phone Recycling Act.

Impacts and Issues

Recycling has limits as a way to reduce the environmental impact of industrialized society. Although most forms of recycling save energy compared to using virgin materials, manufacturing even with recycled materials does use energy. Re-using objects is an even more effective way to reduce harm to Earth. For example, containers can be collected and re-filled rather than being melted down for the manufacture of brand-new containers, and shopping bags can be brought back to the store for re-use rather than being thrown away or even recycled. Reducing how much is used is even more effective than recycling or re-using: that is, buying less stuff to begin with. This hierarchy or ladder of choices is often captured in the slogan "Reduce, Re-Use, Recycle."

Re-using and reducing may involve adjustments in lifestyle that some people consider unpleasant, while others enjoy them. Recycling has the virtue that it requires minimal lifestyle change, so governments can pursue recycling programs without interfering with private consumption choices.

Does Recycling Make Sense?

Critics of recycling have argued that it is an inefficient practice, claiming there is plenty of room for landfills; all U.S. trash for the next 1,000 years, piled up a few hundred yards deep, could fit into a single 35-square-mi (91-square-km) landfill. In a frequently quoted essay first published in the *New York Times Magazine* in 1996, columnist John Tierney argued that "recycling could be America's most wasteful activity." While admitting that recycling saves energy, Tierney argued that "there are much more direct—and cheaper—ways to reduce pollution. Recycling is a messy way to try to help the environment." Some critics of recycling, including those quoted by Tierney, have ideological objections to mandatory recycling because it could interfere with the "free market" approach; among these critics are the Cato Institute, the Reason Foundation, and the Waste Policy Center.

Such arguments have persuaded few governments at the city or national level, and public support for recycling, as measured by polls, has remained high. In defense of recycling, environmentalists, recycling industry representatives, and government scientists cite studies showing that recycling—when properly carried out—does save energy, reduce landfill costs, reduce pollution, and yield other benefits for the environment. Moreover, these benefits increase as a higher percentage of the waste stream is recycled and as recycling programs mature so that methods can be adjusted to achieve the greatest possible benefits. As for landfill space, saving on landfill space is one benefit of recycling, but its main environmental benefit is reducing the harm done by the mining, logging, refining, and manufacturing of virgin materials.

Poisonous Recycling

In the early 2000s, efforts to recycle e-waste were controversial because e-waste materials collected for recycling were often exported to India, China, and other developing countries, where they were recycled using methods that released large amounts of pollution and injured the health of workers and other people. For example, in China, computers from North American are often recycled using crude methods such as burning wiring in heaps to separate its valuable copper from its plastic insulation. The burning plastic releases clouds of toxic chemicals that find their way into people's lungs and into water supplies. Dismantling circuit boards by hand releases lead dust; a 2008 study found that in some family workshops in China, lead levels were 2,800 times higher than those considered safe by normal industrial standards.

SEE ALSO *Electronics Waste; Green Movement; Landfills; Waste Transfer and Dumping*

BIBLIOGRAPHY

Books

Lund, Herbert F. *McGraw-Hill Recycling Handbook.* New York: McGraw-Hill Professional, 2000.

Periodicals

Barnard, Ann. "City Council Passes Bill for Recycling of Plastic Bags." *New York Times* (January 10, 2008).

Barnard, Ann. "Green with Envy: Germany's Green Dot Program Continues Generating Good Collection Numbers." *Recycling Today* (October, 2004).

Fountain, John. "Recycling that Harms the Environment and People." *New York Times* (April 15, 2008).

Tierney, John. "Recycling Is Garbage." *New York Times Magazine* (June 30, 1996).

"The Truth about Recycling." *The Economist* (June 7, 2007).

Web Sites

Container Recycling Institute. "Waste and Opportunity: U.S. Beverage Container Recycling Scorecard and Report." http://www.container-recycling.org/assets/pdfs/reports/2006-scorecard.pdf (accessed May 9, 2008).

National Institutes of Health. "Help Secure Our Future: Recycle Around the Clock." http://nems.nih.gov/outreach/factsheet_recycling.pdf (accessed May 9, 2008).

Select Committee on Energy Independence and Global Warming: U.S. House of Representatives. "New National Bottle Bill Would Cut Heat-Trapping Emissions, Energy Needs" [press release]. http://globalwarming.house.gov/mediacenter/pressreleases?id=0126 (accessed May 9, 2008).

Larry Gilman

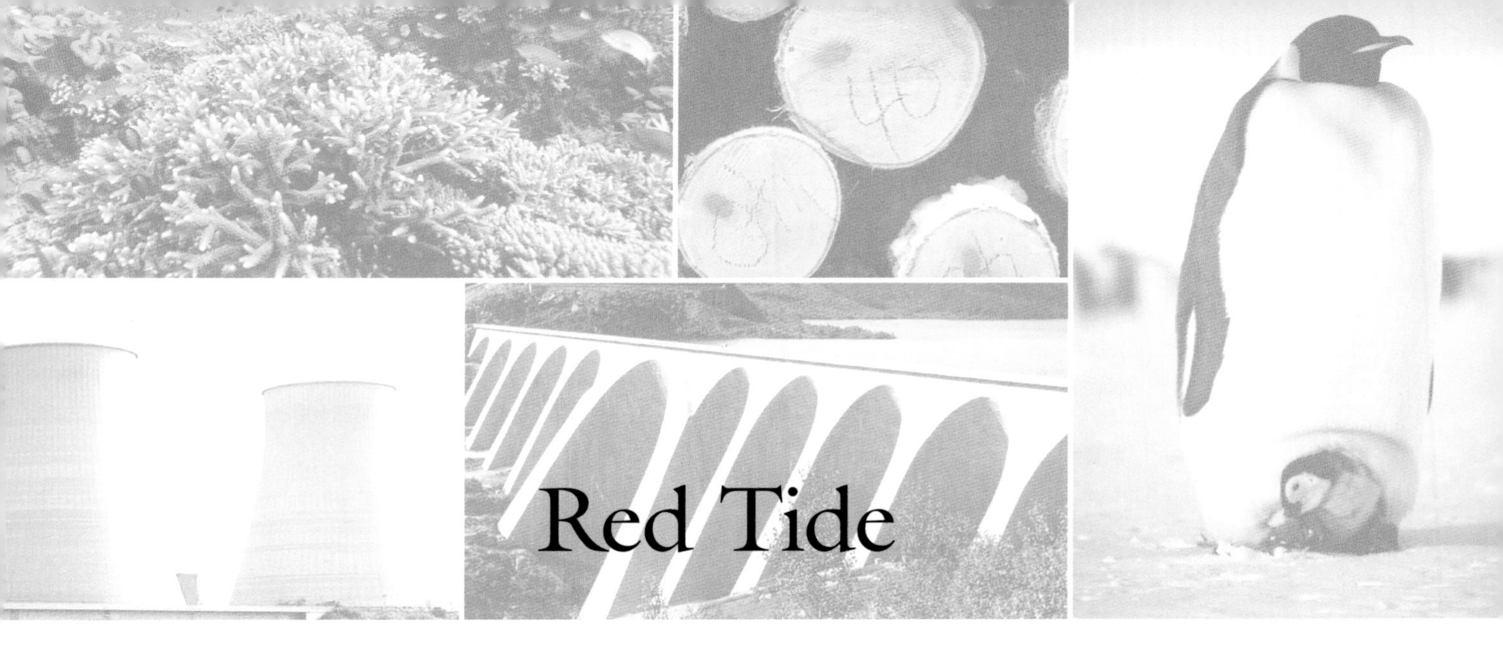

Red Tide

Introduction

Red tide is a type of algal bloom, which is a sudden, large increase in the number of microscopic organisms living in a body of water. These organisms are called dinoflagellates and are a type of phytoplankton that live in marine and freshwater habitats. Sudden large increases in their numbers are thought to be attributable to changes in salinity, temperature, and water depth, and sometimes to human activities such as the addition of nutrients to water from agricultural runoff.

When dinoflagellates that contain various natural neurotoxins—such as brevetoxin, domoic acid, and saxitoxin—occur in large enough concentrations, they can cause heavy and murky spots of reddish, discolored surface waters. This reddish discoloration is the source of the term "red tide." When these toxins are present in the dinoflagellates causing the "red tide," it is termed a harmful algal bloom (HAB). In such cases, these HABs can be eaten by fish, birds, marine mammals, and humans, causing adverse symptoms and, sometimes, death. Toxins can become airborne, as well; when winds set inland during a large red-tide algal bloom off the coast of Florida, for example, asthma attacks send an increased number of people to hospital emergency rooms.

Historical Background and Scientific Foundations

Most species of phytoplankton are not harmful to humans. In fact, they are beneficial as a food source for larger fish and mammals. However, sometimes phytoplankton grow (bloom) so fast that they become visible on the surface of the water. When this happens, the term "algal bloom" is used to describe the various species of phytoplankton that are involved.

Because various phytoplankton species contain pigments of different colors—ranging from green to brown to red—they can turn water different colors depending on which types of phytoplankton have grown to large enough concentrations. The water can appear to be colored from white to nearly black, but usually the water looks reddish or brown, depending on the species. Because of this color variance, scientists usually prefer the term "algal bloom" to the term "red tide." Red tide is used most frequently in the United States, while algal bloom is more often used in other parts of the world.

A particular species of dinoflagellate called *Karenia brevis* is often responsible for the red tides that occur in the waters of the eastern Gulf of Mexico. Red tides also occur in the northeastern part of the United States. The dinoflagellate species *Alexandrium* spp., which are especially common along the coasts of the North Atlantic Ocean and the Pacific Ocean of North America, are associated with red tides in these areas. These dinoflagellate species are eaten by a wide range of organisms, including mussels, soft-shell clams, sea scallops, oysters, lobsters, crabs, salmon, herring, mackerel, whales, sea birds, and sea otters.

Studies by researchers at the Woods Hole Oceanographic Institution and the U.S. National Oceanographic and Atmospheric Administration (NOAA) show that the frequency of algal blooms has increased along the coasts of the United States and other countries since the 1970s. Although the cause of the increased number of blooms is not absolutely certain, a general consensus among scientists is that the documented warming of the coastal oceans has made conditions more favorable for algal growth. If so, a consequence of global warming could be more algal blooms and more cases of marine toxin-related illness.

WORDS TO KNOW

ALGAL BLOOM: Sudden reproductive explosion of algae (single-celled aquatic green plants) in a large, natural body of water such as a lake or sea. Blooms near coasts are sometimes called red tides.

BREVETOXIN: Any of a class of neurotoxins produced by the algae that cause red tide (coastal algae blooms). Brevetoxins can be concentrated by shellfish and poisonous to humans who eat the shellfish.

DINOFLAGELLATE: Small organisms with both plant-like and animal-like characteristics, usually classified as algae (plants). They take their name from their twirling motion and their whip-like flagella.

DOMOIC ACID: A neurotoxin produced by the algae that cause red tide (coastal algae blooms). Domoic acid can be concentrated by shellfish and poisonous to humans who eat the shellfish.

NEUROTOXIN: A poison that interferes with nerve function, usually by affecting the flow of ions through the cell membrane.

PHYTOPLANKTON: Microscopic marine organisms (mostly algae and diatoms) that are responsible for most of the photosynthetic activity in the oceans.

SALINITY: Measurement of the amount of sodium chloride in a given volume of water.

SAXITOXIN: A neurotoxin found in a variety of dinoflagellates. If ingested, it may cause respiratory failure and cardiac arrest.

Impacts and Issues

Scientists are unsure about the cause of red tides, and whether they are natural occurrences or due to human activities. Some parts of the world only experience red tides at certain times of the year, while other regions seem only to have red tides when human activities promote them.

Some scientific studies have found that any pollution that increases the temperature of the water can increase the frequency of red tides. Other studies show that the nitrates and phosphates found in agricultural runoff can cause red tides. Still other studies show an increase in iron concentration as the cause of red tides. However, currently no definitive link has been found between human activities and the incidence of red tides.

Organisms that consume HABs are harmed by the neurotoxin that the dinoflagellates contain. One such neurotoxin—saxitoxin—is found in shellfish that ingest the harmful dinoflagellates. This toxin can produce paralytic shellfish poisoning (PSP) in humans. Patients who eat shellfish containing this toxin experience symptoms—such as amnesia, diarrhea, disorientation, memory loss, nausea, paralysis, and vomiting—that can eventually worsen to severe problems with the respiratory, gastrointestinal, and neurological systems. Death usually comes from respiratory failure. According to the Woods Hole Oceanographic Institution, a person weighing 165 lb (75 kg) can die from digesting just 0.02 oz (0.5 g) of saxitoxin.

Brevetoxin, another neurotoxin, causes neurotoxic shellfish poisoning in humans, while domoic acid produces amnesic shellfish poisoning. Okadaic acid causes diarrhetic shellfish poisoning. Ciguatoxin—from the algae genus *Gambierdiscus*, which lives in tropical waters such as those around coral reefs—produces ciguatera fish poisoning.

The incidence of red tides has increased over the past 30 or so years, possibly implying that human activity is responsible. Scientists have suggested that climate change and increased pollution levels could make red tides more pronounced. The increased incidence of red tides could have an adverse effect on the availability of food sources for marine animals. However, at the present time, scientists do not know conclusively what causes red tides.

Why some dinoflagellate species produce harmful toxins while others do not also is unknown. Scientists continue to study red tides, especially in an attempt to identify the specific genes that are involved in producing red tides.

Scientists urge governments to use better and more frequent monitoring and detection systems for red tides. Although the annual incidence of death and serious illnesses in the United Sates due to fish and shellfish poisoning is relatively low, new toxins appear periodically that make it more difficult to protect the environment and human health.

The artificial creation of non-toxic algal blooms by adding powdered iron to the ocean surface has been proposed as a way of combating global climate change. The idea is that phytoplankton would absorb carbon dioxide from the atmosphere, incorporating its carbon into their tissues; when the phytoplankton died and sank, they would take the carbon with them to the bottom of the sea, sequestering it (isolating it for a long time) from the atmosphere. However, both the efficacy and safety of this technique, called iron fertilization, have been questioned by many scientists. Small-scale tests have seemed to show little carbon sequestration from artificially induced algal blooms.

In 2007, researchers announced that *Karenia brevis* not only produces a number of brevetoxins that can constrict bronchial passages and so interfere with breathing, but also produces an antidote to its own toxins, a substance called brevenal. Brevenal, first discovered in 2004, can be made synthetically and is being evaluated for

Red tide has been a persistent problem on the coast of Coquina Beach, Florida, which prompted a $47 million federal grant to help determine the cause. *AP Images*

treatment of some of the symptoms of cystic fibrosis, a genetic disorder that afflicts about 30,000 people in the United States. Brevenal may also be used in Florida to save manatees (rare water-dwelling mammals found only in certain tropical coastal waters) that are threatened with poisoning by red tides.

SEE ALSO *Algal Blooms; Coral Reefs and Corals; Iron Fertilization*

BIBLIOGRAPHY

Books

Okaichi, Tomotoshi, ed. *Red Tides.* Dordrecht, Netherlands: Kluwer, 2004.

Subba Rao, D. V. *Algal Cultures, Analogues of Blooms and Applications.* Enfield, NH: Science Publishers, 2006.

Periodicals

Potera, Carol. "Florida Red Tide Brews Up Drug Lead for Cystic Fibrosis." *Science* 316 (2007): 1561–1562.

Web Sites

Fathom (Columbia University) and Woods Hole Oceanographic Institution. "Toxic Blooms: Understanding Red Tides." http://www.fathom.com/course/10701012/index.html (accessed March 29, 2008).

Woods Hole Oceanographic Institution. "Harmful Algae and Red Tides." http://www.whoi.edu/page.do?pid=11913 (accessed April 2, 2008).

William Arthur Atkins

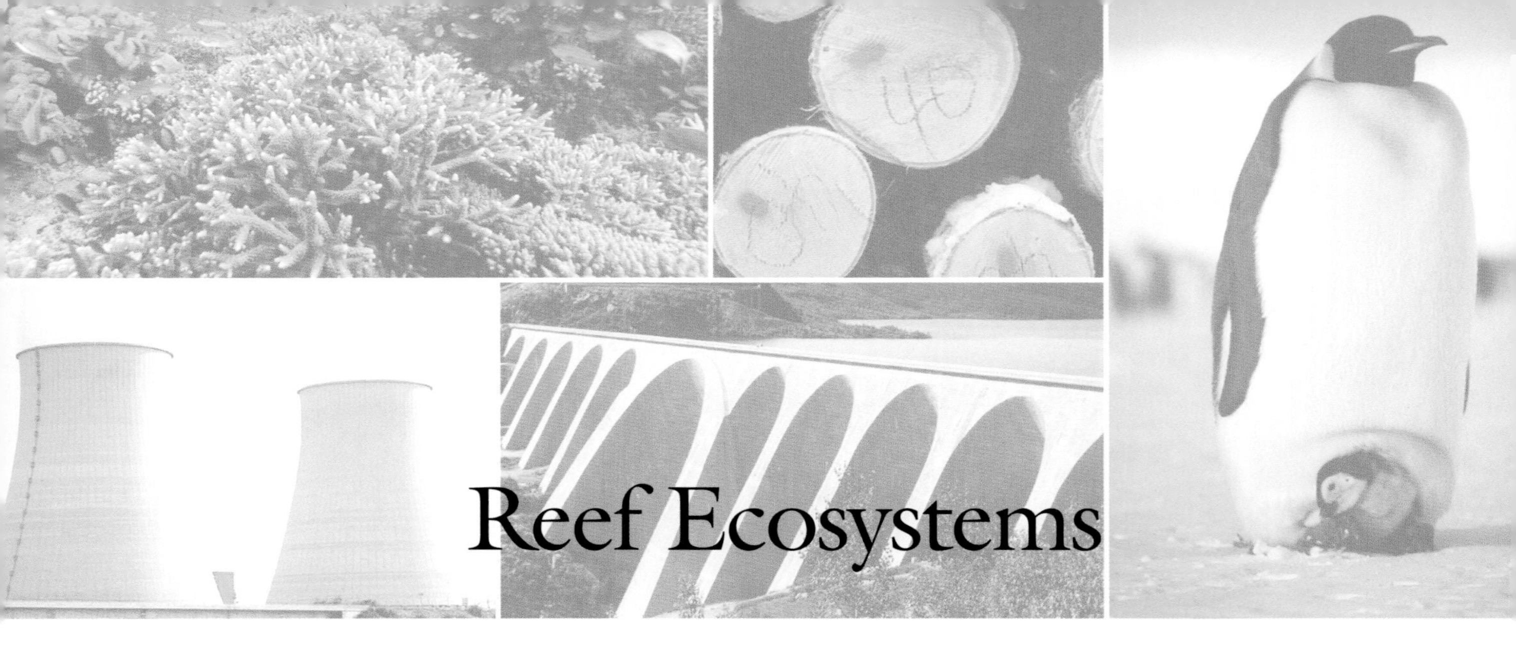

Reef Ecosystems

■ Introduction

Reefs are any rigid, water-resistant structure, usually in shallow water, that is inhabited by aquatic organisms. Most reefs are found in shallow, tropical ocean waters and are built primarily by invertebrates called corals. Corals exude a calcium carbonate skeleton, which over time forms a three-dimensional structure containing many nooks and caves. This reef structure provides abundant habitat for sessile, encrusting, and burrowing animals. Fish, swimming invertebrates, and some reptiles also flourish on coral reefs. As a result, coral reefs are among the most diverse and productive ecosystems in the world.

■ Historical Background and Scientific Foundations

Reefs are shallow water areas that create abundant habitat for marine organisms. Reefs can be formed by some species of worm, oysters, cyanobacteria, and algae. Engineers also create artificial reefs by sinking large machinery in shallow waters. However, the majority of reefs in the ocean result from the accumulation of calcium carbonate skeletons of tiny invertebrates called corals.

Corals are invertebrates, classified in the phylum *Cnidaria*. They are related to jellyfish and sea anemones. Reef-building coral are generally colonial, growing in clusters. Each individual has a shape similar to a flower: a cuplike body with tentacles in the place of petals. Within the center of the tentacles is a structure that serves both as a mouth for ingesting small plankton food particles as well as an excretory structure. Coral secrete a hard, cementlike substance called aragonite that makes up a skeleton in which the coral live. As subsequent generations of coral lay layers of aragonite on top of older skeletons, a reef is formed.

Symbiotic algae called zooxanthellae live within the tissue of coral. As these symbionts perform photosynthesis, they facilitate the conversion of dissolved calcium into the calcium carbonate that forms aragonite. In particular, zooxanthellae create an alkaline environment in which calcium carbonate deposition proceeds easily. In addition, zooxanthellae can provide a very significant portion of a coral's energy requirements. In turn the coral provide a stable environment in which the algae can live and grow as well as a constant source of carbon dioxide that is required for photosynthesis.

Because coral rely heavily on their relationship with zooxanthellae, they must grow in environments in which light is sufficient for photosynthesis. This limits the range of coral reefs to shallow waters where light penetrates. Zooxanthellae also restrict in the temperatures at which coral can live. These algae require water temperatures between about 70 to 79°F (21 to 26°C), which limit corals to tropical waters roughly between latitudes of 25°N and 25°S. Corals themselves are susceptible to changes in salinity. As a result reefs stay below the surface of the water because rain mixing with surface waters can decrease the salinity of ocean water and cause osmotic shock to corals. Sediment can be lethal to corals, as it can clog their feeding appendages, so reefs are rarely found near river mouths or other areas of significant erosion.

Coral reefs create abundant ecological niches for many different types of marine organisms. The three-dimensional structure of the reef provides holes to hide in, surface to grow on, and substrate to burrow through. As animals come to reefs to perform these functions, predators and mates are attracted to the reefs as well. Reefs become places teeming with life from the microscopic diatoms that swarm across the surfaces of encrusting algaes to enormous manta rays that cruise the reef for food. Reefs are home to nearly every phylum of invertebrate as well as many classes of fish and shark, and are often inhabited by sea turtles and ocean snakes.

A typical reef ecosystems might include staghorn corals shaped like deer antlers, green corals shaped like brains, sea stars of blue, red, and orange, and pink and orange sea anemones with orange and white striped clownfish swimming among their tentacles. Large, purple fanlike gorgonians wave in the currents as speckled sea cucumbers crawl along the reef bottom. Spider crabs, mantis shrimp, spiny lobsters, and sea slugs all hunt along the reefs for meals of smaller invertebrates and plant material. Bryozoans, both circus-like and drab, filter water through their elaborate feeding apparatus. Jellyfish and ctenophores float through the tropical waters hunting for small crustaceans that they sting with specialized dart cells. Fish of every color and size from tiny purple gobys to enormous white-tipped reef sharks are found in reef ecosystems.

The enormous diversity in the ecosystem drives intense competition among the members of the reef community. Organisms compete for space, nutrients, light, mates, and hiding spaces from predators. As a result, many reef animals have developed bright colors to warn away predators and competitors and to attract mates. Some reef animals have also developed sophisticated defenses, like poisons, stinging cells, barbs, and spines.

There are three major types of coral reefs, as first classified by Charles Darwin (1809–1882), and still recognized today. Fringing reefs border a coastline. They are common near Hawaii and in the Caribbean. Barrier reefs are found farther offshore than fringing reefs. A large channel usually forms between the reef and the shoreline due to tectonic activity and erosion. Barrier reefs can be found in the Caribbean and along the eastern coast of Australia. Atolls form when a fringing reef forms around a small island and the island subsides below the surface of the ocean, leaving a circular reef formation. Atolls are found most often in the Indo-Pacific. Some of the coral reefs that are in existence today probably formed as early as 10,000 years ago when the most

WORDS TO KNOW

CORAL: Invertebrate organisms in the phylum *Cnidaria* that form reefs in tropical ocean waters.

CYANOBACTERIA: Photosynthetic bacteria, commonly known as blue-green algae.

ECOSYSTEM: The community of individuals and the physical components of the environment in a certain area.

SESSILE: Any animal that is rooted to one place. Barnacles, for example, have a mobile larval stage of life and a sessile adult stage of life.

SYMBIOSIS: A pattern in which two or more organisms of different species live in close connection with one another, often to the benefit of both or all organisms.

ZOOXANTHELLAE: Algae that live in the tissues of coral polyps and, through photosynthesis, supply them with most of their food.

A new study suggests the range of species in the world's oceans is being depleted at such a rapid rate that several stocks will soon become extinct as the water itself becomes more polluted. *AP Images*

Artificial reefs have been made of nearly anything that sinks including shipwrecks and old tires; state and federal regulators are now prohibiting the uses of such items. Reef-building programs around the world are switching to prefabricated structures that are specifically designed as fish shelters, or to mimic natural reefs, and engineered to stay put in storms. *AP Images*

recent glacial epoch ended and sea levels rose to their current height.

■ Impacts and Issues

Corals play an important role in the global ocean and atmosphere system. Because of their intricate symbiosis with zooxanthellae, they act as a sink for carbon dioxide. The zooxanthellae remove carbon dioxide from the atmosphere through photosynthesis and synthesize carbohydrates that provide energy to coral for metabolic processes. Among these processes is the production of aragonite, which contains carbon, thereby removing this important greenhouse gas from the atmosphere for a significant period of time.

Humans directly benefit from reef ecosystems. Reefs themselves act as a physical barrier to the shoreline, creating a buffer from floods, storms, and strong waves. When reefs are destroyed, erosion of shoreline often accelerates. This can result in significant destruction of property during serious storms and extreme tidal events. Reefs provide an important fishery resource. Crabs, lobster, mollusks, shrimp, and fish are often harvested from reef areas. In addition, reefs are popular tourist spots, often creating important economic value in tropical regions.

Because of the great diversity of organisms found in coral reef ecosystems, scientists suspect that inhabitants of the reef could be a potential source of medicine and other chemicals useful to humans. Research has suggested that effective sunscreens could be derived from chemicals found in coral reef organisms. Corals are used in bone replacement therapy because aragonite has been shown to be an effective matrix into which new human bones can grow.

Perhaps the most important problem facing coral reefs is known as bleaching. With ever-increasing frequency, corals go through periods where they eject the zooxanthellae from their tissues. Without their important contribution to the energy requirements of the coral, the coral start to die. Although the exact mechanism that causes bleaching is not completely understood, increases in water temperature likely act as a trigger. Since 1979 there have been six significant bleaching events in the world's reefs. In 1998 during a severe El Niño event, 48% of the reefs in the Western Indian Ocean underwent bleaching. Nearly 16% of the world's reefs died by the end of the year. It is estimated that between 60 and 95% of the coral in the Great Barrier Reef also suffered from bleaching. As water temperatures continue to rise in the future due to changes in the global climate, bleaching events are predicted to become even more severe and frequent.

Another concern regarding the health of coral reef ecosystems is acidification of the ocean due to increases in dissolved carbon dioxide into ocean water. As carbon dioxide accumulates in the atmosphere, some of it dissolves into the ocean. This strips carbonate ions out of the water, leaving it more acidic. Coral cannot produce aragonite in acidic environments. Currently atmospheric carbon dioxide is around 380 ppm (parts per million). At concentrations of 500 ppm, scientists estimate that corals will not be able to lay down calcium carbonate skeletons. At current rates of atmospheric carbon dioxide accumulation, the Intergovernmental Panel on Climate Change (IPCC) estimates that this threshold will be reached in about 2050.

Reef ecosystems are impacted by human activities in other ways as well. Oil, pesticides, heavy metals, and garbage all damage or poison both corals and the organisms that live on them. Because they occur in coastal areas, they are impacted by runoff. In particular, when sediments, soils, and other urban wastes are washed into ocean waters, they can clog the feeding structures of corals and cause them to die. Dredging for harbors, overfishing, fishing with dynamite and cyanide, careless underwater divers, the seashell industry, and the aquarium hobbyists have all contributed to coral reef destruction throughout the world.

In response to these human-induced threats, many preserves have been established to protect reef ecosystems. In 1975, the Great Barrier Reef Marine Park was established by the Australian government, protecting the largest coral reef ecosystem on Earth. Many Caribbean countries, such as the Virgin Islands, Belize, and Panama have also established marine preserves and parks. The Conference on International Trade in Endangered Species of Wild Fauna and Flora (CITES) has developed regulations that oversee the use and trade of reef organisms. The United States, Guam, and Puerto Rico have banned the collection of corals, both living and dead.

See Also *Aquatic Ecosystems; Benthic Ecosystems; Climate Change; Coastal Ecosystems; Conservation; Coral Reefs and Corals; Ecosystem Diversity; Marine Ecosystems; Marine Water Quality; Oceans and Coastlines; Oil Pollution Acts; Runoff; Sea Level Rise; Temperature Records; Water Pollution*

BIBLIOGRAPHY

Books

Garrison, Tom. *Oceanography: An Invitation to Marine Science*, 5th ed. Stamford, CT: Thompson/Brooks Cole, 2004.

Raven, Peter H., Linda R. Berg, and George B. Johnson. *Environment.* Hoboken, NJ: Wiley, 2002.

Web Sites

Reefbase. "A Global Information System for Coral Reefs." http://www.reefbase.org (accessed March 24, 2008).

Sea World. "Corals and Coral Reefs." http://www.seaworld.org/animal-info/info-books/coral/reef-ecosystem.htm (accessed March 24, 2008).

Texas A&M University Ocean World. "Coral Reefs." March 24, 2008. http://oceanworld.tamu.edu/students/coral/index.html (accessed March 24, 2008).

University of the Virgin Islands. "An Introduction to Coral Reefs." http://www.uvi.edu/coral.reefer/ (accessed March 19, 2008).

Juli Berwald

Reforestation

Introduction

Reforestation is the re-growing of forests that have previously been cut down using tree species that are native to the geographic area. Another term for reforestation is afforestation, the practice of restoring forests that used to exist but had been cut down. The resurgent forest can benefit the environment, preserve endangered species and renew valuable resources, and can also remove carbon dioxide from the atmosphere, which benefits the effort to slow global warming.

Reforestation can either occur naturally or it can be managed by people. In natural reforestation, an area is simply left undisturbed by human activity. Seedlings in the ground or carried to the area by wind and water flow germinate and grow. The forest is reestablished in due course according to a succession of plant species that is characteristic of that geographic area.

In managed reforestation, people attempt to re-establish forests. However, managed reforestation can give rise to debate over whether the re-established forest has as much biodiversity as the original forest or a forest that has been naturally reestablished. For example, some forests have been replanted with just a single-tree species, while other tree types are prevented from growing back, giving rise to a forest monoculture that resembles agriculture. Increasingly, reforestation is accomplished by planting seedlings from multiple species native to the area. In clear-cut areas, the natural regeneration of a range of plant and animal species can occur. Thus, in some managed reforestation projects, areas of mature or climax forests in which one species or a natural monoculture predominates, areas of the forest are clear-cut or burned, and the deforested areas are naturally reforested. This results in a more diverse ecosystem than previously existed in the natural climax forest. This procedure is particularly beneficial for old forests on preserved public lands, which have often been protected from the renewing influence of forest fires for many years.

Reforestation is often managed by the lumber and paper industries in order to maintain the resources that make the continuation of wood-dependent industries possible. These industries use the forest as a crop in which trees are replanted to replace those that have been cut. Tree cutting is selective and careful in order to maximize the crop output. These industries are particularly important in Canada, and their forest replanting operations employ large numbers of people during the summer.

Historical Background and Scientific Foundations

The scientific basis for reforestation efforts lies in the preservation of the ecological relationships between the many plant and animal species that depend on forest habitat, including those that have and continue to sustain human communities. As deforestation across the world has changed the environment and resulted in the loss of many of these species as well as the loss of many human livelihoods and the reduction of human quality of life, reforestation efforts have gained ground. The scope and value of these efforts is best presented in a series of examples from the United States and developing nations outlined in the following sections.

The Reforestation of New England

On tours of contemporary New England, travelers are often struck by the extent of evergreen and hardwood forests. It can be difficult to imagine that the primordial New England forest had been almost completely obliterated by the middle of the nineteenth century by settlers and businesses cutting timber for homes, shipbuilding, paper pulp, and the creation of open fields for agriculture and dairy farming.

Most of the European settlement of New England took place during the 1700s. The British Colonial government granted large tracts of land to settler groups who were dubbed proprietors. These groups of 6 to 10 families, or more, were given a certain number of years to develop the land by clearing woodlands for crops and farm animals, and felling trees for homes, fences, and businesses.

This deforestation reached a peak in the mid-1800s, when an estimated 80% of the New England forests were cleared except for parts of Northern Maine and the more mountainous areas. Woodland wildlife species such as wolves, turkeys, beavers, moose, and cougars disappeared from the region, while open-land species such as skunks, meadowlarks, rabbits, and foxes moved in.

With all of the high quality land taken and because much of the remaining land had poor, rocky soil, and because the productivity of the poorer land rapidly declined, many settlers migrated west and abandoned their farms in New England, which provided an opportunity for the return of the forests. The white pine was the most common tree to repopulate the New England forest. However, by the time that the white pine stands reached maturity, commerce created by the building of the Panama Canal and the new national network of railroads concurrently created a demand for wooden transport containers. Movable sawmills depleted the forest re-growth, and this second round of deforestation was further exacerbated by the hurricane of 1938, which devastated central New England.

The first large-scale, concerted effort to re-grow New England's forests began in 1897, when conservationists established the Massachusetts Forests and Parks Association, aiming to tackle some of New England's environmental problems. Though the association's initial focus was wildlife conservation, founding member Harry Reynolds saw the link between woodland habitat and animal conservation and corresponded with state and federal government officials to advocate for land protection measures. His initiative eventually resulted in the formation of the New England Forestry Foundation (NEFF) in 1944.

The NEFF proposed clear principles for harvesting trees, and collaborated with landowners and lumber companies to implement the new guidelines. Although their efforts met with initial resistance, by 1946 the foundation was helping to manage 20 properties with an average area of 150 acres. Some sixty years later, the NEFF is managing over 20,000 acres of forestland in New England.

The NEFF's forest management techniques involve first analyzing the condition of an area, often after a previous manager has already initiated a treatment, or forest management system. NEFF's forest manager decides whether to continue the current treatment, or develop a new system. If a new treatment is advisable, NEFF pursues one of three harvesting methods: Intermediate Thinning, Regeneration, and Allowable Harvests. Going forward, the NEFF manager will monitor and maintain the forest—a complex task.

WORDS TO KNOW

CLEAR-CUT: A parcel of forest that has been denuded of trees. Clear-cutting can be destructive of forests, particularly when the cycle of reforestation is slow and the processes of wind and water erosion of deforested land make it inhospitable to reforestation. However it can also be a tool for increasing the biodiversity of forests that have been protected from forest fires for many years.

DISTURBANCE SEVERITY: The amount of vegetation and root system killed by fire or tree cutting activity, and the type of growing space made available for new plants.

ECOSYSTEM PROCESSES: The dynamic interrelationships among and between these living organisms and their particular habitat elements.

ECOSYSTEM SERVICES: Services that a natural or restored ecosystem provides to human communities, including improving water quality, reducing soil erosion, flooding, and landslides and increasing carbon sequestration. These services can also include the generation of income for people through ecotourism, investment opportunity and local employment.

ECO-TOURISM: Environmentally responsible travel to natural areas that promotes conservation, has a low visitor impact, and provides for beneficially active socio-economic involvement of local peoples.

RETURN INTERVAL: The average time between occurrences of disturbances in a given stand of trees.

Another initiative that helped to revolutionize forest management in New England was the foundation of Harvard Forest, in which a Harvard professor, Richard T. Fisher, and his students developed a comprehensive plan for reforestation in New England. This plan, with its central concept of ecological forestry, took account of factors such as land-use history, human activity, and natural disturbances like hurricanes and electrical storms. Fisher produced a famous set of dioramas depicting the dramatic changes in the New England landscape from pre-settlement times to the twentieth century.

Reforestation efforts in New England have been quite successful; about 80% of the six-state region is now covered with forest. The New England region has effectively combined natural and managed reforestation to create more diverse and ecologically balanced forests. Most of the new forest growth is hardwood oak, maple,

Tourists ride through the middle layer of the rain forest canopy aboard the "Rain Forest Arial Tram" east of the capital of San Jose, Costa Rica, in March 1998. While Costa Rican officials say the country plants 30,000 acres (12,000 hectares) more trees than are cut down in a year, many environmentalists play down the value of replacing forests grown over thousands of years with young trees of limited ecological value. *AP Images*

and ash, with large stands of evergreen pine and spruce in the more northern and mountainous areas.

Impacts and Issues

Reforestation in Rwanda and the Preservation of Chimpanzee Habitat

A reforestation project in Rwanda aims to create a forest corridor that will link an isolated group of chimpanzees to larger areas of habitat in the country's Nyungwe National Park. The Rwandan government backs the Rwandan National Conservation Park initiative, which is receiving funding from the Great Ape Trust of Iowa and other U.S. conservation organizations. It is hoped that the initiative will save the Gishwati chimpanzees from extinction, since each newly planted tree increases their chances of survival by providing food, shelter, and security from people. Linking their habitat to the Nyungwe and the Kibira National Parks in Rwanda offers the additional advantage of bringing them closer to a larger, more secure population that will allow them to avoid inbreeding.

Backers of the initiative say the project will help restore biodiversity and ecosystem services such as improving water quality, reducing soil erosion, flooding, and landslides, and increasing carbon sequestration. The initiative is expected to generate income for Rwandans through ecotourism, investment opportunity, and local employment.

The natural forest cover of Rwanda has been almost completed stripped by decades of subsistence agriculture and fuel wood cutting. The depletion of the environment damaged the land to the point where scarcity precipitated one of the most widespread and brutal genocides in human history. However, since the end of the 1994 genocide, Rwanda has pushed to promote better land practices, including conservation and reforestation, as well as ecotourism. In the past two years, hundreds of millions of dollars in foreign investment for the development of hotels and tourist facilities has flooded into the country.

Gishwati is thought to have great potential for tourism for nature-lovers. The forest originally covered 100,000 hectares (250,000 acres) in the early 1900s but by 1994 it had been reduced to just 1,500 acres (600 hectares). Recent reforestation efforts have increased the forested area to 2,500 acres (1,000 hectares). Only 15 chimps survive in this small area, but conservationists hope that this decimated population will increase as the forest area expands and is connected to Nyungwe National Park.

Reforestation in Mexico and the Preservation of Monarch Butterfly Habitat

Relentless illegal logging and deforestation threatens the species survival of the monarch butterfly because of its singular migration pattern, which concentrates nearly the entire population of insects in the small, mountainous Michoacán jungle area in south central Mexico.

In a migration pattern thought to have been created during the time of advancing ice sheets in North America, monarch butterflies born in Canada and the United States migrate southward in the fall all the way to Mexico, a flight of about 2,800 mi (4,506 km). Their flight begins individually, but the migrating insects are soon joined by tens of thousands of fellow monarchs. They roost together at night in trees and suck nectar from flowers in order to build fat reserves as they head toward Mexico, an unknown destination. They converge by the millions in Mexico's Neo-Volcanic Mountain Range in Mexico on the first of November at altitudes of 9,000 to 12,000 ft (2,740 to 3,660 m) to spend the winter roosting at night in oyamel trees by the hundreds of thousands per tree and flutter around the forest in the daytime in a spectacular display of brilliantly colored, rustling wings moving in ever-changing patterns.

The monarchs mate in the early spring and head north by the end of March. During their flight, they seek out milkweed plants, where the females deposit their eggs and die. The few caterpillars that survive predation develop into butterflies that continue the migration northward. Three or four generations are required to reach the original northern habitat range. In the fall, the migration to Mexico begins again.

The forest canopy of the Michoacán jungle has been degraded by 45% in the monarch butterfly's wintering areas during the past 30 years. This degradation has been so devastating that a severe winter storm in January 2002 killed 75–80% of the monarch butterfly population. Now, an urgent international effort is underway to fund the reforestation and protection of this unique overwintering home of the monarch butterflies. The funds are used not only to return the forests to the mountain slopes, but also to help offer the local population a sustainable way of life in forest management and eco-tourism.

According to Greenpeace, Mexico is one of the world's five leading deforesters, with Brazil in first place and India in second. In 2008, the Mexican government has allocated $500 million, an unprecedented amount of funds, for tree planting. However, overall the Mexican reforestation program has received mixed reviews from Greenpeace and local environmental organizations. In 2007 environmental activists criticized Mexico's national forest policy as being deceptive and inadequate. However, some organizations praise the government's effort, with its goal of planting 280 million trees in 2008, an increase of 30 million over 2007. Greenpeace claims that the program is a failure because few of the saplings survive and there is too much emphasis on planting and not enough on forest management and monitoring. The government counters that the environmental groups do not have sufficient contact with local populations that have been enlisted in reforestation efforts.

Reforestation Practices in the New Jersey Pinelands

The New Jersey Pinelands in southern New Jersey is another example of a major government-backed initiative to restore large tracts of forestland in the eastern United States. The Pinelands National Reserve, the first national reserve in the United States, was created by the

IN CONTEXT: "JOHNNY APPLESEED," SEEDS OF LEGEND

John Chapman is best known by his popular nickname: "Johnny Appleseed." He was born in Massachusetts in 1774. Little verifiable information is known about his early life, however in 1791 he began traveling through Pennsylvania, Ohio, Indiana, and Illinois planting trees, including thousands of apple trees.

Chapman's usual practice was to find a suitable piece of land, clear it, and plant an orchard. Then, over the course of two to three years he would return to tend the orchard until the trees were ready to be sold. By locating his orchards in the path of advancing settlements, Chapman ensured himself a steady supply of ready buyers for his apple saplings, which he sold for about six cents apiece. Estimates place his total plantings at approximately 1,200 acres (486 hectares) of orchards.

Chapman appeared to be part agronomist, part naturalist, and part eccentric philosopher. Although tales of a shaggy man scattering seeds wildly are more fiction than fact, the impact of Johnny Appleseed is undeniable. By spreading apple trees across the frontier, he helped ease the lives of pioneers, and his efforts were responsible for some of the large orchards dotting the Midwest today. More than a century after his death, some of the trees he planted are still bearing fruit.

Like most legends, the man behind the story was probably less exciting than the tall tales about him. But Johnny Appleseed played a unique role in American frontier life. In 1966, the U.S. Postal Service issued a commemorative stamp honoring Johnny Appleseed and his accomplishments.

National Parks and Recreation Act of 1978. The New Jersey legislature's Pinelands Protection Act of 1979 provided for establishment of a Pinelands Commission, which created a Comprehensive Management Plan, approved in 1981 to balance forest protection and development interests. The plan established the Pinelands Commission to oversee a 337,000 acre (136,380 hectare) core preservation district to be maintained in its natural state through strict regulation of development, and a protection area where there are various categories of land use (forest, agriculture, regional growth, rural development, pinelands, towns and villages, military and federal institutions) based on existing natural features and projected need. The commission appointed a Pinelands Forestry Advisory Committee to devise a forest management plan for the region. In 2004 the Pinelands Commission charged this committee increased responsibilities to review, clarify, and expand the forestry provisions of the Pinelands Comprehensive Management Plan.

During the past 300 years, Pinelands forests were almost completely harvested several times over to provide fuel and raw materials for industries ranging from iron forges, charcoal making, glassblowing, and shipbuilding. Hundreds of small sawmills dotted the Pinelands landscape, producing wood products ranging from timber to cedar shakes. This forest activity and its economic potential dwindled during the twentieth century, which helped to create the consensus that led to the passage of the Pinelands Act. The goals of the Pinelands forestry program are to:

- Maintain native pinelands forest types
- Mimic historic influences, and
- Encourage multiple-use forestry.

In maintaining the native Pinelands forest types, three factors contribute to the essential character of the Pinelands: the physical features of the landscape (relief, soils, and hydrology); the living organisms (Pinelands plants and animals); and ecosystem processes, which are the dynamic interrelationships among and between these living organisms and their particular habitat elements. Fire has profoundly influenced the development of contemporary plant and animal species distribution patterns in the Pinelands. Pinelands forestry management program goals specify that forestry practices should maintain general patterns of native species and ecological communities while allowing for some post-disturbance dynamics. Pinelands forestry should maintain these broad patterns of native species and communities by mimicking the effects of historic and prehistoric fires and tree cutting that created these patterns. For example, the clear-cutting of white pines leads to the establishment of a hardwood forest because pines cannot sprout again from roots as do hardwood trees. The Pinelands forestry guidelines allow for conducting forestry for commercial, stewardship, ecological, and hazard reduction goals.

The forestry guidelines encourage proprietor landscape-scale management plans that help maintain biological diversity and ecological processes, including practices that encourage natural regeneration to maintain locally native forest types, species, and genotypes. Importantly, guidelines suggest that forestry techniques should avoid introducing invasive species or facilitating their spread. Forestry practices should also avoid significant permanent conversion from one native forest type to another, and should help to maintain an understory of native plants. In order to maintain the distribution of cover types, species, fuel structure, and soil structure patterns on the landscape, these forestry practices involve taking account of the return interval (the average time between occurrences of disturbances in a given stand); severity (the amount of vegetation and root system killed, and the type of growing space made available for new plants); landscape pattern (distribution of disturbance patch mosaic effects); the size and timing of fire; cover types; age classes; and the pressures of human development both within and adjacent to the managed forest area.

Forest management should promote commercial forestry, wildfire hazard reduction, and forest stewardship while providing for the long-term environmental integrity of the Pinelands, but avoid irreversible adverse affect on habitat critical to the survival of any threatened or endangered plant or animal species. For example, prescribed fire management is encouraged to achieve ecological and wildfire hazard reduction goals as long as it does not threaten rare or endangered species.

See Also *Agricultural Practice Impacts; Deforestation; Rain Forest Destruction*

BIBLIOGRAPHY

Books

Bellemare, Jesse, Motzkin, Glenn, and David R. Foster. *Thoreau's Country: Journey through a Transformed Landscape: "Legacies of the Agricultural Past."* Cambridge, MA: Harvard, 1999.

Florio, James R., Governor, and the Pinelands Forestry Advisory Committee. *Final Report: Recommended Forestry Management Practices.* Trenton, NJ: New Jersey Pinelands Commission, 2006.

Web Sites

Fisher Museum Harvard Forest. "Landscape History of Central New England." 2006. http://harvardforest.fas.harvard.edu/museum/landscape.html#farm (accessed April 6, 2008).

Michoacán Restoration Fund. "Monarchs." 2006. http://michoacanmonarchs.org/ (accessed April 6, 2008).

Mongabay.com. "Rwanda Launches Reforestation Project to Protect Chimps, Drive Ecotourism." March 18, 2008. http://news.mongabay.com/2008/0318-rwanda.html (accessed April 6, 2008).

Kenneth Travis LaPensee

Resource Extraction

Introduction

Resource extraction refers to activities that involve withdrawing materials from the natural environment. Logging is one example of resource extraction. If not done in a sustainable manner, logging extracts trees and their removal causes other changes that can result in soil and nutrient removal from the logged area. Even if it is accomplished sustainably, logging changes an environment.

Mining that involves the creation of an open-air pit is another example of resource extraction. Other examples include the oil sands project (the extraction of oil-laden sediment from regions of the Canadian province of Alberta), and the more conventional processes of oil recovery.

In contrast to logging, where trees (a renewable resource) can be replanted and nurtured to permit lumbering for a long time, oil extraction is permanent. Fossil fuels that have taken millions of years to develop cannot be renewed. Indeed, agencies including the U.S. Geological Survey (USGS) have forecast that the amount of recoverable oil will peak and then begin to decline before the year 2100.

Efforts to minimize the environmental damage of resource extraction can be challenging, as the economic benefits of activities such as logging, mining, and the oil sands are considerable. For example, some critics of the oil sands project in Alberta have decried the minimal restrictions placed on greenhouse gas emissions by the provincial and federal governments.

Historical Background and Scientific Foundations

Resource extraction activities have been practiced for thousands of years. Archaeological evidence indicates that 6,000 years ago Egyptians mined by building a fire against a rock face and then dousing the fire with cold water to split the rock. The use of gunpowder to blast away rock dates back to 1617 in Germany. Open-pit mining is a more recent development, since sophisticated machinery is needed to expose the subsurface material and to haul it away.

Logging dates back to the use of land to raise crops and livestock. In North America, this began in the 1820s on the East Coast and near the end of that century as settlement from Europe expanded westward.

Clear-cutting—the type of logging in which nearly all or all trees in a given area are removed—is a resource extractive practice on several levels. Complete removal of the forest canopy exposes the ground to more sunlight, which can dry the soil. Also, competition for nutrients becomes more intense as fast-growing shrubs and grasses grow and grazing animals move into the area.

On another level, removing trees can increase the surface movement of water, which can increase erosion. Along with directly extracting the soil, erosion removes more nutrients from the environment. Flow of soil into streams, rivers, and lakes occurs more easily, which can stimulate the increased use of oxygen. As a result, the water can become too oxygen-poor to support fish and plant life.

Operations such as open pit mining and the Alberta tar sands project extensively disrupt the surface. In the case of the tar sands, efforts are made to repair the damage. However, the result will not be the same as the environment prior to the extraction of the oil-laden sediment. More importantly, both operations generate toxic compounds that can escape to the environment. The runoff water from mines and the leftover material can be extremely acidic and can contain concentrations of heavy metals that are lethal to many forms of life.

Impacts and Issues

Clear-cutting can be accomplished faster and less expensively than a tree-by-tree survey and harvesting of a forest. However, the consequences of the clear-cut approach to logging includes erosion, reduced diversity of life in the logged region, and loss of the forest canopy—all of which can be environmentally destructive.

The loss of trees reduces the ability of a forest to retain carbon dioxide. As a result, more of this potent greenhouse gas is released into the atmosphere; the increased warming of the atmosphere since the time of the Industrial Revolution, which began in the late-eighteenth century, has been conclusively linked to increased atmospheric levels of greenhouse gases such as carbon dioxide. Thus, resource extraction not only has a local effect, but influences the global climate.

The Brazilian rain forest, for example, which makes up 30% of the total acreage of rain forest on the planet, is being logged at a rate that if continued could eliminate the forest entirely by 2050.

The oil sands also damage the global environment due to the generation of carbon dioxide in the processing of the extracted oil. Although the Canadian government has committed to reducing Canada's greenhouse-gas emissions by 20% by 2020, the oil sands project has been given approval for expansion, and the resulting increased production of greenhouse gases could make the reduction target difficult to achieve.

SEE ALSO *Air Pollution; Clear-cutting; Commercial Fisheries; Cultural Practices and Environmental Destruction; Logging; Oil Spills; Rain Forest Destruction; Superfund Site*

WORDS TO KNOW

EROSION: The wearing away of the soil or rock over time.

RENEWABLE RESOURCE: Any resource that is renewed or replaced fairly rapidly (on human historical time-scales) by natural or managed processes.

RUNOFF: Water that falls as precipitation and then runs over the surface of the land rather than sinking into the ground.

SILVICULTURE: Management of the development, composition, and long-term health of a forest ecosystem. The objective is often to allow logging of the forest over many years.

BIBLIOGRAPHY

Books

Diamond, Jared. *Collapse: How Societies Choose to Fail or Succeed.* New York: Viking, 2004.

Molyneaux, Paul. *Swimming in Circles: Aquaculture and the End of Wild Oceans.* New York: Thunder's Mouth Press, 2006.

Starr, Christopher. *Woodland Management.* Ramsbury, UK: The Crowood Press, 2005.

Wild, Anthony. *Coffee: A Dark History.* New York: Norton, 2005.

Rivers and Waterways

Introduction

A river is a natural, flowing stream of water that provides an avenue for drainage of water from higher elevations to a standing body of water at lower elevations, which is typically a lake or ocean. Some rivers, however, provide avenues for drainage directly into the groundwater table (for example, a river draining into a sinkhole or a river draining into porous materials lying above the groundwater table). In a river, the flow of water is generally confined to a channel or a network of interconnected channels except when the discharge of water exceeds the river's channel capacity. In that instance, the river temporarily spreads out beyond its channel to a surrounding floodplain until the flood subsides.

A waterway, by contrast, is defined as a body of water that is navigable by boat. This is usually taken to mean a commercial vessel or boat, but a broader definition is any type of boat or floating means of conveyance. A waterway can be a natural feature, a human-made feature, or a natural feature modified by humans. A waterway can be a river, but this concept also includes lakes, oceans, and human-made canals. For a waterway to be navigable, it must be deep enough to accommodate the type of boat at issue (waterway draft), it must be wide enough to accommodate the boat (waterway beam width), it must be free of rapids or waterfalls (or have a way around such features, for example, locks and dams), and it must have a water current velocity that allows boats to move easily against the current. In addition to the world's oceans and large lakes such as the North American Great Lakes, other well-known waterways are the Erie Canal (United States); the Intracoastal Waterway (United States); the Panama Canal (Panama); the St. Lawrence Seaway (Canada); the Suez Canal (Egypt); and the Tennessee-Tombigbee Waterway (United States).

Historical Background and Scientific Foundations

Water flows across the surface of Earth in response to the force of gravity. Thus, surface water always flows naturally in a downhill direction. The slope or gradient of Earth's surface can be very steep in mountainous terrains and very low in plains areas or low-lying coastal areas. However, even a very slight gradient, or slope, will cause water flow. Ever since Earth first had a solid, rocky crust, meteoric water (water than falls as rain) has flowed over the surface. We know this because the oldest rocks on Earth are thermally altered sediments once deposited in river channels and on shorelines.

Rivers have had a profound effect upon the landscapes that we see on Earth's surface. From space, we can see that almost all the land surface has been eroded to one degree or another by the action of running water in streams and rivers. Areas with high rainfall, or that had high rainfall in the past, are the most profoundly affected by the action of running water. Nearly all Earth's surface that is not covered by water or ice is divided into drainage basins. A drainage basin is the area drained by a river system. All rainfall within a drainage basin flows out of that basin through a series of interconnected streams and rivers. Smaller streams and rivers feed into progressively larger rivers, which are referred to as higher order rivers. With higher order rivers, the size of the river and amount of water in the river increase downstream, and the gradient of the river decreases as well.

Impacts and Issues

Humans have used rivers as waterways since the time of nomadic tribes and hunter-gatherer groups. Initially, humans used rivers as resource areas for water and food, but rivers also provided means for transportation. With the advent of civilizations, humans commenced making modifications on rivers, including damming and altering the channels of rivers to accommodate various means of water transportation. As land transportation developed, rivers and waterways were spanned by bridges and bypassed by tunnels. In the past, it was common for rivers to be partially blocked by huge masses of downed trees floating in the river, called log jams. Log jams are rare today and, in fact, most rivers that we see today are managed and controlled by dams or other artificial means.

It is unusual to see a completely natural or uncontrolled river, even in remote areas. In many areas, dams on rivers are used to generate power (called hydroelectricity). In hydroelectric power generation, the flow of water is used to turn turbines, which in turn generate electrical power. Rivers have also become waste disposal avenues with the advent of manufacturing processes that generate liquid wastes. For this reason, may rivers in manufacturing areas are highly polluted or have been polluted in the recent past.

SEE ALSO *Floods; Surface Water*

WORDS TO KNOW

EROSION: The wearing away of the soil or rock over time.

HYDROELECTRICITY: Electricity generated by causing water to flow downhill through turbines, which are fan-like devices that turn when fluid flows through them. The rotary mechanical motion of each turbine is used to turn an electrical generator.

SEDIMENT: Solid unconsolidated rock and mineral fragments that come from the weathering of rocks and are transported by water, air, or ice and form layers on Earth's surface. Sediments can also result from chemical precipitation or secretion by organisms.

BIBLIOGRAPHY

Books

Barry, John M. *Rising Tide: The Great Mississippi Flood of 1927 and How It Changed America.* New York, Simon and Schuster, 1998.

Web Sites

Mississippi Development Authority. "The Tennessee-Tombigbee Waterway." 2007. (accessed April 16, 2008).

U.S. Environmental Protection Agency (EPA). "The Great Lakes." March 4, 2008. (accessed April 16, 2008).

David T. King Jr.

Environmentalists and members of American Rivers—a national conservation group—stand on a pier next to the Passaic River in Newark, New Jersey (April 1998), to announce that a 6-mile (10-km) stretch of the Passaic River has been named among the nation's 20 most endangered rivers. *AP Images*

Runoff

Introduction

Runoff occurs when pollutants are washed into a water body by the movement of water. When rates of precipitation are greater than rates at which water percolates into the soil, water begins moving over the land surface. As the water travels, it picks up sediments and pollutants, which are eventually deposited into water bodies. Runoff is the largest contributor to pollution in both freshwater and marine systems. It contributes to increased turbidity, causes eutrophication, adds toxins to the aquatic food web, and increases rates of erosion. Urbanization, which creates impervious surfaces, exacerbates damage caused by runoff.

Historical Background and Scientific Foundations

Runoff forms from water, usually precipitation, that hits Earth's surface. If the rate of precipitation is greater than the rate at which water can infiltrate the ground, or if the ground is already saturated with water, water accumulates in depressions in the ground. Once small depressions are filled, the water flows across the land. As the water travels, it collects residues and wastes that have collected on the surface of the land or in the soils. If the moving water does not evaporate, it eventually runs into a water body, depositing the collected wastes there. In the continental United States, approximately 31% of the precipitation becomes runoff. Two-thirds return to the atmosphere through evaporation or transpiration of plants. Only about 2% of precipitation infiltrates the ground and returns to the aquifer.

Numerous factors affect runoff. Runoff can form from many different types of precipitation: rain, snow, snowmelt, or sleet. Each affects the quantity and movement of runoff. The intensity, quantity, and duration of the precipitation influence the quantity and speed of the runoff and the amount of material it picks up. Other meteorological factors such as the distribution of precipitation in the watershed, storm direction, air temperature, wind speed and direction, and humidity can affect how much runoff actually enters the water. In addition, the conditions of the soil, the drainage basin topography, and the locations of the water reservoirs impact runoff.

Impacts and Issues

There are several forms of runoff. Agricultural runoff refers to water that has run through agricultural lands. This runoff often contains sediment, herbicides, pesticides, and fertilizer. Urban runoff is water that has collected on city streets, parking lots, and residential areas. This runoff most often includes motor oil, gasoline, garbage, animal waste, herbicides, and pesticides from residential yards. River runoff refers to the water that enters the ocean via a river or stream. The river may carry with it pollutants that it has picked up along its entire length. These pollutants vary depending on the health of the river and the pathway along which it travels. Pollution that enters the ocean via river runoff is point-source pollution. Because the river is a known entry point, it can be monitored. This is in contrast to agricultural runoff and urban runoff, which are nonpoint-sources of pollution and much more difficult to manage. Runoff accounts for more than 40% of all marine pollution. Agricultural runoff is the leading source of environmental degradation in surveyed lakes and rivers in the United States.

The pollutants that are deposited into water bodies via runoff decrease water quality in a variety of ways. Sediment, primarily from agricultural runoff, absorbs light in the water decreasing the light energy available for photosynthesis by plants and phytoplankton. It can also clog feeding structures of filter-feeding organisms. Pollutants like oils, herbicides, pesticides, and heavy metals can kill or damage invertebrates and plants in the water, just like they kill or injure organisms on land. These effects are propagated up the food chain and concentrated in the top predators, which are often fished for human consumption. Animal waste and fertilizer add excessive quantities of nutrients to water bodies. These nutrients trigger massive blooms of microscopic plants known as phytoplankton. Cellular respiration by phytoplankton and the bacteria that consume phytoplankton waste deplete a water body of dissolved oxygen, killing fish and invertebrates that require oxygen for respiration. This condition is known as eutrophication.

Urbanization increases runoff because impervious surfaces like roads and parking lots prevent infiltration of water into the ground. This has several negative consequences. Because water from precipitation runs into streams and the ocean instead of percolating through the soil, aquifers receive less water. This slows rates of groundwater recharge and lowers the water table. When water runs over impervious surfaces, its flow rate increases. This means that it moves with more energy and has the potential to carry large amounts of sediment. The movement of sediment by water is known as erosion. Erosion removes coastal land on which delicate coastal ecosystems depend and increases turbidity to the water. Impervious surfaces also increase the quantity of runoff. Excessive runoff exacerbates the frequency and intensity of floods.

Some legislation in the United States has been aimed at monitoring runoff and decreasing the amount of pollution that reaches water bodies as a result of runoff. The Clean Water Act of 1972 created a permit system to regulate runoff from point-source polluters. In 1987, the act was amended to help monitor and regulate nonpoint-sources of runoff. However, nonpoint-source runoff is extremely hard to assess and regulate. The Coastal Zone Act was amended in 1990 in order to regulate development on coastal lands and to decrease pollution by these activities. In 2002, Congress increased

WORDS TO KNOW

AQUIFER: Rock, soil, or sand underground formation that is able to hold and/or transmit water.

EUTROPHICATION: The process whereby a body of water becomes rich in dissolved nutrients through natural or man-made processes. This often results in a deficiency of dissolved oxygen, producing an environment that favors plant over animal life.

WATERSHED: The expanse of terrain from which water flows into a wetland, water body, or stream.

This least bittern searches for food among the lily pads of the Florida Everglades where an abundance of birds, animals, reptiles, fish, and insects gorge themselves daily from the "River of Grass." Environmentalists say that the waters that form this river are being corrupted by land drainage, contamination from phosphorus-laden runoff and the manipulation of natural water flow. *AP Images*

funding for conservation efforts in the Farm Bill in order to decrease pollution in agricultural runoff.

SEE ALSO *Algal Blooms; Aquatic Ecosystems; Bays and Estuaries; Clean Water Act; Coastal Ecosystems; Environmental Protection Agency (EPA); Groundwater; Groundwater Quality; Industrial Pollution; Industrial Water Use; Insecticide Use; Lakes; Marine Ecosystems; Marine Water Quality; Nonpoint-Source Pollution; Oceans and Coastlines; Precipitation; Water Pollution*

BIBLIOGRAPHY

Books

Cunningham, W.P., and A. Cunningham. *Environmental Science: A Global Concern*. New York: McGraw-Hill International Edition, 2008.

Garrison, Tom. *Oceanography: An Invitation to Marine Science*, 5th ed. Stamford, CT: Thompson/Brooks Cole, 2004.

Web Sites

U.S. Environmental Protection Agency (EPA). "Polluted Runoff (Nonpoint Source Pollution)." February 25, 2008. http://www.epa.gov/owow/nps/ (accessed March 25, 2008).

U.S. Environmental Protection Agency (EPA). "Protecting Water Quality from Agricultural Runoff." March 2005. http://www.epa.gov/owow/nps/Ag_Runoff_Fact_Sheet.pdf (accessed March 28, 2008).

U.S. Geological Survey. "Earth's Water: Runoff." August 30, 2005. http://ga.water.usgs.gov/edu/runoff.html (accessed March 25, 2008).

Juli Berwald

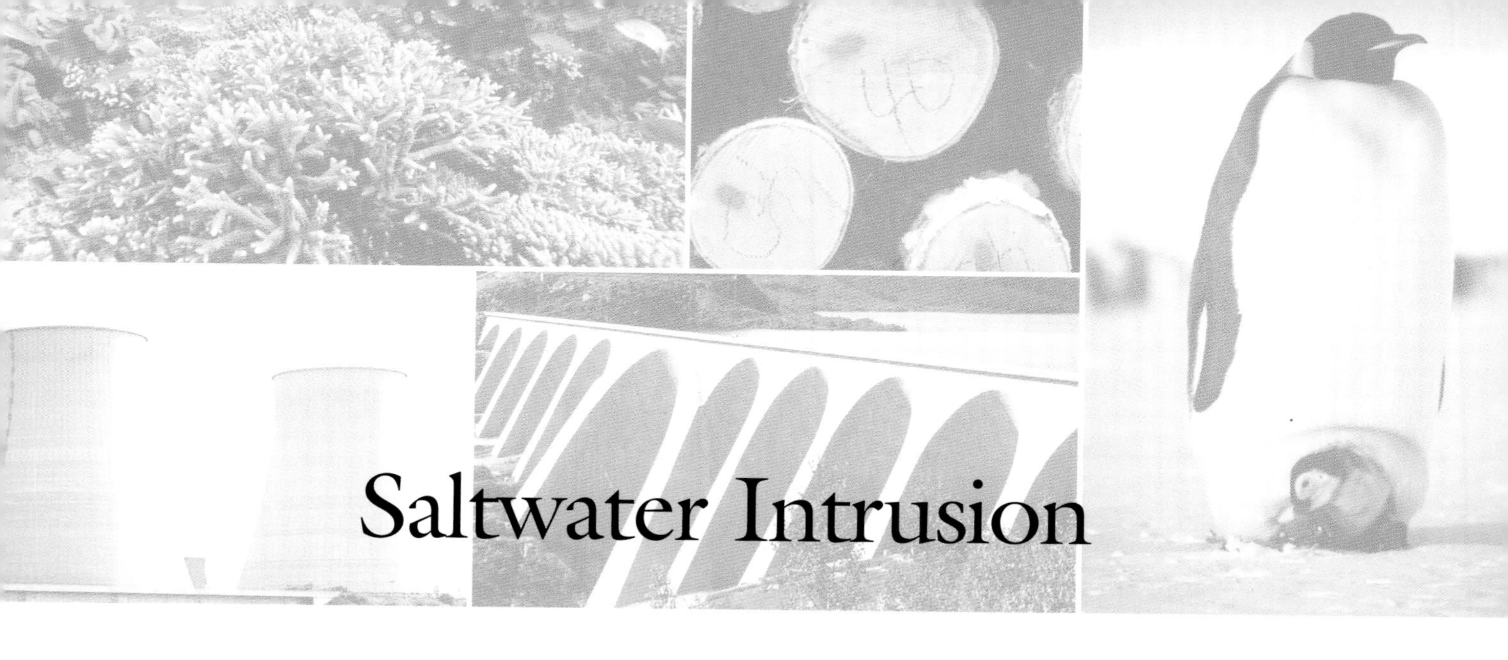

Saltwater Intrusion

Introduction

Saltwater intrusion or encroachment is the movement of saltwater into underground sources (aquifers) of freshwater, which can occur in coastal regions or inland, and the surface movement of saltwater inland from the coast.

Saltwater intrusion is most common in coastal regions, where the freshwater is displaced by the inland movement of saltwater from the ocean. But it can also occur inland, far away from an ocean, as freshwater is pumped out from underground reservoirs and the salt-laden water from surrounding salty layers of the earth flow in.

The most common cause of saltwater intrusion is the pumping of freshwater from wells near coasts. Climate change can increase saltwater encroachment along coastal regions as sea level rises. Increased salinity of coastal freshwater can threaten the plant life and wildlife of coastal areas, destroy habitats such as marshes, and force the abandonment of drinking-water supplies.

Historical Background and Scientific Foundations

Normally, along coastal regions, freshwater flows downhill to the sea through aquifers. This outward flow prevents seawater from moving inland. Also, freshwater contains a lesser content of minerals than saltwater and so will float on top of saltwater. The boundary between the freshwater and saltwater tends to be near the coast.

The most common cause of saltwater encroachment is the pumping of freshwater from wells. Since freshwater runs down slope to the sea, even underground, intercepting and removing some of this flow causes less freshwater to reach the coast; this allows saltwater to migrate inland and can force the abandonment of wells. Another mechanism of saltwater encroachment on wells is termed upconing. This occurs when a well draws freshwater from an aquifer that overlies saltwater. Pulling large amounts of freshwater can draw up a cone-shaped volume of saltwater into the porous rock of the freshwater aquifer; if the peak of this cone of saltwater reaches the draw point of the well, the well ceases to produce freshwater even though most of the freshwater aquifer is undisturbed. Rising coastal populations are accompanied by increased freshwater demand, which results in more water being pumped from wells; this, in turn, increases rates of saltwater intrusion.

An increasingly important cause of saltwater encroachment is sea-level rise due to anthropogenic (human-caused) climate change. Human beings are changing global climate by adding certain gases to the atmosphere, especially carbon dioxide (CO_2) and methane (CH_4), through burning fossil fuels and through agriculture. Destroying forests also contributes to global climate change. All these activities tend to warm Earth as a whole, although this warming is not evenly spread over time or space.

Measurements of tide levels that have been conducted for over a century and satellite measurements since the 1970s have shown a continuing rise in sea level. According to the 2007 Assessment Report of the Intergovernmental Panel on Climate Change (IPCC), the average global sea level rose by 0.07 in (1.8 mm) per year from 1961 to the early 1990s, and, since 1993, has been rising by 0.12 in (3.1 mm) each year.

The cause of the sea level rise is twofold. First, warming ocean waters expand. Second, melting mountain glaciers and polar ice sheets increase the volume of the ocean by adding water to it. Both processes are results of anthropogenic climate change.

As the volume of ocean waters increases, sea level rises, causing the coast to move inland. A rise in sea level can cause saltwater to move farther inland in regions where the coastline is near sea-level in height, simply

WORDS TO KNOW

WATERSHED: The expanse of terrain from which water flows into a wetland, water body, or stream.

WETLANDS: Areas that are wet or covered with water for at least part of the year.

from the increased volume of water present. Also, the tendency of the saltwater to penetrate to underground freshwater reservoirs is increased.

If the involved freshwater reservoir is used as a source of drinking water, the pumping of the freshwater from wells can encourage the movement of the saltwater from lower in the aquifer to the vicinity of the wells. This can contaminate drinking water and lead to abandonment of the drinking water source.

Impacts and Issues

The rising sea levels documented by the IPCC have already affected some coastal locales. One example is the Canadian province of Prince Edward Island. According to the Geological Survey of Canada, rising sea levels are contributing to increased damage caused by coastal storms. Erosion of the beaches is threatening the island's nearly $3 million annual tourism industry. Almost half of the coastline could be submerged by 2100.

Another area being affected by saltwater encroachment is the peninsular state of Florida in the United States. The advancing sea level has already affected freshwater bodies, including coastal wetlands. The low elevation of the state even puts the Florida Everglades at risk. Wetlands such as the Everglades are home to many species of birds and wildlife. A threat to wetlands water quality threatens the entire ecosystem.

IN CONTEXT: SALTWATER INTRUSION

Following the 2005 Indian Ocean tsunami destruction, saltwater intrusion created substantial problems, including saltwater contamination of agricultural areas, freshwater wells, and septic systems. The United Nations Environmental Program (UNEP) estimated that 62,000 freshwater wells in Sri Lanka alone were contaminated by marine water.

The fact that many of Florida's shorelines are gently sloping will increase the height of the sea-level rise, which could exceed 2 ft (0.6 m) by 2100. The state's drinking water supplies will be threatened.

Developed regions such as Florida have the resources and infrastructure to at least attempt to deal with the changing ecology caused by saltwater encroachment. In less developed areas of the world, however, the threat to drinking water supplies and coastal life will be far less easily overcome, and could displace many people.

Freshwater supplies of small islands are particularly threatened by sea-level rise. On small islands, freshwater aquifers exist as relatively small lens-shaped bodies sitting on top of, and surrounded by, saltwater. Even small rises in sea level can drastically shrink, or even eliminate, such lenses of freshwater.

As the global climate warms and drinking water scarcity grows, the threat of saltwater contamination of aquifers will also grow. For example, in the United States, approximately two thirds of the known underground sources of drinking water are surrounded by formations containing saline water. Overuse of the reservoirs to supply drinking water to an increasing number of people and for agricultural use will lead to saltwater contamination.

SEE ALSO *Climate Change; Global Warming; Sea Level Rise; Water Resources*

BIBLIOGRAPHY

Books

Houghton, John. *Global Warming: The Complete Briefing*. New York: Cambridge University Press, 2004.

Parry, M. L., et al, eds. *Climate Change 2007: Impacts, Adaptation and Vulnerability: Contribution of Working Group II to the Fourth Assessment Report of the Intergovernmental Panel on Climate Change*. New York: Cambridge University Press, 2007.

Pugh, David. *Changing Sea Levels: Effects of Tides, Weather and Climate*. New York: Cambridge University Press, 2004.

Valsson, Trausti. *How the World Will Change with Global Warming*. Oxford, UK: Oxbow Books, 2007.

Web Sites

U.S. Geological Survey. "Freshwater-Saltwater Interactions along the Atlantic Coast." http://water.usgs.gov/ogw/gwrp/saltwater/salt.html (accessed April 3, 2008).

Brian D. Hoyle

Sea Level Rise

Introduction

Global sea level, the average height of the ocean's surface apart from the daily changes of the tides, is rising. Both the main causes of this change are linked to anthropogenic (human-caused) global warming. First, warming of the ocean causes it to expand. This effect, confined mostly to the ocean's top 2,300 ft (701 m), is called thermal expansion or thermosteric expansion. Second, melting of glaciers and other bodies of ice lying on land causes the total mass of water in the ocean to increase. The heat energy causing both thermal expansion and ice melting comes from warming of the atmosphere, which has in turn been caused in recent decades primarily by anthropogenic climate change. Retention of liquid water on the continents, as for example by the damming of rivers, makes a small negative contribution to sea-level rise, withholding some water from the oceans.

In the late 1800s, after two or three thousand years of stability, sea level began to rise steadily at about 0.07 in (1.7 mm) per year. From 1993 to 2007, the rate of sea-level rise increased to about 0.12 in/yr (3 mm/yr). Sea level will continue to rise for at least the next few centuries because of global-warming processes already set in motion, but how much more it rises will depend on factors that are difficult to predict, including whether human societies reduce their greenhouse-gas emissions and the degree to which the West Antarctic and Greenland ice sheets melt. The future behavior of these ice sheets is uncertain. Recent data show that melting of both sheets has recently been faster than any computer climate model had predicted. Rising sea levels harm human coastal communities and natural coastal ecosystems.

Historical Background and Scientific Foundations

History of Sea Level Change

Earth's sea levels have risen and fallen due to natural causes many times over the planet's history. During the last half-million years, Earth has gone through about half a dozen ice ages alternating with warmer periods called interglacial periods. During each ice age, water evaporating from the oceans falls as snow on cold regions near the poles and stays there, locked up in the form of land-based ice sheets. This causes sea level to drop. (Ice that forms as floating sheets, on the other hand, such as the sea ice floating over the North Pole today, does not affect sea level directly, either by forming or melting.)

During the last four glacial cycles, each of which has lasted about 100,000 years, sea level has fallen about 400 ft (122 m) when the ice was at its height and then risen again by the same amount as the ice melted. During the warmest part of the last interglacial period, about 130,000–118,000 years ago, Earth's average global air temperature was 3.6–5.4°F (2–3°C) warmer than today and sea level was 13–20 ft (4–6 m) higher than today. During the warm interglacial period of the Middle Pliocene, about three million years ago, global temperature was 6.3–9.9°F (3.5–5.5°C) warmer than today and sea level was 80–115 ft (24–35 m) higher than today.

The patterns of ancient sea-level change are therefore not a matter of merely abstract interest: Scientists study them for clues to the changes that the human race may face in coming decades and centuries. Since most of the water that is available for increasing the mass of the oceans resides in the ice sheets of Antarctica and Greenland, understanding the nature of past changes to these ice sheets is particularly important for predicting future sea-level rise and other aspects of climate change. If the ice sheets can be partly or wholly destabilized by

WORDS TO KNOW

CALVING: Process of iceberg formation in which huge chunks of ice break free from glaciers, ice shelves, or ice sheets due to stress, pressure, or the forces of waves and tides.

EUSTATIC SEA LEVEL: Change in global average sea level caused by increased volume of the ocean (caused both by thermal expansion of warming water and by the addition of water from melting glaciers). Often contrasted to relative sea level rise, which is a local increase of sea level relative to the shore.

FORAMINIFERA: Single-celled marine organisms that inhabit small shells and float free in ocean surface waters. The shells of dead foraminifera sink to the sea bottom as sediment, forming thick deposits over geological time. Because the numbers and types of foraminifera are climate-sensitive, analysis of these sediments gives data on ancient climate changes.

GREENHOUSE-GAS EMISSIONS: Releases of greenhouses gases into the atmosphere.

ICE AGE: Period of glacial advance.

INTERGOVERNMENTAL PANEL ON CLIMATE CHANGE (IPCC): The Intergovernmental Panel on Climate Change (IPCC) was established by the World Meteorological Organisation (WMO) and the United Nations Environment Programme (UNEP) in 1988 to assess the science, technology, and socioeconomic information needed to understand the risk of human-induced climate change.

PALEOCLIMATE: The climate of a given period of time in the geologic past.

RELATIVE SEA LEVEL: Sea level compared to land level in a given locality. Relative sea level may change because land is locally sinking or rising, not because the sea itself is sinking or rising (eustatic sea level change).

THERMOSTERIC EXPANSION: Expansion as a response to a change in temperature, also called thermal expansion.

TREE RINGS: Marks left in the trunks of woody plants by the annual growth of a new coat or sheath of material. Tree rings provide a straightforward way of dating organic material stored in a tree trunk.

TIDE GAUGE: A device, usually stationed along a coast, that measures sea level continuously. Measurements from tide gauges were the main source of sea-level data prior to the beginning of satellite measurements in the 1970s.

global warming and melt relatively rapidly—that is, over decades and centuries rather than over thousands of years—then sea-level rise over the next century might turn out to be significantly larger than the half-meter to a meter predicted by the Intergovernmental Panel on Climate Change (IPCC). Many scientists—for example, Jonathan Overpeck and colleagues (2006)—have argued that standard models of ice-sheet instability and melting are inadequate, and that sea-level rise may happen larger and faster than the IPCC's cautious predictions allow.

For example, warming of the polar regions—especially the north—by the end of the twenty-first century may raise air temperatures to levels not seen for at least 118,000 to 130,000 years, during the last major interglacial period, when sea level was 13 to 20 ft (4–6 m) higher than today. If sea-level rise corresponds to temperature increase, equally large increases in sea level may happen again, though not all in the coming century. This is not, however, a worst-case scenario. In 2007, the IPCC estimated that unless successful efforts are made to curb greenhouse-gas emissions, by 2100 Earth's average temperature may rise by 9.9°F (5.5°C), a temperature associated three million years ago with sea levels up to 115 ft (35 m) higher than today's. At the end of the last interglacial period, sea levels rose at speeds of between 0.43–0.78 in/yr (11–20 mm/yr). Although such an increase would be slow by everyday human standards, it is much faster than the 0.102±0.002 in/yr (2.6±0.04 mm/yr) that was occurring as of 2006. Since many coastal lands are nearly level with the sea, a 0.5 in/yr (1.2 cm/yr) sea-level rise would cause them to retreat by several feet to many feet annually. This would disrupt coastal human settlements and natural coastal ecosystems at a pace that would be hard to adapt to.

The most recent interglacial sea-level rise happened from about 7,000 to 20,000 years ago. After that time, sea levels rose slightly or not at all until the late 1800s. Starting then, the oceans began to rise at a rate of about 0.07 in/yr (1.7 mm/yr). Recently, as a result of global climate change, sea-level rise has accelerated: from 1993 to 2007, the oceans rose at about 0.12 in/yr (3 mm/yr).

Measuring Sea Level Changes

Ancient climate, called paleoclimate from the Greek *palaois*, is revealed by the various marks it left on the land, the sea bottoms, the ice caps, and elsewhere. For example, average global air temperatures up to about 900,000 years ago can be inferred from the atomic composition of air bubbles trapped in ancient ice layers in Antarctica and Greenland. These layers record the date of their formation much as tree rings record the age of a tree. Also, cylinders of layered muck drilled from the ocean floors contain trillions of shells of single-celled organisms called foraminifera, the atomic composition of whose shells records water temperature. Fossil corals found above or below present-day sea level record ancient sea levels, since corals grow only in a narrow range of water depth, and these fossil corals can also be dated by their atomic composition (the ratios of certain atomic isotopes in their skeletons). Sea levels from the present back to about 500,000 years ago can be inferred from fossil corals. More ancient sea levels are shown by other evidence, such as fossils marking ancient inland shorelines.

Modern sea levels are measured both on-site and remotely. On-site measurements are made by tide gauges, instruments attached to posts standing in the water near shore. High-quality tide-gauge data from around the world date back to about 1950. Lower-quality data from a smaller number of gauges, mostly in the Northern Hemisphere, go back to 1870.

Tide gauges have several disadvantages. First, the gauges must be checked regularly to make sure that the pilings they are attached to are not sinking into the ground, giving a false impression of sea-level rise. Second, their readings must be corrected for changes of local altitude due to post-glacial rebound. Third, tide gauges measure sea level only at a series of unevenly distributed points around the edges of the ocean.

Since 1992, data from satellite radar altimeters aboard the TOPEX/Poseidon and Jason satellites have allowed precise global measurement of sea levels. Satellite radar altimeters send pulses of radio energy toward Earth's surface and measure the time until an echo is received. The distance from the satellite to the surface can be calculated from the delay. Satellite data have shown faster sea-level rise since 1993, up to about 0.12 in/yr (3 mm/yr) from 0.07 in/yr (1.7 mm/yr). This observation has been confirmed by the tide-gauge data.

Global Sea Level Budget

Thermal expansion and ocean mass are the two biggest contributors to global sea-level change. Thermal expansion of the ocean (thermosteric expansion) occurs mainly as its top 2,300 ft (700 m) or so swells because of warming. For most of the twentieth century, thermosteric expansion was the main cause of sea-level rise: The ocean's heat content increased by about 2×10^{23} joules, about 20 times as much as the atmosphere's heat content increased during that time. (The ocean can absorb more heat than the atmosphere because it weighs far more.) The expansion caused by this much heat corresponds to a sea-level rise of 1.2 in (3 cm). Today, this warmth-driven, expansive sea-level rise has accelerated with increased warming of the atmosphere to about 2.4 in (6 cm) per century.

The other major cause of sea-level rise is the melting of land-based ice. Ice on land is transported to the land by two processes, namely melting followed by runoff and dumping of glacial ice into the sea. Mountain glacial ice and ice around the edges of Greenland has been melting and running off as water,

Earth layers are seen on the shoreline of the Dead Sea in Jordan, showing the shrinking water level of the sea—the lowest point and the saltiest sea on Earth. The surface level of the saltiest water in the world has been receding by 3.3 ft (1 m) every year for the past 25 years. *AP Images*

IN CONTEXT: SEA LEVEL RISING, BUT BY HOW MUCH?

In early 2008, ongoing independent research projects confirmed that global sea level—the average height of the ocean's surface apart from the daily changes of the tides—is rising. Sea-level rise threatens freshwater supplies, coastal land use, and vital economic activity. For some islands, significant sea-level rise could be devastating. In 2007, the Intergovernmental Panel on Climate Change (IPCC)—an international body of hundreds of government-appointed scientists and economists representing as broadly as possible the world community of climate and weather scientists—foretold that by 2100, sea levels would rise between 7 and 23 in (18 and 59 cm). It was widely reported that the upper end of this estimate's range had been reduced from the IPCC's 2001 estimate of 3.1–34.7 in (8–88 cm). However, no other figure in the IPCC report has been so disputed by scientists as the prediction for sea-level rise. The range calculated by the IPCC for its 2007 report did not include the possibility of increased contributions to sea-level rise from accelerated glacial melting. A number of scientists have argued in the world's leading scientific journals that future sea-level rise may be greater than the IPCC's 2007 forecast, but is unlikely to be less.

while glaciers in southern Greenland and along the edges of the West Antarctic Peninsula have recently accelerated their slide to the sea. As chunks of glacial ice break off into the water, a process called calving, they raise sea level at once; when they finally melt, there is no further change. Since about 1993, because of accelerated loss of mountain glaciers around the world and of parts of the Greenland and West Antarctic ice sheets, land-based ice has been responsible for about half of the observed sea-level rise, more than in the earlier part of the twentieth century.

The Twentieth-Century Sea Level Enigma

Scientists have puzzled over what they call the enigma of twentieth century sea-level rise, first pointed out by American oceanographer Walter Munk in 2002. The enigma or mystery is that the known sources of sea-level rise do not seem to add up to the observed rise. From 1961 to 2003, sea level rose at about 0.017 in/yr (0.43 mm/yr) due to thermal expansion. In many reports, a "±" sign indicates a range of uncertainty around such data (e.g., 0.028 ±0.002 in/yr). For simplicity and readability, the uncertainty figures are often omitted and so such values should be considered approximate. Sea levels rose at approximately 0.020 in/yr (0.51 mm/yr) due to the melting of glaciers and other minor bodies of land-based ice, 0.002 in/yr (0.05 mm/yr) due to melting of the Greenland ice sheet, and 0.006 in/yr (0.15 mm/yr) due to melting of the Antarctic ice sheet. The sum of these contributions to sea-level rise was approximately 0.043 in/yr (1.0 mm/yr), but the observed rise was 0.071 in/yr (1.8 mm/yr), leaving a difference of 0.028 in/yr (0.7 mm/yr) unexplained. For the period 1993 to 2003, the unexplained remainder was smaller, only 0.012 in/yr (0.3 mm/yr).

A number of explanations for the discrepancy have been suggested, including shifts in Earth's rotational axis (polar wander), overestimation of sea-level rise, and underestimation of either the thermal expansion or ice melting. In 2007, European researchers claimed to have resolved the enigma of excess sea-level rise noted by Munk in 2002. According to G. Wöppelmann and colleagues, the apparent shortfall in the sea-level rise budget is whittled away from both ends at once: 1) Since 2002, estimates of the anthropogenic greenhouse contribution to sea-level rise have been upped to 0.06 in/yr (1.4 mm/yr) from Munk's figure of 0.02 in/yr (0.7 mm/yr); and 2) Wöppelmann and colleagues' corrections to tidal-gauge data using the Global Positioning System (GPS) reduce, they say, the amount of rise to be explained from 0.07 in/yr (1.8 mm/yr) to 0.05 in/yr (1.3 mm/yr). Thus, the enigma is resolved. However, the new claim will have to be verified by other scientists before agreement can be reached that the global sea-level budget is closed. Other explanations for the enigma, including polar wander, have also been strongly defended.

Impacts and Issues

Effects of Sea Level Rise

In some parts of the world, the land is still rising after the removal of the weight of the ice sheets at the end of the last glacial period about 10,000 years ago: The land bobs upward after the ice melts off, like a cork pushed partly under water and then released. This effect, termed post-glacial rebound, is causing the relative sea level in some places to shrink faster than eustatic (actual, global) sea-level rise is making it grow. In Stockholm, Sweden, for example, post-glacial rebound is causing the relative sea level to decrease at about 0.16 in/yr (4 mm/yr). Such locations will experience stable sea levels if eustatic sea-level rise increases to an equal rate, and in that case will see no direct negative effects from sea-level rise.

However, some other locations, such as Chesapeake Bay in the eastern United States, are experiencing the opposite effect. There, the southern edge of the area pushed down by the last Ice Age, formerly raised, is sinking again like one end of a see-saw after the rider on the other end gets off. Chesapeake Bay is therefore seeing faster relative sea-level rise (about twice the global average) than the eustatic rise. Most of the planet, including the areas inhabited by most of the 100 million or so people who live within several feet (a meter) of the present-day sea level, are seeing neither form of post-glacial rebound and will experience the effects of eustatic sea-level rise directly. Those effects will be almost universally negative. They include flooding of river deltas, erosion of shores, infiltration of saltwater into island groundwater supplies, disappearance of small, low-lying islands, increased vulnerability of large, low-lying cities to violent storms, and more.

Disputed Projections

In 2007, the IPCC—an international body of hundreds of government-appointed scientists and economists representing as broadly as possible the world community of climate and weather scientists—foretold that by 2100, sea levels would rise between 7 and 23 in (18 and 59 cm). It was widely reported that the upper end of this estimate's range had been reduced from the IPCC's 2001 estimate of 3.1-34.7 in (8–188 cm). It was apparently reassuring news: Sea-level rise might not be as bad as we thought.

However, no other figure in the authoritative IPCC report has been so disputed by scientists as this one. The range calculated by the IPCC for its 2007 report did not include the possibility of increased contributions to sea-level rise from accelerated glacial movement in the Greenland and West Antarctic ice sheets. The West Antarctic ice sheet contains enough water to raise sea level by 16 ft (5 m) and Greenland contains enough to raise it by 23 ft (7 m). During most of the twentieth century, these bodies of ice contributed only a few tenths of a millimeter per year to sea-level rise; as of 2006, however, scientists estimated that due to a combination of surface melting and accelerated glacial flow, Greenland was now contributing over 0.02 in/yr (0.5 mm/yr) to global sea-level rise, more than double previous estimates.

A number of scientists have argued on the basis of paleoclimatic data that the ice sheets are capable of much more rapid change than the IPCC has reported—change that could cause sea-level rise of up to 0.78 in/yr (20 mm/yr). The IPCC acknowledges uncertainties about ice-sheet instabilities. Future sea-level rise may be greater than the IPCC's 2007 forecast, but is unlikely to be less. The group underestimated sea-level rise in its 2001 report, predicting a maximum rise of no more than 0.08 in/yr (2 mm/yr) from 1993 to 2006: The actual rise turned out to be 0.13 in/yr (3.3 mm/yr). German climatologist Stefan Rahmstorf commented in the journal *Science* in 2007, "In a way, it is one of the strengths of the IPCC to be very conservative and cautious and not overstate any climate change risk."

■ Primary Source Connection

The following news article recognizes the Dutch as true innovators of controlling water through dikes, levees, and dam systems, and are now moving forward with "climate-adaptation plans." Such plans are being drafted and acted on in hopes of preventing flooding as a response to climate change, which could cause sea levels to rise significantly each year.

HOW TO FIGHT A RISING SEA

The Dutch enjoy a hard-earned reputation for building river dikes and sea barriers. Over centuries, they have transformed a flood-prone river delta into a wealthy nation roughly twice the size of New Jersey.

If scientific projections for global warming are right, however, that success will be sorely tested. Globally, sea levels may rise up to a foot during the early part of this century, and up to nearly three feet by century's end. This would bring higher tidal surges from the more-intense coastal storms that scientists also project, along with the risk of more frequent and more severe river floods from intense rainfall inland.

Nowhere does this aquatic vise squeeze more tightly than on the world's densely populated river deltas.

So why is one of the most famous deltas—the Netherlands—breaching some river dikes and digging up some of the rare land in this part of the country that rises (barely) above sea level?

In the Biesbosch, a small inland delta near the city of Dordrecht, ecologist Alphons van Winden looks out his car window at a lone excavator filling a dump truck with soil. He considers the question and laughs. "We do have a hard time explaining this to foreigners," he says.

The work here represent a keystone in the country's climate-adaptation plans, Mr. van Winden says. Indeed, nowhere are adaptation planning efforts to address rising sea levels and flooding more advanced than in the Netherlands.

To be sure, the country's economic wealth and long experience dealing with threats from seas and rivers give it an advantage over other low countries that face rising waters, such as Bangladesh, Vietnam, and the tiny tropical island nation of Tuvalu in the South Pacific. But many of the approaches the Netherlands is taking can and are being slowly adopted even in countries far poorer, specialists say.

The excavation work here is one example of what van Winden calls "soft approaches" to flooding in this small nation where competing interests jostle for every square foot of land. By buying out the few farmers remaining in this region, breaching the dikes they built to protect their land, and digging additional water channels, the

Dutch government aims to reduce peak flood flows at Dordrecht and other cities downstream. No longer will tightly constricted river and canal channels hold high water captive. Big floods will overspread the Biesbosch, reducing the threat of water spilling over the top of levees that guard densely populated cities to the west.

The Biesbosch may also be critical to the future of farming on the productive southwest coast. There, most of the area's fresh water sources are close to the coast—and vulnerable to salt-water contamination from a rising North Sea. This could make farming difficult, if not impossible. The Biesbosch, however, hosts three large reservoirs, each surrounded by a 20-foot-high dike. Fresh water piped from these reservoirs, some 50 miles inland, could keep coastal areas supplied.

1.4 billion live near seacoast

Globally, some 21 percent of the world's 6.6 billion people live within 20 miles of a seacoast—and nearly 40 percent within 60 miles, says Robert Nicholls, a professor of civil and environmental engineering at the University of Southampton in England.

Seacoast populations who face the greatest risk from floods, storms, and sea-level rise live on river deltas, says the UN-sponsored Intergovernmental Panel on Climate Change.

In the IPCC's latest set of reports on the impact of global warming, released earlier this year, scientists looked at data from 40 of the globe's river deltas, home to 300 million people. If current trends continue through 2050, flooding in the Nile, Mekong, and Ganges-Brahmaputra river deltas could each displace more than 1 million people. Up to a million more may be forced to head for higher ground in each of another nine deltas, including the Mississippi River delta. Up to 50,000 could be forced to relocate in each of 12 other deltas, including the Rhine River delta—an area known more widely as the Netherlands.

Besides global warming, scientists say the challenges these regions face have other causes as well. Levees, sea walls, drainage canals, dams, and other land-use patterns have taken a toll. Deltas tend to subside (sink) naturally, accentuating the rise in sea level. Past engineering projects can actually limit the ability of natural processes to replenish the landmass of deltas.

A patch of the Netherlands between Rotterdam and Gouda, called Zuidplaspolder, highlights the issue in a way that New Orleans might recognize. The 19-square-mile area is bounded by dikes and the Gouwe River. Face the river, and the landscape looks like a typical river plain. But turn and face Zuidplaspolder, and you see a steep decline dropping more than 20 feet. The huge dimple in the delta stretches as far as the eye can see.

It's the lowest spot in Europe, some 23 feet below sea level.

"And it's all subsidence," says Willemien Croes, a planner with the provincial government of South Holland. Over the centuries, residents dug up thick layers of peat to warm their homes in Gouda, Rotterdam, and Amsterdam, she says. Much of Zuidplaspolder then filled with water. Farmers pumped it dry, grew crops, and raised dairy herds on the rich clay and peat. When the soil settled, farmers ringed the area with dikes for protection.

The area's low elevation and the anticipated increased future risk of floods, combined with development pressures from Rotterdam and Gouda, have turned this area into one of the country's biggest adaptation challenges. But it's hardly alone: Some 60 percent of the country, accounting for 70 percent of its gross domestic product, lies below sea level.

These sinking lowlands are protected along the coast by sand dunes, dikes, and sea barriers that stretch across the mouths of estuaries. These natural and engineered defenses have protected millions from the North Sea since a devastating storm surge hit the country in 1953. But these defenses have come at an ecological cost. Unlike river deltas such as the Mississippi's, which grew as sediment washed downriver from deep in the North American interior, the Dutch delta was built by the sea. Currents swirling through the Strait of Dover since the end of the last ice age eroded the white cliffs and deposited the material along the Dutch coast.

That process has slowed substantially, says van Winden, who works for Stroming, an environmental consulting firm in Nijmegen. Although the delta drains three of Europe's major rivers—the Rhine, Meuse, and Scheldt—the rivers never carried enough sediment to build the delta, and don't carry enough silt to maintain it today. From that standpoint, he says, over the long term "we are living beyond our means."

Dutch humble in face of rising threat

Faced with the twin threats of increased river flooding from inland storms and higher ocean storm surges as the climate warms and sea levels rise, the country aims to meet these challenges with a variety of approaches, ranging from complex engineering to "natural." But it's doing so with increased humility, given the levee failures in New Orleans after hurricane Katrina in 2005.

"If you want a caricature of the Netherlands, it's: 'We have the dikes; we are 100 percent safe. So just go on with your life,'" says Pieter Bloemen, who runs the government's Adaptation Program for Spatial Planning and Climate. But these days, "even we proud Dutch, with climate change in the back of our heads, have to think about broken dikes. That's a big paradigm shift."

Zuidplaspolder is a case in point. As the lowest real estate in one of the Netherlands' most vulnerable provinces, it has become a test bed for factoring water and climate change into zoning and development plans. In the next 20 years, 15,000 to 30,000 new housing units will be built here. Anticipating this growth, in 2004, officials from provincial and local governments joined with non-governmental organizations to develop a master plan for the polder. (A polder is a large tract of land containing farms and villages encircled by dikes. The dikes offer flood protection, but they also turn the polders into enormous bathtubs with bottoms that slowly, inexorably sink.)

The new homes that rise in the polder may look nothing like those in the villages the Dutch are used to, Mr. Bloemen says. To deal with floods, homes on this higher ground could be designed to float in place or built on stilts. They may sport tall ground floors, with living space and utilities placed on higher floors. Entire villages might be built to float in place, linked by buoyant sidewalks and roads.

In addition, he adds, officials may ask developers to use a technique that dates back centuries: building houses, even whole villages, on mounds. That low-tech approach is appearing in other parts of the world, too. Oxfam International is working with villages in Bangladesh to build individual homes and even small villages on flood-resistant mounds.

In the Netherlands, river floods are a top item on the climate-change adaptation must-fix list. To be sure, the country has tried to be forward-looking in tackling flood control and sea-level rise, notes Hans Balvoort, with the Netherlands' Ministry of Public Works, Transport, and Water Management. It typically uses a 50-year planning horizon. But a wake-up call came in the 1990s, "when, for the first time, rainfall was so heavy and intense that our pumping systems could not cope," he says.

Powerful pumps long ago replaced the signature windmills as the way to keep the polders from flooding. "On such a large scale," he says, the inability of pumps to keep pace with rainfall was "something we had not experienced before."

Moreover, for two winters during that decade, flooded rivers rose so high that officials evacuated some 250,000 people out of concern that levees might not hold. Instead of building large numbers of new levees, he continues, scientists, engineers, and officials looked for other ways to store flood waters over the short term to reduce the risk.

The Biesbosch project, with its dike removal, or "depoldering," is one approach. The government also is working on a range of other strategies to give flooded rivers more room to flow. They might spread dikes farther apart, excavate land between river and dikes (to capture overflow), deepen central river channels, remove jettylike groins that now force most of the flow into the center of a river, remove other obstructions, and even add new channels to the flood plain or restore old ones.

Storm surge is biggest coastal worry

The government plans to spend €2.2 billion ($3.2 billion) to make these changes to its rivers. Meanwhile, along the coast, the big worry is not about any average increase in sea level, which scientists project to rise here between 35 and 85 cm (14 to 33 inches) by 2100. Instead, the biggest concern is the change in storm-surge patterns that will ride atop that rise, says Pier Vellinga, who heads the climate program at Wageningen University.

As if to highlight this point, last weekend Britain and the Netherlands closed their sea barriers in the face of a storm in the North Sea that sent a 13-foot surge bearing down on their coasts.

Planners in other countries often design for a once-every-hundred-years storm. While that approach can be useful, the challenge is that climate change may throw those projections out of whack. For example, some researchers say that in the U.S. Northeast, midcentury coastal winter storms could lead to flood levels every three or four years—floods of a severity that used to occur only once every 100 years. Netherlands planners aim for a 10,000-year storm for the country's most vulnerable areas. And even that may be inadequate, Dr. Vellinga says.

"When you do an economic assessment of the damage," he says, "and what you can afford to [spend to] avoid that damage, a better safety level would be a recurrence of 1 in 100,000 years." One storm like that could cost the country up to a year's worth of gross domestic product—€500 billion ($730 billion).

In 1990, the government decided to maintain the country's existing coastline by replenishing its extensive phalanx of coastal dunes using enormous deposits of sand that lie far offshore—another geological gift delivered over millennia from the English Coast to the Netherlands.

Three years ago, the government added that it will strive not only to maintain the coastline at its current position, but also to maintain the shape of the current offshore slope to a depth of about 130 feet. Today, that means dredging and depositing nearly 16 million cubic yards of sand along the coast each year. So, as the sea level rises, the dunes will, too, says Joost Stronkhoorst, with the National Institute for Coastal and Marine Management at The Hague.

Offshore sand deposits are large enough to allow the Dutch to accommodate a rise in sea level up to 16 feet, he says. But the line of coastal dunes is not unbroken. The gaps are spanned by barriers that in some cases will require 20 feet added to their height given sea-level-rise scenarios out to 2100.

In some cases, that's not possible. The northern coastal town of Petten shows why. It's tucked hard against the back of a sea dike that traces its origins to the Middle Ages—and sits 14 feet below the level at which waves crash on the other side.

To build up a dike, you must expand its base, explains Roel Posthoorn, with the Dutch nature trust Natuurmonumenten, as he stands on the crest of the dike on a blustery fall afternoon. The presence of the village eliminates the chance to expand the dike's base inland. And churning North Sea currents already sweep away precious coastal sand from the seaward edge of the dike's base, preventing planners from trying to expand the dike seaward.

Possible solution: artificial reefs

Here, Mr. Posthoorn says, the long-term solutions may lie in building an offshore reef to reduce the height of the waves slamming into the dike. Or, as some are now beginning to suggest, perhaps the large deposits of sand offshore should also be used to build the country's coast westward by nearly a mile.

In the meantime, groups like Natuurmonumenten are working to meet two of the country's adaptation goals by trying to prevent further development behind sea dikes like this one and converting the land to nature reserves. These "climate buffers" are another tool in the Netherlands' kit for coping with global warming.

Adaptation experts generally agree that scientists, engineers, and policymakers already know what needs to be done to adapt to global warming. For the most part, they say, it means doing what they already know how to do to reduce risks from natural hazards—it's just doing more of it and a better job of it.

As if to underscore the point, Henk Wolfort, a researcher at Alterra, an institute at Waganingen University that focuses on sustainable development, shows a set of maps illustrating the evolution of watery areas and polders in the country since the 14th century.

"Our problems are not so very different from the problems the people in the Middle Ages had," he says. Even back then, techniques like building on mounds or widening the space between river dikes to accommodate flooding were well understood. The lesson? In a high-tech age, some of the effective adaptation approaches may come from a decidedly low-tech time.

"I think the people in the Netherlands have forgotten about those old ideas because they have relied on technological solutions," he says. "Now they see that technical solutions don't provide 100 percent safety. So perhaps we should think about the old solutions again."

Peter N. Spotts

SPOTTS, PETER N. "HOW TO FIGHT A RISING SEA." *CHRISTIAN SCIENCE MONITOR* (NOVEMBER 15, 2007).

SEE ALSO *Climate Change; Global Warming; Intergovernmental Panel on Climate Change; IPCC 2007 Report; Oceans and Coastlines*

BIBLIOGRAPHY

Books

Parry, M. L., et al, eds. *Climate Change 2007: Impacts, Adaptation and Vulnerability: Contribution of Working Group II to the Fourth Assessment Report of the Intergovernmental Panel on Climate Change.* New York: Cambridge University Press, 2007.

Solomon, S., et al, eds. *Climate Change 2007: The Physical Science Basis: Contribution of Working Group I to the Fourth Assessment Report of the Intergovernmental Panel on Climate Change.* New York: Cambridge University Press, 2007.

Periodicals

Alley, Richard B., et al. "Ice-Sheet and Sea-Level Changes." *Science* 310 (2005): 456–460.

Dowdeswell, Julian A. "The Greenland Ice Sheet and Global Sea-Level Rise." *Science* 311 (2006): 963–964.

Ericson, Jason P. "Effective Sea-Level Rise and Deltas: Causes of Change and Human Dimension Implications." *Global and Planetary Change* 50 (2006): 63–82.

Gregory, J. M., and P. Huybrechts. "Ice-Sheet Contributions to Future Sea-Level Change." *Philosophical Transactions of the Royal Society A* 364 (2006): 1709–1731.

Meehl, Gerald A., et al. "How Much More Global Warming and Sea Level Rise?" *Science* 307 (2005): 1769–1772.

Munk, Walter. "Twentieth Century Sea Level: An Enigma." *Proceedings of the National Academy of Sciences* 99 (2002): 6550–6555.

Oppenheimer, Michael. "The Limits of Consensus." *Science* 317 (2007): 1505–1506.

Overpeck, Jonathan T. "Paleoclimatic Evidence for Future Ice-Sheet Instability and Rapid Sea-Level Rise." *Science* 311 (2006): 1747–1,750.

Rahmstorf, Stefan. "A Semi-Empirical Approach to Projecting Future Sea-Level Rise." *Science* 315 (2007): 368–370.

Shepherd, Andrew, and Duncan Wingham. "Recent Sea-Level Contributions of the Antarctic and Greenland Ice Sheets." *Science* 315 (2007): 1529–1532.

Web Sites

Goddard Institute for Space Studies, NASA. "Coastal Populations, Topography, and Sea Level Rise." http://www.giss.nasa.gov/research/briefs/gornitz_04 (accessed April 2, 2008).

National Aeronautics and Space Administration (NASA). "NASA Satellites Measure and Monitor Sea Level." http://www.nasa.gov/home/hqnews/2005/jul/HQ_05175_sea_level_monitored.html (accessed April 2, 2008).

Woods Hole Oceanographic Institution. "Rising Sea Levels and Moving Shorelines." http://www.whoi.edu/page.do?pid=12457&tid=282&cid=2484 (accessed April 2, 2008).

Larry Gilman

Seasonal Migration

Introduction

Seasonal migration refers to the movement of various bird, insect, and mammal species from one habitat to another during different times of the year. The migration is necessary because seasonal fluctuation in factors such as the availability of food, sunlight, and the temperature of the air or water become intolerable in one habitat, usually during the winter, making breeding and even survival difficult.

An example is the migration of various whale species from their summer habitat in Arctic or Antarctic waters to their wintertime breeding grounds in tropical waters near the equator. Some bird species also migrate from their summer home in the Arctic to spend winter at warmer latitudes.

For some species, the seasonal migration cycle involves journeys that exceed tens of thousands of miles or kilometers in length. A migration route can involve resting and refueling stops, although some species complete their migration non-stop.

Efforts to protect migratory species include public and private initiatives, routing of shipping away from known migration routes, and international agreements. The latter includes the Convention on Migratory Species (the Bonn Convention) and the Ramsar Convention on Wetlands.

Historical Background and Scientific Foundations

Seasonal migration likely dates back at least 3,000 years, as records of bird migrations written by the ancient philosophers Homer and Aristotle have been recovered. Biblical descriptions of bird behavior consistent with migration can also be found in the Books of Job and Jeremiah.

Seasonal migration can take place over very long distances. The route of the Arctic tern between the Arctic and Antarctic exceeds 13,670 mi (22,000 km), and the Bar-tailed Godwit travels non-stop over 6,835 mi (11,000 km) from Alaska to New Zealand. Among whale species, the blue whale journeys from Arctic or Antarctic waters to its breeding grounds in equatorial waters in a journey that can take four months, and during this time it does not eat. The route of the Eastern Northern Pacific gray whale between the coastal waters off Baja, California, and the northern Okhotsk Sea between Russia and Japan is over 6,200 mi (10,000 km) in length.

Bat, seal, turtle, and insect species all seasonally migrate. For example, the monarch butterfly migrates from southern Canada to winter in central Mexico.

Stopping en route to rest and refuel with food and water is common among migratory bird species. Wetlands and agricultural land that has been harvested are popular rest stops for species such as the Canada Goose. The need for space and resources such as water along the migration route makes seasonal migration susceptible to climate conditions that adversely affect the land. Drying or pollution of wetlands or conversion of land for urban or industrial use can threaten the migratory journey. As one example, dwindling populations of the monarch butterfly have been traced to overlogging of their wintering grounds in central Mexico. Organizations including Ducks Unlimited work to acquire and preserve wetlands to preserve migratory routes.

Impacts and Issues

Climate change can affect seasonal migration. Bird migration has been affected for species whose route takes them over deserts. Regions such as the Sahel region of

Africa have become drier, decreasing the opportunity to stop and refuel. In highly developed areas of the world, light pollution can also pose problems for birds who migrate at night and use the moon and stars for navigation.

Whale species can be threatened when their migration takes them through shipping routes. An example is the northern right whale, whose migration terminates in Canada's Bay of Fundy. Every year several whales are struck and killed by commercial vessels, depleting a population that has been estimated to be only about 300. As well, the pollution of coastal waters with agricultural and urban runoff is a health threat.

Because migration can be over national boundaries, international cooperation is necessary to help protect migratory species. The Convention on Migratory Species, implemented in Bonn, Germany, in 1979 under the United Nations Environment Programme (UNEP) is concerned with the global conservation of migratory wildlife and habitats. While North America and coastal waters are part of the migratory routes for a variety of bird and mammal migratory species, the United States, Canada, and Mexico are not among the 104 nations to have ratified the convention.

SEE ALSO *Habitat Loss; Light Pollution; Migratory Species*

WORDS TO KNOW

HABITAT: The natural location of an organism or a population.

LIGHT POLLUTION: Also known as photopollution and luminous pollution, refers to the presence of excessive amounts of light in the atmosphere.

WETLANDS: Areas that are wet or covered with water for at least part of the year.

BIBLIOGRAPHY

Books

Freyfogle, Eric T. *Why Conservation Is Failing and How It Can Regain Ground*. New Haven: Yale University Press, 2006.

Hermes, Patricia. *Fly Away Home*. New York: Newmarket, 2005.

Web Sites

Journey North. http://www.learner.org/jnorth/ (accessed May 2, 2008).

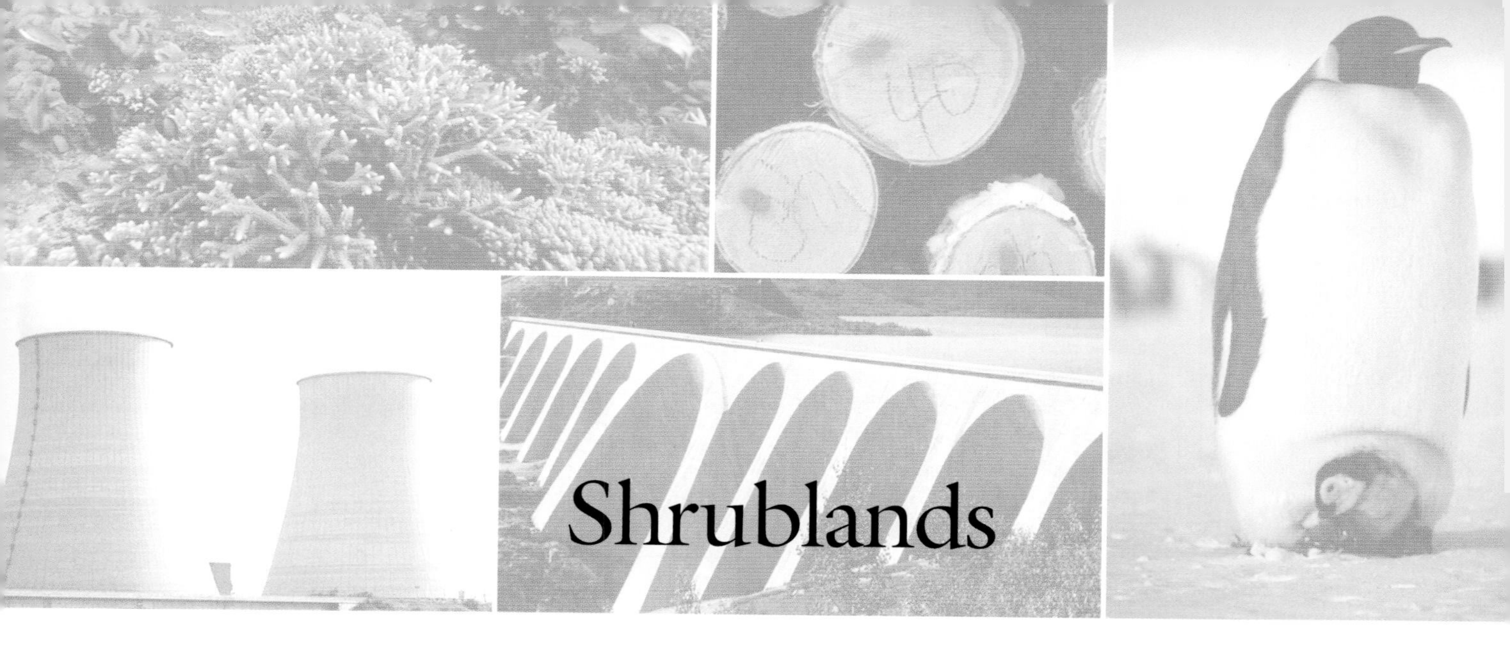

Shrublands

Introduction

Shrublands refer to regions whose main plant life are woody shrubs—low-lying plants that have several stems instead of a single trunk protruding from the base. Woody shrubs are also known as bushes. Woody shrubs are perennial (they bloom every year without the need of planting of new seeds).

Shrublands are located in various regions of the world. They can be permanent or temporary. Temporary shrublands can arise after an event such as a forest fire or clear-cut logging, and represent a transition between the former environment and the re-established mature environment. Other shrublands are created by the overuse of an environment by humans.

The barren appearance of a shrubland can be deceiving, as the environment can support a variety of life. However, the barren appearance can lead to misuse and overexploitation of the land for other purposes, which can result in the encroachment of nearby desert areas (one form of desertification).

The loss of shrubland plant cover may increase the nighttime radiation of heat back to the atmosphere, further fueling atmospheric warming.

Historical Background and Scientific Foundations

The plants that occupy shrubland are small bushes that grow up to a maximum of about 20 ft (6 m) in height. Most often, shrubland plants are lower than 10 ft (3 m) in height. Shrublands are extensive in regions of the globe near the equator (between 32° and 40° north and south of the equator) that do not receive extensive rainfall. This territory includes southwestern regions of North America such as Arizona, New Mexico, and Texas, and the Mediterranean, the central area of Chile, portions of Brazil, the Cape area of South Africa, and regions of Australia.

These regions are characterized by hot, dry summers and cool, moist winters, or by distinct seasons that are very wet and very dry. The shrubs tend to be evergreen plants that are hardy, and able to tolerate fire and sparse nutrients, or non-evergreens that shed their leaves during the dry time of the year. Examples of plants are the chaparral (Mediterranean), caatinga (Brazil), mallee (Australia), and sagebrush (North America).

These naturally occurring shrublands tend to be stable and long lasting. Other shrubland that also persists over time can result from the excessive lumbering of trees and the overgrazing of cattle. An example is the moors of Scotland, which at one time were pine forests. Excessive lumbering altered the ecosystem so drastically that the re-establishment of forests was hindered. Instead, the bushy moor plants became dominant.

Other shrubland is not permanent, but represents a stage between the former environment and the similar environment that will develop over time. This is known as successional shrubland; the shrubland is one of a number of stages (successions) between the original and ultimate environments. Most shrublands in temperate regions of the globe tend to be successional systems, and are an indication of the extensive land use that has occurred in the past.

Although shrublands can appear desolate, they are actually ecosystems that support many species of plant and animal life. The parts of the plant that are above the ground are slower growing that the roots, which can grow far down into the soil to reach water. The soil is stabilized by the roots. The low-lying and dense plants provide protection for birds and animals.

Impacts and Issues

Some shrubs can produce and secrete compounds that are toxic to herbs in the vicinity. This helps the shrub compete for the available nutrients by eliminating some of its competitors. These compounds may be an exploitable source of natural herbicides. Thus, the destruction of shrublands may be depriving scientists of valuable compounds that could reduce the use of synthetic herbicides.

The barren and dull-colored appearance of shrublands can lead to their disregard. Use of the land for other purposes destroys the groundcover that birds and animals rely on for protection and habitat. Furthermore, the absence of soil stabilizing roots can lead to use of soil through wind and water erosion, and the loss of the fire-tolerant bushes can increase the tendency of fires to spread and cause more extensive destruction.

The extensive areas of arid, semi-arid, and temperate regions of Earth that are covered by shrubland make the plants an influential contributor to climate modification. Experiments have shown that the absence of shrubs increases the radiation of heat to the atmosphere, especially at night. The influence of extensive loss of shrubland on the global climate is unclear, as is the effect of global warming on shrublands.

SEE ALSO *Desertification; Ecosystems; Grasslands; Tundra*

WORDS TO KNOW

CLEAR-CUT: A parcel of forest that has been denuded of trees. Clear-cutting can be destructive of forests, particularly when the cycle of reforestation is slow and the processes of wind and water erosion of deforested land make it inhospitable to reforestation. However it can also be a tool for increasing the biodiversity of forests that have been protected from forest fires for many years.

DESERTIFICATION: Transformation of arid or semiarid productive land into desert.

ECOSYSTEM: The community of individuals and the physical components of the environment in a certain area.

SAVANNA: One of Earth's biomes characterized by an extensive cover of grasses with scattered trees. The savanna biome is a transitional biome between those dominated by forests and those dominated by grasses and is associated with climates having seasonal precipitation accompanied with a seasonal drought.

TUNDRA: A type of ecosystem dominated by lichens, mosses, grasses, and woody plants. It is found at high latitudes (arctic tundra) and high altitudes (alpine tundra). Arctic tundra is underlain by permafrost and usually very wet.

BIBLIOGRAPHY

Books

Burnie, David. *Shrublands.* New York: Raintree, 2003.

Kump, Lee R., James F. Kasting, and Robert G. Crane. *The Earth System.* New York: Prentice-Hall, 2003.

Parker, Gary. *Exploring the World Around You: A Look at Nature from Tropics to Tundra.* Green Forest, AR: Master Books, 2003.

Web Sites

World Resources Institute. "Ecosystem Area: Open Shrublands." http://earthtrends.wri.org/searchable_db/index.php?theme=9&variable_ID=756&action=select_countries (accessed April 28, 2008).

Silent Spring

Introduction

Silent Spring, a book about the dangers of pesticide use on the environment and human health, was written by American marine biologist Rachel Carson (1907–1964). It was first published as a series of articles in the *New Yorker* magazine, then as a book in 1962. It had a remarkable impact, perhaps more so than any other environmentalist book of the last century, and it has remained continuously in print since its publication. *Silent Spring* is often credited with having changed the scope and nature of the environmental movement, and helping trigger the creation of the U.S. Environmental Protection Agency (EPA).

Historical Background and Scientific Foundations

In the mid-twentieth century, agriculture in the United States and some other countries became highly industrialized. That is, rather than being carried on by a large population of farmers working many relatively small farms using animals for draft (pulling) power in the fields and fertilizing with manure, it became mechanized and reliant on chemical inputs, especially fertilizers manufactured using fossil fuel, pesticides (chemicals designed to poison pests, especially insects), and herbicides (chemicals designed to poison unwanted plants). By the 1950s, hundreds of millions of pounds of pesticides were applied to crops in the United States, as well as to lawns and gardens. Large profits were generated by sales of these chemicals, but little public attention was paid to the harms they might be doing. Rather, journalists tended to celebrate the wondrous new age of high-productivity in industrial agriculture, where farmers driving air-conditioned tractors and harvesters across vast fields could listen to music while raising history's largest crops of pest-free wheat, corn, and other foods.

Scientists, meanwhile, were gathering a large body of evidence that the uncontrolled application of herbicides and pesticides was causing death and sickness in many bird and animal species, including some humans. Rachel Carson, a marine biologist, gathered together much of this evidence and wrote a passionate appeal advocating controls for poison spraying, *Silent Spring,* that received praise both within the scientific community and the public at large. Carson argued that pesticides were being used recklessly and that this was endangering human health, threatening species of wild animals and birds with extinction, and—if the trend toward greater and greater use of such poisons continued—leading toward a world stripped of much of its life, especially its birds. Her grim vision of a spring with no birds to sing—a silent spring—gripped millions of readers.

The pesticide on which she focused most was DDT (dichloro-diphenyl-trichloroethane), a chemical discovered in 1874, but not used as a pesticide until 1939. DDT and related chemicals are particularly damaging to the environment because they do not break down quickly and can be passed up the food chain, concentrating in the tissues of predators. For example, DDT sprayed on leaves or seeds is first eaten by insects, which are poisoned by the chemical. Before or after the insects die, they are eaten by birds, which build up DDT in their own tissues. At sufficiently high concentrations, the DDT kills the birds; at lower concentrations, it interferes with egg development, causing eggshell thinning and fewer chicks to hatch. Fish that eat DDT-contaminated insects also concentrate DDT in their tissues; eagles eating the fish concentrate the DDT still further. In North America, bald-eagle reproduction rates were depressed by DDT while it was still widely used, but recovered in the years following the banning of the chemical in 1972. The EPA has also classified DDT as a likely human carcinogen (chemical that potentially causes cancer).

Impacts and Issues

Carson's book was an immediate bestseller and triggered widespread public discussion of chemical pollution, pesticides, and the need for regulation. She was called to testify before Congress, and President John F. Kennedy (1917–1963) established a special commission to study whether chlorine-containing insecticides such as DDT should be banned. In 1963, the commission recommended such a ban, and in 1972, DDT was banned by law in the United States.

Carson's book was attacked angrily by supporters of the chemical industry even before its publication and continues to be reviled today. A number of critics make the argument that use of DDT was halted for insufficient scientific reason, allowing malaria—a disease spread by mosquitoes that causes 1.2 to 2.7 million deaths per year—to spread and kill without restraint. In 2007, the hundredth anniversary of Carson's birth, a number of editorials in politically conservative journals and mainstream newspapers were devoted to expressing this view. For example, columnist John Tierney wrote in the *New York Times* that *Silent Spring* was a "hodgepodge of science and junk science, dubious statistics, and anecdotes.... The human costs have been horrific in the poor countries where malaria returned after DDT spraying was abandoned."

In fact, Carson never called for a total ban on DDT, in *Silent Spring* or elsewhere, and no such ban has ever been instituted. The 1972 law that banned DDT in the United States contained an exception for malaria control—never invoked because malaria is rare in the United States. Nor has DDT ever been banned for anti-malarial use internationally (only for agriculture): As of 2007, DDT is again used for indoor spraying in 10 African countries. The global decline in DDT use for outdoor anti-malaria spraying is partially due to the evolution of increased insect resistance to the chemical, a problem with all widely used pesticides. In *Silent Spring*, Carson advised DDT use to be limited to "Spray as little as you possibly can." Thus, Carson never advocated a reckless and inhumane ban on using DDT or any other pesticides for fighting insect-borne disease.

In 1980, Carson was posthumously awarded the Presidential Medal of Freedom.

WORDS TO KNOW

DDT (DICHLORO-DIPHENYL-TRICHLOROETHANE): One of the earliest insecticides, dichloro-diphenyl-trichloroethane, used until banned by many countries in the 1960s after bird populations were decimated by the substance, and other negative environmental consequences occurred. Lately, selective DDT use has returned in targeted areas in Africa in order to eliminate high concentrations of the mosquitoes that carry the parasite that causes malaria.

FOOD CHAIN: A sequence of organisms, each of which uses the next lower member of the sequence as a food source.

FOSSIL FUEL: Hydrocarbon fuel that has been obtained from the death and decay of living matter millions of years ago.

Holding her then controversial book "Silent Spring," Rachel Carson stands in her library in March 1963. She says she "wanted to bring to public attention" her charges that pesticides are destroying wildlife and endangering the environment. *AP Images*

Primary Source Connection

Rachel Carson's book *Silent Spring* was the first meticulously referenced, book-length attack by a scientist on the practices of an entire industry, in this case the pesticide industry. *Silent Spring* provoked a response that was also the first of its kind, as partly described in the 1962 article "'Silent Spring' Is Now Noisy Summer" by John M. Lee. A many-fronted public-relations counterattack on Carson was mounted by corporations displeased by her book.

The full extent of the industry response was not known at the time that Lee wrote his article. Besides the criticisms quoted there (published before the book version of *Silent Spring* appeared), the chemical and pesticide industry took a number of specific actions against

IN CONTEXT: RACHEL CARSON

Rachel Carson (1907–1964) is best known for her 1962 book, *Silent Spring*, which is often credited with beginning the environmental movement in the United States. The book focused on the uncontrolled and often indiscriminate use of pesticides, especially dichloro-diphenyl-trichloroethane (commonly known as DDT), and the irreparable environmental damage caused by these chemicals. The public outcry Carson generated by the book motivated the U.S. Senate to form a committee to investigate pesticide use. Carson's eloquent testimony before the committee altered the views of many government officials and helped lead to the creation of the U.S. Environmental Protection Agency (EPA).

Before beginning her graduate studies at Johns Hopkins, Carson had arranged an interview with the U.S. Bureau of Fisheries for a part-time science writer to work on radio scripts. The only obstacle was the civil service exam, which women were then discouraged from taking. Carson not only did well on the test, she outscored all other applicants. She went on to become only the second woman ever hired by the bureau for a permanent professional post.

At the Bureau of Fisheries, Carson wrote and edited a variety of government publications—everything from pamphlets on how to cook fish to professional scientific journals.

Carson. A group of chemical companies—including DuPont, Monsanto, Shell, Dow Chemical, W.R. Grace, and the members of the Manufacturing Chemists Association—hired public relations experts to question Carson's credibility and even, on occasion, her sanity. The National Agricultural Chemicals Association spent over a quarter of a million dollars to oppose the book through ads, press conferences, and other public-relations methods.

Velsicol Chemical Company, a pesticides manufacturer, threatened to sue Houghton Mifflin for libel. Velsicol also threatened *Audubon* magazine with a lawsuit if it published excerpts from *Silent Spring*, warning that printing "a muckraking article containing unwarranted assertions about Velsicol pesticides [might] jeopardize [the] financial security" of persons employed by the magazine and that of their families. One chemical company threatened to sue the *New Yorker* if the magazine ran the final installment of Carson's book. The editor responded, "Everything in those articles has been checked and is true. Go ahead and sue." In the end, however, no libel lawsuits were actually brought against Carson or anyone else involved in the publication of her book.

'SILENT SPRING' IS NOW NOISY SUMMER

Pesticides Industry Up In Arms Over a New Book

Rachel Carson Stirs Conflict—Producers Are Crying 'Foul'

The $300,000,000 pesticides industry has been highly irritated by a quiet woman author whose previous works on science have been praised for the beauty and precision of the writing.

The author is Rachel Carson, whose "The Sea Around Us" and "The Edge of the Sea" were best sellers in 1951 and 1955. Miss Carson, trained as a marine biologist, wrote gracefully of sea and shore life.

In her latest work, however, Miss Carson is not so gentle. More pointed than poetic, she argues that the widespread use of pesticides is dangerously tilting the so-called balance of nature. Pesticides poison not only pests, she says, but also humans, wildlife, the soil, food and water.

The men who make the pesticides are crying foul. "Crass commercialism or idealistic flag waving," scoffs one industrial toxicologist. "We are aghast," says another. "Our members are raising hell," reports a trade association.

Some agricultural chemicals concerns have set their scientists to analyzing Miss Carson's work, line by line. Other companies are preparing briefs defending the use of their products. Meetings have been held in Washington and New York. Statements are being drafted and counter-attacks plotted.

A drowsy midsummer has suddenly been enlivened by the greatest uproar in the pesticides industry since the cranberry scare of 1959.

Miss Carson's new book is entitled "Silent Spring." The title is derived from an idealized situation in which Miss Carson envisions an imaginary town where chemical pollution has silenced "the voices of spring."

The book is to be published in October by the Houghton Mifflin Company and has been chosen as an October selection of the Book-of-the-Month Club. About half the book appeared as a series of three articles in The New Yorker magazine last month.

A random sampling of opinion among trade associations and chemical companies last week found the Carson articles receiving prominent attention.

Many industry spokesmen preface their remarks with a tribute to Miss Carson's writing talents, and most say that they can find little error of fact.

What they do criticize, however, are the extensions and implications that she gives to isolated case histories of the detrimental effects of certain pesticides used or misused in certain instances.

The industry feels that she has presented a one-sided case and has chosen to ignore the enormous benefits in increased food production and decreased incidence of disease that have accrued from the development and use of modern pesticides.

The pesticides industry is annoyed also at the implications that the industry itself has not been alert and concerned in its recognition of the problems that accompany pesticide use.

Last week, Miss Carson was said to be on "an extended vacation" for the summer and not available for comment on the industry's rebuttal. Her agent, Marie Rodell, said she had heard nothing directly from chemical manufacturers concerning the book.

Houghton Mifflin referred all questions to Miss Rodell. The New Yorker said it had received many letters expressing great interest in the articles and "only one or two took strong objection."

In an interview, E. M. Adams, assistant director of the biochemistry research laboratory of the Dow Chemical Company, said he would be among the first to acknowledge that there were problems in the use or misuse of pesticides.

"I think Miss Carson has indulged in hindsight," he said. "In many cases we have to learn from experience and often it is difficult to exercise the proper foresight."

Emphasizing that he spoke as a private toxicologist, Mr. Adams said that in some procedures, such as large-scale spraying, the possible benefits had to be balanced against the possible ills.

He referred to the extensive testing programs and Federal regulations prevalent in the pesticides industry and said, "What we have done, we have not done carelessly or without consideration. The industry is not made up of money grubbers."

Tom K. Smith, vice president and general manager of agricultural chemicals for the Monsanto Chemical Company, said that "had the articles been written with necessary attention to the available scientific data on the subject, it could have served a valuable purpose-helping alert the public at large to the importance of proper use of pesticide chemicals."

However, he said, the articles suggested that Government officials and private and industrial scientists were either not as well informed on pesticide problems as Miss Carson, not professionally competent to evaluate possible hazards or else remiss in their obligations to society.

P. Rothberg, president of the Montrose Chemical Corporation of California, said in a statement that Miss Carson wrote not "as a scientist but rather as a fanatic defender of the cult of the balance of nature." He said the greatest upsetters of that balance, as far as man was concerned, were modern medicines and sanitation.

Montrose, an affiliate of the Stauffer Chemical Company, is the nation's largest producer of DDT, one of the pesticides that Miss Carson discusses at length. She also discusses the effect of malathion, parathion, dieldrin, aldrin and endrin.

"It is ironic to think," Miss Carson states at one point, "that man may determine his own fixture by something so seemingly trivial as his choice of insect spray." She acknowledges, however, that the effects may not show up in new generations for decades or centuries.

The Department of Agriculture reported that it had received many letters expressing "horror and amazement" at the department's support of the use of potentially deadly pesticides.

The industry had a favorite analogy to use in rebuttal. It conceded that pesticides could be dangerous. The ideal was to use them all safely and effectively.

The public debate over pesticides is just beginning and the industry is preparing for a long siege. The book reviews and publicity attendant upon the book's publication this fall will surely fan the controversy.

John M. Lee

LEE, JOHN M. "'SILENT SPRING' IS NOW NOISY SUMMER." *NEW YORK TIMES* (JULY 22, 1962).

■ Primary Source Connection

The following news article acknowledges the American ornithologist Bridget Stutchbury's book *Silence of the Songbirds*, which, like Rachel Carson's *Silent Spring*, discusses how human actions can negatively impact the environment, particularly in relation to birds. Stutchbury makes the connection between consumer choices and the future of the habitat of these birds. Stuchbury holds the Canada Research Chair in Ecology and Conservation Biology at Yale University.

PICKING UP WHERE 'SILENT SPRING' LEFT OFF

By now, the litany of human-driven environmental problems is probably beginning to sound depressingly familiar. Coral reefs are bleaching, forests and the wildlife they host are disappearing, and humanity, which has doubled in numbers since the 1960s, is stressing Earth's resources. Scientists predict that global warming will only exacerbate these problems. Many worry about imminent ecosystem meltdowns. Civilization is inextricably linked to the natural world and as Jared Diamond and others have pointed out, in the past, ecological shifts have coincided with the collapse of entire civilizations.

To this rather grim picture, ornithologist Bridget Stutchbury brings a book about an often overlooked and important denizen of the natural world: the songbird. In *Silence of the Songbirds*, she explores the reasons behind songbirds' alarming decline in the past 40 years. She tells of how, flying thousands of miles at nighttime in vast flocks that show up on radar, the little birds connect Canada's boreal forests with Central and South America's jungles. She explains how forests rely on them for pollination, seed dispersal, and insect control. With the exhaustive and sometimes plodding style of a good trial lawyer, she documents how habitat loss, forest fragmentation, and pesticides have wreaked havoc on their populations. And she postulates new and fascinating reasons why fragmentation might cause songbird populations to decline: Songbirds avoid forests devoid of other birds. They need habitat large enough to sustain bird communities.

Thankfully, after outlining these problems, Stutchbury doesn't desert the reader. By connecting consumer choices with harmful land practices in the birds' wintering grounds, she gives readers an avenue of action. What coffee you choose—organic and shade-grown versus sun-grown and pesticide-treated—can lead to either more or less bird habitat. Even your choice of toilet paper can mean a tree cut down in Canada's boreal forests, where many songbirds summer. Choose the right products as a consumer, and you can help rather than hurt.

Unfortunately, Stutchbury doesn't explore possible solutions beyond consumer choice. As she's well aware, enlightened consumer choices can't fix everything. The poverty, weak regulation, and nearly nonexistent enforcement of existing laws that drive environmental degradation in Central and South America cannot be remedied only by drinking Fair Trade coffee in Boston. Effective solutions will require both top-down and bottom-up efforts by all those involved. And we are all involved.

So yes: *Silence of the Songbirds* is another book you probably won't feel like reading. Who needs to feel forlorn, dejected, and guilty and responsible? But before you move on to more pleasant and probably less relevant fare, let me explain why the book is, in fact, worth reading.

Extinction and ecosystem collapse are part of Earth's natural history. Species have eaten themselves and others out of existence before and they'll do so again. And yet, in the current human-driven disturbance, there is a new element to this oft-repeated drama. Probably never before has a species possessed both full knowledge of what was happening and the know-how to avoid it.

And this is why, as both observer and participant, it's worth being fully apprised of what's at stake. If we make the wrong decisions, it's simply nature's version of "business as usual." But in a "sustainable" resolution to our environmental dilemmas lies the beginning of something truly unique on earth. For the first time, an organism will have altered its behavior not when catastrophe hit but before. Homo sapiens—Latin for "wise human"—will have truly earned its moniker. This self-adjustment would herald the emergence of a new intelligence in both the human and natural realms.

Moises Velasquez-Manoff

VELASQUEZ-MANOFF, MOISES. "PICKING UP WHERE 'SILENT SPRING' LEFT OFF." *CHRISTIAN SCIENCE MONITOR* (AUGUST 14, 2007).

SEE ALSO *Agricultural Practice Impacts; Environmental Activism*

BIBLIOGRAPHY

Books

Carson, Rachel. *Silent Spring*. Greenwich, CT: Fawcett Crest, 1962.

Murphy, Priscilla Coit. *What a Book Can Do: The Publication and Reception of Silent Spring*. Amherst, MA: University of Massachusetts Press, 2007.

Russell, Edmund. *War and Nature: Fighting Humans and Insects with Chemicals from World War I to Silent Spring*. Cambridge, UK: Cambridge University Press, 2001.

Periodicals

Tierney, John. "Fateful Voice of a Generation Still Drowns Out Real Science." *New York Times* (June 5, 2007).

Web Sites

People's Weekly World Newspaper. "Defending Rachel Carson," by David Pimentel. http://www.pww.org/article/articleprint/11914/ (accessed April 12, 2008).

Larry Gilman

Smog

Introduction

Smog is a particularly potent type of air pollution. It is usually highly visible as a brownish or yellowish haze in the air, and although smog is often associated with large cities, it occurs around the world. Smog occurs when emissions from factories and cars mix with air under certain atmospheric conditions that trap pollutants near to the ground. Heavy traffic, warm weather, and still air tend to contribute to smog build up in cities such as Los Angeles, California. Ground level ozone is often a component of smog, and it is created by a complex series of chemical reactions catalyzed by sunlight.

Smog has a number of damaging impacts on health. The most obvious effects in a thick urban smog would be watering eyes and coughing. More serious impacts include triggering of asthma and heart attacks. Smog also reduces visibility, which can be dangerous when driving. Environmental legislation promoting cleaner air has dramatically reduced the incidence of smog in many countries, but it continues as a worldwide issue.

Historical Background and Scientific Foundations

In 1952, around 4,000 people in London, England, died as a result of the notorious Great Smog. Emissions from factories combined with smoke from people's chimneys to create a great cloud of pollution that enveloped the city. The term smog, which means fog intensified by smoke, was coined in 1905, although the problem of the so-called "pea-souper" fogs occurring in urban areas dates back much further, and is mentioned in the work of nineteenth-century British writer Charles Dickens (1812–1870).

Smog is visible as a brownish haze polluting the air, particularly in cities and especially in the summer. It comprises an aerosol of noxious droplets and particles. One major component of smog is ozone, which is formed when nitrogen oxides and unburned hydrocarbons from motor vehicles react in the presence of sunlight to create a photochemical smog. Ozone in the stratosphere is helpful, as it absorbs ultraviolet (UV) light from the sun. Ground level ozone is a powerful pollutant that affects the eyes and lungs.

Los Angeles is a classic example of a city where climate and geographical factors create conditions for the formation of photochemical smog. The city is surrounded by mountains on three sides and has a dry, sunny climate. At night, the sky is usually clear and the ground cools fast. Air near the surface is cool, while upper layers are relatively warm. Lack of mixing of the layers forms an inversion and traps pollution from traffic close to the ground. Morning sunshine triggers a number of photochemical reactions, creating ozone. This photochemical smog is sometimes clearly visible as a brown haze by the afternoon.

Another example of smog is the Asian Brown Cloud, which is produced by burning of agricultural wastes and increases in the use of fossil fuels. The cloud covers the entire Indian sub-continent for much of the year and drifts out over the Indian Ocean at the end of the monsoon season.

Impacts and Issues

The World Health Organization (WHO) says that air pollution, including smog, claims five to six million lives per year. Fine particles in smog are linked with heart attacks, worsening asthma, lung cancer, and immune suppression. Ozone and formaldehyde cause eye irritation, coughing, chest pain, and attacks of asthma and bronchitis. Meanwhile, smog reduces visibility. If all sources of air pollution were removed from the atmos-

WORDS TO KNOW

AEROSOL: Liquid droplets or minute particles suspended in air.

INVERSION: A type of chromosomal defect in which a broken segment of a chromosome attaches to the same chromosome, but in reverse position.

OZONE: An almost colorless, gaseous form of oxygen, with an odor similar to weak chlorine, that is produced when an electric spark or ultraviolet light is passed through air or oxygen.

PHOTOCHEMICAL SMOG: A type of smog created by the action of sunlight on pollutants.

phere, people in cities could see an estimated ten times farther than they can today.

There are various legislative measures in place in the United States and elsewhere designed to reduce smog and air pollution. The 1990 Clean Air Act covers ozone, nitrogen oxides, and particles in smog. States must meet the criteria that the Environmental Protection Agency (EPA) lays down for these pollutants or take measures to clean up. Generally, this involves applying new, cleaner technology to vehicles and tackling traffic congestion. Some states such as California have gone one step farther by enacting tougher vehicle emission standards than are required by federal law.

SEE ALSO *Air Pollution; Atmospheric Inversions; Industrial Pollution*

BIBLIOGRAPHY

Books

Cunningham, W.P., and A. Cunningham. *Environmental Science: A Global Concern*. New York: McGraw-Hill International Edition, 2008.

Kaufmann, R., and C. Cleveland. *Environmental Science*. New York: McGraw-Hill International Edition, 2008.

Web Sites

Met Office. "The Great Smog of 1952." http://www.metoffice.gov.uk/education/secondary/students/smog.html (accessed March 16, 2008).

U.S. Environmental Protection Agency (EPA). "Smog/Regional Transport of Ozone." http://www.epa.gov/airmarkets/envissues/smog.html (accessed March 16, 2008).

Fishing from Boston's Alford Street Bridge in July 1998, with the Mystic Power Plant in the background. A report by the Northeast Clean Power Campaign blasted 14 of New England's biggest polluters for spewing smog into the atmosphere. *AP Images*

Snow and Ice Cover

Introduction

Snow and ice are both the reflective covers that turn away the sun's warming rays and the thermal blanket under which much biological activity takes place. While plants and animals struggle to adapt to extreme cold, snow and ice cover protects them. In the summer, snow and ice again protect plants and animals by melting into water that nourishes them in times of drought. Climate change as a result of global warming has had an impact on Earth's snow and ice cover.

Historical Background and Scientific Foundations

The snow and ice cover involves much more than snowflakes that have accumulated on the ground. As snowflakes fall to the earth, they are tumbled in the wind, fractured and compacted, then often melted and refrozen. This initial deterioration results in the formation of ice crystals. These grains of ice may then coalesce with others, until all are nearly the same size. Throughout this process, air spaces within the snowpack are reduced in size as individual ice grains pack together and bond at their points of contact. Both the snowpack density and the mechanical strength of the snow increase substantially through this process. Water vapor then moves upward, reducing the size of the ice crystals at the bottom of the snowpack. The subsequent formation of depth hoar facilitates the movement of small animals as they forage under the snow during the winter. Under typical winter conditions, the snowpack is warmest at the bottom and coldest at the top.

As the temperature of the planet warms, the snowpack diminishes. The shrinking snowpack in the Arctic and Antarctic has received much attention as concern about global warming rises. However, climate change has already begun to leave a dramatic mark on more inhabited regions of the planet. In the 1990s, the inhabitants of Shishmaref, a Native American village on the Alaskan island of Sarichef, noticed that sea ice was forming later and melting earlier. The change meant that a protective skirt of ice no longer buffered the small settlement from destructive storm waves.

Numerous mountain systems, including the Canadian Rockies, have also experienced a reduction in snow and ice cover. The loss may disturb the sensitive ecosystem of the areas. The survival of many plants and animals depends on an annual snow cover.

Impacts and Issues

Global warming has clearly already begun to influence the amount of snow and ice cover. Scripps Institution of Oceanography researchers studied climate changes in the American West between 1950 and 1999. They found that winter precipitation fell increasingly as rain rather than snow and that snow melted faster over the years. The scientists concluded that up to 60% of changes in the snowpack could be attributed to human activities that release emissions including carbon dioxide into the atmosphere.

Scientists have developed scenarios that indicate air temperature will increase in high latitude regions in coming decades, causing the period of snow cover to shorten, the growing season to lengthen, and soil temperatures to change during the winter, spring, and early summer. A 2006 Swedish study concluded that a warmer climate by the end of the twenty-first century will increase the growing season by 30 to 43 days and shorten the duration of a consistent snowpack by 73 to 93 days. Although a longer growing season is a benefit, the food chain is likely to be disrupted since variable soil temperatures will adversely affect insects as well as all the plants and animals that depend on insects for their survival.

WORDS TO KNOW

DEPTH HOAR: Brittle, loosely arranged crystals at the base of a snowpack.

ICE CRYSTALS: An arrangement of water molecules in which motion among the molecules slows and the structure takes on a rigid shape, as a consequence of temperatures near freezing. Crystals often form around particulate matter (dust, pollutants, etc.).

SEE ALSO *Antarctic Issues and Challenges; Arctic Darkening and Pack-Ice Melting; Glacial Retreat; Glaciation; Human Impacts; Ice Cores*

BIBLIOGRAPHY

Books

Doesken, Nolan J. *The Snow Booklet: A Guide to the Science, Climatology, and Measurement of Snow in the U.S.*. Fort Collins: Colorado State University, 1997.

Kolbert, Elizabeth. *Field Notes from a Catastrophe: Man, Nature, and Climate Change*. New York: Bloomsbury USA, 2006.

Marchand, Peter J. *Life in the Cold: An Introduction to Winter Ecology*. Hanover, NH: University Press of New England, 1987.

Web Sites

Rutgers University. "Global Snow Lab." April 9, 2008. http://climate.rutgers.edu/snowcover/ (accessed on April 9, 2008).

IN CONTEXT: CLIMATE CHANGE IMPACTS ON SNOW AND ICE

According to the Intergovernmental Panel on Climate Change (IPCC): "With regard to changes in snow, ice and frozen ground (including permafrost), there is high confidence that natural systems are affected. Examples are:"

- "enlargement and increased numbers of glacial lakes;"
- "increasing ground instability in permafrost regions, and rock avalanches in mountain regions;"
- "changes in some Arctic and Antarctic ecosystems, including those in sea-ice biomes, and also predators high in the food chain."

Source: Parry, M. L., et al. *IPCC, 2007: Summary for Policymakers. Climate Change 2007: Impacts, Adaptation and Vulnerability. Contribution of Working Group II to the Fourth Assessment Report of the Intergovernmental Panel on Climate Change*. New York: Cambridge University Press, 2007.

Soil Chemistry

Introduction

Plants depend on soil for their growth because it supplies physical support for their roots and nutrients. Soil is a complex substance that is produced by the ongoing weathering of rocks. Its characteristics vary from place to place. The acidity, texture, and nutrient content of a soil affect what kind of crop can grow in it.

The nutrient content of soil cannot usually keep up with the demands for high productivity imposed by modern agriculture. Therefore, either natural or chemical fertilizers are often added to boost fertility. These can become polluting if they run off into local surface water. Soil is also prone to erosion by wind or rain and also human activities like logging, road building, mining, and overgrazing. It can also absorb pollution from air or water. Soil is a precious resource, and sustainable agriculture and pollution control are the best ways of preserving it.

Historical Background and Scientific Foundations

Soil is a complex mixture of organic and mineral components. The organic content comes mainly from humus, a material produced by the decay of the plants living in the soil. Humus gives soil its characteristic brown color and provides food for the microbial communities that live in the soil. The mineral component comes from the underlying rocks and contains various metals that can also be essential nutrients for plants growing in the soil. Humus coats the mineral content, giving soil a crumblike and spongy texture, holding onto water and nutrients that are taken up by the roots of plants. Soil may have very little organic content, such as sand, or may be almost 100% organic content, like peat.

Soil is a renewable resource. It is produced by the effects of weathering upon the underlying rocks. The weathering process is both physical and chemical. Physical weathering fragments rock into smaller particles. Chemical weathering involves a range of chemical reactions between the rock and surrounding environmental components. A soil is defined according to the size of its particles. Clay comprises the smallest particles, which are 0.002 mm in diameter, while gravel particles are the largest, with a diameter of 2 mm or more. Silt and sand are components with particle diameters that lie between these extremes.

The range of particle size in a soil determines its texture which, in turn, influences the number of pores it contains that can carry water. This is important when it comes to the accumulation of groundwater and the supply of water to plant roots. Soils dominated by clay hold water, while those dominated by sand have good drainage and dry out easily. Another important characteristic of soil is its pH, which is a measure of acidity. Few plants like to grow on an acidic soil, but pH can be corrected by the addition of chemicals.

A mature soil is layered. The topsoil is usually around 4 in (10 cm) deep and it contains material such as fallen leaves and partially decomposed organic matter as well as some minerals and living organisms. Then there is a layer between topsoil and subsoil through which dissolved or suspended matter moves. The subsoil, below this, is where humic compounds, clay, iron, and aluminum may accumulate after leaching from the upper layers. Finally there is the bedrock, which is the source of the soil through weathering processes.

Impacts and Issues

Soil is a renewable resource, for it is constantly being formed from the underlying rocks. However, weathering

WORDS TO KNOW

CLAY: The portion of soil comprising the smallest particles, resulting from the weathering and breakdown of rocks and minerals.

GRAVEL: The most coarse particles in soil.

PEAT: Partially carbonized vegetable matter that can be cut and dried for use as fuel.

SOIL: Unconsolidated materials above bedrock.

SUSTAINABLE AGRICULTURE: Agricultural use that meets the needs and aspirations of the present generation without compromising those of future ones.

WEATHERING: The natural processes by which the actions of atmospheric and other environmental agents, such as wind, rain, and temperature changes result in the physical disintegration and chemical decomposition of rocks and earth materials in place, with little or no transport of the loosened or altered material.

can take many years to produce new soil and, in the meantime, soil can be degraded by various processes. Soil erosion occurs through wind, rain, and activities like mining and agriculture. Some farming practices, such as planting rows of corn and soybean, or the eradication of all weeds with herbicides, can leave the soil very exposed to erosion. The eroded soil often ends up as sediment in surface waters, which may cause problems for aquatic ecosystems. Soil erosion may also lead to desertification, which is the conversion of fertile soil into desert.

The fertility of soil depends on its levels of nitrogen, phosphorus, and organic matter. These may not be enough to support long-term crop productivity. Some crops, like corn, tobacco, and cotton, take a lot of nutrients out of the soil. This can be replenished in various ways. For instance, crop rotation involves planting legumes, whose roots fix nitrogen into the soil. Often, fertilizer is added to enrich the soil. These might be natural, like compost or manure, or chemical, like ammonium nitrate or phosphate. Fertilizer runoff can be a major pollutant for surface waters because it can enrich them and encourage the overgrowth of algae, which can upset aquatic ecosystems.

SEE ALSO *Agricultural Practice Impacts; Desertification; Geochemistry*

BIBLIOGRAPHY

Books

Cunningham, W.P., and A. Cunningham. *Environmental Science: A Global Concern*. New York: McGraw-Hill International Edition, 2008.

Williams, I. *Environmental Chemistry*. Chichester: Wiley, 2001.

Web Sites

U.S. Department of Agriculture Natural Resources Conservation Service. "What Is Soil?" http://soils.usda.gov/education/facts/soil.html (accessed April 10, 2008).

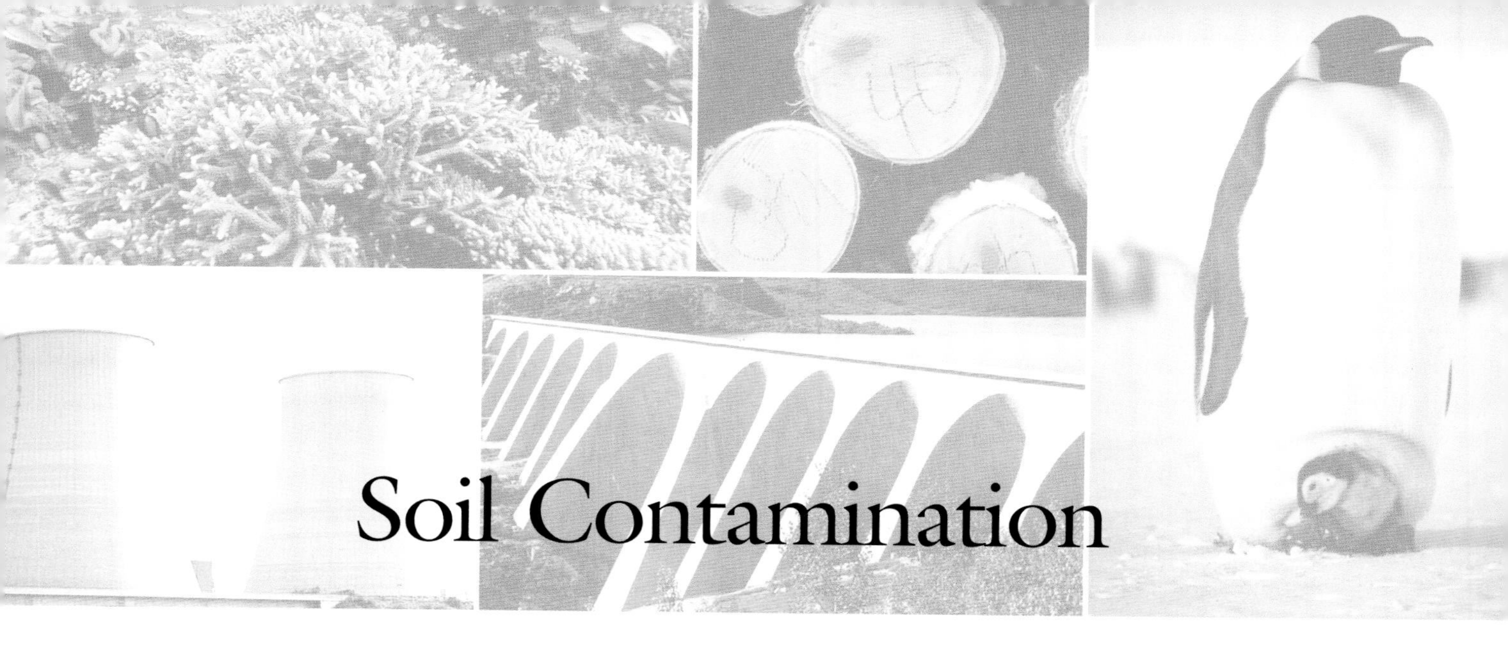

Soil Contamination

Introduction

Soil contamination refers to the presence of un-natural (human-made) chemicals or other substances in the soil. The pollutant can be a solid or liquid. Examples include gasoline released from the rupture of an underground tank at a gas station, runoff of oil from a parking lot onto adjacent ground, deliberate dumping of solid or liquid waste, and the uncontrolled use of pesticides or herbicides.

Oil-based compounds (hydrocarbons), heavy metals, and solvents that do not mix well with water also tend to adhere tightly to soil particles, making them harder to extract. This is also true if they enter the groundwater, as they can then be transported a considerable distance away from the original site of contamination.

With the growth of urban centers and their associated industries beginning in European countries and the United States in the early nineteenth century, the contamination of soil became extensive. Among the many examples in the United States is Love Canal, a neighborhood in Niagara Falls, New York, that was created on the site of land that had been used for the public and private disposal of toxic waste.

A legacy of Love Canal and other sites of soil contamination was the passage of legislation in 1980 that created what is popularly known as Superfund sites—long abandoned, heavily contaminated sites for which the U.S. Environmental Protection Agency (EPA) provides funds for clean up and restoration. As of 2008, nearly 1,300 sites had been identified for clean up under this program.

Cleaning contamination soil can involve the removal of the soil or on-site treatment. Since the promising results obtained with the use of bacteria to digest the oil-coated shoreline in the aftermath of the *Exxon Valdez* oil spill in Alaska in 1989, this strategy, which is known as bioremediation, has become an accepted means of treating contaminated soil. The approach can be beneficial if the added bacteria are able to use the soil contaminant as a food source.

Historical Background and Scientific Foundations

Although natural soil contamination has always occurred (the main example is the movement of salty water into soils in coastal regions), soil contamination is mostly associated with the expansion of human development. This dates back to the growth of industries and urban areas beginning in the late eighteenth century with the Industrial Revolution, and, later, with the use of pesticides and herbicides on agricultural land to increase crop production.

In many developed countries including the United States, the end of World War II (1939–1945) was a time of great economic prosperity and growth of industry and agriculture. The use of chemicals expanded. At the same time, environmental regulations were not nearly as stringent as in 2008. Although not widely recognized at that time, this combination led to the environmental release of chemicals that were both toxic and slow to degrade.

A turning point came in 1962, with the publication of *Silent Spring*. In the book, author Rachel Carson (1907–1964) warned of the consequences of the widespread and extensive use of the pesticide DDT on the natural environment and humans, given that relatively little was known of the interactions of DDT with living creatures. The outcry following the book's publication spurred efforts to understand and deal with environmental contamination including that of soil.

Love Canal is one of the most infamous examples of soil contamination in U.S. history. The area took its name from William Love, who began to develop the land

WORDS TO KNOW

GROUNDWATER: Fresh water that is present in an underground location.

REMEDIATION: A remedy. In the case of the environment, remediation seeks to restore an area to its unpolluted condition, or at least to remove the contaminants from the soil and/or water.

in the early 1890s with the aim of building a canal, first as a means of generating electrical power and then to allow ships to bypass Niagara Falls, New York. The project was abandoned and in the 1920s the land was acquired by the City of Niagara Falls. The city elected to use the area to dispose of chemical waste, since the ground was mainly composed of clay, in which liquids do not penetrate well, and since the area was far from existing neighborhoods. Dumping expanded in the 1940s, when the Hooker Chemical and Plastics Corporation bought the site. Records show that from 1947 to 1952 alone, approximately 22,000 tons of toxic chemical waste were buried by the company. After the site reached capacity in 1952, the waste was overlaid with a thick layer of clay and the site was sold back to the city for one dollar.

In the 1950s, housing was permitted on the site; 100 homes and a school were built. By the 1960s, residents were complaining of health problems, and the presence of potential cancer-causing chemicals including benzene leaching from the soil had been detected. Initial tests did not conclusively link the chemicals to the health problems. By 1979, the EPA had established a link between the chemical waste and various health maladies including genetic damage.

The growing public outcry led U.S. President Jimmy Carter to declare a federal emergency at Love Canal in May 1980 and provide financial aid in relocating residents most at risk. Ultimately, more than 800 families were relocated.

Love Canal is an example that shows some soil contaminations are handled most quickly and economically by abandoning the site, as the scope of the clean up is too large for other options. However, the danger in this strategy is that contaminants can continue to migrate outward underground and spread over a wider area. In 2008, if a site is to be abandoned without remediation, the boundary of the contaminated site needs to contain a plastic liner that prevents escape of the contaminant.

More typically, the soil is treated. This treatment can be done on-site (also refereed to as in-place or *in situ* remediation). This can involve drilling a well to the site of the contaminants and pumping them to the surface. However, since the contaminants may not be that easily extracted, often *in situ* treatment has to take the solution to the site of the problem. One example is bioremediation, where the bacteria and a food source (water can be sufficient) are pumped down into the subsurface. After a time to allow the bacteria to digest the pollutants, the subsurface water is pumped out through another well drilled nearby. The cycle is repeated until tests done on the extracted water show that the level of contaminants have dropped to an acceptable level.

In reality, the process is more difficult. But, bioremediation can be successful, particularly if carried out on contaminated soil that has been transported off-site. When treated away from the original site of contamination, the soil is placed on a liner that prevents leakage. Aside from bioremediation, soil can be exposed to steam or chemicals that disperse the pollutant from the soil.

The off-site treatment strategy has the advantage of better control over the decontamination, since the soil can be spread out to increase the surface area that can be treated, and reduce the depth of the soil. Afterward, the liner can be collected and destroyed.

■ Impacts and Issues

As shown by Love Canal, soil contamination is a serious health concern, causing illness (which can be long lasting) and genetic damage that can lead to cancer, developmental and nervous system damage in children (if lead or heavy metal contamination has occurred), and many other consequences. As two other examples, exposure to mercury-contaminated soil can cause kidney damage that may be permanent, and contamination with polychlorinated biphenyls (PCBs), which were widely used as coolants and insulating fluids in transformers until being banned in the 1970s, can damage the liver. Chemicals such as PCBs and DDT degrade in the environment very slowly. So, once present in soil and if not detected and removed, such chemicals can remain capable of causing illness for decades.

Contaminated soil harms other creatures as well as humans. Since an ecosystem is based on a food chain—where organisms serve as food for other organisms, which in turn can be a food source for another organism—disruption of soil populations can ripple upward through the food chain. Furthermore, some soil pollutants accumulate in animals that feed on contaminated prey (bioaccumulation). Also, as exemplified by DDT, which caused weakened egg shells in some exposed species of birds, a soil contaminant can cause population decrease due to reduced efficiency in producing offspring.

Vegetation can also be harmed by the absorption of toxins through their roots.

Until the 1970s, despite the Love Canal disaster, soil contamination was not regarded as being wide-

In Phoenix, protesters from the Children for a Safe Environment (October 1996) hold up a large poster protesting the toxic trains which have been unloading DDT-tainted soil. Eleven-hundred rail cars filled with the contaminated soil are being dumped at the Butterfield Landfill near Mobile, Arizona. *AP Images*

spread. However, as more sites were identified as a part of the Superfund initiative, the extent of soil contamination in the United States became apparent. As of 2008, the EPA has identified about 1,300 abandoned sites that are contaminated seriously enough to warrant clean-up funding. The actual number of contaminated sites is much higher—some estimates range up to 200,000.

As challenging as the U.S. soil clean-up efforts are, the situation is better than in developing nations, where disposal of toxic wastes can be less regulated, if at all, and where toxic chemicals banned in the United States are still in use (an example is DDT). In some African countries, where DDT has been sanctioned for use by the World Health Organization for the control of mosquito populations, and so as a malaria control, the application of the chemical is supposed to be strictly controlled to avoid contamination of soil and water.

SEE ALSO *Aquifers; DDT; Groundwater Quality; Radioactive Waste; Spill Remediation*

BIBLIOGRAPHY

Books

Atlas, Ronald M., and Jim Philip. *Bioremediation: Applied Microbial Solutions for Real-World Environment Cleanup*. Washington, DC: ASM Press, 2005.

Livingstone, James. *Agriculture and Soil Pollution: New Research*. Hauppauge: Nova Science Publishers, 2006.

Wang, Zhendi, and Scott Stout. *Oil Spill Environmental Forensics: Fingerprinting and Source Identification*. New York: Academic, 2006.

Brian D. Hoyle

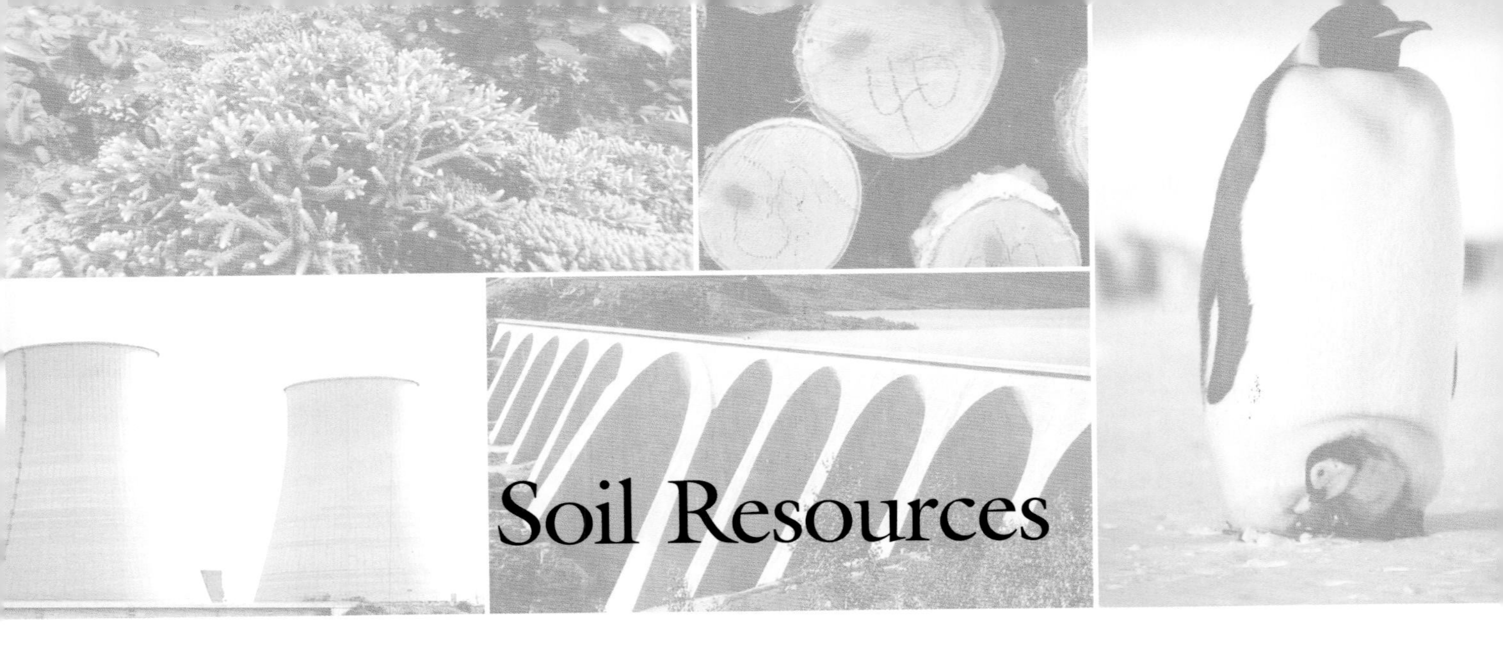

Soil Resources

■ Introduction

The coverage of Earth's surface with soil varies from place to place. Tropical forests have thin soils that are poor in nutrients, while grasslands in temperate regions have soils that are rich and well able to support crops. There are several different classes of soil depending on how the soil is formed and where it is located. An understanding of which class of soil is found in a particular location is an important foundation to obtaining successful crop yields.

Soil resources need conserving as much as water resources do. Erosion is the key process by which soil is created from rock and destroyed. Wind and rain are the main factors that cause erosion of soil from agricultural areas, possibly converting them into new desert. Some modern agricultural practices leave soil exposed to the elements and thereby risk loss of long-term productivity because of soil erosion.

■ Historical Background and Scientific Foundations

Soil is as important as water as a resource. It provides nutrients and an anchor to the roots of plants and is therefore essential to their healthy growth and yield of food. It is a complex mixture of organic and mineral content which is constantly being formed by the weathering of rocks.

Soil has a layered structure, with the topsoil being around 4 in (10 cm) deep and rich in organic material. Then there is a layer between topsoil and subsoil through which dissolved or suspended matter moves. Below this is the subsoil, where humic compounds, clay, iron, and aluminum may accumulate after leaching from the upper layers. Finally there is the actual bedrock, which is the source of the soil through weathering processes. Soil is also classified according to the size of the particles it contains. A soil with a lot of clay has fine particles, while one with a lot of gravel is coarser.

A knowledge of soil resources, which is concerned with the type and distribution of soils around the world, is an important factor in global food supply. Soils are classified according to their type and composition. In the United States, these considerations give rise to twelve soil orders and these can be related to the type of environment where the soil is located. The best soils for farming are known as mollisols and alfisols, both of which are rich in organic matter. Both of them form wherever rainfall and precipitation are moderate. Spodosol is another class of soil; it is formed under pine forests, where the acidic needle litter will form a characteristic white and ashy looking middle layer.

Hot and rainy environments have quite different soils. Oxisols and ultisols are the two classes found in these locations. They are severely depleted of nutrients and they are red in color because they contain a lot of iron-rich minerals. In arid environments, aridosols are the predominant soil class. These are characterized by their low organic content and the presence of accumulated salt. Some classes of soil are defined by how they are formed. Andisol comes from volcanic material and vertisol from clay-rich material, for example.

■ Impacts and Issues

Soil is a renewable resource. It is constantly being formed and destroyed, mainly by erosion processes. Although erosion can spread rich soils by wearing down mountains, it can also lead to removal of top soils from agricultural areas. When erosion removes too much soil, the farmer will need to apply more fertilizer, which can cause pollution problems through runoff. Annual soil loss from agricultural lands amounts to as much as 25 billion metric tons, although this only leads to a loss of about 1% in crop production, because of compensating applications of water and fertilizer.

Wind and water are the main agents of soil erosion. Their impact is accelerated by removal of vegetation, as in deforestation. Soil erosion is not always compensated by the farmer and may therefore lead to desertification, in which productive agricultural land is gradually converted into desert. Desertification is particularly marked in China, Canada, and the United States. The U.S. Department of Agriculture states that 170 million acres of U.S. farmland are eroding their soil at a rate that is sure to reduce long-term productivity. Intensive farming makes a large contribution to soil erosion because practices like sowing row crops leaves soil exposed to wind and rain.

SEE ALSO *Desertification; Soil Chemistry*

BIBLIOGRAPHY

Web Sites

Cunningham, W.P., and A. Cunningham. *Environmental Science: A Global Concern*. New York: McGraw-Hill International Edition, 2008.

Kaufmann, R., and C. Cleveland. *Environmental Science*. New York: McGraw-Hill International Edition, 2008.

Web Sites

National Aeronautics and Space Administration (NASA). "Soil and the Environment." August 15, 2001. http://soil.gsfc.nasa.gov/env.htm (accessed April 23, 2008).

WORDS TO KNOW

ALFISOIL: Rich soil formed under deciduous forest.

ARIDOSOL: Type of soil found in arid environments.

DESERTIFICATION: Transformation of arid or semiarid productive land into desert.

EROSION: The wearing away of the soil or rock over time.

MOLLISOL: Rich soil formed under grasslands.

RENEWABLE RESOURCE: Any resource that is renewed or replaced fairly rapidly (on human historical time-scales) by natural or managed processes.

RUNOFF: Water that falls as precipitation and then runs over the surface of the land rather than sinking into the ground.

SPODOSOL: Acidic soil formed under pine forests.

Solar Power

Introduction

Solar power refers to the use of energy from the sun. This use occurs naturally as a part of life. For example, plants that are capable of photosynthesis use solar power as a source of the energy they need to survive and grow. Without the energy of the sun warming Earth, no life would exist; surface life would die off quickly and, as the atmosphere cooled, even the deepest parts of the ocean would freeze.

More practically, solar power has been harnessed for warmth for thousands of years, and, beginning in the nineteenth century, as a means of generating energy to provide heat, light, and electricity.

Solar power is an increasingly attractive alternative energy source as it is a stand-alone energy source; that is, it does not require a connection to a power grid, and the sun provides an unlimited and virtually endless supply of energy. As well, solar power can be used to supplement electricity generated conventionally using water (hydroelectric power) and burning fossil fuels.

In an era when the human-related generation of greenhouse gases such as carbon dioxide have been acknowledged by the Intergovernmental Panel on Climate Change (IPCC) to be a major reason for the increasing warming of Earth's atmosphere that has been occurring for the past 150 years, and which has accelerated since the mid-twentieth century, the use of solar power reduces the greenhouse-gas emissions associated with conventional generation of electricity. This is one reason why solar power is moving more into the mainstream as an energy source.

Historical Background and Scientific Foundations

Solar power has existed ever since there was material that could be warmed by the rays of the sun. Roman bathhouses built in the first century operated essentially like greenhouses, letting sunlight in through south-facing rooms. The same design was used in the thirteenth century in dwellings made by the Anasazi in Arizona and New Mexico; some of these dwellings still exist. In the fifteenth century, Italian inventor and artist Leonardo da Vinci (1452–1519) made sketches of a device that would concentrate sunlight to weld copper. A device that concentrated solar energy to provide a temperature high enough to cook food was first built in 1767. In 1839, French physicist Edmond Becquerel (1820–1891) discovered the photovoltaic effect (which was experimentally proven in 1916) and in 1891, the first solar water heater was patented in the United States.

Solar power is still used today to generate heat in a more sophisticated form known as concentrated solar thermal systems. These consist of an arrangement of mirrors that track the motion of the sun and reflect the sunlight to a small central area. All this reflected sunlight creates temperatures of up to 932°F (500°C), which is used to heat water or oil, which is, in turn, transferred to a facility to generate power. These facilities can convert up to 40% of the incoming sunlight to usable power.

The first operational photovoltaic cell was developed at Bell Laboratories in New Jersey in 1954. It was capable of converting solar energy to provide enough electricity to run everyday electrical equipment. Within several years the cell's efficiency had been increased from 4% to 11%, which is remarkable considering that the versions in common use in 2008 have only a marginally higher efficiency of about 15%.

The solar panels that power objects ranging from Earth-bound pocket calculators to orbiting satellites are photovoltaic cells. The word photovoltaic, which comes from photo (meaning light) and voltaic (electricity), describes the function of the process, in which sunlight is directly converted into electricity.

A photovoltaic cell is typically composed of silicon. When light contacts silicon, some of the energy is absorbed. This added energy dislodges electrons from the silicon; once freed, the electrons can move. The movement of electrons represents a current. The current can be tapped to power a device like a calculator, or can be stored.

A solar panel is made up of many photovoltaic cells. The collective power produced by an array of solar panels, such as are found on a house roof, can supply at least some of the building's electricity needs.

Silicon is able to function in a solar panel because of its structure. In a silicon crystal, each atom has 14 electrons that are in orbit around it. The electrons tend to arrange to be most energetically stable. The result is a three-tiered arrangement of electrons (each tier is known as a shell). The two shells of electrons that are nearest to the central atom are full and thus are stable. However, the outermost shell has only four electrons, leaving it half-full. As this is not energetically stable, one atom will tend to share its four outermost electrons with a neighboring silicon so that each atom will have an outer shell containing eight electrons.

Because this arrangement produces outer shells that are stable, it becomes more difficult to get electrons to dislodge and flow, which is what is required for a solar photovoltaic cell to operate. So, in reality, the silicon is present along with other atoms such as phosphorus. Phosphorous has five electrons in its outer shell. It can share four electrons with silicon, leaving one unpaired electron. It is these unpaired electrons that can be dislodged to generate the electrical current.

The presence of other atoms in the crystalline arrangement of silicon atoms is termed an impurity. But, this does not imply that impurities are accidental. Rather, impurities are introduced deliberately; the process is referred to as doping. Silicon doped with phosphorus atoms is known as N-type silicon (N stands for negative, because the free electrons produce an overall negative charge). N-type silicon conducts electricity much better than pure silicon does.

A photovoltaic cell also has another component called P-silicon, which is generated by using boron atoms to create the impurity instead of phosphorus atoms. A boron atom has only three electrons in its outermost shell, in effect creating a hole into which one electron can be occupied.

The photovoltaic cell is designed so that not all the electrons become energetically stable. The result is the creation of an electrical field (known as voltage) between the N-silicon and P-silicon regions. When sunlight hits the cell, the flow of electrons provides the current. The combination of the voltage and current is electrical power.

A photovoltaic cell also has a coating that inhibits the reflection of sunlight, since reflection will reduce the amount of light that contacts the panel and the resulting insolation. As well, the panel is covered by glass as a protective layer.

This process is not as efficient as it sounds in theory. Only about 15% of the sunlight striking a cell is converted to electrical energy, because energy is required to dislodge electrons and because doped silicon is not close to being 100% efficient.

Still, when produced on a daily basis (when enough sunlight strikes the panel), storage of the energy in batteries can be a longer-term source of electricity, similar to the powering of a laptop computer by its battery when it is not connected to a wall receptacle.

Some solar panels also have water-filled tubes running though them. The circulating water that is heated during a back-and-forth passage through the array of solar panels can then be used as a source of hot water for laundry, bathing, or as heat in radiant flooring. Often the heated water is not used directly, but is routed to the building's hot water heater to supplement the hot water and so reduce the amount of conventional (and purchased) electricity required to heat the tank of water.

WORDS TO KNOW

GLOBAL WARMING: Warming of Earth's atmosphere that results from an increase in the concentration of gases that store heat, such as carbon dioxide (CO_2).

INSOLATION: Solar radiation received at Earth's surface.

SOLAR RADIATION: Energy received from the sun is solar radiation. The energy comes in many forms, such as visible light (that which we can see with our eyes). Other forms of radiation include radio waves, heat (infrared), ultraviolet waves, and x-rays. These forms are categorized within the electromagnetic spectrum.

Impacts and Issues

The popularity of solar power has long been based on its economy and as a renewable alternative to the nonrenewable use of fossil fuels to generate electricity. With global warming, the use of solar power has become even more attractive, as it does not generate carbon dioxide.

As well, solar power represents a form of energy that can be harnessed rapidly and with little controversy, in contrast to, for example, a nuclear power plant. Establishing a solar power facility requires money and space; once these are available, the solar collectors can be set up with minimal construction. Acquiring the means to rapidly generate electricity may become urgent, according to a 2008 study from the Scripps Institution of Oceanography. The study examined Lake Mead in the Western United States, a key water source for millions of Americans in the Southwest and the main source of power (in the form of hydroelectricity) for Las Vegas, Nevada. Current climate-related change could dry the reservoir by 2021 if water use is not altered.

The amount of energy used globally in a year is contained in the sunlight that strikes Earth for only 40 minutes. The large amount of land in the United States that could be used to install solar arrays means that the country could become energy self-sufficient.

Already several U.S. companies have recognized the potential of large-scale solar power projects as becoming feasible as the need for energy becomes more difficult or environmentally unacceptable to be met by conventional technologies.

In March 2008, the U.S. Department of Energy announced an initiative to invest almost $14 million over the next three years in a number of projects that are aimed at increasing the efficiency of the conversion of sunlight to electricity by solar photovoltaic cells. Using different materials and solar cell designs, the goal is to manufacture a solar cell that could convert 45% of the sun's energy into electricity.

SEE ALSO *Greenhouse Effect; Wind and Wind Power*

BIBLIOGRAPHY

Books

Kemp, William. *The Renewable Energy Handbook: A Guide to Rural Energy Independence, Off-Grid and Sustainable Living*. Tamworth, Ontario, Canada: Aztext Press, 2006.

Thomas, Isabel. *The Pros and Cons of Solar Power*. New York: Rosen Central, 2008.

Web sites

Natural Resources Canada. "Technologies and Applications: About Solar Energy." April 26, 2005. www.canren.gc.ca/tech_appl/index.asp?CaId=5&PgId=121 (accessed March 19, 2008).

U.S. Department of Energy. "DOE to Invest up to $13.7 Million in 11 Solar Cell Projects." March 12, 2008. www.eere.energy.gov/news/news_detail.cfm/news_id=11638 (accessed March 19, 2008).

Brian D. Hoyle

Solid Waste Treatment Technologies

Introduction

Eleven billion tons of solid waste are produced every year in the United States. Most of it is still buried in landfill sites with no prior treatment, other than removal of materials that can be recycled. However, costs of landfill are soaring and the number of suitable sites is decreasing. Therefore, there is increasing interest in treating solid waste to re-use its material or extract energy from it, rather than merely dumping it.

Incineration and composting are traditional approaches to treating solid waste. However, new designs can lead to more widely applicable, safer, and more efficient versions of these treatments. For example, gasification and ash melting technology are being applied to incineration of solid waste to deal with the ash that remains and thereby reduce the potential for dioxin pollution. Meanwhile, treatment of hazardous solid wastes, including toxic wastes, is important to render them harmless and stop widespread environmental contamination.

Historical Background and Scientific Foundations

Domestic, commercial, and industrial human activities produce nearly 11 billion tons of solid waste a year. Around half of this comes from agriculture and includes manure and crop residues, which farmers generally put back into the soil. Mining and metal processing account for another third of solid wastes produced per year. Other industries produce about 400 million metric tons of solid waste, much of which may be dealt with by the company concerned at their own facilities. Around 60 million metric tons of this is hazardous waste, including toxic waste, and this may require special treatments to make it safe. Municipal waste, which is domestic and commercial refuse, amounts to more than 200 million metric tons a year, according to the U.S. Environmental Protection Agency (EPA). Most of this is dealt with by local refuse collection agencies.

Previously, solid wastes were dumped on the ground, wherever space could be found. Dumping has become socially and economically unacceptable in industrialized societies, although vast waste sites can still be seen in some developing countries. Over the last 50 years, it was realized that treating solid waste can lessen our dependence on Earth's resources. Solid waste can yield value in terms of energy and materials, which provides incentives for increasingly sophisticated technologies to be developed for their extraction.

Most solid waste in the United States is still disposed of by landfill. Solid waste receives minimal treatment before being put into a landfill site, other than removal of items that can be recycled. Once in the site, the waste needs to be monitored for leaks. More advanced landfills are managed for production of methane, which can be used as a source of energy. Landfill is increasingly expensive and appropriate sites are becoming scarce. In Japan, where there is little land available for landfill, solid waste is more likely to be treated by recycling or incineration.

The amount of solid waste dealt with by recycling or incineration has increased in the United States. Recycling begins with separating out recyclable materials from solid waste. Recycled material can be processed to produce another version of the same object. For instance, aluminum cans are made into more aluminum cans. Recycled material can also be processed to make something completely different, such as tires that are made into road surfacing material.

Incineration is used to deal with about 20% of solid waste in the United States. The waste may be sorted first to remove any non-combustible material, particularly plastics that may give off toxic emissions. This sorted material is called refuse-derived fuel. It is particularly useful for energy recovery, where the heat from the

WORDS TO KNOW

BIOREMEDIATION: The use of living organisms to help repair damage such as that caused by oil spills.

COMPOSTING: Breakdown of organic material by microorganisms.

DEMANUFACTURING: The disassembly, sorting, and recovering of valuable or toxic materials from electronic products such as televisions and computers.

ENERGY RECOVERY: Incineration of solid waste to produce energy.

REFUSE-DERIVED FUEL: Solid waste from which unburnable materials have been removed.

incinerator is used either directly or to generate electricity. Incinerator facilities are costly to build, but the investment is offset if energy is produced alongside the waste disposal. There is often public resistance to the building of incinerators in an area because the ash remaining after incineration could contain toxins. Incineration of plastics produces dioxins, which are very toxic to humans, and it tends to remain in fine particulate ash, which could enter the surrounding air. Therefore, pollution controls on incinerators need to be very strict if the facility is to be operated safely. Plastics and batteries, which contain toxic heavy metals, must be removed from solid waste prior to incineration.

Composting is the other main method of treating solid waste. Composting involves using microbial action to turn waste with an organic content, such as kitchen and garden waste, into a nutrient-rich addition to soil. Composting can be accomplished by an individual household or on a larger scale. Compost itself does not have a high market value, but the composting process also produces methane, which can be captured and sold as fuel in an advanced composting facility.

Hazardous and toxic waste requires careful treatment to make it safe. One important area is demanufacturing, which is applied to the increasing number of discarded computers, televisions, cell phones, and refrigerators. This involves retrieval of toxic heavy metals, such as mercury, lead, and gallium, from electronic components, along with plastics from the casing. Hazardous wastes produced by industry may be subjected to chemical treatment such as oxidation or neutralization in order to convert them into harmless substances. Sometimes isolating the toxic component is all that is needed, such as trapping it in a charcoal filter, for instance.

Some toxic materials, such as the polychlorinated biphenyls that were widely used in industry before their health hazards were appreciated, can often be treated by bioremediation. Many microbes will feed on unusual carbon-containing compounds, even if they are toxic to humans, breaking them down into carbon dioxide and water. If applied to land contaminated with toxic waste, they can render if safe in the long term. Plants, such as the mustards, reeds, and the water hyacinth, have the ability to detoxify waste by taking up heavy metals and other dangerous substances in their roots. Phytoremediation is an attractive and cheap option for treating toxic waste, although care has to be taken that the plants do not release the toxins back into the environment.

Hazardous solid waste cannot always be treated to make it safe, and in this instance, safe storage may be the only option. Storage locations include deep ground burial, perhaps in a discarded mine, or in a remote, secure building. Often retrievable storage is used so that if the waste begins to leak, it can be rescued before it causes damage to the environment. Vitrification, where the waste is injected into glass, is sometimes used for long-term safe storage of hazardous material including radioactive waste.

Impacts and Issues

An increasing number of new technologies is being applied to solid waste treatment. For instance, the thermal conversion process applies heat and high pressure to a mixture of manure, tires, plastics, and sewage sludge, converting it by a complex series of chemical reactions into gasoline, oil, and methane. Another approach, being developed in Japan, is gasification and ash melting applied to incineration. This technique uses the energy from the waste to treat the ash and reduce its dioxin content to harmless levels in a system that is readily adaptable to municipal incinerators. If waste is seen as a scientific and technical challenge, then its safe disposal and recovery of valuable materials are more likely.

SEE ALSO *Electronics Waste; Hazardous Waste; Industrial Pollution; Recycling; Toxic Waste; Waste Transfer and Dumping*

BIBLIOGRAPHY

Books

Cunningham, W.P., and A. Cunningham. *Environmental Science: A Global Concern*. New York: McGraw-Hill International Edition, 2008.

Web Sites

Virtual Centre for Environmental Technology Exchange. "Future Municipal Solid Waste Treatment Technologies." http://www.apec-vc.or.jp/e/modules/tinyd00/index.php?id=122&kh_open_cid_00=11 (accessed March 27, 2008).

Susan Aldridge

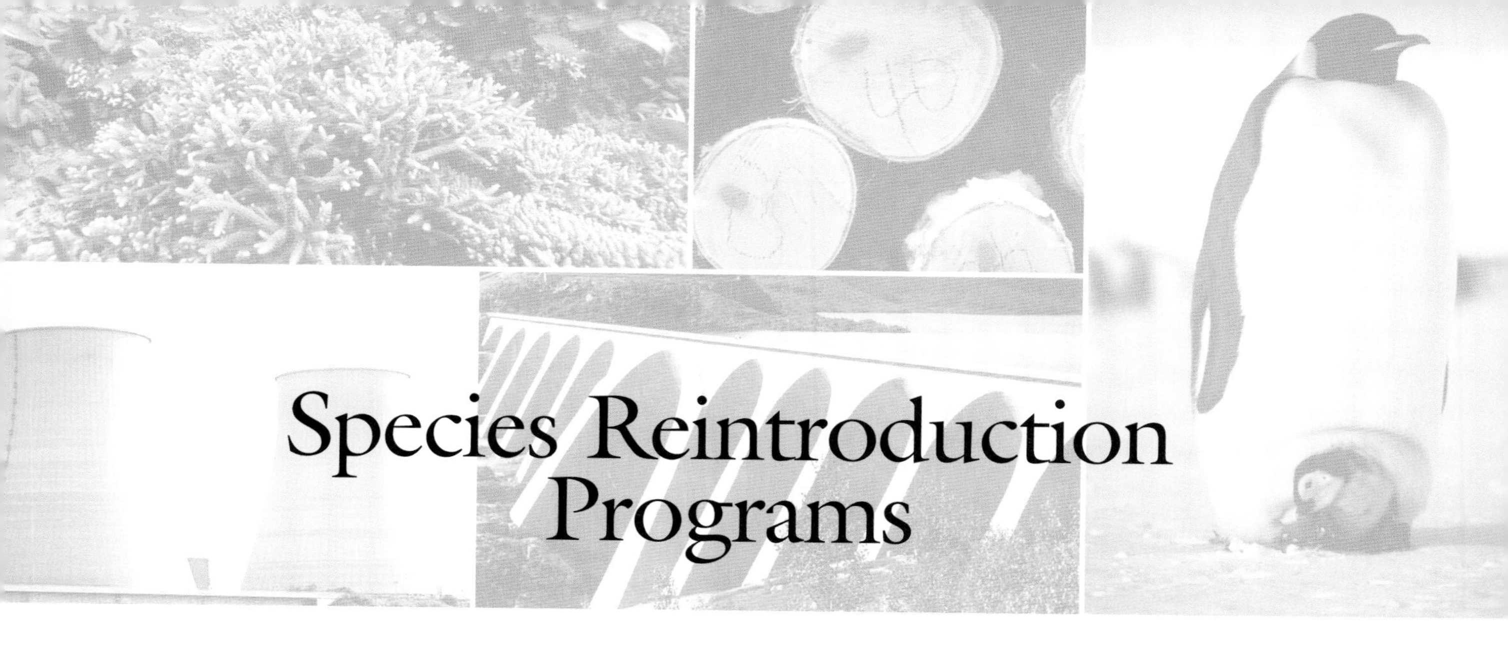

Species Reintroduction Programs

Introduction

The reintroduction of species to the wild is the latest step in the effort to preserve animals, birds, insects, and plants that are endangered or extinct in the wild. Although captive breeding programs increase populations, they do not restore wild environments. Species reintroduction plans attempt to create biodiversity.

Historical Background and Scientific Foundations

Species reintroduction programs are an important feature of global conservation efforts. The reintroduction of native species is essential to the long-term environmental health of a region. In Canada, the Karner Blue butterfly is being reintroduced because the species supports the Prairie thistle, the Blue racer, the Frosted elfin, and the Antenna-waving wasp. The restoration of oak savanna is dependent upon the restoration of the Karner Blue. Each level in the food chain is critical, with the removal of one level resulting in environmental damage or chaos.

Many of the species restoration programs are not controversial. In 2007, the U.S. Fish and Wildlife Service (USFWS) began to restore 21 threatened and endangered species to the Tennessee River system as part of a broader campaign to reverse more than 75 years of damage caused by dams and dredging for navigation. The reintroduced species included fifteen types of mussels, five fish, and a river snail.

In contrast, reintroductions of some predators have proved enormously controversial. The Mexican gray wolf, one of the first species to be reintroduced to the wild, is the southernmost and smallest subspecies of the North American gray wolf. From prehistoric times into the mid-nineteenth century, it ranged from central and northern Mexico to western Texas, southern New Mexico, and central Arizona. When the numbers of native prey species, such as deer and elk, dropped because of human predation, these wolves turned for food to the large numbers of cattle that had newly arrived in the West. The increased predation on livestock by the wolves led to their demise. Wolves were trapped, shot, and poisoned by both private individuals and government agents. Bounties were paid for the corpses of wolves.

By the 1970s, there were few, if any, Mexican gray wolves left in the United States. In 1976, they were listed as endangered under the Endangered Species Act (1973). Such a listing indicates that the peril of extinction is greater and more imminent than for a "threatened" species. In 1977, Mexico teamed with the United States to create a breeding program with five wolves in captivity. According to the USFWS, the 1982 Mexican Gray Wolf Recovery "recommended maintenance of the captive breeding program and re-establishment of a viable self-sustaining population of at least 100 wolves in the wild within the Mexican wolf's historic range." In 1998 wolves that were raised in captivity were reintroduced by release into designated recovery areas within Arizona and New Mexico. By 2007, population estimates indicated that approximately 60 wolves lived in the recovery areas.

Impacts and Issues

When high-level predators, such as Mexican gray wolves or African wild dogs, are reintroduced to an area, human residents typically fear the loss of their livestock. In New Mexico in 2007, Grant County Commissioners sent a resolution to the USFWS calling for the federal government to pay ranchers for livestock losses due to wolves. Later in 2007, the USFWS removed two adult female

WORDS TO KNOW

CAPTIVE BREEDING: A wildlife conservation method in which rare or endangered species are bred in restricted environments such as zoos or wildlife preserves.

FOOD CHAIN: A sequence of organisms, each of which uses the next lower member of the sequence as a food source.

REWILDING: The restoration of a plant or animal to its historic habitat.

ZOONOSIS (ZOONOTIC DISEASE): Any disease that can be transmitted from animals to humans.

Mexican gray wolves from the Gila Forest in New Mexico because of multiple cow depredations by the animals. Some humans attacked the introduced predators. In 2007, USFWS law enforcement officials issued a reward for the person responsible for the disappearance of three members of another wolf pack. The success of the Mexican wolf reintroduction is in doubt.

Additionally, the reintroduction of large mammals after decades of absence may favor the spread of zoonotic diseases and may put ecological communities at risk. The reintroduction of the wild boar to Denmark is possible, but the animal may act as a reservoir of diseases, such as leptospirosis. Failure to prevent the disease consequences of species restoration can negate the conservation benefits and may limit public cooperation with other species reintroduction efforts.

SEE ALSO *Biodiversity; Extinction and Extirpation; Genetic Diversity; Natural Resource Management; Wildlife Population Management; Wildlife Protection Policies and Legislation*

BIBLIOGRAPHY

Books

Akcakaya, H. Resit. *Species Conservation and Management: Case Studies.* New York: Oxford University Press, 2004.

Klyza, Christopher McGrory. *Wilderness Comes Home: Rewilding the Northeast.* Hanover, NH: University Press of New England, 2001.

Norton, Bryan G. et al. *Ethics on the Ark: Zoos, Animal Welfare, and Wildlife Conservation.* Washington, DC: Smithsonian Institution Press, 1995.

Web Sites

U.S. Fish and Wildlife Service. "The Mexican Wolf Recovery Program." March 11, 2008. http://www.fws.gov/southwest/es/mexicanwolf/ (accessed March 15, 2008).

Caryn E. Neumann

A Mexican gray wolf leaves cover at the Seviellta National Wildlife Refuge, New Mexico. After a five-year review, the program to reintroduce the Mexican gray wolf in eastern Arizona and western New Mexico will continue, reported the U.S. Fish and Wildlife Service (July 2006). *AP Images*

Spill Remediation

Introduction

Remediation is the removal of a toxic compound from water or soil, or the containment of a spill so that the area that is contaminated does not increase. Spill remediation refers to the response to an environmental spill that occurs in freshwater and marine water, and on land.

The responses to a spill are varied and depend on a number of factors including the type of spill, and whether people or wildlife are immediately threatened by the spill. A spill may be handled at the site, or the contaminated soil and sand can be trucked to another site to be treated.

Spill remediation can be a task that involves various experts including those familiar with the geology of the area, water flow patterns (hydrology), behavior of chemically diverse compounds, and people who can generate computer models of the spill site.

In developing countries in North America and western Europe, the extent and nature of spills are now well-known. As well, legislation has been passed and penalties imposed to lessen the risk of accidental spills and to make the deliberate dumping of toxic material less likely. But, the situation in developing countries is far different. Dumping can be widespread and deliberate, and can go undetected for decades.

Historical Background and Scientific Foundations

Spill remediation depends on the nature of the spill. For example, when the spill involves a liquid on soil or a beach, the clean up will be both at the surface and underground. This is because the liquid can percolate down into the ground in between the spaces in between the soil and sand particles. As the liquid moves downward, the pollutant will bind to the particles. This binding can be very strong, particularly if the pollutant tends not to mix with water. A spill that occurs on more impermeable terrain, such as rock or concrete, needs to be treated quickly to prevent the spill from spreading out over the surface to adjacent land or watercourses.

Spills involving compounds that tend not to mix with water can be very difficult to contain if the response is delayed. This is because the compound will spread out on the surface of the water. As the spill continues, the surface layer will become thinner and larger in diameter, rather than remaining the same diameter and increasing in thickness. This is why oil spills in freshwater and marine water can quickly grow to cover hundreds of square miles/kilometers in area. Clean up of a spill of such a large area presents difficult challenges.

Oil spills that coat the rocks, soil, and sand beaches of coastal shorelines can be treated in several ways. Hot water sprayed onto the shore can help dislodge pollutants. However, as was found in the analysis of the outcome of the clean up of the *Exxon Valdez* spill off Alaska in 1989, the use of high pressure, steaming water can destroy microbial life in the area being treated. Most bacteria cannot tolerate the boiling water temperature that is used in spill clean up. Although there are a few bacteria that are able to tolerate such temperatures, such as bacteria found naturally in hot springs, these extremely hardy bacteria typically do not degrade the pollutants, whereas the bacteria that are naturally present in the area of a spill may have developed the ability to use the pollutants as nutrients. The loss of the natural microbial population can reduce the efficiency of a spill remediation, since bacteria that are in the soil or sand may be able to use some of the pollutants as sources of food and degrade them. The compounds that are left over following this degradation may not be as toxic, if at all.

Another strategy for remediating a spill, especially if it involves hydrocarbons or other compounds that cling tenaciously to rocks, soil, and sand, is to spray on a solution that contains chemicals that contain dispersants that

WORDS TO KNOW

BIOREMEDIATION: The use of living organisms to help repair damage such as that caused by oil spills.

***EXXON VALDEZ*:** A tanker that spilled almost 11 million gallons of oil in Prince William Sound, Alaska, beginning on March 24, 1989.

GROUNDWATER: Fresh water that is present in an underground location.

SUPERFUND: Legislation that authorizes funds to clean up abandoned, contaminated sites.

gather around the oil and form oil-containing droplets. The dispersant compounds are described as being amphipathic; they have a region of their structure that tends to associate with water (the hydrophilic region) and a different region that tends not to associate with water (the hydrophobic region). When added to water, the hydrophobic portion will want to escape water. The way that this can be achieved is by the formation of a sphere composed of the dispersant compounds, in which the hydrophobic regions are on the inside and the hydrophilic regions are on the outside. Since hydrocarbons also tend to be hydrophobic, they will associate with the hydrophobic regions of the dispersant, and so will tend to gather on the inside of the spheres as they form. In this way the pollutant can be trapped inside the spheres, which aids in the dispersion of the spill from rocks, soil, and sand. The dispersed oil is then collected by vacuuming or, if it has re-entered the water, by gathering it in an ever-decreasing space using a floating barrier known as a boom, or by the addition of sponge-like material that can soak up the liquid. Spills that occur on the surface of the ground or on an artificial surface like concrete can be treated by adding an absorbent, powdery material. The soaked material needs to be collected afterward and properly disposed of, such as by incineration or storage in leak-proof bags and burial in a landfill.

Spill remediation can be less exotic. For example, when a heavy oil spill has washed up on a beach, much of the spill may be removed manually by shoveling or using a grader. This option is not appropriate in gravely areas, however. The choice of strategy depends on the nature of the spill, the urgency of the clean up (for example, a spill that threatens coastal regions may be more urgent than an inland spill), and whether or not a spill is in an urban or a rural area.

The reliance on bacteria to degrade spills has grown since the time of the *Exxon Valdez* disaster. At the time of this oil spill, this strategy, which is known as bioremediation, was known but had not been widely used, out of concern for the possible adverse consequences of the rapid growth of the bacteria. Their sudden and explosive increase in numbers could upset the balance of the particular ecosystem. Those advocating the approach argued that since the ecosystem had already been drastically affected by the spill, this concern was not as great as the need to deal with the spill itself.

In the *Exxon Valdez* response, a spray that contained nutrients was applied to the area of the spill. The intent was to stimulate the growth of the bacteria that were already resident in the soil and sandy beaches, with the hope that their growth would aid in the degradation of the pollutants. Although not completely successful, the strategy worked well enough to earn approval as a spill remediation technique from the U.S. Environmental Protection Agency (EPA).

Another example of bioremediation is the use of bacteria to clean up a site contaminated with polychlorinated biphenyls (PCBs). Over 1.5 million pounds of PCBs were made and used in the United States before being banned in 1977, due to concerns about their adverse health effects and their persistence in the environment. Even given this environmental longevity, bacteria have been successfully used to degrade PCBs.

Bioremediation can also be done by tailoring organisms to be capable of degrading target pollutant compounds. This approach usually involves genetic engineering—an alteration of the genetic material so that either a gene coding for the production of the pollutant degrading protein is introduced into the bacteria, or that a gene already present is made able to function more efficiently.

The genetic engineered approach to bioremediation is still controversial. Critics argue that the release of bacteria that are not naturally present in the particular environment really could disrupt the ecosystem. There is merit in this argument, and so genetically engineered bioremediation is typically done when the polluted material has been collected and moved to a controlled site, where escape of the engineered bacteria does not occur.

Spills that occur in soil can be treated at the site of the spill. Typically, a well is drilled down into the main site of the spill. If bioremediation is being done, a nutrient solution can be pumped into the ground. Often this can be as simple as water, since the water can contain traces of nutrients that are sufficient to stimulate the growth of bacteria dwelling underground. Another well drilled nearby is used to pump up the underground fluid allowing time for degradation to occur. The cycle of nutrient addition and fluid removal can be repeated over and over. The fluid that is recovered after each cycle can be tested for the presence and concentration of the pollutant. Hopefully, with time the level of the contaminant drops. Remediation is complete when the level of the recovered pollutant is at a safe level, according to regulations in place that govern the clean up.

The same strategy can be used for contaminated soil that has been collected and taken to another site. The soil is deposited in an enclosed area. Alternately, the contaminated soil can be spread out over a wide area of leak-proof plastic. This can be advantageous since it increases the amount of contaminated soil that is exposed, which can make it easier to treat.

Spill remediation regulations also govern how the clean up is done. For example, if a spill site is near a watercourse such as a stream or river, the shoreline must be protected so that any accidental release of recovered contaminant does not drain into the surface water. This can be done by the installation of a barrier of synthetic material such as plastic or straw. The site must be monitored during the spill remediation to ensure that the barrier is intact, and the material must be removed and properly disposed of once the job is completed.

Spill remediation also involves caring for the wildlife that has been affected. Marine birds and mammals can be especially affected, since a pollutant such as oil can destroy the insulation ability of fur and the water repellent properties of feathers. Clean up of pollutant-soiled birds and animals has to be done creature by creature, and typically involves trying to wash off as much of the pollutant as possible. Even so, ingestion of the chemical as the bird or animal tries to groom itself can be fatal.

Impacts and Issues

Spill remediation is occurring constantly in developed countries. For example, the EPA estimates that there are nearly 14,000 oil spills each year. Many of these are handled by federal, state, and even local personnel trained in the specialized clean up that is required. However, some spills are so toxic that even clean up is too dangerous. In such cases, the spill site is isolated from development and any use, and the site is abandoned. An example is the Chernobyl nuclear plant accident that occurred in Russia in 1986. The reactor was entombed rather than an attempt made to clean up the toxic levels of radioactivity.

Spills involving oil are an inevitable consequence of the global use of oil and the need to transport the oil long distances from the site of recovery to the refinery. According to the U.S. National Oceanic and Atmospheric Administration (NOAA), the United States uses about 700 million gallons of oil every day. Globally, the daily figure is almost 3 billion gallons.

In the United States, responsibility for a spill clean up varies depending on the location of the spill. Typically, either the U.S. Coast Guard or the EPA takes charge. NOAA can also be called in to assist.

In most countries, the extent and consequences of spills became better known in the 1960s. In the United States, incidents such as New York's Love Canal contamination in the 1970s increased public awareness of chemical spills, after the passage of the National Environmental Policy Act in 1969. The legislation requires stringent analysis of federal projects, in part to reduce the likelihood of a spill. Other legislation directly pertained to spill remediation. The best example in the United States is the Comprehensive Emergency Response Compensation and Liability Act that was passed in 1980. This legislation specifies who has legal responsibility for spill remediation, including the federal government when clean up of an abandoned site was necessary. The federal government's role is through a program popularly known as Superfund. As of 2008, 1,305 Superfund sites have been scheduled for remediation, and over 14 million Americans live within one mile of a Superfund site, according to the National Institute for Environmental Health Sciences.

The actual number of soil-contaminated sites in the United States is much higher, with estimates ranging up to 200,000. In developing and underdeveloped countries the situation is much worse, since legislation governing pollution and clean up is not as extensive. Chemicals that persist in the environment that have been banned in the United States can in some cases still be used in other areas of the world. One example is the pesticide DDT; although its use in the control of malaria has been sanctioned in Africa by the World Health Organization, concerns over its use remain.

The natural impacts of a spill can be disastrous. Images of oil-soaked birds and other wildlife struggling to survive are real and all too frequent. For example, an oil spill that occurred off the coast of Spain in late 2002, when an estimated 64,000 tonnes of oil leaked from the tanker *Prestige*, killed thousands of sea birds and fouled hundreds of miles of coastline.

Clean up of an oil spill can be difficult, time consuming, and expensive. As an example, the 1989 spill from the *Exxon Valdez* released enough oil to fill almost 125 Olympic-sized swimming pools (still, this amount represents only about 2% of the oil used by the United States each day in 2008). It was the location of the spill in an area of Alaska that was rich in fish and wildlife that generated the most concern. The slick, which fouled over 1,000 mi (1,600 km) of Alaskan coastline, took four summers to clean up. The clean up involved 10,000 workers and 1,000 boats. Exxon's bill for the clean up, compensation to local fishermen and other business owners whose jobs were affected by the spill, penalties, and other expenses reached $3.5 billion.

Still, all this time and effort does not necessarily restore an area to its pre-spill state. As of 2008, the region affected by the *Exxon Valdez* spill has not fully recovered from the spill in terms of fish stocks and the health of the coastline. In February 2008, the U.S. Supreme Court considered a civil action filed on behalf of those affected by the 1989 spill, who consider Exxon's compensation of about $15,000 per person too little, arguing that the lasting damage has jeopardized their jobs. An additional $75,000 per person has been sought in the civil suit. A decision has yet to be announced.

SEE ALSO *Chemical Spills; Groundwater Quality*

BIBLIOGRAPHY

Books

Leacock, Elspeth. *The Exxon Valdez Oil Spill.* New York: Facts on File, 2005.

Owens, Peter. *Oil and Chemical Spills.* New York: Lucent Books, 2003.

Wang, Zhendi, and Scott Stout. *Oil Spill Environmental Forensics: Fingerprinting and Source Identification.* New York: Academic, 2006.

Brian D. Hoyle

Streamflow

Introduction

Streamflow, which is also known as channel runoff, refers to the flow of water in natural watercourses such as streams and rivers. Without streamflow, the water in a given watershed would not be able to naturally progress to its final destination in a lake or ocean. This would disrupt the ecosystem.

Streamflow is one important route of water from the land to lakes and oceans. The other main routes are surface runoff (the flow of water from the land into nearby watercourses that occurs during precipitation and as a result of irrigation), flow of groundwater into surface waters, and the flow of water from constructed pipes and channels.

A given watercourse has a maximum streamflow rate that can be accommodated by the channel, and which can be calculated. If the streamflow exceeds this maximum rate, as happens when an excessive amount of water is present in the watercourse, the channel cannot handle all the water and flooding occurs.

Since the streamflow can be calculated, watercourses that are prone to flooding can be modified to reduce the likelihood of flooding, or development of lands in the vulnerable areas can be restricted.

Historical Background and Scientific Foundations

Measurement of streamflow in a small watercourse can be accomplished by releasing a floating object from a designated spot and timing how long it takes the object to travel to another designated spot farther downstream. By knowing the distance between the two points and the travel time, the flow can be calculated (typically as cubic meters per second). The result of at least three determinations is known as a hydrograph.

The measurements should be conducted in the middle of the watercourse, since this is where the water is deepest. The flow toward either side can be slower, due to the increased friction of the shallower water with the bottom.

The result can be made even more realistic of the actual streamflow by factoring in the nature of the bottom of the watercourse. A rough bottom will produce a more turbulent waterflow, as the water will flow around and over the rocks and stones present. This will slow the streamflow from the value calculated using the speed of the floating object. Generally, the streamflow value obtained from midstream is multiplied by 0.8 to approximate the actual flow rate.

A smoother bottom will not impede flow as much. For such watercourses, the calculated mid-stream flow rate is multiplied by 0.9 to obtain the estimate of the actual flow.

Watercourses that are wider and deeper are more complicated to measure. For these, the width, depth, and waterflow must be measured at many spots across the width of the watercourse and at multiple depths. The measurements are combined to yield an average flow at the average depth.

For these measurements, a simple floating object is insufficient. Rather, a piece of equipment called a current meter can be used. The meter consists of a probe attached to a torpedo-shaped object. The object has fins at one end that positions the object in the path of the current. Spinning cups on the probe measure the velocity of the passing water, which is recorded electronically.

Streamflow is not constant. This is especially evident after a heavy rainstorm or during the springtime snowmelt. A stream or river that is gently flowing one day can contain a thundering flow of water only hours later, as the excess water enters the watercourse directly or from surface runoff. The change is especially evident in larger rivers that are receiving the flow of water from

WORDS TO KNOW

EROSION: The wearing away of the soil or rock over time.

RUNOFF: Water that falls as precipitation and then runs over the surface of the land rather than sinking into the ground.

SPRING: The emergence of an aquifer at the surface, which produces a flow of water.

WATERSHED: The expanse of terrain from which water flows into a wetland, water body, or stream.

other watercourses in a watershed, since the combined flow of the various watercourses is combined into the larger watercourse.

By monitoring streamflow, the capacity of the watercourse can be determined at various points. Areas that are narrower or have bends may be more susceptible to flooding than other areas. All this is valuable information, since warnings can be issued when conditions are such that the capacity of the watercourse is likely to be exceeded. Susceptible areas of the watercourse may be altered or barriers installed to lessen the chances of flooding.

■ Impacts and Issues

Streamflow is a vital part of the water cycle, where water is cycled between Earth's surface and the atmosphere, and is vital for the movement of water within a watershed.

Knowledge of the streamflow of a particular watercourse is also important in determining how best to use the land that borders the watercourse. For example, regions that are determined to be susceptible to flooding should not be used for residential, industrial, or commercial development. Use of such regions as parkland, for example, allows the land to function as a floodplain when necessary.

As well, the knowledge of streamflow is important in determining environmentally acceptable activities that are permitted in the lands bordering the river. For example, it would be unwise to permit the release of industrial effluent into a watercourse with a very slow flow, since any toxic compounds in the effluent would have more opportunity to affect life in the water than in a watercourse with a high flow rate.

Localized manipulation of streamflow is exploited as a means of generating power. By damming a watercourse and releasing the water at a controlled rate, the artificially created higher streamflow is used to generate hydroelectricity.

See Also *Aquifers; Floods; Freshwater and Freshwater Ecosystems*

BIBLIOGRAPHY

Books

Bray, R. N. *Environmental Aspects of Dredging*. New York: Taylor & Francis, 2008.

Chiras, Daniel D., John P. Reganold, and Oliver S. Owen. *Natural Resource Conservation: Management for a Sustainable Future*. New York: Prentice-Hall, 2004.

Grover, Velma I. *Water: Global Common and Global Problems*. Enfield, NH: Science Publishers, 2006.

Superfund Site

Introduction

A Superfund site is a location contaminated by hazardous waste that has been designated by the U.S. Environmental Protection Agency (EPA) for management and cleanup. Superfund sites are prioritized in the National Priorities List, which has been maintained since 1983. The Superfund system was established in 1980, when a federal law creating a special fund for cleaning up such sites was passed. As of 2007, there were 1,569 Superfund sites on the National Priorities List.

Historical Background and Scientific Foundations

Thousands of locations throughout the world have been polluted by long-lasting toxic substances, including metals such as mercury and lead, asbestos, and organic pollutants such as dioxins. Many of these sites are in industrialized countries, including the United States. For many decades, industry was free to dump such toxins into the ground or nearby waterways, or to accumulate poorly secured stockpiles of waste in drums or other fragile containers with little legal oversight. In the 1970s, as the environmental movement became strong, several severe pollution disasters made national headlines in the United States.

One of the most notorious of these was the Love Canal neighborhood of the town of Niagara Falls, New York, where 21,000 tons of toxic chemical waste were dumped by the Hooker Chemical and Plastics Company from 1942 to 1952. The waste was covered with dirt, and about 200 homes and a school were built near the site. Gradually, it became clear that the residents of Love Canal were suffering from cancers and other diseases at an unusually high rate. The cause was traced to the chemical waste, and President Jimmy Carter declared a state of emergency for the neighborhood in 1978 and 1980. Another notorious toxic disaster site was the Valley of the Drums in Bullitt County, Kentucky, where tens of thousands of leaking drums of toxic chemicals had been strewn over a 23-acre site.

In response, Congress passed the Comprehensive Environmental Response, Compensation and Liability Act (CERCLA, also known simply as the Superfund Act) in 1980 to deal with abandoned toxic dumps. The law authorized the EPA to deal with threatening releases of "any pollutant or contaminant which may present an imminent and substantial danger to public health or welfare." When possible, the parties responsible for creating dangerous sites would be forced to pay for cleanup. When this could not be done—as, for example, at Love Canal, where the polluting corporation was no longer in business—CERCLA also established a trust fund, the Superfund. Money for the fund came mostly from a tax on crude oil and some industrial chemicals and an income tax on some corporations. The Superfund would pay for cleanup when there was no other recourse. Finally, CERCLA specified which substances would be considered actionable pollutants. This list has since been expanded.

In 1986, CERCLA was followed by SARA (Superfund Amendments and Reauthorization Act). This made several changes to the Superfund system. One of SARA's most important provisions was that it extended CERCLA to federal facilities, not just private property. Both laws apply only to sites in the United States.

Before being dealt with under CERCLA, a toxic site must be deemed officially needy of Superfund treatment. When the EPA is notified of the existence of a possible Superfund site, it adds it to a list called the Comprehensive Environmental Response, Compensation, and Liability Information System. Either the EPA or the state where the candidate site is located investigates the site to deter-

WORDS TO KNOW

REMEDIATION: A remedy. In the case of the environment, remediation seeks to restore an area to its unpolluted condition, or at least to remove the contaminants from the soil and/or water.

mine the level of contamination that exists. If it is severe enough according to the EPA's Hazard Ranking System, it is added to the National Priority List (NPL) and becomes eligible for Superfund money, if needed, for long-term remediation.

IN CONTEXT: FUNDS FOR THE SUPERFUND

The types of remediation that the EPA uses at Superfund sites varies depending on the type of hazard in the location. In emergency situations, the EPA removes hazardous material immediately, often incinerating the waste. In some cases, people are relocated to protect them from exposure to the toxins. In most cases, removal of the toxic material takes much longer, months or even years.

When a site is listed on the Superfund National Priority List (NPL), the EPA attempts to fund the cleanup from those responsible for dumping the chemicals. These groups are known as Potentially Responsible Parties (PRPs). If these groups cannot be identified or are no longer viable entities, then the money for the cleanup comes from the Superfund trust that was initially generated from taxes on petrochemical companies. The fund was exhausted by 2004 and Superfund remediation was then funded from general revenues.

Impacts and Issues

The first NPL was announced in 1983 and contained 406 sites, including Love Canal and the Valley of the Drums. From 1983 to 2007, the EPA placed 1,569 sites on the NPL. Plans for final cleanup at 75% (1,180) of the sites had been adopted by 2007, and construction of remedial solutions had been finished at two thirds (1,030) of NPL sites.

The infamous Love Canal site was added to the National Priority List in September 1983. Federal funds were used to relocate 800 families from the Love Canal neighborhood, one of the most dramatic actions taken at a Superfund site. Similar action was taken in Pensacola, Florida, where in 1991, an emergency cleanup action at a former wood-treatment plant created a heap over 5.4 million cubic ft (153,000 cubic m) of dioxin-contaminated dirt that became known as "Mount Dioxin." Over 400 households were relocated.

SEE ALSO *Landfills; Toxic Waste*

BIBLIOGRAPHY

Books

Macey, Gregg P., and Jonathan Z. Cannon, eds. *Reclaiming the Land: Rethinking Superfund Institutions, Methods and Practices.* New York: Springer, 2007.

Web Sites

U.S. Environmental Protection Agency (EPA). "Superfund." http://www.epa.gov/superfund/ (accessed March 29, 2008).

Larry Gilman

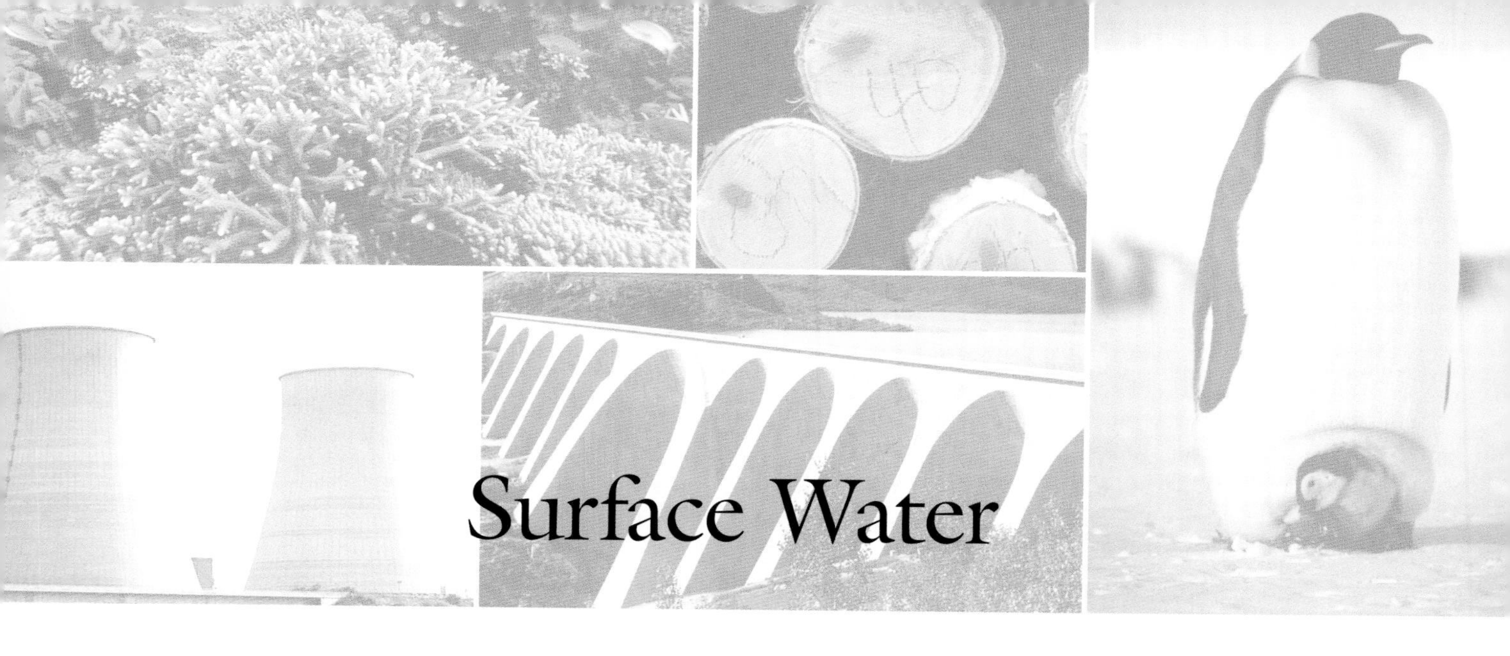

Surface Water

Introduction

Surface water includes water found in streams, rivers, lakes, marshland, snow, ocean water, or any other water found on Earth's surface. Groundwater is located in the subsurface in reservoirs (aquifers).

Approximately 70% of Earth's surface is covered by surface water. Of this estimated 1.4 billion cubic kilometers of water, almost 98% is saltwater and between 2 and 3% is freshwater. Surface water can be replenished by precipitation and can pass to the atmosphere via evaporation, and so is a crucial part of the planet's hydrologic cycle.

The increased temperature and drier climate that are developing in some regions of the world under the influence of global warming—the increasing warming of the atmosphere that is being driven in part by the production of compounds associated with human activity that restrict the dissipation of heat in the atmosphere—is causing some surface water sources to decrease or dry up altogether. In underdeveloped countries, this is a concern to human welfare and survival, where freshwater is scarce.

Surface water is also more susceptible to microbial contamination than is groundwater, and so is an important cause of waterborne illnesses such as cholera.

Historical Background and Scientific Foundations

The surface water that is replenished by precipitation and which can release water vapor to the atmosphere is a small proportion of the total surface freshwater on Earth. Of the estimated 35 million cubic kilometers of fresh surface water, almost 25 million cubic kilometers are unavailable, as they are present as glacial ice, permafrost, or permanently frozen snow.

Evaporation of surface water adds an estimated 580,000 cubic kilometers of water vapor to the atmosphere every year. Without this infusion of water vapor, the planetary water cycle, and so life on Earth, would cease.

Because of its location, surface water is susceptible to contamination by a variety of pollutants and microorganisms. Typically, the disease-causing microbes normally live in the intestinal tract of warm-blooded animals including humans. They enter the surface water in feces. Examples of disease-causing bacteria are those in the genera of *Salmonella*, *Shigella*, *Vibrio*, and *Escherichia*.

Escherichia coli O157:H7 has become prominent. Contamination of drinking water with O157:H7 can be devastating. An infamous example was the contamination of the municipal water supply of Walkerton, Ontario, Canada, in 2000. Several thousand people became ill, and seven people died due to ingestion of the contaminated water that had flowed into an uncapped well from an adjacent cattle field.

Protozoa can also contaminate surface water. The main two examples are Giardia and Cryptosporidium, which are naturally present in the intestinal tract of beavers and deer. Waterborne illness due to these protozoans is increasing in North America, as human settlement encroaches on previously natural areas.

Impacts and Issues

The substantial proportion of fresh surface water that has been inaccessible is becoming more available as polar regions of Earth increase in temperature due to global warming. Large regions of glacial ice in the Antarctic have broken away into the surrounding sea. Eventually the ice will melt, increasing the volume of liquid water present in the global ocean.

WORDS TO KNOW

EROSION: The wearing away of the soil or rock over time.

RECHARGE: Replenishment of an aquifer by the movement of water from the surface into the underground reservoir.

RUNOFF: Water that falls as precipitation and then runs over the surface of the land rather than sinking into the ground.

WATERSHED: The expanse of terrain from which water flows into a wetland, water body, or stream.

Measurements of tide levels have been done throughout the twentieth century and satellite measurements have been ongoing since the 1970s. These measurements have established that the global sea level is rising. According to the 2007 report of the Intergovernmental Panel on Climate Change (IPCC), the average global sea level rose by 1.8 millimeters per year from 1961 to the early 1990s, and has been rising by over 3 millimeters per year since 1993.

Rising ocean waters are already affecting coastal regions. For example, according to the Geological Survey of Canada, rising sea levels are contributing to the increased damage of beach coastline in the maritime province of Prince Edward Island caused by coastal storms, which is threatening the province's nearly $3 million annual tourism industry. If the rising surface waters continue, almost half the coastline could be submerged by 2100.

Florida is another area being affected. The advancing sea level has already affected surface water bodies including coastal wetlands. The low elevation of the state puts even the interior of Florida, including the Everglades, at risk, and so could threaten many species of birds and wildlife.

The fact that many of Florida's shorelines are gently sloping will increase the height of the sea-level rise, which could exceed 2 ft (0.6 m) by 2100. The state's drinking water supplies could be challenged.

In contrast, climate change is causing a decline in surface waters in regions of the globe that are becoming warmer and drier. For the 600 million people who already face water scarcities, the further loss of surface water could be life threatening. The World Health Organization estimates that by the year 2025, almost 3 billion people could be affected by inadequate water supplies.

SEE ALSO *Aquifers; Dams; Drainage Basins; Floods; Groundwater; Oceanography; Rivers and Waterways; Tidal or Wave Power*

BIBLIOGRAPHY

Books

Grover, Velma I. *Water: Global Common and Global Problems.* Enfield, NH: Science Publishers, 2006.

O'Neill, Karen. *Rivers by Design: State Power and the Origins of U.S. Flood Control.* Raleigh, NC: Duke University Press, 2006.

Web Sites

World Health Organization. "Water Resource Quality." http://www.who.int/water_sanitation_health/resources/resquality/en/index.html (accessed April 20, 2008).

Surveying

Introduction

In environmental science, a survey refers to the examination of a site to precisely determine selected characteristics such as the presence of certain wildlife; type and condition of the soil; presence and extent of features such as wetlands, grassland, and lakes; indications of the presence of minerals in the subsurface (by measurement of variations in radiation levels); and information on the surface geology (by measurement of fluctuations in magnetism).

Surveys can be done on foot, but are most easily carried out from the air, typically using equipment onboard the aircraft or towed behind the aircraft.

Historical Background and Scientific Foundations

Environmental science survey techniques are generally known as geophysical surveys, which refer to the physical aspects of Earth's surface and immediate subsurface.

A common environmental science geophysical survey is the aeromagnetic survey. In a typical aeromagnetic survey, an instrument called a magnetometer is carried onboard or towed behind an aircraft. Magnetometers are portable and so can be carried during on-foot surveys. However, the use of an aircraft permits a much broader territory to be examined in a day, while examination of the same area on foot would likely require weeks, even if the terrain is hospitable enough to permit easy access to the survey area.

In an aeromagnetic survey, the aircraft flies in a regular and pre-determined pattern over the region of interest. Typically, the flight path is a grid pattern, with the aircraft flying back and forth, for example north-south followed by south-north, to form one area of the grid, before paths at 90 degrees to the first pattern (for example, west-east followed by east-west) to complete the grid. Geographic information system (GIS) technology allows the location of the aircraft to be determined at any point during the flight, and this geographic information can be correlated with the aeromagnetic findings to precisely map findings of interest.

A magnetometer measures small variations in the magnetic field at Earth's surface. When the contributions of Earth's magnetic field are filtered out, the resulting variations correspond to a difference on local magnetic energy due to the minerals present at the surface or the immediate subsurface. In this way a map of the geological composition can be generated.

Another environment survey is known as an aeroradiometric survey. This can be done in exactly the same way as an aeromagnetic survey, except for the use of a different piece of equipment that measures optical radiation such as the ultraviolet, visible, and infrared portions of the light spectrum. A common use of the survey is to measure variations in surface temperature. This can also be done using radiometers positioned onboard satellites orbiting Earth. The latter surveys permit the examination of huge swaths of the surface during each orbit.

Surveys can also be done by eye or photographically. These are useful when the intent is to monitor an area to roughly gauge the population of a target species such as elk or to gauge the success of a species re-introduction campaign.

Impacts and Issues

Aeromagnetic surveys are important in generating detailed geological maps of Earth's surface, and they are commercially important in mineral exploration. Deposits of minerals such as iron are magnetic, while other nonmagnetic minerals will produce a fluctuation in the local magnetic field when they are present in abundance.

WORDS TO KNOW

ANTHROPOGENIC: Made by humans or resulting from human activities.

GREENHOUSE GAS: A gas that accumulates in the atmosphere and absorbs infrared radiation, contributing to the greenhouse effect.

HABITAT: The natural location of an organism or a population.

PRIMARY POLLUTANT: Any pollutant released directly from a source to the atmosphere.

Aeroradiometric surveys can be conducted over water, which can be useful in revealing locations where colder and deeper water is moving upward (upwelling). This can be helpful in revealing potential fishing sites, since the upwelling water brings nutrients to the surface.

Surveys of the surface temperature of the ocean that have been conducted in satellite studies since the 1970s have revealed the increasing surface temperature of the ocean in certain regions. In tropical regions this change may be influencing the frequency and intensity of extreme weather events such as hurricanes. Additionally, some modeling studies have indicated that the changing ocean temperature could alter currents such as the Gulf Stream.

The data from environmental surveys are also useful in generating models that attempt to predict other effects of climatic change that includes weather patterns, the frequency of heat waves, and rise of sea level.

SEE ALSO *Climate Modeling; Geospatial Analysis*

BIBLIOGRAPHY

Books

Berg, Linda, and Mary Catherine Hager. *Visualizing Environmental Science*. New York: Wiley, 2006.

Pepper, Ian, Charles Gerba, and Mark Brusseau. *Environmental and Pollution Science*. New York: Academic Press, 2006.

Web Sites

Appalachian Mountain Club "Ecological Survey and Land Management Planning on the Katahdin Iron Works Property." http://www.outdoors.org/conservation/wherewework/maine/ecological-inventory.cfm (accessed May 2, 2008).

Sustainable Development

Introduction

Sustainable development is the practice of managing growth and change in ways that meet the needs of the present without damaging future generations' ability to develop further. Although sustainable development is usually thought of as an environmental effort, it also includes the social and economic aspects of development. The concept was established in a series of international policy meetings starting in the 1970s, with the term first being used in the early 1980s.

The United Nations has assumed an important role in promoting and facilitating sustainable development all over the world, including in developing nations. Through its Division for Sustainable Development, the United Nations provides technical expertise, facilitates communication, and evaluates development efforts. For sustainable development to be effective, it must be implemented in both large- and small-scale projects. In pursuit of this goal, governments, charities, international organizations, and private companies are working to provide development benefits to humanity in a sustainable manner.

Historical Background and Scientific Foundations

The idea of sustainable development has its roots in environmentalism, which arose in the second half of the nineteenth century in Great Britain and the United States as a response to the pollution of the Industrial Revolution. To power factories and blast furnaces, early industrialists relied almost exclusively on the burning of coal. This cheap, abundant fuel provided the energy needed to power the era's economic expansion, but it soon became clear that widespread coal use had serious consequences. Without the modern technologies that trap pollutants, huge amounts of soot, toxic gases, and heavy metals were released into the air. Other industries were simply dumping their poisonous refuse into the rivers and oceans, with many trusting that the water would simply wash the waste away.

Furthermore, the methods used to acquire resources to fuel industrial development were becoming more destructive as the technology to exploit them improved in the first half of the twentieth century. Large areas were deforested by clear-cutting without thought to what would become of the land. Huge strip-mines removed whole mountains, creating cavernous pits that leached toxic water. More potent chemical pesticides injured birds and other wildlife, not just the crop pests they were meant to deter. Even chemical fertilizers had unintended consequences, with the excess nutrients causing harmful algae blooms. The damage caused to wildlife by habitat loss, chemical toxins, and unrestrained hunting led to the worrying collapse of the American bison population, and the 1914 extinction of the passenger pigeon, once the most abundant bird in North America.

By the mid-twentieth century, U.S. public policy had long favored unrestrained economic development without heed to the environmental costs. In 1962, Rachel Carson brought environmental issues to the public consciousness with her groundbreaking book *Silent Spring*, which exposed the damage that the pesticide DDT was doing to bird species. In response to public concern, President John F. Kennedy ordered an investigation of the book's claims, leading to a ban on DDT. Carson's book is widely credited with spurring the creation of the environmental movement as it exists today.

The 1970s saw the first steps toward worldwide, coordinated efforts on environmental issues, as well as the first international summits on environment and human development. In 1972, representatives from 113 countries met in Sweden at the United Nations

WORDS TO KNOW

ALTERNATIVE ENERGY: An energy source that is used as an alternative to fossil fuels. Solar, wind, and geothermal power are examples of alternative energies.

CLEAR-CUTTING: A forestry practice involving the harvesting of all trees of economic value at one time.

ECO-TOURISM: Environmentally responsible travel to natural areas that promotes conservation, has a low visitor impact, and provides for beneficially active socio-economic involvement of local peoples.

MICROFINANCING: An economic development strategy in which small loans (microcredit) and other financial services are provided to very low-income individuals.

REFORESTATION: The replanting of a forest that had been cleared by fire or harvesting.

RENEWABLE RESOURCES: Resources that can be renewed or replaced fairly rapidly by natural or managed processes.

SUSTAINABILITY: Practices that preserve the balance between human needs and the environment, as well as between current and future human requirements.

Stockholm Conference on the Human Environment. The conference declared that humans have the ability to improve or destroy the environment, and that the health of the environment is a major factor in humans' quality of life. It also recognized that environmental problems have different causes in developed and undeveloped countries. Although more developed countries usually create pollution because of industrial activities, underdeveloped countries' pollution is most often caused by their very lack of development.

Although the phrase was not yet in use, the Stockholm conference laid down some of the fundamental principles of sustainable development. Its acknowledgement that continued human development is inevitable laid the foundation of practicality that characterizes the current movement. By calling on both developed and undeveloped nations to use better environmental practices, the conference stated that preserving the environment is the responsibility of all humanity. It appealed to individuals, local and national governments, and the international community to adopt policy and behaviors that protect the environment while allowing for sufficient human development.

The rise of the sustainable development and environmental movements coincided with a series of disasters and scandals that highlighted these issues in the public consciousness. In February 1972, a massive coal-waste dam failed in the area of Buffalo Creek, West Virginia. More that a hundred million gallons of black water rushed down the creek and into the towns below, killing 125 people, injuring over 1,000, and leaving almost 4,000 homeless. Around the same time, it was publicly revealed that widely used polychlorinated biphenyl compounds (PCBs) were carcinogens and caused developmental defects in children. The discovery of a hole in the ozone layer of the atmosphere over Antarctica, and it subsequent link to chlorofluorocarbons (CFCs) in refrigerants and aerosol propellants, led to concern about exposure to radiation and the possibility of further damage to the atmosphere. In 1978, outrage erupted in the media over the toxic contamination of the Love Canal neighborhood in Niagara, New York, which had been built over a dumping site for poisonous waste. These incidents and others made it clear to many that human development had, to this point, progressed without heed to human and environmental costs. While some progress was made with the 1970 founding of the U.S. Environmental Protection Agency (EPA), international concern continued to grow surrounding environmental and sustainability issues.

In 1980, the World Conservation Strategy coined the term sustainable development, expanding the ideas laid out in Stockholm and in subsequent conferences on environmental issues. It stated that the problems of environmental protection and meaningful development for underprivileged people are inseparable and must be addressed together. This new perspective, namely that development and conservation are not in opposition to one another, is the heart of sustainable development.

Issues and Impacts

From its roots in environmentalism, the concept of sustainable development has been applied to many other issues. The idea of sustainability is now common and of concern to individuals, development planners, political leaders, and humanitarian organizations. Today, sustainable development is generally viewed to have three main areas of focus, environmental sustainability, social sustainability (including political issues), and economic sustainability. The United Nations also applies sustainable practices more specifically to agriculture, renewable resources, water management, ecotourism, technology, disaster management, trade, waste management, international law, and infrastructure organization.

It is important to note that the three major areas of sustainable development are interdependent; environmentally destructive practices can harm individuals in society, misused social policy can destroy the environment or harm the economy, and poor economic planning can lead to loss of resources and social distress. Conversely, by preserving and improving the environment, humanity can hope to gain tangible economic and social benefits. In many ways, the dangerous mismanagement of sustainability factors facilitated many of the bloody African conflicts of the 1990s, some of which continue today. Decades of poverty and exploitation led to violence as mistrustful ethnic groups competed for natural resources, overpopulation created scarcity of food and jobs, and ineffective governments abused their populations, failed to control crime, and unleashed waves of refugees across unregulated borders.

Although many of these conflicts happened in the developing world, more developed nations were also contributing to instability with their unsustainable practices. A particularly tragic example of this occurred during Sierra Leone's civil war of the 1990s. Multiple parties were seeking control of the country's diamond resources, leading to social conflict and widespread violence. The willingness of western countries to purchase the diamonds, despite their dubious origins, stimulated the conflict. The current state of environmental degradation in Sierra Leone still leaves the population vulnerable to food and resource shortages along with the social conflicts that can follow.

Today's sustainable development tries to address sustainability problems before they erupt into violence or disaster. In the western world, sustainability efforts are often viewed with skepticism for several reasons. One of the most persistent obstacles is the idea that sustainable development will reduce quality of life or decrease the profitability of industry or commerce. While this may sometimes be true in the short term, the argument loses some credibility when viewed against the massive costs associated with environmental contamination or the current mortgage credit crisis caused by unsustainable financial practices.

The development and use of new technology, however, is often the most widely accepted way of improving the environment while benefiting humanity. Currently, green and other alternative energy sources are a popular target for sustainable development at many levels. Individuals, corporations, charities, and governments in both the developed and undeveloped world are all thinking about how humanity's energy needs can be met with more environmentally friendly technologies. Concern about global climate change is the greatest influence driving interest in non-fossil fuels. Although the threats seem abstract or distant to many people, the effects of increased global temperatures can already be observed and are affecting humans, animals, and the planet. By employing energy sources that emit less carbon, there is hope that climate change can be slowed or reversed. At the same time, better alternative energy technologies may someday make electricity available in areas with little infrastructure or to people who are unable to afford the current high prices of fossil fuels. This is a good example of the positive link between development and environmental protection.

Although new technologies are important to sustainable development, new ways of thinking are perhaps even more critical to achieving sustainability. In the past, inflexible thinking has been a barrier to progress toward environmental benefits. A good example of this is the former use of CFCs in propellants and refrigerants. Although many industry groups protested that it would be impossible to reduce or eliminate their use, within a decade they were able to find new chemicals to take their place in most applications. There are many cases in which the development of better industrial processes was motivated by short-term economic benefits. What can be more difficult to see is the long-term, less conspicuous links between environmentally friendly practices and future success. The sustainability movement is currently struggling to convince both industry and its regulatory bodies of the need to broaden their thinking and embrace more sustainable methods.

Members of the Landless Peoples Movement march with other protest groups in August 2002 from the impoverished township of Alexandria to nearby Sandton, Johannesburg, South Africa, where the U.N. World Summit on Sustainable Development was taking place. *AP Images*

Working toward this same goal, the field of industrial ecology seeks to improve industrial operations by the use of more efficient processes, less dangerous chemicals, and the production of longer-lasting materials. When developing new processes, industrial ecologists recognize that sustainable development is not simply being able to continue a process forever, but working in ways that avoid dangerous future consequences. Similarly, the field of green engineering seeks to design better ways to make chemicals, treat wastewater, and decrease the need for raw materials. Where industrial ecology seeks to optimize current processes, green engineering looks to the future with new, more ecologically sound techniques. A new area of interest is in engineering to restore damaged ecosystems. Where previously most efforts in environmental restoration were directed toward removing harmful toxins, environmental engineering seeks to rehabilitate streams, rivers, wetlands, or other fragile habitat to their original function. Environmental engineering can help to increase the effectiveness of traditional restoration projects such as reforestation and eradication of invasive species.

Economic volatility can also be viewed through the lens of sustainability. Although agreement has not been reached on this subject, there is widespread concern that the current deficit spending and large national debt accrued by the United States is not compatible with continuing prosperity. Other problems, not just limited to the United States, include reliance on decreasing supplies of fossil fuels, high consumption of disposable items, and increasing disparity between the rich and the poor. All of these problems, while primarily economic in nature, are also linked to environmental and social issues that could amplify any crises that occur. Proponents of sustainable economies seek a system in which the natural environment is viewed as more than a source of raw materials and a place to dispose of waste. So far, there has been a great deal of experimentation with new economic models, but not a lot of large-scale sustainable economic changes. These methods include carbon trading, microfinancing, product take-back programs, renewable energy credits, and car sharing schemes. While many of these programs can be encouraged by governments, many of them rely on changes in the behavior of individuals, which can be hard to create.

For environmental sustainability to flourish, it must integrate with the needs of a society that successfully provides for its members. Humans have a complex hierarchy of needs ranging from basics like food and shelter to security requirements like safety from crime and protection from exploitation, to psychological necessities like freedom of religion, expression, and the right to choose to have a voice in their own governance. A lack of any one of these necessities can result in hostility in the form of social protest, a resort to crime, or even armed conflict. If environmental sustainability efforts weaken societies by destroying jobs or threatening traditional ways of life, this can also paradoxically threaten sustainable development.

Some social sustainability efforts have therefore focused on the preservation of traditional lifestyles, languages, foods, crafts, occupations, and other cultural capital that is in danger of being subsumed by larger populations. Indeed, the Stockholm Conference on the Human Environment stated that a richness of culture was vital to human vitality. Other projects have sought to decrease the stress and harm associated with modern city life. The "slow food" and "slow city" movements want to improve health and the environment by decreasing the use of harmful technologies like private car use in cities and long-distance transportation of food. They also encourage social cohesiveness by encouraging people to eat with their families and friends, develop relationships with their neighbors, and contribute to their local economies.

National governments must embrace sustainability as core policy if it is to succeed in providing real human benefits. Though many non-governmental organizations, state and local governments, private companies, and enthusiastic individuals are applying sustainability principles at many levels in many parts of the world, national governments still possess the capital and decision-making power to affect the most change. Another important consideration is that even the best sustainability efforts can fail if weakened by larger unsustainable practices. However, that does not mean that all efforts are doomed. With the continuing support of the international community, increasing interest from business and industry, and expanding scholarship on sustainability issues at the university level, incremental gains can form the foundation of a more sustainable future.

Primary Source Connection

Wangari Muta Maathai was awarded the Nobel Peace Prize in 2004, in recognition of her work with the Green Belt Movement, a group that organizes disadvantaged women in Africa to plant trees in order to preserve the environment and improve women' quality of life.

In 2002, Maathai was elected to Kenya's parliament, and was later appointed Kenya's Assistant Minister for Environment and Natural Resources.

In awarding the 2004 Nobel Peace Prize to Maathai, the foundation specifically commended Maathai "for her contribution to sustainable development, democracy and peace." In a press release announcing the prize, the foundation asserted that Maathai had "taken a holistic approach to sustainable development that embraces democracy, human rights and women's rights in particular. She thinks globally and acts locally."

The award of the Nobel Peace Prize to Maathai also helped to focus international attention on the concept of sustainable development and how a range of actions from local to international can foster such development.

WANGARI MAATHAI NOBEL LECTURE

Your Majesties

Your Royal Highnesses

Honourable Members of the Norwegian Nobel Committee

Excellencies

Ladies and Gentlemen

I stand before you and the world humbled by this recognition and uplifted by the honor of being the 2004 Nobel Peace Laureate.

As the first African woman to receive this prize, I accept it on behalf of the people of Kenya and Africa, and indeed the world. I am especially mindful of women and the girl child. I hope it will encourage them to raise their voices and take more space for leadership. I know the honor also gives a deep sense of pride to our men, both old and young. As a mother, I appreciate the inspiration this brings to the youth and urge them to use it to pursue their dreams.

Although this prize comes to me, it acknowledges the work of countless individuals and groups across the globe. ... To all who feel represented by this prize I say use it to advance your mission and meet the high expectations the world will place on us.

This honor is also for my family, friends, partners and supporters throughout the world. ... Because of this support, I am here today to accept this great honour.

I am immensely privileged to join my fellow African Peace laureates, Presidents Nelson Mandela and F.W. de Klerk, Archbishop Desmond Tutu, the late Chief Albert Luthuli, the late Anwar el-Sadat and the UN Secretary General, Kofi Annan.

... I have always believed that solutions to most of our problems must come from us.

In this year's prize, the Norwegian Nobel Committee has placed the critical issue of environment and its linkage to democracy and peace before the world. For their visionary action, I am profoundly grateful. ...

... As I was growing up, I witnessed forests being cleared and replaced by commercial plantations, which destroyed local biodiversity and the capacity of the forests to conserve water.

In 1977, when we started the Green Belt Movement, I was partly responding to needs identified by rural women, namely lack of firewood, clean drinking water, balanced diets, shelter and income.

Throughout Africa, women are the primary caretakers, holding significant responsibility for tilling the land and feeding their families. As a result, they are often the first to become aware of environmental damage as resources become scarce and incapable of sustaining their families.

The women we worked with recounted that unlike in the past, they were unable to meet their basic needs. ...

Tree planting became a natural choice to address some of the initial basic needs identified by women. Also, tree planting is simple, attainable and guarantees quick, successful results within a reasonable amount time. ...

So, together, we have planted over 30 million trees that provide fuel, food, shelter, and income to support their children's education and household needs. The activity also creates employment and improves soils and watersheds. ... This work continues.

Initially, the work was difficult because historically our people have been persuaded to believe that because they are poor, they lack not only capital, but also knowledge and skills to address their challenges. Instead they are conditioned to believe that solutions to their problems must come from "outside." ...

In order to assist communities to understand these linkages, we developed a citizen education program, during which people identify their problems, the causes and possible solutions. They then make connections between their own personal actions and the problems they witness in the environment and in society. ...

On the environment front, they are exposed to many human activities that are devastating to the environment and societies. These include widespread destruction of ecosystems, especially through deforestation, climatic instability, and contamination in the soils and waters that all contribute to excruciating poverty.

In the process, the participants discover that they must be part of the solutions. They come to recognize that they are the primary custodians and beneficiaries of the environment that sustains them.

Entire communities also come to understand that while it is necessary to hold their governments accountable, it is equally important that in their own relationships with each other, they exemplify the leadership values they wish to see in their own leaders, namely justice, integrity and trust.

Although initially the Green Belt Movement's tree planting activities did not address issues of democracy and peace, it soon became clear that responsible governance of the environment was impossible without democratic space. Therefore, the tree became a symbol for the democratic struggle in Kenya. ...

In time, the tree also became a symbol for peace and conflict resolution, especially during ethnic conflicts in Kenya when the Green Belt Movement used peace trees to reconcile disputing communities. ... Using trees as a symbol of peace is in keeping with a widespread African tradition. For example, the elders of the Kikuyu carried a staff from the thigi tree that, when placed between two disputing sides, caused them to stop fighting and seek reconciliation. ...

Such practices are part of an extensive cultural heritage, which contributes both to the conservation of habitats and to cultures of peace. With the destruction of these cultures and the introduction of new values, local biodiversity is no longer valued or protected and as a result, it is quickly degraded and disappears. For this reason, The Green Belt Movement explores the concept of cultural biodiversity, especially with respect to indigenous seeds and medicinal plants.

...

It is 30 years since we started this work. ... Today we are faced with a challenge that calls for a shift in our thinking, so that humanity stops threatening its life-support system. We are called to assist the Earth to heal her wounds and in the process heal our own—indeed, to embrace the whole creation in all its diversity, beauty and wonder. This will happen if we see the need to revive our sense of belonging to a larger family of life, with which we have shared our evolutionary process.

In the course of history, there comes a time when humanity is called to shift to a new level of consciousness, to reach a higher moral ground. A time when we have to shed our fear and give hope to each other.

That time is now.

The Norwegian Nobel Committee has challenged the world to broaden the understanding of peace: there can be no peace without equitable development; and there can be no development without sustainable management of the environment in a democratic and peaceful space. This shift is an idea whose time has come.

... Those of us who have been privileged to receive education, skills, and experiences and even power must be role models for the next generation of leadership. ...

Culture plays a central role in the political, economic and social life of communities. Indeed, culture may be the missing link in the development of Africa. ...

Africans, especially, should re-discover positive aspects of their culture. In accepting them, they would give themselves a sense of belonging, identity and self-confidence.

There is also need to galvanize civil society and grassroots movements to catalyze change. I call upon governments to recognize the role of these social movements in building a critical mass of responsible citizens, who help maintain checks and balances in society. On their part, civil society should embrace not only their rights but also their responsibilities.

Further, industry and global institutions must appreciate that ensuring economic justice, equity and ecological integrity are of greater value than profits at any cost. The extreme global inequities and prevailing consumption patterns continue at the expense of the environment and peaceful co-existence. The choice is ours.

I would like to call on young people to commit themselves to activities that contribute toward achieving their long-term dreams. They have the energy and creativity to shape a sustainable future. To the young people I say, you are a gift to your communities and indeed the world. You are our hope and our future.

The holistic approach to development, as exemplified by the Green Belt Movement, could be embraced and replicated in more parts of Africa and beyond. It is for this reason that I have established the Wangari Maathai Foundation to ensure the continuation and expansion of these activities. Although a lot has been achieved, much remains to be done.

As I conclude I reflect on my childhood experience when I would visit a stream next to our home to fetch water for my mother. I would drink water straight from the stream. Playing among the arrowroot leaves I tried in vain to pick up the strands of frogs' eggs, believing they were beads. But every time I put my little fingers under them they would break. Later, I saw thousands of tadpoles: black, energetic and wriggling through the clear water against the background of the brown earth. This is the world I inherited from my parents.

Today, over 50 years later, the stream has dried up, women walk long distances for water, which is not always clean, and children will never know what they have lost. The challenge is to restore the home of the tadpoles and give back to our children a world of beauty and wonder.

Thank you very much.

Wangari Muta Maathai

MAATHAI, WANGARI MUTA. NOBEL LECTURE. DELIVERED BEFORE MEMBERS OF THE NOBEL FOUNDATION, DECEMBER 10, 2004, OSLO, NORWAY. AVAILABLE ONLINE AT <HTTP://NOBELPRIZE.ORG/PEACE/LAUREATES/2004/MAATHAI-LECTURE-TEXT.HTML> (ACCESSED MARCH 10, 2006).

SEE ALSO *Corporate Green Movement; Forest Resources; Green Movement; Human Impacts; Natural Resource Management*

BIBLIOGRAPHY

Books

Brown, Oli. *The Environment and Our Security: How Our Understanding of the Links Has Changed.* Winnipeg, Manitoba, Canada: International Institute for Sustainable Development, 2005.

Gardner, Gary, and Thomas Prugh. *State of the World 2008: Seeding the Sustainable Economy.* Washington, DC: Worldwatch Institute, 2008.

Web Sites

Encyclopedia of Earth. "United Nations Conference on the Human Environment." November 9, 2007. http://www.eoearth.org/article/United_Nations_Conference_on_the_Human_Environment_(UNCHE)_Stockholm_Sweden (accessed March 4, 2008).

National Science Foundation. "Environmental Sustainability." February 14, 2008. http://www.nsf.gov/funding/pgm_summ.jsp?pims_id=501027 (accessed March 4, 2008).

Sustainability Reporting Program. "A Brief History of Sustainable Development." 2000. http://www.sustreport.org/background/history.html (accessed March 2, 2008).

United Nations. "Declaration of the United Nations Conference on the Human Environment." http://www.unep.org/Documents.multilingual/Default.asp?DocumentID=97&ArticleID=1503 (accessed March 4, 2008).

United Nations Department for Social and Economic Affairs. "United Nations Division for Sustainable Development." February 2008. http://www.un.org/esa/sustdev/index.html (accessed March 4, 2008).

West Virginia State Archives. "The Buffalo Creek Disaster." http://www.wvculture.org/hiStory/buffcreek/bctitle.html (accessed March 4, 2008).

Kenneth Travis LaPensee

Tellico Dam Project (Snail Darters) and Supreme Court Case (*TVA v. Hill*, 1977)

Introduction

In the mid-1970s, a well-publicized legal battle pitted environmental groups and an endangered species of fish against a $100 million federal construction project. The Tennessee Valley Authority (TVA) was constructing Tellico Dam on the Little Tennessee River when a new freshwater fish species, the snail darter, was discovered upstream. The U.S. Fish and Wildlife Service listed the snail darter as an endangered species under the Endangered Species Act of 1973.

Numerous individuals and environmental groups filed suit to stop construction of the Tellico Dam, arguing that damming the Little Tennessee River would destroy the snail darter's natural habitat. The court case, *Tennessee Valley Authority v. Hill*, went to the Supreme Court of the United States. The Supreme Court decided to uphold the mandate of the Endangered Species Act to protect the snail darter and its habitat. After further political maneuvering, however, the Tellico Dam was completed, and the snail darter's natural habitat was destroyed.

Historical Background and Scientific Foundations

In 1972, U.S. President Richard M. Nixon (1913–1994) called on Congress to enact comprehensive legislation to protect endangered species. Congress responded and passed the Endangered Species Act of 1973 (ESA). The ESA provides "a means whereby the ecosystems upon which endangered species and threatened species depend may be conserved, ... 'and provides' a program for the conservation of such endangered species and threatened species."

The U.S. Fish and Wildlife Service (FWS) and the National Oceanic and Atmospheric Administration (NOAA) administer the ESA. NOAA, through the National Marine Fisheries Service, manages all endangered and threatened marine species. The FWS manages all other endangered or threatened species, including all freshwater species.

Under the ESA, plant and animal species may be classified as endangered or threatened. The ESA defines an endangered species as being in "danger of extinction throughout all or a significant portion of its range." A threatened species is any species "likely to become an endangered species within the foreseeable future throughout all or a significant portion of its range."

A species may be listed as endangered or threatened in one of two ways. First, the FWS or NOAA may decide to list a species directly. Second, any individual or organization may petition the FWS or NOAA to conduct a scientific review of any species threatened or endangered. The FWS or NOAA then determines whether to list the species. Once FWS or NOAA lists an endangered or threatened species, they are to identify the critical habitat of the species. All federal agencies are then required to work with FWS or NOAA to ensure that the critical habitat of that species is not destroyed or modified.

Section 11 of the ESA contains a citizen suit clause that allows any citizen to sue "to enjoin any person, including the United States and any other governmental instrumentality or agency ... who is alleged to be in violation of any provision of this Act." In 1976, Hank Hill, a student at the University of Tennessee, and others filed suit in federal court seeking an injunction to stop the construction of a dam that would destroy the habitat of the snail darter.

The snail darter is a small fish, about 2 in (10 cm) in length that was discovered in the Little Tennessee River in 1973. In 1975, the FWS listed the snail darter as an endangered species under the ESA. When the snail darter was listed on the endangered species list, the secretary of the U.S. Department of the Interior, which oversees the FWS, stated that the snail darter resided only in a small portion of the Little Tennessee River. Furthermore, the secretary stated that the construction of the nearby Tellico Dam would destroy the critical habitat of the snail darter. The secretary then stated that "all Federal agencies must take such action as is necessary to insure that actions authorized, funded, or carried out by them do not result in the destruction or modification of this critical habitat area."

The secretary's findings troubled the Tennessee Valley Authority, the agency responsible for the construction of Tellico Dam. The TVA began the Tellico Dam project in 1967. Many homes and farms were relocated because Tellico Reservoir would flood 16,500 acres of land. The snail darter had not been discovered when the Tellico Dam project began. Tellico Dam was 70% to 80% complete (a cost of tens of millions of dollars) by the time the snail darter was listed as an endangered species.

In the lawsuit to enjoin the Tellico Dam project, now called *Tennessee Valley Authority v. Hill*, the TVA argued that the ESA's protection of endangered species and their critical habitats did not extend to government projects that were authorized, funded, and underway at the time Congress passed the ESA. Essentially, the TVA argued that Congress did not intend for the ESA to interfere with projects that Congress had approved. The TVA also pointed to the fact that in 1975 Congress appropriated an additional $29 million for the Tellico Dam project after the TVA informed Congress that the project could threaten the snail darter.

In May 1976, the U.S. District Court for the Eastern District of Tennessee denied relief for the snail darter advocates and dismissed their complaint. The court determined that the completion of Tellico Dam would probably jeopardize the continued existence of the snail darter. The court concluded, however, that "at some point in time, a federal project becomes so near completion and so incapable of modification that a court of equity should not apply a statute enacted long after inception of the project to produce an unreasonable result."

In January 1977, the U.S. Court of Appeals for the Sixth Circuit reversed the lower court's decision. The Court of Appeals remanded, or sent back, the case to the lower court "with instructions that a permanent injunction issue halt[ed] all activities incident to the Tellico Project which may destroy or modify the critical habitat of the snail darter." The Court of Appeals directed that the injunction "remain in effect until Congress, by appropriate legislation, exempts Tellico from compliance with the Act or the snail darter has been deleted from the list of endangered species or its critical habitat materially redefined." The Court of Appeals noted that the near completion of the Tellico Dam project and congressional appropriation of money to the project were irrelevant. The only relevant legal consideration for the court was the language of the ESA, "the meaning and spirit" of which was "clear on its face."

The TVA appealed the Court of Appeals ruling to the Supreme Court of the United States. The Supreme Court issued its opinion in June 1978, affirming the ruling of the Court of Appeals. The Supreme Court found that the ESA prohibited the construction of Tellico Dam, because the project would threaten the snail darter and its critical habitat. The court noted that the language of the ESA is plain and does not make any exception for projects underway at the time Congress passed the ESA. If Congress had intended to exempt such projects from the requirements of the ESA, they could have simply added language to that effect to the ESA. The court refused to read exemption into the ESA in the absence of such language.

The court also noted that the "plain intent of Congress in enacting this statute was to halt and reverse the trend toward species extinction, whatever the cost." If Congress wanted to exempt Tellico Dam or other projects from the requirements of the ESA, then Congress would have to do so clearly and unequivocally. The court stated that continued congressional appropriation for the Tellico Dam project and a few comments by congressional representatives on the Appropriations Committee were not enough to repeal or modify existing law.

WORDS TO KNOW

ECOSYSTEM: The community of individuals and the physical components of the environment in a certain area.

EXTINCT: No longer in existence. In geology, it can be used to mean a process or structure that is permanently inactive (e.g., an extinct volcano).

HABITAT: The natural location of an organism or a population.

THREATENED: When a species is pressured, but technically not yet endangered.

Issues and Impacts

One month after the Supreme Court decided *Hill*, Congress amended the ESA to include a process by which economic impacts could be reviewed by the Endangered Species Committee. The committee had the power to exempt projects from ESA restrictions. Congressional representatives sought an exemption for the Tellico Dam project through the Endangered Species Committee. After weighing the scientific and economic evidence, the Endangered Species Committee denied an exemption for Tellico Dam in early 1979.

In September 1979, Congress passed an amendment that exempted the Tellico Dam project from the Endangered Species Act. The TVA closed the gates of Tellico Dam in November 1979, flooding the native habitat of the snail darter. In an effort to save the snail darter, the TVA had relocated many snail darters to the nearby Hiwassee River prior to damming the Little Tennessee River. Scientists also discovered snail darters in other areas of the Tennessee River Valley watershed after the completion of Tellico Dam. In 1984, the Fish and Wildlife Service delisted the Tellico Dam area as a critical habitat for the snail darter because the species died off in that area. Because of the newly discovered snail darter populations, the FWS changed the classification of the snail darter from endangered to threatened in 1985.

The snail darter case changed the face of environmental law in the United States, especially in regard to the Endangered Species Act. The decision of the Supreme Court demonstrated the importance of the ESA and other environmental laws. The court rejected the view that environmental laws should be enforced only when convenient. The court sided with the ESA over a $100 million federal project. Despite this victory in court, however, the snail darter case illustrated the political nature of environmental law when Congress voted to exempt the Tellico Dam project from the ESA.

SEE ALSO *Endangered Species; Environmental Protection Agency (EPA); Extinction and Extirpation*

BIBLIOGRAPHY

Web sites

Chattanooga Times Free Press. "Knoxville: Little Fish, Big Fight." April 19, 2008. http://www.timesfreepress.com/news/2008/apr/19/knoxville-little-fish-big-fight/ (accessed May 2, 2008).

Findlaw. "Tennessee Valley Authority v. Hill." http://caselaw.lp.findlaw.com/scripts/getcase.pl?court=US&vol=437&invol=153 (accessed May 2, 2008).

Forest History Society. "1979: Snail Darter Exemption Case." November 1, 2004. http://www.foresthistory.org/research/usfscoll/policy/northern_spotted_owl/1979owl.snaildarter.html (accessed May 2, 2008).

U.S. Environmental Protection Agency (EPA). "Summary of the Endangered Species Act." March 6, 2008. http://www.epa.gov/regulations/laws/esa.html (accessed May 2, 2008).

U.S. Fish and Wildlife Service. "Endangered Species Act of 1973." http://www.fws.gov/endangered/esa/content.html (accessed May 2, 2008).

Joseph P. Hyder

Temperature Records

Introduction

Temperature records are information about air and water temperatures at specific times and places. Without such records, little could be said about the history of Earth's climate.

Most temperature records are either proxy records or instrumental records. A proxy is something that stands for something else: Proxy records are natural deposits or features that record information about temperature at the time that they were formed. Past temperatures are not directly recorded by proxies, but can be deduced from them. For example, the amounts of different isotopes of oxygen and hydrogen in snow that fell thousands of years ago in Antarctica and Greenland record the temperature of the air at the time the snow fell, since heavier isotopes fall in snow at a higher rate when the air is colder. By studying ancient snow layers, climatologists can gather information about ancient temperatures.

Instrumental records are readings from instruments that have been written down. Since about 1880, instrumental records from thermometers have been kept in many parts of the world. Since the 1970s, the instrumental record includes satellite observations of the entire planet.

Historical Background and Scientific Foundations

The temperature at any one place on Earth's surface at any given time is the result of many causes, including short-term weather events and long-term climate trends. Earth's average, long-term surface temperature depends on how much energy the planet receives from the sun and how much it radiates back out into space. Since the mid-1800s, human beings have been altering this energy balance, mainly by adding greenhouse gases (carbon dioxide, methane, and others) to the atmosphere and by destroying forests. This has caused Earth's average temperature to rise, the phenomenon known as global warming.

Global warming is known through a combination of proxy temperature records and instrumental records. Instrumental records reveal Earth's average temperature with high precision over recent decades. For years prior to about 1850, historical non-instrumental records of temperature-related events such as freezing rivers are also a source of information about temperatures. The main source of pre-1850 temperature information, however, is proxy records. These include tree rings, layers of sediment (muck) on ocean and lake bottoms, snow layers preserved in ice sheets, and others. By far the longest continuous proxy records are found in ice cores, cylindrical samples of ice drilled from the depths of the ice sheets of Greenland and Antarctica, which are several miles deep in places. Each year's snowfall has been preserved as a thin, distinct layer of ice in these ice sheets. In 2004, a record of snowfall layers was recovered from Antarctica that goes back 800,000 years.

Proxies must be interpreted, and interpretation is not perfect. Therefore, temperature records derived from different proxies do not always agree exactly. However, temperature records derived from various proxies all agree on the general temperature history of Earth over the last millennium or two: The warming seen in the last century is unprecedented, with high probability. An earlier warming period around the year 1000, called the Medieval Warm Period, was neither as sudden nor as intense as today's warming.

The temperature record over the last one or two thousand years, if plotted with temperature on the vertical axis and time on the horizontal axis, is a wiggling but more or less level line from at least 1,300 years ago until about 100 years ago, with a sudden, unprecedented rise

WORDS TO KNOW

HOCKEY STICK GRAPH: A chart of global average air temperatures indicating that recent climate warming is unprecedented for at least the last 1100 years or so. Critics argued the graph was flawed and misleading, but in 2006, after a careful review of the evidence, the US National Academy of Sciences affirmed that the graph is essentially accurate, though its shape is less certain for earlier dates.

IN CONTEXT: TEMPERATURE RECORDS

Recent World Meteorological Organization (WMO) reports provide an extensive portrait of the increasing severity of weather activity around the globe. The report noted not only higher average temperatures worldwide, but wide variations in rainfall, with some regions experiencing flooding and others plagued with drought. The WMO report from 2005 echoed similar themes, with higher-than-average temperatures resulting in what was then the second warmest year on record. The year 2005 also brought a record number of strong Atlantic hurricanes, which some scientists claim are due to higher ocean temperatures brought on by global warming and climate change.

at the very end of the graph—recent, observed, global warming. Since the overall shape of this curve resembles the outline of a hockey stick on its side with its blade pointing up, such a curve is often called a hockey-stick graph.

Impacts and Issues

Throughout the 1990s, skeptics of global warming noted that surface instrumental temperature records did not agree well with satellite records measuring temperatures higher up in the atmosphere. How, they asked, if the satellite and surface records did not agree, could climatologists say with any confidence that the world was warming? As the U.S. Climate Change Science Program stated in 2005, "previously reported discrepancies between the amount of warming near the surface and higher in the atmosphere have been used to challenge the reliability of climate models and the reality of human-induced global warming. Specifically, surface data showed substantial global-average warming, while early versions of satellite and radiosonde data showed little or no warming above the surface." However, the scientists of the Climate Change Science Program concluded that the discrepancy "no longer exists because errors in the satellite and radiosonde data have been identified and corrected. New data sets have also been developed that do not show such discrepancies."

Another climate controversy erupted in 1998, when Michael E. Mann, Raymond S. Bradley, and Malcolm K. Hughes published an article in the journal *Nature*. The authors reconstructed average air temperatures in the Northern Hemisphere from 1500 to the present from various proxies and presented a hockey-stick graph of average temperatures in the Northern Hemisphere from 1500 to the present day. A version of the graph appeared in the Intergovernmental Panel on Climate Change's third Assessment Report in 2001 and in former U.S. Vice President Al Gore's Academy Award-winning documentary *An Inconvenient Truth* (2006). Scientific critics of the graph, including W. Soon, S. Baliunas, S. McIntyre, R. McKitrick, and Hans von Storch, argued that it was deeply flawed and that the Medieval Warm Period was probably just as warm as our present period. However, the hockey-stick shape is typical of all published reconstructions of average surface temperature over the last 1,300 to 2,000 years.

Moreover, in early 2007, the Intergovernmental Panel on Climate Change reviewed the hockey-stick controversy in detail. The criticisms of Soon, Baliunas, and others were dismissed as having been adequately answered. The panel published a graph overlaying a dozen different reconstructions of average Northern Hemisphere air surface temperature for the last 1,300 years. All the reconstructions, created using various proxy temperature records, agreed that it was at least 66% probable that the twentieth century was the warmest century in the last 1,300 years, and at least 90% probable that the second half of the twentieth century was the warmest 50-year period in the last 500 years. A 2006 study by the National Academy of Sciences also affirmed the basic validity of the hockey-stick temperature record.

SEE ALSO *Climate Change; Global Warming; Ice Cores*

BIBLIOGRAPHY

Books

North, Gerald R., et al. *Surface Temperature Reconstructions for the Last 2,000 Years.* Washington, DC: National Academy of Sciences, 2006.

Parry, M. L., et al, eds. *Climate Change 2007: Impacts, Adaptation and Vulnerability: Contribution of Working Group II to the Fourth Assessment Report of the Intergovernmental Panel on Climate Change*. New York: Cambridge University Press, 2007.

Solomon, S., et al, eds. *Climate Change 2007: The Physical Science Basis: Contribution of Working Group I to the Fourth Assessment Report of the Intergovernmental Panel on Climate Change*. New York: Cambridge University Press, 2007.

Periodicals

Brumfiel, Geoff. "Academy Affirms Hockey-Stick Graph." *Nature* 441 (2006): 1032–1033.

Kerr, Richard A. "Politicians Attack, But Evidence for Global Warming Doesn't Wilt." *Science* 313 (2006): 421.

Kerr, Richard A. "Yes, It's Been Getting Warmer in Here Since the CO_2 Began to Rise." *Science* 312 (2006): 1854.

Mann, Michael E., et al. "Global-Scale Temperature Patterns and Climate Forcing Over the Past Six Centuries." *Nature* 392 (1998): 779–105.

McIntyre, S., and R. McKitrick. "Corrections to the Mann et al. (1998) Proxy Database and Northern Hemispheric Average Temperature Series." *Annual Review of Energy and the Environment* 14 (2003): 751–771.

Web Sites

National Oceanographic and Atmospheric Administration. "Global Surface Temperatures over the Past Two Millennia." http://www.ncdc.noaa.gov/paleo/pubs/mann2003b/mann2003b.html (accessed May 12, 2008).

U.S. Climate Change Science Program. "Temperature Trends in the Lower Atmosphere—Understanding and Reconciling Differences [Executive Summary]." http://www.climatescience.gov/Library/sap/sap1-1/finalreport/sap1-1-final-execsum.pdf (accessed May 12, 2008).

Larry Gilman

IN CONTEXT: TEMPERATURE RECORDS AND INCREASES IN GREENHOUSE GASES

Globally, all of the warmest years since instrumental measurements of surface temperature began to be recorded (around 1880) have occurred since the late 1980s. Typically, these warm years have averaged about 1.5–2.0°F (0.8–1.0°C) warmer than those during the decade of the 1880s. As of 2007, the average temperature of Earth's atmosphere near the surface had risen about 1.33°F (0.74°C) since 1906. This warming closely matches increases in greenhouse gases emitted by human activities.

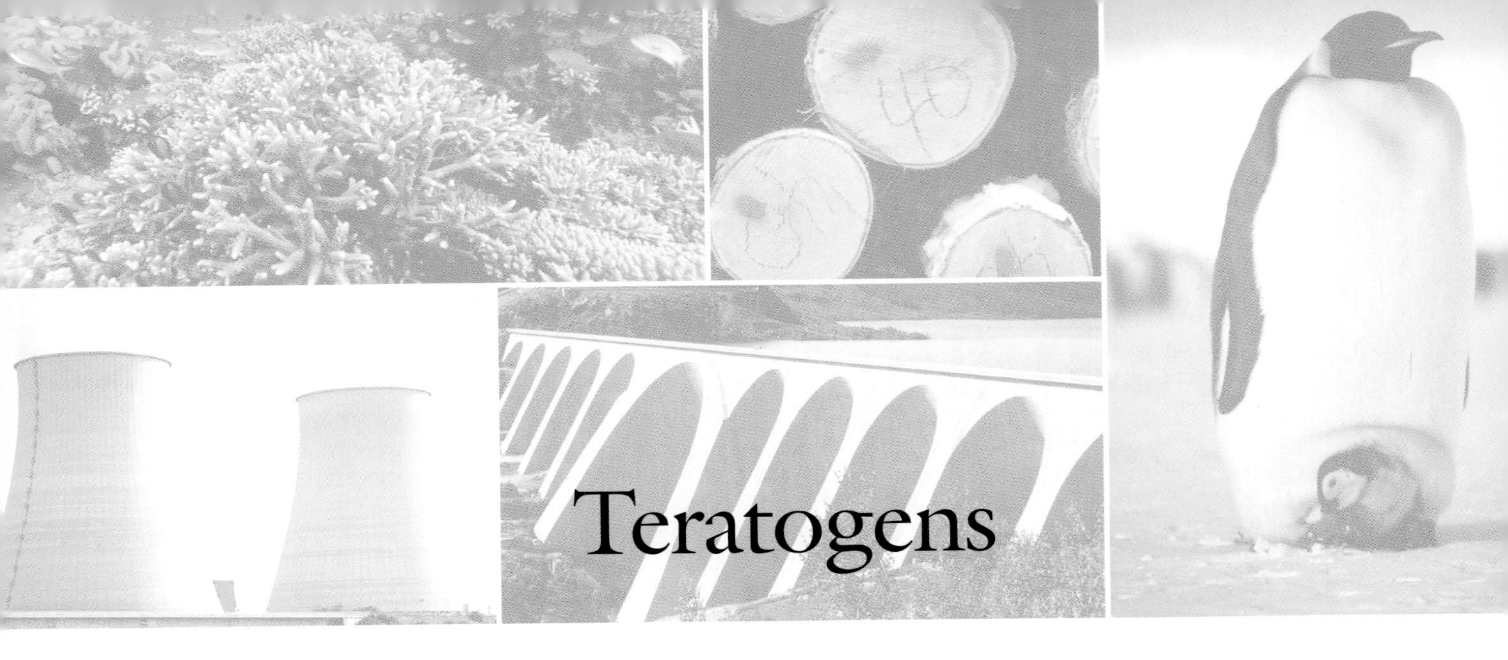

Teratogens

Introduction

The word teratogen derives from the Greek word *teratos,* which had a double meaning of both prodigy and demon. The Greek term also referred to any natural phenomenon that evoked fascination and horror. Today, it has a much more precise medical meaning, describing any environmental agent, contaminant, or chemical that causes fetal abnormalities during pregnancy.

Historical Background and Scientific Foundations

Fetal abnormalities were traditionally thought to be a sign of divine displeasure, that all was not right with the world. The Tablet of Nineveh written in Babylonia nearly 4,000 years ago listed 62 malformations and interpretations of different birth defects. A child born without a right hand, for example, presaged an earthquake. This connection between birth defects and messages from God became particularly pronounced during the Wars of Religion in the sixteenth and seventeenth centuries. For example, the theologian Martin Luther (1483–1546) asserted that a deformed child was the sign of the imminent collapse of the Roman Catholic Church.

On the other hand, the Bible also focused upon maternal responsibility for fetal viability. Thus by the medieval era, it was commonly believed that fetal deformities could not only be a sign from God, but could also be due to maternal failings. Birth defects were assumed to result from a variety of causes, such as coitus between people of different races, possession by the devil, or sex with animals. It was also postulated that the thoughts of pregnant women could be imprinted on the developing fetus. So, if a woman unduly thought about cherries, perhaps due to a food craving, the baby could be born with cherry shaped and colored birthmarks. The Roman naturalist and physician Pliny the Elder (23–79) thus advised women that during pregnancy, they should not look at animals in order to avoid the abnormal development of their developing child.

It was not until systematic study of embryology took place in the nineteenth century that the true causes of birth defects and the role of teratogens were beginning to be understood. In 1822, Étienne Geoffroy Saint-Hilaire (1772–1844) published the second volume to his major work, *Philosophie Anatomique.* Utilizing studies in comparative anatomy, he argued that birth defects were due to an interruption to fetal development. Geoffrey, in a later collaboration with his son, reproduced deformities in chicken eggs with mixed success, but their work did help to put an end to the idea of material imagination.

Researcher Camille Dareste (1822–1899) also argued that birth defects were due to incomplete embryonic development, and he realized that external agents—teratogens—could be a factor in deformities. He thus used different teratogenic agents to reproduce the abnormalities in chicks that he had observed in nature. As the historian Jean-Louis Fischer indicated, to transform embryos experimentally, Dareste exposed them to teratogens such as varnish, electricity, or different chemicals, as well as lowering incubation temperatures for several hours during early development.

Charles R. Stockard (1879–1939), a zoologist at Cornell Medical College, continued Dareste's research in the same spirit, demonstrating how different ionic concentrations could produce abnormalities in minnow embryos. Stockard later did experimental tests to reveal the effects of alcohol on the fetal development of guinea pigs. His results seemed to indicate that maternal exposure to alcohol fumes could cause fetal abnormality and an increase in stillbirths. Stockard's research indirectly led to the discovery of Fetal Alcohol Syndrome (FAS) in 1973. American physician Sterling Clarren confirmed the teratogenic affects of alcohol on fetal development, and as additional scientists have demonstrated, FAS is the leading known cause of mental retardation in the western world.

The latter part of the twentieth century saw the discovery of many teratogens. In 1941, Sir Norman Gregg (1892–1968) discovered the human rubella virus as the first human teratogen; exposure *in utero* led to heart defects and congenital cataracts. Later experimental findings connecting exposure to mercury to teratogenic effects raised awareness about *in utero* exposure to chemicals in the environment.

Though some of the effects of teratogens were discovered through animal studies in the laboratory, most teratogens are discovered empirically when there is a sudden or increased occurrence of a particular birth defect. The spike in occurrences of phocomelia, a previously rare disorder resulting in shortened limbs due to the lack or malformation of long bones, led to the identification of a chemical isomer of thalidomide as a human teratogen. Thalidomide was given to expecting mothers to quell morning sickness during the first trimester when the embryo was vulnerable to limb abnormalities. Though the compound thalidomide in itself is safe, the medicine given to mothers was contaminated with its chemical isomer, its molecular mirror image.

The following postulates, a modern modification of Robert Koch's postulates of disease, are generally used to determine if an agent is a human teratogen. These postulates include the necessity of demonstrating a direct statistical correlation between the occurrence of teratogens and fetal abnormality. The correlation must be reproduced in experiments using animals. Increased doses or exposure to the teratogen should cause an increase in teratogenic effects, and the suggested effect must be scientifically and biologically plausible.

WORDS TO KNOW

EMBRYOLOGY: The study of early development in living things.

PHOCOMELIA: A birth defect in which the upper portion of a limb is absent or poorly developed, so that the hand or foot attaches to the body by a short, flipperlike stump.

People gather frogs along the edge of Porter Pond in Sterling, Connecticut (July 1998). They were participating in a program run by the Connecticut State Department of Environmental Protection to survey the frog population in the pond for deformities. *AP Images*

Impacts and Issues

Increased awareness of teratogens has also led to several epidemiological techniques to identify them in the environment and in medications.

Monitoring birth defects and developmental disabilities in a given population can lead to a "case-control" study in which parents of the children with abnormalities as well as control groups are interviewed. Drugs given in pregnancy are also now subject to continuous monitoring to identify teratogens, and as a result, the U.S. Food and Drug Administration (FDA) in the 1980s created five drug categories to designate the safety of medications for use during pregnancy. Despite the fact that few drugs have been demonstrated to cause birth defects, it is important for a pregnant woman to discuss any medication she is taking with her physician, including over-the-counter medications, herbal remedies, or vitamins.

Longitudinal studies involving a large population, in which children and their parents are interviewed from birth until adulthood, can make connections between environment and potential teratogens. In Denmark, investigators studied the risks of mumps-measles-rubella (MMR) vaccines causing autism. Cross-referencing immunization records of children with those later diagnosed with autism demonstrated that the rate of autism was the same in vaccinated and non-vaccinated groups.

There was thus no increased risk of autism involved with MMR vaccination, and the MMR vaccine was therefore determined not to be a teratogen. As José F. Cordero indicated, in the United States, the large population and privacy concerns would have made such a study impossible. Epidemiologists must always consider patient confidentiality when engaging in studies to identify teratogens.

In 2004, the U.S. Environmental Protection Agency (EPA) issued recommendations for women of childbearing age concerning the levels of methylmercury in fish and shellfish. Methylmercury is a known teratogen, and occurs naturally and also makes its way into rivers and oceans when mercury is released as an industrial pollutant. Once in the water, mercury converts to methylmercury and is absorbed by fish. Although the EPA suggested that fish are an important part of a nutritious diet, the agency recommends that pregnant women and young children avoid eating fish such as shark, king mackerel, swordfish, and tilefish that contain higher amounts of methylmercury. Salmon, shrimp, canned light tuna, and catfish are among the suggested alternatives that have low levels of methylmercury, although the suggested amount of consumption is six ounces per week for tuna and 12 ounces of these fish overall per week. Finally, the EPA recommends that only six ounces of fish caught by family or friends from local waters be consumed per week, and that children receive smaller portions.

SEE ALSO *Clean Air Mercury Rule of 2005; Inland Fisheries; Marine Fisheries; Marine Water Quality*

BIBLIOGRAPHY

Books

Persaud, T. A., and S. R. Chudley. *Basic Concepts in Teratology*. New York: Alan R. Liss, 1985.

Shardein, James. *Chemically Induced Birth Defects*, 3rd ed. New York: Informa Healthcare, 2000.

Warkany, J., ed. *Issues and Reviews in Teratology*. New York: Plenum Press, 1983.

Periodicals

Abel, E. L., and Sokol, R. J. "Incidence of Fetal Alcohol Syndrome and Economic Impact of FAS-Related Anomalies: Drug Alcohol Syndrome and Economic Impact of FAS-Related Anomalies." *Drug and Alcohol Dependency* 19 (1987): 51–70.

Betterton, Rosemary. "Promising Monster: Pregnant Bodies, Artistic Subjectivity, and Maternal Imagination." *Hypatia* 2 (2006): 80–100.

Brent, R. L. "Reproductive and Teratologic Effects of Low-Frequency Electromagnetic Fields: A Review of In Vivo and In Vitro Studies Using Animal Models." *Teratology* 59 (1999): 261–286.

Cooper, Melinda. "Regenerative Medicine: Stem Cells and the Science of Monstrosity." *Medical Humanities* 30 (2004): 12–22.

Cordero, José F. "A New Look at Behavioral Outcomes and Teratogens: A Commentary." *Birth Defects Research (Part A)* 67 (2003): 900–902.

Fischer, Jean-Louis. "The Embryological Oeuvre of Laurent Chabry." *Development Genes and Evolution* 201 (1992): 125–127.

Madsen, K. M., Hviid A., Vestergaard M., et al. "A Population-Based Study of Measles, Mumps, and Rubella Vaccination and Autism." *New England Journal of Medicine* 347 (2002): 1477–1482.

Pauly, Philip J. "How Did the Effects of Alcohol on Reproduction Become Scientifically Uninteresting?" *Journal of the History of Biology* 29 (1996): 1–28.

Web Sites

University of Washington, TERIS (Teratogen Information System) Program. "Clinical Teratology Web." http://depts.washington.edu/terisweb/ (accessed February 17, 2008).

Anna Marie E. Roos

Tidal or Wave Power

Introduction

Large amounts of energy are involved in the motions of ocean water. These motions include steady ocean currents, the repetitive motions of the tides, and the irregular motions of surface waves. For decades, machines have been in development to harvest energy in the form of electricity from these motions. Although a few projects have been in operation since the 1960s, a commercial market for ocean-power harvesting devices has only opened up in the early 2000s. Ocean power is poised to make a significant contribution to a future portfolio of renewable energy sources that will also include wind and solar power.

Historical Background and Scientific Foundations

Twice a day, over much of the world, sea level rises and falls, anywhere from a few inches to 56 ft (a few centimeters to 17 m). The most extreme variation occurs at the Bay of Fundy in coastal Canada. This rise and fall entails the movement of trillions of tons of water, involving an immense amount of energy. Tidal energy ultimately derives from the Earth-moon system: The energy dissipated as friction by the tides is subtracted from Earth's rotational kinetic energy, slowing the planet's spin. Because of the tides, Earth's day is getting longer at the rate of about 0.0017 seconds per century.

Earth's seas also circulate in currents and are covered everywhere by the short-lived, small-scale, up-and-down motions of waves. Some of the energy in currents and waves is derived from the tides, but most is from the sun, which drives air and ocean movements around the planet by heating it unevenly.

French engineer Pierre Girard (1765–1836) patented an early device for gathering wave power in 1799. In the centuries since this time, many schemes have been proposed for harvesting tidal and wave power, but until the second half of the twentieth century none produced large amounts of power. Development efforts are underway to find less expensive means of harvesting these energy flows. Challenges remain, but commercial sales of a variety of sea-power systems were ongoing in the early 2000s, and coming decades may see greatly increased output from ocean energy.

Tidal and Current Power

In some locations, bays are connected to the open ocean through narrow openings, forcing tidal waters to flow back and forth like a reversing river. Energy is particularly easy to gather in such locations: all that is needed is to place a windmill-like turbine or other energy-capturing device in the tidal current. A prototype of this kind of device was installed near Hammerfest, Norway, in 2003, generating 300 kilowatts (300,000 watts, 300 kW), with the Norwegian government planning to install a larger facility after gaining experience with the prototype.

In April 2008, a similar but larger system was installed in Strangford Lough, Ireland, a SeaGen dual-turbine device capable of generating a peak power output of 1.5 megawatts (1.5 million watts, five times more than the Hammerfest system). The SeaGen machine consists of an anchored column tall enough to protrude from the sea's surface with two wing-like underwater supports, one on each side, each supporting a two-bladed turbine resembling a windmill. Each turbine is 50 ft (15 m) in diameter. The system, due to begin producing power later in 2008, was to be monitored by an independent team of government scientists to see what effects it had on seals, fish, and other aspects of the local environment. The blades turn too slowly to be a threat to most marine life.

WORDS TO KNOW

RENEWABLE ENERGY: Energy that can be naturally replenished. In contrast, fossil fuel energy is nonrenewable.

Such systems involve little modification of coastlines and are therefore environmentally relatively low-impact. A larger, but higher-impact, type of system is the tidal dam. A tidal-dam energy project has been operating at La Rance, France, since 1966, generating a round-the-clock average of 65 MW (240 MW peak). At La Rance, the rising tide is allowed to fill a reservoir behind a dam; when the tide begins to fall, floodgates are closed and the water in the reservoir is allowed to run out through turbines, generating electricity. Such systems will probably never be built in large numbers, both because of their great expense and because they involve large-scale modifications of coastlines.

Another low-impact system is the hydrofoil or pulse generator. A hydrofoil is a wing-like object designed to experience an up or down force when water flows over it. Hydrofoil generators being tested in the early 2000s were large structures that sat on the ocean floor with a 33-ft (10-m) hydrofoil blade held out on a horizontal arm. The current flowing over the hydrofoil forced it up and down, like the handle on a car jack or a dolphin's tail being wagged. This motion operated a water pump, and water pressurized by the pump flowed through a generator, producing electricity. In April 2008, the government of the United Kingdom, a world leader in tidal and wave-power research, gave its go-ahead for the installation of a 0.1-MW pulse generator supplying power to the national grid (network of power lines supplying homes and businesses with electricity). Experience in operating the prototype would, if all went according to plan, lead to the design and installation of units 10 times larger.

Wave Power

Tidal power has the advantage of being completely predictable: The tides occur with perfect regularity every day. In contrast, waves vary in strength, from a mere ripple to large storm waves. Yet sites with deep, concentrated tidal currents are not common, and waves wash all shores. There are at least six different technologies for harvesting wave power:

1. Attenuator: Long, tubelike floats linked end-to-end by flexible hoses rock as waves pass under them. The motion pressurizes oil, which drives generators. The world's first commercial wave-power farm, along the coast of Portugal, uses this technology and was being installed as of mid-2008.
2. Point absorber: A float bobs up down on the surface of the waves. A vertical shaft is worked by the motion, pressurizing generators.
3. Oscillating wave surge converter: A buoylike arm is tethered to the bottom and waved back and forth like an upside-down pendulum as waves surge past.
4. Oscillating water column: A sealed column with water in its lower portion and air in its upper portion is connected to the sea. Wave action causes the water part of the column to rise and fall, raising and lowering the air and causing air-driven turbines to rotate.
5. Overtopping device: Waves slosh over the top of a barrier and the water flows back to the sea through a turbine.
6. Submerged pressure differential: A drum or tube along which a piston moves is anchored to the ocean floor not far below the surface. Water pressure on the piston varies as waves pass above, causing it to move back and forth in the shaft.

Impacts and Issues

Why not put up windmills instead of tidal and wave power machines? There are two answers. First, windmills produce power irregularly, only when the wind blows. Tidal machines produce it on a reliable schedule, and even though waves are also irregular, they are easier to predict than winds. Second, water is 800 times denser than air (i.e., each cubic foot of water weighs 800 times as much as a cubic foot of air). This means that even at lower velocities, intercepting a relatively small amount of water in motion can yield a large amount of energy. Tidal and wave machines should, therefore, be able to harvest more energy for their size than wind machines, and so ultimately compete with them in cost.

However, wind power will never be displaced by sea power. Winds blow over continental interiors, far from the ocean, and are a very large, well-distributed energy resource. Also, windmills face fewer mechanical design challenges, due in part to the low density of air: They are subject to less extreme forces (for example, during storms). Also, sea-power systems are more difficult to access for maintenance and must survive immersion in a metal-corroding salt solution, namely seawater. It is likely that in the long term, renewable-energy systems harvesting energy from many sources, including geothermal, photovoltaic, solar thermal, passive solar, tidal and wave, wind, and other, will be used to supply some or all of society's energy needs.

Wave power can also be used to drive oceangoing ships, gathering energy by using horizontal fins that rock up and down as the vessel encounters waves. A vessel demonstrating the technology set out from Hawaii in March 2008, bound for Japan. By mid-May 2008, it had reached the halfway point to its destination.

SEE ALSO *Tides*

BIBLIOGRAPHY

Books

Cruz, Joao. *Ocean Wave Energy: Current Status and Future Perspectives.* New York: Springer, 2008.

Periodicals

Bachtold, Daniel. "Britain to Cut CO_2 without Relying on Nuclear Power." *Science* 299 (2003): 1291.

Callaway, Ewen. "Energy: To Catch a Wave." *Nature* 4 (1979): 823–831.

Geoghegan, John. "Long Ocean Voyage Set for Vessel that Runs on Wave Power." *New York Times* (March 11, 2008).

Stone, Richard. "Norway Goes with the Flow to Light Up Its Nights." *Science* 299 (2003): 339.

Voss, Alfred. "Waves, Currents, Tides—Problems and Prospects." *Energy* 4 (1979): 823–831.

Web Sites

U.S. Department of Energy. "Ocean Wave Power." http://www.eere.energy.gov/consumer/renewable_energy/ocean/index.cfm/mytopic=50009 (accessed May 14, 2008).

Larry Gilman

Tides

Introduction

A tide is the alternating rise and fall in sea level that occurs at marine coastal regions each day. The sea level change is caused by the gravitational attraction between Earth and the moon that causes the ocean to swell and recede at different regions of Earth at the same time.

The influence of tides on coastal regions could increase in the twenty-first century with the increase in sea level that has been predicted to accompany global warming. Low lying coastal regions, such as the Florida Keys, could be especially affected, as the effect of rising sea level will cause the tide to submerge more on the land than at present.

Depending on the geography of the coast, tides can move tremendous amounts of water back and forth each day. For example, the Bay of Fundy on the Gulf of Maine, which is narrow and increasingly shallow, can experience differences between high and low tides of over 50 ft (15 m). Harnessing this energy on a large scale, which is feasible with currently available technology, would be a source of renewable energy that could power thousands of homes with no greenhouse-gas emissions.

The biggest tidal power system in the world is located at St. Malo, France. The dam-based system generates almost 240 megawatts of electricity, enough to power about 100,000 homes for a year. The expanded use of the renewable, emission free generation of electricity from tides would reduce the amount of carbon dioxide released to the atmosphere, which would aid in curbing the warming of the atmosphere that has been liked to carbon dioxide and other greenhouse gases.

Historical Background and Scientific Foundations

Tides have occurred as long as the ocean and the moon have existed. Both Earth and moon have their own gravitational fields that interact with each other. This interaction keeps the moon in orbit about Earth. Without this attraction the moon would drift off into space.

As the moon's gravity pulls on Earth, it also pulls on the surface water. The ocean bulges out toward the moon on the side of Earth that faces the moon. At the same time, the ocean water on the opposite side of Earth also is pulled out by a type of force known as inertia, which arises due to the orbit of the moon about Earth and the rotation of Earth.

As Earth rotates on its axis, any single point on the planet's surface will experience a high tide, low tide, another high tide, and finally another low tide in any 24-hour period. This cycle creates what is termed semidiurnal tides.

In reality, regions in the Northern and Southern Hemisphere experience only one high tide and one low tide each day (a diurnal pattern). The reason is that the moon's orbit is inclined to Earth—the moon does not rotate just around Earth's equator but instead the orbit is inclined so that the moon's monthly path takes it over both hemispheres. This orbit does change in a cycle of 19 years, so there will be times when many regions of Earth do experience the semidiurnal pattern of tides.

If the moon remained orbiting above the same spot on Earth, the tides in any one spot point on the planet would always occur at the same time. But the moon moves around Earth in an orbit that takes 30 days to complete. Earth is also spinning in the same direction as the moon's rotation. As a result, the twice-daily high-low tide cycle is not exactly 24 hours long, but 24 hours and 50 minutes. This means that at any point on the planet's coasts, the high and low tides occur 50 minutes later each day.

The gravitational attraction between Earth and the sun also influences tides depending on the position of the moon and sun. As the moon orbits around Earth, there will be a time in the month when the moon is between Earth and the sun (the "new moon"). Then, there will be combined pull on Earth from both the moon and sun, and so the high tides will be higher and the low tides lower than they are when just the moon's gravity is being considered. The same effect occurs when the moon and sun are oriented on the opposite sides of Earth (the "full moon"). The resulting tides are known as spring tides.

When the moon's orbit around Earth is one-quarter and three-quarters complete, the moon is perpendicular to the sun. The sun's gravity pulls water away from the regions of high tide toward the regions of low tide, making the high less high and the low tides higher than is the case if just the moon's gravity was the influencing factor. These tides are known as neap tides.

The tidal movement of water periodically exposes areas of the shoreline each day. The community that lives in this region has adapted to times of exposure to air during low tides and high tide, when they are submerged. The result is known as an intertidal ecosystem, which is unique from the neighboring ecosystems that are always submerged and which are above the high tide line.

Impacts and Issues

Ships that are traveling near the coast, or are moving in the mouth of a coastal river need to account for tidal patterns to ensure that their paths stay true. If a navigational path is determined without taking tides into consideration, a ship can end up off course. If the coastline is shallow or has reefs or rocks in the vicinity, this misdirection could be disastrous. For example, if the ship in distress is carrying a load of toxic material such as oil, leakage or rupture of the ship's hull can carry the oil to shore. Currents near the shore can distribute the spill over an even wider area, adversely affecting shore vegetation and life.

A spill can damage a special ecosystem. The organisms found in the intertidal zone are very different from those that live farther in from the water and those that are always submerged. The intertidal organisms face more environmental stresses than their counterparts and their adaptations have created a unique environment. Some intertidal areas have been set aside as protected regions to try and preserve their unique character for its own sake and to allow the study of the ecosystem. Research that revealed, for example, how an organism copes with a lack of water (desiccation) could be useful in helping other organisms survive in the coming century, when global warming is predicted to increase drought in some regions of the globe.

Tides can be harnessed as a source of power. For example, in the Bay of Fundy on the east coast of Canada, the tide-related change in water level can be up to 52 ft (16 m). A small tidal power project, in which tidal water passes through a dam and is held in place before being passed back through a generator, has been operating since 1984, and is one of only three concerted tidal power projects in the world. In 2008, plans were being finalized to position larger turbines deeper in the tidal flow. The result would be enough electricity, about 300 megawatts, to power over 200,000 homes for a year.

Pending funding, the initiative could be complete in 2009. A similar approach is being tested at Orkney, Scotland.

WORDS TO KNOW

DIURNAL: Performed in twenty-four hours, such as the diurnal rotation of Earth; also refers to animals and plants that are active during the day.

GRAVITY: An attractive force that exists between all mass in the universe such as the moon and Earth.

RENEWABLE RESOURCE: Any resource that is renewed or replaced fairly rapidly (on human historical time-scales) by natural or managed processes.

STORM SURGE: Rise of the sea at a coastline due to the effect of storm winds.

TSUNAMI: A series of ocean waves that result because of an undersea disturbance such as an earthquake.

SEE ALSO *Bays and Estuaries; Coastal Zones; Marine Ecosystems; Sea Level Rise; Tidal or Wave Power*

BIBLIOGRAPHY

Books

Garrison, Tom S. *Oceanography: An Invitation to Marine Science*. New York: Brooks Cole, 2007.

McCully, James Greig. *Beyond the Moon: A Conversational, Common Sense Guide to Understanding the Tides*. Singapore: World Publishing, 2006.

Brian D. Hoyle

Toxic Waste

Introduction

Toxic substances, such as heavy metals and organic solvents, are often present in the waste stream produced by human activities. The chemical, petroleum, and mining industries are the most likely industries to produce toxic wastes, but they are also generated by domestic users, hospitals, and dry cleaners. Toxic waste can contaminate groundwater and possibly enter drinking water or the food chain.

Toxins have many different adverse effects on the body, from acute poisoning to the slow accumulation of cell defects over many years, which can lead to cancer. Environmental legislation has been enacted following instances of toxic waste harming human health, such as the famous incident at Love Canal in the 1970s. Those producing or handling toxic waste are now required to prevent its entering the public domain. But there are still many sites needing to be cleaned of toxic waste, and problems with generating, storing, and exporting it to developing countries.

Historical Background and Scientific Foundations

People often use the term toxic waste and hazardous waste interchangeably. But the U.S. Environmental Protection Agency (EPA) defines toxic waste as a type of hazardous waste, which is a material dangerous or potentially harmful to human health or the environment. The other types of hazardous waste are flammable, corrosive, or reactive waste. A toxic waste contains components that poison body cells or tissues, thereby producing adverse effects on health.

The petrochemical, chemical, metal, and mining industries produce most toxic wastes, but homes produce some too—garden pesticides and cleaning materials are an important source. Toxic wastes can be present in the form of a solid, liquid, or gas and they may enter air, water, or soil. There are, therefore, a variety of ways in which humans can be exposed and a wide range of toxins. The United States Agency for Toxic Substances & Disease Registry lists arsenic, lead, mercury, vinyl chloride, and polychlorinated biphenyls (PCBs) as the most hazardous toxic wastes.

Impacts and Issues

The Industrial Revolution and the growth of the chemical industry brought many toxins into everyday use, long before medical research revealed their potential for harm to the cells of the body. Toxins often cause mutations in cells that can lead to birth defects, miscarriage, and cancer. Cancer may show up only many years after exposure. Chemicals found in toiletries, pesticides, and many other everyday products are now known to mimic the effect of the female hormone estrogen. Recent research has shown that these environmental estrogens cause developmental abnormalities in fish and may be linked to infertility and other human health problems.

The potential of toxic waste to harm was highlighted in events at Love Canal, Niagara Falls, New York, in the 1970s. In the 1920s, this area had been a landfill site for industrial chemicals. In the 1950s, around 100 homes and a school were built on the land. Twenty-five years later, the *New York Times* reported how the waste containers had rotted, allowing toxic waste to leach into the backyards of these homes. A higher than expected incidence of birth defects and miscarriages was noted in the local population, and many had higher than expected white cell counts, a possible precursor of leukemia. In the end, President Jimmy Carter declared a state of emergency and had the residents evacuated.

The Love Canal incident led to the Comprehensive Environmental Response, Compensation, and Liability Act, also known as the Superfund Act, in 1980. This legislation deals with the treatment and liability for these contaminated sites, the most seriously polluted of which are designated as Superfund sites.

Toxic waste continues to pose problems, but these are more likely to occur in developing nations that do not yet have tough environmental legislation. They may also be the recipients of toxic wastes from elsewhere. These exports continue, albeit illegally since a ban in 1989. In 2006, ten people died and thousands were made ill after exposure to petroleum waste exported to the Ivory Coast. In 2007, the Mattel toy company recalled almost two million toys from stores in the United States that were contaminated with lead-based paint; the source of the paint was traced to factories in China. China has also been accused of exporting toxic waste intended for use as components in fertilizer; Australian officials seized shipments of raw materials shipped from China and meant for fertilizers containing dangerously high levels of cadmium in 2003. Inside China, recent discharges of toxic waste into the Beijiang, Songhua, and Sungari Rivers all resulted in environmental emergencies that required the suspension of local water supplies.

SEE ALSO *Chemical Spills; Electronics Waste; Industrial Pollution; Hazardous Waste; Superfund Site; Waste Transfer and Dumping*

WORDS TO KNOW

ENVIRONMENTAL ESTROGEN: Compounds in toxic waste that mimic estrogen in their effect on humans and other animals.

SUPERFUND: Legislation that authorizes funds to clean up abandoned, contaminated sites.

TOXIN: A substance which is harmful to the body.

BIBLIOGRAPHY

Books

Cunningham, W.P., and A. Cunningham. *Environmental Science: A Global Concern.* New York: McGraw-Hill International Edition, 2008.

Web Sites

U.S. Agency for Toxic Substances & Disease Registry. "2007 CERCLA Priority List of Hazardous Substances." http://www.atsdr.cdc.gov/cercla/07list.html (accessed July 10, 2008).

U.S. Agency for Toxic Substances and Disease Registry. "Toxic Substances Portal." April 13, 2007.

Activists of the environmental organization Greenpeace protest to end the dumping of toxic ships in Asia and sub-standard working conditions in front of the U.N. headquarters in Geneva, Switzerland, 2005. *AP Images*

Environmental activists picket the Japanese Embassy in Manila in October 2006, to protest the recent signed agreements by both countries known as Japan-Philippines Economic Partnership Agreement. The protesters alleged that the Phillippine president allowed into the country—through zero tariff rates—Japanese toxic and hazardous wastes. *AP Images*

http://www.atsdr.cdc.gov/substances/index.html (accessed April 7, 2008).

U.S. Environmental Protection Agency (EPA). "The Love Canal Tragedy: EPA History." January 1979. http://www.epa.gov/history/topics/lovecanal/01.htm (accessed March 17, 2008).

Tsunami Impacts

Introduction

A tsunami can be described as a series of long, powerful ocean waves produced by a disturbance at or close to the ocean. This disturbance can result from earthquakes, volcanic eruptions, submarine slides, meteorites hitting the ocean, or the explosion of nuclear devices near the ocean. The word tsunami, a Japanese term, was adopted by an international scientific conference in 1963. It is made up of two characters: *tsu* and *nami*, meaning harbor wave. Tsunami are also known as seismic sea waves as they are mostly caused by seismic activity, or earthquakes.

Earthquakes that impact the ocean floor give rise to tsunami. The quake deforms the floor, causing a displacement of water. This triggers a series of waves that move toward the shore. It is not possible to spot the buildup to a tsunami from outside, and it usually cannot be felt on board a ship or a vessel at sea. Documented records of tsunami have existed for years, but there is a renewed interest in researching this natural disaster in the twenty-first century. This can be attributed to the large-scale devastation caused by the Indian Ocean tsunami in 2004.

Historical Background and Scientific Foundations

Over the years, the world has witnessed several instances of tsunami of varying magnitude. The Pacific Ocean has been the most vulnerable to tsunami. This is because of the earthquake-prone regions along its periphery. However, tsunami have also occurred in the coastal regions along the Indian Ocean, the Mediterranean Sea, the Caribbean region, and even the Atlantic Ocean. Countries that are earthquake prone and have populated coastal regions, such as Japan and Indonesia, have suffered huge losses due to tsunami. History records also show tsunami occurring in Alaska, the Hawaiian Islands, and the coastal regions of northwestern Europe. There is historical evidence indicating that a tsunami hit the Minoan civilization in about the fifteenth century BC. It destroyed a major portion of the coastal Minoan settlements on the Aegean Sea Islands. This tsunami is attributed to a volcanic eruption on the Greek island of Santorini. The oldest recorded tsunami dates back to AD 365 when an earthquake in the sea ravaged the city of Alexandria, Egypt. Casualties mounted to thousands of people.

A series of earthquakes took place in Lisbon, Portugal, on November 1, 1755. Soon after the earthquake, a 50-ft (15-m) tsunami occurred and took the lives of 60,000 people. Krakatoa, a volcanic island in the Indian Ocean, experienced a massive tsunami on August 26 and 27, 1883, with a death toll of more than 36,000 in Java and Sumatra. Waves measuring up to 115 ft (35 m) wiped out several villages situated along the coast.

Japan is one of the worst hit regions due to this phenomenon. In 1896, a series of tsunami caused the deaths of 27,000 people and destroyed more than 10,000 houses. Sixty eight tsunami have been recorded in Japan between 684 and 1984. In April 1946, Hawaii and Alaska bore the brunt of a tsunami. The Hawaiian Islands saw frequent tsunami in the 1950s and 1960s. Several tsunami occurred in 1952, 1956, 1957, 1960, 1964, and 1975. In 1960, Chile experienced a series of earthquakes that generated tsunami killing 1,500 people. The damage caused by these tsunami extended to Hawaii and Japan.

A powerful tsunami occurred in the Moro Gulf in the Philippines on August 16, 1976, taking more than 8,000 human lives, injuring 10,000 people, and rendering 90,000 people homeless. Another tsunami occurred in August 1977 in the Lesser Sunda Islands, Indonesia. On December 12, 1979, a destructive tsunami occurred

WORDS TO KNOW

BENTHIC: Living on, or associated with, the ocean floor.

EUTROPHICATION: The process whereby a body of water becomes rich in dissolved nutrients through natural or man-made processes. This often results in a deficiency of dissolved oxygen, producing an environment that favors plant over animal life.

RUN-UP HEIGHT: The vertical distance between the mean-sea-level surface and the maximum point attained on the coast.

SHALLOW-WATER WAVES: Waves that have a greater wave-length and in which the ratio between the depth of water and the wavelength is very small.

SHOALING EFFECT: Transformation that takes place when a wave travels from deep water to shallow water resulting in the decrease in wavelength and increase in the height of the wave.

SUBMARINE SLIDES: Marine landslides that can transport subsurface rock and sediment down the continental slope.

in the southwest region of Columbia killing hundreds of people and severely impacting the economy. During the 1990s, 10 tsunami were recorded in Indonesia resulting in thousands of deaths and property damages running to $1 billion.

A huge tsunami occurred in December 2004 in the Indian Ocean. It hit several coastal regions in Thailand, India, Sri Lanka, Indonesia, Malaysia, Myanmar, Bangladesh, Maldives, Somalia, and other regions. This tsunami was triggered by an earthquake of magnitude 9.0, which displaced the ocean floor off the Indonesian Island of Sumatra. This had a far-reaching impact affecting the eastern coastal regions of India and Sri Lanka and even Africa. It killed almost 200,000 people and destroyed property worth billions of U.S. dollars.

A breakthrough was made in the detection of tsunami when the Tsunami Warning System (TWS) was invented. A TWS can monitor seismological and tidal stations and forecast impending tsunami. It can be used for determining tsunamigenic earthquakes, monitoring tsunami, and alerting people in the coastal area of an impending tsunami. With the help of this system, the geological society can predict the occurrence of any earthquake of high magnitude. Using this technology, the meteorological agencies can evaluate changes in the sea level. Subsequently, the warning center can combine this information with data pertaining to the depth and features of the ocean floor. The output aids in calculating the path, magnitude, and the arrival time of a tsunami. Government agencies can use this time to evacuate the coastal areas and thus lessen the destructive impact of the tsunami.

The first TWS was established in the Pacific Region on April 1, 1946, after the tsunami hit the Hawaiian Islands. From October 1953, this warning information was made available to other regions such as California, Oregon, and Washington. It was the Chilean tsunami of May 1960 that motivated several countries to join the TWS. In 1965, the United Nations Educational Scientific and Cultural Organization's Intergovernmental Oceanographic Commission expanded its Tsunami Center in Honolulu; thus, the Pacific Tsunami Warning Center (PTWC) was established. Twenty-six countries from the Pacific region are members of the Pacific Tsunami Warning System (PTWS) network established by the PTWC in 1968. The International Coordination Group (ICG/ITSU) and the International Tsunami Information Center (ITIC) subsequently emerged. They were primarily established to evaluate and organize the activities of the International Tsunami Warning System for the Pacific (ITWS). Apart from the ITWS, several Regional Warning Systems have been established in tsunami-prone regions including the Soviet Union, Japan, Alaska, and Hawaii. UNESCO recognized the need for installing Tsunami Warning Systems in other regions particularly in the Indian Ocean after the tsunami of December 2004. Soon after the disaster, work commenced on installing these systems in the Indian Ocean, the Caribbean Sea, the Atlantic Ocean, and the Mediterranean Sea.

Impacts and Issues

Tsunami waves are extremely powerful ocean waves that can travel at a speed of 400 to 500 mph (650 to 800 km/h). The crests of major tsunami waves striking the coast are generally between 33 and 100 ft (10 and 30 m) in height. Owing to its enormity, the impact it has on the coastal regions is immense. Further, a tsunami is a consequence of other destructive phenomena such as earthquakes, volcanic eruptions, or submarine slides.

In general, any disturbance capable of displacing a huge volume of water from its equilibrium position can cause tsunami. As mentioned earlier, earthquakes are the primary cause of tsunami. When an earthquake occurs, there is a sudden deformation of the sea floor. The size of deformation depends on various factors such as the earthquake's magnitude, depth, and fault characteristics. This movement causes the overlying water to become displaced from its equilibrium state. When the displaced water tries to get back to its initial equilibrium state, gravitational force acts on it, giving rise to waves. The extent of vertical sea floor deformation is the prime determinant of the size of tsunami along the coast at the outset. Apart from this, there are other factors influencing the size of tsunami such as the seashore and bathymetric pattern, the speed of the sea floor deformation, and the depth of water near the earthquake source.

Earthquakes can cause submarine landslides, which in turn can lead to tsunami. In the event of a submarine landslide, the sediment moves along the sea floor. This disturbs the equilibrium position of the sea level. Gravitational forces further act on the sea level giving rise to huge sea waves. Similarly, a powerful volcanic eruption occurring in the sea can displace the water column thereby generating a tsunami. Further, subarial landslides and meteorite impacts on the sea can cause a tsunami. The tsunami produced by nonseismic mechanisms are less powerful than those generated by earthquakes; hence, the impact is less.

Offshore, tsunami have shorter wavelengths and often pass unnoticed. However, as they approach the coast, the waves gain momentum and as they enter the land, the waves become powerful enough to destroy life and exterminate property, flora and fauna, and anything else that comes in their way. Additionally, the receding waves pull people, buildings, trees, and sand into the ocean. Tsunami are shallow-water waves because their wavelengths are long, about 300 mi (500 km). The energy of a wave primarily depends on its wavelength. Because a tsunami has a large wavelength, it loses little energy as it advances. In an ocean, a tsunami can easily cover a huge distance without losing much energy. In fact, the speed of a raging tsunami when it travels in a very deep ocean can easily match that of a jet plane.

As soon as a tsunami approaches the coast, its speed decreases considerably. This is because the speed of a tsunami depends on the depth of the water; the lesser the depth of the water, the lesser the speed of an approaching tsunami. However, there is no change in the overall impact of a tsunami. As the speed of a tsunami decreases in shallow water, the shoaling effect takes place whereby the waves grow in height and assume a gigantic form. A tsunami may assume different forms such as a bore, a swiftly rising or declining tide, or a series of breaking waves. A number of factors can alter a tsunami as it approaches the sea shore—reefs, bays, entrances to rivers, undersea features, and the beach slope.

In recent times, the severity of tsunami impact has been greatly augmented because of human activities that harm the ecosystem. Urban development, aquaculture, and tourism are some of the factors that have severely damaged the coastal ecosystem. The impact of tsunami can be direct as well as indirect. The direct impact primarily depends on the ecosystem. If it is well protected by natural barriers, such as a coral reef, coconut palms, or mangroves, the impact is less. In fact, coastal ecosystems experience the worst impact of tsunami. Along with the physical structure of the coastal regions, it destroys the flora and fauna, the corals, and sea grass, and it affects several marine species. As the tsunami waves advance, they damage whatever obstructs their way, including bridges, seawalls, and buildings.

The indirect impacts of a tsunami are often more harmful than the tsunami. A noteworthy indirect impact of a tsunami is sedimentation due to tremendous runoff of coastal silt, sand, and organic matter. In addition, the floating debris such as buildings, vehicles, boats, and large electrical appliances cause further damage. They damage power lines that may lead to an outbreak of fire. Fires resulting from ravaged ships, oil storage tanks, or refineries can also lead to severe damage. The floating debris can adversely affect corals and other benthic substrates. Spilled harmful chemicals, including oils, paints, freons, and cleansers, are deposited and affect nearshore marine ecosystems. This can lead to disease in corals, algae, fish, and other invertebrates. However, such damages and their impacts do not become evident immediately; it may take months or even years to substantiate the consequences.

Destruction from a tsunami can be in the form of inundation, the impact of sea waves on structure, and erosion. Apart from physical changes, it also brings about chemical changes such as intrusion of salty ocean water, eutrophication of water due to high runoff, raw sewage, and decay of dead plants and animals. Toxic wastes including plastics and other nonbiodegradable wastes give rise to marine garbage.

Tsunami have caused grave destruction of life, property, and ecosystems. Besides, the survivors of this tragedy also suffer from the aftermath for a long time. They are often devoid of the basic necessities of life—

IN CONTEXT: TSUNAMI IMPACTS

Following the 2004 Indian Ocean tsunami, most environmental experts expected that the environmental toll of the event would be devastating. Although the human toll is fundamentally and tragically incalculable, studies following the tsunami show that the environmental damage caused by the tsunami was severe but uneven.

Places that were already suffering from environmental damage were the places most affected by the tsunami. On the other hand, places that had healthy coastal ecosystems sustained substantially less human and environmental damage.

Healthy coastal ecosystems such as mangroves, coral reefs, and vegetated sand dunes acted as buffers to the tsunami, often protecting both structures as well as human life. For example in the Yala and Bundala National Parks in Sri Lanka, where sand dunes were completely vegetated, the tsunami had minimal environmental impact. Conversely, in places where the coral reefs had been mined, damage by the waves was most destructive.

A fisherman rows his boat amid dead trees destroyed due to waterlogging by waters brought in by the tsunami at India's southeastern Andaman and Nicobar Islands archipelago, December 2005. *AP Images*

clean drinking water, food, and shelter. A large number suffer from physical injuries as well as mental trauma. Floodwater resulting from a tsunami is also a cause of major health problems. Contamination of water and food brings about food-borne and waterborne illnesses. Furthermore, there is a noticeable increase in infectious and insect-transmitted diseases. With their homes destroyed by the tsunami, the survivors are exposed to harsh weather and other environmental hazards. The Indian Ocean tsunami of 2004, which greatly affected several regions in Thailand, India, Sri Lanka, Myanmar, Indonesia, Africa, and the Maldives, gave impetus to several communicable diseases such as cholera, malaria, dengue, tuberculosis, influenza, and rabies.

Natural disasters such as earthquakes and tsunami cannot be prevented. However, preventive measures may be taken to reduce their impacts. One of the measures is to protect the coastal ecosystem. Some of the methods are promoting bio-shields; restoration of natural barriers, such as mangroves, casuarinas plantations, sand dunes, and coral reefs; and establishing artificial barriers like sea walls and embankments. Another way is conserving natural coastal habitats, for example, by discouraging aquaculture farms. Coastline development should be done only through proper management and in keeping with the Coastal Regulation Zone (CRZ) norms. Further, human activities should be controlled to protect marine ecosystems, and sustainable use of natural resources should be encouraged.

Several Tsunami Warning Systems (TWS) have been set up at various locations around the globe for detecting potential tsunami. However, the technology is not infallible and does not guarantee security. Predictions through a TWS can be inaccurate, causing false alarms or failure to detect a tsunami. Repeated false alarms could lead people to not respond to a warning. Though a TWS operates in real-time, it cannot always be relied upon to save lives. This is because the ferocity of a tsunami gives people little time to react or escape. The real-time operation also makes it difficult to predict a tsunami run-up resulting from an earthquake. Other hurdles include inadequate data and lack of proper communication. The TWS technology is yet to become widespread, and as a result, several vulnerable locations across the world remain unguarded.

SEE ALSO *Aquatic Ecosystems; Coastal Ecosystems; Earthquakes; Ecodisasters; Volcanoes*

BIBLIOGRAPHY

Books

Bryant, Edward. *Tsunami: The Underrated Hazard.* Cambridge: Cambridge University Press, 2001.

Kusky, Timothy M. *Geological Hazards: A Sourcebook.* Westport, CT: Greenwood Publishing Group, 2003.

Web Sites

Centers for Disease Control and Prevention (CDC). "Emergency Preparedness & Response."

http://www.bt.cdc.gov/disasters/tsunamis/healtheff.asp (accessed February 29, 2008).

Indian Concrete Journal. "An Introduction to Tsunami in the Indian Context." http://www.nicee.org/ICJ_TsunamiSpecialReortl.pdf (accessed February 29, 2008).

International Tsunami Information Center (ITIC). "Tsunamis on the Move." http://ioc3.unesco.org/itic/contents.php?id=205 (accessed February 29, 2008).

Office of Climate, Weather, and Water Services. "Tsunami: The Great Waves." http://www.weather.gov/om/brochures/tsunami.htm (accessed February 29, 2008).

UN Atlas of the Oceans. "Impact of Tsunami on Ecosystems." http://www.oceansatlas.com/servlet/CDSServlet?status=ND03MTY4NyZjdG5faW5mb192aWV3X3NpemU9Y3RuX2luZm9fdmlld19mdWxsJjY9ZW4mMzM9KiYzNz1rb3M~ (accessed February 29, 2008).

UNESCO. "Tsunami: The Great Waves." http://ioc3.unesco.org/itic/files/great_waves_en_2006_small.pdf (accessed February 29, 2008).

University of Wollongong. "Introduction to Tsunami." http://www.uow.edu.au/science/eesc/research/UOW002909.html (accessed February 29, 2008).

WWF-India & Wetlands International-South Asia. "Post Tsunami Conversation Issues and Challenges Consultative Meeting for Coordinated Action." http://assets.wwfindia.org/downloads/final_report_may2005.doc (accessed February 29, 2008).

Amit Gupta

Tundra

Introduction

Tundra are areas in far northern and southern regions of Earth where tree growth is limited due to low temperatures and a growing season that is only weeks long. This arctic tundra contains shrubs, mosses, and lichens. Another kind of treeless zone that can occur near the top of the world's highest mountains is called an alpine tundra.

Despite its barren appearance, the tundra that occupies northern Canada and Russia is inhabited by about 50 species including large herds of caribou.

The word tundra is derived from a Finnish word *tunturia,* meaning barren land. The subsurface of the arctic tundra is permanently frozen, making it impossible for tree roots to penetrate more than a few feet into the ground.

In the winter, when daytime temperatures average –18°F (–27.7°C) and nighttime temperatures can be below –90°F (–67.7°C) and sunlight is almost absent, the tundra's surface is cold and snow-covered. As the snow melts and the sun shines almost constantly in the short summer season, the tundra becomes soggy and a breeding ground for insects. The summer tundra is also a popular nesting ground for some species of migrating birds.

Tundra are becoming an important concern for global warming because they are a major reserve of carbon dioxide and methane. The increased warming of the north is melting the tundra permafrost, which is releasing the greenhouse gases to the atmosphere. Continued thaw of the permafrost could lead to the release of large levels of the greenhouse gases. These consequences for global warming have not been modeled and so are unknown.

Historical Background and Scientific Foundations

At first glance, the tundra appears inhospitable. The barren landscape supports the growth of only low-lying shrubs, grasses, moss, and lichens. As well, with little to block the wind, the tundra is a windy region, with winds of 60–90 mph (97–145 km/h) not uncommon.

The permanently frozen subsurface does not support the growth of trees. Neither does the lack of moisture. Most tundra are deserts, receiving only around 10 in (25 cm) of precipitation annually. In the summer, the ground remains frozen, so the melting snow cannot drain. This produces a boggy surface with many small lakes and marches, which provides a resting place for migratory birds.

The harshness of the tundra environment is reflected in its low biodiversity. Only about 1,700 plant species and 50 animal species have been documented. However, some of the animal species can be present in huge numbers; one example is the caribou. Other animals that reside in the tundra include the arctic hare, arctic fox, snowy owl, and, in the far northern reaches of the tundra, polar bear.

Although some aboriginal people live in the northern tundra of Canada and Russia, most tundra are relatively unpopulated and undeveloped. However, some development that has occurred has had adverse effects on the environment. For example, an oil pipeline built in Alaska crosses a caribou migration route. The pipeline has been raised off the ground at some points so that the caribou have an unimpeded north-south route. In developed areas, the hoards of insects present during the summer have been controlled using pesticides for the comfort of workers. The presence of the pesticides in surface waters poses a threat to birds who use the same environment as a summer migration resting place.

Oil exploration and mining can damage the fragile tundra ecosystem. For example, toxic by-products from nickel mining in Russia have caused severe pollution in some areas. As well, the tracks left by vehicles can remain for decades, becoming even more prominent as erosion occurs.

Impacts and Issues

The fragility of the tundra environment makes it prone to great damage from development. This threat may grow, as oil and mining deposits continue to be discovered. As the world's current supplies of non-renewable fossil fuel are depleted (predictions are that current supplies could be exhausted during the twenty-first century), the tundra will become more attractive for oil and mining.

The tundra has also become a concern as the north continues to warm. The permafrost contains large amounts of carbon dioxide that are trapped in plants that have collected over thousands of years but which have not been able to completely decompose. Also, the permafrost contains methane. Both compounds are greenhouse gases—their presence in the atmosphere increases the retention of solar radiation, which causes the atmosphere to heat up. This is especially so for methane.

With the increased warming of the north, some polar scientists have cautioned that massive plant decomposition and the abundant release of methane could seed the atmosphere with huge quantities of carbon dioxide and methane that have not been accounted for in computer projections of the progression of global warming. The effect could be an unanticipated acceleration of atmospheric warming.

The effects of Earth's changing climate are already evident in the tundra, where boreal forests are appearing.

SEE ALSO *Antarctic Issues and Challenges; Arctic Darkening and Pack-Ice Melting; Deserts*

WORDS TO KNOW

BIOME: A well-defined terrestrial environment (e.g., desert, tundra, or tropical forest) and the complex of living organisms found in that region.

BOREAL FORESTS: A forest biome of coniferous trees running across northern North America and Eurasia; its high northern latitudes are often referred to as taiga.

DESERT: A land area so dry that little or no plant or animal life can survive.

PERMAFROST: Perennially frozen ground that occurs wherever the temperature remains below 0°C for several years.

TAIGA: The part of the boreal forest occurring in high northern latitudes, consisting of open woodland of coniferous trees growing in a rich floor of lichen.

BIBLIOGRAPHY

Books

Moore, Peter. *Tundra*. New York: Facts on File, 2007.

Parker, Gary. *Exploring the World Around You: A Look at Nature from Tropics to Tundra*. Green Forest, AR: Master Books, 2003.

Woodford, Chris. *Arctic Tundra and Polar Deserts*. Chicago: Raintree, 2003.

United Nations Conference on the Human Environment (1972)

Introduction

In response to the growing environmental movement of the 1960s, many nations began to take actions to protect the environment within their borders. By the early 1970s, however, governments began to realize that pollution did not stop at their borders. International consensus and cooperation were required to tackle environmental issues, which affected the entire world.

In 1972, the United Nations Conference on the Human Environment (UNCHE) was convened to address issues concerning the environment and sustainable development. UNCHE, also known as the Stockholm Conference, linked environmental protection with sustainable development. The Stockholm Conference also produced concrete ideas on how governments could work together to preserve the environment. The concepts and plans developed by the Stockholm Conference have shaped every international conference and treaty on the environment over the last 35 years.

Historical Background and Scientific Foundations

The United Nations Conference on the Human Environment (UNCHE), held in Stockholm, Sweden, in 1972, was the first major international conference on the environment. The United Nations General Assembly convened the UNCHE at the request of the Swedish government. Representatives from 113 nations and over 400 non-governmental organizations (NGOs) attended the Stockholm Conference.

The gathering produced the Declaration of the Conference on the Human Environment and an action plan. The declaration noted that many factors harm the environment, including population growth, developing economies, and technological and industrial advancements. Despite the pressure placed on the environment, the declaration proffered 26 principles "to inspire and guide the peoples of the world in the preservation and enhancement of the human environment."

The Declaration of the Conference stated that every human has the right to enjoy a clean and healthy environment. With this right, however, comes the responsibility to preserve the environment for future generations. The document noted that humans must properly manage wildlife and their ecosystems to ensure their continued survival, and it sought an end to the discharge of pollution into the environment. The declaration also called on industrialized nations to provide financial and technological assistance to developing nations to enable them to develop their economies in an environmentally responsible manner.

The declaration was the first major international document to recognize that both developing and industrialized economies contribute to environmental problems, and it noted that most environmental problems in developing economies occur because of underdevelopment. Poverty in developing nations leads to poor health, poor sanitation, and release of toxic chemicals. These conditions release harmful human, animal, and chemical products into the environment. Developing economies also often seek advancement of the economy with little regard for environmental regulation. Industrialized nations contribute to environmental problems through technological advancements and industrialization. Energy production, automobile emissions, and factory production release greenhouse gases and other chemicals into the environment.

Whereas the Declaration of the Convention contained many lofty ideals, the action plan of the Stockholm Conference contained 109 specific recommendations for achieving these goals. The action plan presented 69 recommendations on how governments, intergovernmental agencies, and NGOs could work together to implement environmental protection strategies. The action plan also contained 16 proposals for dealing with pollution in general. Recommendation 70 contains one of the first references to global climate change contained in an international document. It recommends that governments be "mindful of activities in which there is an appreciable risk of effects on the climate."

The action plan also called for establishing international standards for pollutants after scientific research into the effect of certain pollutants on the environment. The action plan then recommended the creation of a network of national and international pollution monitoring agencies. The United Nations founded the United Nations Environment Programme (UNEP) in 1972 to coordinate its environmental initiatives and to provide support to developing nations on environmental issues.

WORDS TO KNOW

ECOSYSTEM: The community of individuals and the physical components of the environment in a certain area.

GREENHOUSE GASES: A gas whose accumulation in the atmosphere increases heat retention.

NON-GOVERNMENTAL ORGANIZATION (NGO): A voluntary organization that is not part of any government; often organized to address a specific issue or perform a humanitarian function.

SUSTAINABLE DEVELOPMENT: Development (i.e., increased or intensified economic activity; sometimes used as a synonym for industrialization) that meets the cultural and physical needs of the present generation of persons without damaging the ability of future generations to meet their own needs.

Impacts and Issues

The objectives and action plans produced by the Stockholm Conference have inspired every subsequent international conference on the environment. In 1983, the United Nations convened the World Commission on Environment and Development (WCED), also called the Brundtland Commission. The Brundtland Commission discussed and devised international and national strategies for protecting the environment and promoting sustainable development. The Brundtland Commission published its final report, *Our Common Future*, in 1987. *Our Common Future* states that governments could not address environmental protection separately from the related crises of economic development and energy production. *Our Common Future* outlined a plan for dealing with these interlocking crises.

The Stockholm Conference also laid the foundation for the United Nations Conference on Environment and Development (UNCED), commonly called the Earth Summit. In June 1992, representatives from 172 nations convened in Rio de Janeiro, Brazil, for the unprecedented Earth Summit, which included 108 heads of state, 2,400 representatives from various non-governmental organizations (NGOs), and nearly 10,000 journalists. An additional 17,000 NGO representatives attended a parallel NGO forum that provided recommendations to the Earth Summit.

The massive interest and participation in the Earth Summit indicated a shift in global attitudes toward the environment. Scientific evidence gathered in the second half of the twentieth century indicated that human activity affected the environment and climate. The scientific evidence also revealed that pollution and depletion of natural resources that occurred in one country could have a profound effect on the environment of other nations or even the entire planet. At the Earth Summit, world leaders devised plans and policies to protect the environment by involving national and local governments and NGOs in the process. Earth Summit 1992 produced two key documents: Agenda 21 and the Rio Declaration on Environment and Development. Earth Summit 1992 also produced the United Nations Framework Convention on Climate Change (UNFCCC), which seeks to combat global climate change by reducing global greenhouse-gas emissions.

Earth Summit 2002, a ten-year follow-up to Earth Summit 1992, produced the Johannesburg Declaration, which reiterated many of the points contained in the Rio Declaration and Agenda 21. The Johannesburg Declaration, however, contains a more general statement about the environment and sustainable development. It calls for an end to all conditions that threaten sustainable development, including drug use, corruption, terrorism, ethnic intolerance, and natural disasters. However, the declaration does not contain many specific proposals for addressing many of these issues.

The Stockholm Conference also inspired the Kyoto Protocol, one of the best known and far-reaching plans undertaken by the UNFCCC. The Kyoto Protocol is an international treaty that seeks to stabilize greenhouse-gas emissions by committing countries to specific greenhouse-gas emissions goals. The Kyoto Protocol, which went into effect in February 2005, requires developed signatory countries to reduce their greenhouse-gas emissions to 5% below 1990 levels by 2012.

SEE ALSO *Earth Summit (1992); Earth Summit (2002); Sustainable Development; United Nations World Commission on Environment and Development (WCED) Our Common Future Report (1987)*

BIBLIOGRAPHY

Web Sites

Secretariat of the United Nations Framework Convention on Climate Change. "Essential Background: The United Nations Framework Convention on Climate Change." http://unfccc.int/essential_background/convention/items/2627.php (accessed April 21, 2008).

Secretariat of the United Nations Framework Convention on Climate Change. "Kyoto Protocol." http://unfccc.int/kyoto_protocol/items/2830.php (accessed April 21, 2008).

United Nations. "Documents: Agenda 21." December 15, 2004. http://www.un.org/esa/sustdev/documents/agenda21/index.htm (accessed April 21, 2008).

United Nations. "U.N. Conference on Environment and Development (1992)." May 23, 1997. http://www.un.org/geninfo/bp/enviro.html (accessed April 21, 2008).

United Nations Environment Programme. "Stockholm 1972: Report of the United Nations Conference on the Human Environment." http://www.unep.org/Documents.Multilingual/Default.asp?documentID=97 (accessed April 21, 2008).

Joseph P. Hyder

United Nations Convention on the Law of the Sea (UNCLOS)

Introduction

Technological advancements during the last century led nations to compete for and exploit the ocean and its resources at an unprecedented level. Large fishing vessels that could remain at sea for months at a time depleted fish stocks. The discovery of oil, gas, and minerals under the seabed led to a competition to claim areas of the sea that held this wealth. The magnitude of these activities also had a profound effect on the environment as dwindling fish populations and increased pollution destroyed coastal ecosystems.

Over 160 nations met in the Third United Nations Conference on the Law of the Sea to address the issues surrounding use of the oceans. After nine years of negotiations, the conference adopted the United Nations Convention on the Law of the Sea (UNCLOS). UNCLOS addressed the territorial limits of nations over the ocean, economic rights to the ocean's resources, and rights of transit. UNCLOS also addressed pollution and other environmental concerns. Article 192 of UNCLOS states that all nations have a general obligation "to protect and preserve the environment." Even though UNCLOS was not strictly intended to be an environmental treaty, it has had a profound effect on environmental regulation in the oceans.

Historical Background and Scientific Foundations

Until the mid-twentieth century, most nations adhered to the freedom-of-the-seas doctrine, which originated in the seventeenth century. The freedom-of-the-seas doctrine granted nations jurisdiction over the oceans for an area extending three nautical mi (5.5 km) from shore. Nations considered the waters beyond this three-mile limit to be international waters that were open to all nations but not belonging to any one nation in particular.

By the mid-twentieth century, deep-sea discoveries and technological advancements led several nations to question the continued utility of the freedom-of-the-seas doctrine. In the early twentieth century, valuable metals and diamonds were discovered in the oceans, which raised the possibility of unrestricted deep-sea mining. Offshore petroleum and gas reserves were also discovered on continental shelves in the North Sea and off the coast of North America. The first offshore oil drilling occurred in the Gulf of Mexico in 1947. By the late 1960s, oil production in the Gulf of Mexico had grown to nearly 400 million tons of oil per year.

In 1945, in order to protect oil, gas, and mineral reserves, U.S. President Harry Truman declared that the United States had jurisdiction over the entire continental shelf. President Truman's declaration made the United States the first nation to abandon the freedom-of-the-seas doctrine.

Other nations followed suit and abandoned the freedom-of-the-seas doctrine in order to protect mineral deposits, petroleum reserves, and rapidly depleting fish stocks. Many nations also became concerned about the environmental damage caused by pollution and overfishing. Nations soon became involved in numerous territorial and sovereignty claims as they competed for the oceans' vast resources and debated the issue of the rights of ships to pass through their waters.

In 1967, Arvid Pardo, Malta's ambassador to the United Nations, called on all nations to address the environmental, economic, legal, and political issues surrounding the exploration and use of the oceans and seabed. In 1973, the Third United Nations Conference on the Law of the Sea convened in New York City. Nine years later, in 1982, the conference adopted the United Nations Convention on the Law of the Sea (UNCLOS).

WORDS TO KNOW

CONTIGUOUS ZONE: A maritime zone extending 24 nautical miles (44 km) from the outer edge of the territorial sea, in which a coastal state can exert limited control of its laws.

CONTINENTAL SHELF: A gently sloping, submerged ledge of a continent.

DEEP SEA MINING: The extraction of valuable mineral deposits from the ocean floor; not yet practiced due to legal, environmental, and monetary concerns.

EXCLUSIVE ECONOMIC ZONE (EEZ): A maritime zone extending 200 nautical miles (370 km) from the outer edge of the territorial sea, in which a coastal state has special rights over the exploration and use of marine resources.

FREEDOM-OF-THE-SEAS DOCTRINE: An eighteenth-century agreement establishing the right of neutral shipping in international waters.

INNOCENT PASSAGE: The right of all ships to pass through the territorial waters of another state subject to certain restrictions.

TERRITORIAL WATER: A maritime zone of coastal waters extending up to 12 nautical miles (22 km) from the baseline of a coastal state, within which a state can exert control of its laws and regulations.

TRANSIT PASSAGE: The right of free navigation and overflight for the purpose of continuous transit through a strait between one part of the high seas and another.

UNCLOS replaced the three-nautical-mile limit that existed under the freedom-of-the-seas doctrine with a tiered approach. UNCLOS expanded each nation's territorial waters to 12 nautical mi (22 km). Within territorial waters, a nation can set and enforce laws, regulate use, and exploit resources. UNCLOS also allowed for an additional 12-mile contiguous zone where a nation could enforce smuggling and immigration laws. Under UNCLOS, ships from one nation may pass through the territorial waters of another nation under the right of innocent passage. Under the right of transit passage, military vessels may pass through the territorial waters of another country when other opportunities do not exist, such as passing through a strait.

The establishment of exclusive economic zones (EEZs) was one of the most important features of UNCLOS. Under UNCLOS, each nation has the exclusive right to use, exploit, or develop any resource that lies within 200 nautical mi (370 km) of its coastline. This placed 38 million square nautical miles of ocean and seabed under the exclusive control of individual countries. Virtually all fish stocks, most deep-sea minerals, and nearly 87% of all offshore oil and gas reserves now lie within a specific EEZ.

Although EEZs carry an obvious economic benefit, the establishment of EEZs has also had an impact on environmental regulation. Under UNCLOS, a nation has the ability to conserve the natural resources that lie within its EEZ. Nations can now prevent foreign nations from depleting fish stocks, extracting minerals, and drilling for oil within its EEZ. The drafters of UNCLOS envisioned that a nation would be more likely to take precautions to prevent environmental accidents when the consequences of a disaster would affect its own coastal ecosystem.

Impacts and Issues

Although UNCLOS is not expressly an environmental treaty, the scope and relatively universal acceptance of UNCLOS make it one of the most influential treaties regarding the environment. Part XIII of UNCLOS addresses the protection and preservation of the marine environment, including the following sources of pollution: land-based pollution, deep-sea mining, dumping, vessel-source pollution, continental shelf drilling, and pollution from or through the atmosphere. UNCLOS permits states to enforce their anti-pollution and other environmental standards throughout their EEZs, except those concerning vessel-source pollution.

Each nation has the right to exploit all mineral resources that lie within its EEZ. Although most minerals are located within the EEZ of a particular nation, Part XI of UNCLOS addresses mineral resources that lie in international waters. Currently, the technology does not exist to remove minerals located at such great depths. Part XI established the International Seabed Authority (ISA) to regulate mining in international waters. The United States objected to the establishment of the ISA. As a result, the U.S. Senate has not ratified UNCLOS. The United States is a signatory to UNCLOS, however, and regards all of its remaining provisions as binding customary international law.

Finally, global climate change has greatly increased the importance of the UNCLOS provisions regarding the right of foreign ships to pass through another nation's waters. As noted, UNCLOS states that a nation's vessels may pass through the territorial waters of another nation under the right of innocent passage. In 2007, the loss of Arctic sea ice due to global climate change led to the opening of the Northwest Passage for the first time in recorded history. The Northwest Passage provides a shortcut from the North Atlantic to the Pacific Ocean by passing through a chain of Canadian islands known as the Canadian Arctic Archipelago. Scientists predict that the Northwest Passage will open regularly during summer months because of global climate change.

A dispute over international waters erupted over transit through the Northwest Passage. Most nations, including the United States and most European nations, regard the Northwest Passage as a strait through which foreign vessels have the rights of innocent passage and transit passage of military vessels. Canada claims that the Northwest Passage is part of its internal waters and not part of its territorial waters. A nation may prohibit all use of its internal waters, including the right of innocent passage. A similar international dispute is likely to occur when the Northeast Passage, a passage through the Russian Arctic, is ice-free in the near future.

SEE ALSO *Coastal Ecosystems; Oceans and Coastlines; Overfishing; Water Conservation; Water Pollution; Water Resources*

BIBLIOGRAPHY

Books

Churchill, R. R., and A. V. Lower. *The Law of the Sea*. Huntington, NY: Juris, 1999.

Periodicals

Halfar, Jochen, and Rodney M. Fujita. "Danger of Deep-Sea Mining: Plans for Deep-Sea Mining Could Pose a Serious Threat to Marine Ecosystems." *Science* 316 (2007): 987.

Web Sites

Canadian Broadcasting Corporation. "Arctic Sovereignty: Drawing a Line in the Water." August 2, 2007. http://www.cbc.ca/news/background/cdnmilitary/arctic.html (accessed January 31, 2008).

National Geographic. "Arctic Melt Opens Northwest Passage." September 17, 2007. http://news.nationalgeographic.com/news/2007/09/070917-northwest-passage.html (accessed January 31, 2008).

United Nations. "United Nations Convention on the Law of the Sea." http://www.un.org/Depts/los/convention_agreements/texts/unclos/closindx.htm (accessed January 31, 2008).

United Nations. "The United Nations Convention on the Law of the Sea: A Historical Perspective." http://www.un.org/Depts/los/convention_agreements/convention_historical_perspective.htm (accessed January 31, 2008).

U.S. Department of State. "The UN Convention on the Law of the Sea: Written Testimony Before the Senate Foreign Relations Committee of John D. Negroponte, Deputy Secretary of State." September 27, 2007. http://www.state.gov/s/d/2007/92921.htm (accessed January 31, 2008).

Joseph P. Hyder
Adrienne Wilmoth Lerner

United Nations Framework Convention on Climate Change (UNFCCC)

Introduction

The United Nations Framework Convention on Climate Change (UNFCCC) is an international environmental treaty. Agreement on the treaty was reached at the Earth Summit in Rio de Janeiro, Brazil, in June 1992. The treaty sets out to reduce emissions of greenhouse gases from human activities in order to combat global climate change.

Historical Background and Scientific Foundations

Held over June 3–14, 1992, the United Nations Conference on Environment and Development—better known as the Earth Summit—was unprecedented for a U.N. conference, in terms of its size and breadth of its concerns. It came exactly twenty years after the U.N.'s Conference on the Environment and followed the resurgence of environmentalism in the late 1980s.

The Earth Summit was less a formal intergovernmental meeting, however, than a vast gathering in which tens of thousands of environmentalists, campaigners, and ordinary people sought to make their voices heard. Their overall message was that world leaders needed to make difficult decisions to protect Earth's environment.

There was a universal acknowledgement at the Earth Summit of the need to ensure that future national and international policy should take into account environmental impact. There was a commitment from national governments to scrutinize and regulate local industrial pollution; an affirmation to explore the development of alternative and renewable energy sources; and a recognition of the growing scarcity of essential water sources.

The key document to emerge from Rio was the United Nations Framework Convention on Climate Change (UNFCCC), signed by 154 nations on June 12. The treaty represented—for the first time—international recognition that atmospheric concentrations of greenhouse gases needed to be reduced in order to counteract climate change. Its stated objective was "to achieve stabilization of greenhouse gas concentrations in the atmosphere at a low enough level to prevent dangerous anthropogenic interference with the climate system." The UNFCCC set in motion the period of negotiation that would culminate in binding targets for emissions reductions set in Kyoto, Japan, five years later.

Impacts and Issues

The UNFCCC reflected the spirit of the Earth Summit, but while it was the clearest international commitment to counteract climate change produced up to that time, it was legally non-binding on the parties. The UNFCCC set no mandatory limits on greenhouse-gas emissions for individual nations. Instead, the treaty included provisions for updates and amendments—known as "protocols"—that could set future mandatory emission limits and reduction targets.

No mention was made of the creation of an international body to monitor and enforce reductions of greenhouse-gas emissions. The implication was that any future targets would be voluntary and self-monitored. The actual formula for calculating a country's greenhouse emissions was also vague. This deliberate omission was remedied by the 1977 Kyoto Protocol to the UNFCCC.

Environmental advocates saw the UNFCCC as a good beginning, but too weak, and called for more urgent action. For example, Maurice Strong, the Earth Summit's secretary general, lobbied unsuccessfully to include a target of 60% in the reduction in greenhouse-gas emissions. Another concern for several environmental organizations and participating nations was the reluctance of some high-polluting nations, such as the

United States, to discuss binding emissions limits or to codify target reductions.

The UNFCCC entered into force on March 21, 1994. A year later, its signatories met in Berlin, Germany, for a general meeting, the first in an annual series called for by the terms of the UNFCCC. At that meeting, there was an acknowledgement that the existing convention did not go far enough in its efforts to counteract global warming and a mandate was agreed, which initiated a period of negotiation to establish binding greenhouse-gas emissions reduction targets. From this, the Kyoto Protocol emerged in December 1997, and entered into force in February 2005. The Kyoto Protocol sets binding limits on emissions for developed countries, requires other countries (including, controversially, India and China) only to monitor their greenhouse emissions, and sets up mechanisms for trading in carbon credits and for transferring money from wealthy nations to poorer nations for greenhouse-mitigation projects such as the setup of more efficient manufacturing plants.

A successor agreement to the Kyoto Protocol, which might remedy some of that agreement's admitted shortcomings, was due to enter into force in 2012. Negotiations on the terms of that successor agreement began in late 2007. However, diplomats did not expect much progress to be made until after the U.S. presidential election of 2008, since the administration of George W. Bush had since 2001 been resolutely opposed to U.S. cooperation with the Kyoto Protocol (which was never ratified by the United States, alone of industrialized nations). The Bush administration opposed any climate treaty that would entail U.S. promises to abide by specific greenhouse-emissions caps.

Although it did not ratify the Kyoto Protocol, the United States remains a member of the UNFCCC. The United States also cooperates in the operations of the IPCC, a separate UN group dedicated to monitoring the state of global scientific knowledge about climate change.

Primary Source Connection

The United Nations Framework Convention on Climate Change was an international treaty created based on UN recognition that climate change is a global issue and relevant to all countries. It was also created based on recognizing that human activities would increase the concentration of greenhouse gases, especially carbon dioxide, which would then, in turn, cause a further increase in global temperature.

The United Nations also recognized that the management of carbon dioxide emissions needs to be based on a global agreement. It is not suitable for every country to develop their own standards because the decisions made by one country can impact other countries. This especially applies to those countries more vulnerable to changes in temperature, including developing countries in Africa, small island countries, and countries with low-lying areas vulnerable to flooding.

Based on these recognitions, the treaty requires that all countries provide information on their carbon dioxide emissions, implement plans to manage emissions and mitigate climate change, promote and cooperate in activities related to managing climate change, and take climate change into account when developing social, economic, and environmental plans.

In 1997, the Framework Convention on Climate Change was amended with the addition of the Kyoto Protocol. This addition required countries to reduce their emissions of greenhouse gases and also set greenhouse-gas emission limits.

WORDS TO KNOW

ANTHROPOGENIC: Made by humans or resulting from human activities.

GREENHOUSE GAS: A gas whose accumulation in the atmosphere increases heat retention.

KYOTO PROTOCOL: Extension in 1997 of the 1992 United Nations Framework Convention on Climate Change (UNFCCC), an international treaty signed by almost all member countries with the goal of mitigating climate change.

RENEWABLE ENERGY: Energy that can be naturally replenished. In contrast, fossil fuel energy is nonrenewable.

FRAMEWORK CONVENTION ON CLIMATE CHANGE

ARTICLE 2

OBJECTIVE

The ultimate objective of this Convention and any related legal instruments that the Conference of the Parties may adopt is to achieve, in accordance with the relevant provisions of the Convention, stabilization of greenhouse gas concentrations in the atmosphere at a level that would prevent dangerous anthropogenic interference with the climate system. Such a level should be achieved within a time-frame sufficient to allow ecosystems to adapt naturally to climate change, to ensure that food production is not threatened and to enable economic development to proceed in a sustainable manner.

ARTICLE 3

PRINCIPLES

In their actions to achieve the objective of the Convention and to implement its provisions, the Parties shall be guided, inter alia, by the following:

1. The Parties should protect the climate system for the benefit of present and future generations of humankind, on the basis of equity and in accordance with Their common but differentiated responsibilities and respective capabilities. Accordingly, the developed country Parties should take the lead in combating climate change and the adverse effects thereof.
2. The specific needs and special circumstances of developing country Parties, especially those that are particularly vulnerable to the adverse effects of climate change, and of those Parties, especially developing country Parties, that would have to bear a disproportionate or abnormal burden under the Convention, should be given full consideration.
3. The Parties should take precautionary measures to anticipate, prevent or minimize the causes of climate change and mitigate its adverse effects. Where there are threats of serious or irreversible damage, lack of full scientific certainty should not be used as a reason for postponing such measures, taking into account that policies and measures to deal with climate change should be cost-effective so as to ensure global benefits at the lowest possible cost. To achieve this, such policies and measures should take into account different socio-economic contexts, be comprehensive, cover all relevant sources, sinks and reservoirs of greenhouse gases and adaptation, and comprise all economic sectors. Efforts to address climate change may be carried out cooperatively by interested Parties.
4. The Parties have a right to, and should, promote sustainable development. Policies and measures to protect the climate system against human-induced change should be appropriate for the specific conditions of each Party and should be integrated with national development programmes, taking into account that economic development is essential for adopting measures to address climate change.
5. The Parties should cooperate to promote a supportive and open international economic system that would lead to sustainable economic growth and development in all Parties, particularly developing country Parties, thus enabling them better to address the problems of climate change. Measures taken to combat climate change, including unilateral ones, should not constitute a means of arbitrary or unjustifiable discrimination or a disguised restriction on international trade.

ARTICLE 4

COMMITMENTS

1. All Parties, taking into account their common but differentiated responsibilities and their specific national and regional development priorities, objectives and circumstances, shall:

(a) Develop, periodically update, publish and make available to the Conference of the Parties, in accordance with Article 12, national inventories of anthropogenic emissions by sources and removals by sinks of all greenhouse gases not controlled by the Montreal Protocol, using comparable methodologies to be agreed upon by the Conference of the Parties;

(b) Formulate, implement, publish and regularly update national and, where appropriate, regional programmes containing measures to mitigate climate change by addressing anthropogenic emissions by sources and removals by sinks of all greenhouse gases not controlled by the Montreal Protocol, and measures to facilitate adequate adaptation to climate change;

(c) Promote and cooperate in the development, application and diffusion, including transfer, of technologies, practices and processes that control, reduce or prevent anthropogenic emissions of greenhouse gases not controlled by the Montreal Protocol in all relevant sectors, including the energy, transport, industry, agriculture, forestry and waste management sectors;

(d) Promote sustainable management, and promote and cooperate in the conservation and enhancement, as appropriate, of sinks and reservoirs of all Greenhouse gases not controlled by the Montreal Protocol, including biomass, forests and oceans as well as other terrestrial, coastal and marine Ecosystems;

(e) Cooperate in preparing for adaptation to the impacts of climate change; develop and elaborate appropriate and integrated plans for coastal zone management, water resources and agriculture, and for the protection and rehabilitation of areas, particularly in Africa, affected by drought and desertification, as well as floods;

(f) Take climate change considerations into account, to the extent feasible, in their relevant social, economic and environmental policies and actions, and employ appropriate methods, for example impact assessments, formulated and determined nationally, with a view to minimizing adverse effects on the economy, on public health and on the quality of the environment, of projects or measures undertaken by them to mitigate or adapt to climate change;

(g) Promote and cooperate in scientific, technological, technical, socio-economic and other research, systematic observation and development of data archives related to the climate system and intended to further the understanding and to reduce or eliminate the remaining uncertainties regarding the causes, effects, magnitude and timing of climate change and the economic and social consequences of various response strategies;

(h) Promote and cooperate in the full, open and prompt exchange of relevant scientific, technological, technical, socio-economic and legal information related to the climate system and climate change, and to the economic and social consequences of various response strategies;

(i) Promote and cooperate in education, training and public awareness related to climate change and encourage the widest participation in this process, including that of non-governmental organizations; and

(j) Communicate to the Conference of the Parties information related to implementation, in accordance with Article 12.

United Nations

UNITED NATIONS. *EPA GLOBAL WARMING PUBLICATIONS.* "FRAMEWORK CONVENTION ON CLIMATE CHANGE." HTTP://YOSEMITE.EPA.GOV/OAR/GLOBALWARMING.NSF/CONTENT/RESOURCECENTERPUBLICATIONSREFERENCEFRAMEWORKCONVENTIONONCLIMATECHANGECONVENTION.HTML (ACCESSED APRIL 10, 2008).

SEE ALSO *Climate Change; Global Warming; Intergovernmental Panel on Climate Change; IPCC 2007 Report; Kyoto Protocol*

BIBLIOGRAPHY

Periodicals

Forneri, Claudio, et al. "Keeping the Forest for the Climate's Sake: Avoiding Deforestation in Developing Countries Under the UNFCCC." *Climate Policy* 6 (2006): 275–294.

Web Sites

NewScientist. "Leading Nations Find Agreement on Climate Change." http://environment.newscientist.com/channel/earth/dn11199-leading-nations-find-agreement-on-climate-change.html (accessed April 4, 2008).

United Nations Framework Convention on Climate Change. http://unfccc.int/2860.php (accessed April 4, 2008).

United Nations Intergovernmental Panel on Climate Change. "IPCC Reports." http://www.ipcc.ch/ipccreports/assessments-reports.htm (accessed March 29, 2008).

James Corbett

United Nations Policy and Activism

Introduction

The United Nations (UN) is an international organization that works toward promoting democracy, preserving world peace, and tackling pressing socioeconomic issues globally. Representatives from 51 countries officially established the UN on October 24, 1945. In 1972, the first ever UN Conference on the Human Environment was held in Stockholm, Sweden. The issues focused on during the conference included depletion of the ozone layer and global warming. What resulted from the conference was a declaration that contained 26 principles. One of the key recommendations during the meeting was to set up a global environmental organization. This led to the creation of the United Nations Environment Programme (UNEP) by the UN General Assembly in the same year. In the twenty-first century, UNEP is at the forefront of environmental activism, policy making, and recognition of commendable environmental initiatives. UNEP has its footprints across the world through its offices and centers. It is headquartered in Nairobi, Kenya.

Historical Background and Scientific Foundation

When UNEP was established, climate change was yet to receive the worldwide attention it receives in the twenty-first century. In 1980, the UNEP Governing Council brought forth the issue of the depleting ozone layer and suggested measures to combat it. In 1985, the Vienna Convention for the Protection of the Ozone Layer focused on the impact of human activities on the ozone layer. A multilateral agreement was created during the convention. The convention laid out the responsibilities of states in protecting human health and the environment from the consequences of ozone depletion. This convention was thrown open for signatures in 1985 and was formally adopted in 1988.

In 1987, the Montreal Protocol on Substances That Deplete the Ozone Layer was created. This treaty was intended to phase out the use of substances that could severely impact the ozone layer. The treaty was formally adopted in 1989. In 1987, the UN General Assembly also took up a framework—Environmental Perspective to the Year 2000 and Beyond—to direct the policies and programs to attain sustainable development.

In 1988, a forum called the Intergovernmental Panel on Climate Change (IPCC) was established. It was the outcome of the joint effort of the World Meteorological Organization (WMO) and the United Nations Environment Programme (UNEP). It serves as an information repository for anyone interested in learning about climate change. It has a global network of authors, scientists, and decision makers who collaborate to sensitize the world toward climate change. In 1989, the Helsinki Declaration on the Protection of the Ozone Layer encouraged states to adopt the phasing out of chlorofluorocarbons (CFCs), as outlined in the earlier Montreal Protocol.

At the second World Climate Conference (October 29–November 7, 1990), it was realized that climate change is a global problem and requires global response. Soon the importance of a more potent international action on the environment was realized. The United Nations Conference on Environment and Development, also known as The Earth Summit, was held in 1992 in Rio de Janeiro, Brazil. This summit came up with a new framework for protecting global environments—the Rio Declaration and Agenda 21. This conference created the United Nations Framework Convention on Climate Change (UNFCCC), a major step in dealing with global warming. The most prominent climate change action taken so far is the Kyoto Protocol on December 11, 1997, in Japan. Its target is to decrease the emanation of carbon dioxide and other greenhouse gases, by the industrialized countries, by at least 5% below the 1990 levels in the period from 2008 to 2012. However, it was in 2005 that the Kyoto Protocol on climate change came into force.

The year 2000 saw the Malmö Declaration come into force. Over 100 environment ministers from across the world brainstormed to come up with this declaration. The issues discussed include the environmental challenges in the twenty-first century, the role of the industrial sector, and the responsibility of citizens in general. The following year saw the Stockholm Convention on Persistent Organic Pollutants (POPs). Also, in 2005, UNEP adopted the Bali Strategic Plan for Technology Support and Capacity Building. The Millennium Ecosystem Assessment was held in the same year, which focused on the degradation of the environmental ecosystems globally.

In December 2007, the United Nations Climate Change Conference was hosted by the government of Indonesia in Bali. During this conference, participants agreed on setting up an Adaptation Fund to help high-risk developing countries alleviate the threat of climate change. This fund is reported to be worth up to $500 million annually by 2012.

Impacts and Issues

The IPCC predicts that if the emission of harmful gases is not checked, global temperature will rise from 3.2°F to 7.1°F (1.8°C to 4°C) throught the world by the end of the twenty-first century, thus impacting the environment, economy, and society. Similarly, sea levels could rise by as much as 7 to 19.6 in (18 to 59 cm) by 2100. Besides being an environmental challenge, climate change is also an economic and political challenge.

UNEP has led several initiatives in spreading awareness about the impact of climate change and in taking corrective measures. Failing to control climate change would result in food scarcity, natural disasters, and the threat of epidemics.

In order to cater to the information needs about climate change, the Division of Environmental Law and Conventions (DELC) of UNEP has created an outreach program. The aim of this outreach program is to support governments, NGOs, and the general public in understanding the issue better. The program reaches out to people through visual aids, brochures, radio programs, and television.

However, controversy has dogged the assessment of climate change impact by the IPCC. In its third report, the IPCC had created a graph of the Northern Hemisphere temperature. The graph, shaped like a hockey stick, came to be known as the Hockey Stick representation. However, this method was widely disputed by scientists. The primary argument against this method was that it did not consider documented evidence of the Medieval Warm Period and the Little Ice Age. This was followed by a reexamination of the statistical methods employed in creating the Hockey Stick representation. As a result, the IPCC used more than one technique in creating its temperature graphs for its fourth Assessment Report.

Another criticism of the climate change movement has been that human industrialization is not the only factor leading to global warming. It was argued that other planets such as Neptune and Mars have also experienced global warming even without the presence of any proven life forms.

There was also a controversy surrounding the IPCC Working Group I report in 1990. Some of the participating scientists complained about being pressured to highlight the results that were supportive of the prevailing global warming theory and to downplay results that suggested otherwise. Many of the participants disagreed with the contents of the final report, and the UN's claims of having a consensus among scientists were disputed.

WORDS TO KNOW

CHLOROFLUOROCARBONS (CFCS): A family of chemical compounds consisting of carbon, fluorine, and chlorine that were once used widely as propellants in commercial sprays but regulated in the United States since 1987 because of their harmful environmental effects.

CLIMATE NEUTRAL: The process of reducing greenhouse emissions so as to create a neutral impact on climate change.

OZONE LAYER: The layer of ozone that begins approximately 15 km above Earth and thins to an almost negligible amount at about 50 km, and which shields Earth from harmful ultraviolet radiation from the sun. The highest natural concentration of ozone (approximately 10 parts per million by volume) occurs in the stratosphere at approximately 25 km above Earth. The stratospheric ozone concentration changes throughout the year as stratospheric circulation changes with the seasons. Natural events such as volcano eruptions and solar flares can produce changes in ozone concentration, but man-made changes are of the greatest concern.

SUSTAINABLE DEVELOPMENT: Development (i.e., increased or intensified economic activity; sometimes used as a synonym for industrialization) that meets the cultural and physical needs of the present generation of persons without damaging the ability of future generations to meet their own needs.

Even before the IPCC had released its three-volume report on climate in 2007, there was a debate regarding the relevance of the findings. Some of the scientists working with the IPCC reportedly criticized the gap of five to six years between the IPCC's climate change review. Additionally, the report did not consider several important phenomena such as the melting of glaciers in Greenland, melting Arctic ice cap, and an increase in the carbon dioxide levels on Earth.

See Also *United Nations Conference on the Human Environment (1972); United Nations Framework Convention on Climate Change (UNFCCC)*

BIBLIOGRAPHY

Web Sites

allAfrica.com. "South Africa: Bali Break Through in Climate Change Talks" http://allafrica.com/stories/200712110610.html (accessed April 1, 2008).

Our Planet. "Spirit of Optimism" http://www.unep.org/OurPlanet/imgversn/102/trittin.html (accessed March 12, 2008).

United Nations Environment Programme. "About DELC." http://www.unep.org/dec/About/index.asp (accessed April 1, 2008).

United Nations Environment Programme. "About UNEP: The Organization." http://www.unep.org/Documents.Multilingual/Default.asp?DocumentID=43 (accessed March 12, 2008).

United Nations Environment Programme. "Identifying and Responding to Emerging Issues." http://www.roap.unep.org/region/emerge.cfm (accessed March 12, 2008).

United Nations Publication. "The United Nations: An Introduction for Students." http://www.un.org/cyberschoolbus/unintro/unintro.asp (accessed March 12, 2008).

U.S. Senate Committee on Environment & Public Works. "U.S. Senate Report: Over 400 Prominent Scientists Disputed Man-Made Global Warming Claims in 2007." http://epw.senate.gov/public/index.cfm?FuseAction=Minority.Blogs&ContentRecord_id=f80a6386-802a-23ad-40c8-3c63dc2d02cb (accessed April 1, 2008).

Amit Gupta

United Nations World Commission on Environment and Development (WCED) *Our Common Future* Report (1987)

Introduction

The World Commission on Environment and Development (WCED), also called the Brundtland Commission after its Chairman Gro Harlem Brundtland (a Norwegian physician), was an international commission that discussed and devised strategies for protecting the environment and promoting sustainable development. The Brundtland Commission published its final report, *Our Common Future,* in 1987. *Our Common Future* stated that governments could not address environmental protection separately from related crises, such as economic development and energy production. *Our Common Future* also outlined a blueprint for dealing with these interlocking crises simultaneously. The findings and proposals of *Our Common Future* have shaped international environmental policy for the last two decades.

Historical Background and Scientific Foundations

The World Commission on Environment and Development was not the first United Nations conference to address environmental issues. In 1972, the United Nations convened the United Nations Conference on the Human Environment in Stockholm, Sweden. Representatives from 113 nations and over 400 non-governmental agencies (NGOs) attended the Stockholm Conference. This conference, often called the Stockholm Conference, was the first international conference to address environmental problems directly.

The Stockholm Conference produced the Declaration of the Conference on the Human Environment, which stated that every person deserves a clean, healthy environment. The declaration recognized that unchecked technological and scientific advancements permitted humans to "transform the environment in countless ways and on an unprecedented scale." The Stockholm Conference declaration also stated that environmental protection is one of the major humanitarian and economic issues facing the world.

The Declaration of the Conference on the Human Environment was the first major international document to recognize that environmental problems originate from both developing and developed economies. The World Commission on Environment and Development and every subsequent United Nations conference on the environment have sought to address these seemingly contradictory sources of environmental degradation. The Stockholm Conference Declaration noted that most environmental problems in developing economies occur because of underdevelopment. Poverty in these nations leads to poor health, poor sanitation, and toxic cleanup, which place harmful human, animal, and chemical products into the environment. Governments with developing economies also often seek advancement of the economy with little regard for environmental regulation. Industrialized nations contribute to environmental problems through technological advancements and industrialization.

The Stockholm Conference also produced an action plan, which contained 109 specific recommendations for improving the environment, including limiting the use of ozone-depleting chlorofluorocarbons (CFCs). The action plan also called for a reduction of marine pollution. Finally, the action plan called on industrialized nations to provide economic and technological assistance to developing nations so the developing nations could grow their economies in an environmental responsible manner.

In 1983, the United Nations General Assembly established the United Nations World Commission on Environment and Development (WCED), also called the Brundtland Commission, with the passage of Resolution

WORDS TO KNOW

DEVELOPING NATION: A country that is relatively poor, with a low level of industrialization and relatively high rates of illiteracy and poverty.

NON-GOVERNMENTAL ORGANIZATION (NGO): A voluntary organization that is not part of any government; often organized to address a specific issue or perform a humanitarian function.

SUSTAINABILITY: Practices that preserve the balance between human needs and the environment, as well as between current and future human requirements.

38/161 (*Process of Preparation of the Environmental Perspective to the Year 2000 and Beyond*). The resolution called for the creation of a commission to propose international environmental strategies that would lead to sustainable development for the year 2000 and beyond. The Brundtland Commission first met in 1984.

Under the requirements of Resolution 38/161, the Brundtland Commission served as an independent body outside control of the United Nations and national governments. The Brundtland Commission addressed three major environmental issues, as mandated by Resolution 38/161. First, the commission examined critical environmental and developmental issues and formulated proposals for dealing with these issues. Second, the commission proposed new forms of international cooperation on these issues. Finally, the commission addressed ways to raise awareness of environmental issues and commitments to address those issues from individuals, NGOs, governments, and intergovernmental agencies.

In 1987, after three years of information gathering and debate, the Brundtland Commission issued *Our Common Future*, a report containing the commission's findings and recommendations. *Our Common Future* asserted that any international environmental initiative must address sustainable development. *Our Common Future* defined sustainable development as "development that meets the needs of the present without compromising the ability of future generations to meet their own needs." Every United Nations environmental convention since the release of *Our Common Future*, including Earth Summit 1992 and Earth Summit 2002, has embraced the idea of sustainable development as a vital aspect of environmental policy.

Our Common Future asserted that sustainable development was the only solution to the interlocking crises occurring in environmental preservation, economic development, and energy production. The report asserted that any attempt to address problems occurring in only one of these areas would further exacerbate problems in the other two areas. *Our Common Future* stressed that governments cannot manage these interlocking crises on a local or national basis. The report stated that the only effective remedy is an international approach that simultaneously addresses all three crises.

The report stated that large-scale farming in industrialized nations harmed the environment through increased pesticide use, destruction of ecosystems through clearing land, and overuse of the soil. Industrialized nations often donate this food surplus to developing nations. This practice creates insecurity in the food supply by inhibiting the development of agriculture in the developing nations. *Our Common Future* argued that a better practice would be technological assistance to developing nations that would allow those nations to develop their own environmentally responsible food supplies.

Our Common Future addressed other sustainable development and environmental preservation methods. Nations could reduce energy consumption by using technology to create more energy-efficient appliances, automobiles, and machines. Cleaner forms of energy production, such as wind and solar energy, would also reduce the pollution and greenhouse gases emitted by fossil fuels. The report also suggested that developing nations design and implement urban planning initiatives to deal with their growing populations. Rapid urban population growth will place an enormous strain on the environment and the economies of developing nations in the twenty-first century.

Issues and Impacts

Our Common Future has served as the blueprint for international action on environmental issues. Every major international convention on the environment since 1987, including the United Nations Conference on Environment and Development (Earth Summit 1992) and the World Summit on Sustainable Development (Earth Summit 2002), has drawn heavily from the principles espoused in *Our Common Future*.

Earth Summit 1992 produced the Rio Declaration on Environment and Development and Agenda 21. The Rio Declaration on Environment and Development states the rights and responsibilities of nations toward environmental protection and sustainable development. The Rio Declaration states that nations have the right to exploit natural resources within their borders if their actions do not affect the environment in other nations. Furthermore, the Rio Declaration calls on all governments to develop plans to preserve the environment and natural resources for future generations.

Agenda 21, like *Our Common Future*, addresses environmental issues through detailed social and economic proposals. Agenda 21 proposes addressing environmental issues through poverty reduction, conservation of natural resources, deforestation prevention, promotion of sustainable agriculture, modification of production and consumption patterns, and protection of the atmosphere and oceans.

Earth Summit 2002 produced the Johannesburg Declaration, which reiterated many of the points contained in the Rio Declaration and Agenda 21. The Johannesburg Declaration contains a more general statement about the environment and sustainable development, and it calls for an end to all conditions that threaten sustainable development, including drug use, corruption, terrorism, ethnic intolerance, and natural disasters. The Johannesburg Declaration, however, does not contain specific proposals for addressing many of these issues.

SEE ALSO *Earth Summit (1992); Earth Summit (2002); Sustainable Development*

BIBLIOGRAPHY

Web Sites

Organization for Economic Co-operation and Development. "Sustainable Development: Our Common Future." http://Hww.oecdobserver.org/news/fullstory.php/aid/780/Sustainable_development:_Our_common_future.html (accessed April 27, 2008).

United Nations. "Documents: Agenda 21." December 15, 2004. http://www.un.org/esa/sustdev/documents/agenda21/index.htm (accessed April 21, 2008).

United Nations. "Our Common Future: Report of the World Commission on Environment and Development." http://www.un-documents.net/ocf-ov.htm#I (accessed April 27, 2008).

United Nations. "Report of the World Summit on Sustainable Development." hhtp://www.un.org/jsummit/html/documents/summit_docs/131302_wssd_report_reissued.pdf (accessed April 27, 2008).

United Nations. "Res. 38/161: Process of Preparation of the Environmental Perspective to the Year 2000 and Beyond." http://www.un.org/documents/ga/res/38/a38r161.htm (accessed April 27, 2008).

United Nations. "U.N. Conference on Environment and Development (1992)." May 23, 1997. http://www.un.org/geninfo/bp/enviro.html (accessed April 21, 2008).

United Nations Environment Programme. "Stockholm 1972: Report of the United Nations Conference on the Human Environment." http://www.unep.org/Documents.Multilingual/Default.asp?documentID=97 (accessed April 21, 2008).

Joseph P. Hyder

Vegetation Cycles

Introduction

All of life on Earth depends on plants, whether they are tall trees or tiny mosses on the forest floor. The plant life of an ecosystem is known as its vegetation. It can be seen from space as the green cover of an area. Vegetation plays many roles in the biosphere. It is a primary source of food and it controls the characteristics of soils. Vegetation is also part of the great biogeochemical cycles that provide water, carbon, and nitrogen to living things.

Vegetation varies in its composition and structure. Its nature characterizes a terrestrial biome, whether forest or grassland. The study of vegetation was originally concerned with identifying the species that made it up. Modern ecology is more concerned with the functions and interactions of vegetation with other environmental factors. A better understanding of vegetation cycles may help combat climate change and preserve biodiversity.

Historical Background and Scientific Foundations

Few regions of the world, even deserts, are completely free of vegetation. The term refers to the plant cover of a biome and includes trees, shrubs, wild flowers, and grasses. It can vary from dense tropical rain forest to wheat fields or patches of weeds in a garden. Vegetation varies over an area and over time. Vegetation has a three-dimensional structure with height, width, and length. Natural vegetation is not usually uniform. For instance, a forest consists of tall trees, shorter trees, flowering plants, and shrubs, maybe interspersed with clearings. Commercial tree plantations, however, may be uniformly spaced giving an even density of vegetation cover.

Vegetation also varies in a temporal way. There are the natural germination, growth, and death cycles, whose length varies with the nature of the plant. For areas with many deciduous species, there are dramatic changes in vegetation cover with the seasons. Human activities, such as deforestation and urban development, can also cause changes in vegetation over time. It is rare for vegetation to disappear completely. Changes in land use tend to lead to succession, which is the growth of a plant community that probably differs from the original one.

Vegetation plays an important role in the three main biogeochemical cycles, which are the hydrological or water cycle, the nitrogen cycle and the carbon cycle. The fact that vegetation itself has temporal cycles means that these interactions can be complex. They have an impact upon local and global climate and on soil characteristics, all of which have a feedback effect on the structure and productivity of the vegetation.

Vegetation takes part in the hydrologic or water cycle. It contributes about 10% of the water vapor in the atmosphere through transpiration, which is evaporation through leaves. The rest of the water vapor comes from evaporation from the seas and oceans. The water vapor forms clouds, then falls as precipitation onto the land, then runs through into bodies of water, eventually ending up back in the ocean.

Nitrogen fixation is the ability of the roots of leguminous plants, such as soybean or clover, to convert atmospheric elemental nitrogen into nitrate. These plants have nodules in their roots, which house nitrogen-fixing bacteria that carry out this conversion. The plant then converts nitrate into proteins and nucleic acids. These more complex nitrogen-containing molecules are supplied to animals when they eat plant foods. Eventually, these organisms die and decay, returning their nitrogen-containing compounds to the soil, where they are processed into nitrates by nitrifying bacteria. The final step in the nitrogen cycle is conversion of nitrate into nitrogen by denitrifying bacteria.

Finally, vegetation also plays a key role in the carbon cycle through sequestration, which is the absorption of carbon dioxide from the atmosphere through photosynthesis. Fixing carbon dioxide through photosynthesis makes glucose, which provides biochemical fuel for animals eating plants and for plants themselves. Carbon dioxide is thus released back into the atmosphere quite quickly, when animals respire. However, some of the carbon sequestered in plants becomes buried when plants die and decay, forming fossil fuels, thereby removing carbon dioxide from the atmosphere for a long period of time.

Impacts and Issues

Vegetation is a dynamic system whose interaction with biogeochemical cycles is complex and changing all the time. Activities such as deforestation need to be understood in terms of how they will affect climate through these interactions. Loss of transpiration may lead to decreased precipitation and higher risk of drought. Reforestation, however, may provide a useful redress against global warming through carbon sequestration. Vegetation also protects against soil erosion and provides habitats for many species. A better understanding of the type and structure of vegetation which provides maximum benefit to the environment both globally and locally is therefore highly desirable.

WORDS TO KNOW

BIOGEOCHEMICAL CYCLE: The chemical interactions that take place among the atmosphere, biosphere, hydrosphere, and geosphere.

BIOME: A well-defined terrestrial environment (e.g., desert, tundra, or tropical forest) and the complex of living organisms found in that region.

DECIDUOUS: Plants that shed leaves or other foliage after their growing season.

NITROGEN CYCLE: Biochemical cycling of nitrogen by plants, animals, and soil bacteria.

NITROGEN FIXATION: Conversion of atmospheric nitrogen into nitrate by the roots of leguminous plants.

Bemidji State University (Minnesota) environmental science professor Dr. Drago Bilanovic holds a glass bottle containing xanthan, a product extracted from fermented potato waste, January 2006. Bilanovic and other BSU researchers are looking for ways to convert potato waste into valuable and environmentally friendly materials. *AP Images*

SEE ALSO *Carbon Sequestration; Deforestation; Forests; Soil Resources*

BIBLIOGRAPHY

Books

Cunningham, W.P., and A. Cunningham. *Environmental Science: A Global Concern*. New York: McGraw-Hill International Edition, 2008.

Web Sites

NASA. "Image of the Day: Global Vegetation." April 7, 2008. http://www.nasa.gov/multimedia/imagegallery/image_feature_1056.html (accessed May 2, 2008).

Volcanoes

Introduction

A volcano is a vent in Earth's crust through which molten rock, gas, or ash flow onto the surface or are injected into the atmosphere. Volcanoes vary in size and violence, with some flowing steadily for centuries and others, termed supervolcanoes, capable of exploding with a force equal to that of thousands of nuclear weapons.

For billions of years, volcanism has reshaped the surface of Earth and several other bodies in the solar system. Eruptions have destructive short-term environmental and human effects, and can change global climate for years, but their long-term effects are not destructive. The history of life on Earth has been altered by periods of intense volcanism that contributed to mass extinctions.

Historical Background and Scientific Foundations

The molten rock or magma that is expelled by most volcanoes is either forced up through cracks from the mantle (the layer of Earth's structure just below the crust) or, working its way upward through the denser, solid rocks around it, accumulates in pockets within the crust. An underground pocket of magma is termed a magma chamber. When a magma chamber becomes pressurized enough to force some of its contents to the surface, a volcanic eruption occurs. Often, volcanologists can predict a volcanic eruption by measuring the swelling of Earth's surface above a growing magma chamber. For example, the center of Yellowstone National Park, whose central region resides inside the caldera (crater) of a gigantic volcano 40 mi (70 km) across, has been rising at 3 in (7.5 cm) per year since 2004 as a large magma chamber beneath it swells.

The molten rock expelled by a volcanic eruption—called lava as soon as it emerges—begins as rock in Earth's mantle, about 40 to 120 mi (70 to 200 km) underground. For material to travel through the crust to the surface and emerge from a volcano, it must be mobile—namely, gas, liquid, or a mixture of both. The heat that melts the rock comes from the decay of radioactive elements over geological time. There is not enough radioactive material inside Earth to make lava significantly more radioactive than surface rock, but inside the planet, heat from radioactivity has nowhere to go, and so accumulates.

There are several kinds of volcano. The most common type is the subduction-zone volcano. A subduction zone is a linear (long, narrow) region where one large, semi-rigid piece of Earth's rocky crust—termed a plate—is being pushed against another so that its leading edge is forced down into the mantle. At a typical subduction zone, crust forming an ocean floor is forced down into the mantle at an inch or two per year. Some water is carried down with the subducted rock.

Mixing rock with even a small amount of water lowers its melting point significantly, so some of the descending crust turns into magma at a depth where the surrounding, hot, dry rocks of the upper mantle are still solid. These pockets or droplets of liquid rock also contain gases such as water vapor and carbon dioxide. Being less dense than the surrounding solid rock, they migrate upward through the crust. These ascending masses of magma are termed plutonic diapirs. Most do not reach the surface, but cool and solidify while still deep within the crust, forming large masses of igneous rock. If a diapir gets close enough to the surface, however, it may form a magma chamber and be joined by other diapirs, becoming pressurized. This pressure may build until a narrow channel to the surface is forced open or the overlying rock is blown away in a massive explosion. Volcanic eruptions of magmas containing water tend to be violent: Without steam (hot water vapor) or large amounts of some other gas, magma does not explode, but flows.

Because subduction zones more or less surround the edges of the Pacific Ocean, volcanoes are common along much of the ocean's rim. This area is often termed the Ring of Fire.

Another kind of volcano is the shield volcano. This type includes the largest volcanoes on Earth and in the solar system: Earth's largest volcano, Mauna Loa in Hawaii, is a shield volcano, as is Olympus Mons on Mars. A shield volcano is formed by hotspot volcanism, a "hotspot" being a place in the mantle where an upwelling of liquid rock persists for geologic time (millions of years). A runny form of lava is usually produced by these volcanoes, causing the resulting mountains to have a wide, round, spread-out shape reminiscent of an ancient hand-carried shield.

As an oceanic or continental plate moves slowly above a hotspot, a series of volcanoes may be formed above it as the hotspot burns a series of holes in the moving plate. A live volcano will usually be found right above the hotspot and inactive (extinct) volcanoes trailing away from it in the direction of the plate's motion. The Hawaiian Island chain has been formed by this process. On Mars, which does not have moving plates, the now-extinct Olympus Mons volcano sat motionless upon its hot spot for millions of years and grew to be the largest mountain in the solar system. Shield volcanoes, because their lava is so runny, tend not to become plugged up and build gas pressure, so they erupt steadily and calmly, rather than explosively. However, the Yellowstone supervolcano is a hotspot volcano that erupts explosively due to the presence of steam in its magma. Yellowstone explodes every 20,000 years or so but has undergone three especially large explosions, the most recent of which, about 640,000 years ago, spread ash over most of North America.

■ Impacts and Issues

Violent venting or outright explosion pulverizes rock around the volcanic vent, turning it to a powder termed ash. This ash, lofted into the atmosphere, can settle around a volcano and be carried far downwind from it, smothering animals and vegetation. Volcanic explosions can cause great damage by direct force: in the vicinity of Mt. St. Helens, a volcano in the state of Washington that erupted in 1980, the initial explosion killed 62 people and destroyed about 150 square mi (390 square km) of surrounding forests, a loss to the timber industry of about $1 billion.

Volcanoes also emit gases, including carbon dioxide, water vapor, sulfur dioxide, and hydrogen sulfide. Over billions of years, these emissions have helped form Earth's atmosphere. Hot gases mixed with pulverized rock can sweep down the slopes of a volcano at many miles an hour, often traveling far from the caldera: This is called a pyroclastic flow, and can be one of the most dangerous volcanic events. On the Caribbean island of Martinique, a pyroclastic flow from the volcano Mt. Pelee completely wiped out the town of Saint-Pierre in 1902, killing about 30,000 people. About 640,000 years ago, a pyroclastic eruption of the Yellowstone volcano blanketed the surrounding landscape with 240 cubic mi (1,000 cubic km) of ash. Another hazard from volcanic eruptions is the sudden melting of snow and ice or the mixing of heavy rain with ash, which can produce floods or lahars (hot mudflows).

Over the middle term—several years, as opposed to the hours or days of the eruption itself—individual volcanoes can affect the climate of the entire world. The eruption of Mt. Pinatubo in the Philippines in 1991 ejected about 20 million tons of sulfur dioxide aerosol particles into the upper atmosphere. These bright particles reflect sunlight into space and infrared light back toward Earth, tending to make summers cooler and winters warmer. In the Northern Hemisphere, summer was about 3.6°F (2°C) cooler the year after the Pinatubo eruption and winter was about 5.4°F (3°C) warmer. The main effect of bright aerosols such as sulfur dioxide, as predicted by theory and observed following the Pinatubo eruption, is to cool climate. After a few years, the climatic effect of such an eruption fades away.

Particularly massive and prolonged volcanic eruptions can alter the history of life, wiping out thousands of species. In the largest mass extinction in Earth's history, the Permian-Triassic extinction event about 251 million years ago, about 70% of all land-dwelling plants and animals and 96% of all sea-dwelling creatures died. Geologists state that the extinction coincided and may have been caused by the most massive volcanic eruption in Earth's history, the Siberian Traps, which covered tens of thousands of square miles with lava and darkened the skies of the world with ash. Several other mass extinctions have coincided with large volcanic eruptions, as well as with strikes by asteroids. The extinction of the dinosaurs was probably caused by an asteroid impact, not volcanism.

WORDS TO KNOW

CALDERA: Volcanic crater that has collapsed to form a circular depression greater than 1 mi (1.6 km) in diameter.

MANTLE: The thick, dense layer of rock that underlies Earth's crust and overlies the core.

IN CONTEXT: ACTIVE AFRICA

Given the intense activity and frequent eruptions related to the Pacific rim "Ring of Fire," most people do not think of Africa as volcanically active. Yet much of the African landscape and rich mineral deposits are related to volcanic and tectonic activity. Both Kilimanjaro and Africa's second highest peak, Mount Kenya (17,058 ft; 5,117 m) sitting astride the equator, are actually composite volcanos, part of the vast volcanic field associated with the East African rift valley.

The most distinctive and dramatic geological feature in Africa is undoubtedly the East African rift system. The rift opened up in the Tertiary period, approximately 65 million years ago, shortly after the dinosaurs became extinct. Seismically, the rift valley is very much alive. Lava flows and volcanic eruptions occur about once a decade in the Virunga Mountains north of Lake Kivu along the western stretch of the rift valley. One volcano in the Virunga area in eastern Zaire that borders Rwanda and Uganda actually dammed a portion of the valley formerly drained by a tributary of the Nile River, forming Lake Kivu as a result.

Despite volcanoes' destructive effects, the long-term environmental results of volcanic eruptions are usually beneficial. Forests quickly re-colonize ash-covered ground and lava flows, especially in tropical regions. Lava and ash are rich in minerals, and when weathered produce rich soils that are excellent for farming.

The Pinatubo eruption of 1991 shot so much sulfurous debris into the stratosphere that it is believed it cooled Earth by 0.9 °F (–17°C) for about a year. *AP Images*

SEE ALSO *Climate Change; Earthquakes; Extinction and Extirpation; Tsunami Impacts*

BIBLIOGRAPHY

Books

de Boer, Jelle Zeilinga *Volcanoes in Human History: The Far-Reaching Effects of Major Eruptions.* Princeton, NJ: Princeton University Press, 2004.

Schmincke, Hans-Ulrich. *Volcanism.* New York: Springer, 2004.

Solomon, Susan, et al, eds. *Climate Change 2007: Physical Science Basis. Contribution of Working Group I to the Fourth Assessment Report of the Intergovernmental Panel on Climate Change.* New York: Cambridge University Press, 2007.

Periodicals

Kerr, Richard A. "Did Volcanoes Drive Ancient Extinctions?" *Science* 289 (2000): 1130–1131.

McEwen, Alfred S. "Active Volcanism on Io." *Science* 297 (2002): 2220—2221.

Robock, Alan. "Pinatubo Eruption: The Climatic Aftermath" *Science* 295 (2002): 1242–1244.

Web Sites

U.S. Geological Survey. "Volcano Hazards Program." http://volcanoes.usgs.gov/ (accessed May 12, 2008).

Larry Gilman

Walden

■ Introduction

Walden is a book by Henry David Thoreau (1817–1862) that was first published in 1854. It describes the two years (telescoped, in the book, into one) that Thoreau spent living in a one-room house that he built by Walden Pond in Concord, Massachusetts. The book sold poorly during Thoreau's lifetime and was little read for decades after his death. In the early twentieth century, however, its fame began to grow. There was a particularly strong *Walden* boom in the 1960s, when the book was read by many young people seeking wisdom from outside the cultural mainstream.

Walden's fame has continued to grow: There were six editions in print in 1948, eleven in 1958, 23 in 1968, and over 100 in 2008. Unlike the writings of Thoreau's friend Ralph Waldo Emerson (1803–1882), who owned the land on which Thoreau built his cabin and was a much more successful author when the two men were alive, *Walden* is now read every year by millions of people seeking their own encounter with Thoreau's challenging thoughts about how to live. Thoreau was not a nature writer in the sense of writing long, inspiring descriptions of natural scenes; rather, he wrestled with how to live rightly as a free individual in a social group and as a human animal in a wider environment. These questions have become even more pressing, if possible, since *Walden*'s appearance.

■ Historical Background and Scientific Foundations

Thoreau was born in Concord, Massachusetts, in 1817, lived all his life in that small town, and died there in 1862. He was educated at Harvard but earned his living in Concord by land-surveying, making pencils, and hiring out for odd jobs. As a youth he was greatly influenced by the writings of his fellow townsman Ralph Waldo Emerson, the most famous proponent of the philosophy called Transcendentalism. Transcendentalism taught that people can transcend the dreary round of daily routine and make ecstatic contact with the universe—what Emerson sometimes called the "Oversoul"—through engagement with nature. In some ways, Transcendentalism was a precursor of the New Age movement that arose in the late twentieth century, but Thoreau, at least, was an intensely practical man not given to magical thinking. He could build, garden, survey, hike, or paddle for many miles through rugged wilderness, and he was an expert naturalist, recording detailed observations of the fauna and flora of New England. He preached against slavery, helping to move escaped slaves north along the Underground Railroad, and was jailed briefly for refusing to pay a poll tax (tax levied on all adults, regardless of income or land ownership) to support a war with Mexico.

Although Thoreau lived most of his life with his parents and siblings (he never married), from July 4, 1845 to September 6, 1847, he lived in a small house that he built on the shore of Walden Pond. He described his motive for doing so in *Walden*: "I went to the woods because I wished to live deliberately, to front only the essential facts of life, and see if I could not learn what it had to teach, and not, when I came to die, discover that I had not lived."

The pond was only semi-wild at the time; a railroad had recently been built a few yards from its southern tip and the pond and its surroundings were subject to logging, ice-harvesting, fishing, and other commercial uses. Thoreau grew beans in a field, sallied forth for long exploratory walks and boating trips, worked on a book about earlier experiences, and journaled about his daily life. He would later re-shape and supplement his journal writings to produce *Walden: Or, Life in the Woods.* While living at Walden he was jailed for refusing to pay a tax

WORDS TO KNOW

ENVIRONMENTAL MOVEMENT: A diverse social, political, and scientific movement revolving around the preservation of Earth's environment.

NATURALIST: One who studies or is an expert in natural history, especially in zoology or botany.

TRANSCENDENTALISM: A literary and philosophical movement asserting the existence of an ideal spiritual reality that transcends the empirical and scientific and is knowable through intuition.

supporting the U.S. war with Mexico (1846–1848) and wrote, in response, the essay "On Civil Disobedience" (1849), which was later to influence Indian spiritual leader Mohandas Gandhi (1869–1948) and American civil rights leader Martin Luther King Jr. (1929–1968), among many others.

Impacts and Issues

Although millions of readers have loved *Walden*, some have found its style irritating. It is written as a deliberate challenge to the reader, using paradox, exaggeration, and self-contradiction to provoke thought. Thoreau says in the book that he wants his readers to live authentically, not mimic his own actions. The book has also annoyed some readers because it criticizes standard patterns of behavior that are often admired—hard work at a high-paying job, the acquisition of a large house and many possessions, and so forth. Such a life, no matter how apparently wealthy, is spiritually poverty-stricken, Thoreau urges. "Most men, even in this comparatively free country," he writes, "through mere ignorance and mistake, are so occupied with the factitious cares and superfluously coarse labors of life that its finer fruits cannot be plucked from them."

Thoreau saw nature as a standard against which human economic life might be compared. Nature is patient, self-renewing, self-regulating, as spiritually bottomless as Walden Pond was rumored (falsely) to be. In contrast, "The mass of men lead lives of quiet desperation." Thoreau counsels that it need not be so: "I learned this, at least, by my experiment: that if one advances confidently in the direction of his dreams, and endeavors to live the life which he has imagined, he will meet with a success unexpected in common hours."

Walden's provoking qualities have encouraged stubborn myths about Thoreau. Many people assume that he was born rich, or concealed from readers the fact that his hut was within an easy walk of his family's house. In fact, Thoreau was not rich, but had to work for a living, and makes clear throughout *Walden* that his hut was not isolated. Thoreau never sought to leave society behind, but rather to withdraw from it just far enough to better discover his own place in it and its relationship (and his own) to the natural world. For these goals, a spot on the edge of town, rather than in a remote wilderness, was best.

Walden has been a basic text of the environmental movement, though Thoreau wrote it before public parklands even existed in the United States (the National Park Service, for example, was not established until 1916, 54 years after his death). His essays and books have helped shape the environmentalist conviction that our human identity, spiritual well-being, and ultimate survival depend not only on clearing and occupying land but on using that land wisely and on preserving wilderness as well: "In wildness is the preservation of the world."

Thoreau made careful notes of various natural events, such as the first appearance of migratory birds or of native wildflowers in spring. In the early 2000s, researchers from Boston University combed Thoreau's journals for such information and combined it with similar data from other sources (including dated photographs by Herbert Gleason (1855–1937), an early-twentieth-century photographer who took pictures of many scenes described by Thoreau). Their goal was to compare these precise nineteenth-century notes to present-day patterns of leafing and migration. They found that by 2005, due to global climate change, spring was arriving about 3 weeks earlier than it had in Thoreau's day.

Today, Walden Pond is part of a state park, where tens of thousands of people visit yearly to swim under the eye of a lifeguard, explore the trails in the park, and view a replica of Thoreau's hut.

Primary Source Connection

Walden is one of the most influential essays written in nineteenth-century America. Regarded as a practical utopian tract and as a guide to meaningful living, it has stood and continues to stand as a touchstone for a way of thinking that values the authority of the individual against the state, the morality of self-reliance, and the primacy of nature. In *Walden* nature is presented as the proper guide for how people ought to live and as the model for how society ought to be designed, organized, and run. In contact with nature, Thoreau asserts, people experience themselves authentically and can find the strength to act on principle rather than by conforming to social rules or traditions that may actually be in violation of those principles. Thoreau chronicled the operation of principle in action in his other classic work, *On Civil Disobedience*

Walden has become a primary text for environmentalists and an inspirational source for people disaffected from the established routines or values of organized societies. Along with *On Civil Disobedience*, it has provided inspiration for world figures, too, like the Indian leader Mohandas Gandhi and the American civil rights leader Martin Luther King Jr.

WALDEN

In the savage state every family owns a shelter as good as the best, and sufficient for its coarser and simpler wants; but I think that I speak within bounds when I say that, though the birds of the air have their nests, and the foxes their holes, and the savages their wigwams, in modern civilized society not more than one half the families own a shelter. In the large towns and cities, where civilization especially prevails, the number of those who own a shelter is a very small fraction of the whole. The rest pay an annual tax for this outside garment of all, become indispensable summer and winter, which would buy a village of Indian wigwams, but now helps to keep them poor as long as they live. I do not mean to insist here on the disadvantage of hiring compared with owning, but it is evident that the savage owns his shelter because it costs so little, while the civilized man hires his commonly because he cannot afford to own it; nor can he, in the long run, any better afford to hire. But, answers one, by merely paying this tax, the poor civilized man secures an abode which is a palace compared with the savage's. An annual rent of from twenty-five to a hundred dollars (these are the country rates) entitles him to the benefit of the improvements of centuries, spacious apartments, clean paint and paper, Rumford fire-place, back plastering, Venetian blinds, copper pump, spring lock, a commodious cellar, and many other things. But how happens it that he who is said to enjoy these things is so commonly a poor civilized man, while the savage, who has them not, is rich as a savage? If it is asserted that civilization is a real advance in the condition of man—and I think that it is, though only the wise improve their advantages—it must be shown that it has produced better dwellings without making them more costly; and the cost of a thing is the amount of what I will call life which is required to be exchanged for it, immediately or in the long run. An average house in this neighborhood costs perhaps eight hundred dollars, and to lay up this sum will take from ten to fifteen years of the laborer's life, even if he is not encumbered with a family—estimating the pecuniary value of every man's labor at one dollar a day, for if some receive more, others receive less; —so that he must have spent more than half his life commonly before his wigwam will be earned. If we suppose him to pay a rent instead, this is but a doubtful choice of evils. Would the savage have been wise to exchange his wigwam for a palace on these terms?...

The Fitchburg Railroad touches the pond about a hundred rods south of where I dwell. ... The whistle of the locomotive penetrates my woods summer and winter, sounding like the scream of a hawk sailing over some farmer's yard, informing me that many restless city merchants are arriving within the circle of the town, or adventurous country traders from the other side... Here come your groceries, country; your rations, countrymen! ...And here's your pay for them! screams the countryman's whistle; timber like long battering-rams going twenty miles an hour against the city's walls, and chairs enough to seat all the weary and heavy-laden that dwell within them... All the Indian huckleberry hills are stripped, all the cranberry meadows are raked into the city. Up comes the cotton, down goes the woven cloth; up comes the silk, down goes the woollen; up come the books, but down goes the wit that writes them.

When I meet the engine with its train of cars moving off with planetary motion... with its steam cloud like a banner streaming behind in golden and silver wreaths, like many a downy cloud which I have seen, high in the heavens, unfolding its masses to the light—as if this traveling demigod, this cloud-compeller, would ere long take the sunset sky for the livery of his train; when I hear the iron horse make the hills echo with his snort like thunder, shaking the earth with his feet, and breathing fire and smoke from his nostrils, ...it seems as if the earth had got a race now worthy to inhabit it. If all were as it seems, and men made the elements their servants for noble ends! If the cloud that hangs over the engine were the perspiration of heroic deeds, or as beneficent as that which floats over the farmer's fields, then the elements and Nature herself would cheerfully accompany men on their errands and be their escort.

I watch the passage of the morning cars with the same feeling that I do the rising of the sun, which is hardly more regular. Their train of clouds stretching far behind and rising higher and higher, going to heaven while the cars are going to Boston, conceals the sun for a minute and casts my distant field into the shade, a celestial train beside which the petty train of cars which hugs the earth is but the barb of the spear. The stabler of the iron horse was up early this winter morning by the light of the stars amid the mountains, to fodder and harness his steed. Fire, too, was awakened thus early to put the vital heat in him and get him off. If the enterprise were as innocent as it is early!

Henry David Thoreau

THOREAU, HENRY DAVID. *WALDEN*. BOSTON: 1854.

SEE ALSO *Climate Change; Environmental Activism; Global Warming*

BIBLIOGRAPHY

Books

Thoreau, Henry David. *The Illustrated Walden.* Princeton, NJ: Princeton University Press, 1973.

Web Sites

National Wildlife Federation. "Walden Warming." http://www.nwf.org/nationalwildlife/article.cfm?issueID=117&articleID=1510 (accessed April 12, 2008).

Larry Gilman

War and Conflict-Related Environmental Destruction

Introduction

Armed conflict often causes negative environmental impacts. This can occur on local and regional levels. Impacts of war on ecosystems depend on the magnitude and duration of a conflict, as well as the types of weapons used. For example, a conflict over an isolated water source may impact only a few villages, but large international conflicts that employ modern weaponry could harm many regions. Recent environmental effects of armed conflict have included soil degradation, radioactive pollution, acid rain, and diminution of air quality.

Historical Background and Scientific Foundations

By definition, war is destructive, not just on people and infrastructure, but also on the surrounding natural environment. Throughout history, from the Roman sacking of Carthage to the 2003 U.S.-led invasion of Iraq, conflict has caused large-scale environmental damage. Yet only in recent decades have researchers measured the environmental costs of war. The media have paid increasing attention to issues of environmental degradation caused by human conflict in recent years. When Israel invaded Lebanon in 2006, the media reported on a conflict-caused release of 100,000 tons of oil into the Mediterranean Sea.

Modern warfare has particularly severe environmental impacts. Scorched earth tactics have been deployed for almost as long as war is old. Although the initial effect is devastating, the environmental effects have for most of history been short term: trees and crops will grow again, towns and cities can be rebuilt, wildlife will usually return. Yet in the Gulf War (1990–1991) between a U.S.-led coalition and Iraq, the retreating Iraqi army deployed scorched earth tactics on a different scale, by destroying the Kuwaiti oil field it had recently seized.

In total, 736 oil wells were set ablaze, consuming 6 million barrels of oil daily and dousing significant parts of the Persian Gulf in impenetrable black clouds. The fires burned until November 1991, eight months after a ceasefire. Although initial fears proved unfounded that the smoke would penetrate the upper atmosphere and influence weather patterns—similar to the effects of major volcanic eruptions—the sabotage had a significant regional effect. Released sulfur contributed to acid rain and local air quality was significantly diminished. Non-ignited oil formed vast lakes on the Arabian Desert, contaminating huge amounts of soil and freshwater.

Gulf War alliance forces used depleted uranium (DU)—a dense metal used to enhance the penetrative quality of munitions. Because DU is essentially a low-level nuclear waste, it is toxic. Scientific opinion remains split on the degree of harm caused by exposure, but DU use can contaminate its very immediate surroundings.

After fighting in France and Belgium in World War I (1914–1918), and through Poland and the Ukraine in World War II (1939–1945), most of the major battlefields were successfully turned back to agricultural land with the commencement of peace—despite thousands of tons of munitions lying unexploded in the ground. By contrast, modern battlefields lie contaminated not just by DU, but often by very sensitive weapons like unexploded cluster bombs and landmines—both of which are difficult to detect but remain deadly for decades.

Cheap, easy to deploy, and deadly (and just as difficult and costly to deactivate), landmines are a problem in large areas of the former Yugoslavia, West Africa, and parts of East Asia. Landmines render land an uninhabitable environmental disaster zone. In Cambodia and Bosnia there is approximately one landmine for every person; in Afghanistan, Iraq, Croatia, Eritrea, and Sudan, the ratio is one mine for every two persons.

Other effects of war on local ecosystems are less immediately obvious. Changes in land use, abandon-

WORDS TO KNOW

ACID RAIN: A form of precipitation that is significantly more acidic than neutral water, often produced as the result of industrial processes.

ATOMIC BOMB: A highly destructive weapon that derives its explosive power from the fission of atomic nuclei.

DEPLETED URANIUM (DU): A byproduct of spent nuclear fuel, DU is a dense metal with a variety of civilian and military uses. It is often used to enhance the armor piercing qualities of munitions, although its deployment is riddled with controversy and it has been linked to increased incidence of cancer rates and birth defects.

ENVIRONMENTAL DEGRADATION: The overall deterioration of environmental quality due to a range of issues, such as deforestation, desertification, pollution, and climate change.

LANDMINE: A bomb planted on or near the surface of the ground that is triggered by something passing over it.

NATO: The North Atlantic Treaty Organization is a military alliance comprising the United States and 25 other members states, along with 14 major allies. It was involved in the bombing of Serbia in 1999 and most recently in Afghanistan.

NUCLEAR WEAPON: A military device whose explosive power is derived from nuclear fission or fusion.

SCORCHED EARTH: A military policy involving widespread destruction of property and resources, especially by burning, so that an advancing enemy cannot use them.

This NOAA satellite image from 520 miles (836 k) in space shows an undetermined number of oil well fires in Iraq, the plumes of which are visible, indicated by the arrow, March 20, 2003. *AP Images*

Bedouin Arabs stand next to their camel during a protest organized by environmental groups at the future site of Israel's separation barrier, on the Good Samaritan Hill in the Judean Desert, near the West Bank town of Jericho, March 2007. The Friends of the Earth organization says the proposed route of Israel's separation barrier will carve up parts of the Judean Desert around the Israeli settlement of Ma'aleh Adumim. Israel says the barrier is necessary for security. *AP Images*

ment of cultivated land, or significantly increased demands on natural resources can have long-term consequences on an environment. During the war between Eritrean and Ethiopia (1961–1991), the country's forest cover was reduced from nearly one third of its overall landmass to less than 1%, resulting in high rates of soil erosion, reduced wildlife populations, sedimentation of rivers and reservoirs, and crop failures.

Impacts and Issues

Nuclear weapons remain controversial because of their capacity for destruction. It is estimated that such a conflict could result in more than 1 billion deaths, with a similar number injured due to the combined effects of blast, fire, and radiation. Scientists predict an appalling array of environmental consequences too: a tripling of UV-B radiation; increased doses of ionizing radiation; atmospheric pollution; chemical pollution of surface waters; and release of toxic chemicals from secondary fires.

In August 1945, during World War II, the United States used atomic bombs against the Japanese cities of Hiroshima and Nagasaki. The human and environmental effects were devastating. The blasts caused air pollution from dust particles and radioactive debris. Fires burned for some six weeks, destroying plants and wildlife. The region's water table was polluted by radioactive particles. Radioactive precipitation contaminated humans and agriculture far beyond the blast zone. Birth defects and incidence of cancer were significantly enhanced over following decades. Several nations have since developed exponentially more powerful nuclear weapons than the Hiroshima and Nagasaki bombs, though these weapons were used in tests or never detonated. Even testing of nuclear weapons has caused serious environmental damage. Some researchers have studied test sites for clues about the possible impacts of wide-scale use of nuclear weapons in future conflicts. Some assert that nuclear weapons pose a considerable threat to the environment and increasing global instability could lead to increased use of nuclear devices. However, others counter that modern tactical nuclear weapons allow for more accurate targeting and concentrated explosions, limiting the scope of human and environmental impacts when compared to the earliest atomic bombs.

As the effects of climate change are increasingly felt, global environmental degradation is also seen as a major contributing factor to future conflicts. The United Nations and the Intergovernmental Panel on Climate Change (IPCC) have addressed the possibility that climate-change induced food and resource scarcity could spark regional conflict. In the Middle East, disputes over water access are already a significant impediment to lasting peace settlements between some nations. The armed conflict in the Sudan in the 1980s is an example of conflict initiated by food insecurity and famine. Elsewhere in East Africa deforestation and soil erosion combined with spiraling populations and uncontrolled urban migration have diminished food self-sufficiency and heightened the danger of civil conflict.

SEE ALSO *Human Impacts; International Environmental Law; Oil Spills*

BIBLIOGRAPHY

Books

Gleditsch, Nils Petter, ed. *Conflict and the Environment, NATO Advanced Research Workshop on Conflict and the Environment.* Dordrecht: Kluwer, 1997.

Gleditsch, Nils Petter, and Paul F. Diehl, eds. *Environmental Conflict.* Boulder: Westview, 2001.

Schneider, Gerald, Katherine Barbieri, and Nils Petter Gleditsch, eds. *Globalization and Armed Conflict.* New York: Rowman and Littlefield, 2003.

Periodicals

"Special Issue on Environmental Conflict." *Journal of Peace Research* 35, no.3 (May 1998).

James Corbett

Waste Transfer and Dumping

Introduction

Waste is an inevitable and increasing byproduct of modern life, whether it is in solid, liquid, or gaseous form. Despite advances in awareness of ways of reducing, recycling, or reusing waste materials, most of it is still discarded by dumping or dropping it somewhere. Dumping of waste needs to be carefully controlled, if it is not to create health and environmental hazards. In most places where there are regulations controlling waste, this means creating a landfill site to put the waste in.

There are increasing pressures on landfill, however, and waste is sometimes dumped illegally, either in the place where it originated, in the ocean, or exported to another country, where the technology to deal with the waste, especially hazardous waste, may not be readily available. For instance, electronics waste is often exported to China where the local population strips it of valuable components, but often without protecting themselves from the impact of heavy metal pollution.

Historical Background and Scientific Foundations

The United States produces around 600 million metric tons of waste every year, of which about two thirds comes from industry, with the rest being municipal waste, issuing from homes and businesses. Some of this waste is dumped which, at its most basic, means just dropping refuse somewhere. Uncontrolled, dumping results in a mountain of litter or waves of refuse eventually deposited upon the beach.

Over the last 50 years, it has become socially and economically unacceptable to dump refuse in many countries. Environmental legislation requires people to deal with their refuse by making it available for collection by local authorities or specialized disposal services. Uncontrolled dumping has been replaced by landfill, with ordinary and hazardous wastes being separated.

Municipal solid waste, less any material for recycling, is taken to a facility called a waste transfer station. Here it is held before being reloaded onto larger long-distance transport vehicles, which convey it to a landfill site. This reduces labor and transport costs. A waste transfer station should be designed and located with regard to the health and sensibilities of the local population, for solid municipal waste produces odors and attracts vermin.

A landfill site is basically a hole in the ground with a lining into which wastes are put and covered. It needs to be located away from water, to reduce the risk of contamination of the water supply, and on rocks that are reasonably leakproof. The geology of the area should be well understood so that any leaks that do occur can be easily tracked and corrected. Landfills may be lined with clay, plastic, or a mixture of both materials. The landfill also contains a method of collecting and dealing with leachate, which is the liquid that accumulates in the landfill as the waste ages. The cover, which is usually a mixture of clay and soil, keeps the landfill secure from water, vermin, and human interference. The site must be monitored continuously for leaks.

Impacts and Issues

Illegal dumping causes a variety of health and environmental hazards. Refuse is blown by the wind and rain leaches heavy metals and other toxins into the ground, where it could reach the water supply. Many countries where environmental legislation is weak have large areas of piled rubbish, where local residents affected by poverty often put themselves at risk by scavenging. Mexico City and Manila are just two examples of such

WORDS TO KNOW

INCINERATION: The burning of solid waste as a disposal method.

LANDFILL SITE: Solid waste disposal site consisting of a lined, covered hole in the ground.

MUNICIPAL WASTE: Waste that comes from households or is similar to household waste.

WASTE TRANSFER STATION: A structure in which wastes are temporarily held and sorted before being transferred to larger facilities.

dumping hotspots. Illegal export of waste is a problem too. For instance, petroleum waste was dumped in the Ivory Coast in 2006, where toxic byproducts killed ten people and made thousands of others ill.

Landfills were filled beyond capacity in Naples, Italy, in mid-2007, and photographs of streets blocked with heaps of rubbish were sent via the media around the world, tarnishing the image of the southern Italian city's charm. Blamed on years of weak public infrastructure and organized crime, the government finally appointed a task force to clear the streets of tons of refuse and divert it to temporary storage facilities and landfills in other parts of Europe. By April 2008, more than a year later, the downtown streets and waterfront were cleared, but tourism, which is Naple's main source of income, remained hard hit, as large tourist hotels on the Bay of Naples waterfront were only 30% occupied on average. Stories-high piles of rotting trash along the outskirts of the city remained, and citizens in the outlying areas of Naples awaited a workable waste disposal plan.

There is increasing pressure on landfills because transport and land costs are rising. Furthermore, as people become more environmentally aware, they may campaign against having a landfill site in their neighborhood. This means that reducing, recycling, and reusing are more important than ever so that the amount of waste requiring landfill use can be controlled. Incineration or burning of non-toxic waste materials found in landfills can generate heat and energy, and is employed in some areas to deal with bulging landfills. Landfills can even be useful, as they generate methane, which can be tapped off and used as an energy resource.

SEE ALSO *Electronics Waste; Hazardous Waste; Industrial Pollution; Recycling; Solid Waste Treatment Technologies; Toxic Waste*

A Chinese sailor inspects a cargo of plastic and domestic waste in Hong Kong's Victoria Harbor, 1996. The cargo, which originated in Atlanta, Georgia, is at the center of controversy after China and Hong Kong officials rejected it as unwanted waste. Environmentalists say the shipment, thousands of plastic bags from Florida-based U.S. supermarket chain Winn-Dixie, was proof U.S. companies are dumping waste in Asia. *AP Images*

IN CONTEXT: HIGH INCOME ELASTICITY AND INTERNATIONAL WASTE TRANSFERS

Environmental standards are lax in developing countries because the people there are desperate for employment. This reasoning was laid out in an extraordinary 1991 memo signed by Larry Summers (1954–), then chief economist of the World Bank. The memo asked, "shouldn't the World Bank be encouraging more migration of the dirty industries to the [less developed countries]?" and went on to say that "the demand for a clean environment for aesthetic and health reasons is likely to have very high income elasticity"—that is, the poorer you are, the more ugliness and disease you can be forced to put up with. "I think," the document went on to say, "the economic logic behind dumping a load of toxic waste in the lowest wage country is impeccable and we should face up to that." Summers later became deputy secretary of the Treasury Department under President Bill Clinton and was president of Harvard University from 2001 to 2006. He maintained that the memo was meant as a "sardonic counterpoint," not to be taken seriously. In any case, it describes the reasoning operative in actual waste transfer to poor countries.

BIBLIOGRAPHY

Books

Cunningham, W.P., and A. Cunningham. *Environmental Science: A Global Concern.* New York: McGraw-Hill International Edition, 2008.

Web Sites

Environmental Research Foundation. "The Basics of Landfills." http://www.ejnet.org/landfills/ (accessed March 21, 2008).

U.S. Environmental Protection Agency (EPA). "Solid Waste Landfills." http://www.epa.gov/epaoswer/non-hw/muncpl/landfill/landfills.htm (accessed March 21, 2008).

U.S. Environmental Protection Agency (EPA). "Waste Transfer Stations." http://www.epa.gov/epaoswer/non-hw/transfer.htm (accessed March 21, 2008).

Greenpeace members point to a discarded battery (August 1996), symbolic of waste which they say ends up in a Philippine dumpsite after being imported from other countries under the guise of recycling. *AP Images*

Wastewater Treatment Technologies

Introduction

Domestic and industrial activities produce billions of gallons of wastewater every year. This contains a wide range of impurities, ranging from human waste and bits of food, to detergents and industrial chemicals. If discharged straight into the environment, these quantities of wastewater would soon overwhelm and poison ecosystems, and create a serious human health hazard. Therefore, wastewater must be treated before it is allowed to rejoin the water supply.

Most wastewater is collected by the sewage system. Some industrial wastewater may be treated first to remove specific chemical pollutants. Afterward, the wastewater is filtered to remove solids and treated by chemical or biological processes to reduce levels of other pollutants to the safe limits specified by local environmental legislation. Pollution incidents arise when wastewater treatment is inadequate, or non-existent. There is an ongoing need for lower-cost treatment methods to meet the needs of developing countries.

Historical Background and Scientific Foundations

Sewage is the name usually given to the wastewater that is discharged from homes, offices, and industrial premises into the public sewerage system. Often, the terms sewage and wastewater are used interchangeably. In a home or office, wastewater is everything that goes down the drain, so it comprises water from the bath, shower, toilet, and washbasin, as well as that from the sink, washing machine, and dishwasher. Much industrial wastewater has been used for washing or cooling in various processes.

Wastewater contains human waste, with accompanying microbes, some of which are pathogenic (disease-causing), detergents, oil, food particles, grease, and industrial chemicals of various kinds. Wastewater is about 99.94% water and it looks cloudy because of its suspended solid content. Storm water is an additional type of wastewater arising from rain or melting snow, which runs along the ground. Storm water issuing from industrial premises, car parks, streets, and rooftops may contain oil, tars, metals, pesticides, fertilizers, and various other undesirable substances. Wastewater needs to be treated to remove these various impurities, or at least reduce them down to acceptable levels, before it is discharged back into the environment. Raw sewage, as untreated wastewater is often called, poisons fish and other aquatic organisms, and may cause human disease, as it contains pathogenic microorganisms. It also looks unsightly in waters used for recreational purposes.

Wastewater enters the sewage system, which is a series of pipes and pumps that carries it away from its place of origin to a facility where it can be treated to clean it up. Sewage systems date back to the nineteenth century, when the role that contaminated water plays in disease was first realized. Before that, people would just discard wastewater in the street.

At the sewage works, preliminary treatment consists of using a coarse screen on the incoming wastewater to remove larger particles such as rags, stones, and sand. It then undergoes a stage called primary treatment, where it is left to settle in a tank so that solids either sink to the bottom or float on the top of the wastewater. These bottom and top layers are skimmed off to form a sludge, which undergoes a drying treatment before being shipped as a fertilizer, which will return organic matter to the soil.

The clarified wastewater emerging from primary treatment is known as effluent. It then undergoes a biological treatment stage, known as secondary treatment, in which sewage microorganisms break down the organic matter in the effluent. There are various technologies for secondary treatment. In fixed film technology, microorganisms are grown on a surface such as rock, plastic, or sand where they create a film. The effluent then flows past the microbial film. One fixed film system is the trickling filter bed, which is a bed of stones coated with microbial film over which effluent drips via a series of perforated pipes or an overhead sprinkler.

Another system includes the suspended film system for secondary treatment in which microbes are suspended within the effluent in a tank and aerated. This system is sometimes also known as activated sludge. The third type of secondary treatment is the lagoon. This consists of a shallow basin in which the effluent is held for several months to allow natural microbial degradation approaches to work on it. Finally, the effluent undergoes chlorine or ultraviolet treatment to ensure that pathogenic organisms have been reduced to a safe level. Generally, the amount of pathogens in sewage is assessed using the count of coliform bacteria present within a sample. If chlorine treatment is used, then a neutralizing chemical will need to be added afterward, because chlorine can have an adverse impact on aquatic ecosystems once the treated wastewater is finally discharged.

Wastewater often contains high levels of nitrogen and phosphorus from human waste and detergents. Both are nutrients and, as such, enrich surface water when they are discharged. This enrichment, known as eutrophication, encourages the growth of algae and decomposer microorganisms to the detriment of other organisms, such as fish, and may upset the local ecosystem. Therefore, nitrogen and phosphorus may be removed from wastewater by adding chemicals that will precipitate them out.

WORDS TO KNOW

COLIFORM: Bacteria present in the environment and in the feces of all warm-blooded animals and humans; useful for measuring water quality.

EFFLUENT: Wastewater from which solids have been removed.

SLUDGE: Semisolid material formed as a result of wastewater treatment or industrial processes.

STORM WATER: Water that flows over the ground when it rains or when snow melts.

WASTEWATER: Water that carries away the waste products of personal, municipal, and industrial operations.

Steam from cooling towers rises above the rotary kiln as wastewater from VX nerve agent is incinerated at Veolia Environmental Services in Port Arthur, Texas, April 2007. Some residents are protesting the incineration of the wastewater, saying it is dangerous. *AP Images*

Impacts and Issues

Standardized sewage treatment is usually provided for under environmental legislation such as the U.S. Clean Water Act of 1972. Those discharging wastewater into any kind of surface water need a permit and must adhere to limits of certain pollutants that have been laid down in law.

Setting up a sewer infrastructure can be costly and there is a need for lower-cost technologies for the developing world, so that clean water is available to all. One option is effluent sewerage, which is a mixture of a septic tank and a full sewerage system. The septic tank is a container used to allow sewage to settle and be digested by bacteria. It is intended for small-scale wastewater treatment. In effluent sewerage, a tank is used by each dwelling and its effluent is pumped to a central facility. Because only liquids are being treated here, smaller pipes, pumps, and treatment beds can be used, which lowers the cost.

IN CONTEXT: WASTEWATER TREATMENT STANDARDS

Wastewater treatment is usually subject to local and national standards of operational performance and quality in order to ensure that the treated water is of sufficient quality so as to pose no threat to aquatic life or settlements downstream that draw the water for drinking.

Chlorination remains the standard method for the final treatment of wastewater. However, the use of the other systems is becoming more popular. Ozone treatment is popular in Europe, and membrane-based or ultraviolet treatments are increasingly used as a supplement to chlorination.

Another approach is to use the natural remedial action of natural or artificial wetlands to treat wastewater. For instance, poplar plantations in India have been shown to be capable of carrying out secondary treatment on wastewater at a cost of a third, or less, of conventional mechanical treatment. Such constructed wetlands are now being used in American cities and in developing countries. Another low-tech option is to use duckweed on the surface of lagoons to purify wastewater, at a cost of one tenth of a conventional system. The duckweed can then be harvested and used as high-protein feed or as a fuel.

SEE ALSO *Surface Water; Water Resources; Water Supply and Demand*

BIBLIOGRAPHY

Books

Cunningham, W.P., and A. Cunningham. *Environmental Science: A Global Concern.* New York: McGraw-Hill International Edition, 2008.

Web Sites

Ohio State University. "Wastewater Treatment Principles and Regulations." http://ohioline.osu.edu/aex-fact/0768.html (accessed March 25, 2008).

World Health Organization. "Water, Sanitation, and Health: Wastewater Use." http://www.who.int/water_sanitation_health/wastewater/en/ (accessed March 10, 2008).

Susan Aldridge

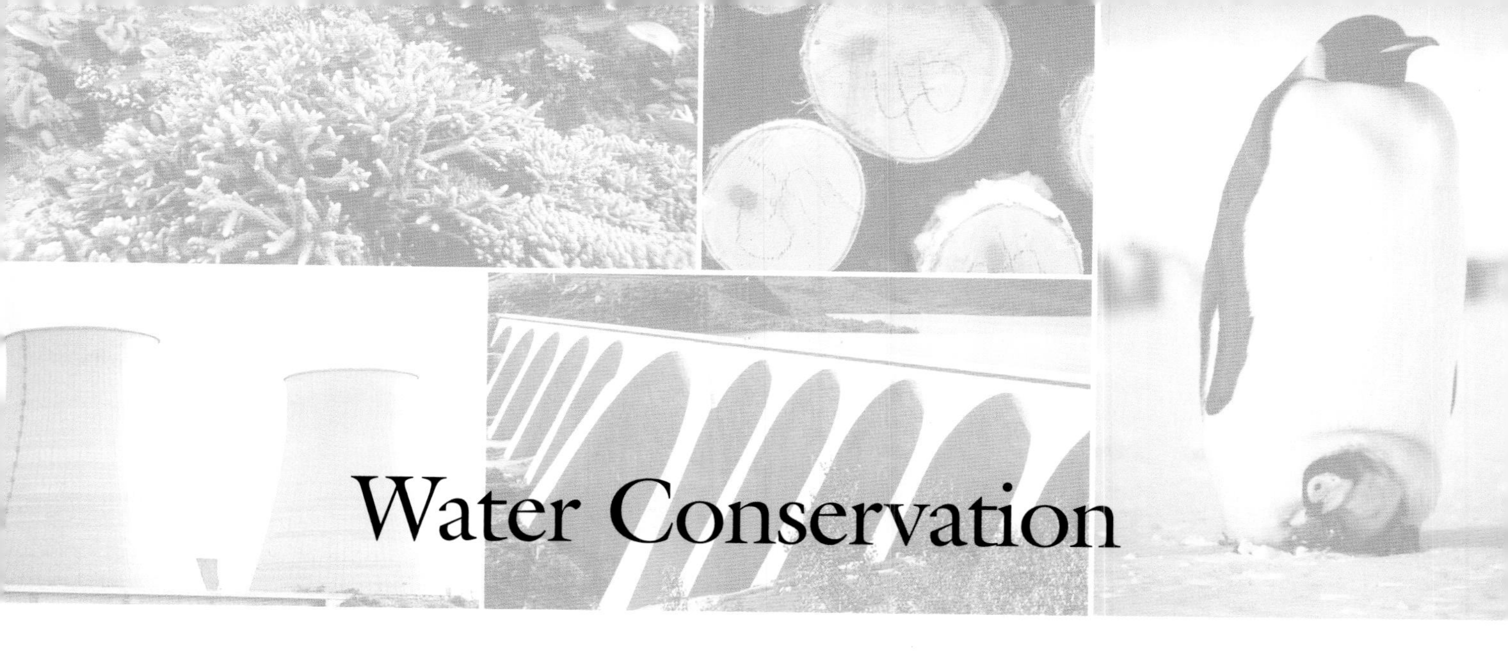

Water Conservation

Introduction

It is easy for some people in developed nations to take their water supply for granted. But water is essential to life and, according to the United Nations, one billion people around the world lack access to safe drinking water and 2.6 billion people lack adequate sanitation. Therefore, there is not enough water in circulation to even meet people's basic needs. Added to this, agriculture and industry rely upon an adequate water supply for productivity. Therefore, water is a precious resource and its conservation is important.

There are many simple ways in which water can be conserved by the individual. Authorities, such as the U.S. Environmental Protection Agency (EPA), have various schemes to encourage a responsible attitude towards water consumption. Technical advances can aid water conservation, such as the low-volume toilet and new ways of irrigating crops. Water conservation, like energy efficiency, brings big benefits in terms of improving quality of life and protecting the environment.

Historical Background and Scientific Foundations

Forty-five countries around the world suffer from frequent water shortages, and in many other places drought can make water supplies uncertain. Much water is wasted though leaks in pipes, toilets, taps, and irrigation systems. Meanwhile, the growing world population and increasing industrialization increase the demand for freshwater. Without some kind of controlled reduction in this demand through water conservation, pressure on water supplies could have a widespread effect on human health and productivity.

Water conservation relies upon a mixture of behavioral change and technical advances. There are many simple ways in which individuals can save water, from fixing leaks and having shorter showers, to collecting and reusing water for the garden. Water conservation efforts should be driven by an awareness of which activities use the most water. Toilet flushing accounts for about 47% of typical household water use in the United States, followed by 31% for bathing, and 20% for laundry and dishwashing.

Therefore, low-volume toilets can make a difference to domestic water consumption. Prior to 1992, when these were introduced, a typical flush took six gallons of water but now it takes only 1.6 gallons. There are also water-efficient dishwashers and washing machines as well as low-volume shower heads. Wastewater can be reclaimed and recycled, with several cities now using purified sewage effluent. Agriculture is the greatest consumer of water worldwide and some practices are very wasteful. In standard irrigation, much water is lost through evaporation or runoff. But drip irrigation helps conserve water, by releasing controlled amounts of water just above plant roots, so that nearly all of the water is used by the plant.

Awareness of the need for water conservation is being increased by the water footprint concept, which has been developed by researchers at the University of Twente in the Netherlands. The water footprint can be applied to an individual, a business, or a nation, and it refers to the volume of water used in the production of goods and services used. For instance, it takes about 4,227 gallons (16,000 liters) of water to produce 2.2 pounds (1 kilogram) of beef, and 37 gallons (140 liters) of water to produce a cup of coffee.

Impacts and Issues

Wasting water not only makes shortages worse, it also contributes to water pollution. For example, diverting

WORDS TO KNOW

DRIP IRRIGATION: Slow, localized application of water just above the soil surface.

SEWAGE: Waste and wastewater discharged from domestic and commercial premises.

WATER FOOTPRINT: The total volume of water used to produce goods and services consumed by an individual, business or nation.

natural water supplies by building dams traps sediment, which concentrates pollutants and reduces dissolves oxygen. The more water people use, the more dirty wastewater is produced, which flows into natural water supplies, compromising their purity.

Water conservation improves the overall quality of the water supply and the environment in many ways. The sewage system is less liable to failure through overloading. More efficient irrigation means less polluted runoff from agricultural land. Natural pollution filters like wetlands are less likely to be dried up or depleted, while reduced need to build dams and reservoir preserves species' habitats. Moreover, using less hot water means using less electricity, which benefits both the consumer and the environment.

Due to a lack of water facilities at her home, a woman drinks the water that melted from the snow of a mountain in Bolivia, in March 2003. According to a U.N. World Water Development Report, by the middle of this century, as many as seven billion people in 60 countries could be faced with water scarcity. *AP Images*

SEE ALSO *Drought; Industrial Water Use; Irrigation; Runoff; Water Resources; Water Supply and Demand*

BIBLIOGRAPHY

Books

Cunningham, W.P., and A. Cunningham. *Environmental Science: A Global Concern*. New York: McGraw-Hill International Edition, 2008.

Web Sites

U.S. Environmental Protection Agency (EPA). "WaterSense®." January 25, 2008. http://www.epa.gov/watersense/water/index.htm (accessed March 23, 3008).

Waterfootprint.org. "Water Footprint." http://www.waterfootprint.org/?page=files/home (accessed March 23, 2008).

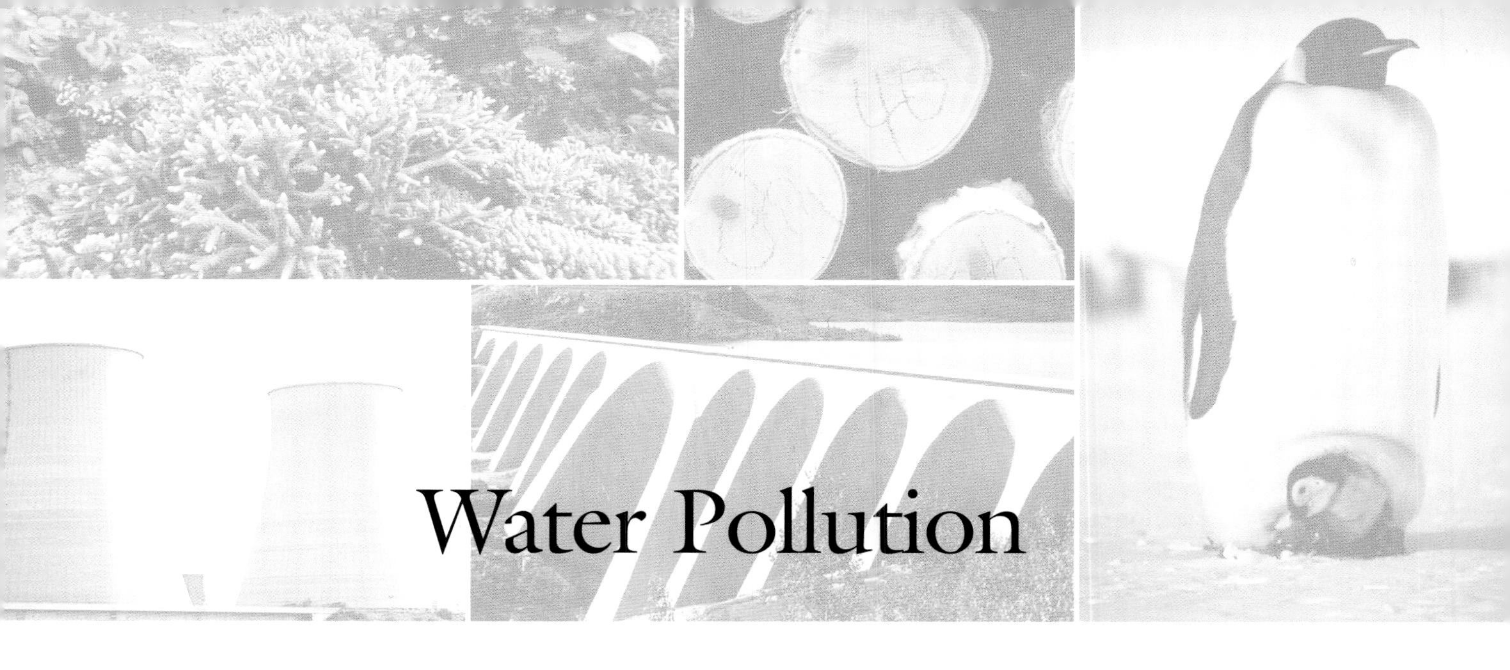

Water Pollution

Introduction

Water pollution refers to the presence of compounds that decrease the quality of fresh or marine water. This is a broad definition that takes into account a variety of water sources including lakes, streams, rivers, oceans, and groundwater. As well, the sources of water pollution can be natural, such as animal waste, or more commonly, due to human activity.

There are many causes of human-mediated water pollution, including runoff of pesticides, herbicides, and feces from agricultural land; runoff of gasoline and other chemicals from concrete in urban settings; and household chemicals discarded by flushing down the toilet (such as medicines or household cleaners).

Water pollution is a major global concern. In developing countries, waterborne diseases such as cholera, typhoid fever, and hepatitis A are among the leading causes of illness and death, killing over 3.4 million people every year (including 4,000 children every day), according to the World Health Organization (WHO).

In the environment, water pollution can alter or destroy a watercourse directly by the addition of the pollutants, and indirectly when the pollutant can be used as food source by microorganisms, whose explosive increase in growth and total numbers depletes the water of oxygen.

Historical Background and Scientific Foundations

As humans began to shift from a nomadic way of living to settlements, the communities were virtually always located near water. The importance of clean water to health was recognized thousands of years ago, even if the reason was not understood. For example, in ancient Rome, the Tiber River became so polluted with the direct discharge of human sewage that separate channels were built for drinking water. Throughout the next two millennia, waterborne illness outbreaks occurred; from the descriptions provided, it is likely that many were cholera and typhoid. Even today, cholera remains a problem in areas of the world that do not have access to clean water.

It was not until the 1850s that the connection was made between water contaminated with human feces and disease. Then, during a cholera outbreak in London, physician John Snow (1813–1858) deduced that fecal-contaminated water from a particular well was implicated in the illness. Stopping the use of the well stopped the spread of the disease.

Similarly, in the United States, rivers were used to dispose of human waste for centuries. With the growth of industries, waters also became the disposal system for industrial waste, a practice that continued into the twentieth century. An example is the Cuyahoga River in Cleveland, Ohio, which periodically caught fire from the 1930s into the 1970s due to oil and other flammable industrial waste discharged into the water. Finally, a fire in 1969 that was immortalized in a song ("Burn On") by American recording artist Randy Newman drew attention to the deplorable condition of the Cuyahoga and other American waterways, which spurred passage of the Clean Water Act in 1972. The legislation banned the release of pollutants in waterways used by craft, although it has subsequently been criticized for a lack of standards of toxicity and difficulty in identifying pollution that does not come from a single source (non-point source pollution).

In the decades of prosperity following World War II (1939–1945), many chemicals were synthesized and commercialized. Antibiotics and DDT, for example, were spectacularly successful at stopping the spread of a variety of diseases (at least until resistance developed). At the same time, these and other compounds were finding

WORDS TO KNOW

AQUIFER: Rock, soil, or sand underground formation that is able to hold and/or transmit water.

COLIFORMS: Bacteria that live in the intestinal tract of warm-blooded animals. The presence of coliforms in water indicates fecal pollution.

DISINFECTION BY-PRODUCT: A compound created by the interaction of a disinfectant chemical and organic compounds in the water.

EUTROPHICATION: The process whereby a body of water becomes rich in dissolved nutrients through natural or man-made processes. This often results in a deficiency of dissolved oxygen, producing an environment that favors plant over animal life.

GROUNDWATER: Fresh water that is present in an underground location.

HYDROLOGY: The study of the distribution, movement, and physical-chemical properties of water in Earth's atmosphere, surface, and near-surface crust.

RUNOFF: Water that falls as precipitation and then runs over the surface of the land rather than sinking into the ground.

their way into water by runoff. In 1962, the book *Silent Spring* was published. Written by Rachel Carson, the book was a condemnation of the environmental release of pesticides and especially of DDT.

Preventing Waterborne Disease

Microorganisms are an important source of water pollution. Some microorganisms that are of concern live in the intestinal tract of warm-blooded animals, including humans. Examples of disease-causing (pathogenic) bacteria are *Salmonella*, *Shigella*, *Vibrio*, and *Escherichia*. With the latter, *Escherichia coli* O157:H7 is particularly dangerous. Contamination of drinking water with O157:H7 can cause severe illness, permanent kidney damage, and even death. In an infamous example, contamination of the municipal water supply of Walkerton, Ontario, Canada, in the summer of 2000 because of runoff from adjacent cattle farms sickened over 2,000 people and killed 7.

The intestinal tract of warm-blooded animals also harbors pathogenic (disease-causing) viruses that can contaminate water, in particular rotavirus, enterovirus, norovirus, and coxsackievirus. A number of protozoan microorganisms can also affect water quality. The two most prominent protozoans are Giardia and Cryptosporidium, which normally live in the intestinal tracts of wild animals such as beaver and deer.

Municipal drinking water is usually treated with disinfectants in order to minimize the risk of microbial contamination. One common disinfectant is chlorine. Other treatments that kill bacteria include the use of a gaseous form of ozone (a compound containing three oxygen atoms, which is also present in Earth's atmosphere), and exposure of shallow water to ultraviolet light. The ultraviolet radiation breaks apart the bacterial genetic material so much so that it cannot be repaired, and the microbe dies. Filters can also be used in water treatment; they contain openings of some filters that are so tiny that even viruses that are 10 nanometers in diameter are blocked from passing through (a nanometer is one-billionth of a meter).

Water polluted with protozoans can be more difficult to treat, as both Giardia and Cryptosporidium form chlorine-resistant structure cysts that can pass through the water treatment filters. In 1993, the contamination of the water supply of Milwaukee, Wisconsin, sickened over 400,000 people and at least 47 people died.

Non-point Source Pollution

Non-point source pollution enters water from many different sites, rather than from a single site. Examples of non-point source pollution are acid rain and runoff of polluted water from the urban and rural land. The nitrogen and phosphorus in fertilizers can be a source of food to microorganisms such as algae. Their resulting rapid growth and increase in numbers can deplete the oxygen in the water. In some areas of North America, huge factory farms contain tens of thousands of poultry or pigs. The waste material and huge amounts of water used in the farm operation are typically stored in large lagoons. Rupture of a lagoon wall can release the toxic water into nearby watercourses or into the groundwater.

Because of the many possible sites of non-point source pollution, tracking down the origin of the water pollution can be difficult. In contrast, an example of a point source is polluted material flowing from one factory into a river. This type of pollution can be much easier to detect and to stop.

According to the U.S. Environmental Protection Agency (EPA), as of 2008, about 40% of all the U.S. freshwater sources that have been mapped have been so affected by non-point source pollution that they are not safe for swimming or fishing.

There are a variety of organic (carbon-containing) water pollutants in addition to microorganisms. There are many types of chemicals that kill insects (insecticides) and weeds (herbicides). Runoff of moisture from agricultural land or urban lawns can carry these chemicals into streams, rivers, and lakes. Similarly, leakage of industrial chemicals, oil, and gasoline can occur both above the ground and from underground storage tanks. The latter can be especially serious since it may escape detection for a long time, and the pollutant can move outward from the source, even into groundwater.

The treatment of water can generate disinfection byproducts such as trihalomethanes, when chlorine reacts with organic matter. This tends to occur with surface waters, as more organic material can be present in the water. Trihalomethanes are considered a pollutant since they affect the smell and taste of the drinking water.

A more recent concern are the various drugs that people discard by flushing them down the toilet or the ones that can end up in landfills. With the development of technologies that can detect very low levels of drugs (parts per billion, equivalent to a drop of water in an Olympic sized swimming pool), the presence of (as a few examples) cosmetics, detergents, toiletries, painkillers, tranquilizers, anti-depressants, antibiotics, birth control pills, estrogen replacement therapy drugs, chemotherapy drugs, and epilepsy treatment drugs have all been detected in the environment. Some pharmaceuticals and drugs, including caffeine and acetaminophen (Tylenol), have also been detected in drinking water.

Although concentrations of most of these chemicals are not a concern in the short term, it is unknown if longer-term exposure is harmful. In 1999–2000, the most recent survey of the U.S. Geological Survey tested 139 streams from 30 states for 95 contaminants. Some 75% contained two or more of these contaminants, with 13% of the streams containing 20 or more.

Inorganic Pollutants

Inorganic water pollutants include heavy metals, acidic materials, nitrogen and phosphorus that are part of fertilizers, and silt from construction sites and clear-cut lands that is washed into watercourses.

A well-known inorganic water pollutant is Dichloro-Diphenyl-Trichloroethane (DDT). DDT exposure causes adverse effects in humans, animals, and birds in the short-term and more chronically. Possible longer-term effects include difficulties in reproduction, altered development of a fetus, and both liver and lung cancer. Prior to the ban on DDT use in the United States, bald eagle populations had declined to levels that threatened the species due to the accumulation of the pesticide in the bird's tissue, which caused production of eggs with thinner, more fragile shells. As the compound has gradually declined in the environment in the decades following the ban, bald eagle populations have rebounded.

One major reason for the chronic effects of DDT is bioaccumulation, the accumulation of a compound in tissues of one species that is passed on to species higher in the food chain. As a result, creatures including the bald eagle ingest a much higher level of the toxin than they otherwise would and store the poison in fatty tissues, where it can persist.

The structure of DDT also makes it difficult to degrade. It can persist in water for months. Degradation is not a solution, since two breakdown products also persist in the environment and can cause adverse effects.

Compounds that can leak out of landfills can also degrade slowly, as do petroleum and petroleum-based products; polychlorinated biphenyls (PCBs); dioxins and polyaromatic hydrocarbons (PAHs); metals such as lead, mercury, and cadmium; and some kinds of radioisotopes. As an example, one of the considerations in selecting Yucca Mountain in Nevada as an underground garbage site for radioactive waste was the geology of the volcanic rock, which lacked cracks that could allow the waste to move from the site of storage in the event of a leak, and the fact that groundwater was very far beneath the storage area. Thus, the likelihood of radioactive contamination of the groundwater would be remote. This is good, since radioisotopes can have a half-life (the time for half the radioactivity to be degraded) of thousands of years, which would irreversibly contaminate water. Although the concern over water pollution may have been met, other concerns have pushed back the planned 1998 opening to 2017.

Water pollution due to chemicals is an inevitable consequence of modern society. Chemicals are used for a variety of everyday activities; only a very few examples are dishwashing, house cleaning, lawn care, and interior and exterior painting. Environment Canada estimates that 100,000 chemicals are commonly used.

Some such as pesticides, PCBs (which were used as a coolants in electrical transformers), and polychlorinated phenols (PCPs; which are present in wood preservatives) are toxic even when present in very low amounts. While PCBs were banned in the 1970s because of their environmental toxicity, they can still be found in older transformers. These compounds are designed and used because of their toxicity and long-lasting property—the same properties that make them so detrimental to the environment.

Acid rain is a serious form of water pollution, particularly in the northeastern United States. The term comes from the presence of nitric and sulfuric acids in the atmosphere close to the surface. Acid rain can be produced naturally, but the majority is generated by human activities, mainly from electricity generation that relies on the burning of fossil fuels such as coal. The burning releases sulfur dioxide and nitrogen oxides, which react with atmospheric water, oxygen, and other compounds to form a mild solution containing the nitric and sulfuric acids. The small droplets can stay suspended in the air and can be blown for hundreds of miles. As time goes on, the continued deposition of the droplets in water sources makes the water more acidic, which can kill vegetation and fish. As species lower in the food chain are affected, the entire ecosystem of a lake can be affected. In areas where acid rain is especially pronounced, the result can be lakes that are devoid of life. A lake that has been affected by acid rain can be sparkling clear in appearance. This beauty is deceptive, since the reason the water is so clear is that there is no microscopic of other life present.

In contrast, the phenomenon of eutrophication occurs because of the opposite reason; the presence of too much life. The overgrowth of the many organisms depletes the oxygen from the water. Although eutrophication is a natural process, it is greatly accelerated by pollution with compounds such as nitrogen and phosphorus, which are components of fertilizer. Drainage of fertilizer from surrounding land into a waterway boosts the amount of nutrients in the water, which can cause the explosive growth of microorganisms. The oxygen-starved water becomes inhospitable for fish. In turn, the decay of the dead fish uses up even more oxygen.

Impacts and Issues

One recent example of the effects of eutrophication and of the recovery that is possible even in severely polluted water is Lake Erie. In the 1960s, the addition of nitrogen and phosphorus to Lake Erie, the shallowest of the five Great Lakes, had depleted the oxygen to such an extent that massive deaths of fish were occurring. In 1972, legislation banning the use of phosphate in laundry detergent was passed. This reduced the amount of phosphorus entering Lake Erie by about 90%. The quality of the water recovered. In 2008, Lake Erie supports an abundance of life.

Groundwater pollution represents an even more serious threat to human health, because the pollutant can remain in the water; water contaminated with a radioactive substance, for example, can be unusable for thousands of years. In the United States, about half the population relies on groundwater as a drinking water source, for industrial purposes, or for crop irrigation.

Pollution of groundwater is not rare. For example, the EPA has records of over 400,000 cases of petroleum-based fuel spills from underground storage tanks. Although detection and remediation of groundwater pollution is a priority in the United States as mandated by the Resource Conservation and Recovery Act (RCRA) and the Comprehensive Environmental Response, Compensation, and Liability Act (CERCLA, which is also known as the Superfund), it is often a less important priority in developing countries where resources are limited. Developing countries also bear the brunt of waterborne diseases, due to the presence of pathogenic bacteria and viruses in water.

Recent studies indicate that global warming might create a more favorable environment for *V. cholerae*, the bacteria that causes cholera, and increase the incidence of the disease in vulnerable areas. A majority of cholera deaths occur in children under five years of age. Aside from the human tragedy of this loss, the effect on the economy is felt for decades. The World Health Organization has estimated that, in the poorest of developing nations, every dollar spent to remedy water pollution and inadequate sanitation would produce an economic return of $3 to $34.

The United Nations has set a goal of reducing by half the number of people who do not have access to safe drinking water and adequate sanitation by the year 2015. This goal will require the participation of developed countries and comes at a time when the changing global climate is beginning to affect the availability of water, and the prevalence of waterborne microbial diseases such as malaria and cholera.

IN CONTEXT: WATER POLLUTION IMPACTS ON MICROORGANISMS

There is a microbiological aspect of water pollution that experts fear reduces our ability to treat some bacterial infections: the presence in water of agents used to treat bacteria in other environments. For example, in the household a number of disinfectant compounds are routinely employed in the cleaning of household surfaces. In the hospital, the use of antibiotics to kill bacteria is an everyday occurrence. Such materials have been detected in water both before and after municipal wastewater treatment. The health effect of these compounds is not known at the present time. However, by analogy with other systems, the low concentration of such compounds might provide selective pressure for the development of resistant bacterial populations.

Primary Source Connection

The Water Quality Act of 1965 and the amendments that became law in 1977 represent two of several initiatives by the U.S. government to protect and ensure the quality of surface and ground waters.

Legislative concern over water quality began in 1948, with the passage of the Water Pollution Control Act. The act was essentially an adoption of principles to be followed in the pursuit of water quality. It was the Water Quality Act of 1965 that put some legislative teeth to these principles.

The 1965 legislation directed the states to develop water-quality standards. A federally directed initiative was deemed necessary since many watersheds and waterways crossed state boundaries. By the early 1970s, water quality standards had been developed and enacted by all the states. Since then, revisions have occurred to reflect changing scientific information and new testing procedures.

Environmentalist groups charge that brine discharged by the Mexican salt plant, ESSA, has killed animals in the waters off the Baja, California, coast. ESSA and the Mexican government deny these charges. *AP Images*

THE WATER QUALITY ACT OF 1965

An Act

To amend the Federal Water Pollution Control Act to establish a Federal Water Pollution Control Administration, to provide grants for research and development, to increase grants for construction of sewage treatment works, to require establishment of water quality criteria, and for other purposes.

Be it enacted by the Senate and House of Representatives of the United States of America in Congress assembled, That (a) (1) section 1 of the Federal Water Pollution Control Act (33 U.S.C. 466) is amended by inserting after the words "Section 1." a new subsection (a) as follows:

(a) The purpose of this Act is to enhance the quality and value of our water resources and to establish a national policy for the prevention, control, and abatement of water pollution.

Federal Water Pollution Control Administration

Sec. 2. Effective ninety days after the date of enactment of this section there is created within the Department of Health, Education, and Welfare a Federal Water Pollution Control Administration (hereinafter in this Act referred to as the "Administration").

Grants for Research and Development

Sec. 6. (a) The Secretary is authorized to make grants to any State, municipality, or intermunicipal or interstate agency for the purpose of assisting in the development of any project which will demonstrate a new or improved method of controlling the discharge into any waters of untreated or inadequately treated sewage or other waste from sewers which carry storm water or both storm water and sewage or other wastes, and for the purpose of reports, plans, and specifications in connection therewith. The Secretary is authorized to provide for the conduct of research and demonstrations relating to new or improved methods of controlling the discharge into any waters of untreated or inadequately treated sewage or other waste from sewers which carry storm water or both storm water and sewage or other wastes, by contract with public or private agencies and institutions and with individuals without regard to sections 3648 and 3709 of the Revised Statutes, except that not to exceed 25 per centum of the total amount appropriated under authority of this section for any fiscal year may be expended under authority of this sentence during such fiscal year.

(b) Federal grants under this section shall be subject to the following limitations: (1) No grant shall be made for any project pursuant to this section unless such project shall have been approved by an appropriate State water pollution control agency or agencies and by the Secretary; (2) no grant shall be made for any project in an amount exceeding 50 per centum of the estimated reasonable cost thereof as determined by the Secretary; (3) no grant shall be made for any project under this section unless the Secretary determines that such project will serve as a useful demonstration of a new or improved method of controlling the discharge into any water of untreated or inadequately treated sewage or other waste from sewers which carry storm water or both storm water and sewage or other wastes.

(c) There are hereby authorized to be appropriated for the fiscal year ending June 30, 1966, and for each of the next three succeeding fiscal years, the sum of $20,000,000 per fiscal year for the purposes of this section. Sums so appropriated shall remain available until expended. No grant or contract shall be made for any project in an amount exceeding 5 per centum of the total amount authorized by this section in any one fiscal year.

(3) Standards of quality established pursuant to this subsection shall be such as to protect the public health or welfare, enhance the quality of water and serve the purposes of this Act. In establishing such standards the Secretary, the Hearing Board, or the appropriate State authority shall take into consideration their use and value for public water supplies, propagation of fish and wildlife, recreational purposes, and agricultural, industrial, and other legitimate uses.

(5) The discharge of matter into such interstate waters or portions thereof, which reduces the quality of such waters below the water quality standards established under this subsection (whether the matter causing or contributing to such reduction is discharged directly into

such waters or reaches such waters after discharge into tributaries of such waters), is subject to abatement in accordance with the provisions of paragraph (1) or (2) of subsection (g) of this section, except that at least 180 days before any abatement action is initiated under either paragraph (1) or (2) of subsection (g) as authorized by this subsection, the Secretary shall notify the violators of other interested parties of the violation of such standards. In any suit brought under the provision of this subsection the court shall receive in evidence a transcript of the proceedings of the conference and hearing provided for in this subsection, together with the recommendations of the conference and Hearing Board and the recommendations and standards promulgated by the Secretary, and such additional evidence, including that relating to the alleged violation of the standards, as it deems necessary t a complete review of the standards and to a determination of all other issues relating to the alleged violation. The court, giving due consideration to the practicability and to the physical and economic feasibility of complying with such standards, shall have jurisdiction to enter such judgment and orders enforcing such judgment as the public interest and the equities of the case may require.

(6) Nothing in this subsection shall (A) prevent the application of this section to any case to which subsection (a) of this section would otherwise be applicable, or (B) extend Federal jurisdiction over water not otherwise authorized by this Act.

(7) In connection with any hearings under this section no witness or any other person shall be required to divulge trade secrets or secret processes.

Public Law 89-235

Joint Resolution

Authorizing and requesting the President to extend through 1966 his proclamation of a period to "See the United States," and for other purposes.

Resolved by the Senate and House of Representatives of the United States of America in Congress assembled, That the president is authorized and requested (1) to extend through 1966 the period designated pursuant to the joint resolution approved August 11, 1964 (Public Law 88-416), as a period to see the United States and its territories; (2) to encourage private industry and interested private organizations to continue their efforts to attract greater numbers of the American people to the scenic, historical, and recreational areas and facilities of the United States of America, its territories and possessions, and the Commonwealth of Puerto Rico; and (3) to issue a proclamation specially inviting citizens of other ceremonials to be celebrated in 1966 in the United States of America, its territories and possessions, and the Commonwealth of Puerto Rico.

Sec. 2. The President is authorized to publicize any proclamations issued pursuant to the first section and otherwise to encourage and promote vacation travel within the United States of America, its territories and possessions, and the Commonwealth of Puerto Rico, both by American citizens and by citizens of other countries, through such departments or agencies of the Federal Government as he deems appropriate, in cooperation with State and local agencies and private organizations.

Sec. 3. For the purpose of the extension provided for by this joint resolution, the President is authorized during the period of such extension to exercise the authority conferred by section 3 of the joint resolution approved August 11, 1964 (Public Law 88-416), and for such purpose may extend for such period the appointment of any person serving as National Chairman pursuant to such section.

Approved October 2, 1965.

U.S. Congress

U.S. CONGRESS. "THE WATER QUALITY ACT OF 1965." 79 STAT. 903, 70 STAT. 498. WASHINGTON, DC: U.S. CONGRESS, OCTOBER 2, 1965.

SEE ALSO *Agricultural Practice Impacts; Chemical Spills; Groundwater; Insecticide Use; Marine Water Quality; Oil Spills; Runoff; Toxic Waste; Wastewater Treatment Technologies*

BIBLIOGRAPHY

Books

Barlow, Maude. *Blue Covenant: The Global Water Crisis and the Coming Battle for the Right to Water.* New York: New Press, 2008.

Davis, Devra Lee. *When Smoke Ran Like Water: Tales of Environmental Deception and the Battle Against Pollution.* Oshkosh, WI: Basic Books, 2004.

Morris, Robert. *The Blue Death: Disease, Disaster, and the Water We Drink.* New York: Harper Collins, 207.

Web Sites

U.S Environmental Protection Agency (EPA). "Water Pollution." http://www.epa.gov/ebtpages/watewaterpollution.html (accessed May 11, 2008).

World Heath Organization. "Water, Sanitation, and Hygiene." http://www.who.int/water_sanitation_health/en/ (accessed May 11, 2008).

Brian D. Hoyle

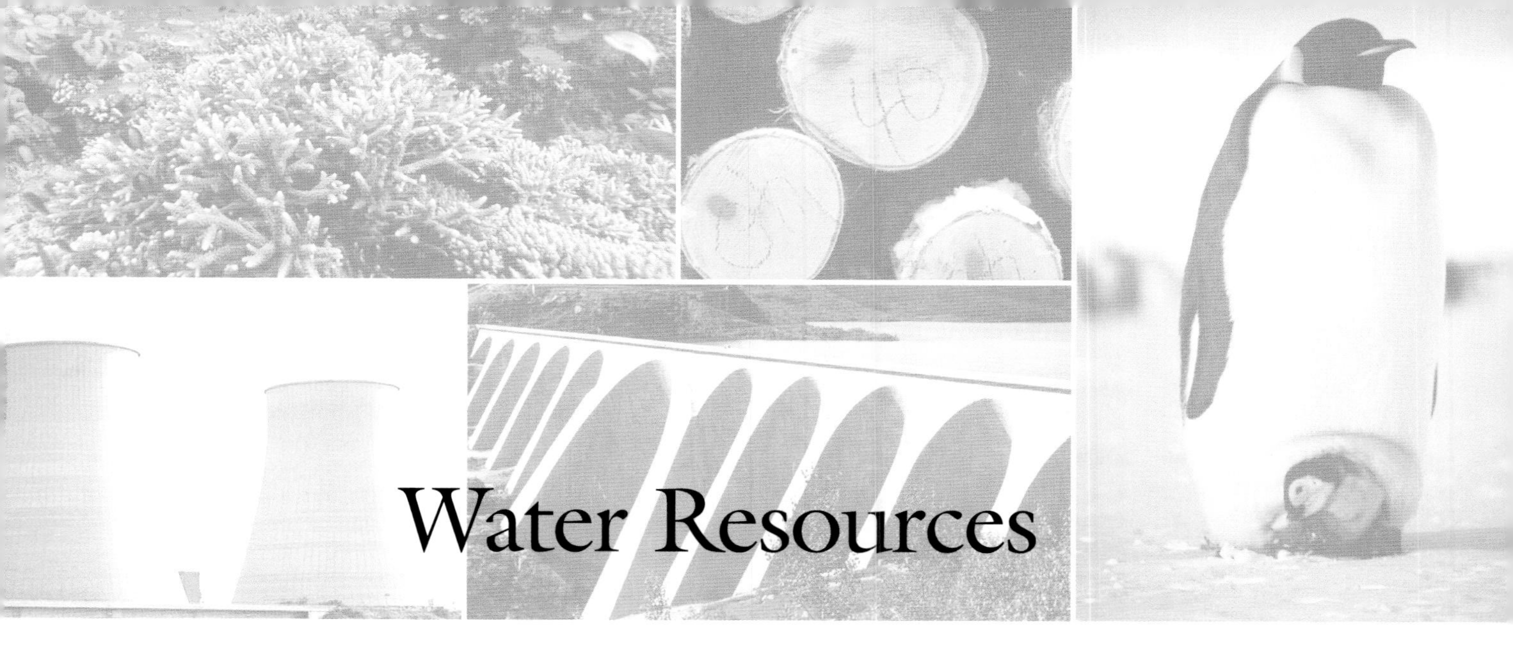

Water Resources

Introduction

Water is the world's most precious resource because the life of animals and plants depends on it. Most industries also require water for various applications, so the global economy depends on it as well. Most of the water on Earth is saltwater, which cannot be used by terrestrial organisms. Glaciers are the major freshwater resource, while the most important resource for human use is the surface runoff found in lakes and rivers.

Water is a renewable resource through the hydrologic cycle whereby water from the ocean moves onto the land and back again. Sometimes human intervention in the form of dams and pipelines diverts natural water resources to meet local needs. As need for water grows, tensions over water resources are likely to increase. Conservation measures and smarter technologies may help to ensure more equitable distribution of water around the world

Historical Background and Scientific Foundations

The oceans contain about 0.3 billion cubic miles of liquid water, which is around 97% of all water on Earth. Saltwater contains more than one gram per liter of dissolved solids, of which the most significant is sodium chloride or common salt. This renders it unfit for use by terrestrial animals, including humans, and plants, and for most industrial applications. Freshwater contains less than one gram per liter of dissolved solids, and is the main resource for human use. Most of it is, however, inaccessible because it is locked in glaciers, icecaps, and snow cover in the polar regions and elsewhere. Ice and snow account for about 90% of all freshwater. Varying amounts are released into nearby streams at various times during the year where they then become available as a resource. Around 1% of all freshwater is available for human use, amounting to only around 0.007% of the total amount of water on Earth.

Fortunately, water is a renewable resource through the hydrologic or water cycle, which allows the vast body of water in the oceans to be tapped for human use. Water molecules at the surface of the oceans evaporate into the atmosphere and move over the land as droplets of freshwater in clouds. These fall as rain or snow, known as precipitation. Some precipitation evaporates and the rest moves either vertically or horizontally on land. The vertical portion fills up pores in rocks, clay, sand, or soil which, when saturated, are known as aquifers. The water in an aquifer is called groundwater. When an aquifer is trapped between two layers of impermeable rocks, the resulting pressure may create an Artesian well that brings the water up to the surface. Pumps can also bring groundwater to the surface through a well. Groundwater is generally of high quality but care should be taken not to draw off too much, as it renews itself only slowly, particularly in dry regions.

Precipitation that moves along horizontally is called surface runoff and it is carried, by gravity, to the nearest body of surface water, which could be a stream, river, pond, or lake. Eventually it arrives back in the ocean, completing the hydrological cycle. Surface water is the most important water resource because it is often readily accessible. A stream is a small channel of water that eventually runs into a river. Streams and rivers account for about nearly 500 billion gallons of water, which is about 0.0001% of the total water on Earth, including the oceans. Yet they are probably the most important water source. If rivers and streams were not replenished by precipitation, melting snow and ice, and seeping groundwater, they would probably run dry in a matter of weeks because of human withdrawals.

A pond is a small body of water that is shallow enough for plants to root there. Lakes are larger bodies

WORDS TO KNOW

AQUIFER: Rock, soil, or sand underground formation that is able to hold and/or transmit water.

GROUNDWATER: Fresh water that is present in an underground location.

SURFACE WATER: Water collecting on the ground or in a stream, river, lake, wetland, or ocean, as opposed to groundwater.

of water whose depths may vary from a few feet to over a mile, as in Lake Baikal in Siberia. Their areas vary from around an acre to hundreds of thousands of acres like Lake Superior, which is really an inland sea. Reservoirs are natural or artificial ponds or lakes used for storing water. Lakes and reservoirs account for around 0.28% of the total water on Earth and they are also an important resource for human use.

Impacts and Issues

Although water is a renewable resource, the amount available for human use is affected by various threats. These include pollution, urban growth, landscape changes, drought, and climate change. Farming, deforestation, mining, and road-building can all impair the quality of water by allowing too much soil and pollutants to enter local rivers, streams, and lakes.

Care must also be taken not too overexploit a water resource. Twenty million people in Chad, Niger, Nigeria, and Cameroon depend on Lake Chad for water. But the lake has shrunk drastically in recent years, and shortages have caused conflicts between the local populations. Meanwhile, the Aral Sea between Kazakhstan and Uzbekistan in Central Asia has been shrinking since the 1960s because two of the rivers that feed it are diverted for irrigation. The water is also heavily polluted because of sewage dumping. Many major rivers, such as the Nile, Ganges, and Rio Grande, are showing signs of drying up. One major factor is that water withdrawals are still being poorly managed and controlled. There has also been a trend to increase groundwater withdrawals, where depletions are less obvious than in surface water. But signs of poor management are there in terms of subsidence, poor water quality, and a sinking water table, which is the top part of an aquifer.

The relationship between climate change and water resources is currently little understood, although it is more likely than not to lead to water shortages. For instance, while climate change tends to shrink glaciers, the effect is not to enrich nearby water resources. Most of the water released tends to evaporate long before it reaches any drought-stricken areas that need it. Global warming increases the incidence of drought, increasing the pressure on water supplies in dry areas. Meanwhile, extreme weather events stemming from global warming, such as floods, tend to degrade the quality of water resources.

Clearly there is a need to develop water resources in a more sustainable manner, taking account of the various pressures on them. Organizations such as the World Health Organization stress that water policy should be driven more by scientific understanding of the consequences of a lack of adequate freshwater for all peoples, rather than by short-term economic or political goals. This includes applying what is known of local water resources and how they interact with the water cycle. Traditionally, rising demand has been met by storing surface water in a reservoir, diverting flows to drier areas, and using increasing amounts of groundwater. Other techniques, such as rainwater collection, desalination, and water reuse can be added to help protect water resources so they can continue to meet local needs without becoming depleted or degraded.

The United Nation's Millennium Goal number seven—to ensure environmental sustainability—includes the following additional goal: "to reduce by half the proportion of people without sustainable access to safe drinking water." The UN declared the years 2005–2015 the "Water for Life" decade, an International Decade for Action where all nations should aim to end the unsustainable exploitation of water resources.

Primary Source Connection

The rivers and waterways of the United States have always been used for transportation, irrigation, drinking water, and for the removal of waste. In the twentieth century, many rivers were dammed to prevent flooding and to be harnessed for hydroelectric power. Rivers were diverted or channeled so that their flow could be controlled. Levees were built to contain flooding. Development of the rivers and their adjacent land was seen as a means to increase productivity and economic growth. The Wild and Scenic Rivers Act was signed into the U.S. Code on October 2, 1962 by President Lyndon B. Johnson. The act established the Wild and Scenic Rivers System that oversees the preservation of all rivers that are designated under the act. In order to be designated, a river must be free-flowing and contain an "outstandingly remarkable" feature, such as scenery, historical value, geological features, or particular fish and wildlife. The first three sections and the seventh section of the act are excerpted below.

The Wild and Scenic Rivers Act is considered an important piece of environmental regulation. Since its inception, 156 rivers comprising nearly 11,000 mi

(17,700 km) of waterways have been designated part of the Wild and Scenic Rivers System. Although this is an impressive achievement, it contrasts with the more than 600,000 mi (966,000 km) of once free-flowing rivers in the United States that are contained by more than 60,000 dams.

The Wild and Scenic Rivers Act has been criticized for not doing enough to protect rivers from pollution or development (especially along critical feeder streams and banks). However, most environmentalists consider the act to be a valuable conservation tool. No river that has been designated by Congress has ever been removed from the system, indicating that the benefits of the act outweigh any detriments. Part of the reason that the act has been successful is because its intent is not to completely block the use of a designated river, but instead to require that the river be managed so that the fundamental character of the river is preserved. Any development on the river must respect the free flow of water and protect its natural features. The Wild and Scenic Rivers System must co-exist with development so as to create a management plan that protects natural values as well as property values.

WILD AND SCENIC RIVERS ACT

P.L. 90-542, as amended

16 U.S.C. 1271–1287

AN ACT

To provide for a National Wild and Scenic Rivers System, and for other purposes.

Be it enacted by the Senate and House of Representatives of the United States of America in Congress assembled, that

SECTION 1.

(a) this Act may be cited as the "Wild and Scenic Rivers Act."

(b) It is hereby declared to be the policy of the United States that certain selected rivers of the Nation which, with their immediate environments, possess outstandingly remarkable scenic, recreational, geologic, fish and wildlife, historic, cultural, or other similar values, shall be preserved in free-flowing condition, and that they and their immediate environments shall be protected for the benefit and enjoyment of present and future generations. The Congress declares that the established national policy of dam and other construction at appropriate sections of the rivers of the United States needs to be complemented by a policy that would preserve other selected rivers or sections thereof in their free-flowing condition to protect the water quality of such rivers and to fulfill other vital national conservation purposes.

(c) The purpose of this Act is to implement this policy by instituting a national wild and scenic rivers system, by designating the initial components of that system, and by prescribing the methods by which and standards according to which additional components may be added to the system from time to time.

SECTION 2.

(a) The national wild and scenic rivers system shall comprise rivers

(i) that are authorized for inclusion therein by Act of Congress, or

(ii) that are designated as wild, scenic or recreational rivers by or pursuant to an act of the legislature of the State or States through which they flow, that are to be permanently administered as wild, scenic or recreational rivers by an agency or political subdivision of the State or States concerned, that are found by the Secretary of the Interior, upon application of the Governor of the State or the Governors of the States concerned, or a person or persons thereunto duly appointed by him or them, to meet the criteria established in this Act and such criteria supplementary thereto as he may prescribe, and that are approved by him for inclusion in the system, including, upon application of the Governor of the State concerned, the Allagash Wilderness Waterway, Maine; that segment of the Wolf River, Wisconsin, which flows through Langlade County; and that segment of the New River in North Carolina extending from its confluence with Dog Creek downstream approximately 26.5 miles to the Virginia State line.

Upon receipt of an application under clause (ii) of this subsection, the Secretary shall notify the Federal Energy Regulatory Commission and publish such application in the Federal Register. Each river designated under clause (ii) shall be administered by the State or political subdivision thereof without expense to the United States other than for administration and management of federally owned lands. For purposes of the preceding sentence, amounts made available to any State or political subdivision under the Land and Water Conservation 'Fund' Act of 1965 or any other provision of law shall not be treated as an expense to the United States. Nothing in this subsection shall be construed to provide for the transfer to, or administration by, a State or local authority of any federally owned lands which are within the boundaries of any river included within the system under clause (ii).

(b) A wild, scenic or recreational river area eligible to be included in the system is a free-flowing stream and the related adjacent land area that possesses one or more of the values referred to in Section 1, subsection (b) of this Act. Every wild, scenic or recreational river in its free-flowing condition, or upon restoration to this condition, shall be considered eligible for inclusion in the national wild and scenic rivers system and, if included, shall be classified, designated, and administered as one of the following:

1. Wild river areas—Those rivers or sections of rivers that are free of impoundments and generally inaccessible except by trail, with watersheds or shorelines essentially primitive and waters unpolluted. These represent vestiges of primitive America.
2. Scenic river areas—Those rivers or sections of rivers that are free of impoundments, with shorelines or watersheds still largely primitive and shorelines largely undeveloped, but accessible in places by roads.
3. Recreational river areas—Those rivers or sections of rivers that are readily accessible by road or railroad, that may have some development along their shorelines, and that may have undergone some impoundment or diversion in the past.

SECTION 7.

(a) The Federal Power Commission shall not license the construction of any dam, water conduit, reservoir, powerhouse, transmission line, or other project works under the Federal Power Act (41 Stat. 1063), as amended (16 U.S.C. 791a et seq.), on or directly affecting any river which is designated in section 3 of this Act as a component of the national wild and scenic rivers system or which is hereafter designated for inclusion in that system, and no department or agency of the United States shall assist by loan, grant, license, or otherwise in the construction of any water resources project that would have a direct and adverse effect on the values for which such river was established, as determined by the Secretary charged with its administration. Nothing contained in the foregoing sentence, however, shall preclude licensing of, or assistance to, developments below or above a wild, scenic or recreational river area or on any stream tributary thereto which will not invade the area or unreasonably diminish the scenic, recreational, and fish and wildlife values present in the area on the date of designation of a river as a component of the national wild and scenic rivers system. No department or agency of the United States shall recommend authorization of any water resources project that would have a direct and adverse effect on the values for which such river was established, as determined by the Secretary charged with its administration, or request appropriations to begin construction of any such project, whether heretofore or hereafter authorized, without advising the Secretary of the Interior or the Secretary of Agriculture, as the case may be, in writing of its intention so to do at least sixty days in advance, and without specifically reporting to the Congress in writing at the time it makes its recommendation or request in what respect construction of such project would be in conflict with the purposes of this Act and would affect the component and the values to be protected by it under this Act. Any license heretofore or hereafter issued by the Federal Power Commission 'FERC' affecting the New River of North Carolina shall continue to be effective only for that portion of the river which is not included in the national wild and scenic rivers system pursuant to section 2 of this Act and no project or undertaking so licensed shall be permitted to invade, inundate or otherwise adversely affect such river segment.

(b) The Federal Power Commission 'FERC' shall not license the construction of any dam, water conduit, reservoir, powerhouse, transmission line, or other project works under the Federal Power Act, as amended, on or directly affecting any river which is listed in section 5, subsection (a), of this Act, and no department or agency of the United States shall assist by loan, grant, license, or otherwise in the construction of any water resources project that would have a direct and adverse effect on the values for which such river might be designated, as determined by the Secretary responsible for its study or approval

(i) during the ten-year period following enactment of this Act or for a three complete fiscal year period following any Act of Congress designating any river for potential addition to the national wild and scenic rivers system, whichever is later, unless, prior to the expiration of the relevant period, the Secretary of the Interior and where national forest lands are involved, the Secretary of Agriculture, on the basis of study, determine that such river should not be included in the national wild and scenic rivers system and notify the Committees on Interior and Insular Affairs of the United States Congress, in writing, including a copy of the study upon which the determination was made, at least one hundred and eighty days while Congress is in session prior to publishing notice to that effect in the Federal Register: Provided, That if any Act designating any river or rivers for potential addition to the national wild and scenic rivers system provides a period for the study or studies which exceeds such three complete fiscal year period the period provided for in such Act shall be substituted for the three complete fiscal year period in the provisions of this clause (i); and

(ii) during such interim period from the date a report is due and the time a report is actually submitted to the Congress; and

(iii) during such additional period thereafter as, in the case of any river the report for which is submitted to the President and the Congress for inclusion in the national wild and scenic rivers system, is necessary for congressional consideration thereof or, in the case of any river recommended to the Secretary of the Interior for inclusion in the national wild and scenic rivers system under section 2(a)(ii) of this Act, is necessary for the Secretary's consideration thereof, which additional period, however, shall not exceed three years in the first case and one year in the second.

Nothing contained in the foregoing sentence, however, shall preclude licensing of, or assistance to, developments below or above a potential wild, scenic or recreational river area or on any stream tributary thereto which will not invade the area or diminish the scenic, recreational, and fish and wildlife values present in the potential wild, scenic or recreational river area on the date of designation of a river for study as provided in section 5 of this Act. No department or agency of the United States shall, during the periods hereinbefore specified, recommend authorization of any water resources project on any such river or request appropriations to begin construction of any such project, whether heretofore or hereafter authorized, without advising the Secretary of the Interior and, where national forest lands are involved, the Secretary of Agriculture in writing of its intention so to do at least sixty days in advance of doing so and without specifically reporting to the Congress in writing at the time it makes its recommendation or request in what respect construction of such project would be in conflict with the purposes of this Act and would affect the component and the values to be protected by it under this Act.

(c) The Federal Power Commission 'FERC' and all other Federal agencies shall, promptly upon enactment of this Act, inform the Secretary of the Interior and, where national forest lands are involved, the Secretary of Agriculture, of any proceedings, studies, or other activities within their jurisdiction which are now in progress and which affect or may affect any of the rivers specified in section 5, subsection (a), of this Act. They shall likewise inform him of any such proceedings, studies, or other activities which are hereafter commenced or resumed before they are commenced or resumed.

(d) Nothing in this section with respect to the making of a loan or grant shall apply to grants made under the Land and Water Conservation Fund Act of 1965 (78 Stat. 897; 16 U.S.C. 460l-5 et seq.).

U.S. Code

U.S. CODE. "WILD AND SCENIC RIVERS ACT." TITLE 16, CHAPTER 28, SECTIONS 1271, 1272, 1277.

SEE ALSO *Groundwater; Natural Resource Management; Surface Water; Water Conservation; Water Supply and Demand*

BIBLIOGRAPHY

Books

Cunningham, W.P., and A. Cunningham. *Environmental Science: A Global Concern*. New York: McGraw-Hill International Edition, 2008.

Kaufmann, R., and C. Cleveland. *Environmental Science*. New York: McGraw-Hill International Edition, 2008.

Web sites

Green Facts: Facts on Health and Environment. "Scientific Facts on Water Resources." http://www.greenfacts.org/en/water-resources/index.htm#2 (accessed May 2, 2008).

United Nations. "International Decade for Action: Water for Life, 2005–2015." http://www.un.org/waterforlifedecade/background.html (accessed May 2, 2008).

Susan Aldridge

Water Supply and Demand

Introduction

Water is essential to human life and to many industries. While by far the largest body of water on Earth is the saltwater of the oceans, freshwater is required for most human activities. However, most freshwater is locked up in glaciers and polar ice caps. Water supplies are drawn mainly from rivers, lakes, and groundwater. The hydrologic or water cycle renews water supplies by moving water from the oceans, as freshwater, to land.

Water consumption varies widely from place to place, but is increasing everywhere as global population rises. Sometimes supply cannot meet local demand, which creates conflict. Climatic factors, such as drought and global warming, water management practices, and over-exploitation place pressure upon water supplies. Diversion of water resources to increase water supplies often has an adverse impact on water quality and the local ecology. Water conservation measures and water management based on sound scientific principles are needed to avert a global water crisis.

Historical Background and Scientific Foundations

Around 2.5% of the water on Earth is freshwater and therefore suitable for human use. Of this, only 1% is readily available as a water supply, mainly from lakes, rivers, and groundwater. When rain or snow fall to the ground, as precipitation, some of the water forms surface runoff, moving horizontally under gravity toward the nearest body of water, such as a stream or lake. This surface runoff is the most important source of water supplies. The rest moves vertically into the ground where it fills pores in soil, sand, clay, or rocks known as aquifers. This is known as groundwater, and it is increasingly important as a water supply, although it may require pumping to bring it to the surface.

The availability of freshwater supplies varies between countries. The United Nations keeps an ongoing database that calculates the amount of water available per person in 193 countries. Australia, a dry but large and sparsely populated country, has 6,600,000 gallons per person per year, compared to just over 2,000 gallons per person per year in Kuwait, which is also dry but much smaller. Nearby Bahrain has practically no freshwater of its own and relies on desalination of sea water and imports. The figure for the United States is 2,640,000 gallons per person per year. Many African countries have a lot of water available, such as the Democratic Republic of Congo, which has over 57 million gallons per person per year, while others are short of water, such as Rwanda, which has 161,000 gallons per person per year. Overall, South America, West Central Africa, and South and Southeast Asia have plenty of rainfall, along with Canada and Russia. Egypt, even though it has the River Nile, is relatively water poor, with only 11,000 gallons per person per year.

Most water used for human activities returns to rivers and streams in the form of wastewater. In water calculations, water withdrawal refers to amounts taken from the water supply, some of which will be returned into circulation eventually. Water consumption is the water that is withdrawn and used in such a way that it is not returned to circulation. For instance, some water is lost through leaks, evaporation, or some kind of chemical transformation. Some water that is withdrawn, but not consumed, becomes degraded by pollution or heating so it is no longer suitable for reuse. Water is a renewable resource, through the hydrologic cycle, and the amount of water in the oceans is vast. This fact has led people to take water for granted. But the natural hydrologic cycle is slow and water supplies do not renew themselves overnight. Therefore, it is important to return as much of water withdrawals to the supply as possible by minimizing consumption.

Water use has been increasing as the global population has risen over the last century. It has leveled off in industrialized nations, which tend to have small birth rates, but will continue to increase in developed countries. Global average water withdrawal per person per year is 170,544 gallons. Some countries can meet their population's demands with ease. Canada, Brazil, and the Democratic Republic of Congo take less than 1% of their water resources for their populations. But Libya and Israel do not have enough surface water to meet demand and are forced to rely on extracting groundwater, which will not be sustainable in the long run.

Water demand comes from the agricultural, industrial, and domestic sectors. Agriculture accounts for the greatest use and consumption of water worldwide. Crop irrigation accounts for two-thirds of water withdrawals and 85% of consumption worldwide. Industry accounts for 20% of water withdrawals worldwide. Industrial water withdrawals vary from up to 70% in countries with a lot of industry, to less than 5% in some less-developed nations. Power production and mining account for the most industrial water use, with water being required for cleaning and cooling processes. Unlike agricultural water, most industrial water can be returned to the water supply, only it will be degraded if it is not treated first.

Domestic water use accounts for around 20% of water withdrawals worldwide. People in wealthy countries will use 100 to 200 gallons of water per day, compared to only 7 to 40 gallons a day in developing countries. The major domestic use of water is in toilet flushing. In the United States, the average individual will use 13,000 gallons of water a year in flushing the toilet. Bathing comes next, followed by laundry and dishwashing. Around 6% of domestic water is used for drinking and cooking, and miscellaneous activities like brushing teeth account for the remaining 5%. Since 1960, domestic use of water has grown by about 50% as urban populations have increased.

WORDS TO KNOW

DESALINATION: Removal of salt from saltwater to produce freshwater.

WATER CONSUMPTION: Use of water in such a way that it cannot be re-used.

WATER STRESS: Inability to provide enough water to meet basic needs.

WATER WITHDRAWAL: Removal of water from a water supply, some of which will later be returned.

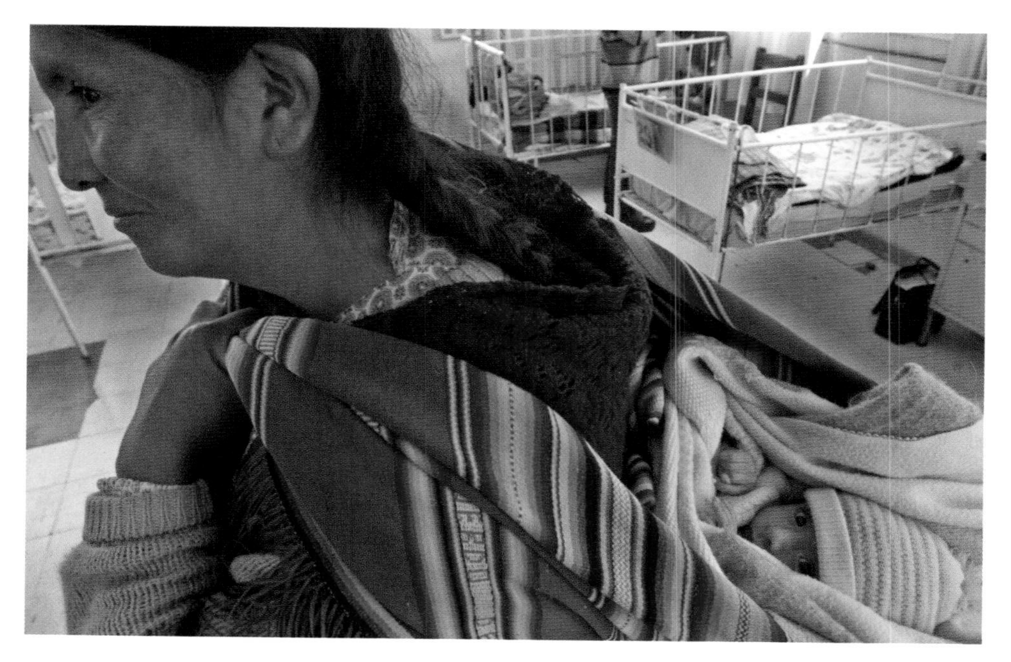

The consumption of contaminated water in Bolivia is the probable cause of a chronic diarrhea resulting in the baby girl weighing less than half of what she should weigh (2003). According to a UNICEF report, 1.6 million children die every year due to the consumption of contaminated water. *AP Images*

With streams drying up, croplands turning to ravines and ambitious government supply projects failing to keep up with demand, Indian environmentalists are suggesting the country return to age-old methods to save itself from a water crisis. *AP Images*

Impacts and Issues

According to the United Nations, at least one billion people around the world lack access to safe drinking water and 2.6 billion do not have adequate sanitation. Global warming and population growth could exacerbate these shortages in years to come unless water conservation and management practices around the world improve.

At least 45 countries around the world, mainly in Africa or the Middle East, have severe water stress in that they cannot provide enough clean water to meet their citizens's basic needs. Sometimes a country may have enough water, but it is not clean enough. This is often more of a problem for rural residents than those living in the city. In some developing countries, there is no infrastructure to deliver water so people have to spend many hours fetching water, instead of doing productive work.

Wars have been fought over oil. Water is, arguably, an even more precious resource and scarcities can lead to conflict. Water has not, as yet, been the sole cause of a major war, but is often a factor. For instance, water supplies may be hit as a military target. Most water-centered conflicts have been local. For instance, farmers and traditional herders have clashed over allocations of water in the Ivory Coast, Burkina Faso, and Ghana. There has also been recent conflict between police and farmers in the Hirakud dam area of India over diversion of water to industrial needs. However, water can also promote cooperation, if both sides work together to protect a water resource that spans their countries. Thus, the Indus River Commission continues to function despite ongoing conflict between India and Pakistan.

Primary Source Connection

The Water Sourcebook Series is a project of the Office of Ground Water and Drinking Water (OGWDW). Developed in 2000, the Water Sourcebook Series is the result of a partnership among the EPA Region Four (based in Atlanta, Georgia), the Alabama Department of Environmental Regulation, and Legacy, Inc.

Water Sourcebooks: K–12 is an environmental education program that contains over 300 activities for schoolchildren from kindergarten to twelfth grade—with the reoccurring environmental message, "Use What You Need and Don't Pollute." The Water Sourcebook program describes the water management cycle and its effect on all parts of the environment. The curriculum involves science and mathematics, along with subject areas in the social studies, language arts, reading, and other educational areas. The activities involved with the program vary from fact sheets, reference materials, hands-on investigations, and a glossary of terms. Each activity is organized by objectives, materials needed, background information, advance preparation, procedures, and resources.

WATER SOURCEBOOKS: K-12

INTRODUCTION

The value of clean, safe water for individuals, communities, businesses, and industries can't be measured. Every living thing depends on water. The economy requires it. Water issues should be everyone's concern, but most people take water quality and availability for granted. After all, clean, safe water is available to most Americans every time they turn on the tap. Water issues do not become a concern until there is a crisis such as a drought or wastewater treatment plant failure. Educating citizens who must make critical water resource decisions in the midst of a crisis rarely results in positive change. Developing awareness, knowledge, and skills for sound water use decisions is very important to young people, for they will soon be making water resource management decisions. Properly equipping them to do so is essential to protect water resources.

WATER SOURCEBOOK PROGRAM

The Water Sourcebook educational program is directed specifically toward the in-school population. The program consists of supplemental activity guides targeting kindergarten through high school. Water Sourcebooks are available for primary (K-2), elementary (3-5), middle (6-8), and secondary (9-12) levels. Materials developed in the program are compatible with existing curriculum standards established by State Boards of Education throughout the United States as well as national standards in science, social studies and geography. Concepts included in these standards are taught by using water quality information as the content. The Water Sourcebooks include five chapters—Introduction, Drinking Water and Wastewater Treatment, Groundwater, Surface Water, and Wetlands.

DEVELOPMENT

The Water Sourcebooks are developed in three stages. First, classroom teachers are selected to write the activities with assistance of education specialists. Teams of teachers are given the task of developing and writing the activities for each of the five instructional chapters. The second step involves testing activities in the classroom and technical reviews by water experts. From the evaluations provided by the testing teachers and technical reviewers, revisions are made. Finally, editing, and illustrations are complete and the Water Sourcebook is published.

ACTIVITY DESIGN

All of the activities include "hands-on" components and are designed to blend with existing curricula in the areas of general sciences, language arts, math, social studies, art, and in some cases, reading or other areas. Each activity details (1) objectives, (2) subjects(s), (3) time, (4) materials, (5) background information, (6) advance preparation, (7) procedure (including activity, follow-up, and extension), and (8) resources. Fact sheets and a glossary section are included at the end of the guide to help equip teachers to deal with concepts and words used in the text which may be unfamiliar.

U.S. Environmental Protection Agency

U.S. ENVIRONMENTAL PROTECTION AGENCY. *WATER SOURCEBOOKS: K–12.* WASHINGTON, DC: U.S. ENVIRONMENTAL PROTECTION AGENCY, 2000.

■ Primary Source Connection

The following news article reports on how the 2007 California water crisis has resulted in enforced water conservation for both California citizens and businesses. Tim Quinn, executive director of the Association of California Water Agencies, explains the difficulty of California's water issue as, "the clash between the environment, the California economy, and the population, which is pouring in at more than 600,000 per year." Water conservation measures are also affecting the economy of the state, especially the farmers and manufacturers, and water supply measures are sometimes at odds with environmentalists.

WATER CRISIS SQUEEZES CALIFORNIA'S ECONOMY

California farmers, who produce half the nation's fruits and vegetables, say they will idle fields and cut back on planting lettuce, cotton, rice, and more.

Silicon Valley computer-chip makers and other industrial/commercial users say they will rethink manufacturing processes that use water, or dramatically raise the price of products they sell.

Cities from Sacramento to San Diego say drought-era practices of rationed water—low-use toilets and washers, designated water days for lawns and cars—are back, including stiff fines for those who don't follow the rules.

After 35 years of hemming and hawing over how to fix the largest estuary in the Western Hemisphere—the sprawl of canals, levees, and flood plains that join the Golden State's two river systems—the state has been told by a federal judge that business-as-usual is now illegal.

A new ruling to stop pumping up to 37 percent of the water that flows through the delta to protect endangered fish species has sent shock waves of concern into the three main sectors that have long competed for it: cities, farms, environment.

The estuary provides water to 23 million Californians and about 5 million acres of farmland. Overused and under maintained for years, the delta and its water are at the heart of the state's economic vitality, its wildlife habitat, shipping, transportation, drinking water, and recreation.

"In the water business we are facing the biggest challenges here in over half a century … there is no way any knowledgeable person could contest that," says Tim Quinn, executive director of the Association of California Water Agencies, which represents more than 450 of the state's water agencies that provide water to 95 percent of the state's farms and cities. "The state has had some success in better managing this problem for the last decade, but we have hid ourselves from the biggest issue … and Mother Nature is telling us there is no more hiding."

The biggest issue, say Mr. Quinn and others, is the clash between the environment, the California economy, and the population, which is pouring in at more than 600,000 per year.

In recent years, the use of water in the delta has been crippled as a result of drought as well as age, with deteriorating levees that are vulnerable to flood, earthquake, and subsidence. It was environmental groups who most recently challenged water use. The ruling by U.S.

District Court Judge Oliver Wanger in Fresno Aug. 31 came after a suit against state and federal water officials by the Natural Resources Defense Council (NRDC) and three other environmental groups.

The agricultural community in the Central Valley appears to be the most worried about the consequences of the judge's action.

Stephen Patricio, chairman of the board of directors for the Western Growers Association and a 30-year farmer of 2,000 acres of cantaloupe near Firebaugh, Calif., says the impact of a 30 percent or more reduction of water to his region will have a domino effect on other jobs.

He expects his $6 million payroll, employing about 600 people during the harvest, will drop to about $1.5 million this year and force him to cut 400 jobs. Those losses will contribute to another 2,400 layoffs in related industries: truck drivers, tractor operators, seed operations, warehousing, repair, and fuel, he says.

The announcement is already having an effect on the loans farmers receive to operate their farms during 2008.

"Ninety percent of these farms need to be financed, and lenders have made it very clear that without a water plan, there is no money for 2008 crops," Mr. Patricio says.

Eighty percent of the water in California moves from above the delta to farms and communities in the south of it via pumps. The environmental groups said that current use of water pumping through the delta endangered several species of fish, including two kinds of smelt (long fin and delta), steel head, green sturgeon, winter and spring salmon, and split tail.

"This ruling was essentially an agreement that we need to protect habitat in the delta more than we have been, and what state and federal agencies have been doing is likely to drive the smelt to extinction," says Barry Nelson of the NRDC. "For all the serious concern about how the state is now going to meet its water needs, no one is saying that the court got it wrong. Everyone has known for a long time that this was coming."

Water agencies, farmers, and scientists agree, saying the ruling will force a much-needed opportunity to fine-tune the use of water to avoid waste.

"This alarm as been sounding about the delta for 30 years, and we've been pushing the snooze alarm," says Peter Gleick, president of the Pacific Institute, a non-profit think tank in Oakland, Calif. "This may lead to some of the first discussions ever on how we manage growth in the state, how people live, how much we waste, and what crops we farm."

He notes that farming just three common crops—rice, cotton, and alfalfa—as well as irrigated pastures for cows use about half of the agricultural sector's allotment but earn a fraction of agricultural income. "We can continue to have a healthy economy with less water, but there has been no demand to do that yet," he says. "This ruling may drive us to do things we ought to be doing anyway."

Anticipating the reduced water spigot from the delta as of January 2008, water agencies north of it are telling their clients to cut back on water use. They are already spending money in new ad campaigns to remind users to cut back or face the possibility of mandatory laws with fines.

"We are spending millions to get the conservation message out that we need to conserve as if we are in a crunch," says Jeff Kightlinger, of the Metropolitan Water District of Southern California, which serves agencies and 18 million residents in six counties. He says the new rationing will affect 2 of every 3 Californians.

"Farmers across the state know this will be very tough and not pleasant," says Dave Kranz, spokesman for the California Farm Bureau Federation. "To the extent that you take farmland out of production for whatever reason, it increases another problem, which is providing enough American-grown food to serve the US population as well as demand from other countries."

Solutions are now in the works. A commission appointed by Gov. Arnold Schwarzenegger is planning to make recommendations next month, including ideas for dams and more storage capacity.

State Senate President pro tem Don Perata (D) is pushing two measures. One provides $200 million for immediate safeguards of freshwater flows from north to south. The second is a $5 billion bond that includes $2 billion to fix the water supply, improve flood protection, and boost fisheries in the delta, and $2 billion for water storage projects such as dams.

Daniel B. Wood

WOOD, DANIEL B. "WATER CRISIS SQUEEZES CALIFORNIA'S ECONOMY." *CHRISTIAN SCIENCE MONITOR* (SEPTEMBER 12, 2007).

SEE ALSO *Groundwater; Surface Water; Water Conservation; Water Resources*

BIBLIOGRAPHY

Books

Cunningham, W.P., and A. Cunningham. *Environmental Science: A Global Concern*. New York: McGraw-Hill International Edition, 2008.

Kaufmann, R., and C. Cleveland. *Environmental Science*. New York: McGraw-Hill International Edition, 2008.

Web Sites

Green Facts: Facts on Health and Environment. "Scientific Facts on Water Resources." http://www.greenfacts.org/en/water-resources/index.htm#2 (accessed July 15, 2008).

Pacific Institute: The World's Water. "Water and Conflict." February 2008. http://www.worldwater.org/conflict.html (accessed March 25, 2008).

Susan Aldridge

Watershed Protection and Flood Prevention Act of 1954

Introduction

The Watershed Protection and Flood Prevention Act of 1954 (WPFPA) is a law that protects watersheds from erosion, sedimentation, and flooding. Under WPFPA, federal agencies work with local organizations to develop and implement flood control and watershed runoff plans. Flooding and poor watershed runoff management both damage the environment by carrying sediment and pollutants into streams and rivers. Sedimentation and pollution in water systems harms ecosystems and makes rivers and lakes unsuitable for fishing, swimming, or drinking. Federal and local agencies have also implemented numerous flood control plans to prevent property damage and loss of life that can occur from flooding.

Historical Background and Scientific Foundations

Within a watershed, all surface and underground water drains to the same place. Watersheds are an important adjunct to the hydrological cycle. Precipitation that falls in an isolated mountain area may feed into a small stream, which then feeds into larger river systems. Precipitation that falls hundreds of miles inland may eventually feed back into the ocean. The watershed of an area determines where all water in that area goes. Almost every place in the world is located within a watershed. The Mississippi River Basin is the largest watershed in the United States. Approximately 41% of the land in the continental United States drains through the Mississippi River to the Gulf of Mexico.

Restricting pollution within watersheds is important, because watersheds play a vital role in ecosystems and provide humans with water for irrigation and drinking. Water draining through watersheds can also be destructive. Floods are the result of too much water trying to drain through a watershed at once. By controlling the amount of water that drains through a watershed, humans can ensure a steady supply of water while also preventing, or lessening, flooding.

The U.S. Congress passed the Watershed Protection and Flood Prevention Act of 1954 (WPFPA) to support watershed flood control projects throughout the United States. The goal of WPFPA is to preserve, protect, and improve the land and water resources and the quality of the environment in the United States. The WPFPA seeks to achieve the following: prevent flood, sediment, and erosion damage; further the conservation, development, use, and disposal of water; and further the conservation and use of land.

The WPFPA sought cooperation between U.S. Department of Agriculture and local organizations in preparing and implementing flood control plans. The WPFPA defines local organizations as a state or local government; soil or water conservation district; flood prevention or control district; other agencies with authority under state law to carry out works of improvement; nonprofit irrigation or reservoir company, water users' association or similar organization; or tribal organization.

The WPFPA authorized the Secretary of Agriculture to construct flood protection measures below a certain volume limit. Such initiatives were to be cost-shared and localities were required to contribute rights-of-way. The law also required that the Secretary of the Interior be consulted regarding plans that affect reclamation, irrigation, or public lands controlled by the Department of the Interior.

Congress amended the WPFPA in 1956 to give Congress greater control over large flood control projects. The 1956 Amendments required the Secretary of Agriculture to submit to Congress any plan for works of improvement in watershed areas where the federal contribution exceeded $250,000 ($5 million under recent amendments) or the plan included a structure with a capacity greater than 2,500 acre-feet (3.1 million square meters, or about 108 million square feet).

The 1972 Amendments to the WPFPA stated that conservation of water and preservation of the environment were permissible general purposes for authorized projects. This change allowed the U.S. Department of Agriculture and local organizations to undertake watershed projects for the sole purpose of protecting an ecosystem.

Impacts and Issues

The U.S. Department of Agriculture and local organizations have implemented numerous watershed protection and flood prevention plans over the last 50 years. Although these projects have greatly improved the environment and water supply, a 1994 report by the U.S. Environmental Protection Agency (EPA) found that nearly 40% of all waters in the United States were too polluted for fishing, swimming, or other uses. According to the report, the primary sources of pollution were silt, sewage, fertilizers, disease-causing bacteria, toxic metals, and oil and grease. Continued pollution of watersheds could take a large toll on the U.S. economy. The EPA estimates that more that $450 billion of the U.S. economy relies on healthy watersheds, primarily food production, manufacturing, and tourism.

The Mississippi River Basin and the Gulf of Mexico hypoxic zone highlight the need for continued watershed protection in the United States. Hypoxia is the absence of oxygen in reaching living tissue. An aquatic hypoxic zone, or dead zone, is an area of water that contains too little dissolved oxygen in the water to support aquatic life. Aquatic hypoxia zones usually occur when there is an overabundance of nutrients, such as nitrogen or phosphorous, in the water. The presence of these nutrients promotes excessive growth of algae, which reduce oxygen levels in the water. Excessive nutrients come from soil erosion, agricultural fertilizers, and runoff from developed land. The Gulf of Mexico hypoxic zone reached a record area of 8,481 square mi (22,000 square km), an area the size of New Jersey.

WORDS TO KNOW

EROSION: The wearing away of the soil or rock over time.

HYDROLOGICAL CYCLE: Hydrological cycle - natural water recycling; water evaporates from lakes, ponds, streams or wet earth; forms clouds; then precipitates as rain or snow back to Earth.

WATERSHED: The expanse of terrain from which water flows into a wetland, water body, or stream.

SEE ALSO *Floods; Hurricanes: Katrina Environmental Impacts; Water Resources; Water Supply and Demand; Watersheds*

BIBLIOGRAPHY

Web sites

Institute of Public Law, University of New Mexico School of Law. "Watershed Protection and Flood Prevention Act." http://ipl.unm.edu/cwl/fedbook/wpfpa.html (accessed May 1, 2008).

National Resources Conservation Service, U.S. Department of Agriculture. "Watershed Protection and Flood Prevention Act." http://www.nrcs.usda.gov/programs/watershed/pl56631705.pdf (accessed May 1, 2008).

U.S. Environmental Protection Agency (EPA). "Watersheds: A Watershed Approach." March 28, 2008. http://www.epa.gov/owow/watershed/approach.html (accessed May 1, 2008).

U.S. Environmental Protection Agency (EPA). "Watersheds: Introduction." May 8, 2007. http://www.epa.gov/owow/watershed/framework/ch1.html (accessed May 1, 2008).

U.S. Fish and Wildlife Service. "Digest of Federal Resource Laws of Interest to the U.S. Fish and Wildlife Service: Watershed Protection and Flood Prevention Act." http://www.fws.gov/laws/lawsdigest/WATRSHD.HTML (accessed May 1, 2008).

Joseph P. Hyder

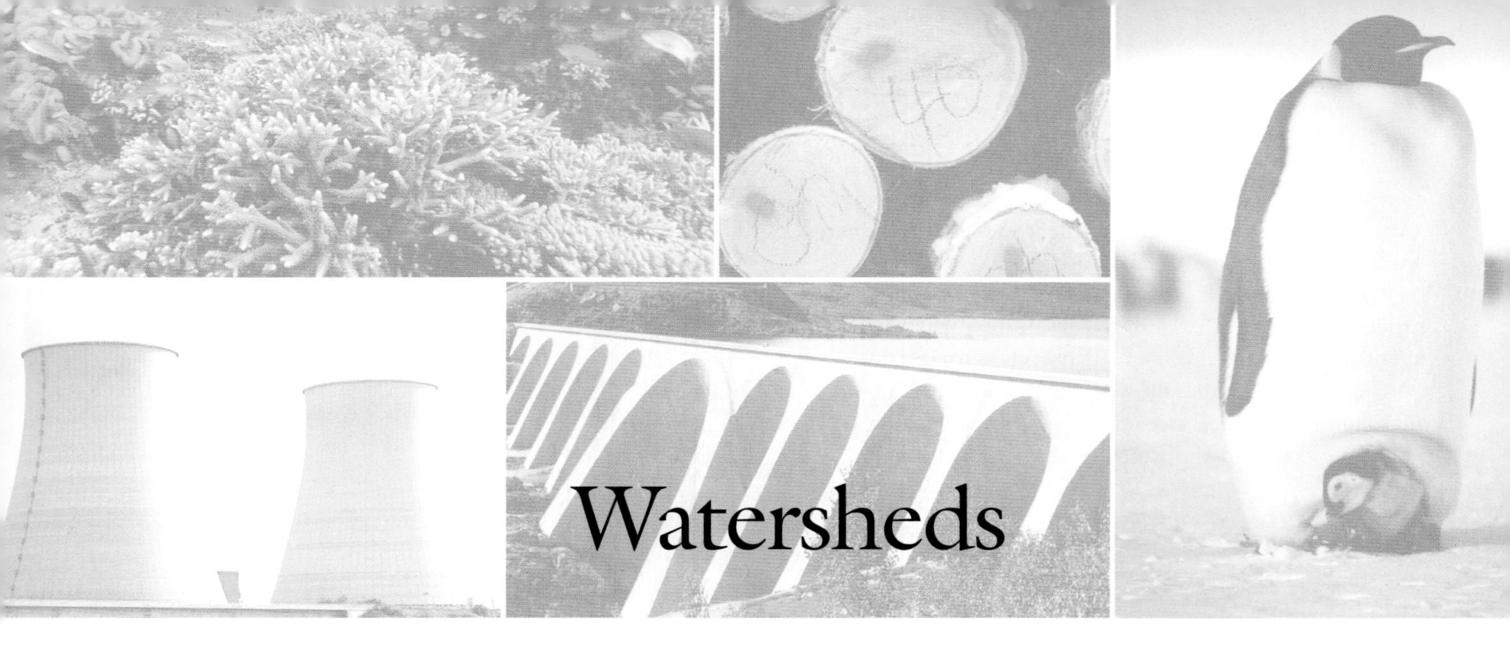

Watersheds

Introduction

A watershed refers to an entire area that drains into a stream, river, or lake, or which is the source of water for an underground water source (an aquifer). Everything in a particular watershed—watercourses, land, vegetation, and human-made structures—can contribute to the water drainage. For example, a shopping center parking lot can be an important part of an urban watershed, as water can easily drain off the concrete into drainage pipes that empty into an adjacent watercourse, or can directly run off into the watercourse.

Pollutants can also be carried in the runoff. A watershed can be very large; the Mississippi River watershed is over 1 million square mi (2.6 million sq km) in area. A polluted environment that is located at the upper reaches of the watershed can affect the waters through the remainder of the system.

The various watercourses in a watershed contribute water that ends up emptying into a larger body of water. For example, while the Mississippi River watershed begins near the border between the United States and Canada, the collective waters that drain over 40% of the continental United States empty into the Gulf of Mexico.

If the watercourses in a watershed are polluted, the effect can be magnified at the site of drainage. Continuing with the Mississippi River watershed example, the tremendous amount of organic material that is continually carried into the Gulf of Mexico has fueled eutrophication, an explosive growth of life that has depleted the oxygen from regions of the Gulf of Mexico.

Historical Background and Scientific Foundations

Watersheds that include urban areas are especially prone to contamination with runoff of various chemicals from concrete. For example, in many communities, some storm drains have a label warning that dumping is not allowed. This is because the drain empties directly into a nearby watercourse. Nonetheless, entry of polluted water can occur during a heavy rainstorm or because of run off of polluted water from driveways after a car is washed, as two examples. As a result, the polluted water continues downstream.

A crude analogy of a watershed is a funnel. The area at the receiving end of the funnel is much greater than the area at the outflow, so the capacity of the funnel to receive water is great. In reality, a watershed is more complicated, since not only can water enter the system at the area corresponding to the top of the funnel, but also at many places throughout the watershed. A watershed is indeed funnel-like, however, since the water ultimately exits from a common site at the mouth of the watershed's main river.

In a visual analog, if all of the watercourses in a watershed could be highlighted when viewed from far overhead, the entire structure would resemble a tree, with multiple and smaller branches leading to successively bigger branches that in turn fed into the main truck. The final outflow of a watershed corresponds to the intersection of the trunk with the ground.

If there are a number of urban centers in the area of a watercourse, the water can be unacceptably polluted by the time it empties into the final receiving body of water. Some pollution occurs inadvertently. As one example, leakage of oil and gas from vehicles while parked can be a source of contaminants to an adjacent watercourse, especially during a rainstorm, as the water drains to the lowest point of land, which is often the watercourse.

Watersheds in rural areas can also become contaminated. A typical example is the runoff of pesticides and herbicides from agricultural land. This has increased in frequency since the 1950s, as family farms have been replaced by very large farms that rely more heavily on mechanical tending, chemical control of pests, and chemical supplementation of soil fertility.

Another example of the adverse effects of farming practices on watershed degradation are factory farms—large indoor or outdoor facilities that house thousands of livestock or poultry. A feature of most factory farms is a very large holding pond that stores wastewater. This is necessary since a factory farm generates huge amounts of waste—a facility of 10,000 hogs can generate as much waste in one day as 25,000 people. Leakage or spillage from a factory farm waste lagoon can compromise a watershed. In 1999, for example, North Carolina was struck by two hurricanes within seven days. The first storm damaged some waste lagoons that subsequently leaked or ruptured during the second storm. One spill released 25 million gallons of feces-laden water into an adjacent river. Up to 10 million fish were killed as the massive glut of pollution moved downriver.

The health of a watershed can be maximized by knowing the capacity of the various watercourses within the watershed for contaminants. The technical term for this measurement is total maximum daily load (or TMDL). The TMDL is the sum of all the sources of pollution that a watershed receives, either from a single source (point source) such as from a drainage pipe or from non-point sources (runoff from fields is an example). Scientists take into account factors such as watercourse volume, flow, oxygen level, and uses of the water to calculate a TMDL.

WORDS TO KNOW

AQUIFER: An underground source of water.

EUTROPHICATION: The process whereby a body of water becomes rich in dissolved nutrients through natural or man-made processes. This often results in a deficiency of dissolved oxygen, producing an environment that favors plant over animal life.

RECHARGE: Replenishment of an aquifer by the movement of water from the surface into the underground reservoir.

RUNOFF: Water that falls as precipitation and then runs over the surface of the land rather than sinking into the ground.

SPRING: The emergence of an aquifer at the surface, which produces a flow of water.

Scientists aren't sure the bay grasses—which provided habitat for blue crabs and spawning fish—can withstand the pressure of more than 15 million people living in the Chesapeake Bay watershed, with more to come. *AP Images*

Nonpoint-source pollution of a watershed occurs when rainfall, melting snow, or irrigation of agricultural land causes water to run over land or, in urban areas, pavement. Because many compounds are able to dissolve in water or can become suspended, pollutants present on the surface can be carried along with the water flow to nearby watercourses. As well, the contaminated water can percolate down into the ground, contaminating the soil and even groundwater.

In a watershed, vegetation and other forms of life can also be affected by the contaminated water. For example, plants can take up heavy metals through their roots as they acquire moisture from the soil. If the plants are consumed, the toxic heavy metals can be transferred.

The most common source of watershed contaminants are soil that has been eroded and nutrients such as nitrogen and phosphorus that are components of fertilizer. Other pollutants include bacteria and viruses (particularly those that normally live in the intestinal tract of warm blooded animals, and which are deposited in water as part of fecal contamination), salt, oil, grease, gasoline, industrial chemicals, and heavy metals.

IN CONTEXT: WATERSHED RESTORATION

The restoration of watershed ecosystems is increasing, as the realization grows of the importance of the watersheds to the health of animals and plants that are part of the ecosystem and to the communities that depend on the watershed for their drinking water. Although ecologists claim the goals are too modest, various U.S. agencies and, in sum, committed to improve or restore 100,000 acres of wetlands each year and 25,000 mi (40,235 km) of stream shoreline.

Impacts and Issues

Almost everyone lives in a watershed. Different cities, states, and even countries can lie within large watersheds. This can complicate watershed preservation, as differing legislation from local or national governments can hinder consensus.

An example is the Amazon watershed, the world's largest. The Amazon River itself is 4,000 mi (6,437 km) long. Combined with the hundreds of tributaries, the watershed drains almost 3 million square mi (7 million square km), and includes the countries of Brazil, Peru, Ecuador, Bolivia, and Venezuela. The priorities of all these governments may not be the same. As watershed quality can be compromised from the headwaters to the mouth, differing environmental priorities can become evident as a deteriorating watershed.

Watersheds are economically important to their respective countries. For example, according to the National Oceanic and Atmospheric Administration (NOAA), watersheds that drain into U.S. coastal regions drive local economies that collectively generated 60 million jobs in 2006.

Economic prosperity can also degrade a watershed. As cities and industrial development has grown around major rivers in watersheds, the resulting development has contributed to the decline in watershed quality. Removal of vegetation can encourage erosion, which adds more sediment to the water. As well, habitat for birds and animals can be altered or destroyed.

In the United States and elsewhere, the importance of watershed preservation has been recognized by governments for decades. In the United States, for example, the Watershed Protection and Flood Prevention Act was passed in 1954. More recently, the Natural Resources Conservation Service of the U.S. Department of Agriculture undertook a watershed program that helps state, local, and aboriginal governments, and other organizations in conservations campaigns.

SEE ALSO *Freshwater and Freshwater Ecosystems; Groundwater Quality; Industrial Water Use; Nonpoint-Source Pollution; Superfund Site*

BIBLIOGRAPHY

Books

Grover, Velma I. *Water: Global Common and Global Problems.* Enfield, NH: Science Publishers, 2006.

Midkiff, Ken, and Robert F. Kennedy Jr. *Not a Drop to Drink: America's Water Crisis (and What You Can Do).* Novato, New World Library, 2007.

World Water Assessment Programme. *Water: A Shared Responsibility.* New York: Berghahn Books, 2006.

Periodicals

Gardner, Gary. "From Oasis to Mirage: The Aquifers that Won't Replenish." *World Watch* 8 (2005): 30–37.

Web Sites

U.S. Environmental Protection Agency (EPA). "What Is a Watershed?" April 3, 2007.http://www.epa.gov/owow/watershed/whatis.html (accessed April 10, 2008).

Brian D. Hoyle

Weather and Climate

Introduction

Weather is the behavior of the atmosphere—wind, temperature, humidity, air pressure, precipitation—at any given point on a planet's surface at a given time. Winds, calms, storms, clouds, droughts, and rains—all are weather events. Weather events last anywhere from seconds to months. Climate is weather averaged over some area (a single spot to the whole globe) for at least a few years. Often, the climate of a region is viewed as its long-term average pattern of seasonal weather. Any change in an average seasonal pattern is a climate change, even if several such changes happen to cancel each other when the seasons are averaged together. For example, if over a period of some years New England were to get less snow in winter and more rain in summer, its climate would have changed even if the total average amount of yearly precipitation remained the same.

Meteorologists (scientists who study weather) and climatologists (scientists who study climate) stress that climate sets the background conditions of weather. As climate changes, the overall chances or probabilities of various weather events also change. Droughts may become more frequent, or heavy rains; cold snaps less frequent, and heat waves more prevalent during periods of climate change. However, no single weather event is a climate event, or proves or disproves the existence of changes in climate. No particular hurricane or a drought can be specifically attributed to climate change, and no particular snowstorm or cold snap contradicts the reality of climate change. It takes many storms, many droughts, many weather events to add up to a climate, or to a change in climate.

The energy driving all weather and thus, all climate, is derived from the sun. It arrives in the form of light and is stored, moved, transformed, radiated, absorbed, and re-radiated inside Earth's complex climate machine. At this time, Earth radiates to space slightly less energy than it receives from the sun.

Historical Background and Scientific Foundations

For tens of thousands of years, human societies feared weather, welcomed it, sought to predict it, and, if only through prayer and sacrifice, to control it. The development of modern physics in the 1600s and 1700s gave scientists the tools they needed to begin deciphering the mechanisms of weather, but scientific recording of large-area weather patterns did not begin until the mid-nineteenth century. In the 1960s and 1970s, with the advent of affordable, powerful computers, scientists began to produce mathematical models of weather, that is, systems of equations describing the various physical forces that produce it. Today, governments around the world spend billions of dollars studying weather and predicting it days in advance. Satellites ring Earth, observing its weather, and as of 2008 space probes were constantly observing the weather on Mars and Saturn from orbits around those planets. Some of the largest supercomputers in existence are devoted to weather modeling. Weather prediction is of high economic value. Agriculture, shipping, emergency response, and planning of outdoor events from picnics to battles all rely on weather forecasting.

Despite all the effort expended on predicting weather, however, weather forecasts are not very reliable for more than about a week into the future. The reason is fundamental: The weather system is chaotic. That is, a very small change in conditions today will produce large changes in the exact weather pattern tomorrow, larger changes the day after that, and so on. Any slight uncertainty in today's conditions (and precise knowledge of all the relevant conditions is impossible) leads to large and

WORDS TO KNOW

CLIMATE MODEL: A quantitative method of simulating the interactions of the atmosphere, oceans, land surface, and ice. Models can range from relatively simple to quite comprehensive.

DEFORESTATION: A reduction in the area of a forest resulting from human activity.

GLOBAL WARMING: Warming of Earth's atmosphere that results from an increase in the concentration of gases that store heat, such as carbon dioxide (CO_2).

growing uncertainty about weather conditions at increasingly future times. Weather cannot be predicted very far in advance.

Climate

The idea that scientists can predict climate changes decades into the future is often countered with the argument, "If they can't predict next week's weather, how can they predict climate 10 years from now?" This argument is mistaken because climate, unlike weather, is not a chaotic process. The climate system is complex but relatively well-behaved: Small changes to the climate system's inputs are unlikely to have effects that get larger and larger over time, whereas large changes are likely to have large effects. In particular, greatly increasing the amounts of greenhouse gases in the atmosphere (carbon dioxide, methane, nitrous oxide, chlorofluorocarbons), as human beings have been doing for several centuries, is highly likely to have effects on climate that can be predicted, though with some range of uncertainty around the predictions. Computer climate modeling is therefore a different game from weather modeling. Today's climate models have a good, and steadily improving, ability to hindcast climate changes—that is, knowing the climate record up to a certain point in the past, they can predict with reasonable accuracy climate changes from that time forward, which are already on record and so can be compared with the predictions.

Impacts and Issues

Climate forecasting, like weather forecasting, is a major concern of earth scientists today. The stakes are high. Since climate is not chaotic, the long-range future is not a blank. All but a few climatologists are now agreed that the global climate is warming because of human activities, especially deforestation and the burning of fossil fuels, and that it is going to continue to warm. Global warming does not mean that every place on Earth is becoming hotter and drier: The same mathematical models of global climate change that predict overall warming also predict cooling in central Antarctica, which is being observed. Some places will get more precipitation as climate changes, others less. Earth's climate system is complex, its changes are complex, and many of its features are still poorly understood.

As the non-chaotic future of climate can be predicted at least in outline, and since climatologists are now confident that human activities are shaping that future, human beings face choices about climate. We cannot control specific weather events, but we can, at least in theory, choose a specific climate future, within a certain range of real-world possibilities. How much fuel we burn, how we choose to alter or not alter Earth's ecosystems, what lifestyles we adopt (that is, those of us lucky enough to have such choices)—all these have affected climate, are affecting it, and will affect it. Climate, in turn, sets the background for all of the billions of particular weather events and conditions, day by day, that determine the livability of the planet.

SEE ALSO *Climate Change; Climate Modeling; El Niño and La Niña; Global Warming; Hurricanes; Weather Extremes*

BIBLIOGRAPHY

Books

Ahrens, C. Donald. *Meteorology Today: An Introduction to Weather, Climate, and the Environment.* Belmont, CA: Brooks Cole, 2006.

Solomon, S., et al, eds. *Climate Change 2007: Physical Science Basis. Contribution of Working Group I to the Fourth Assessment Report of the Intergovernmental Panel on Climate Change.* New York: Cambridge University Press, 2007.

Web Sites

National Aeronautic and Space Administration (NASA). "Weather." http://www.nasa.gov/worldbook/weather_worldbook.html (accessed May 12, 2008).

National Oceanic and Atmospheric Administration (NOAA). "National Weather Service." http://www.nws.noaa.gov/ (accessed May 12, 2008).

Larry Gilman

Weather Extremes

Introduction

Extreme weather refers to weather that is unusually violent in nature or that is abnormal in its frequency of occurrence or length. Tornadoes and hurricanes are well-known forms of weather whose force and destructive power is extreme.

Hurricanes may also be an example of weather that is extreme from the point of view of frequency. Although still debatable, evidence has been published by agencies including the World Meteorological Organization (WMO) and the U.S. Environmental Protection Agency (EPA) that supports the view that hurricanes spawned in the equatorial region between North and South America are becoming more frequent in both their occurrence and ferocity as a result of global warming—the warming of Earth's atmosphere that has been present for the past 150 years, which has been accelerating since the mid-twenty-first century.

Extreme weather also includes droughts, flooding, high winds, blizzards, and colder than normal winter time temperatures. In various regions of the globe, all these types of extreme weather are occurring more often. Whether this is directly due to the atmospheric changes occurring because of global warming is debatable, but scientific evidence to support this view is accumulating.

Historical Background and Scientific Foundations

Determining whether extreme weather events such as hurricanes are increasing in frequency and intensity requires knowledge of the past trends. Since records of hurricanes, cyclones, and typhoons have been in existence for centuries beginning with mariners, and satellite monitoring has been used since the 1970s, there is a great deal of data on these extreme weather events. As well, other atmospheric parameters such as temperatures have been recorded for a long time.

What is also necessary is to model future weather trends based on the incorporation of the atmospheric changes that are taking place due to global warming and the influence of the warming ocean. Several models have been developed; these are collectively termed Atmosphere-Ocean Global Climate Models. These models tend to accurately portray the present climate, which increases confidence that their predictions of future weather will be reasonably accurate, as long as all factors influencing weather have been accounted for.

Models developed by agencies including the U.S. National Oceanic and Atmospheric Administration (NOAA) and the Canadian Center for Climate Modeling and Analysis support the notion that the frequency and severity of hurricanes could increase as the atmosphere and tropical ocean continues to warm. The basis is the warmer air overlying the increasingly warming tropical ocean. The more intense rising of the air into the atmosphere and subsequent return of colder air from aloft, combined with wind, begins the spiral pattern of the storm, whose ferocity is fueled by the continued supply of warmer air from the ocean's surface.

Ocean measurements conducted by Woods Hole Oceanographic Institution and other agencies since the 1950s have demonstrated the warming of the tropical ocean at the surface and to depths of over 1,000 ft (304 m). Indeed, measurements of ocean and atmospheric warming indicate that the amount of heat absorbed by the ocean exceeds that absorbed by the atmosphere by about 20 times. Because warm air can hold more water vapor than colder air, a result has been the increased content of warm water vapor over the tropical ocean. These conditions are prime for the establishment of extreme weather.

A paper published in the journal *Science* in 2005 that examined satellite data from the 1970s onward con-

WORDS TO KNOW

ATMOSPHERE: The air surrounding Earth, described as a series of layers of different characteristics. The atmosphere, composed mainly of nitrogen and oxygen with traces of carbon dioxide, water vapor, and other gases, acts as a buffer between Earth and the sun.

GAIA HYPOTHESIS: The hypothesis that Earth's atmosphere, biosphere, and its living organisms behave as a single system striving to maintain a stability that is conductive to the existence of life.

GLOBAL WARMING: Warming of Earth's atmosphere that results from an increase in the concentration of gases that store heat, such as carbon dioxide (CO_2).

cluded that the number of category 4 and 5 hurricanes, the most severe and destructive forms, had increased. At the same time, the number of less intense hurricanes decreased. The increase matches with data showing a warming tropical ocean and atmosphere. Surface-based studies that charted surface wind and temperature have confirmed the satellite observations, and have indicated that an annual increased ocean surface temperature of about 0.45°F (0.25°C) boosts the likelihood of severe weather by 60%.

Climate models also support the link between climate change and other aspects of extreme weather. For example, a downpour can be considered as extreme weather, as it is rainfall that exceeds the normal pattern of rainfall over time. Short but very heavy rain can overwhelm the capability of drainage systems such as sewers in urban settings. The result can be flooding. Similarly, the runoff of excessive amounts of water into a watercourse during and shortly after a heavy rain can exceed the capacity of the watercourse to contain the water, creating a flood. Monitoring in the United States and Canada through the last century has shown that short duration heavy rains have increased in frequency; one-day rainfalls of more that 2 inches (50 mm) have increased in the United States by 20% since the beginning of the twenty-first century, for example. Predictions from Canadian climate models are that such storms that currently tend to occur about once every 20 years will occur once every 10 years by the end of this century.

Flooding in tropical regions brings the added possibility of the development of diseases of microbial origin such as cholera, since the bacteria are able to grow and multiply in the warm water, and infect people whose resistance to infection has been lowered by injury.

Another related type of extreme weather in northern climates is excessive snowfall. As for extreme rainstorms, the frequency of blizzards, which can bring cities to a standstill and which are hazardous for road and air transportation, have increased since the mid-1970s.

Another type of extreme weather is known as a heat wave, where the temperature becomes much hotter than normal for the time of year and lasts longer than normal. Many regions of the world including Africa, the United Kingdom, Europe, and North America have experienced more frequent heat waves since the mid-twentieth century. In regions that have shortages of water, heat waves are very serious and can be lethal.

■ Impacts and Issues

The possibility of more frequent and severe weather could put coastal human populations in affected regions more at risk of personal harm and property damage. Secondarily, the economies of the affected countries would suffer, as damage claims soared and government-funded and maintained relief efforts were put into action more often. As of 2008, for example, the costs to the U.S. government associated with 2005's Hurricane Katrina are estimated at over $80 billion.

Climate scientists generally agree that one important influencing factor in extreme weather is the regular oscillating weather patterns known as El Niño and La Niña. These oscillations can trigger extreme weather events that include cyclones, droughts, and increased precipitation in different regions of the Pacific. As well, the influence of a La Niña can extend to the Caribbean, and may influence development of tropical storms.

The influences of El Niño and La Niña may increase still further if the frequency of the weather oscillations is affected by global warming. Some climate scientists assume that this is occurring, but this view is not universal.

Analysis of the ocean conditions at the time of Hurricane Katrina indicates that the temperature of the Gulf of Mexico was crucial in fueling the increased intensity of the storm as it crossed the open water of the gulf en route to the Gulf Coast region of the United States. Normally, as a tropical storm builds, deeper ocean water is stirred up to the surface. Because this water tends to be colder, it acts to diminish the storm. However, because the gulf water temperature has increased in the past 50 years, the deeper waters did not diminish Hurricane Katrina, but instead made the storm even more powerful. The results for the Gulf Coast were devastating.

Elsewhere in the world, extreme weather such as droughts and monsoon flooding are already affecting people especially in undeveloped regions, where an already tenuous existence is becoming even more difficult.

SEE ALSO *Dust Storms; Floods; Global Warming; Landslides; Tsunami Impacts; Volcanoes*

IN CONTEXT: EXTREME WEATHER

Increased extreme weather is a predicted consequence of global climate change. In August 2007, scientists at the World Meteorological Organization (WMO), an agency of the United Nations, announced that during the first half of 2007, Earth showed significant increases above long-term global averages in both high temperatures and frequency of extreme weather events (including heavy rainfalls, cyclones, and wind storms). Prior to the release of the findings, the global average land temperatures for January and April of 2007 were the warmest recorded for those two months since records began in 1880.

BIBLIOGRAPHY

Books

Banta, John. *Extreme Weather Hits Home: Protecting Your Buildings from Climate Change.* Gabriola Island, British Columbia, Canada: New Society Publishers, 2007.

Burt, Christopher. *Extreme Weather.* New York: W.W. Norton, 2007.

Mogil, Michael. *Extreme Weather: Understanding the Science of Hurricanes, Tornadoes, Floods, Heat Waves, Snow Storms, Global Warming and Other Atmospheric Disturbances.* New York: Black Dog & Leventhal Publishers, 2007.

Periodicals

Emanuel, K. "Increasing Destructiveness of Tropical Cyclones over the Past 30 Years." *Nature* 436 (2005): 686–688.

Webster, P. J., et al. "Changes in Tropical Cyclone Number, Duration, and Intensity in a Warming Environment." *Science* 309 (2005): 1844–1846.

Web Sites

NASA: Goddard Institute for Space Studies. "NASA Study Predicts More Severe Storms with Global Warming." August 30, 2007. http://www.giss.nasa.gov/research/news/20070830/ (accessed April 20, 2008).

Brian D. Hoyle

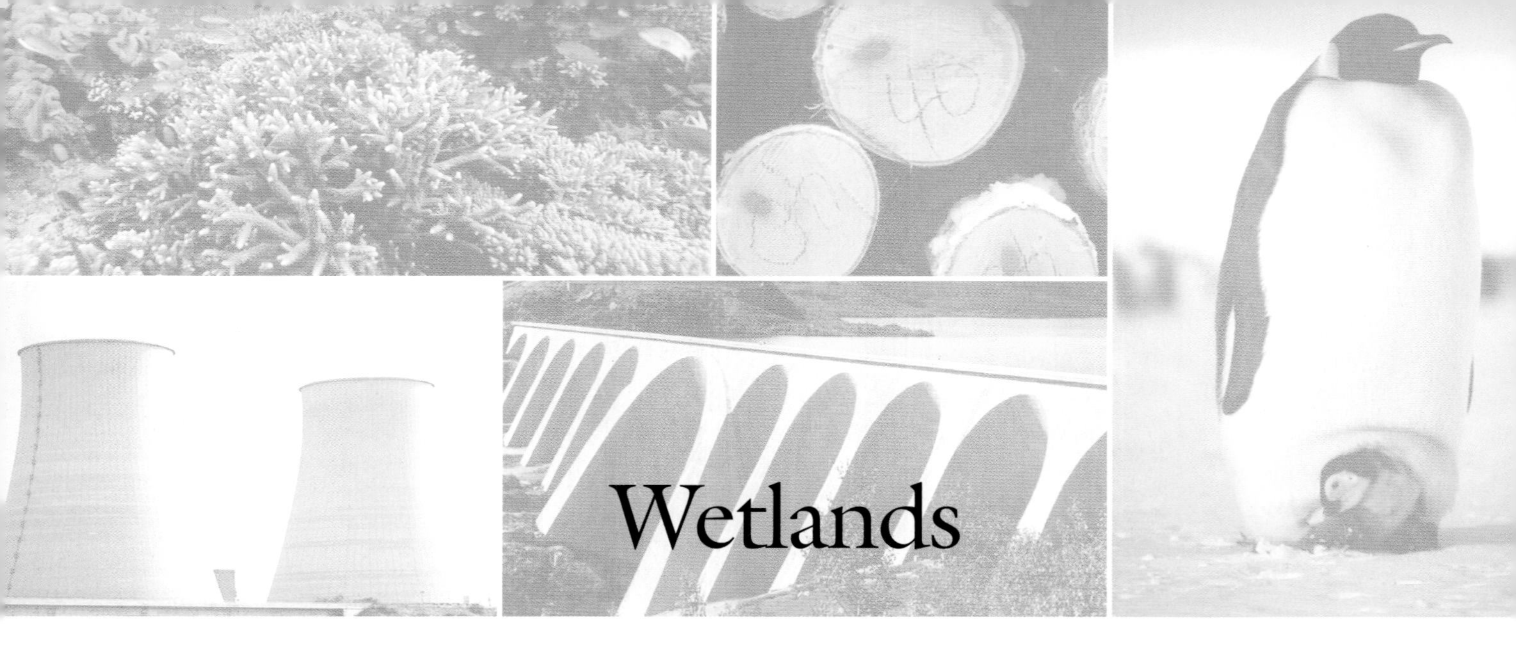

Wetlands

Introduction

Wetlands are areas in which the soil is completely saturated with water. The plants growing in a wetland have adapted to be able to survive and grow in water-soaked soil. Swamps, bogs, peat lands, and marshes are examples of wetlands.

A wetland is located between two very different environments, one located more inland (a terrestrial ecosystem) and the other being the waters of the lake or ocean (an aquatic ecosystem). The wetland is an interface or ecotone between the terrestrial ecosystem and the aquatic ecosystem. The area where one ecosystem blends in with another is referred to as an ecotone.

Although once regarded as wasteland, wetlands have vital environmental functions and, when located on a flyway of migratory birds, are rest and refueling stops for the birds during their seasonal journey. Today, many wetlands are nurtured and, where necessary, restored.

Historical Background and Scientific Foundations

In wetlands, the water table (the underground zone where the underlying soil is completely saturated with water) is very near or at the surface. The saturated soil of a wetland lacks oxygen. Wetlands may not always be saturated with water, however. Depending on the season, tide cycle, presence of a drought, the type or amount of precipitation, and its location, a wetland can be a flooded zone sometimes, and a grassy area another time.

There are additional stresses in wetlands located on the ocean coast (intertidal wetlands). At high tide, the wetland is inundated with saltwater, while at low tide freshwater predominates. A type of tree called the mangrove has adapted to the high salt level by the structure of the roots and leaves, which makes it more difficult for salty water to enter and for the moisture to escape.

Generally, wetlands become wet in three ways: precipitation in the form of rain or snow, the emergence of groundwater, and the flow of freshwater into the wetland from surrounding areas via watercourses or flow of marine water from the nearby ocean. The location and topography of the area is influential in determining the movement of water into the wetland and the speed at which the water moves through the wetland. For example, an area that lies lower than the surrounding land will tend to receive water flowing downhill.

The periodically saturated and oxygen-free soil represents an environment that is too harsh for many varieties of plant to grow. Those that have become adapted to growth in oxygen-free soil are referred to as hydrophytes or wetland plants. Examples include cattails, bulrushes, water lilies, and duckweed. When trees are present, they can have roots raised above the waterline, which increases the exchange of gas between the roots and the air. In tropical regions, mangrove trees contribute sulfur, making a wetland more acidic and increasing the rate at which organic material is decomposed. This helps provide more food for other life in the wetland.

The plants in wetlands can be used as a food source for animals and, if uneaten, also provide nutrients when they decompose. The resulting flow of energy creates what is termed a food web. Although other environments such as prairie land and temperate forests also create food webs, wetland associated food webs are especially productive and support a large and varied population.

Wetlands are also an area where waterfowl can hide. This is important for breeding. Wetlands that are located near watercourses can also be a form of flood control. If the watercourse is breached during spring runoff or during a heavy rain, the excess water can gather in the wetland, to be gradually returned to the surrounding region. Near the coast, wetlands can buffer the effects of storms, providing a place where surging water can collect and dissipate. However, these benefits no longer exist in many regions, where wetlands were filled in to create land for agriculture or housing.

Wetlands can be located in an area where water is abundant. Intertidal wetlands are one example. Wetlands can also be located in a region where surface water is otherwise sparse. Prairie wetlands are an example. Such wetlands are important for wildlife, commercial livestock, and migrating species especially during dry periods.

A well-known wetland is the Florida Everglades, which is actually a very slow moving river almost 62 mi (100 km) wide and over 100 mi (160 km) in length.

WORDS TO KNOW

BOG: Area of wet, spongy ground consisting of decayed plant matter.

PEAT: Partially carbonized vegetable matter that can be cut and dried for use as fuel.

TIDE CYCLE: A period that includes a complete set of tide conditions or characteristics, such as a tidal day or a lunar month.

■ Impacts and Issues

Wetlands have been described as environmental kidneys, because they function to filter out sediment and pollution as the water moves slowly through the wetland vegetation. Usually the water emerging from a wetland is cleaner than the water that entered. This principle has been applied to sewage treatment that utilizes wetland-like regions as a step in the treatment process. Nutrients such as phosphorus and nitrogen, which are commonly present in fertilizer and which can enter the water in runoff, can be retained by wetland vegetation, reducing the nutrient level in the water and so reducing the likelihood of nutrient-fueled explosive growth of microorganisms that can deplete oxygen for the water.

Furthermore, wetlands can retain nutrients, which makes them a source of great biodiversity and a haven for migrating species. Indeed, estimates are that North American wetlands located along migratory pathways are used by upward of 500,000 migrating birds each spring and fall, representing 30% of the continent's migrating bird population. Internationally, wetlands are used by at least 100,000 birds each year.

Wetlands also benefit humans as a source of food and fuel. For example, in Thailand, the rice that is grown

A systems engineer checks the water quality in a wetland area in the restoration project at Richie Woods in Fishers, Indiana. *AP Images*

IN CONTEXT: WETLANDS FILTER WATER

Natural wetlands also contribute to the purification of water. Wetlands can serve as a depositional sump and provide biological filtering.

and harvested in the rich wetland soil is one of the staple foods. In northeastern regions of the United States and maritime Canada, wetland regions are used to grow cranberries. In Russia and some regions of Europe and the United Kingdom, peat has provided a fuel source for millennia.

Although in many places emphasis is on the protection and restoration of wetlands, this was not always the case. The Environmental Protection Agency (EPA) estimates that in the 1600s there were more than 22 million acres of wetlands in the continental United States. However, by the mid-1950s, more than half of these had been drained and the land converted to other uses, with wetlands tending to be viewed as wasteland with little natural value or function. In North America, many wetlands were drained to provide more land for development. Coastal wetlands were particularly vulnerable because their drainage provides land for development of oceanfront housing. According to the EPA, about 60,000 acres of U.S. wetlands are still lost every year.

Everglades National Park, formed in 1947 by President Harry Truman, exists to protect the remaining portion of the Florida Everglades. Over half of the original area of the everglades has been lost to agricultural and urban development.

In the everglades alone restoration efforts by the U.S. Army Corps of Engineers as part of the Comprehensive Everglades Restoration Plan have already cost billions of dollars. For increasing water-strapped Florida, however, the efforts are worthwhile, since the everglades are of crucial influence to the availability of freshwater.

The EPA and Army Corps of Engineers jointly establish environmental standards for wetlands and monitor discharges into wetlands. If necessary, the EPA has the legal authority to intervene to stop unacceptable discharges.

Wetlands are also threatened by air and water pollution. For example, insecticides and herbicides carried into a wetland in the inflowing water can be toxic to plants and wildlife.

In the United States, some wetlands are federally regulated as part of the Clean Water Act. Wetlands protection is also important internationally. One example is the Ramsar Convention on Wetlands, which was signed in Ramsar, Iran, in 1971. As of 2008, the convention had 158 parties and included almost 400 million acres (161 million hectares) of wetland. The convention also addresses the influence of climate change on migratory species.

SEE ALSO *Clean Water Act; Coastal Zones; Conservation; Wildlife Population Management; Wildlife Protection Policies and Legislation*

BIBLIOGRAPHY

Books

Grandin, Temple, Terry Williams, and Rosalie Winard. *Wild Birds of the American Wetlands.* New York: Welcome Books, 2008.

Lockwood, C., and Rhea Gary. *Marsh Mission: Capturing the Vanishing Wetlands.* Baton Rouge: Louisiana State University Press, 2005.

Mitsch, William, and James Gosselink. *Wetlands.* New York: Wiley, 2007.

Web Sites

Sierra Club. "Wetlands Overview." http://www.sierraclub.org/wetlands/ (accessed May 2, 2008).

Brian D. Hoyle

Whaling, International Convention for the Regulation of

Introduction

The International Convention for the Regulation of Whaling is an international agreement in effect since 1948 designed to make commercial whaling a sustainable practice. It established the International Whaling Commission, a British-based international regulatory body, which governs the conduct of whaling throughout the world.

Historical Background and Scientific Foundations

In December 1946, longstanding efforts by the world's whaling nations to exert control over the whaling industry and protect against over-hunting produced the International Convention for the Regulation of Whaling (ICRW). Originally signed by 15 nations in Washington, D.C., on December 2, it came into effect nearly two years later, on November 10, 1948.

The purpose of the convention was "to provide for the proper conservation of whale stocks and thus make possible the orderly development of the whaling industry." But the interests of the whaling industry lay at the heart of the convention. As Article V of the agreement stipulates, while the convention's intent is "to provide for the conservation, development, and optimum utilization of the whale resources," it must take into consideration "the interests of the consumers of whale products and the whaling industry."

The convention established the 14-nation International Whaling Commission (IWC), which was empowered to regulate the whaling industry. But from the outset, the International Whaling Commission was undermined. For the first quarter century of its existence, it tolerated whaling at unsustainable levels, allowing many of the largest species to decline precipitously. As such, by the late 1960s whaling had started to become a campaign issue for environmentalists and creep into wider public consciousness in western nations. Indeed, whale numbers fell so dramatically that in 1972, the United States called for a 10-year moratorium on all commercial whaling. When the moratorium was voted down in 1972, and again a year later, activists called on a boycott of goods from the principal whaling nations—Norway, Iceland, Japan, and the Soviet Union.

In 1974, the IWC endorsed the New Management Procedure—a United States-backed Australian plan—which banned whaling of all over-exploited stocks but permitted commercial catches of other species at sustainable levels. This was heralded as a great compromise, at once satisfying IWC members from non-whaling nations against the excessive size of earlier whale quotas, as well as whaling nations opposed to a moratorium on commercial whaling. It took effect from the 1975–1976 whaling season.

The success of the New Management Procedure, however, did not allay the efforts of environmentalists and non-whaling nations, who continued to seek a complete commercial moratorium. Exploiting the fact that any nation can accept the 1946 convention and become an equal voting member of the IWC, anti-whaling nations tipped the IWC balance in their own favor by recruiting additional non-whaling nations to the commission—increasing its membership from the original 14 to 40. In 1982, the newly enlarged IWC passed a moratorium on all commercial whaling, which came into effect in 1986.

Nevertheless, Norway, the Soviet Union, and later, Iceland registered objections to the moratorium and so were not bound by it. Japan also registered an objection, but this was withdrawn in 1987 on the threat of U.S. sanctions. All these nations continued to fish for limited numbers of whales under the semblance of scientific

WORDS TO KNOW

INTERNATIONAL WHALING COMMISSION: An international body established ICRW to provide for the proper conservation of whale stocks and make possible the orderly development of the whaling industry.

SUSTAINABLE WHALING: The process by which a limited number of whales from thriving species, such as the minke, are culled each year.

WHALE MEAT: A traditional food in Japan as well as in northern high-latitude nations such as Norway.

research, while continuing to sell whale products on the commercial market.

Finally, in June 2006, the International Whaling Commission meeting backed a resolution calling for the eventual return of commercial whaling by a majority of just one vote. Although this did not signal the overturning of the 20-year-old moratorium—for that a three-quarters majority would have been necessary—it instilled the confidence in Japan, Norway, and Iceland to resume commercial whaling in direct contravention of it.

■ Impacts and Issues

The IWC moratorium on commercial whaling has led to repeated accusations from pro-whaling nations that the organization's founding principles—to regulate commercial whaling—have been deviated from and undermined. Moreover, pro-whaling nations have accused the IWC of basing its refusal to allow sustainable whaling upon political and emotional factors, rather than scientific knowledge. They argue the IWC has swayed from its original purpose and is utilizing the guise of conservation to grant whales absolute protection.

Just as the ranks of the IWC were filled—and thus influenced—in the 1970s with nations sympathetic to anti-whaling, so pro-whaling nations have enlisted allies on the IWC with a view to overturning the moratorium. This has led to accusations of vote-buying—where aid has been offered to poorer countries in return for them joining the IWC and supporting pro-whaling positions—particularly by Japan. Indeed, a look at recent pro-whaling IWC members-nations such as the Solomon Islands and Mongolia with no ostensible whaling interests (or even a coastline, as in landlocked Mongolia's case, would seemingly attest to this. The Japanese Fisheries Ministry has paid a reputed $320 million in overseas aid to developing countries sympathetic to its whaling interests.

Seven of the 13 species of great whale, including the blue whale and the bowhead, are considered endangered. But several species—notably Minke whales—are abundant and whaling nations contend that they want to catch only such species. Taken in the context of existing whale stocks, they make a persuasive case: Norway counts 107,000 Minke whales in its economic waters, and Iceland 43,000. From this profusion, whose numbers grow at about 2 to 3% a year, Iceland takes just 38 whales and Norway 711.

Allegations that whaling is an inherently cruel form of hunting—as a minority of governments and most environmentalist groups maintain—is also open to question. For example, every Norwegian whaling expedition must by law include a qualified veterinarian and according to Norway's government, these vets report that 90% of the whales killed by its fishermen die instantly (a grenade attached to the harpoon blows the whale's brain apart), while the remaining 10% are dispatched swiftly with the use of a rifle.

Nevertheless, the actual need to hunt whales is also open to question. The whaling industry has been in decline for some 150 years as demand for its products—which once ranged from candles to cosmetics and engine lubricants—evaporated with the appearance of petrochemicals. All that is left of the commercial whaling business is the market for whale meat, but even Norwegians and Japanese are losing their appetite for what is considered an unfashionable meat, and both countries are left with surpluses of the meat each year.

■ Primary Source Connection

The International Convention for the Regulation of Whaling is an international agreement in effect since 1948 designed to make commercial whaling sustainable. It established the International Whaling Commission (IWC), a British-based international regulatory body, which governs the conduct of whaling throughout the world.

In addition to protecting species and biodiversity, the commission seeks to permit only sustainable whaling, the process by which a limited number of whales from thriving species, such as the minke, are culled each year.

Recent international controversies regarding whaling have the IWC searching for solutions that balance cultural practice with species protection.

INTERNATIONAL CONVENTION FOR THE REGULATION OF WHALING

WASHINGTON, 2ND DECEMBER, 1946

The Governments whose duly authorised representatives have subscribed hereto,

An anti-whaling group's boat, left, collides with a Japanese whale-spotting vessel in the waters of the Antarctic, February 2007. The two collided twice during clashes over a pod of whales. *AP Images*

Recognizing the interest of the nations of the world in safeguarding for future generations the great natural resources represented by the whale stocks;

Considering that the history of whaling has seen over-fishing of one area after another and of one species of whale after another to such a degree that it is essential to protect all species of whales from further over-fishing;

Recognizing that the whale stocks are susceptible of natural increases if whaling is properly regulated, and that increases in the size of whale stocks will permit increases in the number of whales which may be captured without endangering these natural resources;

Recognizing that it is in the common interest to achieve the optimum level of whale stocks as rapidly as possible without causing widespread economic and nutritional distress;

Recognizing that in the course of achieving these objectives, whaling operations should be confined to those species best able to sustain exploitation in order to give an interval for recovery to certain species of whales now depleted in numbers;

Desiring to establish a system of international regulation for the whale fisheries to ensure proper and effective conservation and development of whale stocks on the basis of the principles embodied in the provisions of the International Agreement for the Regulation of Whaling, signed in London on 8th June, 1937, and the protocols to that Agreement signed in London on 24th June, 1938, and 26th November, 1945; and

Having decided to conclude a convention to provide for the proper conservation of whale stocks and thus make possible the orderly development of the whaling industry;

Have agreed as follows:

Article III

1. The Contracting Governments agree to establish an International Whaling Commission, hereinafter referred to as the Commission, to be composed of one member from each Contracting Government. Each member shall have one vote and may be accompanied by one or more experts and advisers.
2. The Commission shall elect from its own members a Chairman and Vice-Chairman and shall determine its own Rules of Procedure. Decisions of the Commission shall be taken by a simple majority of those members voting except that a three-fourths majority of those members voting shall be required for action in pursuance of Article V. The Rules of Procedure may provide for decisions otherwise than at meetings of the Commission.
3. The Commission may appoint its own Secretary and staff.

4. The Commission may set up, from among its own members and experts or advisers, such committees as it considers desirable to perform such functions as it may authorize.
5. The expenses of each member of the Commission and of his experts and advisers shall be determined by his own Government.
6. Recognizing that specialized agencies related to the United Nations will be concerned with the conservation and development of whale fisheries and the products arising therefrom and desiring to avoid duplication of functions, the Contracting Governments will consult among themselves within two years after the coming into force of this Convention to decide whether the Commission shall be brought within the framework of a specialized agency related to the United Nations.
7. In the meantime the Government of the United Kingdom of Great Britain and Northern Ireland shall arrange, in consultation with the other Contracting Governments, to convene the first meeting of the Commission, and shall initiate the consultation referred to in paragraph 6 above.
8. Subsequent meetings of the Commission shall be convened as the Commission may determine.

Article IV

1. The Commission may either in collaboration with or through independent agencies of the Contracting Governments or other public or private agencies, establishments, or organizations, or independently; (a) encourage, recommend, or if necessary, organize studies and investigations relating to whales and whaling; (b)collect and analyze statistical information concerning the current condition and trend of the whale stocks and the effects of whaling activities thereon; (c)study, appraise, and disseminate information concerning methods of maintaining and increasing the populations of whale stocks.
2. The Commission shall arrange for the publication of reports of its activities, and it may publish independently or in collaboration with the International Bureau for Whaling Statistics at Sandefjord in Norway and other organizations and agencies such reports as it deems appropriate, as well as statistical, scientific, and other pertinent information relating to whales and whaling.

ARTICLE V

1. The Commission may amend from time to time the provisions of the Schedule by adopting regulations with respect to the conservation and utilization of whale resources, fixing (a) protected and unprotected species; (b) open and closed seasons; (c) open and closed waters, including the designation of sanctuary areas; (d) size limits for each species; (e) time, methods, and intensity of whaling (including the maximum catch of whales to be taken in any one season); (f) types and specifications of gear and apparatus and appliances which may be used; (g) methods of measurement; and (h) catch returns and other statistical and biological records.
2. These amendments of the Schedule (a) shall be such as are necessary to carry out the objectives and purposes of this Convention and to provide for the conservation, development, and optimum utilization of the whale resources; (b) shall be based on scientific findings; (c) shall not involve restrictions on the number or nationality of factory ships or land stations, nor allocate specific quotas to any factory ship or land station or to any group of factory ships or land stations; and (d) shall take into consideration the interests of the consumers of whale products and the whaling industry.
3. Each of such amendments shall become effective with respect to the Contracting Governments ninety days following notification of the amendment by the Commission to each of the Contracting Governments, except that (a) if any Government presents to the Commission objection to any amendment prior to the expiration of this ninety-day period, the amendment shall not become effective with respect to any of the Governments for an additional ninety days; (b) thereupon, any other Contracting Government may present objection to the amendment at any time prior to the expiration of the additional ninety-day period, or before the expiration of thirty days from the date of receipt of the last objection received during such additional ninety-day period, whichever date shall be the later; and (c) thereafter, the amendment shall become effective with respect to all Contracting Governments which have not presented objection but shall not become effective with respect to any Government which has so objected until such date as the objection is withdrawn. The Commission shall notify each Contracting Government immediately upon receipt of each objection and withdrawal and each Contracting Government shall acknowledge receipt of all notifications of amendments, objections, and withdrawals.

INTERNATIONAL WHALING COMMISSION (IWC). "INTERNATIONAL CONVENTION FOR THE REGULATION OF WHALING." HTTP://WWW.IWCOFFICE.ORG/COMMISSION/CONVENTION.HTM (ACCESSED APRIL 5, 2008).

SEE ALSO *Endangered Species; Environmental Activism; Environmental Protests*

BIBLIOGRAPHY

Web Sites

Economist. "A Bloody War: Whaling—A Special Report." December 30, 2003. "http://www.economist.com/science/displaystory.cfm?story_id=E1_NPTPDRN (accessed March 5, 2008.).

International Whaling Commission. http://www.iwcoffice.org/index.htm (accessed July 15, 2008).

Time. "The Hunt, the Furor." August 2, 1993. "http://www.time.com/time/magazine/article/0,9171,978989,00.html (accessed March 6, 2008).

James Corbett

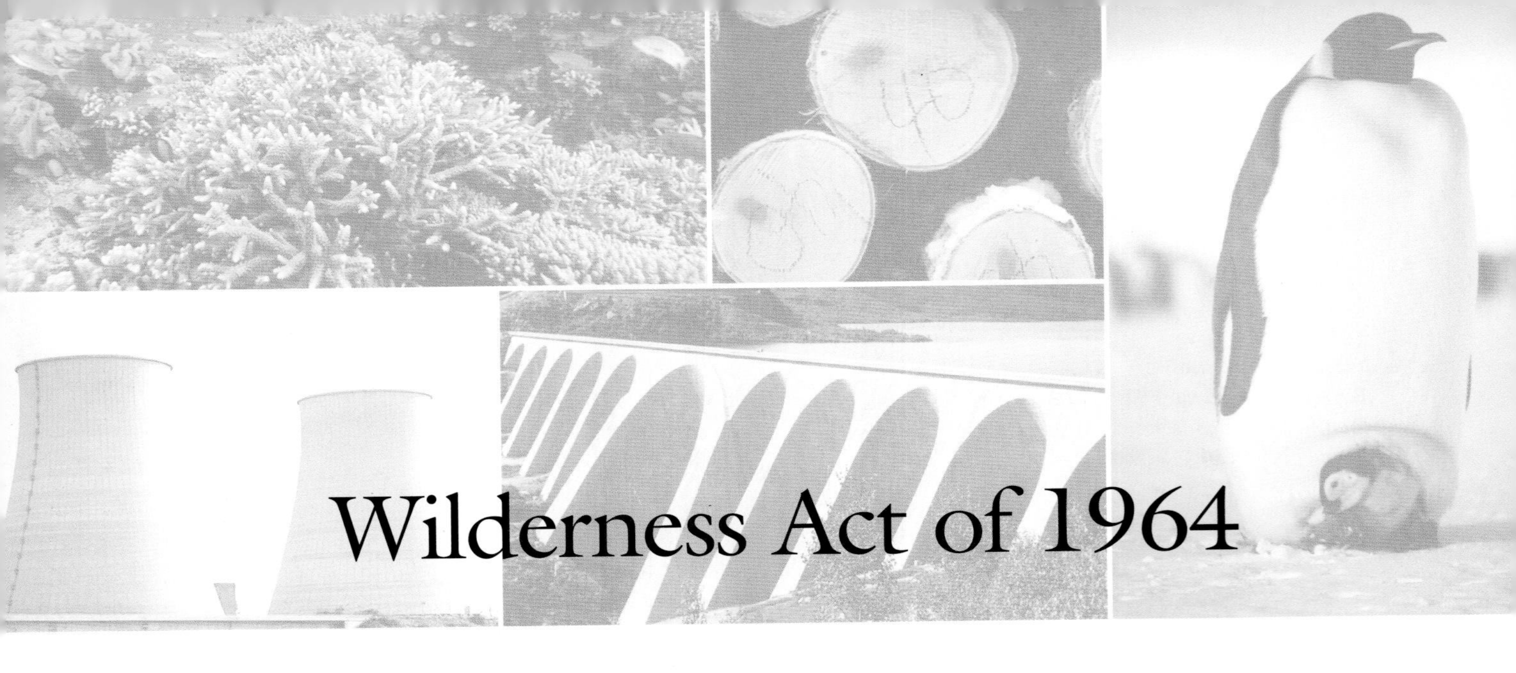

Wilderness Act of 1964

Introduction

The Wilderness Act of 1964 is an environmental law that protects pristine natural lands in the United States. The act contains specific guidelines for setting aside and protecting undeveloped federal land. The Wilderness Act sets aside designated wilderness areas for recreational or scientific use or use as wildlife habitats. Wilderness areas may not be used for timber, grazing, mining, or other destructive purposes. Today, the National Wilderness Preservation System comprises over 100 million acres of federal land.

Historical Background and Scientific Foundations

When European settlers first moved into the area that would become the present-day United States, they encountered over 2 billion acres of relatively pristine wilderness. Settlers destroyed huge swathes of untouched land as they pushed west in the eighteenth and nineteenth centuries to tame the American frontier.

The completion of the transcontinental continental railroad in 1869 connected the West Coast with the East Coast and allowed more Americans to explore, and exploit, the wilderness. With the rise of the automobile in the mid-twentieth century, Americans could travel anywhere in the United States in a matter of days, not months. Companies could also exploit America's vast natural resources and move goods to distant markets quickly and cheaply.

The stress that America's westward expansion placed on the wilderness had become apparent by the mid-nineteenth century. By the 1860s, conservationist called for a halt to the destruction of pristine ecosystems and irresponsible exploitation of natural resources. The efforts of these early conservationists led President Abraham Lincoln (1809–1865) to sign a bill to protect Yosemite Valley for the enjoyment of all Americans. In 1890, Congress passed the Yosemite Reserves Act, which created Yosemite National Park.

In 1897, Congress passed the Organic Act, which established a procedure to set aside federal forest reserves that could not be logged. Although the ostensible purpose of the Organic Act was to ensure future timber supplies, the act was the first comprehensive plan to set aside large tracts of federal land. In 1916, Congress passed the National Park Service Organic Act, which established the National Park Service. Many of the lands that had been set aside under the Organic Act fell under the protection of the newly formed National Park Service. The primary responsibility of the National Park Service was to conserve the natural scenery and wildlife for Americans to enjoy.

Despite nearly a century of conservation efforts, the United States regressed in forest conservation with the passage of the Multiple Use, Sustained Yield Act (MUSY) of 1960. The United States experienced a major period of economic growth following World War II (1939–1945), greatly increasing the demand for timber. Other interests, including recreational opportunities and grazing land for large Western farms, also competed for use of national forests. In MUSY, Congress mandated that the National Forest Service balance the following interests: timber, recreation, grazing, watershed, and wildlife habitats. MUSY stated that a national forest could not be set aside exclusively for one use.

In response to the continued depletion of the American wilderness, Congress passed the Wilderness Act of 1964, after eight years of negotiations and 66 revisions. President Lyndon Johnson (1908–1973) signed the bill into law on September 3, 1964. The Wilderness Act states a legal definition of "wilderness" and provides strong legal protection to any area designated a wilderness area. Under the Wilderness Act, a designated wilderness area may not be used for timber, mining, grazing, or any other activity that would harm its pristine condition. The act also bans all motorized vehicles from designated wilderness areas.

The Wilderness Act contains a famous poetic definition of wilderness written by Howard Zahniser: "A wilderness, in contrast with those areas where man and his own works dominate the landscape, is hereby recognized as an area where the earth and its community of life are untrammeled by man, where man himself is a visitor who does not remain." The act continues with a more specific definition for a wilderness area. Such lands must be at least 5,000 acres, have little noticeable alteration by humans, provide solitude, and may contain "ecological, geological, or other features of scientific, educational, scenic, or historical value."

Under the terms of the Wilderness Act, only Congress has the authority to designate a wilderness area. Congress sets specific boundaries for each wilderness area in statutory law. A separate act of Congress is required to remove a wilderness area from the list of designated wilderness areas.

The Wilderness Act did not place any new land under federal control. The act merely provided protection for some federal lands already under the control of the National Park Service, U.S. Forest Service, and the U.S. Fish and Wildlife Service (FWS). When passed in 1964, the Wilderness Act designated 9.1 million acres of land controlled by these agencies as wilderness areas. The Federal Land Policy and Management Act of 1976 also permitted the Bureau of Land Management (BLM) to recommend land held by the BLM for designation as a wilderness area. Every designated wilderness area falls under the control of the National Wilderness Preservation System (NWPS).

■ Impacts and Issues

The total acreage of designated wilderness areas in the United States has increased tremendously since the passage of the Wilderness Act in 1964. The act directed the secretary of the interior to examine every parcel of land in the National Park System and National Wildlife Refuge. The act directed the secretary of agriculture to do the same with all lands held in the National Forest System. Within ten years, the secretaries of the interior and agriculture were to recommend to the president land that could be included in the National Wilderness Preservation System.

At the request of federal agencies or conservation groups, Congress considers several proposals every year to add new land to the Wilderness System. When the Wilderness Act became law in 1964, the Wilderness System comprised 9.1 million acres of land. Today, the Wilderness System comprises over 107 million acres in over 700 separate wilderness areas. The National Wilderness Preservation System contains wilderness areas in 44 states and Puerto Rico. Wilderness areas are open for recreational use, and millions of Americans visit wilderness areas every year.

WORDS TO KNOW

CONSERVATION: The act of using natural resources in a way that ensures that they will be available to future generations.

ECOSYSTEM: The community of individuals and the physical components of the environment in a certain area.

WATERSHED: The expanse of terrain from which water flows into a wetland, water body, or stream.

In addition to providing scenic and recreational opportunities, wilderness areas possess great ecological value. Most wilderness areas contain thousands of acres. The largest wilderness complex, Noatak and Gates of the Arctic Wildernesses in Alaska, contains over 12 million acres. Such large, contiguous areas of land allow local ecosystems to develop undisturbed by human interference. Many wilderness areas also protect watersheds, thereby preserving the water supply upon which humans and other animals rely.

SEE ALSO *Conservation; National Park Service Organic Act; Natural Reserves and Parks*

BIBLIOGRAPHY

Web Sites

National Wilderness Preservation System. "Wilderness Fast Facts." http://www.wilderness.net/index.cfm?fuse=NWPS&sec=fastFacts (accessed April 22, 2008).

National Wilderness Preservation System. "Wilderness Legislation: Overview." http://www.wilderness.net/index.cfm?fuse=NWPS&sec=legisoverview (accessed April 22, 2008).

U.S. Department of the Interior, National Park Service. "The Idea of Wilderness: Timeline." http://wilderness.nps.gov/idea.cfm (accessed April 22, 2008).

U.S. Fish and Wildlife Service. "Digest of Federal Resource Laws of Interest to the U.S. Fish and Wildlife Service: Wilderness Act." http://www.fws.gov/laws/lawsdigest/wildrns.html (accessed April 22, 2008).

Joseph P. Hyder

Wildfire Control

Introduction

Wildfires attracted comparatively little attention for centuries. Although such blazes destroyed plants, they typically were of small size and had little negative impact on humans. As people have moved closer to wildlands and as the climate has warmed, controlled burns have become more difficult to stage, while uncontrolled burns threaten more homes and people.

Wildfire control does not entail stopping all fires. Some blazes bring benefits to the natural environment by creating vegetative diversity, such as a mixture of wildlife habitats, while destroying the brush that threatens to become a heavy fuel accumulation. Periodic small fires prevent large, massively destructive fires. Wildfires are a natural part of the ecosystem.

Historical Background and Scientific Foundations

Wildfires typically occur when weather conditions increase the risk of fire activity. Unusually wet weather produces abundant vegetation that sets up a region for catastrophic wildfire. Low humidity and dry, windy conditions then provide the conditions for fire potential. Fires propelled by strong winds can move as fast as 60 mi (96 km) per hour. In 2008, Indian scientists reported that biomass density and average precipitation of the warmest quarter of forest had the highest connection to the incidence of forest fires. Other factors included the amount of forest area, rural population density, elevation, and mean annual temperature. Downed power lines, lightning strikes, carelessly tossed cigarettes, sparks from welding equipment, and arson are major causes of wildfires.

Wildfires have been deliberately set in the past to benefit humans. English colonists in the southern colonies of America in the seventeenth century were interested in the Native American practice of burning extensive sections of the forest once or twice a year. The burning consumed all the underwood and brush that would make passage through the forest difficult and complicate hunting. The result was a forest of large, widely spaced trees, few shrubs, and much grass and herbage. Additionally, the removal of brush and fallen trees reduced the total accumulated fuel at ground level. With only small non-woody plants to consume, the annual fires moved quickly, burned with relatively low temperatures, and soon extinguished themselves. These fires rarely grew out of control. Fires of this kind could be used to drive game for hunting, to clear fields for planting, and, on at least one occasion, to fend off European invaders. On the plains, Indians often ignited fires every spring to encourage the new grass growth that would benefit horses and buffalo. In October 1804, explorers Meriwether Lewis and William Clark observed a deliberately set prairie fire that went out of control, killing two people and injuring three more.

Modern governments have used controlled burns to manage natural resources. Firebreaks, sometimes combined with controlled burning, top the list of wildfire control measures. The Natural Resources Conservation Service recommends that firebreaks of 10-foot (3-meter) wide strips of bare soil be set up around structures and fields adjacent to roadways. In large areas, where structures are vulnerable, or where large tracts of land may be affected, a combination of firebreaks and fireguards are used. Firebreaks are constructed during cooler weather with high humidity. The land between firebreaks is then burned to create a black area with no vegetation to support a fire as a fireguard. Some rangeland requires even more protection. Since juniper trees can send bits of burning tree as far as 200 ft (60 m) or more downwind, areas with heavy concentrations of junipers often have 500-foot (152-meter) firebreaks. Such measures reflect changes in federal policy. In July 1972, fire suppression activities used at the Moccasin Mesa blaze in New Mexico destroyed many archeological sites. The resulting public furor led to a national policy to include cultural resource oversight in wildland fires on federal lands.

Although the most prominent fires of the nineteenth and twentieth centuries occurred in the Midwest and West, climate change has brought devastating fires to the East. The August 1995 Sunrise fire in Long Island, New York, that destroyed 5,000 acres alerted Easterners to the fact that a major blaze could happen in their region. Fires in Florida in 1998 forced the evacuation of thousands of people while damaging over 200,000 acres.

WORDS TO KNOW

CONTROLLED BURN: A forest management technique in which small, controlled fires are set to clear brush and prevent larger wildfires in the future.

FIREBREAK: A strip of cleared or plowed land that acts as a barrier to slow or stop the progress of a wildfire.

WILDLAND FIRE USE: A fire management technique that uses naturally ignited fires to benefit natural resources.

Impacts and Issues

As a general rule, U.S. agencies assume responsibility for managing and financing fires that break out on federal lands, while states take charge of blazes on state trust lands or private property. As a result, some states have shouldered massive expenses. On many fires, state and local government crews are often the first to respond, provide most of the fire engines, and carry out most of the fire line work, at least until escaping flames ignite a fast-spreading conflagration. For those events, the U.S. Forest Service and Department of the Interior work closely with state natural resource agencies and rural fire departments through cooperative agreements, combining fire engines, aircraft, and firefighting crews to fight major fires regardless of where they are burning.

The debate within the United States about global warming has harmed efforts to control wildfires. The development of mega-fires, like Colorado's Hayman Fire in 2002 that burned 136,000 acres and 600 structures, means that firefighting agencies need to develop strate-

Environmentalist Carlos Ibero examines the charred remains of pine trees in the mountains of Somosierra, north of Madrid, in August 1995. Over 2,471 acres (1,000 ha) of pine forest were destroyed by fire, which was likely sparked by a bolt of lightning. Contributing to the loss was the lack of firebreaks and excess trees and foliage in the forest. *AP Images*

gies that go beyond fighting fires and fuel management to include variables like the effects of climate change. Since 1999, the U.S. Government Accountability Office (GAO) has recommended the development of a cohesive fire strategy that addresses long-term options and associated funding for reducing hazardous fuels and combating fires. In 2007, a GAO report stated that resource managers in the Agriculture, Interior, and Commerce departments have received limited guidance about whether or how to address climate change. This lack of guidance reduces the ability of these agencies to effectively manage natural resources.

Another GAO report in 2007 found that agencies have identified weaknesses in the management of fire cost-containment efforts but have yet to clearly define cost-containment goals and objectives. In 2007, the U.S. Forest Service spent $1.4 billion fighting fires. The government has been criticized for using too high a percentage of its funds on fire suppression instead of preventive measures such as fuels reduction. State and local governments generally use federal assistance dollars to help fund fuel reduction and the training of state firefighters. These state and local programs face cutbacks in light of federal budget woes and the inability of cash-strapped state and local government to pick up the federal slack. As the risk of fires increases because of climate change, the resources to control wildfires are diminishing.

SEE ALSO *Forest Resources; Forests*

BIBLIOGRAPHY

Books

Krauss, Erich. *Wall of Flame: The Heroic Battle to Save Southern California.* Indianapolis: Wiley, 2006.

Patent, Dorothy Hinshaw. *Fire: Friend or Foe?* New York: Clarion Books, 1998.

Schwab, James. *Planning for Wildfires.* Chicago: American Planning Association, 2005.

Web Sites

National Interagency Fire Center. http://www.nifc.gov (accessed March 14, 2008).

U.S. Forest Service. "Fire and Aviation Management." http://www.fs.fed.us/fire/ (accessed March 14, 2008).

Caryn E. Neumann

Wildfires

Introduction

Wildfires are uncontrolled fires that begin in forests or other wilderness areas. Wildfires often spread to neighboring urbanized areas or to agricultural land. They can be caused by a lightning strike, surface contact of flaming embers from an existing fire set for another purpose such as the burning of trash, by volcanoes, and by human involvement that can be accidental or deliberate (arson).

An example of an accidental wildfire occurred in 1965, when a malfunction caused a truck to ignite that was being driven through a forested area, triggering a wildfire that consumed nearly 510 acres in California's Los Padres National Forest. The driver, singer Johnny Cash (1932–2003), was sued for damage by the federal government and eventually paid approximately $80,000 in an out of court settlement.

Wildfires occur worldwide in regions that are forested and can experience dry weather. For example, areas of Europe, Canada, and the United States that are cool (or cold) and moist during parts of the year can have extended hot and dry periods during the summer and fall. The dried forest floor and falling of dead branches provides the fuel for the ignition of a fire.

The increased warming of the atmosphere related to human activities that has occurred since the mid-nineteenth century and that has been accelerated since the 1950s has been suggested as being influential in the increased prevalence and intensity of wildfires, particularly in southern California and other areas of the western United States. However, as of 2008, a definite link between global warming and wildfires has yet to be established, although the increased frequency and severity of wildfires is consistent with some climate models that have predicted a warmer and drier climate in hard-hit areas such as southern California.

Historical Background and Scientific Foundations

Regions of the world including portions of the United States, Canada, Europe, Australia, and South Africa are prone to wildfires. All experience the protracted periods of hot and dry weather that parch the forest and forest floor or the grassland. Wind also plays an important role in the spread of a wildfire. In California, for example, wildfires that occur during later summer and fall can be fanned by Santa Ana winds—a hot and dry wind that blows westward from the California desert to the coast.

As devastating as wildfires can be to urban areas that border on wilderness regions, and even though they can pose great danger to those caught in a rapidly advancing blaze, wildfires are a natural part of the ecosystem. Vegetation that is aging or dead will be readily burned, providing more space for the growth of a new generation of plants. Indeed, some plants have adapted to be more fire tolerant and capable of rapid regeneration of shoots following a fire, or which produce seeds that are less affected by heat. Research has shown that as some plants burn in a wildfire, the smoke produced is a signal for the subsequent germination of other plants after the fire has passed. This strategy allows the plants to gain a competitive advantage in the charred aftermath of a wildfire and immediately begins the rehabilitation of the area.

The giant redwood trees that grow in regions of northern California have been able to attain their tremendous height because of the periodic lack of competition for nutrients in the aftermath of forest fires.

However, the increased frequency of wildfires that has occurred in southern California through the 1990s and to 2008 has altered such a natural balance of the ecosystem. The repeated fires have eliminated some native plant species and the successors have proven to be more prone to catching fire.

WORDS TO KNOW

CLIMATE MODEL: A quantitative method of simulating the interactions of the atmosphere, oceans, land surface, and ice. Models can range from relatively simple to quite comprehensive.

HABITAT: The natural location of an organism or a population.

SANTA ANA WIND: A warm and dry wind that blows coastward from the desert in southern California.

One reason that a dry forest, grassland, or shrubland is more flammable is the altered chemistry of the vegetation that occurs during hot and dry periods. As many plants dry, they release a flammable gas called ethylene that gathers in the air above the vegetation, increasing the likelihood of a blaze.

The amount of material that is in the vicinity of a fire (the fuel load, typically measured as tons per acre) determines how fiercely the fire will burn. A smaller fuel load will allow a smaller fire than will a high fuel load. Also, a high fuel load can create a hotter fire, which can cause a quicker combustion of wood because the wood releases hydrocarbons that mix with the oxygen in the air to cause an explosive ignition.

A wildfire is no different from a campfire in that the air emerging from the flames is hotter than the surrounding air and so tends to rise. This rising air can contain embers—small bits of still burning material—that can be blown by the current of rising air and any wind present. As the particles settle some distance away, they can ignite new fires, and this cycle can be repeated. As a result, wildfires can jump from location to location, which can make them hard to extinguish.

Santa Ana winds can be influential in spreading a wildfire. Southern California wildfires have covered up to 40 mi (64 km) in a single day during times when this wind is blowing. The advancing fire throws up embers that are spread ahead of the flames by the wind. Some of these wildfires have burned over 1,000 acres per hour.

Complicating the fight to extinguish a wildfire can be a phenomenon called smoldering. This occurs when underbrush is slowly combusted without bursting into flame. An example of smoldering can be seen when a campfire or fire in a wood stove is in the process of being lit. A smoking piece of flammable material can be encouraged to burst into flame when air is supplied by blowing on the material. Similarly, debris in the aftermath of an active wildfire may smolder for days before igniting. In the frenzy of battling an ongoing blaze, these dangerous remnants may be overlooked and can be the cause of the re-emergence of a wildfire in an area that has previously been affected.

Topography also influences the spread of wildfires. A fire can advance more easily uphill, as the flames carry embers upward. Once a fire has reached the peak of a hill, it can become more difficult for it to spread farther, unless embers are dispersed by wind. This behavior can be helpful in fighting a fire; if a fire is contained to the base of a hill, then it is encouraged to travel uphill and so can become easier to extinguish.

Wildfires are battled on the ground and from the air. Fire-fighting crews popularly known as hotshots can remove surface vegetation ahead of the fire in an effort to reduce the amount of available fuel. This is sometimes done for very large wildfires that cannot be safely battled up close by deliberately setting and, hopefully, controlling a fire to consume the surface vegetation. In smaller fires, the ground crews can also carry portable pumps and canisters of water to douse small fires or smoldering areas. In some cases, specially trained fire-fighting personnel dubbed smokejumpers will parachute down near a fire located far from roads.

Often, the effort of the ground personnel is accompanied by the dropping of water or fire retarding chemicals onto the blaze from overhead planes or helicopters (although the latter have to be carefully used, since their blades may fan the fire).

Battling a wildfire is a coordinated process. The locations and activities of the ground and airborne personnel must be known at all times, as any changes in weather such as wind direction or the appearance of a thunderstorm can alter where and how a fire will be dealt with. Rain usually helps firefighters control even the largest wildfire.

Impacts and Issues

On average, about five million acres of forest, grassland, and shrubland are burned each year in the United States. Controlled wildfires, which are also called controlled burns, can be helpful in eliminating underbrush that could be a potential source for more disastrous fires in the future, and can create conditions that are more favorable for the growth of the trees in the area.

In areas where urban settlement is near wilderness, controlled burns are a wise preventative strategy. Ironically, one reason for the preventative approach is the influence of urban growth on the increased frequency of wildfires. This has been particularly true in southern California and in Colorado. As people seek to live in a natural setting and homes are constructed in fire-prone environments and are not as fire-retardant as they could be, the homes become an excellent fuel that spurs the spread of a fire.

In southern California, some wildfires that have occurred since 2000 have been fought using a newly designed fire retardant chemical. The gel-like material is saturated with water, which resides in tiny bubbles dispersed throughout the material. This in effect creates layers of fire retardant in the gel that is sprayed onto a fire or, more commonly, on buildings that are under threat of fire. The layers of the gel burn over time, which increases the time that the material can protect a structure. Although it does not curb the spread of a wildfire, the gel has been effective in reducing property damage and saving homes.

Aside from direct danger and damage, the soot released by large wildfires can alter local climate. Airborne soot from some fires has been tracked in the atmosphere for over 3,100 mi (5,000 km). Atmospheric models have shown that when aloft, the soot can reduce incoming sunlight by up to 15%. Another effect of a wildfire is the sudden release of the carbon dioxide that had been sequestered in the vegetation. A study of an Indonesia forest fire that occurred during 1997 and 1998 demonstrated a release of over 2.5 gigatonnes of carbon dioxide into the atmosphere. Another study conducted at the U.S. National Center for Atmospheric Research has shown that the carbon dioxide released by a large wildfire that burns for several weeks is equal to that released over an entire year by the millions of vehicles in California. As carbon dioxide is a potent greenhouse gas, wildfires likely are a contributor to the warming of the atmosphere that has been accelerating since the 1950s.

SEE ALSO *Ecodisasters; Reforestation; Wildfire Control*

BIBLIOGRAPHY

Books

Halsey, Richard. *Fire, Chaparral, and Survival in Southern California*. San Diego, CA: Sunbelt Publications, 2008.

Reinhart, Kare. *Yellowstone's Rebirth by Fire: Rising from the Ashes of the 1988 Wildfires.* Helena, MT: Farcountry Press, 2008.

Wuerthner, George. *Wildfire Reader: A Century of Failed Forest Policy.* Washington: Island Press, 2006.

Brian D. Hoyle

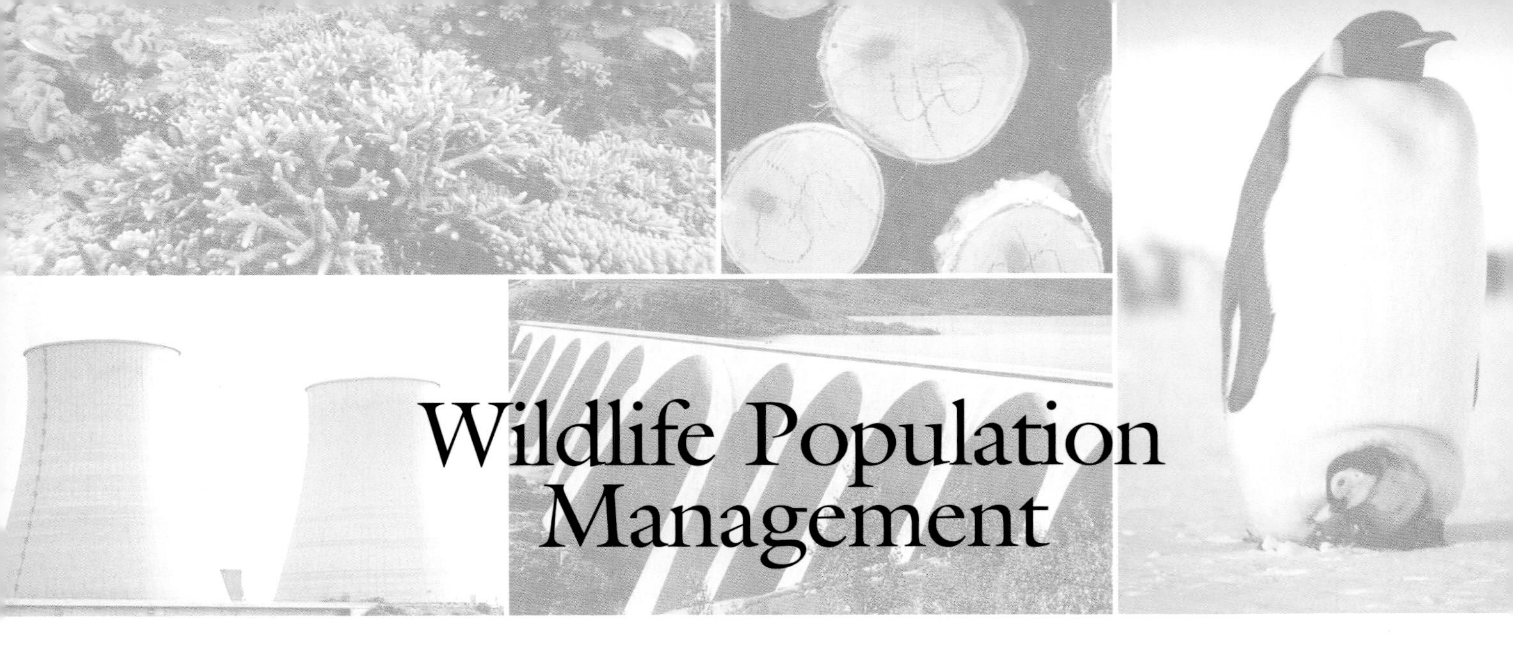

Wildlife Population Management

Introduction

Wildlife population management refers to any strategy that seeks to maintain a target population at a level that can be supported by the ecosystem. This can involve protecting a threatened population from declining further in numbers, or even re-stocking a population. Conversely, when the numbers of a target population have become too great to be sustained by the food or territory available, then predators can be introduced, or a human-mediated cull can be done. Culls have also been done when an infectious disease is present in a population; the deliberate killing of the infected animals can help protect other members of the population as well as other species in the same habitat. Put another way, the management strategies focus on the habitats of the species of concern.

Wildlife population management can also use the legal system to prosecute those found responsible for illegally hunting or trapping species.

Historical Background and Scientific Foundations

In the United States, there were few laws or regulations on the hunting of wild animals before 1900. Virtually all animal and bird species were hunted throughout the year.

Although regulations were introduced to control the timing of hunting of species for commercial purposes and to assign quotas on species killed, and, subsequently, similar controls on recreational hunting, various species came under threat from the changing land use in the country. Territory was lost as land was converted to agricultural, industrial, and urban uses. Furthermore, in agriculture-intensive states, predators such as foxes were actively hunted and killed, allowing their natural prey to increase tremendously in numbers.

By the mid-twentieth century, the need to more actively deal with wildlife populations had been recognized, and the discipline of wildlife biology had been created.

Successful wildlife population management adopts what is known as a landscape perspective. This view takes into account the biological diversity of the particular ecosystem, not just the species of concern. Although in the case of a species at risk the immediate goal is to stabilize the population, the landscape approach recognizes that by addressing the stability of the entire ecosystem, the threatened species also benefits.

Increased numbers of a particular species can upset the balance of a ecosystem, as the burgeoning population requires more food and, if individuals of the species each require substantial territory, can exceed the available space. The limit of an ecosystem on the numbers of any particular species is known as the carrying capacity. There are differing carrying capacities for different species.

Carrying capacity is determined by whatever factor limits the particular population. For wolves, the limiting factor can be territorial size, while for deer the limiting factor can be the available food.

A wildlife population management strategy that can be very effective is hunting. A well-known example is the hunting of deer that takes place in the autumn. Hunters must possess a current license. The number of licenses issued in any year is governed by estimates of the deer population in the state. In years when the deer population is high, more licenses will be issued.

The predator-prey balance is an important influence on wildlife populations. In the absence of wolves, for example, the number of deer can rapidly increase. The high number can exceed the carrying capacity of its habitat, leading to a die-off and the sudden decline in the population.

There are various approaches to managing wildlife. One approach involves the restoration of habitats that have been damaged. One example is the restoration of a wetland. Controlled hunting such as annual deer hunts is another example. A third example is the monitoring of species and the use of the estimated numbers to decide if the species is under threat. Another approach is to review development planned for a region to determine if it could damage vulnerable land or disrupt migratory routes, as two examples, and, if so, to modify or cancel the development.

WORDS TO KNOW

CULL: The selection, often for destruction, of a part of an animal population.

ECOSYSTEM: The community of individuals and the physical components of the environment in a certain area.

POACHING: Illegal hunting.

Impacts and Issues

Wildlife population management is a human-mediated endeavor that involves the work of wildlife biologists, politicians, and volunteers. As well, natural and climatic factors operate. The efforts of non-profit organizations such as Ducks Unlimited, Nature Conservancy, and the Sierra Club have and continue to be tremendously helpful in preserving a variety of bird and animal species.

The near extinction of the bison and the extinction of the passenger pigeon attests to the destructive influence that humans can have in the absence of protective legislation. The implementation and enforcement of wildlife protective laws continues to be an important aspect of population management, especially in areas of the world where poaching is decimating populations such as the tiger and rhinoceros. But, the increased numbers of bison and the rescue of the Whooping Crane from the brink of extinction also attests to the success of population management campaigns.

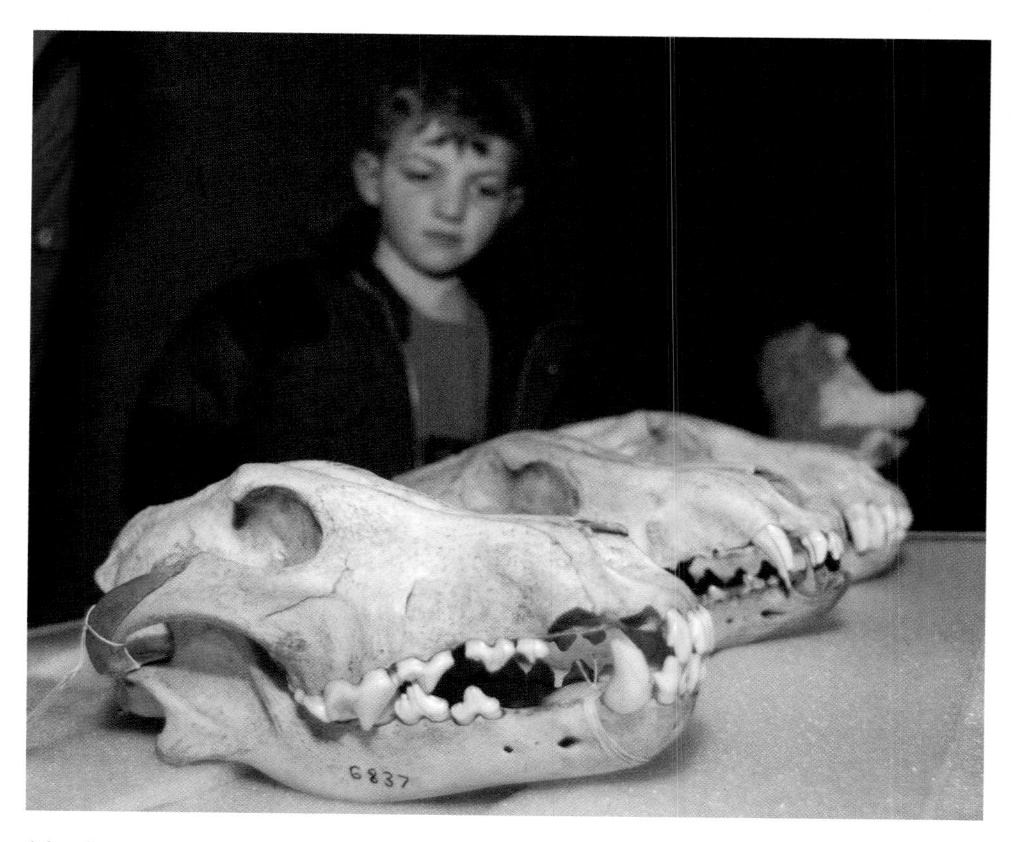

A boy looks at three skulls of wolves found in western Colorado in the early 1900s, on display at the Denver Museum of Science and Nature (2001). A coalition of environmental groups is promoting the return of the gray wolf to Colorado, arguing that the predator would help restore the region's ecological balance. *AP Images*

In February 2006, the U.S. Fish and Wildlife Service (FWS) announced that it was approving Connecticut's Wildlife Conservation Strategy Plan, which provided for the protection of a number of species, including the barred owl. *AP Images*

See Also *Biodiversity; Ecosystem Diversity; Sustainable Development*

BIBLIOGRAPHY

Books

Adams, Clark, Kieran Lindsey, and Sarah Ash. *Urban Wildlife Management.* Boca Raton, FL: CRC, 2005.

Bolen, Eric, and William Robinson. *Wildlife Ecology and Management.* New York: Benjamin Cummings, 2008.

Sinclair, Anthony, John Fryxell, and Graeme Caughley. *Wildlife Ecology, Conservation and Management.* New York: Wiley-Blackwell, 2006.

Wildlife Protection Policies and Legislation

Introduction

Wildlife protection involves policies and legislation to govern land use; hunting and trapping of animals; illicit trapping and trade of animals, reptiles, and birds or products; and other goals that are intended to preserve or restore habitats, protect species that are threatened or in danger of extinction, and enable restoration of populations at risk.

Wildlife protection policies and legislation spans a wide range, from the local level to the international level. Locally, for example, municipal committees can consider development projects in terms of the potential for degradation of watercourses and land as part of an environmental assessment. Such assessments are also required for federal government projects in many countries including the United States. Also, at the federal level, a variety of environmental legislation helps safeguard wildlife.

Policies and legislation also extend to the international level, reflecting the multinational movement of some migratory species and the increasing recognition that environmental preservation, including protection of wildlife, is a global responsibility.

Historical Background and Scientific Foundations

In the United States, few laws or regulations on the hunting of wild animals existed at the beginning of the twentieth century. Hunting of virtually all animal and bird species had no regard for numbers killed or any seasonal timing of hunting, even during breeding periods.

In the United States and other industrialized countries, land use patterns changed from predominantly rural and locally based agriculture to larger, corporate controlled farming and expanding urban centers. Previous pristine territory was lost. By the mid-twentieth century, the need to protect wildlife populations had been recognized.

An important influence in wildlife protection efforts was the 1962 publication of *Silent Spring.* American author Rachel Carson (1907–1964) warned of the consequences on the natural environment and humans of the widespread and extensive use of chemicals such as the pesticide DDT. The book was prophetic, with DDT exposure subsequently being linked to adverse effects in humans, animals, and birds.

In the intervening decades, a variety of policies and laws to protect wildlife were implemented. One example is the environmental assessment process. The multi-step assessment helps identify areas of environmental concern in a project, which can lead to modification of the project or withholding of approval.

In Canada, the Species at Risk Act (SARA) is federal legislation that targets bird, animal, reptile, fish, mollusk, and plant species whose existence is threatened. The act, which was passed by the House of Commons in 2002 as part of the Canadian Biodiversity Strategy and the government's response to the United Nations Convention on Biological Diversity (CBD), is intended to prevent extinction of the species and to do whatever is deemed necessary to ensure their recovery including appropriating land and compensating those affected by the changed land use. As part of the legislation, a committee of experts who are independent of government was established. It is the committee's task to identify species at risk. As of 2008, there are over 200 species on the list including the grizzly bear, Arctic walrus, swift fox, burrowing owl, piping plover, and spotted turtle.

Internationally, the CBD was signed in 1992 at the Earth Summit held in Rio de Janeiro, Brazil. It represented the first global agreement focusing broadly on biological diversity, and the convention contains provi-

WORDS TO KNOW

BIODIVERSITY: Literally, "life diversity": the wide range of plants and animals that exist within any given geographical region.

ECO-TOURISM: Environmentally responsible travel to natural areas that promotes conservation, has a low visitor impact, and provides for beneficially active socio-economic involvement of local peoples.

HABITAT: The natural location of an organism or a population.

sions pertaining to conservation, environmental sustainability, and the open and fair use of benefits of genetic resources. The convention is coordinated and administrated from a United Nation's office in Montreal, Quebec, Canada.

The CBD assists nations in identifying and implementing strategies to protect habitats and preserve biodiversity through scientific research, economic incentives to encourage land use protection, and the legal means to prosecute if necessary.

Another international agreement is the Convention on Migratory Species, which was struck in Bonn, Germany, in 1979 under the United Nations Environment Programme (UNEP). Also known as the Bonn Convention, the agreement is concerned with preserving habitats that are vital for migratory species as well as protecting the species themselves. As of 2008, 104 nations have signed the Bonn Convention.

The Ramsar Convention on Wetlands was signed in Ramsar, Iran, in 1971. The treaty is intended to facilitate national and international actions to preserve wetlands. The 158 signatories represent nearly 1,750 wetland sites that total over 620,000 square mi (161 million hectares).

Impacts and Issues

The many policies and laws concerned with wildlife protection that are in effect globally help preserve habitats, sustain the populations of species at risk, and help in efforts to increase the numbers of species in decline. These efforts have broader benefits. A healthy habitat is better able to support human activities such as agriculture, reduces the development and spread of infectious diseases, and can encourage eco-tourism.

In 2002, the member nations of the CBD committed "to achieve by 2010 a significant reduction of the

World Wildlife Fund activists demand political consequences from the "Pallas" (1998, Germany) oil pollution disaster; the duck represents only one of many animals that died as a direct result of the disaster. *AP Images*

current rate of biodiversity loss at the global, regional and national level as a contribution to poverty alleviation and to the benefit of all life on Earth." Although progress at the global level can be hard to gauge, the CBD 2010 biodiversity target has increased awareness of the importance of wildlife protection and has spurred protective efforts.

SEE ALSO *Biodiversity; Ecosystem Diversity; Sustainable Development*

BIBLIOGRAPHY

Books

Adams, Clark, Kieran Lindsey, and Sarah Ash. *Urban Wildlife Management.* Boca Raton, FL. CRC, 2005.

Bolen, Eric, and William Robinson. *Wildlife Ecology and Management.* New York: Benjamin Cummings, 2008.

Sinclair, Anthony, John Fryxell, and Graeme Caughley. *Wildlife Ecology, Conservation and Management.* New York: Wiley-Blackwell, 2006.

IN CONTEXT: UNITED STATES LISTS POLAR BEARS AS THREATENED SPECIES

Complying with a federal judge's order to make a decision, in May 2008 the United States designated the polar bear as a threatened species and entitled it to Endangered Species Act protections. Federal officials asserted that a two-thirds reduction of the current population level (estimated at 25,000) was possible by 2050. Among the reasons cited was loss of Arctic sea ice that constitutes an essential part of the bear's habitat. The U.S. action further pressured Canada, which as of May 2008 had not listed the polar bear as a threatened species.

In 2001 Greenpeace activists, dressed up as animals, demonstrate for better protection of wild animals in front of the German Science Ministry in Berlin, where a report on environmental policy was published. *AP Images*

Wildlife Refuge

Introduction

A wildlife refuge is land that has been set aside to help protect wild animals, birds, reptiles, and plant species that reside in that area, or that use the area during migration.

Although some wildlife refuges or portions of the land remain undeveloped and in a natural state, other portions may be altered. Alterations are usually made to assist certain species of wildlife. An example is the planting of rice in areas that contain populations of water birds.

Other alterations are permitted for commercial purposes such as lumbering, mining, and oil exploration. These uses are contentious and must be carefully monitored and controlled to prevent overexploitation and land degradation.

Historical Background and Scientific Foundations

The first wildlife refuge in the United States was Pelican Island National Wildlife Refuge. The refuge located on the Atlantic coast of Florida was created by American President Theodore Roosevelt (1858–1919) in 1903.

The idea of setting land aside from development as a haven for wildlife began over 50 years earlier, when explorers returning from the western United States reported on the mass slaughter of species such as the buffalo. Even then the realization was growing that without human intervention in the form of protected lands, animal and bird species could become threatened or extinct.

In 1864, an act of Congress transferred Yosemite Valley to the State of California with the condition that fish and animal species within the allotted area be protected from harm. The area subsequently became a federal government responsibility. In 1872, the first national park was established (Yellowstone National Park), although no provisions were made for the protection of wildlife until passage of the Yellowstone Protective Act in 1894.

Following the establishment of the Pelican Island refuge, wildlife refuges were soon established on many other parcels of land and water. By 1905, the Bureau of Biological Survey (one of the predecessors of the U.S. Fish and Wildlife Service, which now manages federal wildlife refuges) had been created in the U.S. Department of Agriculture, with the responsibility for managing the refuges.

Passage of the Migratory Bird Act in 1918 spurred the creation of refuges specifically for the protection of waterfowl; the first was the Upper Mississippi River Wild Life and Fish Refuge established in 1924. Passage of the Fish and Wildlife Coordination Act in 1934 authorized the acquisition of land specifically for the protection of fish and wildlife.

The National Wildlife Refuge System Administration Act passed in 1966 has been very influential in guiding the development of wildlife refuges in the United States; the act specified the guidelines and policies for the creation of refuges.

As of 2008, there are over 500 sites designated as a National Wildlife Refuge, comprising over 96 million acres in all 50 states and in U.S. territories including Puerto Rico and the U.S. Virgin Islands, almost 46 million acres of which are wetlands. In addition, the various refuges contain undisturbed wilderness (almost 21 million acres), islands, lakes, forests, deserts, and mountainous regions.

More than 200 turtle hatchlings (2-in/5-cm) born in 2003 at a Drexel University incubation lab were released around the Heinze Refuge. *AP Images*

WORDS TO KNOW

EXTINCTION: The total disappearance of a species or the disappearance of a species from a given area.

LAND DEGRADATION: Gradual land impoverishment caused primarily by human activities such as agriculture.

WETLAND: A shallow ecosystem where the land is submerged for at least part of the year.

■ Impacts and Issues

Wildlife refuges have played an influential role in the conservation of land and natural resources in the United States. Their role in the protection of non-migrating wildlife and migratory species has been invaluable in protecting many species from the assaults of land development and pollution that has accompanied the growth of the United States.

For example, the Blackwater Refuge located in eastern Maryland, which was established in 1933 as a haven for ducks and geese during their migrations along the Atlantic Flyway, is used as a rest and refueling stop by an estimated 35,000 geese and 15,000 ducks during their November southern migration. Without the availability of this region, the massive development that has taken place along the northeastern seaboard of the United States would hamper the success of duck and geese migration.

The state of Alaska alone contains 77 million acres of refuge land. The Alaska National Wildlife Reserve, which lies entirely north of the Arctic Circle, has become a flashpoint between environmentalists and those who feel that such reserves can be used both for wildlife protection and for commercial gain. In this case, the commodity is oil; interested companies have estimated that upward of 16 million barrels lies beneath the tundra of the refuge. In 1995, Congress approved oil drilling in the refuge, only to have the initiative vetoed by President Bill Clinton. In 2005, approval for drilling was passed by Congress during budget proceedings, but an outcry led to blockage of the bill by the Senate.

Although no further movement to authorize drilling has been made as of 2008, some environmentalists caution that the growing need for new oil reserves, along with the rising price of oil, will see the return of pressure to drill in the Alaska reserve.

SEE ALSO *Conservation; Wildlife Population Management; Wildlife Protection Policies and Legislation*

BIBLIOGRAPHY

Books

Bass, Rick. *Caribou Rising: Defending the Porcupine Herd, Gwich-'in Culture, and the Arctic National Wildlife Refuge.* San Francisco: Sierra Club Books, 2004.

Butcher, Russell. *America's National Wildlife Refuges: A Complete Guide.* Lanham, MD: Roberts Rinehart Publishers, 2003.

Waterman, Jonathan. *Where Mountains Are Nameless: Passion and Politics in the Arctic Wildlife Refuge.* New York: W. W. Norton, 2007.

Wind and Wind Power

Introduction

Wind is the movement of air from an area of higher air pressure to an area of lower air pressure. The movement of air can be harnessed to provide power. Centuries ago, wind power was used to grind grain and to pump water. In modern times, wind has been harnessed to generate electricity. In the twentieth century, wind power has become more popular, with large arrays of windmills (wind farms) being established worldwide. While in 2008, wind power generated electricity accounts for only about 1% of the world's electricity production, it is becoming increasing attractive as the consequences of fossil fuel generated electricity on atmospheric warming have become recognized.

Historical Background and Scientific Foundations

The earliest documented record of windmills dates back to the late twelfth century in England. However, windmills may have been in use as long ago as AD 500 in Persia, and perhaps even 2,000 years ago in China.

The windmills used in England were designed to replace the use of animals to turn grinding wheels. Windmill-driven grinding of grain soon became very widely used in England and in Europe. By the mid-fifteenth century, there were some 10,000 windmills in the south and east regions of England where wheat was cultivated.

At about the same time, people began using windmills in the Netherlands as a means of draining regions called polders—land that had been created by walling off a section of the sea and draining the land. Because the land was below sea level, draining had to be continued to maintain the land for farming and the raising of livestock. Windmills, sometimes in a series of three or four, pumped water from the lowest lying areas to a higher elevation river or lake.

In North America, windmills were built for the same purposes beginning in the late seventeenth century as colonization to present-day Canada and the eastern United States occurred. As the Midwest became settled and ranging began, windmills also became used as pumps to bring water to the surface from wells. These small windmills took advantage of the near constant winds that are still a feature of prairie regions. From 1850 to 1970, over six million windmills were built on U.S. farms.

In 1888, a large windmill built in Cleveland, Ohio, is the first known version constructed specifically to generate electricity. Within a decade, wind-powered electricity had spread to Europe.

The use of wind power as a source of electricity declined in the United States in the 1930s and 1940s as the electrical demands of farmers increased and also due to the Great Depression. A federal government initiative to spur rural economies during the Depression was to extend the conventional supply of electricity, which replaced the need for an on-site windmill as a power source.

The modern era of wind power began following World War II (post 1945), when European countries such as Denmark and Germany began development of large three-bladed turbines to produce electricity. This design is still in wide use today in wind farms—facilities that house hundreds of turbines that all feed to a common collection point.

Although wind power faded again in the 1960s due to an abundance of inexpensive fossil fuel for the conventional generation of electricity, subsequently it has regained popularity due to fluctuations in fossil fuel availability and price, which increased demand for a supply of renewable energy, and the recognition of the influence on greenhouse gases such as carbon dioxide on the warming of the atmosphere.

Wind can result from the uneven heating of the land and water. Land tends to absorb the heat of the sun more quickly than does water. As well, heat is emitted back to the atmosphere more readily from land than from water. Because warm air rises, cooler air will move in to displace it. This creates wind.

Different regions of land can also heat up and cool down unevenly. For example, a sandy desert will heat up and cool down more quickly than a forested area. As a result, wind can be generated over land, as air moves from a relatively cooler area to areas where warmer air is rising.

This behavior can occur locally over small distances and, in the case of weather systems, over hundreds or thousands of miles/kilometers.

The conversion of wind into electricity begins when the moving air contacts the blades of the wind machine (turbine). The angle of the blades slows the movement of the air. The design is similar to the wings of an aircraft. The air passes more slowly over one surface of a blade than over the opposing surface. This causes a difference in air pressure, and the movement of air from the higher-pressure surface of the blade to the lower pressure area. The blades do not rise into the air as do airplane wings, but rotate about the fixed central point.

The rotating blades are attached to a shaft. As the shaft turns, an electrical generator positioned at the top of the wind machine produces electricity. A cable that runs down the center of the machine conveys the electricity to the ground, where it feeds into a transmission line. The electricity can then flow into the conventional electrical grid to supplement the conventionally generated electricity, or can be directly used on-site.

Traditionally, both the first windmills and modern electricity-generating turbines have had one of two designs. In the horizontal axis design, the blades are oriented horizontal to the ground and perpendicular to the central shaft. In the vertical axis design, the blades are positioned vertical to the ground and parallel to the central shaft.

The horizontal axis version is the most widely used. The turbines that are used in wind farms can be very large, as tall as a 20-story building with three blades that are hundreds of feet long. Even larger versions exist; the blades on the largest turbines are over 300 ft (90 m) in length, longer than a football field.

The electrical output from a wind farm can be considerable. For example, the Horse Hollow Wind Energy Center in Texas contains over 420 wind turbines. The collective electricity generated is enough to power almost one-quarter million homes for a year.

WORDS TO KNOW

ELECTRICAL GRID: Network of power lines that carry electricity from the source of generation to where the power can be used.

GIGAWATT: A unit of power that is equal to one billion watts.

RENEWABLE ENERGY SOURCE: An energy resource that is naturally replenished, such as sunlight, wind, or geothermal heat.

TURBINE: An engine that moves in a circular motion when force, such as moving water, is applied to its series of baffles (thin plates or screens) radiating from a central shaft.

Impacts and Issues

Wind power is a renewable source of electricity that does not generate carbon dioxide. In contrast, the generation of electricity by the burning of coal is non-renewable and emits carbon dioxide to the atmosphere. These attributes of wind power are advantageous both in terms of the economy of power generation and in reducing the emissions of greenhouse gas. It is now recognized that human activities such as electricity generation are the major driver of the atmospheric warming that has been occurring since the mid-nineteenth century and that has been accelerating since the mid-twentieth century. Thus, any technology that can reduce carbon dioxide emissions is potentially beneficial.

However, the advantages of wind power come with some drawbacks. The rotating blades can be lethal to birds. Besides considering the strength and regularity of prevailing winds, wind farm design also needs to consider the migration pattern of birds and avoid bird migration flyways. As well, the visual impact of dozens or hundreds of turbines can be undesirable to some people. This is subjective, however, as the same farm can be viewed approvingly by others.

Turbines do generate noise. Although wind farms tend to be in sparsely populated areas (some are even placed offshore), an individual turbine can be a disruption to those in the immediate vicinity. Whether the frequency of sound produced is merely inconvenient or has some health effects is contentious and unclear.

Despite the disadvantages, wind power is an increasingly popular technology. As an example, an offshore turbine initiative by the United Kingdom approved construction of almost 600 turbines in 2007. When these turbines become operational, combined with existing facilities, the aim is to harness enough wind energy to power the United Kingdom by the year 2020.

See Also *Alternative Fuel Impacts; Nuclear Power; Tidal or Wave Power*

BIBLIOGRAPHY

Books

Gipe, Paul. *Wind Power.* White River Junction, VT: Chelsea Green, 2004.

Spilsbury, Richard, and Louise Spilsbury. *The Pros and Cons of Wind Power.* New York: Rosen Central, 2007.

Williams, Wendy, and Robert Whitcomb. *Cape Wind: Money, Celebrity, Class, Politics, and the Battle for Our Energy Future on Nantucket Sound.* New York: PublicAffairs, 2007.

Web Sites

U.S. Department of Energy. "Wind Powering America." January 29, 2008. http://www.eere.energy.gov/windandhydro/windpoweringamerica/ (accessed April 7, 2008).

Brian D. Hoyle

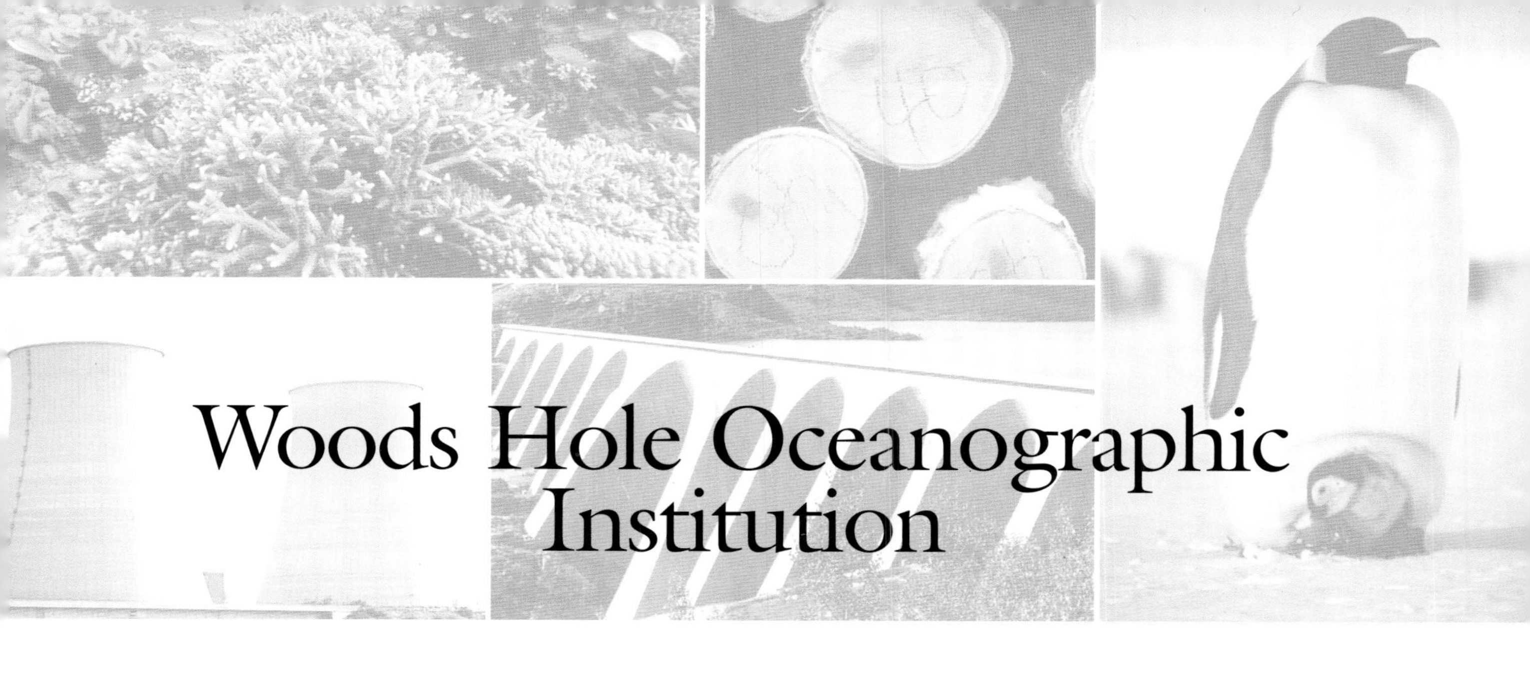

Woods Hole Oceanographic Institution

Introduction

Woods Hole Oceanographic Institution (WHOI) is a private, not-for-profit oceanographic research and educational facility located in Woods Hole, Massachusetts.

WHOI was founded in 1930 primarily as a research institution for the study of the ocean. Later, in collaboration with the Massachusetts Institute of Technology (MIT), it also became an educational institution where students can earn graduate degrees and undertake post-graduate research in a variety of oceanography-related sciences.

WHOI research is both basic and applied. The institution has been, and continues to be, a pioneer in the development of undersea probes that are piloted by onboard personnel (the best-known example is *Alvin*, which was used in the discovery of the first undersea vent and the *Titanic*), probes that are tethered to the ship, and robotic probes capable of independent undersea flight. In addition, WHOI is participating in the establishment of instrumentation arrays that can be deployed on the sea surface and seafloor. All of these technologies are revealing a great deal of information about the composition, behavior, and change of the ocean.

The latter has become very important as the influence of the ocean on Earth's climate has become recognized.

Historical Background and Scientific Foundations

As oceanography became an established multidisciplinary science in the nineteenth century, the need for dedicated research facilities was recognized. By the first decade of the twentieth century, oceanographic research was flourishing. The Scripps Institution of Oceanography was established in San Diego, California, in 1903. The need for an East Coast facility was recognized and recommended in 1927 by an Oceanography Committee of the National Academy of Sciences. On the basis of this recommendation, WHOI was established on January 6, 1930.

The founding mission of WHOI, which remains the same in 2008, was "research and education to advance understanding of the ocean and its interaction with the Earth system, and communicating this understanding for the benefit of society."

By the following year, two research ships had been delivered and research was underway. WHOI expanded considerably during World War II (1939–1945), when many war-related research projects were undertaken (defense-related research, including U.S. Navy-sponsored research on robotic probes, continues in 2008). From 1940 to 1945, WHOI expanded from a scientific and support staff of 92 to over 300.

In the 1950s, while WHOI remained a private institution, internal funds became insufficient to cover the increasing coats of oceanographic research. This led WHOI to compete for public funding. The success of funding led to further expansion, and by 1968 the institute housed over 600 scientists and staff. That year, WHOI became an educational institution when it established a joint graduate program with MIT.

The growing diversity of oceanographic research was recognized in 2000 by the formation of four "ocean institutes" within WHOI: the Coastal Ocean Institute, Deep Ocean Exploration Institute, Ocean and Climate Change Institute, and Ocean Life Institute.

In 2008, WHOI entered its 77th year of operation. In February of that year, Dr. Susan Avery became the ninth director, and the first woman appointed to the position.

Among WHOI's more public accomplishments have been the discovery of hydrothermal vents on the sea floor of the Pacific Ocean in 1977 (new vents continue

WORDS TO KNOW

HYDROSPHERE: The total amount of liquid, solid, and gaseous water present on Earth.

MARIANAS TRENCH: A canyon almost 36,000 ft (11,000 m) in depth located in the floor of the Pacific Ocean; it is the deepest point in the ocean.

to be discovered, the latest in 2007) and the discovery of the sunken ocean liner *Titanic* in 1985.

These accomplishments are a very small part of WHOI's research, which has grown to include the physical, chemical, ecological, and biological (including microbiological) sciences. Modeling of ocean dynamics, currents, and the relationship between the ocean and Earth's atmosphere and climate has grown in importance since the 1980s.

As of 2008, WHOI operated a fleet of three ships and numerous underwater manned and robotic vehicles.

■ Impacts and Issues

WHOI has, and continues to be, at the forefront of oceanographic research. The basic and applied research has led to, as just a few examples, the formulation of paints that decrease the fouling of ships' hulls with biological growth, increased knowledge about practices that help sustain fisheries, the discovery of new forms of life, and increased understanding of ocean currents, and other aspects of ocean composition and behavior.

Refinement of robotic probes to accommodate more and increasingly sensitive instrumentation and to allow probes to descend to greater depths has now made virtually any portion of the ocean accessible. In the past, a limitation to undersea exploration was the supporting cable used to tether some deep ocean probes to the ship. A cable tens of thousands of feet in length would be too heavy to be supported by a ship's winch. In 2006, however, researchers at Woods Hole successfully developed fiber optic cables that were capable of both supporting more instrumentation and that weighed a fraction of the existing cable. This has made it feasible to manufacture a tethering cable capable of sending undermanned probes to the Marianas Trench, which at a depth of 6.8 mi (11 km) is the deepest-known part of the ocean.

Research from WHOI and other institutions has also revealed the deterioration of the ocean due to human activities, and has shown that human-related activities are changing the chemistry of the ocean. For example, since the 1980s, the pH (a measure of the acidity of the water) has dropped. Given the immense volume of the global ocean, the finding that the ocean chemistry can be changed is very significant.

An engineering assistant assembles REMUS, one of the remote environmental monitoring units at the Woods Hole Oceanographic Institution's Smith Laboratory in May 2003. *AP Images*

A changing ocean is fundamentally important for climate. As of 2008, climate research has become a very important focus of WHOI research.

The increasing link between the ocean and Earth's climate is increasing the pressure on WHOI to play a more prominent role in U.S. environmental policy. This is creating challenges for an institution whose foundation and existence is based on scientific research.

SEE ALSO *Climate Modeling; Iron Fertilization; Marine Ecosystems; Oceanography*

BIBLIOGRAPHY

Books

Cullen, Vicky. *Down to the Sea for Science: 75 Years of Ocean Research Education & Exploration at the Woods Hole Oceanographic Institution*. Woods Hole, MA: Woods Hole Oceanographic, 2005.

Parsons, Tim. *The Sea's Enthrall: Memoirs of an Oceanographer*. Victoria, British Columbia, Canada: Trafford Publishing, 2007.

Sverdrup, Keith A. *An Introduction to the World's Oceans*. New York: McGraw-Hill, 2008.

IN CONTEXT: DISCOVERING A NEW ECOSYSTEM

A vibrant community of bacteria, tubeworms that are unique to the geothermal vent environment and other creatures, exists around hydrothermal vents. The entire ecosystem is possible because of the activity of the bacteria. These bacteria have been shown, principally through the efforts of Holger Jannasch (1927–1998) of the Woods Hole Oceanographic Institution, to accomplish the conversion of sulfur to energy in a process that does not utilize sunlight called chemosynthesis. The energy is then available for use by the other life forms, which directly utilize the energy, consume the bacteria, or consume the organisms that rely directly on the bacteria for nourishment. For example, the tubeworms have no means with which to take in or process nutrients. Their existence relies entirely on the bacteria that live in their tissues.

Sources Consulted

BOOKS

Acquaah, George. *Horticulture: Principles and Practices.* New York: Prentice Hall, 2004.

Adams, Clark, Kieran Lindsey, and Sarah Ash. *Urban Wildlife Management.* Boca Raton, FL: CRC, 2005.

Ahrens, C. Donald. *Meteorology Today: An Introduction to Weather, Climate, and the Environment.* Belmont, CA: Brooks Cole, 2006.

Akcakaya, H. Resit. *Species Conservation and Management: Case Studies.* New York: Oxford University Press, 2004.

Allen, John. *The Biospheres: A Memoir by the Inventor of Biosphere II.* Santa Fe, NM: Synergetic Press, 2008.

Anderson, Stephen O., and K. Madhava Sarma. *Protecting the Ozone Layer: The United Nations History.* London: Earthscan Publications, 2005.

Atlas, Ronald M., and Jim Philip. *Bioremediation: Applied Microbial Solutions for Real-World Environment Cleanup.* Washington, DC: ASM Press, 2005.

Audubon, John James. *Birds of America.* Philadelphia: J. B. Chevalier, 1842.

Baillie, M.G.L. *A Slice Through Time: Dendrochronology and Precision Dating.* New York, Cambridge University Press, 1995.

Balfour, Eve. *The Living Soil.* London: Faber and Faber, 1948.

Banta, John. *Extreme Weather Hits Home: Protecting Your Buildings from Climate Change.* Gabriola Island, British Columbia, Canada: New Society Publishers, 2007.

Barlow, Maude. *Blue Covenant: The Global Water Crisis and the Coming Battle for the Right to Water.* New York: New Press, 2008.

Barry, John M. *Rising Tide: The Great Mississippi Flood of 1927 and How It Changed America.* New York, Simon and Schuster, 1998.

Barry, Roger G. *Atmosphere, Weather and Climate.* Oxford, United Kingdom: Routledge, 2003.

Bass, Rick. *Caribou Rising: Defending the Porcupine Herd, Gwich-'in Culture, and the Arctic National Wildlife Refuge.* San Francisco: Sierra Club Books, 2004.

Belk, Colleen, and Virginia Borden. *Biology: Science for Life with Physiology.* New York: Benjamin Cummings, 2006.

Bellemare, Jesse, Glenn Motzkin, and David R. Foster. *Thoreau's Country: Journey through a Transformed Landscape: "Legacies of the Agricultural Past."* Cambridge, MA: Harvard, 1999.

Berg, Linda, and Mary Catherine Hager. *Visualizing Environmental Science.* New York: Wiley, 2006.

Bernstein, Aaron. *Sustaining Life: How Human Health Depends on Biodiversity.* New York: Oxford USA, 2008.

Berry, Wendell. *The Unsettling of America.* San Francisco: Sierra Club Books, 1977.

Bigg, Grant R. *The Oceans and Climate.* Cambridge: Cambridge University Press, 2004.

Bolen, Eric, and William Robinson. *Wildlife Ecology and Management.* New York: Benjamin Cummings, 2008.

Bray, R.N. *Environmental Aspects of Dredging.* New York: Taylor & Francis, 2008.

Brinkley, Douglas. *The Great Deluge: Hurricane Katrina, New Orleans, and the Mississippi Gulf Coast.* New York: Harper Perennial, 2007.

Brown, Oli. *The Environment and Our Security: How Our Understanding of the Links Has Changed.* Winnipeg, Manitoba, Canada: International Institute for Sustainable Development, 2005.

Bryant, Edward. *Tsunami: The Underrated Hazard.* Cambridge: Cambridge University Press, 2001.

Burchell, Jon. *The Evolution of Green Politics: Development and Change within European Green Parties.* London: Earthscan, 2002.

Bureau of Land Management, U.S. Department of the Interior. *Land Use Planning Handbook.* March 11, 2005.

Burnie, David. *Shrublands.* New York: Raintree, 2003.

Burns, Ronald G., and Michael J. Lynch. *Environmental Crime: A Sourcebook.* New York: LFB Scholarly Publishing, 2004.

Burt, Christopher. *Extreme Weather.* New York: W.W. Norton, 2007.

Butcher, Russell. *America's National Wildlife Refuges: A Complete Guide.* Lanham: Roberts Rinehart Publishers, 2003.

Carson, Rachel. *Silent Spring.* Greenwich, CT: Fawcett Crest, 1962.

Centner, Terence J. *Empty Pastures: Confined Animals and the Transformation of the Rural Landscape.* Champaign: University of Illinois Press, 2004.

Chiras, Daniel D., John P. Reganold, and Oliver S. Owen. *Natural Resource Conservation: Management for a Sustainable Future.* New York: Prentice-Hall, 2004.

Chivian, Eric, and Aaron Bernstein, eds. *Sustaining Life: How Human Health Depends on Biodiversity.* Cambridge, MA: Oxford, 2008.

Churchill, R.R., and A.V. Lower. *The Law of the Sea.* Huntington, NY: Juris, 1999.

Clifford, Mary. *Environmental Crime: Enforcement Policy and Social Responsibility.* Boston: Jones and Bartlett, 1998.

Clover, Charles. *The End of the Line: How Overfishing Is Changing the World and What We Eat.* Berkeley: University of California Press, 2008.

Collin, Robert W. *The Environmental Protection Agency: Cleaning Up America's Act.* Westport, CT: Greenwood Press, 2005.

Committee on the Future of Rainfall Measuring Missions. *NOAA's Role in Space-Based Global Precipitation Estimation and Application.* Washington, DC: National Academies Press, 2007.

Committee on the Science of Climate Change, National Academy of Sciences. *Climate Change Science: An Analysis of Some Key Questions.* Washington, DC: National Academies Press, 2001.

Conard, Rebecca, and Wayne Franklin. *Places of Quiet Beauty: Parks, Preserves, and Environmentalism.* Iowa City: University of Iowa Press, 1997.

Cornforth, Derek. *Landslides in Practice: Investigation, Analysis, and Remedial/Preventative Options in Soils.* New York: Wiley, 2005.

Corzine, John S., Governor, and the Highlands Water Protection and Planning Council. *Highlands Regional Master Plan* State of New Jersey: November 2007.

Cruz, Joao. *Ocean Wave Energy: Current Status and Future Perspectives.* New York: Springer, 2008.

Cullen, Vicky. *Down to the Sea for Science: 75 Years of Ocean Research Education & Exploration at the Woods Hole Oceanographic Institution.* Woods Hole, MA: Woods Hole Oceanographic Institution, 2005.

Cullet, Philippe. *Differential Treatment in International Environmental Law.* Aldershot, UK: Ashgate, 2003.

Cunningham, W.P., and M.A. Cunningham. *Environmental Science: A Global Concern.* New York: McGraw-Hill International Edition, 2008.

Davis, B.E. *Geographic Information Systems: A Visual Approach.* Clifton Park, NY: OnWord Press, 1996.

Davis, Devra Lee. *When Smoke Ran Like Water: Tales of Environmental Deception and the Battle Against Pollution.* Oshkosh, WI: Basic Books, 2004.

de Boer, Jelle Zeilinga *Volcanoes in Human History: The Far-Reaching Effects of Major Eruptions.* Princeton, NJ: Princeton University Press, 2004.

DeLong, James. *Out of Bounds and Out of Control: Regulatory Enforcement at the EPA.* Washington, DC: Cato Institute, 2002.

DeMers, M.N. *Fundamentals of Geographic Information Systems*, 2nd ed. Hoboken, NJ: Wiley, 2000.

Desonie, Dana. *Oceans: How We Use the Seas.* New York: Chelsea House Publications, 2007.

Diamond, Jared. *Collapse: How Societies Choose to Fail or Succeed.* New York: Viking, 2004.

DiMento, Joseph F.C., and Pamela M. Doughman. *Climate Change: What It Means for Us, Our Chil-*

dren, and Our Grandchildren. Boston: MIT Press, 2007.

DiPippo, Ronald. *Geothermal Power Plants: Principles, Applications, Case Studies and Environmental Impact.* Portsmouth, NH: Butterworth-Heinemann, 2007.

Doesken, Nolan J. *The Snow Booklet: A Guide to the Science, Climatology, and Measurement of Snow in the U.S..* Fort Collins: Colorado State University, 1997.

Dombeck, Michael P., Christopher A. Wood, and Jack E. Williams. *From Conquest to Conservation: Our Public Lands Legacy.* Washington, DC: Island Press, 2003.

Dryzek, John S., et al. *Green States and Social Movements: Environmentalism in the U.S., U.K., Germany and Norway.* Oxford: Oxford University Press, 2003.

Dutta, Subijoy. *Environmental Treatment Technologies for Hazardous and Medical Wastes: Remedial Scope and Efficacy.* New York: McGraw-Hill, 2007.

Ellis, Richard. *The Empty Ocean.* Washington, DC: Island Press, 2004.

Enger, Eldon, and Bradley Smith. *Environmental Science: A Study of Interrelationships.* Boston: McGraw-Hill, 2006.

Environmental Careers Organization. *The ECO Guide to Careers that Make a Difference: Environmental Work for a Sustainable World.* Washington, DC: Island Press, 2007.

Erjavec, Jack, and Jeff Arias. *Hybrid, Electric and Fuel-Cell Vehicles.* Clifton Park, NJ: Thomson Delmar Learning, 2007.

Fasulo, Mike, and Paul Walker. *Careers in the Environment.* New York: McGraw-Hill, 2007.

Flannery, Tim. *The Great Lakes: The Natural History of a Changing Region.* Vancouver, British Columbia, Canada: Greystone Books, 2007.

Flannery, Tim. *The Weather Makers: How Man Is Changing the Climate and What It Means for Life on Earth.* New York: Atlantic Monthly Press, 2005.

Fletcher, Thomas. *From Love Canal to Environmental Justice: The Politics of Hazardous Waste on the Canada-U.S. Border.* Peterborough, Ontario, Canada: Broadview Press, 2003.

Flint, Anthony. *This Land: The Battle over Sprawl and the Future of America.* Baltimore: Johns Hopkins University Press, 2006.

Florio, James R., Governor, and the Pinelands Forestry Advisory Committee. *Final Report: Recommended Forestry Management Practices.* Trenton, NJ: New Jersey Pinelands Commission, 2006.

Flynn, Ann Marie. *Health Safety and Accident Management in the Chemical Process Industries.* New York: Marcel Dekker, 2002.

Freyfogle, Eric T. *Why Conservation Is Failing and How It Can Regain Ground.* New Haven, CT: Yale University Press, 2006.

Fujita, Rodney. *Heal the Ocean: Solutions for Saving Our Seas.* Gabriola Island, British Columbia, Canada: New Society Publishers, 1998.

Gardner, Gary, and Thomas Prugh. *State of the World 2008: Seeding the Sustainable Economy.* Washington, DC: Worldwatch Institute, 2008.

Gipe, Paul. *Wind Power.* White River Junction: Chelsea Green, 2004.

Gleditsch, Nils Petter, and Paul F. Diehl, eds. *Environmental Conflict.* Boulder, CO: Westview, 2001.

Gleditsch, Nils Petter, ed. *Conflict and the Environment, NATO Advanced Research Workshop on Conflict and the Environment (1996 Bolkesjo, Norway).* Dordrecht, Netherlands: Kluwer, 1997.

Goodell, Jeff. *Big Coal.* New York: Mariner Books.

Goodwin, Lei. *Discover Your World with NOAA.* Washington, DC: NOAA, 2007.

Gore, Al. *An Inconvenient Truth: The Planetary Emergency of Global Warming and What We Can Do About It.* Emmaus, PA: Rodale Press, 2006.

Gould, Stephen Jay. *Time's Arrow, Time's Cycle: Myth and Metaphor in the Discovery of Geological Time.* Cambridge, MA: Harvard University Press, 1987.

Governments of Argentina, Australia, Belgium, Chile, the French Republic, Japan, New Zealand, Norway, the Union of South Africa, The Union of Soviet Socialist Republics, the United Kingdom of Great Britain and Northern Ireland, and the United States of America. *The Antarctic Treaty.* United Nations, December 1, 1959.

Gradstein, F.M., et al. *A Geologic Time Scale.* Cambridge: Cambridge University Press, 2004.

Grady, Wayne. *The Weather Makers: How Man Is Changing the Climate and What It Means for Life on Earth.* Jackson TN: Atlantic Monthly Press, 2006.

Grandin, Temple, Terry Williams, and Rosalie Winard. *Wild Birds of the American Wetlands.* New York: Welcome Books, 2008.

Grant, William, and Todd Swannack. *Ecological Modeling.* Malden, MA: Blackwell Publishing, 2007.

Grover, Velma I. *Water: Global Common and Global Problems.* Enfield, NH: Science Publishers, 2006.

Gunn, Angus. *Unnatural Disasters: Case Studies of Human-Induced Environmental Catastrophes.* Westport: Greenwood Press, 2003.

Gunter, Valerie Jan. *Volatile Places: A Sociology of Communities and Environmental Controversies.* Thousand Oaks, CA: Pine Forge Press, 2007.

Gupta, Harsh, and Roy Sukanta. *Geothermal Energy: An Alternative Resource for the 21st Century.* Boston: Elsevier Science, 2006.

Hadlock, Charles R. *Mathematical Modeling in the Environment.* Boca Raton, FL: CRC Press, 2007.

Halderman, James, Tony Martin, and Craig Van Batenburg. *Hybrid and Alternative Fuel Vehicles.* Clifton Park, NJ: Cengage Delmar Learning, 2006.

Haller, Rebecca, and Christine Kramer. *Horticulture Therapy Methods.* Boca Raton: CRC, 2006.

Halsey, Richard. *Fire, Chaparral, and Survival in Southern California.* El Cajon: Sunbelt Publications, 2008.

Hansen, Keith A. *The Comprehensive Nuclear Test Ban Treaty: An Insider's Perspective.* Stanford, CA: Stanford University Press, 2006.

Hanson, J.R. *Chemistry in the Garden.* Cambridge: Royal Society of Chemistry, 2007.

Harley, John Brian, and David Woodward. *The History of Cartography.* Chicago: University of Chicago Press, 1987.

Hermes, Patricia. *Fly Away Home.* New York: Newmarket, 2005.

Heywood, I., S. Carver, and S. Cornelius. *An Introduction to Geographical Information Systems.* Boston: Longman, 1998.

Hillman, Mayer, Tina Fawcett, and Sudhir Chella Rajan. *The Suicidal Planet: How to Prevent Global Climate Catastrophe.* New York: Thomas Dunne Books, 2007.

Ho, Mun S., and Chris P. Nielsen. *Clearing the Air: The Health and Economic Damages of Air Pollution in China.* Boston: MIT Press, 2007.

Hocking, Colin. *Global Warming and the Greenhouse Effect.* Berkeley, CA: GEMS, 2002.

Horn, Tammy. *Bees in America: How the Honey Bee Shaped a Nation.* Lexington: University of Kentucky, 2006.

Hough, Susan Elizabeth. *Earthshaking Science: What We Know (and Don't Know) about Earthquakes.* Princeton, NJ: Princeton University Press, 2004.

Hough, Susan Elizabeth. *Richter's Scale: Measure of an Earthquake, Measure of a Man.* Princeton, NJ: Princeton University Press, 2006.

Houghton, John. *Global Warming: The Complete Briefing.* Cambridge, MA: Cambridge University Press, 2004.

House of Commons. *Environment Audit Committee. Environmental Crime and the Courts. Sixth Report of Session 2003-4.* London: HM Stationary Office, 2004.

Hulme, Mike. *Nature's Revenge?: Hurricanes, Floods and Climate Change.* New York: Hodder, 2003.

Hulot, Nicholas. *One Planet: A Celebration of Biodiversity.* New York: Abrams, 2006.

Hyne, Norman J. *Nontechnical Guide to Petroleum Geology, Exploration, Drilling and Production*, 2nd ed. Tulsa, OK: PennWell Books, 2001.

Ingrahm, Scott. *The Chernobyl Nuclear Disaster.* New York: Facts on File, 2005.

Jackson, Patrick Wyse. *The Chronologers' Quest: The Search for the Age of the Earth.* Cambridge, UK: Cambridge University Press, 2006.

Jang, Hwee-Yong. *Gaia Project: 2012; The Earth's Coming Great Changes.* Woodbury: Llewellyn Publications, 2007.

Jarrell, Melissa. *Environmental Crime and the Media: News Coverage of Petroleum Refining Industry Violations.* New York: LFB Scholarly Publishing LLC, 2007.

Johnsen, Carolyn. *Raising a Stink: The Struggle over Factory Hog Farms in Nebraska.* Winnipeg, Canada: Bison Books, 2003.

Kahn, Matthew, E. *Green Cities: Urban Growth and the Environment.* Washington, DC: Brookings Institution Press, 2006.

Kallen, Stuart A. *Is Factory Farming Harming America?* San Diego, CA: Greenhaven Press, 2006.

Kaufmann, R., and C. Cleveland. *Environmental Science.* New York: McGraw-Hill International Edition, 2008.

Kemp, William. *The Renewable Energy Handbook: A Guide to Rural Energy Independence, Off-Grid and Sustainable Living.* Tamworth, Ontario, Canada: Aztext Press, 2006.

Kidd, J.S., and R.A. Kidd. *Air Pollution: Problems and Solutions.* New York: Facts on File, 2005.

Klyza, Christopher McGrory. *Wilderness Comes Home: Rewilding the Northeast.* Hanover, NH: University Press of New England, 2001.

Kolbert, Elizabeth. *Field Notes from a Catastrophe: Man, Nature, and Climate Change.* New York: Bloomsbury USA, 2006.

Krauss, Erich. *Wall of Flame: The Heroic Battle to Save Southern California.* Indianapolis: Wiley, 2006.

Kruger, Paul. *Alternative Energy Resources: The Quest for Sustainable Energy.* New York: Wiley, 2006.

Kump, Lee R., James F. Kasting, and Robert G. Crane. *The Earth System.* New York: Prentice-Hall, 2003.

Kurlansky, Mark. *Cod: A Biography of the Fish that Changed the World.* New York: Penguin, 1997.

Kurlansky, Mark. *Salt: A World History.* New York, Walker and Company, 2002.

Kurtzman, Dan. *A Killing Wind: Inside Union Carbide and the Bhopal Catastrophe.* New York: McGraw-Hill, 1987.

Kusky, Timothy M. *Geological Hazards: A Sourcebook.* Westport, CT: Greenwood Publishing Group, 2003.

Lal, Rattan, et al. *Climate Change and Global Food Security.* New York: CRC, 2005.

Leacock, Elspeth. *The* Exxon Valdez *Oil Spill.* New York: Facts on File, 2005.

Liddick, Donald R. *Eco Terrorism: Radical Environmental and Animal Liberation Movements.* Westport CT: Praeger Publishers, 2006.

Livingstone, James. *Agriculture and Soil Pollution: New Research.* Hauppauge: Nova Science Publishers, 2006.

Llindstrom, Matthew, and Zachary A. Smith. *The National Environmental Policy Act: Judicial Misconstruction, Legislative Indifference, & Executive Neglect.* College Station, TX: Texas A&M University Press, 2002.

Lockwood, C., and Rhea Gary. *Marsh Mission: Capturing the Vanishing Wetlands.* Baton Rouge: Louisiana State University Press, 2005.

Lockwood, Michael, et al., eds. *Managing Protected Areas: A Global Guide.* Sterling, VA: Earthscan Publications, 2006.

Lomolino, Mark, and Lawrence Heaney. *Frontiers of Biogeography.* Sunderland, MA: Sinauer Associates, 2004.

Louka, Elli. *International Environmental Law: Fairness, Effectiveness, and World Order.* Cambridge: Cambridge University Press, 2003.

Lovelock, James. *The Revenge of Gaia: Earth's Climate Crisis and the Fate of Humanity.* New York: Basic Books, 2006.

Luken, Ralph, Frank van Rompaey, and Kandeh Yumkella. *Environment and Industry in Developing Countries.* Northampton, MA: Edward Elgar, 2007.

Lund, Herbert F. *McGraw-Hill Recycling Handbook.* New York: McGraw-Hill Professional, 2000.

Lurie, Edward. *Louis Agassiz: A Life in Science.* John Hopkins University Press, Baltimore, MD: 1996.

Macey, Gregg P., and Jonathan Z. Cannon, eds. *Reclaiming the Land: Rethinking Superfund Institutions, Methods and Practices.* New York: Springer, 2007.

Mackenzie, Fred T. *Carbon in the Geobiosphere: Earth's Outer Shell.* Dordrecht, Netherlands: Springer, 2006.

Maher, Neil M. *Nature's New Deal: The Civilian Conservation Corps and the Roots of the American Environmental Movement.* New York: Oxford University Press, 2007.

Marchand, Peter J. *Life in the Cold: An Introduction to Winter Ecology.* Hanover, NH: University Press of New England, 1987.

McCaffrey, Paul. *Global Climate Change.* Minneapolis: H.W. Wilson, 2006.

McCully, James Greig. *Beyond the Moon: A Conversational, Common Sense Guide to Understanding the Tides.* Singapore: World Publishing, 2006.

McEvoy, T.J. *Positive Impact Forestry: A Sustainable Approach to Managing Woodlands.* Washington, DC: Island Press, 2004..

McFarlane, Allison, and Rodney C. Ewing, eds. *Uncertainty Underground: Yucca Mountain and the Nation's High-Level Nuclear Waste.* Cambridge, MA: MIT Press, 2006.

McGowran, Brian. *Biostratigraphy: Microfossils and Geological Time.* New York, Cambridge University Press, 2005.

McGranahan, Gordon, and Frank Murray. *Air Pollution and Health in Rapidly Developing Countries.* London: Earthscan Publications, 2003.

McNeely, J.A., ed. *The Great Reshuffling: Human Dimensions of Invasive Alien Species.* Cambridge, MA: World Conservation Union, 2001.

Metz, B., et al, eds. *Climate Change 2007: Mitigation of Climate Change: Contribution of Working*

Group III to the Fourth Assessment Report of the Intergovernmental Panel on Climate Change. New York: Cambridge University Press, 2007.

Midkiff, Ken, and Robert F. Kennedy, Jr. *Not a Drop To Drink: America's Water Crisis (and What You Can Do)*. Novato: New World Library, 2007.

Miller, Bruce G. *Coal Energy Systems*. San Diego, CA: Academic Press, 2004.

Miller, G. Tyler. *Environmental Science: Working with the Earth*. New York: Brooks Cole, 2005.

Mitchell, Nora, et al, eds. *The Protected Landscape Approach: Linking Nature, Culture and Community*. Gland, Switzerland: World Conservation Union, 2004.

Mitsch, William, and James Gosselink. *Wetlands*. New York: Wiley, 2007.

Mogil, Michael. *Extreme Weather: Understanding the Science of Hurricanes, Tornadoes, Floods, Heat Waves, Snow Storms, Global Warming and Other Atmospheric Disturbances*. New York: Black Dog & Leventhal Publishers, 2007.

Molyneaux, Paul. *Swimming in Circles: Aquaculture and the End of Wild Oceans*. New York: Thunder's Mouth Press, 2006.

Mooney, Chris. *Storm World: Hurricanes, Politics, and the Battle over Global Warming*. New York: Harcourt, 2007.

Mooney, H.A., and R.J. Hobbs, eds. *Invasive Species in a Changing World*. Washington, DC: Island Press, 2000.

Moore, Peter. *Tundra*. New York: Facts on File, 2007.

Morgan-Griffiths, Lauris. *Ansel Adams: Landscapes of the American West*. London: Quercus, 2008.

Morganstein, Stanley. *The Greenhouse Effect*. Cheshire, UK: Trafford, 2003.

Morris, Robert. *The Blue Death: Disease, Disaster, and the Water We Drink*. New York: Harper Collins, 207.

Muhn, James. *Opportunity and Challenge: The Story of the Bureau of Land Management*. U.S. Department of Interior, Bureau of Land Management, 1988.

Muir, John. *Our National Parks*. Boston: Houghton-Mifflin, 1901.

Murphy, Priscilla Coit. *What a Book Can Do: The Publication and Reception of Silent Spring*. Amherst: University of Massachusetts Press, 2007.

Nash, Madeline. *El Niño: Unlocking the Secrets of the Master Weather-Maker*. New York: Grand Central Publishing, 2003.

National Academy of Engineering, National Research Council of the National Academies. *The Carbon Dioxide Dilemma: Promising Technologies and Policies*. Washington, DC: National Academies Press, 2003.

Nelson, Gaylord, et al. *Beyond Earth Day: Fulfilling the Promise*. Madison, WI: University of Wisconsin Press, 2002.

Newman, William, and Wilfred E. Houlton. *Boston's Back Bay: The Story of America's Greatest Nineteenth-Century Landfill Project*. Holliston, MA: Northeastern, 2007.

North, Gerald R., et al. *Surface Temperature Reconstructions for the Last 2,000 Years*. Washington, DC: National Academy of Sciences, 2006.

Norton, Bryan G. et al. *Ethics on the Ark: Zoos, Animal Welfare, and Wildlife Conservation*. Washington, DC: Smithsonian Institution Press, 1995.

Odum, Eugene, and Gary W. Barrett. *Fundamentals of Ecology*, 5th ed. Detroit: Brooks Cole, 2004.

Okaichi, Tomotoshi, ed. *Red Tides*. Dordrecht, Netherlands: Kluwer, 2004.

O'Neill, Karen. *Rivers by Design: State Power and the Origins of U.S. Flood Control*. Raleigh, NC: Duke University Press, 2006.

Owens, Peter. *Oil and Chemical Spills*. New York: Lucent Books, 2003.

Parker, Gary. *Exploring the World Around You: A Look at Nature from Tropics to Tundra*. Green Forest, AR: Master Books, 2003.

Parker, Lee. *Environmental Communication: Messages, Media, and Methods*. Dubuque: Kendall/Hunt Publishing, 2005.

Parry, M.L., et al, eds. *Climate Change 2007: Impacts, Adaptation and Vulnerability. Contribution of Working Group II to the Fourth Assessment Report of the Intergovernmental Panel on Climate Change*. New York: Cambridge University Press, 2007.

Parson, Edward A. *Protecting the Ozone Layer: Science and Strategy*. Oxford: Oxford University Press, 2003.

Parsons, Tim. *The Sea's Enthrall: Memoirs of an Oceanographer*. Victoria, British Columbia, Canada: Trafford Publishing, 2007.

Pastorok, Robert A., Steven M. Bartell, Scott Ferson, and Lev R. Ginzberg, eds. *Ecological Modeling in Risk Assessment: Chemical Effects on Populations, Ecosytems, and Landscapes*. Boca Raton, FL: CRC Press, 2002.

Patent, Dorothy Hinshaw. *Fire: Friend or Foe?* New York: Clarion Books, 1998.

Payne, Daniel G. *Voices in the Wilderness: American Nature Writing and Environmental Politics.* Hanover: University Press of New England, 1996.

Pepper, David. *Modern Environmentalism: An Introduction.* London: Routledge, 1996.

Pepper, Ian, Charles Gerba, and Mark Brusseau. *Environmental and Pollution Science.* New York: Academic Press, 2006.

Perrings, C., et al., eds. *The Economics of Biological Invasions.* Northampton, MA: Edward Elgar, 2000.

Persaud, T.A., and S.R. Chudley. *Basic Concepts in Teratology.* New York: Alan R. Liss, 1985.

Pisano, Gary P. *Science Business: The Promise, The Reality, and the Future of Biotech.* Boston: Harvard Business School Press, 2006.

Poynter, Jane. *The Human Experiment: Two Years and Twenty Minutes Inside Biosphere 2.* New York: Thunder's Mouth Press, 2006.

Preece, John, and Paul Read. *The Biology of Horticulture.* New York: Wiley, 2005.

Pugh, David. *Changing Sea Levels: Effects of Tides, Weather and Climate.* New York: Cambridge University Press, 2004.

Raven, Peter H., Linda R. Berg, and George B. Johnson. *Environment.* Hoboken, NJ: Wiley, 2002.

Ravenel, Ramsey, Ilmi Granoff, and Carrie Magee. *Illegal Logging in the Tropics.* Boca Raton, FL: CRC Press, 2005.

Reeve, Rosalind. *Policing International Trade in Endangered Species: The CITES Treaty and Compliance.* London: Royal Institute of International Affairs, 2004.

Reinhart, Kare. *Yellowstone's Rebirth by Fire: Rising from the Ashes of the 1988 Wildfires.* Helena: Farcountry Press, 2008.

Rice, Tony. *Deep Ocean.* London: Natural History Museum, 2000.

Robbins-Roth, Cynthia. *Alternative Careers in Science: Leaving the Ivory Tower.* New York: Academic Press, 2005.

Roberts, Callum. *The Unnatural History of the Sea.* Washington, DC: Island Press, 2007.

Roberts, Rodger. *The Green Killing Fields: The Need for DDT to Defeat Malaria and Reemerging Diseases.* Washington, DC: AEI Press, 2007.

Robinson, Arthur H., et al. *Elements of Cartography*, 6th ed. Hoboken, NJ: Wiley, 1995.

Robinson, Darren. *Hurricane Katrina: The Destruction of New Orleans.* Charleston: Booksurge Publishing, 2005.

Rodgers, Heather. *Gone Tomorrow: The Hidden Life of Garbage.* New York: The New Press, 2006.

Rootes, Christopher, ed. *Environmental Protest in Western Europe.* Oxford: Oxford University Press, 2003.

Royte, Elizabeth. *Garbage Land: On the Secret Trail of Trash.* Boston: Back Bay Books, 2006.

Rubin, Charles T. *The Green Movement: Rethinking the Roots of Environmentalism.* New York: Free Press, 1994.

Russell, Edmund. *War and Nature: Fighting Humans and Insects with Chemicals from World War I to Silent Spring.* Cambridge, UK: Cambridge University Press, 2001.

Sax, Dov, John Stachowicz, and Steven Gaines. *Species Invasions: Insights into Ecology, Evolution, and Biogeography.* Sunderland, MA: Sinauer Associates, 2005.

Schmincke, Hans-Ulrich. *Volcanism.* New York: Springer, 2004.

Schneider, Gerald, Katherine Barbieri, and Nils Petter Gleditsch, eds. *Globalization and Armed Conflict.* New York: Rowman and Littlefield, 2003.

Schwab, James. *Planning for Wildfires.* Chicago: American Planning Association, 2005.

Schwartz, Joel. *Air Quality in America: A Dose of Reality on Air Pollution Levels, Trends, and Health Risks.* Washington: AEI Press, 2008.

Seinfeld, John H., and Spyros N. Pandis. *Atmospheric Chemistry and Physics: From Air Pollution to Climate Change.* New York: Wiley Interscience, 2006.

Shardein, James. *Chemically Induced Birth Defects*, 3rd ed. New York: Informa Healthcare, 2000.

Sinclair, Anthony, John Fryxell, and Graeme Caughley. *Wildlife Ecology, Conservation and Management.* New York: Wiley-Blackwell, 2006.

Sliggers, Johan, and Willem Kakebeeke, eds. *Clearing the Air: 25 Years of the Convention on Long-range Transboundary Air Pollution.* New York: United Nations, 2005.

Smith, W. Brad, Patrick D. Miles, John S. Vissage, and Scott A. Pugh. *Forest Resources of the United States.* Washington, DC: U.S. Department of Agriculture Forest Service, 2002.

Solomon, S., et al, eds. *Climate Change 2007: The Physical Science Basis: Contribution of Working Group I to the Fourth Assessment Report of the*

Intergovernmental Panel on Climate Change. New York: Cambridge University Press, 2007.

Spilsbury, Richard, and Louise Spilsbury. *The Pros and Cons of Wind Power*. New York: Rosen Central, 2007.

Starr, Christopher. *Woodland Management*. Ramsbury, UK: Crowood Press, 2005.

Stein, Bruce, et al. *Precious Heritage: The Status of Biodiversity in the United States*. Cambridge: Oxford, 2000.

Stokke, Olav Schram, and Davor Vidas, eds. *Governing the Antarctic: The Effectiveness and Legitimacy of the Antarctic Treaty System*. Cambridge, UK: Cambridge University Press, 1997.

Subba Rao, D.V. *Algal Cultures, Analogues of Blooms and Applications*. Enfield, NH: Science Publishers, 2006.

Svensmark, Henrik, and Nigel Calder. *The Chilling Stars: The New Theory of Climate Change*. London: Totem Books, 2007.

Sverdrup, Keith A. *An Introduction to the World's Oceans*. New York: McGraw-Hill, 2008.

Tacconi, Luca. *Illegal Logging: Law Enforcement, Livelihoods and the Timber Trade*. London: Earthscan Publications, 2007.

Tarbuck, E.J., F.K. Lutgens, and D. Tasa. *Earth: An Introduction to Physical Geology*. Upper Saddle River, NJ: Prentice Hall, 2005.

Thomas, Isabel. *The Pros and Cons of Solar Power*. New York: Rosen Central, 2008.

Thoreau, Henry David. *The Illustrated Walden*. Princeton, NJ: Princeton University Press, 1973.

Trefil, Calvo. *Earth's Atmosphere*. Geneva, IL: McDougal Littell, 2005.

Triggs, Gillian, and Anna Riddell. *Antarctica: Legal and Environmental Challenges for the Future*. London: British Institute of International and Comparative Law, 2007.

U.S Environmental Protection Agency. *The Water Sourcebooks: K–12*. Washington, DC: U.S. EPA, 2000.

Vaillant, John. *The Golden Spruce: A True Story of Myth, Madness, and Greed*. New York: W.W. Norton, 2005.

Valsson, Trausti. *How the World Will Change with Global Warming*. Oxford, UK: Oxbow Books, 2007.

van Santen, Christa. *Light Zone City: Light Planning in the Urban Context*. Basel: Birkhäuser Basel, 2006.

Walker, M.J.C. *Quaternary Dating Methods*. Chichester, UK: Wiley, 2005.

Walker, Sharon. *Biotechnology Demystified*. New York: McGraw-Hill Professional, 2006.

Wang, Zhendi, and Scott Stout. *Oil Spill Environmental Forensics: Fingerprinting and Source Identification*. New York: Academic, 2006.

Warkany, J., ed. *Issues and Reviews in Teratology*. New York: Plenum Press, 1983.

Wasser, Nickolas, and Jeff Ollerton. *Plant-Pollinator Interactions: From Specialization to Generalization*. Chicago: University of Chicago Press, 2006.

Waterman, Jonathan. *Where Mountains Are Nameless: Passion and Politics in the Arctic Wildlife Refuge*. New York: W.W. Norton, 2007.

Weart, Spencer. *The Discovery of Global Warming*. Cambridge, MA: Harvard University Press, 2004.

Welborne, Victor I. *Biofuels in the Energy Supply System*. New York: Novinka Books, 2006.

White, Christopher P. *Chesapeake Bay: Nature of the Estuary*. Ithaca, NY: Cornell/Tidewater Publishers, 1989.

Wild, Anthony. *Coffee: A Dark History*. New York: Norton, 2005.

Williams, I. *Environmental Chemistry*. Chichester: Wiley, 2001.

Williams, Wendy, and Robert Whitcomb. *Cape Wind: Money, Celebrity, Class, Politics, and the Battle for Our Energy Future on Nantucket Sound*. New York: PublicAffairs, 2007.

Woodford, Chris. *Arctic Tundra and Polar Deserts*. Chicago: Raintree, 2003.

World Water Assessment Programme. *Water: A Shared Responsibility*. New York: Berghahn Books, 2006.

Wright, Gerald, ed. *National Parks and Protected Areas: Their Role in Environmental Protection*. Cambridge, MA: Blackwell Science, 1996.

Wuebbles, Donald J., and Jae Edmonds. *Primer on Greenhouse Gases*. Boca Raton, FL: CRC, 1991.

Wuerthner, George. *Wildfire: A Century of Failed Forest Policy*. Washington, DC: Island Press, 2006.

PERIODICALS

Abel, E.L., and R.J. Sokol. "Incidence of Fetal Alcohol Syndrome and Economic Impact of FAS-Related Anomalies: Drug Alcohol Syndrome and Economic Impact of FAS-Related Anomalies." *Drug and Alcohol Dependency* 19 (1987): 51–70.

"Air Pollution Prevention and Control." *U.S. Code* Title 42, Chapter 85, Subchapter I, Part A, Section 7401.

Alley, Richard B. "Abrupt Climate Change." *Scientific American* (November 2004).

Alley, Richard B. "'C'ing Arctic Climate with Black Ice." *Science* 317 (2007): 1333–1334.

Alley, Richard B., et al. "Ice-Sheet and Sea-Level Changes." *Science* 310 (2005): 456–460.

Allison, Ian. "Antarctic Ice and the Global Climate System." *Australian Antarctic Magazine* (Autumn 2002): 3–4.

Anderson, Theodore L., et al. "Climate Forcing by Aerosols—A Hazy Picture." *Science* 300 (2003): 1103–1104.

Araúo, Miguel B., and Carsten Rahbek. "How Does Climate Change Affect Biodiversity?" *Science* 313 (2006): 1396–1397.

Arnoldy, Ben. "Erosion from Tahoe Fire May Hurt Lake's Health." *Christian Science Monitor* (July 5, 2007).

Asner, Gregory P. et al. "Condition and Fate of Logged Forests in the Brazilian Amazon." *Proceedings of the National Academy of Sciences* 103 (2006): 12947–12950.

Asner, Gregory, et al. "Selective Logging in the Brazilian Amazon." *Science* 310 (2005): 480–482.

Associated Press. "Coal Use Grows Despite Warming Worries." *New York Times* (October 28, 2007).

Bachtold, Daniel. "Britain to Cut CO_2 without Relying on Nuclear Power." *Science* 299 (2003): 1291.

Barnard, Ann. "City Council Passes Bill for Recycling of Plastic Bags." *New York Times* (January 10, 2008).

Barnard, Ann. "Green with Envy: Germany's Green Dot Program Continues Generating Good Collection Numbers." *Recycling Today* (October, 2004).

Barnola, J.M., et al. "Vostok Ice Core Provides 160,000-year Record of Atmospheric CO_2." *Nature* 329 (1987): 408–413.

Barringer, Felicity. "Park Service to Emphasize Conservation in New Rules." *New York Times* (August 31, 2006).

Beck, Eckardt C. "The Love Canal Tragedy." *EPA Journal* (January 1979).

Berg, Rebecca. "Environmental Health and the Media, Part 3: Make Noise, Make the News." *Journal of Environmental Health* 68: 73–78 (2006).

Berstein, Lenny, et al. "Carbon Dioxide Capture and Storage: A Status Report" *Climate Policy* 6 (2006): 241–246.

Betterton, Rosemary. "Promising Monster: Pregnant Bodies, Artistic Subjectivity, and Maternal Imagination." *Hypatia* 2 (2006): 80–100.

Biello, David. "Conservative Climate: Consensus Document May Understate the Climate Change Problem." *Scientific American* (March 18, 2007).

Biello, David. "No Arctic Oil Drilling? How About Selling Parks?" *San Francisco Chronicle* (September 24, 2005).

Biesmeijer, J.C., et al. "Parallel Declines in Pollinators and Insect-Pollinated Plants in Britain and the Netherlands." *Science* 313 (2006): 354–357.

Blaine, Stéphane, et al. "Effect of Natural Iron Fertilization on Carbon Sequestration in the Southern Ocean." *Nature* 446 (April 26, 2007): 1070–1074.

Boyd, P.W., et al. "Mesoscale Iron Enrichment Experiments 1993–2005: Synthesis and Future Directions." *Science* 315 (February 2, 2007): 612–617.

Bradsher, Keith. "Clean Power that Reaps a Whirlwind." *New York Times* (May 9, 2007).

Brent, R.L. "Reproductive and Teratologic Effects of Low-Frequency Electromagnetic Fields: A Review of In Vivo and In Vitro Studies Using Animal Models." *Teratology* 59 (1999): 261–286.

Brown, Barbara E., and John C. Ogden. "Coral Bleaching." *Scientific American* (January 1993): 64–70.

Browne, Malcolm W. "Folk Remedy Demand May Wipe out Tigers." *New York Times* (September 22, 1992).

Brumfiel, Geoff. "Academy Affirms Hockey-Stick Graph." *Nature* 441 (2006): 1032–1033.

Buesseler, Ken O., and Philip W. Boyd. "Ocean Iron Fertilization—Moving Forward in a Sea of Uncertainty." *Science* 319 (2008): 162.

Buesseler, Ken O., and Philip W. Boyd. "Will Iron Fertilization Work?" *Science* 300 (April 4, 2003): 67–68.

Bumiller, Elisabeth. "Energy Politics on Earth Day as Bush Tours California." *New York Times* (April 23, 2006).

Caldeira, Ken, et al. "Climate Sensitivity, Uncertainty and the Need for Energy without CO_2 Emissions." *Science* 299 (2003): 2052–2054.

Callaway, Ewen. "Energy: To Catch a Wave." *Nature* 4 (1979): 823–831.

Cazenave, Anny. "How Fast Are the Ice Sheets Melting?" *Science* 314 (2006): 1250–1252.

Chew, G.L., J. Wilson, F.A. Rabito, et al. "Mold and Endotoxin Levels in the Aftermath of Hurricane Katrina: A Pilot Project of Homes in New Orleans Undergoing Renovation." *Environmental Health Perspectives* 114 (2006): 1883–1889.

"Chronological FBI Summary of Terrorist Incidents, 1980–2004." *Seattle Times* (May 7, 2006).

Clayton, Mark "EPA's Record Settlement with Utility Could Lead to Other Deals." *Christian Science Monitor* (October 11, 2007).

Clayton, Mark. "Climate-Change Paradox: Greenhouse Gas Is Big Oil Boon." *Christian Science Monitor* (September 11, 2007).

Clayton, Mark. "Earth Too Warm? Bury the CO2." *Christian Science Monitor* (July 31, 2007).

Clayton, Mark. "New Tool to Fight Global Warming: Endangered Species Act?" *Christian Science Monitor* (September 7, 2007).

Clayton, Mark. "The Politics of Ethanol Outshine Its Costs." *Christian Science Monitor* (November 15, 2007).

Clayton, Mark. "Problem Dams on the Rise in US." *Christian Science Monitor* (September 13, 2007).

Clayton, Mark. "States Are Closer to Trimming Autos' CO2 Emissions." *Christian Science Monitor* (September 14, 2007).

"A Clear Direction" (editorial). *Nature* 447 (2007): 115.

"Coastal Zone Renewable Energy Act of 2003." *U.S. Congress.* Washington, DC: U.S. Congress, 2003.

Cochran, Elizabeth S., et al. "Earth Tides Can Trigger Shallow Thrust Fault Earthquakes." *Nature* 306 (2004): 1164–1166.

Cohen, Mark A. "Environmental Crime and Punishment: Legal/Economic Theory and Empirical Evidence on Enforcement of Federal Environmental Statues." *Journal of Criminal Law and Criminology* 82, 4 (1992): 1054–1108.

Collins, William, et al. "The Physical Science Behind Climate Change." *Scientific American* (August 2007).

Cooper, Melinda. "Regenerative Medicine: Stem Cells and the Science of Monstrosity." *Medical Humanities* 30 (2004): 12–22.

Cordero, José F. "A New Look at Behavioral Outcomes and Teratogens: A Commentary." *Birth Defects Research (Part A)* 67 (2003): 900–902.

Cressey, Daniel. "Advanced Biofuels Face an Uncertain Future." *Nature* 452 (2008): 670–671.

Cross, Michael. "Antarctica: Exploration or Exploitation?" *New Scientist* (June 22, 1991).

Dean, Cornelia. "Coral Reefs and What Ruins Them." *New York Times* (February 26, 2008).

Dean, Cornelia. "Even Before Its Release, World Climate Report Is Criticized as Too Optimistic." *New York Times* (February 2, 2007).

DePalma, Anthony. "EPA Sued Over Mercury in the Air." *New York Times* (March 30, 2005).

Dugger, Celia W. "As Prices Soar, U.S. Food Aid Buys Less." *New York Times* (September 29, 2007).

Eilperin, Juliet. "188 More Species Listed as Near Extinction." *Washington Post* (September 13, 2007).

Eilperin, Juliet. "U.S., China Got Climate Warnings Toned Down." *Washington Post* (April 7, 2007).

Eilperin, Juliet. "Yes, the Water's Warm ... Too Warm." *Washington Post* (July 15, 2007).

Elsner, James B. "Tempests in Time." *Nature* 447, no. 7145 (June 7, 2007): 647–648.

Emanuel, Kerry. "Increasing Destructiveness of Tropical Cyclones over the Past 30 Years." *Nature* 436, no. 7051 (August 4, 2005): 686–688.

"Energy Policy Act of 2005." *U.S. Congress* Sec. 1342, Sec. 30B. Washington, DC: U.S. Congress, 2005.

EPICA Community Members. "Eight Glacial Cycles from an Antarctic Ice Core." *Nature* 429 (2004): 623–628.

Ericson, Jason P. "Effective Sea-Level Rise and Deltas: Causes of Change and Human Dimension Implications." *Global and Planetary Change* 50 (2006): 63–82.

Etheridge, D.M., et al. "Natural and Anthropogenic Changes in Atmospheric CO_2 Over the Last 1,000 Years from Air in Antarctic Ice and Firn." *Journal of Geophysical Research* 101 (1996): 4115–4128.

Fargione, Joseph. "Land Clearing and the Biofuel Carbon Debt." *Science* 319 (2008): 1235–1237.

Farrell, Alexander E. "Ethanol Can Contribute to Energy and Environmental Goals." *Science* 311 (2006): 506–508.

Fearnside, Philip M. "Do Hydroelectric Dams Mitigate Global Warming? The Case of Brazil's

Curuá-Una Dam." *Mitigation and Adaptation Strategies for Global Change* 10 (2005): 675–691.

"Federal Insecticide, Fungicide, and Rodenticide Act (FIFRA)." *U.S. Code* Title 7, Chapter 6, Subchapter II. June 25, 1947.

Fischer, Jean-Louis. "The Embryological Oeuvre of Laurent Chabry." *Development Genes and Evolution* 201 (1992): 125–127.

Flemming, James R. "The Climate Engineers." *Wilson Quarterly* (Spring 2007).

Ford, Peter. "China Ready to Leap from Industrial to Information-Age Economy." *Christian Science Monitor* (September 5, 2007).

Forero, Juan. "As Andean Glaciers Shrink, Water Worries Grow." *New York Times* (November 24, 2002).

Forneri, Claudio, et al. "Keeping the Forest for the Climate's Sake: Avoiding Deforestation in Developing Countries Under the UNFCCC'." *Climate Policy* 6 (2006): 275–294.

Fountain, Henry. "More Acidic Ocean Hurts Reef Algae as Well as Corals." *New York Times* (January 8, 2008).

Fountain, John. "Recycling that Harms the Environment and People." *New York Times* (April 15, 2008).

Gardner, Gary. "From Oasis to Mirage: The Aquifers that Won't Replenish." *World Watch* 8 (2005): 30–37.

Geoghegan, John. "Long Ocean Voyage Set for Vessel that Runs on Wave Power." *New York Times* (March 11, 2008).

Gibbs, Walter. "Gore and U.N. Panel Win Peace Prize for Climate Work." *New York Times* (October 13, 2007).

Giles, Jim. "Europe Set for Tough Debate on Curbing Aircraft Emissions." *Nature* 436 (2005): 764–765.

Glanz, James. "Panel Finds No Major Flaws in Nuclear Treaty." *New York Times* (August 1, 2002).

Gregory, J. M., and P. Huybrechts. "Ice-Sheet Contributions to Future Sea-Level Change." *Philosophical Transactions of the Royal Society A* 364 (2006): 1709–1731.

"The Greening of Nuclear Power." *New York Times* (May 13, 2006).

Guo, Jerry. "The Galápagos Islands Kiss Their Goat Problem Goodbye." *Science* 313 (2006): 1567.

Halfar, Jochen, and Rodney M. Fujita. "Danger of Deep-Sea Mining: Plans for Deep-Sea Mining Could Pose a Serious Threat to Marine Ecosystems." *Science* 316 (2007): 987.

Halpern, Benjamin S., et al. "A Global Map of Human Impact on Marine Ecosystems." *Science* 319 (2008): 948–952.

Hansen, James, and Larissa Nazarenko. "Soot Climate Forcing Via Snow and Ice Albedos." *Proceedings of the National Academy of Sciences* 101 (2004): 423–428.

Hansen, Keith. "CTBT: Forecasting the Future." *Bulletin of the Atomic Scientists* (March/April 2005).

"The Heat Is On—Survey: Climate Change." *Economist* (September 7, 2006).

Higgins, Paul A.T. "Biodiversity Loss Under Existing Land Use and Climate Change: An Illustration Using Northern South America." *Global Ecology and Biogeography* 16 (2007): 197–204.

Hirn, Alfred, and Mireille Laigle. "Silent Heralds of Megathrust Earthquakes?" *Nature* 305 (2004): 1917–1918.

Hobbs, Peter V., and Lawrence F. Radke. "Airborne Studies of the Smoke from the Kuwait Oil Fires." *Science* 256, no. 5059 (May 15, 1992): 987–991.

Hoffert, M., et al. "Advanced Technology Paths to Global Climate Stability: Energy for a Greenhouse Planet." *Science.* 298 (2002): 981–987.

Howat, Ian M. "Rapid Changes in Ice Discharge from Greenland Outlet Glaciers." *Science* 315 (2007): 1559–1561.

Hurrell, James W. "The North Atlantic Oscillation." *Science* 291 (2001): 605–605.

Jenkins, Martin. "Prospects for Biodiversity." *Science* 302 (2003): 1175–1177.

Jonsson, Patrik. "Katrina Rated Largest U.S. Ecodisaster." *Christian Science Monitor* (November 19, 2007).

Jouzel, J., et al. "Vostok Ice Core: A Continuous Isotope Temperature Record Over the Last Climatic Cycle (160,000 Years)." *Nature* 329 (1987): 403–408.

Kahn, Joseph, and Jim Yardley. "As China Roars, Pollution Reaches Deadly Extremes." *New York Times* (August 26, 2007).

Kanter, James. "Radioactive Nimby: No One Wants Nuclear Waste." *New York Times* (November 7, 2007).

Karl, Thomas R., and Kevin E. Trenberth. "Modern Global Climate Change." *Science* 302 (2003): 1719–1723.

Katz, Jonathan M. "Haiti Deforestation a Means for Survival." *Contra Costa Times* (February 17, 2008).

Keefer, David K., and Matthew C. Larsen. "Assessing Landslide Hazards." *Science* 316 (2007): 1136–1138.

Kerr, Richard A. "A Tempestuous Birth for Hurricane Climatology." *Science* 312, no. 5774 (May 5, 2006): 676–677.

Kerr, Richard A. "Did Volcanoes Drive Ancient Extinctions?" *Science* 289 (2000): 1130–1131.

Kerr, Richard A. "Politicians Attack, But Evidence for Global Warming Doesn't Wilt." *Science* 313 (2006): 421.

Kerr, Richard A. "Yes, It's Been Getting Warmer in Here Since the CO_2 Began to Rise." *Science* 312 (2006): 1854.

Knickerbocker, Brad. "Accidents Dim Hopes for Green Nuclear Option." *Christian Science Monitor* (July 19, 2007).

Knickerbocker, Brad. "Clean Energy vs. Whales: How to Choose?" *Christian Science Monitor* November 28, 2007.

Knickerbocker, Brad. "King Coal's Crown Is Losing some Luster." *Christian Science Monitor* August 2, 2007.

Kniemeyer, O., et al. "Anaerobic Oxidation of Short-chain Hydrocarbons by Marine Sulphate-reducing Bacteria." *Nature* 449 (2007): 898–901.

Krkosek, Martin, Jennifer S. Ford, Alexandra Morton, Subhash Lele, Ransom A. Myers, and Mark A. Lewis. "Declining Wild Salmon Populations in Relation to Parasites from Farm Salmon." *Science* 318 (2007): 1772–1775.

"Kyoto Challenge Has Just Begun." *Nature* 431 (2004): 613.

LaFranchi, Howard. "UN Revs Up Over Global Warming." *Christian Science Monitor* (September 24, 2007).

Lashof, D.A., and Dilip R. Ahuja. "Relative Contributions of Greenhouse Gas Emissions to Global Warming." *Nature* 344 (1990): 529–531.

Law, Kathy S., et al. "Arctic Air Pollution: Origins and Impacts." *Science* 315 (2007): 1537–1540.

Lee, John M. "'Silent Spring' Is Now Noisy Summer." *New York Times* (July 22, 1962).

Lee, Joosung J., et al. "Historical and Future Trends in Aircraft Performance, Cost, and Emissions." *Annual Review of Energy and Environment* 26 (2001): 167–200.

Lima, Ivan B.T., et al. "Methane Emissions from Large Dams as Renewable Energy Resources: A Developing Nation Perspective." *Mitigation and Adaptation Strategies for Global Change* (February 2007).

Liptak, Adam. "Power to Build Border Fence Is Above U.S. Law." *New York Times* (April 8, 2008).

Livingstone, Daniel A. "Global Climate Change Strikes a Tropical Lake." *Science* 301 (2003): 468–469.

Lu, Chensheng, et al. "Organic Diets Significantly Lower Children's Dietary Exposure to Organophosphorus Pesticides." *Environmental Health Perspectives* 114 (2006): 260–263.

Lund, H. "The Kyoto Mechanisms and Technological Innovation." *Energy* 31 (2006): 2325–2332.

Mäder, Paul, et al. "Soil Fertility and Biodiversity in Organic Farming." *Science* 296 (2002): 694–1697.

Madsen, K.M., A. Hviid, M. Vestergaard, et al. ".A Population-Based Study of Measles, Mumps, and Rubella Vaccination and Autism." *New England Journal of Medicine* 347 (2002): 1477–1482.

Mahony, Dianne E. "The Convention on International Trade in Endangered Species of Flora and Fauna: Addressing Problems in Global Wildlife Trade Enforcement." *New England International and Comparative Law Annual* 3 (1997).

Mann, Michael E., et al. "Global-Scale Temperature Patterns and Climate Forcing Over the Past Six Centuries." *Nature* 392 (1998): 779–105.

Marquand, Robert. "Swiss Glacier Retreats at a Rapid Clip." *Christian Science Monitor* (July 17, 2007).

Marshall, Eliot. "Is the Friendly Atom Poised for a Comeback?" *Science* 309 (2005): 1168–1169.

Masood, Ehsan. "Kyoto Agreement Creates New Agenda for Climate Research." *Nature* 390 (1997): 390.

McEwen, Alfred S. "Active Volcanism on Io." *Science* 297 (2002): 2220—2221.

McGinn, Anne P. "Malaria, Mosquitoes, and DDT: The Toxic War against a Global Disease." *World Watch* 15 (2002): 10–17.

McIntyre, S., and R. McKitrick. "Corrections to the Mann et al. (1998) Proxy Database and Northern Hemispheric Average Temperature Series." *Annual Review of Energy and the Environment* 14 (2003): 751–771.

"Medical Waste Tracking Act of 1988. An Amendment to the Solid Waste Disposal Act." *U.S. Code*

Title 42, Chapter 82, Subchapter X, Sections 6992, 6992a and 6992b.

Meehl, Gerald A., et al. "How Much More Global Warming and Sea Level Rise?" *Science* 307 (2005): 1769–1772.

Metz, B., et al. "IPCC, 2007: Summary for Policymakers." In *Climate Change 2007: Mitigation of Climate Change. Contribution of Working Group III to the Fourth Assessment Report of the Intergovernmental Panel on Climate Change.* New York: Cambridge University Press, 2007.

Milly, P.C.D., R.T. Wetherald, K.A. Dunne, and T.L. Delworth. "Increasing Risk of Great Floods in a Changing Climate." *Nature* 15 (2002): 514–517.

Muir, John. "Yosemite Glaciers." *New York Tribune* (December 5, 1871).

Munk, Walter. "Twentieth Century Sea Level: An Enigma." *Proceedings of the National Academy of Sciences* 99 (2002): 6550–6555.

Narvaez, Alfonso. "New York City to Pay Jersey Town $1 Million Over Shore Pollution." *New York Times* (December 8, 1987).

"National Contingency Plan." *U.S. Code* Title 42, Chapter 103, Subchapter I, Sections 9605(a) and 9605(b).

"National Park Service Organic Act." *U.S. Congress* Washington, DC: U.S. Congress, 1916.

Nielsen, Rolf Haugaard. "The Climate Conundrum: Chill Warnings from Greenland." *New Scientist* (August 28, 1993).

Nyberg, Johan, et al. "Low Atlantic Hurricane Activity in the 1970s and 1980s Compared to the Past 270 Years." *Nature* 447, no. 7415 (June 7, 2007): 698–702.

Oerlemans, J. "Extracting a Climate Signal from 169 Glacier Records." *Science* 308 (2005): 675–677.

O'Hear, Michael. "Sentencing the Green-Collar Offender: Punishment, Culpability, and Environmental Crime." *Journal of Criminal Law and Criminology* 95 (2004): 133–276.

"Oil Pollution Act." *U.S. Congress* 33 U.S.C.A. 40. II, sec. 2731–8. Washington, DC: U.S. Congress, 1990.

Oliveira, Paulo, et al. "Land Use Allocation Protects the Peruvian Amazon." *Science* 317 (2007): 1233–1236.

Oppenheimer, Michael, et al. "The Limits of Consensus." *Science* 317 (2007): 150–151.

Oreskes, Naomi. "The Scientific Consensus on Climate Change." *Science* 306 (2004): 1686.

Overpeck, Jonathan T. "Paleoclimatic Evidence for Future Ice-Sheet Instability and Rapid Sea-Level Rise." *Science* 311 (2006): 1747–1750.

Palmer, T.N., and J. Räisänen. "Quantifying the Risk of Extreme Seasonal Precipitation Events in a Changing Climate." *Nature* 415 (2002): 512–514.

Panofsky, Wolfgang K.H. "Nuclear Insecurity." *Foreign Affairs* (September/October 2007).

"Parks to Study Snowmobiles' Effect on Bison." *New York Times* (September 28, 2007).

Pauly, Philip J. "How Did the Effects of Alcohol on Reproduction Become Scientifically Uninteresting?" *Journal of the History of Biology* 29 (1996): 1–28.

Pimentel, David. "Ethanol Fuels: Energy Balance, Economics, and Environmental Impacts Are Negative." *Natural Resources Research* 12 (2003): 127–134.

Potera, Carol. "Florida Red Tide Brews Up Drug Lead for Cystic Fibrosis." *Science* 316 (2007): 1561–1562.

Pounds, J. Alan, et al. "Widespread Amphibian Extinctions from Epidemic Disease Driven by Global Warming." *Nature* 439 (2006): 161–167.

Raghu, S., et al. "Adding Biofuels to the Invasive Species Fire?" *Science* 3131 (2006): 1742.

Rahmstorf, Stefan. "A Semi-Empirical Approach to Projecting Future Sea-Level Rise." *Science* 315 (2007): 368–370.

Raupach, Michael R., et al. "Global and Regional Drivers of Accelerating CO_2 Emissions." *Proceedings of the National Academy of Sciences* 104 (2007): 10288–10293.

Rayburn, Ed. "Overgrazing Can Hurt Environment, Your Pocketbook." *West Virginia Farm Bureau News* (November, 2000).

Regalado, Antonio, and Jim Carlton. "NOAA Scientists Say Hurricanes, Warming Linked." *Wall Street Journal* (February 16, 2006): A4.

Reilly, J., et al. "Multi-gas Assessment of the Kyoto Protocol." *Nature* 401 (1999): 549–555.

Revkin, Andrew C. "Analysts See 'Simply Incredible' Shrinking of Floating Ice in the Arctic." *New York Times* (August 10, 2007).

Revkin, Andrew C. "Scientists Report Severe Retreat of Arctic Ice." *New York Times* (September 21, 2007).

Righelato, Renton, and Dominick V. Spracklen. "Carbon Mitigation by Biofuels or by Saving and Restoring Forests?" *Science* 317 (2007): 902.

"Rising to the Climate Challenge." *Nature* 449 (2007): 755.

Robock, Alan. "Pinatubo Eruption: The Climatic Aftermath" *Science* 295 (2002): 1242–1244.

Rosenthal, Elisabeth, and Andrew C. Revkin. "Science Panel Calls Global Warming 'Unequivocal.'" *New York Times* (February 3, 2007).

Ruddiman, William F. "The Anthropogenic Greenhouse Era Began Thousands of Years Ago." *Climatic Change* 61 (2003): 261–293.

Sala, Enric, and Nancy Knowlton. "Global Marine Biodiversity Trends." *Annual Review of Energy and the Environment* 231 (2006): 93–122.

Sala, Osvaldo E., et al. "Global Biodiversity Scenarios for the Year 2100." *Science* 287 (2000): 1770–1774.

Sanderson, Katharine. "King Coal Constrained." *Nature* 449 (2007): 14–15.

Saunders, Mark A., and Adam S. Lea. "Large Contribution of Sea Surface Warming to Recent Increase in Atlantic Hurricane Activity." *Nature* 451 (2008): 557–561.

Scharlemann, Jörn P.W., and William F. Laurance. "How Green Are Biofuels?" *Science* 319 (2008): 43–44.

Scherer, Ron. "Delivery Companies Switch to Hybrids." *Christian Science Monitor* (November 30, 2007).

Scherer, Ron. "Even as Economy Lags, Corporate 'Green' Push May Advance." *Christian Science Monitor* (October 15, 2007).

Schiermeier, Quirin. "The Oresmen." *Nature* 421 (January 9, 2003): 109–110.

Schmidt, Charles W. "Biodiesel: Cultivating Alternative Fuels." *Environmental Health Perspectives* 115 (February 2007): A86–A91.

Schrope, Mark. "Treaty Caution on Plankton Plans." *Nature* 447 (June 28, 2007): 1039.

Schubert, Siegfried D., et al. "On the Causes of the 1930s Dust Bowl." *Science* 303, no. 5665 (March 19, 2004): 1855–1859.

Scott, Joni. "From Hate Rhetoric to Hate Crime: A Link Acknowledged Too Late." *The Humanist* (January 1, 1999).

Searchinger, Timothy. "Use of U.S. Croplands for Biofuels Increases Greenhouse Gases Through Emissions from Land-Use Change." *Science* 319 (2008): 1598–1600.

Seelye, Katharine Q. "Approval of Snowmobiles Contradicts Park Service Study." *New York Times* (January 31, 2003).

Sengupta, Somini. "Glaciers in Retreat." *New York Times* (July 17, 2007).

Shackleton, Nicolas J. "The 100,000-Year Ice-Age Cycle Identified and Found to Lag Temperature, Carbon Dioxide, and Orbital Eccentricity." *Science* 289 (2000): 1897–1902.

Shepherd, Andrew, and Duncan Wingham. "Recent Sea-Level Contributions of the Antarctic and Greenland Ice Sheets." *Science* 315 (2007): 1529–1532.

Smith, Jessica. "Park Destruction Blamed on ATVs." *East Brunswick [NJ] Sentinel* (January 25, 2007).

"Special Issue on Environmental Conflict." *Journal of Peace Research* 35, no. 3 (May 1998).

Spotts, Peter N. "How to Fight a Rising Sea." *Christian Science Monitor* (November 15, 2007).

Stevenson, Richard W. "Bad Weather Forces Change in Bush's Earth Day Plans." *New York Times* (April 23, 2005).

Stokstad, Erik. "Feared Quagga Mussel Turns Up in Western United States." *Science* 315 (2007): 453.

Stokstad, Erik. "Organic Farms Reap Many Benefits." *Science* 296 (2002): 1589.

Stone, Andrea. "Eco-Terror Suspected in Seattle Blazes." *USA Today* (March 3, 2008).

Stone, Richard. "Norway Goes with the Flow to Light Up Its Nights." *Science* 299 (2003): 339.

Stone, Richard. "Too Late, Earth Scans Reveal the Power of a Killer Landslide." *Science* 311 (2006): 1844–1845.

Stuber, Nicola, et al. "The Importance of the Diurnal and Annual Cycle of Air Traffic for Contrail Radiative Forcing." *Nature* 441 (2006): 864–867.

Thomas, Chris D. "Extinction Risk from Climate Change." *Nature* 427 (2004): 145–148.

Thomas, Karl, and Kevin Trenberth. "Modern Global Climate Change." *Science* 302 (2003): 1719–1723.

Thompson, Lonnie G., et al. "Tropical Glacier and Ice Core Evidence of Climate Change on Annual to Millennial Time Scales." *Climate Change* 59 (2003): 137–155.

Tierney, John. "Fateful Voice of a Generation Still Drowns Out Real Science." *New York Times* (June 5, 2007).

Tierney, John. "Recycling Is Garbage." *New York Times Magazine* (June 30, 1996).

Tilman, David, et al. "Carbon-Negative Biofuels from Low-Input High Diversity Grassland Biomass." *Science* 314 (2006): 1598–1600.

Trenberth, Kevin. "Uncertainty in Hurricanes and Global Warming." *Science* 308, no. 5279 (June 17, 2005): 1753–1754.

Trenberth, Kevin E. "Warmer Oceans, Stronger Hurricanes." *Scientific American* 297, no. 1 (July 2007): 44–51.

Trewaves, Anthony. "Urban Myths of Organic Farming." *Nature* 410: 409–410.

"The Truth about Recycling." *The Economist* (June 7, 2007).

Tsui, Bonnie. "Trophies in a Barrel: Examining 'Canned Hunting.'" *New York Times* (April 9, 2006).

Vekataraman, C., et al. "Residential Biofuels in South Asia: Carbonaceous Aerosol Emission and Climate Impacts." *Science* 307 (2005): 1454–1456.

Velasquez-Manoff, Moises. "Can 'Green Chic' Save the Planet?" *Christian Science Monitor* (July 26, 2007).

Velasquez-Manoff, Moises. "Climate Warming Skeptics: Is the Research Too Political?" *Christian Science Monitor* (October 4, 2007).

Velasquez-Manoff, Moises. "Picking Up Where 'Silent Spring' Left Off." *Christian Science Monitor* (August 14, 2007).

Venter, J.C., et al. "Environmental Genome Shotgun Sequencing of the Sargasso Sea." *Science* 304 (2004): 66–174.

Visbeck, Martin H., et al. "The North Atlantic Oscillation: Past, Present, and Future." *Proceedings of the National Academy of Sciences* 98 (2001): 12876–12877.

Voss, Alfred. "Waves, Currents, Tides—Problems and Prospects." *Energy* 4 (1979): 823–831.

Vuorinen, H.S., P.S. Juuti, and T.S. Katko. "History of Water and Health from Ancient Civilizations to Modern Times." *Water Science and Technology: Water Supply* 7, no. 1 (2007): 49–57.

Warner, Melanie. "What Is Organic? Powerful Players Want a Say." *New York Times* (November 1, 2005).

"The Water Quality Act of 1965." *U.S. Congress* 79 Stat. 903, 70 Stat. 498. Washington, DC: U.S. Congress, October 2, 1965.

Weaver, Andrew J., and Claude Hillaire-Marcel. "Global Warming and the Next Ice Age." *Science* 304 (2004): 400–402.

Webster, P.J., et al. "Changes in Tropical Cyclone Number, Duration, and Intensity in a Warming Environment." *Science*. 309: 1844–1846 (2005).

"Wild and Scenic Rivers Act." *U.S. Code* Title 16, Chapter 28, Sections 1271, 1272, 1277.

Willis, K.J., and H.J.B. Birks. "What Is Natural? The Need for a Long-Term Perspective in Biodiversity Conservation." *Science* 314 (2006): 1261–1265.

Withgott, Jay. "Are Invasive Species Born Bad?" *Science* 305 (2004): 1100–1101.

Witze, Alexandra. "Bad Weather Ahead." *Nature* 441, no. 7093 (June 1, 2006): 564–566.

Wood, Daniel B. "Clean-Air Rule Targets Existing Diesel-Truck Fleet." *Christian Science Monitor* (October 1, 2007).

Wood, Daniel B. "More Restaurants Are Going Green by Going Local." *Christian Science Monitor* (July 30, 2007).

Wood, Daniel B. "Progress in California on Curbing Emissions." *Christian Science Monitor* (November 7, 2007).

Wood, Daniel B. "Water Crisis Squeezes California's Economy." *Christian Science Monitor* (September 12, 2007).

Worden, John, et al. "Importance of Rain Evaporation and Continental Convection in the Tropical Water Cycle." *Nature* 445 (2007): 528–532.

Wyman, Charles E. "Biomass Ethanol: Technical Progress, Opportunities, and Commercial Challenges." *Annual Review of Energy and Environment* 24 (1999): 189–226.

Yamaguchi, Mitsutsune. "CDM Potential in the Power-Generation and Energy-Intensive Industries of China." *Climate Policy* 5 (2005): 167–184.

Yardley, William. "Climate Change Adds Twist to Debate Over Dams." *New York Times* (April 23, 2007).

"Yellowstone National Park Act." *U.S. Congress* 17 Stat. 32. March 1, 1872.

Zegras, P. Christopher. "As If Kyoto Mattered: The Clean Development Mechanism and Transportation." *Energy Policy* 35 (2007): 5136–5150.

Zimmer, Carl. "Predicting Oblivions: Are Existing Models Up to the Task?" *Science* 317 (2007): 892–893.

WEB SITES

allAfrica.com. "South Africa: Bali Break Through in Climate Change Talks." http://allafrica.com/stories/200712110610.html (accessed April 1, 2008).

AMBIOTEK. "Approaches to Developing a Web-based GIS Modelling Tool: For Application to

Hydrological Nowcasting." March 2003. http://www.ambiotek.com/theses/waleed_thesis_final.pdf (accessed February 26, 2008).

American Civil Liberties Union. "ACLU Criticizes Pennsylvania Eco-Terrorism Legislation as Part of Government's Crackdown on Political Dissent." http://www.aclu.org/freespeech/protest/11263prs20050606.html (accessed April 30, 2008).

American Institute of Biological Sciences. "Searching for Nature's Medicines." http://www.actionbioscience.org/biodiversity/plotkin.html (accessed May 2, 2008).

American Meteorological Society. "History of the Clean Air Act: A Guide to Clean Air Legislation Past and Present." http://www.ametsoc.org/sloan/cleanair/index.html (accessed April 20, 2008).

American Museum of Natural History. "Center for Biodiversity and Conservation." 2007. http://cbc.amnh.org/ (accessed February 18, 2008).

American Society of Plant Biologists. "Gatehouse J, Biotechnological Prospects for Engineering Insect Resistant Plants." 2008. http://www.plantphysiol.org/cgi/content/full/146/3/881 (accessed March 26, 2008).

American University. "TED Case Studies: Rhino." http://www.american.edu/projects/mandala/TED/RHINOBLK.HTM (accessed March 9, 2008).

AmphibiaWeb. "Habitat Destruction, Alteration, and Fragmentation." September 23, 2003. http://amphibiaweb.org/declines/HabFrag.html (accessed April 13, 2008).

Antarctic Treaty Secretariat. http://www.ats.aq/index_e.htm (accessed April 8, 2008).

Antarctica New Zealand. "History of the Antarctic Treaty." http://www.antarcticanz.govt.nz/downloads/information/infosheets/Antarctic-Treaty.pdf (accessed April 8, 2008).

Antarctica New Zealand. "Mining Issues in Antarctica." http://www.antarcticanz.govt.nz/downloads/information/infosheets/mining.pdf (accessed April 8, 2008).

Appalachian Mountain Club "Ecological Survey and Land Management Planning on the Katahdin Iron Works Property." http://www.outdoors.org/conservation/wherewework/maine/ecological-inventory.cfm (accessed May 2, 2008).

Austin-American Statesman. "Let's Talk About Hybrids." September 9, 2006. http://www.statesman.com/life/content/life/stories/cars/09/09/9hybrids.html (accessed April 22, 2008).

Australian Museum. "Australia's Biodiversity." 2005. http://www.amonline.net.au/biodiversity/index.htm (accessed February 18, 2008).

Australian Research Council Centre of Excellence, Coral Reef Studies. "Reef at Risk in Climate Change.'" http://www.coralcoe.org.au/news_stories/climatechange.html (accessed April 1, 2008).

BBC GCSE Biology Revision. "The Impact of Humans." http://www.bbc.co.uk/schools/gcsebitesize/biology/livingthingsenvironment/3impactofhumansrev1.shtml (accessed April 25, 2008).

BBC Weather Centre. "The Kyoto Protocol." http://www.bbc.co.uk/climate/policies/kyoto.shtml (accessed March 1, 2008).

BEEF Magazine. "What Is Overgrazing?" May 1, 1999. http://beefmagazine.com/mag/beef_overgrazing (accessed April 21, 2008).

Biobasics: Government of Canada. "Bioremediation." February 9, 2006. http://biobasics.gc.ca/english/View.asp?x=741 (accessed March 25, 2008).

Bioethics Education Project. "Pollution: Mining and Quarrying." http://www.beep.ac.uk/content/231.0.html (accessed April 30, 2008).

Black, Richard. "Gorillas Head Race to Extinction." *BBC News*, September 12, 2007. http://news.bbc.co.uk/go/pr/fr/-/2/hi/science/nature/6990095.stm (accessed May 22, 2008).

Blacksmith Institute. "The World's Worst Polluted Places 2007." http://www.blacksmithinstitute.org/ten.php (accessed April 30, 2008).

Bureau of Land Management. "About the Bureau of Land Management." February 15, 2008. http://www.blm.gov/wo/st/en/info/About_BLM.html (accessed February 21, 2008).

Bureau of Oceans and International Environmental and Scientific Affairs, U.S. Department of State. "Coral Bleaching, Coral Mortality, and Global Climate Change." http://www.state.gov/www/global/global_issues/coral_reefs/990305_coralreef_rpt.html (accessed April 1, 2008).

California Environmental Protection Agency Air Resources Board. "Low-Emission Vehicle (LEV II) Program." January 23, 2007. http://www.arb.ca.gov/msprog/levprog/levii/levii.htm (accessed April 22, 2008).

Canadian Broadcasting Corporation. "Arctic Sovereignty: Drawing a Line in the Water." August 2, 2007 http://www.cbc.ca/news/background/cdnmilitary/arctic.html (accessed January 31, 2008).

Canadian Broadcasting Corporation. "In Depth: Forces of Nature—Flooding." June 20, 2005. http://www.cbc.ca/news/background/forcesofnature/flooding.html (accessed March 24, 2008).

Canadian Environmental Assessment Agency. "Basics of Environmental Assessment." March 16, 2007. http://www.ceaa.gc.ca/010/basics_e.htm (accessed May 1, 2008).

Carlton, Jim. "Is Global Warming Killing the Polar Bears?" *Wall Street Journal Online*, December 14, 2005. http://online.wsj.com/public/article_print/SB113452435089621905-vnekw47PQGtDyf3iv5XEN71_o5I_20061214.htm (accessed May 22, 2008).

Catalina Island Conservancy. http://www.catalina-conservancy.org/ (accessed April 3, 2008).

Center for Liquefied Natural Gas. "About LNG." http://www.lngfacts.org/ (accessed May 2, 2008).

Center for Nonproliferation Studies. "Summary of the Comprehensive Nuclear Test-Ban Treaty." http://cns.miis.edu/pubs/inven/pdfs/aptctbt.pdf (accessed May 11, 2008).

Centers for Disease Control and Prevention (CDC). "Emergency Preparedness & Response." http://www.bt.cdc.gov/disasters/tsunamis/healtheff.asp (accessed February 29, 2008).

Centers for Disease Control and Prevention (CDC). "Landslides and Mudslides." http://www.bt.cdc.gov/disasters/landslides.asp (accessed May 10, 2008).

Charles Darwin Research Station. "Invasive Rat Eradication." http://www.darwinfoundation.org/files/species/pdf/rats-en.pdf (accessed May 5, 2008).

Chattanooga Times Free Press. "Knoxville: Little Fish, Big Fight." April 19, 2008. http://www.timesfreepress.com/news/2008/apr/19/knoxville-little-fish-big-fight/ (accessed May 2, 2008).

Chesapeake Bay Program. "About the Bay." March 2008. http://www.chesapeakebay.net/aboutbay.aspx?menuitem=13953 (accessed March 31, 2008).

Christian Science Monitor. "Bush Water Follies." April 23, 2003. http://www.csmonitor.com/2003/0423/p11s03-coop.html (accessed April 28, 2008).

CITES Secretariat. "Convention Text." http://www.cites.org/eng/disc/text.shtml (accessed February 1, 2008).

CITES Secretariat. "What Is CITES?" http://www.cites.org/eng/disc/what.shtml (accessed May 5, 2008).

City of West Springfield, Massachusetts. "ATV Use on Public Lands: Who's Having the Fun? And Who's Paying the Price?" http://www.west-springfield.ma.us/Public_Documents/WSpringfieldMA_DPW/ENVIRONMENTAL/ATV_damage.pdf (accessed April 16, 2008).

Climate Research Unit, School of Environmental Sciences, University of East Anglia, UK. "North Atlantic Oscillation." http://www.cru.uea.ac.uk/cru/info/nao/ (accessed April 12, 2008).

CNN.com "Prescription Drugs Found in Drinking Water Across U.S." March 10, 2008. http://www.cnn.com/2008/HEALTH/03/10/pharma.water1.ap/ (accessed May 10, 2008).

CNN.com. "U.N.: Glaciers Shrinking at Record Rate." March 17, 2008. http://www.cnn.com/2008/TECH/03/17/un.climate.ap/index.html?iref=newssearch (accessed May 2, 2008).

Confederation of European Paper Industries. "Paper-online." http://www.paperonline.org/ (accessed March 19, 2008).

Container Recycling Institute. "Waste and Opportunity: U.S. Beverage Container Recycling Scorecard and Report." http://www.container-recycling.org/assets/pdfs/reports/2006-scorecard.pdf (accessed May 9, 2008).

Council on Environmental Quality, Executive Office of the President. "A Citizen's Guide to the NEPA." http://ceq.eh.doe.gov/nepa/Citizens_Guide_Dec07.pdf (accessed April 16, 2008).

CTV Television Network. "Walkerton Chronology." December 20, 2004. http://www.ctv.ca/servlet/ArticleNews/story/CTVNews/1103559265883_98968465 (accessed April 19, 2008).

Denver Museum of Nature and Science. "Ice Age in Depth." http://www.dmnh.org/main/minisites/iceage/ia_indepth/depth1.html (accessed April 1, 2008).

Denver Post. "Congress Revisits the Clean Water Act." April 15, 2008. http://www.denverpost.com/headlines/ci_8934105 (accessed April 28, 2008).

Department of State. "Treaty Banning Nuclear Weapon Tests in the Atmosphere, in Outer Space and Under Water." http://www.state.gov/t/ac/trt/4797.htm (accessed May 11, 2008).

Division of Crop Sciences, University of Illinois. "How Herbicides Work." http://web.aces.uiuc.edu/

vista/pdf_pubs/HERBWORK.PDF (accessed March 25, 2008).

Duke University School of the Environment. "Neuse River & Pamlico Sound." http://moray.ml.duke.edu/faculty/crowder/research/neuse (accessed February 7, 2008).

Earth Observatory. "Tropical Deforestation." March 30, 2007. http://earthobservatory.nasa.gov/Library/Deforestation/ (accessed April 13, 2008).

Earth Restoration Service. "Habitat Loss." http://www.earthrestorationservice.org/page/76/habitat-loss.htm (accessed April 20, 2008).

Ecology.com. "Paper Chase." http://www.ecology.com/feature-stories/paper-chase/index.html (accessed March 20, 2008).

The Economist. "A Bloody War: Whaling—A Special Report." December 30, 2003. "http://www.economist.com/science/displaystory.cfm?story_id=E1_NPTPDRN (accessed March 5, 2008.).

The Economist. "Nobel or Savage?" December 19, 2007. http://www.economist.com/displaystory.cfm?story_id=10278703 (accessed March 27, 2008).

Encyclopedia of Earth. "United Nations Conference on the Human Environment." November 9, 2007. http://www.eoearth.org/article/United_Nations_Conference_on_the_Human_Environment_(UNCHE),_Stockholm,_Sweden (accessed March 4, 2008).

Encyclopedia Romana, University of Chicago. "Dogs in Ancient Greece and Rome." http://penelope.uchicago.edu/~grout/encyclopaedia_romana/miscellanea/canes/canes.html (accessed March 6, 2008).

Energy Information Administration, U.S. Department of Energy. "Petroleum." http://www.eia.doe.gov/oil_gas/petroleum/info_glance/petroleum.html (accessed April 16, 2006).

EnviroLink. "How the First Earth Day Came About." http://earthday/envirolink.org/history.html (accessed April 14, 2008).

Environment Canada. "Aquatic Ecosystems." http://www.ec.gc.ca/water/en/nature/aqua/e_ecosys.htm (accessed April 17, 2008).

Environment Canada. "Causes of Flooding." 2004. http://www.ec.gc.ca/water/en/manage/floodgen/e_cause.htm (accessed March 24, 2008).

Environment Canada. "Mounting Concerns over Electronic Waste." June 26, 2003. http://www.ec.gc.ca/envirozine/english/issues/33/feature1_e.cfm (accessed April 20, 2008).

Environment News Service. "Appeals Court Rejects EPA Mercury Cap-and-Trade Rule." http://www.ens-newswire.com/ens/feb2008/2008-02-08-01.asp (accessed May 2, 2008).

Environment News Service. "U.S. EPA Sued for Ignoring Supreme Court Greenhouse Gas Ruling." April 2. 2008. http://www.ens-newswire.com/ens/apr2008/2008-04-02-01.asp (accessed April 30, 2008).

Environment News Service. "U.S. Sues Property Rights Ranchers over Grazing on Federal Lands." August 30, 2007. http://www.ens-newswire.com/ens/aug2007/2007-08-30-092.asp (accessed April 21, 2008).

Environmental Defense Fund. "Eco-Friendly Seafood Selector." 2008. http://www.edf.org/page.cfm?tagID=1521 (accessed March 31, 2008).

Environmental Literacy Council. "Coastal Ecosystems." August 3, 2007. http://www.enviroliteracy.org/subcategory.php/9.html (accessed February 1, 2008).

Environmental Literacy Council. "Ecosystems." May 14, 2007. http://www.enviroliteracy.org/category.php/3.html (accessed February 21, 2008).

Environmental News Network. "Loss of Genetic Diversity Threatens Species Diversity." September 26, 2007. http://www.enn.com/wildlife/article/23391 (accessed May 2, 2008).

Environmental Science and Technology Online. "Marine Waters." January 19, 2005. http://pubs.acs.org/subscribe/journals/esthag-w/2005/jan/science/jp_nutrient.html (accessed February 28, 2008).

EnvironmentalChemistry.com. "A Brief History of Asbestos Use and Associated Health Risks." October 2004. http://environmentalchemistry.com/yogi/environmental/asbestoshistory2004.html (accessed February 16, 2008).

ESRI. "What Is GIS?" http://www.gis.com/whatisgis/ (accessed February 26, 2008).

Estuaries.gov. "About Estuaries." August 8, 2006. http://www.estuaries.gov/about.html (accessed March 31, 2008).

European Green Party. http://www.european-greens.org/cms/default/rubrik/9/9034.htm (accessed May 4, 2008).

FAO. "Global Forest Resources Assessment, 2005: 15 Key Findings." 2005. http://www.fao.org/forestry/foris/data/fra2005/kf/common/GlobalForestA4-ENsmall.pdf (accessed April 15, 2008).

Fathom (Columbia University) and Woods Hole Oceanographic Institution. "Toxic Blooms: Understanding Red Tides." http://www.fathom.com/course/10701012/index.html (accessed March 29, 2008).

Federal Emergency Management Administration (FEMA). "Landslide and Debris Flow." http://www.fema.gov/hazard/landslide/index.shtm (accessed May 10, 2008).

Findlaw. "Tennessee Valley Authority v. Hill." http://caselaw.lp.findlaw.com/scripts/getcase.pl?court=US&vol=437&invol=153 (accessed May 2, 2008).

"Fish Farming's Growing Dangers." *Time* September 19, 2007. www.time.com/time/health/article/0,8599,1663604,00.html (accessed March 5, 2008).

Fisher Museum Harvard Forest. "Landscape History of Central New England." 2006. http://harvardforest.fas.harvard.edu/museum/landscape.html#farm (accessed April 6, 2008).

Food and Agricultural Organization of the United Nations. "Fisheries and Aquaculture Department." 2008. http://www.fao.org/fishery/ (accessed February 25, 2008).

Food and Agriculture Organization of the United Nations. "Aquastat Review of Agricultural Water Use Per Country." http://www.fao.org/nr/water/aquastat/water_use/index.stm (accessed April 25, 2008).

Food and Agriculture Organization of the United Nations. "The State of World Fisheries and Aquaculture." 2004. http://www.fao.org/docrep/007/y5600e/y5600e00.htm (accessed March 24, 2008).

Forest History Society. "1979: Snail Darter Exemption Case." November 1, 2004. http://www.foresthistory.org/research/usfscoll/policy/northern_spotted_owl/1979owl.snaildarter.html (accessed May 2, 2008).

Friends of the Earth. "The Environmental Consequences of Paper and Pulp Manufacture." 2001. http://www.foe.co.uk/resource/briefings/consequence_pulp_paper.html (accessed April 20, 2008).

Fujifilm. "The Fujifilm Group Green Policy." Revised on April 1, 2006. www.fujifilm.com/about/sustainability/green_policy/outline/index.html (accessed March 2, 2008).

GeoCommunity. http://www.geocomm.com/ (accessed April 10, 2008).

GIS Development. "Mapping GIS Milestones." http://www.gisdevelopment.net/history/1960–1970.htm (accessed February 26, 2008).

Global Invasive Species Program. "Global Strategy on Invasive Alien Species." http://www.gisp.org/publications/brochures/globalstrategy.pdf (accessed May 5, 2008).

Goddard Institute for Space Studies, NASA. "Coastal Populations, Topography, and Sea Level Rise." http://www.giss.nasa.gov/research/briefs/gornitz_04 (accessed April 2, 2008).

Government of Canada. "Biobasics: Bioremediation." February 9, 2006. http://biobasics.gc.ca/english/View.asp?x=741 (accessed February 7, 2008).

Great Lakes Information Network. "Coastal Zone in the Great Lakes Region." November 1, 2006. http://www.great-lakes.net/envt/air-land/cstzone.html (accessed February 8, 2008).

Green Facts. "Desertification: Scientific Facts on Desertification." January 23, 2008. http://www.greenfacts.org/en/desertification/index.htm (accessed February 27, 2008).

Greenland Ice Sheet Project 2. "Ice Cores that Tell the Past." http://www.gisp2.sr.unh.edu/MoreInfo/Ice_Cores_Past.html (accessed March 24, 2008).

Greenpeace. http://www.greenpeace.org/usa/ (accessed May 4, 2008).

Greenpeace International. "The E-Waste Problem in China." http://www.greenpeace.org/china/en/campaigns/toxics/e-waste/the-e-waste-problem/ (accessed April 22, 2008).

Guardian.co.uk. "Report Reveals 'Alarming' Rate of Mangrove Habitat Loss." February 1, 2008. http://www.guardian.co.uk/environment/2008/feb/01/endangeredhabitats.conservation (accessed April 20, 2008).

Heal the Bay. "Beach Report Card." January 16, 2007. http://www.healthebay.org/brc/ (accessed February 28, 2008).

The Humane Society of the United States. "The Hunting Campaign." http://www.hsus.org/wildlife/hunting/ (accessed March 6, 2008).

Hybridcars.com. "Research." 2008. http://www.hybridcars.com/ research.html (accessed April 14, 2008).

Illinois State Museum. "Ice Ages." http://www.museum.state.il.us/exhibits/ice_ages (accessed April 1, 2008).

"In China, Farming Fish in Toxic Waters." *New York Times* December 15, 2007. www.nytimes

.com/2007/12/15/world/asia/15fish.html (accessed March 5, 2008).

Indian Concrete Journal. "An Introduction to Tsunami in the Indian Context." http://www.nicee.org/ICJ_TsunamiSpecialReortl.pdf (accessed February 29, 2008).

Institute of Public Law, University of New Mexico School of Law. "Watershed Protection and Flood Prevention Act." http://ipl.unm.edu/cwl/fedbook/wpfpa.html (accessed May 1, 2008).

Inter Press Service News. "Researchers Highlight Overgrazing." http://ipsnews.net/fao_magazine/environment.shtml (accessed April 21, 2008).

Intergovernmental Panel on Climate Change. "Home Page." http://www.ipcc.ch/ (accessed March 4, 2008).

Intergovernmental Panel on Climate Change. "16 Years of Scientific Assessment in Support of the Climate Convention." December 2004. http://www.ipcc.ch/pdf/10th-anniversary/anniversary-brochure.pdf (accessed March 25, 2008).

Intergovernmental Panel on Climate Change. "Climate Change and Biodiversity." April 2002. http://www.ipcc.ch/pub/tpbiodiv.pdf (accessed November 8, 2007).

Intergovermental Panel on Climate Change. "Contributions of Working Group III to the Fourth Assessment Report of the Intergovermental Panel on Climate Change." May 4, 2007. http://www.ipcc.ch/SPM040507.pdf (accessed November 5, 2007).

Intergovernmental Panel on Climate Change. "Intergovernmental Panel on Climate Change." 2008 .http://www.ipcc.ch (accessed March 24, 2008).

International Commission on Stratigraphy. "International Commission on Stratigraphy." January 2008. (accessed March 26, 2008).

International Lake Environment Committee. "World Lake Database." http://www.ilec.or.jp/database/database.html (accessed May 4, 2008).

International Mine Water Association. "Geochemistry by William M. White, Cornell University." August 25, 2005. http://www.imwa.info/geochemistry/ (accessed March 25, 2008).

International Paper Company. "IP Forest Resources Facts." February, 2006. http://www.internationalpaper.com/PDF/PDFs_for_Our_Company/FRFactsSheet06_FINALQ1.pdf (accessed April 15, 2008).

International Rivers Network. "Citizen's Guide to the World Commission on Dams." http://www.irn.org/wcd/wcdguide.pdf (accessed March 29, 2008).

International Society for Ecological Modeling. "Home Page." October 25, 2007. http://www.isemna.org/ (accessed March 2, 2008).

International Tsunami Information Center (ITIC). "Tsunamis on the Move." http://ioc3.unesco.org/itic/contents.php?id=205 (accessed February 29, 2008).

The International Whaling Commission (IWC). "International Convention for the Regulation of Whaling, 1946." http://www.iwcoffice.org/commission/convention.htm (accessed April 5, 2008).

Iowa State University. "A Brief History of Environmentalism." http://www.public.iastate.edu/~sws/enviro%20and%20society%20Spring%202006/HistoryofEnvironmentalism.doc (accessed March 27, 2008).

IUCN-The World Conservation Union. http://www.iucn.org (accessed May 10, 2008).

Jarboe, James F. "The Threat of Eco-Terrorism." *Federal Bureau of Investigation* February 12, 2002. http://www2.fbi.gov/congress/congress02/jarboe021202.htm (accessed January 5, 2008).

Journey North http://www.learner.org/jnorth/ (accessed May 2, 2008).

Kennedy, John F. "Treaty Banning Nuclear Weapon Tests in the Atmosphere, in Outer Space, and Under Water (Limited Test Ban Treaty)." August 5, 1963. http://www.ucsusa.org/assets/documents/global_security/limited_test_ban_treaty.pdf (accessed April 12, 2008).

Lenfest Ocean Program. June 23, 2006. http://www.lenfestocean.org/publications/Lotze-etal_06_Science.pdf (accessed May 22, 2008).

Los Angeles County Department of Public Health. "State Ocean Water Quality Standards." http://www.lapublichealth.org/eh/progs/envirp/rechlth/ehrecocstand.htm (accessed February 28, 2008).

Lynch, Margaret.*The Geographer's Craft Project, Department of Geography, University of Colorado at Boulder.* "Ethical Issues in Electronic Information Systems." 1994. http://www.colorado.edu/geography/gcraft/notes/ethics/ethics_f.html (accessed February 26, 2008).

MapForum Ltd. "The Earliest Atlases: From Ptolemy to Ortelius." http://www.mapforum.com/01/atlas.htm (accessed April 20, 2008).

Marietta College. "Competition." January 11, 2008. http://www.marietta.edu/~biol/biomes/competition.htm (accessed February 11, 2008).

Marietta College. "Environmental Biology—Ecosystems." http://www.marietta.edu/~biol/102/ecosystem.html (accessed February 21, 2008).

MarionBio.org. "Sustainable Fisheries." February 17, 2008. http://marinebio.org/ (accessed March 24, 2008).

Massachusetts Institute of Technology. "The Future of Coal." http://web.mit.edu/coal/The_Future_of_Coal.pdf (accessed April 30, 2008).

Met Office. "The Great Smog of 1952." http://www.metoffice.gov.uk/education/secondary/students/smog.html (accessed March 16, 2008).

Michigan Energy Center Network. "Fossil Fuels: A Short Blip in History." Michigan Renewable Energy Success Stories." 2003–2004. http://www.urbanoptions.org/RenewableEnergy/FossilFuelsAShortBlip.htm (accessed February 29, 2008).

Michigan Energy Center Network. "A Possible Roman Tide Mill." Kent Archaeological Society." 2003–2004. http://www.kentarchaeology.ac/authors/005.pdf (accessed February 29, 2008).

Michoacán Restoration Fund. "Monarchs." 2006. http://michoacanmonarchs.org/ (accessed April 6, 2008).

Mississippi Development Authority. "The Tennessee-Tombigbee Waterway." 2007. (accessed April 16, 2008).

Monarch Watch. http://www.monarchwatch.org/ (accessed April 21, 2008).

Mongabay.com. "China Drives Elephant Poaching for Ivory Trade." February 27, 2008. http://news.mongabay.com/2007/0226-elephants.html (accessed March 9, 2008).

Mongabay.com. "Rwanda Launches Reforestation Project to Protect Chimps, Drive Ecotourism." March 18, 2008. http://news.mongabay.com/2008/0318-rwanda.html (accessed April 6, 2008).

Mount Shasta Herald. "Growers Hear About New Roundup Ready Alfalfa." February 16, 2005. http://www.monsanto.co.uk/news/ukshowlib.phtml?uid=8599 (accessed March 25, 2008).

MSNBC. "Final Border Fence Impact Report Bypassed with Waiver." http://www.msnbc.msn.com/id/23927996/ (accessed April 21, 2008).

MSNBC. "Not So Fast with Biofuels, U.N. Warns," May 8, 2007. http://www.msnbc.msn.com/id/18551000/ (accessed April 10, 2008).

Narragansett Bay. "Estuarine Science." (accessed March 26, 2008).

National Academy of Sciences. "Health Effects of Exposure to Low-Level Ionizing Radiation: BEIR V." http://www.nap.edu/openbook.php?isbn=0309039959 (accessed April 16, 2008).

National Aeronautics and Space Administration (NASA). "A Chilling Possibility." http://science.nasa.gov/headlines/y2004/05mar_arctic.htm (accessed April 2, 2008).

National Aeronautic and Space Administration (NASA). "Image of the Day: Global Vegetation." April 7, 2008 http://www.nasa.gov/multimedia/imagegallery/image_feature_1056.html (accessed May 2, 2008).

National Aeronautics and Space Administration (NASA). "NASA Satellites Measure and Monitor Sea Level." http://www.nasa.gov/home/hqnews/2005/jul/HQ_05175_sea_level_monitored.html (accessed April 2, 2008).

National Aeronautic and Space Administration (NASA). "Ozone Hole Watch." December 4, 2007. http://ozonewatch.gsfc.nasa.gov/ (accessed May 10, 2008).

National Aeronautics and Space Administration (NASA). "Recent Warming of Arctic May Affect Worldwide Climate." October 23, 2003. http://www.nasa.gov/centers/goddard/news/topstory/2003/1023esuice.html (accessed March 1, 2008).

National Aeronautics and Space Administration (NASA). "Soil and the Environment." August 15, 2001. http://soil.gsfc.nasa.gov/env.htm (accessed April 23, 2008).

National Aeronautics and Space Administration (NASA). "Weather." http://www.nasa.gov/worldbook/weather_worldbook.html (accessed May 12, 2008).

National Aeronautic and Space Administration (NASA): Goddard Institute for Space Studies. "NASA Study Predicts More Severe Storms With Global Warming." August 30, 2007http://www.giss.nasa.gov/research/news/20070830/ (accessed April 20, 2008).

National Aeronautic and Space Administration (NASA): Goddard Space Flight Center. "NASA Keeps Eye on Ozone Layer Amid Montreal Protocol's Success." September 18, 2007. http://www.sciencedaily.com /releases/2007/09/070913181730.htm (accessed January 30, 2008).

National Center for Appropriate Technology. "Smart Communities Network." May 31, 2005. http://www.smartcommunities.ncat.org/landuse/lukey.shtml (accessed March 22, 2008).

National Center for Atmospheric Research. "Climate and the Water Cycle." http://www.ncar.ucar.edu/research/earth_system/watercycle.php (accessed March 29, 2008).

National Earthquake Information Center, U.S. Geological Survey. "NEIC: An Interview with Charles F. Richter." http://neic.usgs.gov/neis/seismology/people/int_richter.html (accessed April 7, 2008).

National Environment Research Council. "What Will Climate Change Mean for Our World?" http://www.nerc.ac.uk/research/issues/climatechange/globalimpact.asp (accessed March 6, 2008).

National Geographic News. "Dozens of Sea Lions Found Massacred in Galápagos." February 4, 2008. http://news.nationalgeographic.com/news/2008/02/080204-sea-lions.html (accessed May 4, 2008).

National Geographic Society. "Arctic Melt Opens Northwest Passage." September 17, 2007 http://news.nationalgeographic.com/news/2007/09/070917-northwest-passage.html (accessed January 31, 2008).

National Geographic Society. "Climate Change Caused Extinction of Big Ice Age Mammals, Scientist Says." http://news.nationalgeographic.com/news/2001/11/1112_overkill.htmll (accessed May 1, 2008).

National Geographic Society. "Deforestation and Desertification." http://www.nationalgeographic.com/eye/deforestation/effect.html (accessed April 13, 2008).

National Geographic Society. "Humans Caused Australia's Ice Age Extinctions, Tooth Study Says." http://news.nationalgeographic.com/news/2007/01/070124-iceage-fossils.html (accessed May 1, 2008).

National Geographic Society. "Intraspecies Diversity Helps Ecosystems, Study Says." August 21, 2002. http://news.nationalgeographic.com/news/2002/08/0821_020821_diversity.html (accessed February 18, 2008).

National Geographic Society. "Satellites Can Warn of Floods, Landslides Worldwide, Scientists Say." May 25, 2006. http://news.nationalgeographic.com/news/2006/05/060525-flood-warn.html (accessed May 10, 2008).

National Geographic Society. "Wolves' Leftovers Are Yellowstone's Gain, Study Says." December 4, 2003. http://news.nationalgeographic.com/news/2003/12/1204_031204_yellowstone wolves.html (accessed April 12, 2008).

National Ground Water Association. "Ground Water Sustainability." February 24, 2004. http://www.ngwa.org/PROGRAMS/government/issues/possustain.aspx (accessed April 19, 2008).

National Highway Traffic Safety Administration. "CAFE Overview—Frequently Asked Questions." http://www.nhtsa.dot.gov/CARS/rules/CAFE/overview.htm (accessed April 22, 2008).

National Institutes of Health. "Help Secure Our Future: Recycle Around the Clock." http://nems.nih.gov/outreach/factsheet_recycling.pdf (accessed May 9, 2008).

National Interagency Fire Center. "National Interagency Fire Center." http://www.nifc.gov (accessed March 14, 2008).

National Oceanic and Atmospheric Administration (NOAA). "Coastal Ecosystems." June 20, 2005. http://www.ncddc.noaa.gov/ecosystems (accessed February 1, 2008).

National Oceanographic and Atmospheric Administration (NOAA). "Frequently Asked Questions About El Niño and La Niña." www.pmel.noaa.gov/tao/elnino/faq.html (accessed February 7, 2008).

National Oceanographic and Atmospheric Administration (NOAA). "Global Surface Temperatures over the Past Two Millennia." http://www.ncdc.noaa.gov/paleo/pubs/mann2003b/mann2003b.html (accessed May 12, 2008).

National Oceanic and Atmospheric Administration (NOAA). "Greenhouse Gases: Frequently Asked Questions." December 1, 2005. http://www.ncdc.noaa.gov/oa/climate/gases.html (accessed March 21, 2008).

National Oceanic and Atmospheric Administration (NOAA). "National Weather Service." http://www.nws.noaa.gov/ (accessed May 12, 2008).

National Oceanic and Atmospheric Administration (NOAA). "NOAA Aquaculture Program." http://aquaculture.noaa.gov/ (accessed April 29, 2008).

National Park Service. http://awww.nps.gov/ (accessed May 3, 2008).

National Park Service. "Codifying Tradition: The National Park Service Act of 1916." http://www.nps.gov/history/history/online_books/sellars/chap2.htm (accessed April 12, 2008).

National Park Service. "Developing Conceptual Models of Relevant Ecosystem Components." February 20, 2008. http://science.nature.nps.gov/im/monitor/ConceptualModels.cfm (accessed March 2, 2008).

National Park Service. "National Park Service Organic Act." http://www.nps.gov/legacy/organic-act.htm (accessed April 12, 2008).

National Research Council of the National Academy of Sciences. "Radiative Forcing of Climate Change: Expanding the Concept and Addressing Uncertainties." http://www.nap.edu/openbook.php?isbn=0309095069 (accessed April 11, 2008).

National Resources Conservation Service, United States Department of Agriculture. "Watershed Protection and Flood Prevention Act." http://www.nrcs.usda.gov/programs/watershed/pl56631705.pdf (accessed May 1, 2008).

National Response Center. http://www.nrc.uscg.mil/ (accessed February 22, 2008).

National Science Foundation. "Antarctic Conservation Act." http://www.nsf.gov/od/opp/antarct/aca/aca.jsp (accessed April 8, 2008).

National Science Foundation. "The Antarctic Treaty." http://www.nsf.gov/od/opp/antarct/anttrty.jsp (accessed April 8, 2008).

National Safety Council. "Background on Air Pollution." March 6, 2006. http://www.nsc.org/EHC/mobile/acback.htm (accessed April 20, 2008).

National Science Foundation. "Environmental Sustainability." February 14, 2008. http://www.nsf.gov/funding/pgm_summ.jsp?pims_id=501027 (accessed March 4, 2008).

National Shooting Sports Foundation. "The Ethical Hunter." http://www.nhfday.org/index.php?option=com_content&task=view&id=105&Itemid=87 (accessed March 7, 2008).

National Snow and Ice Data Center. "All About Glaciers." 2007. http://www.nsidc.org/glaciers (accessed March 25, 2008).

National Snow and Ice Data Center. "Repeat Photography of Glaciers." http://nsidc.org/data/glacier_photo/repeat_photography.html (accessed May 2, 2008).

National Weather Service Forecast Office. "What Is Meant by the Term Drought?" http://www.wrh.noaa.gov/fgz/science/drought.hph?wfo=fgz (accessed February 22, 2008).

National Wilderness Preservation System. "Wilderness Fast Facts." http://www.wilderness.net/index.cfm?fuse=NWPS&sec=fastFacts (accessed April 22, 2008).

National Wilderness Preservation System. "Wilderness Legislation: Overview." http://www.wilderness.net/index.cfm?fuse=NWPS&sec=legisoverview (accessed April 22, 2008).

National Wildlife Federation. "Walden Warming." http://www.nwf.org/nationalwildlife/article.cfm?issueID=117&articleID=1510 (accessed April 12, 2008).

Natural Resources Canada. "Technologies and Applications: About Solar Energy." April 26, 2005. www.canren.gc.ca/tech_appl/index.asp?CaId=5&PgId=121 (accessed March 19, 2008).

Natural Resources Defense Council. "Defending NEPA from Assault." http://www.nrdc.org/legislation/fnepa.asp (accessed April 17, 2008).

Nature Reports Climate Change. "Post-Kyoto Pact: Shaping the Successor." June 7, 2007. http://www.nature.com/climate/2007/0706/full/climate.2007.12.html (accessed April 2, 2008).

Nearctica: The Natural World of North America. "Island Biogeography." 1999. http://www.nearctica.com/ecology/habitats/island.htm (accessed March 31, 2008).

New Hampshire Public Television. "NatureWorks—Ecosystems." 2007. http://stort.unep-wcmc.org/imaps/gb2002/book/viewer.htm (accessed February 18, 2008).

New Scientist. "Biofuel Production May Raise Price of Food." May 9, 2007. http://www.newscientist.com/article/dn11811-biofuel-production-may-raise-price-of-food-.html (accessed March 7, 2008).

New Scientist. "Hydroelectric Power's Dirty Secret Revealed." http://www.newscientist.com/article/dn7046.html (accessed March 29, 2008).

New Scientist. "Leading Nations Find Agreement on Climate Change." http://environment.newscientist.com/channel/earth/dn11199-leading-nations-find-agreement-on-climate-change.html (accessed April 4, 2008).

New York Times. "Companies Pressed to Define Green Policies." February 13, 2007. http://www.nytimes.com/2007/02/13/business/13climate.html (accessed March 2, 2008).

NOAA (National Oceanic and Atmospheric Administration) Magazine. "Dust Storms, Sand Storms and Related NOAA Activities in the Middle East." April 7, 2003. http://www.magazine.noaa.gov/stories/mag86.htm (accessed March 29, 2008).

NOAA Coastal Services Center. "Benthic Habitat Mapping." July 24, 2007. http://www.csc.noaa.gov/benthic/start/what.htm (accessed on April 7, 2008).

Nobel Foundation. "The Nobel Peace Prize 2007." October 2007. http://nobelprize.org/nobel_prizes/peace/laureates/2007/ (accessed March 25, 2008).

Oak Ridge National Laboratory. "Global, Regional, and National Fossil Fuel CO_2 Emissions." http://cdiac.ornl.gov/trends/emis/em_cont.htm (accessed April 23, 2008).

Oasis. "What Is Desertification?" http://www.oasis-global.net/what_is.htm (accessed February 26, 2008).

Office for Disarmament Affairs. "Comprehensive Nuclear-Test-Ban Treaty." http://disarmament.un.org/TreatyStatus.nsf (accessed May 11, 2008).

Office of Climate, Weather, and Water Services. "Tsunami: The Great Waves." http://www.weather.gov/om/brochures/tsunami.htm (accessed February 29, 2008).

Office of Naval Research. "Habitats: Beaches." http://www.onr.navy.mil/focus/ocean/habitats/beaches1.htm (accessed April 3, 2008).

Office of the Press Secretary, The White House. "Fact Sheet: Twenty in Ten: Strengthening Energy Security and Addressing Climate Change." May 14, 2007. http://www.whitehouse.gov/news/releases/2007/05/20070514-2.html (accessed April 30, 2008).

Ohio State University. "Wastewater Treatment Principles and Regulations." http://ohioline.osu.edu/aex-fact/0768.html (accessed March 25, 2008).

The Organic Center. "New Evidence Confirms the Nutritional Superiority of Plant-Based Organic Foods." http://www.organic-center.org/science.latest.php?action=view&report_id=126 (accessed March 29, 2008).

Organization for Economic Co-operation and Development. "Sustainable Development: Our Common Future." http://Hww.oecdobserver.org/news/fullstory.php/aid/780/Sustainable_development:_Our_common_future.html (accessed April 27, 2008).

Our Planet. "Spirit of Optimism." http://www.unep.org/OurPlanet/imgversn/102/trittin.html (accessed March 12, 2008).

Overfishing.org. "Overfishing—A Global Disaster." 2007. http://overfishing.org/ (accessed March 24, 2008).

Overseas Development Institute. "Inland Fisheries." http://www.odi.org.uk/publications/keysheets/green_9_inlandfish.pdf (accessed April 29, 2008).

Oxford University. "Oxford Dendrochronology Laboratory." December 2007. (accessed March 26, 2008).

Pacific Institute: The World's Water. "Water and Conflict." February 2008 http://www.worldwater.org/conflict.html (accessed March 25, 2008).

Palomar College. "Early Human Evolution: Early Human Culture." November 28, 2007. http://anthro.palomar.edu/homo/homo_3.htm (accessed March 6, 2007).

Peace Parks Foundation. "Facilitating Peace Parks." http://www.peaceparks.org (accessed April 20, 2008).

People's Weekly World Newspaper. "Defending Rachel Carson." by David Pimentel. http://www.pww.org/article/articleprint/11914/ (accessed April 12, 2008).

Pew Center on Global Climate Change. "Coral Reefs and Global Climate Change." http://www.pewclimate.org/docUploads/Coral_Reefs.pdf (accessed April 1, 2008).

Pew Center on Global Climate Change. "Hurricanes and Global Warming FAQs." http://www.pewclimate.org/hurricanes.cfm (accessed March 29, 2008).

PhysicalGeography.net. "Introduction to the Atmosphere." http://www.physicalgeography.net/fundamentals/7a.html (accessed February 12, 2008).

PhysicalGeography.net. "Precipitation and Fog." http://www.physicalgeography.net/fundamentals/8f.html (accessed April 16, 2008).

Popular Mechanics. "Las Vegas Tries to Prevent a Water Shortage." February 2007. http://www.popularmechanics.com/science/earth/4210244.html (accessed April 19, 2008).

Public Broadcasting Service. "Adaptive Radiation: Darwin's Finches." http://www.pbs.org/wgbh/evolution/library/01/6/l_016_02.html (accessed February 11, 2008).

Reefbase. "A Global Information System for Coral Reefs." http://www.reefbase.org (accessed March 24, 2008).

Renewable Energy and AEoogle. "Tidal Power." http://www.alternative-energy-news.info/technology/hydro/tidal-power/ (accessed March 7, 2008).

Rice University. "What Are Maps?" http://math.rice.edu/~lanius/pres/map/mapdef.html (accessed April 20, 2008).

Rocky Mountain Institute. "Nuclear Power: Economics and Climate-Change Potential." https://www.rmi.org/images/PDFs/Energy/E05-08_NukePwrEcon.pdf (accessed April 15, 2008).

Royal Aeronautical Society. "Greener by Design: Dedicated to Sustainable Aviation." 2007. http://www.greenerbydesign.org.uk/home/index.php (accessed March 19, 2008).

Royal Botanic Gardens Kew. "Paper Information Sheet." http://www.kew.org/ksheets/paper.html (accessed March 19, 2008).

Rushlimbaugh.com. "Record Cold Raises No Questions; Polar Bear Photo a Fraud." http://www.rushlimbaugh.com/home/daily/site_020507/content/0205071.guest.html (accessed March 29, 2008).

Rutgers University. "Global Snow Lab." April 9, 2008. http://climate.rutgers.edu/snowcover/ (accessed on April 9, 2008).

Rutgers University, Bill Kovarik. "Renewable Energy History Project." http://www.runet.edu/~wkovarik/envhist/RenHist/1.biofuels1.html (accessed March 8, 2008).

Sandia National Laboratories. "Guidance on Risk Analysis and Safety of a Large Liquefied Natural Gas (LNG) Spill Over Water, SAND2004-6258." December 2004. http://www.fossil.energy.gov/programs/oilgas/storage/lng/sandia_lng_1204.pdf (accessed May 2, 2008).

Save the Rainforest. http://www.savetherainforest.org/index.htm (accessed April 7, 2008).

Sea World. "Corals and Coral Reefs." http://www.seaworld.org/animal-info/info-books/coral/reef-ecosystem.htm (accessed March 24, 2008).

Secretariat of the Convention on Biological Diversity. "How the Convention on Biological Diversity Promotes Nature and Human Well-Being." http://www.cbd.int/doc/publications/cbd-sustain-en.pdf (accessed April 21, 2008).

Secretariat of the United Nations Framework Convention on Climate Change. "Kyoto Protocol." http://unfccc.int/kyoto_protocol/items/2830.php (accessed April 21, 2008).

See-the-Sea. "Habitat Alteration." http://see-the-sea.org/topics/habitat/habitat_alteration.htm (accessed April 13, 2008).

Select Committee on Energy Independence and Global Warming: U.S. House of Representatives. "New National Bottle Bill Would Cut Heat-Trapping Emissions, Energy Needs." [press release]. http://globalwarming.house.gov/mediacenter/pressreleases?id=0126 (accessed May 9, 2008).

Sierra Club. "Nuclear Waste." http://www.sierraclub.org/nuclearwaste/nucw.asp (accessed April 16, 2008).

Sierra Club. "Wetlands Overview." http://www.sierraclub.org/wetlands/ (accessed May 2, 2008).

Smithsonian Environmental Research Center. "Ecological ModelingLab." http://www.serc.si.edu/labs/ecological_modeling/index.jsp (accessed March 2, 2008).

Supreme Court of the United States. "*Massachusetts v. Environmental Protection Agency.*" April 2, 2007. http://www.supremecourtus.gov/opinions/06pdf/05-1120.pdf (accessed April 30, 2008).

Surfrider Foundation. "Water Quality." http://www.surfrider.org/waterquality.asp (accessed February 28, 2008).

Sustainability Reporting Program. "A Brief History of Sustainable Development." 2000. http://www.sustreport.org/background/history.html (accessed March 2, 2008).

Telegraph.co.uk. "Beetle Adds to Canada's CO_2 Emissions." http://www.telegraph.co.uk/earth/main.jhtml?view=DETAILS&grid=&xml=/earth/2008/04/23/scibeetle123.xml (accessed April 23, 2008).

Telegraph.co.uk. "Tourism 'Threatens Antarctic.'" http://www.telegraph.co.uk/travel/734551/Tourism-%27threatens-Antarctic%27.html (accessed April 8, 2008).

Texas A&M University Ocean World. "Coral Reefs." March 24, 2008. http://oceanworld.tamu.edu/students/coral/index.html (accessed March 24, 2008).

Think Quest. "Drought." http://library.thinkquest.org/16132/html/drought.html (accessed February 22, 2008).

Time Magazine. "The Hunt, The Furor." August 2, 1993. "http://www.time.com/time/magazine/article/0,9171,978989,00.html (accessed March 6, 2008).

The Times Online. "Bengal Big Cats Face Oblivion as Medicine Trade Fuels Poaching." February 14, 2008. http://www.timesonline.co.uk/tol/news/world/asia/article3366053.ece (accessed March 9, 2008).

The Times, South Africa. "WFP warns EU about Biofuels." March 7, 2008. http://www.thetimes.co.za/News/Article.aspx?id=722306 (accessed March 8, 2008).

Transportation Research Board, National Academies of Science. "Critical Issues in Aviation and the Environment 2005." http://onlinepubs.trb.org/onlinepubs/circulars/ec089.pdf (accessed March 19, 2008).

Tulane University. "Coastal Zones." April 9, 2007. http://www.tulane.edu/~sanelson/geol204/coastalzones.htm (accessed February 8, 2008).

U.K. CITES Authorities. "Convention on Trade in Endangered Species: UK." http://www.ukcites.gov.uk/default.asp (accessed February 1, 2008).

The University Corporation for Atmospheric Research. "Drought's Growing Reach: NCAR Study Points to Global Warming as Key Factor." January 10, 2005. http:www.ucar.edu/news/releases/2005/drought_research.shtml (accessed February 22, 2008).

UN Atlas of the Oceans. "Impact of Tsunami on Ecosystems." http://www.oceansatlas.com/servlet/CDSServlet?status=ND03MTY4NyZjdG5faW5mb192aWV3X3NpemU9Y3RuX2luZm9fdmlld19mdWxsJjY9ZW4mMzM9KiYzNz1rb3M~ (accessed February 29, 2008).

UNEP. "About DELC." http://www.unep.org/dec/About/index.asp (accessed April 1, 2008).

UNEP. "About UNEP: The Organization." http://www.unep.org/Documents.Multilingual/Default.asp?DocumentID=43 (accessed March 12, 2008).

UNEP. "Identifying and Responding to Emerging Issues." http://www.roap.unep.org/region/emerge.cfm (accessed March 12, 2008).

UNESCO. "Tsunami: The Great Waves." http://ioc3.unesco.org/itic/files/great_waves_en_2006_small.pdf (accessed February 29, 2008).

Union of Concerned Scientists. "Biodiesel Basics." December 20, 2007. http://www.ucsusa.org/clean_vehicles/big_rig_cleanup/biodiesel.html (accessed March 7, 2008).

Union of Concerned Scientists. "Coal Power: Air Pollution." August 18, 2005. http://www.ucsusa.org/clean_energy/coalvswind (accessed March 1, 2008).

United Kingdom Health and Safety Executive. "General Information about Aasbestos." http://www.hse.gov.uk/asbestos (accessed February 12, 2008).

United Nations. "The Biosphere Conference: 25 Years Later." http://unesdoc.unesco.org/images/0014/001471/147152eo.pdf (accessed May 5, 2008).

United Nations. "Declaration of the United Nations Conference on the Human Environment." http://www.unep.org/Documents.multilingual/Default.asp?DocumentID=97&ArticleID=1503 (accessed March 4, 2008).

United Nations. "Documents: Agenda 21." December 15, 2004. http://www.un.org/esa/sustdev/documents/agenda21/index.htm (accessed April 21, 2008).

United Nations. "Earth Summit 2002." March 24, 2003. http://www.un.org/jsummit/ (accessed May 1, 2008).

United Nations. "Intergovernmental Panel on Climate Change." 2008. http://www.ipcc.ch (accessed March 19, 2008).

United Nations "International Decade for Action: Water for Life, 2005–2015." http://www.un.org/waterforlifedecade/background.html (accessed May 2, 2008).

United Nations. "Johannesburg Declaration on Sustainable Development." http://www.un-documents.net/jburgdec.htm (accessed May 1, 2008).

United Nations. "Our Common Future: Report of the World Commission on Environment and Development." http://www.un-documents.net/ocf-ov.htm#I (accessed April 27, 2008).

United Nations. "Overfishing: A Threat to Marine Biodiversity." http://www.un.org/events/tenstories_2006/story.asp?storyID=800 (accessed March 24, 2008).

United Nations. "Plan of Implementation of the World Summit on Sustainable Development." http://www.un-documents.net/jburgpln.htm (accessed May 1, 2008).

United Nations. "Report of the World Summit on Sustainable Development." http://www.un.org/jsummit/html/documents/summit_docs/131302_wssd_report_reissued.pdf (accessed April 27, 2008).

United Nations. "Res. 38/161: Process of Preparation of the Environmental Perspective to the Year 2000 and Beyond." http://www.un.org/documents/ga/res/38/a38r161.htm (accessed April 27, 2008).

United Nations. "Rethinking Policies to Cope with Desertification." June 28, 2007. http://news.bbc.co.uk/2/shared/bsp/hi/pdfs/28_06_07unreportdesert.pdf (accessed February 27, 2008).

United Nations. "The Road from Johannesburg: What Was Achieved and the Way Forward." http://www.un.org/esa/sustdev/media/Brochure.doc (accessed April 21, 2008).

United Nations. "United Nations Conference on Environment and Development (1992)." May 23, 1997 http://www.un.org/geninfo/bp/enviro.html (accessed April 21, 2008).

United Nations Convention to Combat Desertification. "Dust and Sandstorms from the World's Drylands." http://www.unccd.int/publicinfo/duststorms/menu.php (accessed March 29, 2008).

United Nations. "United Nations Convention on the Law of the Sea." http://www.un.org/Depts/los/convention_agreements/texts/unclos/closindx.htm (accessed January 31, 2008).

United Nations. "United Nations Framework Convention on Climate Change." 2008. http://unfccc.int/2860.php (accessed March 19, 2008).

United Nations Department for Social and Economic Affairs. "United Nations Division for Sustainable Development." February 2008. http://www.un.org/esa/sustdev/index.html (accessed March 4, 2008).

United Nations Department of Public Information. "Montreal Protocol on Ozone-Depleting Substances Effective, But Work Still Unfinished, Says Secretary-General in Message for International Day." September 7, 2006. http://www.un.org/News/Press/docs/2006/sgsm10620.doc.htm (accessed May 6, 2008).

United Nations Economic Commission for Europe. "Convention on Long-range Transboundary Air Pollution." 2008 http://www.unece.org/env/lrtap/ (accessed March 21, 2008).

United Nations Environment Programme. "Climate Change and Dams: An Analysis of the Linkages Between the UNFCCC Legal Regime and Dams." http://www.dams.org/docs/kbase/contrib/env253.pdf (accessed March 29, 2008).

United Nations Environmental Programme.. "Climate Change and Biodiversity." http://www.cbd.int/climate/default.shtml (accessed May 10, 2008).

United Nations Environment Programme. "E-Waste Management." http://www.unep.fr/pc/pc/waste/e_waste.htm (accessed April 22, 2008).

United Nations Environment Programme. "Meltdown in the Mountains." http://www.unep.org/Documents.Multilingual/Default.asp?DocumentID=530&ArticleID=5760 (accessed May 2, 2008).

United Nations Environment Programme. "The Montreal Protocol on Substances that Deplete the Ozone Layer." 2000. http://www.unep.org/ozone/pdfs/Montreal-Protocol2000.pdf (accessed January 28, 2008).

United Nations Environment Programme. "Stockholm 1972: Report of the United Nations Conference on the Human Environment." http://www.unep.org/Documents.Multilingual/Default.asp?documentID=97 (accessed April 21, 2008).

United Nations Environment Programme. "Tourism and Biodiversity: Mapping Tourism's Global Footprint." http://www.unep.fr/pc/tourism/library/mapping_tourism.htm (accessed May 3, 2008).

United Nations Environment Programme. "United Nations List of Protected Areas." http://www.unep-wcmc.org/protected_areas/UN_list/ (accessed May 3, 2008).

United Nations Environment Programme. "The Vienna Convention for the Protection of the Ozone Layer." 2001. http://www.unep.org/Ozone/pdfs/viennaconvention2002.pdf (accessed May 6, 2008).

United Nations Environmental Programme. "World Atlas of Biodiversity." 2007. http://stort.unep-wcmc.org/imaps/gb2002/book/viewer.htm (accessed February 18, 2008).

United Nations Framework Convention on Climate Change Secretariat. "Clean Development Mechanism." June 22, 2007. http://unfccc.int/kyoto_protocol/mechanisms/clean_development_mechanism/items/2718.php (accessed March 21, 2008).

United Nations Framework Convention on Climate Change. "Kyoto Protocol Status of Ratification." http://unfccc.int/files/essential_background/kyoto_protocol/application/pdf/kpstats.pdf (accessed April 2, 2008).

United Nations Framework Convention on Climate Change. "Kyoto Protocol to the United Nations Framework Convention on Climate Change." http://unfccc.int/resource/docs/convkp/kpeng.html (accessed April 2, 2008).

United Nations General Assembly. "Report of the Secretary-General's High-Level Panel on Threats, Challenges, and Change." http://www.un.org/secureworld/ (accessed May 4, 2008).

United Nations Intergovernmental Panel on Climate Change. "IPCC Reports." http://www.ipcc.ch/ipccreports/assessments-reports.htm (accessed March 29, 2008).

United Nations Publication. "Brief The United Nations: An Introduction for Students." http://www.un.org/cyberschoolbus/unintro/unintro.asp (accessed March 12, 2008).

United Nations Statistics Division. "CO_2 Emissions." http://unstats.un.org/unsd/environment/air_co2_emissions.htm (accessed April 23, 2008).

United Nations University. "Two Billion People Vulnerable to Floods by 2050." June 13, 2004. http://www.unu.edu/news/ehs/floods.doc (accessed March 24, 2008).

United Nations World Water Development Report. "World Water Assessment Programme." http://www.unesco.org/water/wwap/facts_figures/water_industry.shtml (accessed April 20, 2008).

University College Dublin and European Union. "CODTRACE." http://www.ucd.ie/codtrace/index.htm (accessed February 25, 2008).

University Corporation for Atmospheric Research. "Abrupt Ice Retreat Could Produce Ice-Free Arctic Summers by 2040." http://www.ucar.edu/news/releases/2006/arctic.shtml (accessed April 8, 2008).

University Corporation for Atmospheric Research. "Ecosystems." January 15, 2008. http://www.windows.ucar.edu/tour/link=/earth/ecosystems.html&edu=elem (accessed February 21, 2008).

University Corporation for Atmospheric Research. "Hurricanes Are Getting Stronger, Study Says." http://www.ucar.edu/news/releases/2005/hurricanestudy.shtml (accessed March 29, 2008).

University of Arizona. "Anthropogenic Desertification vs. 'Natural' Climate Trends." August 10, 1997. http://ag.arizona.edu/~lmilich/desclim.html (accessed April 21, 2008).

University of Arizona. "Biosphere 2." http://www.b2science.org/ (accessed April 28, 2008).

University of California Berkeley. "Biostratigraphy: William Smith." 2006. (accessed March 26, 2008).

University of California Berkley: College of Natural Resources. "Can Organic Farming Feed the World?" http://www.cnr.berkeley.edu/˜christos/articles/cv_organic_farming.html (accessed April 1, 2008).

University of California Museum of Paleontology. "The Desert Biome." http://www.ucmp.berkeley.edu/exhibits/biomes/deserts.php (accessed February 14, 2008).

University of California Museum of Paleontology. "The Forest Biome." http://www.ucmp.berkeley.edu/exhibits/biomes/forests.php (accessed March 4, 2008).

University of California Museum of Paleontology. "The Freshwater Biome." http://www.ucmp.berkeley.edu/exhibits/biomes/freshwater.php (accessed April 17, 2008).

University of California Museum of Paleontology. "The Grassland Biome." http://www.ucmp.berkeley.edu/exhibits/biomes/grasslands.php (accessed April 2, 2008).

University of California Museum of Paleontology. "Tour of Geologic Time." http://www.ucmp.berkeley.edu/exhibits/geologictime.php (accessed March 29, 2008).

University of California Museum of Paleontology. "What Killed the Dinosaurs?" http://www.ucmp.berkeley.edu/diapsids/extinction.html (accessed May 1, 2008).

University of Florida. "What Is Environmental Horticulture?" http://hort.ifas.ufl.edu/aboutus/whatis.htm (accessed April 20, 2008).

University of Michigan. "Ecological Communities: Networks of Interacting Species." November 2, 2005. http://www.globalchange.umich.edu/globalchange1/current/lectures/ecol_com/ecol_com.html (accessed February 11, 2008).

University of Minnesota. "Ware GW, Whiteacre GM, An Introduction to Insecticides, 4th ed.

(extracted from The Pesticide Book).” 2004 http://ipmworld.umn.edu/chapters/ware.htm (accessed March 27, 2008).

University of South Dakota. “The Dust Bowl.” http://www.usd.edu/anth/epa/dust.html (accessed March 27, 2008).

University of the Virgin Islands. “An Introduction to Coral Reefs.” http://www.uvi.edu/coral.reefer/ (accessed March 19, 2008).

University of Washington, TERIS (Teratogen Information System) Program. “Clinical Teratology Web.” http://depts.washington.edu/terisweb/ (accessed February 17, 2008).

University of Wollongong. “Introduction to Tsunami.” http://www.uow.edu.au/science/eesc/research/UOW002909.html (accessed February 29, 2008).

Uranium Information Centre Limited. Melbourne, Australia. “Clean Coal” Technologies.” February 2008. http://www.uic.com.au/nip83.htm (accessed March 1, 2008).

U.S. Agency for Toxic Substances & Disease Registry. “2007 CERCLA Priority List of Hazardous Substances.” http://www.atsdr.cdc.gov/cercla/07list.html (accessed July 10, 2008).

U.S. Climate Change Science Program. “Temperature Trends in the Lower Atmosphere—Understanding and Reconciling Differences [Executive Summary].” http://www.climatescience.gov/Library/sap/sap1-1/finalreport/sap1-1-final-execsum.pdf (accessed May 12, 2008).

U.S. Congress. “Alaska National Interest Lands Conservation Act.” December 2, 1980. http://www.r7.fws.gov/asm/anilca/title01.html#101 (accessed April 20, 2008).

U.S. Congress. “Marine Debris Research Prevention and Reduction Act,” February 10, 2005. http://www.govtrack.us/congress/billtext.xpd?bill=s109-362 (accessed April 10, 2008).

U.S. Congress. “National Environmental Policy Act of 1969.” Washington, DC: U.S. Congress, 1969. http://ceq.eh.doe.gov/nepa/regs/nepa/nepaeqia.htm (accessed April 10, 2008).

U.S. Department of Agriculture. “Agricultural Chemical Usage.” http://www.nass.usda.gov/Statistics_by_State/Pennsylvania/Publications/Annual_Statistical_Bulletin/2006_2007/nur_chemuse.pdf (accessed March 25, 2008).

U.S. Department of Agriculture. “National Invasive Species Information Center.” http://www.invasivespeciesinfo.gov/ (accessed May 5, 2008).

U.S. Department of Agriculture Natural Resources Conservation Service. “What Is Soil?” http://soils.usda.gov/education/facts/soil.html (accessed April 10, 2008).

U.S. Department of Agriculture and U.S. Department of Energy. “Biomass as Feedstock for a Bioenergy and Bioproducts Industry: The Technical Feasibility of a Billion-Ton Annual Supply.” www1.eere.energy.gov/biomass/ pdfs/final_billionton_vision_report2.pdf (accessed April 10, 2008).

U.S. Department of Energy. “Carbon Sequestration.” http://cdiac2.esd.ornl.gov/ (accessed April 14, 2008).

U.S. Department of Energy “DOE to Invest up to $13.7 Million in 11 Solar Cell Projects.” March 12, 2008. www.eere.energy.gov/news/news_detail.cfm/news_id=11638 (accessed March 19, 2008).

U.S. Department of Energy. “Ocean Wave Power.” http://www.eere.energy.gov/consumer/renewable_energy/ocean/index.cfm/mytopic=50009 (accessed May 14, 2008).

U.S. Department of Energy. “Yucca Mountain Repository.” http://www.ocrwm.doe.gov/ym_repository/index.shtml#skiptop (accessed April 16, 2008).

U.S. Department of the Interior. “Forest Resources of the United States.” October 2, 2007.http://nationalatlas.gov/articles/biology/a_forest.html (accessed April 15, 2008).

U.S. Environmental Protection Agency. “About EPA.” January 25, 2008. http://www.epa.gov/epahome/aboutepa.htm (accessed March 2, 2008).

U.S. Environmental Protection Agency. “Agriculture: Clean Water Act.” March 27, 2008. http://www.epa.gov/oecaagct/lcwa.html#Summary (accessed April 28, 2008).

U.S. Environmental Protection Agency. “Air Trends: Basic Information.” April 8, 2008. http://www.epa.gov/airtrends/sixpoll.html (accessed April 20, 2008).

U.S. Environmental Protection Agency. “Aircraft Contrails Factsheet.” September 2000. http://www.epa.gov/otaq/regs/nonroad/aviation/contrails.pdf (accessed March 19, 2008).

U.S. Environmental Protection Agency. “Aquatic Biodiversity.” http://www.epa.gov/bioindicators/aquatic (accessed April 17, 2008).

U.S. Environmental Protection Agency. “Asbestos.” February 4, 2008. http://www.epa.gov/asbestos

(accessed February 13, 2008).

U.S. Environmental Protection Agency. "Atmospheric Changes." 2007. http://www.epa.gov/climatechange/science/recentac.html (accessed March 26, 2008).

U.S. Environmental Protection Agency. "Carbon Dioxide: Greenhouse Gas Emissions." http://www.epa.gov/climatechange/emissions/co2.html (accessed April 23, 2008).

U.S. Environmental Protection Agency. "Carbon Sequestration in Agriculture and Forestry." http://www.epa.gov/sequestration/faq.html (accessed April 14, 2008).

U.S. Environmental Protection Agency. "The Clean Air Act of 1970." http://www.epa.gov/history/topics/caa70/11.htm (accessed April 20, 2008).

U.S. Environmental Protection Agency. "Clean Air Mercury Rule." http://www.epa.gov/camr/index.htm (accessed May 2, 2008).

U.S. Environmental Protection Agency. "Clean Energy, Basic Information." December 28, 2007. http://www.epa.gov/solar/basic-information.html (accessed March 1, 2008).

U.S. Environmental Protection Agency. "Climate Change." March 14, 2008. http://epa.gov/climatechange/index.html (accessed March 24, 2008).

U.S. Environmental Protection Agency. "Coastal Zones and Sea Level Rise." December 20, 2007. http://www.epa.gov/climatechange/effects/coastal/index.html (accessed February 8, 2008).

U.S. Environmental Protection Agency. "Ecosystems." February 20, 2008. http://www.epa.gov/ebtpages/ecosystems.html (accessed February 21, 2008).

U.S. Environmental Protection Agency. "EPA Establishes Hazardous Waste Enforcement and Emergency Response System; Names 60 New Sites." July 11, 1979. http://www.epa.gov/history/topics/hazard/01.htm (accessed April 15, 2008).

U.S. Environmental Protection Agency. "Fact Sheet—EPA Proposes Options for Significantly Reducing Mercury Emissions from Electric Utilities." http://www.epa.gov/oar/mercuryrule/hg_factsheet1_29_04.htm (accessed May 2, 2008).

U.S. Environmental Protection Agency. "The Great Lakes." March 4, 2008. (accessed April 16, 2008).

U.S. Environmental Protection Agency. "Greenhouse Gas Impacts of Expanded Renewable and Alternative Fuels Use." April, 2007. http://earth1.epa.gov/otaq/renewablefuels/420f07035.htm (accessed March 8, 2008).

U.S. Environmental Protection Agency. "Ground Water Rule." November 21, 2006. http://www.epa.gov/safewater/disinfection/gwr/index.html (accessed April 19, 2008).

U.S. Environmental Protection Agency. "Introduction to the Clean Water Act." October 28, 2002. http://www.epa.gov/watertrain/cwa (accessed April 20, 2008).

U.S. Environmental Protection Agency (EPA). "The Love Canal Tragedy: EPA History." January 1979. http://www.epa.gov/history/topics/lovecanal/01.htm (accessed March 17, 2008).

U.S. Environmental Protection Agency. "Medical Waste." http://www.epa.gov/epaoswer/other/medical/ (accessed May 10, 2008).

U.S. Environmental Protection Agency. "Medical Waste Tracking Act (1988)." http://www.epa.gov/epaoswer/other/medical/mwpdfs/mwta.pdf (accessed May 10, 2008).

U.S. Environmental Protection Agency. "National Estuaries Program." January 16, 2008. http://www.estuaries.gov/about.html (accessed March 31, 2008).

U.S. Environmental Protection Agency. "Oil Program." http://www.epa.gov/region09/waste/sfund/oilpp (accessed April 21, 2008).

U.S. Environmental Protection Agency. "Oil Spills." 2008. http://www.epa.gov/oilspill/ (accessed March 24, 2008).

U.S. Environmental Protection Agency. "Pharmaceuticals and Personal Care Products (PPCPs)." December 14, 2007. http://epa.gov/ppcp/faq.html (accessed April 19, 2008).

U.S. Environmental Protection Agency. "The Plain English Guide to the Clean Air Act." June 11, 2007. http://www.epa.gov/air/caa/peg (accessed April 20, 2008).

U.S. Environmental Protection Agency. "Polluted Runoff (Nonpoint Source Pollution)." February 25, 2008. http://www.epa.gov/owow/nps/ (accessed March 25, 2008).

U.S. Environmental Protection Agency. "Protecting Water Quality from Agricultural Runoff." March 2005. http://www.epa.gov/owow/nps/Ag_Runoff_Fact_Sheet.pdf (accessed March 28, 2008).

U.S. Environmental Protection Agency. "Smog/Regional Transport of Ozone." http://www.epa.gov/airmarkets/envissues/smog.html (accessed March 16, 2008).

U.S. Environmental Protection Agency. "Summary of the Endangered Species Act." March 6, 2008. http://www.epa.gov/regulations/laws/esa.html (accessed May 2, 2008).

U.S. Environmental Protection Agency. "Superfund." http://www.epa.gov/superfund/l (accessed March 29, 2008).

U.S. Environmental Protection Agency. "Threats from Oil Spills." http://www.epa.gov/emergencies/content/learning/effects.htm (accessed April 21, 2008).

U.S. Environmental Protection Agency. "Waste Transfer Stations." http://www.epa.gov/epaoswer/non-hw/transfer.htm (accessed March 21, 2008).

U.S. Environmental Protection Agency. "Water (from 'The Challenge of the Environment: A Primer on EPA's Statutory Authority')." September 21, 2007. http://www.epa.gov/history/topics/fwpca/05.htm (accessed April 28, 2008).

U.S. Environmental Protection Agency. "WaterSense®." January 25, 2008 http://www.epa.gov/watersense/water/index.htm (accessed March 23, 3008).

U.S. Environmental Protection Agency. "Watershed Academy Web: Introduction to the Clean Water Act." March 26, 2008. www.epa.gov/watertrain/cwa/ (accessed April 28, 2008).

U.S. Environmental Protection Agency. "What Is Acid Rain?" 2007 http://www.epa.gov/acidrain/what/index.html (accessed August 16, 2007).

U.S Environmental Protection Agency. "Water Pollution" http://www.epa.gov/ebtpages/watewaterpollution.html (accessed May 11, 2008).

U.S. Fish and Wildlife Service. "Coastal Program." September 27, 2006. http://ecos.fws.gov/coastal/viewContent.do?viewPage=home (accessed February 1, 2008).

U.S. Fish and Wildlife Service. "Digest of Federal Resource Laws of Interest to the U.S. Fish and Wildlife Service: Federal Water Pollution Control Act (Clean Water Act)." http://www.fws.gov/laws/lawsdigest/FWATRPO.HTML (accessed April 28, 2008).

U.S. Fish and Wildlife Service. "Endangered Species Act of 1973." http://www.fws.gov/endangered/esa/content.html (accessed May 2, 2008).

U.S. Fish and Wildlife Service. "The Endangered Species Act and What We Do." http://www.fws.gov/endangered/whatwedo.html (accessed May 10, 2008).

U.S. Fish and Wildlife Service. "The Mexican Wolf Recovery Program." March 11, 2008. http://www.fws.gov/southwest/es/mexicanwolf/ (accessed March 15, 2008).

U.S. Forest Service. "Fire and Aviation Management." http://www.fs.fed.us/fire/ (accessed March 14, 2008).

U.S. Geological Survey. "Bioremediation: Nature's Way to a Cleaner Environment." June 28, 2006. http://water.usgs.gov/wid/html/bioremed.html (accessed February 7, 2008).

U.S. Geological Survey. "Decades Required for Natural Processes to Clean Wastewater-Contaminated Ground Water." December 3, 2007. http://toxics.usgs.gov//highlights/gw_cessation.html (accessed April 19, 2008).

U.S. Geological Survey."Desertification." http://pubs.usgs.gov/gip/deserts/desertification/ (accessed March 29, 2008).

U.S. Geological Survey. "Earth's Water: Runoff." August 30, 2005. http://ga.water.usgs.gov/edu/runoff.html (accessed March 25, 2008).

U.S. Geological Survey. "Earthquake Hazards Program." http://earthquake.usgs.gov/ (accessed April 7, 2008).

U.S. Geological Survey. "Endangered Ecosystems of the United States: A Preliminary Assessment of Loss and Degradation." http://biology.usgs.gov/pubs/ecosys.htm (accessed February 27, 2008).

U.S. Geological Survey. "Energy Research and News." April 16, 2008. http://energy.usgs.gov (accessed April 16, 2006).

U.S. Geological Survey. "Flood Hazards: A National Threat." January 2006. http://pubs.usgs.gov/fs/2006/3026/2006-3026.pdf (accessed April 7, 2008).

U.S. Geological Survey. "Freshwater-Saltwater Interactions along the Atlantic Coast." http://water.usgs.gov/ogw/gwrp/saltwater/salt.html (accessed April 3, 2008).

U.S. Geological Survey. "Geographic Information Systems." http://erg.usgs.gov/isb/pubs/gis_poster/#what (accessed February 26, 2008).

U.S. Geological Survey. "Geologic Time: Online Edition." http://pubs.usgs.gov/gip/geotime (accessed March 29, 2008).

U.S. Geological Survey. "Ground Water and Surface Water a Single Resource." September 1, 2005.

http://pubs.usgs.gov/circ/circ1139 (accessed April 19, 2008).

U.S. Geological Survey. "Industrial Water Use." August 30, 2005. http://ga.water.usgs.gov/edu/wuin.html (accessed April 20, 2008).

U.S. Geological Survey. "Irrigation Techniques." http://ga.water.usgs.gov/edu/irmethods.html (accessed April 25, 2008).

U.S. Geological Survey. "Landslides Hazard Program." http://landslides.usgs.gov/ (accessed May 10, 2008).

U.S. Geological Survey. "New Polar Bear Findings." http://www.usgs.gov/newsroom/special/polar_bears/default.asp (accessed May 10, 2008).

U.S. Geological Survey. "Recent Highlights: Natural Resources." January 29, 2003. http://www.usgs.gov/themes/FS-187-97/ (accessed April 20, 2008).

U.S. Geological Survey. "Sustainability of Ground-Water Resources." October 30, 2007. http://pubs.usgs.gov/circ/circ1186 (accessed April 19, 2008).

U.S. Government Accountability Office. "Natural Gas Flaring and Venting: Opportunities to Improve Data and Reduce Emmissions." August 13, 2004. http://www.gao.gov/htext/d04809.html (accessed May 2, 2008).

U.S. Government Printing Office. "National Environmental Policy Act of 1969." http://frwebgate.access.gpo.gov/cgi-bin/getdoc.cgi?dbname=browse_usc&docid=Cite:+42USC4321 (accessed April 16, 2008).

U.S. National Forest Service. "Success Stories: ATVs and the Forest Service." May 18, 2004. http://www.fs.fed.us/plan/par/2003/success/stories/atv_fs.shtml (accessed April 16, 2008).

U.S. News and World Report. "Green Isn't Gold for MBAs." January 15, 2008. http://www.businessweek.com/bschools/content/jan2008/bs20080115_911253.htm?chan=bschools_bschool+index+page_finding+a+job (accessed March 2, 2008).

U.S. State Department. "Senior Officials' Meeting of the Coral Triangle Initiative on Coral Reefs, Fisheries, and Food Security." December 7, 2007. http://www.state.gov/g/oes/rls/rm/2007/96747.htm (accessed February 25, 2008).

U.S. Supreme Court. *United States v. Standard Oil.* May 23, 1966. http://caselaw.lp.findlaw.com/scripts/getcase.pl?court=us&vol=384&invol=224 (accessed January 16, 2008).

Victoria Experimental Network Under the Sea (VENUS). April 8, 2008. http://www.venus.uvic.ca (accessed April 21, 2008).

Virginia Tech Civil Engineering Department. "Groundwater Primer." June 7, 1998. http://www.cee.vt.edu/ewr/environmental/teach/gwprimer/gwprimer.html (accessed April 19, 2008).

Virtual Centre for Environmental Technology Exchange. "Future Municipal Solid Waste Treatment Technologies." http://www.apec-vc.or.jp/e/modules/tinyd00/index.php?id=122&kh_open_cid_00=11 (accessed March 27, 2008).

Wall Street Journal. "The Price of Going Green." February 29, 2008. http://online.wsj.com/article/SB120424591916201491.html (accessed March 2, 2008).

Washington Post. "Corporate Green." May 11, 2005. http://www.washingtonpost.com/wp-dyn/content/article/2005/05/10/AR2005051001182.html (accessed March 2, 2008).

Washington Post. "The Equator Principles." Updated January 25, 2008. http://www.equator-principles.com/index.html (accessed March 2, 2008).

Washington State Department of Ecology. "Marine Waters." http://www.ecy.wa.gov/programs/eap/mar_wat/mwm_intr.html (accessed February 28, 2008).

WaterCare. "Algal Blooms." http://www.water-care.net/wll/himp-algalblooms.html (accessed February 7, 2008).

Waterfootprint.org. "Water Footprint." http://www.waterfootprint.org/?page=files/home (accessed March 23, 2008).

We Campaign. "We Can Solve the Climate Crisis." http://www.wecansolveit.org/ (accessed May 2, 2008).

West Virginia State Archives. "The Buffalo Creek Disaster." http://www.wvculture.org/hiStory/buffcreek/bctitle.html (accessed March 4, 2008).

Windmill World. "History of Windmills." February 24, 2004. http://www.windmillworld.com/windmills/history.htm (accessed February 29, 2008).

Wisconsin Paper Council. "Paper in Wisconsin." http://www.wipapercouncil.org/fun3.htm (accessed March 20, 2008).

Wiser Earth. "Petroleum in the Environment." May 31, 2007. http://www.wiserearth.org/aof/114 (accessed April 16, 2006).

Woods Hole Oceanographic Institution. "Common Misconceptions About Abrupt Climate Change." http://www.whoi.edu/page.do?pid=12455&tid=282&cid=10149#ocean_9 (accessed March 29, 2008).

Woods Hole Oceanographic Institution. "Harmful Algae and Red Tides." http://www.whoi.edu/page.do?pid=11913 (accessed April 2, 2008).

Woods Hole Oceanographic Institution. "News Release: Effects of Ocean Fertilization with Iron to Remove Carbon Dioxide from the Atmosphere Reported." http://www.whoi.edu/page.do?pid=9779&tid=282&cid=886&ct=162 (accessed April 1, 2008).

Woods Hole Oceanographic Institution. "Rising Sea Levels and Moving Shorelines." http://www.whoi.edu/page.do?pid=12457&tid=282&cid=2484 (accessed April 2, 2008).

Woods Hole Oceanographic Institution. "A Scientific Critique of Iron Fertilization as Climate Change Mitigation Strategy." http://www.whoi.edu/cms/files/final_iron_fertilisation_critique_27223.pdf (accessed April 1, 2008).

World Commission on Dams. "Dams and Development: A New Framework for Decision-Making." http://www.dams.org//docs/report/wcdreport.pdf (accessed March 29, 2008).

World Health Organization. "Asbestos." http://www.euro.who.int/document/aiq/6_2_asbestos.pdf (accessed February 12, 2008).

World Health Organization. "Medical Waste." http://www.who.int/topics/medical_waste/en/ (accessed May 10, 2008).

World Health Organization. "Safe Management of Wastes from Health-Care Activities." http://www.healthcarewaste.org/en/documents.html?id=1 (accessed May 10, 2008).

World Health Organization. "Water Resource Quality." http://www.who.int/water_sanitation_health/resources/resquality/en/index.html (accessed April 20, 2008).

World Health Organization. "Water, Sanitation, and Health: Wastewater Use." http://www.who.int/water_sanitation_health/wastewater/en/ (accessed March 10, 2008).

World Health Organization. "Water, Sanitation, and Hygiene" http://www.who.int/water_sanitation_health/en/ (accessed May 11, 2008).

World Resources Institute. "Ecosystem Area: Open Shrublands." http://earthtrends.wri.org/searchable_db/index.php?theme=9&variable_ID=756&action=select_countries (accessed April 28, 2008).

World Wildlife Fund. "Endangered Species." http://www.worldwildlife.org/endangered/ (accessed May 10, 2008).

World Wildlife Fund. "Grasslands." October 19, 2006. http://www.panda.org/news_facts/education/middle_school/habitats/grasslands/index.cfm (accessed April 2, 2008).

World Wildlife Fund. "Problems: Poorly Managed Fishing." February 29, 2008. http://www.panda.org/about_wwf/what_we_do/marine/problems/problems_fishing/index.cfm (accessed March 24, 2008).

World Wildlife Fund. "Wildlife Trade: About CITES." http://www.worldwildlife.org/trade/cites/about.cfm (accessed February 1, 2008).

WWF-India & Wetlands International-South Asia. "Post Tsunami Conversation Issues and Challenges Consultative Meeting for Coordinated Action." http://assets.wwfindia.org/downloads/final_report_may2005.doc (accessed February 29, 2008).

Index

Page numbers in **boldface** indicate the main essay for a topic, and primary source page numbers are in ***boldface and italics***. An *italicized* page number indicates a photo, illustration, chart, or other graphic.

A

Abatement, greenhouse gas. *See* Greenhouse gases
Ablation, glacial. *See* Glacial retreat
Abyssal zone, 2:542–543, 632
Accidents, nuclear power plant, 2:610, 611, 612–614
    See also Ecodisasters
Accidents Dim Hopes For Green Nuclear Option, 2:***613–614***
Accumulation, glacial. *See* Glaciation
Acid rain, 1:**1–4,** *2*
    aquatic ecosystems, 1:32
    Clean Air Act of 1970, 1:109, 110
    ecodisasters, 1:231
    fossil fuel combustion impacts, 1:327, *328*
    freshwater and freshwater ecosystems, 1:330
    habitat loss, 1:410
    industrial pollution, 1:455, 457
    lakes, 1:515–516
    mining and quarrying impacts, 2:572
    nonpoint-source pollution, 2:601
    precipitation, 2:674, 675
    war and conflict-related environmental destruction, 2:825
    water pollution, 2:839–840
Acidity
    acid rain, 1:1, 2
    bays and estuaries, 1:53
    biodiversity, 1:60, 62
    carbon dioxide, 1:83
    carbon dioxide emissions, 1:86
    coral reefs and corals, 1:170, 172
    real-time monitoring and reporting, 2:688
    reef ecosystems, 2:704
    soil chemistry, 2:743
    Woods Hole Oceanographic Institution, 2:892
Action Plan for the Human Environment, 1:212, 215, 478, 2:800–801
Activism, environmental. *See* Environmental activism
Activism, United Nations. *See* United Nations policy and activism
Adams, Ansel, 1:167, 2:669, 670
Adams, E. M., 2:736
Adaptation
    coastal ecosystems, 1:151, 153
    coastal zones, 1:154
    drought, 1:205
    ecological competition, 1:237
    ecosystems, 1:242–243, 244, 245
    endangered species, 1:262
    genetic diversity, 1:335
    habitat alteration, 1:406
    habitat loss, 1:411
    ocean tides, 2:627
    predator-prey relations, 2:677
Advectional inversions, 1:47
    See also Atmospheric inversions
Aeolian dust. *See* Particulates
Aeration, 1:77
Aeromagnetic surveys, 2:767
Aeroradiometric surveys, 2:767, 768
Aerosol, Radiation, and Cloud Processes Affecting Arctic Climate Change Project, 1:37–38
Aerosols
    atmosphere, 1:*42*
    aviation emissions, 1:49, 50
    climate change, 1:130
    radiative forcing, 2:680, 681
    smog, 2:739
    volcanoes, 2:819
Afforestation. *See* Reforestation
Afghanistan, 2:825
Africa
    biodiversity, 1:61, 64
    biofuels, 1:68
    Clean Development Mechanism, 1:115, 117
    DDT, 1:188
    deforestation, 1:189
    desertification, 1:195
    deserts, 1:197
    dust storms, 1:207
    ecodisasters, 1:232–233
    electronics waste, 1:253
    floods, 1:315
    forests, 1:324
    hazardous waste, 1:414
    inland fisheries, 1:467
    invasive species, 1:482, 483
    irrigation, 1:494
    sustainable development, 2:770–771, 772–774
    volcanoes, 2:820
    war and conflict-related environmental destruction, 2:825, 828
    water supply and demand, 2:850

Agar, 1:509–510
Agarose, 1:509–510
Agassiz, Louis, 1:362
Agenda 21
  Earth Summit (1992), 1:212, 213
  Earth Summit (2002), 1:215, 216
  international environmental law, 1:478, 479
  Our Common Future, 2:814–815
Agent Orange, 1:418
Ages, geologic. *See* Geologic history
Agribusiness. *See* Agricultural practice impacts
Agricultural practice impacts, 1:**5–7**
  alternative fuel impacts, 1:15, 16, 19
  aquifers, 1:34–35
  atmosphere, 1:42
  atmospheric inversions, 1:48
  biodiversity, 1:62, 63
  biofuels, 1:68, 70–71, 72
  bioremediation, 1:78
  Bureau of Land Management, 1:81–82
  carbon dioxide emissions, 1:86
  carbon sequestration, 1:90, 91
  clear-cutting, 1:125, 127
  climate change, 1:130, 133, 134, 135
  coastal ecosystems, 1:152
  coastal zones, 1:155
  cultural practices and environment destruction, 1:179
  DDT, 1:186–188
  deforestation, 1:189
  desertification, 1:194, 195, 196
  deserts, 1:*198*
  drainage basins, 1:201
  drought, 1:205, 206
  dust storms, 1:208
  ecodisasters, 1:232
  ecological competition, 1:236, 238
  ecosystem diversity, 1:241
  ecosystems, 1:245–246
  El Niño and La Niña, 1:251
  endangered species, 1:259, 260
  estuaries, 1:297
  factory farms, 1:302–305
  forests, 1:324
  global warming, 1:367
  grasslands, 1:372, 373
  greenhouse effect, 1:387
  groundwater, 1:397, 398
  groundwater quality, 1:402, *402*
  habitat alteration, 1:404, 405
  herbicides, 1:418–419
  human impacts, 1:423, 425
  hunting practices, 1:427
  ice ages, 1:450
  industrial water use, 1:459
  insecticide use, 1:469–471
  irrigation, 1:493–495
  landslides, 1:528
  natural reserves and parks, 2:593
  organic and locally grown foods, 2:645–648
  organic farming, 2:649–651
  Our Common Future, 2:814
  pollinators, 2:673
  precipitation, 2:674–675
  rain forest destruction, 2:686–687
  red tide, 2:700
  reforestation, 2:706–707, 708
  runoff, 2:716, 718
  saltwater intrusion, 2:720
  Silent Spring, 2:734
  soil chemistry, 2:743, 744
  soil resources, 2:748–749
  solid waste treatment technologies, 2:753
  water conservation, 2:835
  water pollution, 2:838
  water supply and demand, 2:849, 851, 852
  watersheds, 2:856–857
Agricultural Research Service, 1:512
An Agricultural Testament (Howard), 2:650
Agronomy, 1:420
Air. *See* Atmosphere
Air pollution, 1:**8–10,** *9*
  acid rain, 1:1–2
  agricultural practice impacts, 1:6
  Arctic, 1:36–38
  atmospheric inversions, 1:47, *48*
  aviation emissions, 1:49–51
  Clean Air Act of 1970, 1:108–111
  coal resource use, 1:147
  dust storms, 1:207–208
  ecodisasters, 1:231–232
  emissions standards, 1:254–258
  endangered species, 1:259
  Environmental Protection Agency, 1:286–287
  geospatial analysis, 1:352
  human impacts, 1:*424*
  Hurricane Katrina, 1:438
  hybrid vehicles, 1:441
  industrial pollution, 1:455, 456–457
  industrialization in emerging economies, 1:462–463
  international environmental law, 1:477
  mining and quarrying impacts, 2:570, 571
  nonpoint-source pollution, 2:601
  recreational use and environmental destruction, 2:690
  smog, 2:739–740
Air Pollution Control Act of 1955, 1:108
Air Quality Act of 1967, 1:108
Airborne dust. *See* Particulates
Aircraft, 1:49–51
Airplanes. *See* Aircraft
Alabama, 1:148, 437, 438, *440*
Alabama Power, 1:3
Alaska
  biodegredation, 1:57
  bioremediation, 1:76
  climate change, 1:134
  conservation, 1:168–169
  earthquakes, 1:218, 219
  ecodisasters, 1:230
  glacial retreat, 1:*358*
  green movement, 1:376
  lakes, 1:517
  oil spills, 2:*642,* 643, *644*
  tsunami impacts, 2:793
  tundra, 2:798
  wildlife refuge, 2:887
Alaskan National Interest Lands Conservation Act (ANILCA), 1:168–169, ***169***
Albedo, 1:37, 383, 388
Albrecht, Laudo, 1:360
Alcmaeon of Croton, 1:397
Aletsch Glacier, 1:359–360
ALF (Animal Liberation Front), 1:224, 225, 227–229
Alfisols, 2:748
Algae
  agricultural practice impacts, 1:5
  algal blooms, 1:11–13
  aquatic ecosystems, 1:32
  biodiversity, 1:61
  carbon sequestration, 1:90, *92*
  closed ecology experiments, 1:142
  coastal ecosystems, 1:151
  coral reefs and corals, 1:170
  fish farming, 1:307
  genetic diversity, 1:336
  greenhouse effect, 1:388
  habitat loss, 1:411
  marine ecosystems, 2:543, 544
  oceans and coastlines, 2:634
Algal blooms, 1:**11–13**
  agricultural practice impacts, 1:5
  coastal ecosystems, 1:152
  coastal zones, 1:155
  fish farming, 1:307
  genetic diversity, 1:336
  industrial water use, 1:460
  irrigation, 1:495
  nonpoint-source pollution, 2:602
  oceanography, 2:630
  oceans and coastlines, 2:634
  photography, 2:669
  predator-prey relations, 2:678

red tide, 2:699, 700
runoff, 2:717
Alien species (ecology). *See* Invasive species
All-terrain vehicles (ATVs), 2:690–691
Alliance of Automobile Manufacturers, 1:88
Alps, 1:359–360
Alternative fuel impacts, 1:**14–20,** *15, 17*
Energy Policy Act, 1:16–20
history, 1:14
laboratory methods in environmental science, 1:513
scientific importance, 1:14–16
See also Biodiesel; Biofuels
Altitude, 1:47, 51, 63
Aluminum
acid rain, 1:1
grasslands, 1:373
recycling, 2:693, 694, 695, 696
Amazon region, 1:*201*
agricultural practice impacts, 1:5
clear-cutting, 1:125, 126, 127
flood control and floodplains, 1:*312*
forests, 1:321
geospatial analysis, 1:*352*
logging, 1:537
mining and quarrying impacts, 2:*572*
photography, 2:669
rain forest destruction, 2:686
watersheds, 2:858
Ambrose, Frank, 1:228
Amchitka, 1:265–266
American Academy of Pediatrics, 1:113
American Civic Association, 2:585
American Electric Power, 1:2, 3, 4
American Nurses Association, 1:113
American Petroleum Institute, 1:177
American Public Health Association, 1:113
American Rivers, 2:*715*
American Society for Civil Engineers, 1:523
American Trader, 2:642
Amnesic shellfish poisoning, 2:700
AMO (Atlantic multidecadal oscillation), 1:432, 433
Amoco Cadiz (ship), 1:230, 2:636, 641
Amosite, 1:39
Amphibians, 1:30, 406, 497
Anadarko Petroleum Corporation, 1:*176*
Anastas, Paul, 2:667
Ancient climate. *See* Paleoclimate
Anderson, A. Scott, 1:395
Anderson, Julia, 1:234, 235
Andes Mountains, 1:359, 360, 363
Andirondack State Park, 2:590
Andisols, 2:748
ANILCA (Alaskan National Interest Lands Conservation Act), 1:168–169
Animal Liberation Front (ALF), 1:224, 225, 227–229, 378
Animal protection policies. *See* Wildlife protection policies and legislation
Animal refuge. *See* Wildlife refuge
Animal rights
eco-terrorism, 1:224, 225, 227–228
factory farms, 1:302, 303, *303*
hunting practices, 1:428–429
Animal waste
agricultural practice impacts, 1:5–6
biofuels, 1:67, 68, 69
factory farms, 1:302, 303
fish farming, 1:307, 308
groundwater, 1:397
groundwater quality, 1:*403*
surface water, 2:765
Animals
biodiversity, 1:60, 63
carbon dioxide, 1:83
CITES, 1:105–107
climate change, 1:133
coastal ecosystems, 1:150, 151, 152
coastal zones, 1:154–155
coral reefs and corals, 1:170–171
deserts, 1:197, 198
Earth Day, 1:*210*
ecosystems, 1:243, 244, 245
endangered species, 1:259–263
estuaries, 1:294
extinction and extirpation, 1:298–300
factory farms, 1:302–305
forests, 1:322, 323, 325
freshwater and freshwater ecosystems, 1:330, 331
global warming, 1:369
greenhouse effect, 1:383, 386
greenhouse gases, 1:390
habitat alteration, 1:404, 405, 406
habitat loss, 1:408, 409, 411–412
hunting practices, 1:427–429
ice ages, 1:448
invasive species, 1:481, 482, 483
IPCC 2007 Report, 1:486
marine ecosystems, 2:541, 542, 543, 544
migratory species, 2:568–569
North Atlantic Oscillation, 2:608
oceans and coastlines, 2:633
oil spills, 2:641, 643
overgrazing, 2:655–657
pharmaceutical development resources, 2:666, 667
pollinators, 2:672
reef ecosystems, 2:702–703
snow and ice cover, 2:741
spill remediation, 2:759
tundra, 2:798
wetlands, 2:864, 865
wildlife protection policies and legislation, 2:883–885, *884, 885*
wildlife refuge, 2:886–887
See also specific types of animals
Annan, Kofi, 2:575
Antarctic and Southern Ocean Coalition, 1:23
Antarctic Conservation Act of 1978, 1:26
Antarctic issues and challenges, 1:**21–24,** *23, 27*
Antarctic Treaty, 1:21, 22, 25–29, *27*
Arctic *vs.,* 1:36
atmospheric circulation, 1:44
biodiversity, 1:61, 63
carbon dioxide emissions, 1:85
chlorofluorocarbons, 1:103
deserts, 1:197
ecological competition, 1:238
ecosystem diversity, 1:240
glacial retreat, 1:357, 358–359
glaciation, 1:362, 363
global warming, 1:368, *370*
greenhouse effect, 1:386
greenhouse gases, 1:392
hydrologic cycle, 1:445
ice ages, 1:448, 449, 450
ice cores, 1:452
IPCC 2007 Report, 1:486
mining and quarrying impacts, 2:573
Montreal Protocol, 2:574, 575
natural reserves and parks, 2:591, 593
ozone hole, 2:658, 659, *659,* 660
ozone layer, 2:661
precipitation, 2:674
sea level rise, 2:721–722, 724, 725
snow and ice cover, 2:741, 742
tundra, 2:798–799
Antarctic Treaty, 1:21, 22, **25–29,** *27, 27–29*
Antarctica, 1:21–26, *349*
Anthers, 2:673
Anthracite, 1:145
Anthropogenic impacts. *See* Human impacts

Antibiotics
agricultural practice impacts, 1:5, 6
biodegradation, 1:57
factory farms, 1:302, 304
fish farming, 1:306, 308
genetic diversity, 1:336
industrialization in emerging economies, 1:463
marine fisheries, 2:547
water pollution, 2:840
Antoci, Mike, 2:648
ANWR. *See* Arctic National Wildlife Refuge (ANWR)
Apex predators, 2:677, 678
Aposematism, 2:677
Apple trees, 2:709
Aquaculture. *See* Fish farming
Aquatic ecosystems, 1:**30–32**
acid rain, 1:1, 2
algal blooms, 1:12
Antarctica, 1:21, 26
benthic ecosystems, 1:55–56
conservation, 1:167–168
ecosystems, 1:242, 245
habitat alteration, 1:405
industrial pollution, 1:457
industrial water use, 1:460, 461
invasive species, 1:481–482
lakes, 1:514–517
nonpoint-source pollution, 2:602
wetlands, 2:864
Aqueducts, 1:397
Aquifers, 1:**33–35,** *34*
carbon sequestration, 1:92
ecosystems, 1:245
greenhouse gases, 1:394–395
groundwater, 1:396–400
groundwater quality, 1:401, 403
runoff, 2:716, 717
saltwater intrusion, 2:719–720
water resources, 2:843
watersheds, 2:856
AR4. *See* IPCC 2007 Report
Arabian Desert, 2:825
Aragonite, 2:702, 704
Aral Sea, 1:232, 494, 2:844
Arboriculture, 1:420
Archaea, 1:56
Archer Daniels Midland, 1:68
Archimedes' screw, 1:310–311
Arctic darkening and pack-ice melting, 1:**36–38**
atmospheric circulation, 1:44, *45*
biodiversity, 1:61, 63, 64
climate change, 1:132
deserts, 1:197
ecosystem diversity, 1:240
global warming, 1:368
greenhouse effect, 1:386, 388
ice ages, 1:449
insecticide use, 1:471
IPCC 2007 Report, 1:486
ozone hole, 2:660
radiative forcing, 2:680, 681
sea level rise, 2:722
snow and ice cover, 2:741, 742
tundra, 2:798–799
United Nations Convention on the Law of the Sea, 2:804–805
wildlife protection policies and legislation, 2:885
Arctic National Wildlife Refuge (ANWR), 1:81, 166, 2:586
Arctic Treaty, 1:26
Argentina, 1:22, 25, 27
Argo Merchant (ship), 2:641
Argon, 1:41
Arid regions, 1:207, 261, 495
Aridosols, 2:748
Aristotle, 2:628
Arizona, 1:88, 257, 2:*747*
Arms control, nuclear. *See* Nuclear test ban treaties
Arms Control Association, 2:616–617
Army Corps of Engineers, 1:221, 313, 2:866
Arrhenius, Svante, 1:83, 130, 474
Arsenic, 1:35, 327, 401, 403, 2:790
Arson. *See* Fires
Artesian springs and wells, 1:33, 397, 403, 2:843
Asbestos contamination, 1:**39–40,** 464
Asbestosis, 1:39
ASDSO (Association of State Dam Safety Officials), 1:184
Aseismic creep, 1:220
Ash, 1:8, 9
Ash, volcanic. *See* Volcanoes
Ash-borer beetles, 1:82
Ash melting, 2:753, 754
Asia
biodiversity, 1:64
biofuels, 1:68, 69, 71
deforestation, 1:189
desertification, 1:195
hazardous waste, 1:414
ice cores, 1:453
industrialization in emerging economies, 1:462
irrigation, 1:494
landslides, 1:528
North Atlantic Oscillation, 2:609
organic farming, 2:649
soot, 1:148
war and conflict-related environmental destruction, 2:825
Asian Brown Cloud, 2:739
Asphyxiates, 1:273
Aspirin, 2:666, 667
Asplund, Ilse Washington, 1:229
Assessment Report 4 (2007). *See* IPCC 2007 Report
Association of California Water Agencies, 2:851
Association of State Dam Safety Officials (ASDSO), 1:184
Asteroids, 1:61, 298, *300,* 2:819
At-risk species. *See* Endangered species
Atacama Desert, 1:197
Athens, Greece, 1:*187,* 522, *525*
Atlantic conveyor belt. *See* Thermohaline circulation
Atlantic multidecadal oscillation (AMO), 1:432, 433
Atlantic Ocean
atmospheric circulation, 1:44
biodiversity, 1:63
commercial fisheries, 1:160–161
El Niño and La Niña, 1:250
geologic history, 1:349
hurricanes, 1:430, 432, 433
hydrologic cycle, 1:447
ice ages, 1:451
mathematical modeling and simulation, 2:556–557
North Atlantic Oscillation, 2:608–609
ocean circulation and currents, 2:620–621, 622
ocean salinity, 2:624
overfishing, 2:652, 653
red tide, 2:699
tsunami impacts, 2:793, 794
Atlases. *See* Maps and atlases
Atmosphere, 1:**41–43**
acid rain, 1:1–2
aerosols, 1:*42*
air pollution, 1:8, 9
alternative fuel impacts, 1:14
biodegradation, 1:57
biodiversity, 1:62
carbon dioxide, 1:83, 84
carbon dioxide emissions, 1:85–86, 88
chlorofluorocarbons, 1:100, 104
climate modeling, 1:139–140
El Niño and La Niña, 1:247–251
Gaia hypothesis, 1:332–334
geochemistry, 1:340
glaciation, 1:363
global warming, 1:366–370
greenhouse effect, 1:383–389
greenhouse gases, 1:390–394
hurricanes, 1:430–432
hydrologic cycle, 1:445
ice ages, 1:449
IPCC 2007 Report, 1:486

light pollution, 1:529, 530
Montreal Protocol, 2:574, 575
National Oceanic and Atmospheric Administration, 2:582–584
ozone hole, 2:658–660
ozone layer, 2:661–663
radiative forcing, 2:680–681
smog, 2:*739*
Atmospheric circulation, 1:**44–46,** *45*
global warming, 1:369
precipitation, 2:674
Atmospheric inversions, 1:**47–48,** *48,* 2:*739*
Atmospheric pollution. *See* Air pollution
Atmospheric pressure
atmospheric circulation, 1:45, 46
atmospheric inversions, 1:47
El Niño and La Niña, 1:247–248, 249, 251
hurricanes, 1:430
North Atlantic Oscillation, 2:608–609
precipitation, 2:674
wind and wind power, 2:888, 889
Atmospheric warming. *See* Global warming
Atolls, coral. *See* Coral atolls
Atomic bombs. *See* Nuclear weapons
ATVs (All-terrain vehicles), 2:690–691
Audubon (periodical), 2:736
Audubon, John James, 2:605–607
Audubon Society for the Protection of Birds. *See* National Audubon Society
Aurora australis, 1:42
Aurora borealis, 1:42
Australia
Antarctic Treaty, 1:25, 27
biodiversity, 1:60
coal resource use, 1:146
desertification, 1:195
El Niño and La Niña, 1:248, 249, 250
Kyoto Protocol, 1:500, 501
predator-prey relations, 2:678
reef ecosystems, 2:705
toxic waste, 2:790
Autism, 2:783–784
AUTOMAP (Automatic Mapping System), 1:342
Automatic Mapping System (AUTOMAP), 1:342
Automobiles. *See* Vehicles
Autotrophs, 1:142
Avalanches, 2:742
Avery, Susan, 2:891
Avian influenza, 1:302, 304
Aviation emissions, 1:**49–52**
Azolla, 1:483, *483*
Azores Islands, 2:608

B

Babbie, Jason, 1:443
Bacillus thuringiensis, 1:470
Bacteria
algal blooms, 1:11, 12
benthic ecosystems, 1:55, 56
biodegredation, 1:57–58
biofuels, 1:69, 70
bioremediation, 1:76–77, 78
carbon sequestration, 1:90
closed ecology experiments, 1:142
coastal ecosystems, 1:152
coastal zones, 1:155
ecological competition, 1:238
ecosystems, 1:242
factory farms, 1:303, 304
fish farming, 1:308
laboratory methods in environmental science, 1:509–510, 511, 512
marine water quality, 2:550, *550,* 551, 552
oceanography, 2:630
oil spills, 2:643
pharmaceutical development resources, 2:666
predator-prey relations, 2:676, 677
soil contamination, 2:745
spill remediation, 2:757–758
water pollution, 2:840
Bactrachochytrium, 1:62
Bags, shopping, 1:*378,* 379
Bahrain, 2:*627*
Baiji, 1:61
Baird's Beaked whales, 1:*428*
Baja, California, 1:*238*
Baker, Andrew, 1:263
Baker, Marc Andre, 1:229
Balcomb, Ken, 1:121
Bald eagles, 1:187, *187,* 188, 336–337, 2:643, 839
Baleen whales, 1:238, 259
Balfour, Eve, 2:649, 650
Bali, 1:*392*
Baliunas, S., 2:780
Balog, James, 2:670
Baltimore, Maryland, 1:113
Ban Ki Moon, 1:64, 65
Band of Mercy, 1:224
Bangladesh
aquifers, 1:35
floods, 1:*311*
groundwater quality, 1:401, 403
hurricanes, 1:434
inland fisheries, 1:467
marine fisheries, 2:*548*
Bans. *See* Government regulation
Bar-built estuaries, 1:294
Barred owls, 2:*882*
Barrier reefs, 2:703
Barros, Francisco Anselmo de, 1:*69*
Barroso, José Manuel, 1:360
Basel Convention, 1:253
Basin Electric Power Cooperative, 1:395
Basins, drainage. *See* Drainage basins
Batteries, 1:441
Bay grasses, 2:*857*
Bay of Fundy, 2:598
Bays and estuaries, 1:**53–54**
algal blooms, 1:11, 13
dredging, 1:202
ocean salinity, 2:624, 625
See also Estuaries
Beaches
coastal ecosystems, 1:150–151
hurricanes, 1:*432*
marine water quality, 2:*550,* 552
medical waste, 2:562
migratory species, 2:569
oceans and coastlines, 2:633
oil spills, 2:641
red tide, 2:*701*
Bean, Michael, 1:263
Becker, Dan, 1:380
Becquerel, Edmond, 2:750
Bedrock, 2:743, 748
Bedstraw, 1:237
Bees, 2:672, 673, *673*
Beetles, 1:82
Beijing, China, 1:9, 10, 462–463, *521*
Belgium, 1:25, 27
Bell, Alexander Graham, 1:14
Benett, Etheldred, 1:80
Bentgrass, 1:82
Benthic ecosystems, 1:**55–56,** *56*
freshwater and freshwater ecosystems, 1:330
marine ecosystems, 2:541–543, 544
pharmaceutical development resources, 2:667
tsunami impacts, 2:793, 794–795
United Nations Convention on the Law of the Sea, 2:804
Benthos. *See* Benthic ecosystems
Benzene, 1:416
Bergeson, Lynn, 1:3

Bering Sea, 1:*453*
Berry, Wendell, 2:645
Besley, Fred W., 1:317
Betsiboka River, 1:*297*
Beverage containers, 2:694, 696
Bhopal, India, 1:100, 231
Biesbosch delta, 2:725–728
Big horned sheep, 1:*82*
Bilanovic, Drago, 2:*817*
Bilharzia, 1:483
Binary plants, 1:355
Bio-shields, 2:796
Bioaccumulation, 1:187, 460, 469, 2:547, 839
Bioaugmentation, 1:77
Biochemical oxygen demand (BOD), 1:11, 12
Biochemistry, 1:510–511
Biochronostratigraphy, 1:79
Biodegredation, 1:**57–59**
- DDT, 1:187
- laboratory methods in environmental science, 1:509, 511

Biodiesel, 1:15, 67, 68, 70
Biodiversity, 1:**60–66**
- aquatic ecosystems, 1:30, 31–32
- biogeography, 1:74, 75
- carbon sequestration, 1:91
- clear-cutting, 1:125, 126, 127
- climate change, 1:60, 61–65, 133
- coastal ecosystems, 1:150, 152
- coastal zones, 1:154, 156
- conservation, 1:166, 168
- coral reefs and corals, 1:170
- deforestation, 1:189, 190
- desertification, 1:194, 195, 196
- deserts, 1:197–198
- Earth Summit (1992), 1:213
- ecosystem diversity, 1:239–241
- ecosystems, 1:243–244, 245
- endangered species, 1:261–262
- estuaries, 1:295
- extinction and extirpation, 1:299, 300
- forest resources, 1:319, 321, 323, 325
- forests, 1:318
- freshwater and freshwater ecosystems, 1:329, 330
- genetic diversity, 1:335
- global warming, 1:369
- habitat alteration, 1:404, 406
- habitat loss, 1:408
- human impacts, 1:60–61, 62, 423, 426
- invasive species, 1:481–482
- island ecosystems, 1:497
- lakes, 1:516
- logging, 1:536
- marine ecosystems, 2:541–542, 543
- natural reserves and parks, 2:590–591, 592
- natural resource management, 2:599
- oceans and coastlines, 2:633
- organic and locally grown foods, 2:645, 646
- pharmaceutical development resources, 2:666
- pollinators, 2:673
- predator-prey relations, 2:678
- rain forest destruction, 2:687
- reef ecosystems, 2:702–703, 704
- reforestation, 2:706, 708, 710
- resource extraction, 2:713
- species reintroduction programs, 2:755
- tundra, 2:798
- wetlands, 2:865
- wildlife population management, 2:880
- wildlife protection policies and legislation, 2:883–885

Biodynamic agriculture, 2:649
Bioengineering. *See* Genetic engineering
Biofuels, 1:**67–73**
- air pollution, 1:10
- alternative fuel impacts, 1:14, 15, 16, 19
- controversy, 1:67
- ethanol, 1:69–72
- history, 1:67–68
- technologies, 1:68–69

Biogas. *See* Methane
Biogeochemical cycles, 1:243, 2:557, 816
Biogeographical zones. *See* Ecozones
Biogeography, 1:**74–75,** 496–497
Bioinformatics, 1:511, 512
Biological cleanup. *See* Bioremediation
Biological communities, 1:239, 242
Biological diversity. *See* Biodiversity
Biological pump, 1:490
Biology
- computational, 1:511, 512
- molecular, 1:509–511

Bioluminescence, 1:32, 2:543, 632
Biomass
- Clean Development Mechanism, 1:116
- clean energy, 1:119
- forests, 1:325
- fossil fuel combustion impacts, 1:328
- genetic diversity, 1:336
- greenhouse effect, 1:386
- habitat loss, 1:410

Biomass methane. *See* Methane
Biomes
- benthic ecosystems, 1:55, 56
- biodiversity, 1:60, 62, 64
- deserts, 1:197
- ecosystems, 1:243
- forests, 1:321
- genetic diversity, 1:335
- natural reserves and parks, 2:593
- oceans and coastlines, 2:634
- vegetation cycles, 2:816

Biomolecules, 1:510
Bioremediation, 1:**76–78**
- biodegredation, 1:57–58
- Comprehensive Environmental Response, Compensation, and Liability Act, 1:163
- hazardous waste, 1:414
- laboratory methods in environmental science, 1:512–513
- Oil Pollution Acts, 2:637
- oil spills, 2:641, 643
- soil contamination, 2:745, 746
- solid waste treatment technologies, 2:754
- spill remediation, 2:757–758

BIOS 3, 1:142–143
Biosphere, 1:83, 332, 340, 449, 478, 2:816
Biosphere 2, 1:142, 143
Biostimulation, 1:77
Biostratigraphy, 1:**79–80,** 85
Biotic exchange, 1:63
Bird flu. *See* Avian influenza
Birds
- aquatic ecosystems, 1:30
- biodiversity, 1:63
- biogeography, 1:75
- coastal ecosystems, 1:150, 151
- ecological competition, 1:237, *238*
- ecosystems, 1:246
- endangered species, 1:260
- flood control and floodplains, 1:313
- insecticide use, 1:471
- light pollution, 1:528, 529
- marine ecosystems, 2:542
- migratory species, 2:568
- non-scientist contributions, 2:604, *605,* 605–607, *606*
- oil spills, 2:641, 642, 643
- predator-prey relations, 2:677–678
- seasonal migration, 2:730–731
- *Silent Spring,* 2:734–735
- spill remediation, 2:759
- tundra, 2:798
- wetlands, 2:864, 865
- wind and wind power, 2:889
- *See also specific types of birds*

Birds of America (Audubon), 2:605, *606–607*
Birth defects, 1:416, 471, 2:782–784
Biscayne Aquifer, 1:34
Bison, 1:*374,* 498, 2:593, 691, 881
Bivalves, 1:62
Bjerknes, Jacob, 1:251
Black, Joseph, 1:83
Black Blizzards, 1:208
Black carbon, 1:36
Black Sea, 1:237
Blacksmith Institute, 2:573
Blackwater Refuge, 2:887
Bleaching, coral. *See* Coral bleaching
Blizzards, 2:862
BLM. *See* Bureau of Land Management (BLM)
Bloeman, Peter, 2:726, 727
Bloomberg, Michael, 1:380, 443
Blue asbestos. *See* Crocidolite
Blue Source, LLC, 1:395
Blue whales, 2:544, 730
Bluefin tuna, 2:*547*
Blum, Barbara, 1:283–284
Boars, wild, 2:756
BOD. *See* Biochemical oxygen demand (BOD)
Bodman, Samuel, 1:*415*
Body waves (seismology), 1:220
Bogs, 1:330
Bolivia, 1:359, 2:*836*
Bombardier, Joseph-Armand, 2:690
Bombay, India. *See* Mumbai, India
Bonn Convention. *See* Convention on Migratory Species
Boreal forests
  biodiversity, 1:62
  ecodisasters, 1:234, 235
  as forests, 1:321, 323
  reforestation, 2:707, 708, 709–710
  tundra, 2:799
Borneo, 1:125, *190, 323, 324*
Bosnia, 2:825
Boston, Massachusetts, 1:522, 2:*740*
Bottle bills, 2:694, 696
Bottom trawling, 1:404–405, 2:547, 670
Bourke-White, Margaret, 2:669
Boutros-Ghali, Boutros, 1:*214*
Bovine spongiform encephalopathy (BSE), 1:305
Bowhead whales, 2:*677*
Brackish water, 1:294
Bradley, Raymond S., 2:780
Brand, Stewart, 2:612
Brazil
  agricultural practice impacts, 1:6
  alternative fuel impacts, 1:15
  atmospheric inversions, 1:48
  biodiversity, 1:64, 65
  biofuels, 1:68, 69, *69,* 70, 71
  clear-cutting, 1:125
  dams, 1:182
  deforestation, 1:190
  desertification, 1:195
  forests, 1:324
  invasive species, 1:483
  island ecosystems, 1:*497*
  logging, 1:537
  medical waste, 2:564
  natural resource management, 2:598, 599
  ocean circulation and currents, 2:*622*
  rain forest destruction, 2:686
Brevenal, 2:700–701
Brevetoxin, 2:699, 700–701
Britain. *See* United Kingdom
British Antarctic Survey, 1:456, 2:574
British Columbia, Canada, 1:125, 308, 536
Brittany, France, 1:230
Brock, John, 1:443
Bromine, 1:100, 152
Brown asbestos, 1:39
Brown tree snakes, 1:336, 410
Brownfields, 1:414
Brundtland, Gro Harlem, 2:813
Brundtland Commission. *See* United Nations World Commission on Environment and Development (WCED)
BSE (Bovine spongiform encephalopathy), 1:305
Buckland, William, 1:80, 345–346
Buffon, Georges Louis Leclerc de, 1:345
Building design, 1:220–221
Building materials, 1:39
Bureau of Biological Survey. *See* Fish and Wildlife Service (FWS)
Bureau of Fisheries, 2:736
Bureau of Land Management (BLM), 1:**81–82,** *82*
  natural reserves and parks, 2:590
  Wilderness Act of 1964, 2:873
Bureau of the Census, 1:341–342
Buried carbon emissions. *See* Carbon sequestration
Burnett, Sterling, 1:258
Burning. *See* Fires
Bush, George W.
  air pollution, 1:10
  alternative fuel impacts, *1:17*
  biodiversity, 1:64, 65
  carbon dioxide emissions, 1:87
  Clean Water Act, 1:122, 123
  Earth Day, 1:210
  Environmental Protection Agency, 1:287
  IPCC 2007 Report, 1:487
  Kyoto Protocol, 1:501
  National Environmental Policy Act, 2:578
  nuclear power, 2:611
  pollution control, 1:3
  recreational use and environmental destruction, 2:691
  United Nations Framework Convention on Climate Change, 2:807
Bushes, 2:732–733
Bushmeat, 2:593
Butterflies
  Karner Blue, 2:755
  Little Wood Satyr, 2:*569*
  Monarch, 2:568, *571,* 672, 673, 709, 730
Bycatch, 2:652, 653
Byrd-Hagel Resolution, 1:501

C

CAA. *See* Clean Air Act of 1970
Cacti, 1:*195,* 197, 198
Cadmium, 1:512, 2:791
Cairo, Egypt, 1:48
Calcifiers, 1:86
Calcium carbonate ($CaCO_3$)
  acid rain, 1:3
  biodiversity, 1:62
  carbon dioxide emissions, 1:86
  carbon sequestration, 1:91
  coral reefs and corals, 1:172
  reef ecosystems, 2:702, 704
Caldera, 2:819
California
  algal blooms, 1:13
  atmospheric inversions, 1:48
  Bureau of Land Management, 1:*82*
  carbon dioxide emissions, 1:87–88
  Clean Air Act of 1970, 1:*110*
  dams, 1:184
  drought, 1:205
  ecological competition, 1:*238*
  ecosystem diversity, 1:239–240
  El Niño and La Niña, 1:*249*
  emissions standards, 1:254, 255–258

geospatial analysis, 1:352
green movement, 1:380
groundwater, 1:398–399, *399*
hybrid vehicles, 1:442
industrial pollution, 1:*457*
irrigation, 1:494
landfills, 1:522
landslides, 1:527, *529*
logging, 1:*538*
marine water quality, 2:*550,* 552
National Environmental Policy Act, 2:578
oil spills, 2:642
organic and locally grown foods, 2:645, 647
smog, 2:739, 740
water supply and demand, 2:851–852
wildfires, 2:877, 878–879
California Air Resources Board (CARB), 1:256
California Clean Air for Life Campaign, 1:255
Callendar, Guy Stewart, 1:130
Calves. *See* Cattle
Calving (glaciers), 2:724
Cambodia, 2:825
Cambridge University, 1:51
The Camera (Adams), 2:670
Cameroon, 1:324
Camouflage, 2:677
Campylobacter, 1:304
CAMR (Clean Air Mercury Rule of 2005), 1:**112–114**
Can 'Green Chic' Save the Planet?, 1:*379–381*
Canada
acid rain, 1:1
air pollution, 1:10
algal blooms, 1:13
Arctic, 1:26
biofuels, 1:70
Clean Development Mechanism, 1:117
El Niño and La Niña, 1:251
emissions standards, 1:255, 257
endangered species, 1:262
environmental activism, 1:265–266
fish farming, 1:306, 308
Geographic Information Systems (GIS), 1:341
geospatial analysis, 1:351
green movement, 1:375–376
groundwater, 1:397
insecticide use, 1:471
international environmental law, 1:477
Kyoto Protocol, 1:500, 503
lakes, 1:514, 517
landfills, 1:526
migratory species, 2:569
natural resource management, 2:599
North Atlantic Oscillation, 2:608
ocean circulation and currents, 2:620
oceanography, 2:630
oil spills, 2:642
overfishing, 2:653
real-time monitoring and reporting, 2:688–689
recreational use and environmental destruction, 2:690
recycling, 2:697
reforestation, 2:706
resource extraction, 2:712, 713
species reintroduction programs, 2:755
surface water, 2:765, 766
tidal or wave power, 2:789
tundra, 2:798
United Nations Convention on the Law of the Sea, 2:804–805
water pollution, 2:838
weather extremes, 2:862
wetlands, 2:866
wildlife protection policies and legislation, 2:883, 885
Canadian Center for Climate Modeling, 2:861
Canadian lynx, 2:*579*
Cancer
chlorofluorocarbons, 1:102, 103
DDT, 1:187
ecodisasters, 1:232
lung, 1:36, 40
ozone hole, 2:658, 659–660
ozone layer, 2:662
toxic waste, 2:790
Canola oil, 1:68
Canopies, tree. *See* Forests
Cap-and-trade programs, 1:112, 257
Capacity, carrying. *See* Carrying capacity
Cape Cod, Massachusetts, 1:398
Captive breeding, 2:755
Capture fisheries, 1:467
CARB (California Air Resources Board), 1:256
Carbamates, 1:470
Carbofuran, 2:593
Carbohydrates, 1:242
Carbon
atmosphere, 1:41, *42*
aviation emissions, 1:49
biodegredation, 1:57–58
biofuels, 1:68
carbon dioxide, 1:83
carbon dioxide emissions, 1:85
carbon sequestration, 1:90
coal resource use, 1:145
ecosystems, 1:243
iron fertilization, 1:490–491
Carbon banking, 1:502
Carbon capture and storage (CCS). *See* Carbon sequestration
Carbon compounds. *See* Organic compounds
Carbon credits. *See* Pollution credits
Carbon cycle, 1:83, 90, 2:816, 817
Carbon dioxide (CO_2), 1:**83–84,** *84*
acid rain, 1:1, 2
agricultural practice impacts, 1:6
air pollution, 1:8, 9
alternative fuel impacts, 1:15, 16
atmosphere, 1:41
aviation emissions, 1:49
biodegredation, 1:57
biodiversity, 1:60, 62, 63
biofuels, 1:69
carbon sequestration, 1:90
Clean Air Act of 1970, 1:109
Clean Development Mechanism, 1:115–117
climate change, 1:128, 129, 130
climate modeling, 1:140
ecosystems, 1:243, 245
emissions, 1:85–88, 385
emissions standards, 1:254, 255
Environmental Protection Agency, 1:286–287
glaciation, 1:363
global warming, 1:366, 367, 368, 369
greenhouse effect, 1:383, 384, 386, 388
greenhouse gases, 1:390, 391, 392, 393, 394–395
horticulture, 1:420–421
hurricanes, 1:430
ice ages, 1:449, 450, 451
ice cores, 1:452, 454
Intergovernmental Panel on Climate Change, 1:474
IPCC 2007 Report, 1:486
iron fertilization, 1:490–491
Kyoto Protocol, 1:500
logging, 1:536–537
mining and quarrying impacts, 2:573
radiative forcing, 2:681
rain forest destruction, 2:687
reef ecosystems, 2:704
reforestation, 2:706
resource extraction, 2:713
tundra, 2:798, 799
vegetation cycles, 2:817
wildfires, 2:879
wind and wind power, 2:889
Carbon dioxide (CO_2) emissions, 1:**85–89**
acid rain, 1:2
air pollution, 1:9
alternative fuel impacts, 1:14, 15, 16

aviation emissions, 1:49, 51
biodiversity, 1:64
biofuels, 1:67, 68
carbon dioxide, 1:83
carbon sequestration, 1:90, 91–93
Clean Development Mechanism, 1:115–116
clean energy, 1:119, 121
climate change, 1:133
coal resource use, 1:147–148
coral reefs and corals, 1:170, 172
dams, 1:182
deforestation, 1:189, 190
dust storms, 1:207
ecodisasters, 1:234–235
emissions standards, 1:257–258
Environmental Protection Agency, 1:286–287
fossil fuel combustion impacts, 1:326, 327–328
geothermal resources, 1:354, 355
greenhouse effect, 1:388
greenhouse gases, 1:393–395
habitat alteration, 1:406
human impacts, 1:425–426
United Nations Framework Convention on Climate Change, 2:807

Carbon monoxide (CO)
air pollution, 1:9, 10
Clean Air Act of 1970, 1:109
emissions standards, 1:254

Carbon sequestration, 1:**90–94**
acid rain, 1:1–2
carbon dioxide emissions, 1:86
Clean Development Mechanism, 1:115
clean energy, 1:119
clear-cutting, 1:127
coal resource use, 1:147–148, 149
deforestation, 1:190
forests, 1:323, *323*
fossil fuel combustion impacts, 1:326
global warming, 1:367, 370
greenhouse effect, 1:388
greenhouse gases, 1:394–395
iron fertilization, 1:490–491
Kyoto Protocol, 1:501
logging, 1:537
oceanography, 2:630
ozone layer, 2:663
paper and wood pulp, 2:665
rain forest destruction, 2:687
red tide, 2:700
reef ecosystems, 2:704
reforestation, 2:706
resource extraction, 2:713
tundra, 2:798, *799*
vegetation cycles, 2:817
wildfires, 2:879

Carbon sinks. *See* Carbon sequestration

Carbon trading. *See* Pollution credits

Carbonic acid, 1:62

Cardboard. *See* Paper and wood pulp

Careers in environmental science, 1:**95–99**, *98*
media, 2:559–560, 561
non-scientist contributions, 2:604
oceanography, 2:628
photography, 2:669, 670
sustainable development, 2:772

Carey, Hugh, 1:416

Caribbean area, 1:195, 247, 250, 263

Caribbean Sea, 1:*171*, 2:793, 794

Caribou, 2:798

Carnivores, 1:242

Carpels, 2:673

Carrying capacity, 1:423, 467, 2:657, 678, 880

Cars. *See* Vehicles

Carson, Rachel
agricultural practice impacts, 1:5
DDT, 1:186, 187
environmental activism, 1:265, 289
environmental assessments, 1:269
environmental science careers, 1:95
green movement, 1:375
insecticide use, 1:471
media, 2:559
organic and locally grown foods, 2:645
Silent Spring, 2:734–737, *735*
soil contamination, 2:745
sustainable development, 2:769
wildlife protection policies and legislation, 2:883

Carter, Jimmy, 1:414, 416, 417, 2:746, 790

Cartography, 1:341

Cash, Connor, 1:228

Cash, Johnny, 2:877

Castillo de Bellver (ship), 2:641–642

Cats, feral, 1:481

Catskills State Park, 2:590

Cattle
agricultural practice impacts, 1:5, 6, *7*
DDT, 1:186
forests, 1:324
organic farming, 2:*650*
overgrazing, 2:655
species reintroduction programs, 2:755–756

Cattle farm impacts. *See* Factory farms, adverse effects of

CCC (Civilian Conservation Corps), 1:317–318

CCS. *See* Carbon sequestration

CCSM3 (Community Climate System Model, version 3), 1:140–141

CDC (Centers for Disease Control and Prevention), 1:303–304, 352, 2:563

CDM. *See* Clean Development Mechanism (CDM)

Cells (atmospheric circulation), 1:45

Cellulose, 1:15, 16

Cellulosic fermentation, 1:70

Census Bureau, 1:341–342

Center for American Progress, 1:395

Center for Automotive Research, 1:88

Center for Biological Diversity, 1:263, *405*

Centers for Disease Control and Prevention (CDC), 1:303–304, 352, 2:563

Central Intelligence Agency (CIA), 1:342

Centrifugation, 1:510

CERCLA. *See* Comprehensive Environmental Response, Compensation, and Liability Act (CERCLA)

Certified Emission Reductions (CERs), 1:115–116

CFCs. *See* Chlorofluorocarbons (CFCs)

CH_4. *See* Methane

Chains, food. *See* Food chains and webs

Chandeleur Islands, 1:*435*

Channel runoff. *See* Streamflow

Channels (estuaries), 1:31, 295

Chapman, Frank, 2:604

Chapman, John, 2:709

Chemical elements, 1:338, 2:624

Chemical industry
Comprehensive Environmental Response, Compensation, and Liability Act, 1:162, 164
industrial pollution, 1:455
industrial water use, 1:455–456
Silent Spring, 2:735–737
toxic waste, 2:790

Chemical spills, 1:**100–101**
aquatic ecosystems, 1:*31*
biodegredation, 1:58
groundwater, 1:398
Hurricane Katrina, 1:437
industrialization in emerging economies, 1:*464*
insecticide use, 1:*471*

spill remediation, 2:757–758, 759
toxic waste, 2:790
tsunami impacts, 2:795
water pollution, 2:840
watersheds, 2:857
Chemicals, toxic. *See* Toxins
Chemistry. *See specific types of chemistry*
Chen Kejian, 1:465
Chernobyl, Ukraine
ecodisasters, 1:231–232
environmental protests, 1:292
nuclear power, 2:610, 611, *612*
Cherry Springs State Park, 1:*532*
Chertoff, Michael, 2:578
Chesapeake Bay
acid rain, 1:4
algal blooms, 1:13
estuaries, 1:295, 296, *296,* 297
nonpoint-source pollution, 2:602
sea level rise, 2:724
watersheds, 2:*857*
Chesser, Michael, 1:177
Chicken farm impacts. *See* Factory farms, adverse effects of
Chickens, 1:238, 302–305, *304*
Children for a Safe Environment, 2:*747*
Chile
Antarctic Treaty, 1:25, 27
Antarctica, 1:22
deserts, 1:197
earthquakes, 1:219
invasive species, 1:483
tsunami impacts, 2:793
Chimpanzees, 2:708
China
air pollution, 1:9, 10
atmospheric inversions, 1:48
aviation emissions, 1:51
biodiversity, 1:61, 64, 65
carbon dioxide emissions, 1:86
Clean Development Mechanism, 1:115, 116
coal resource use, 1:146, 147
commercial fisheries, 1:160
cultural practices and environment destruction, 1:180
dams, 1:181
desertification, 1:195, 196
deserts, 1:197, *198*
earthquakes, 1:218, 219
electronics waste, 1:152, 253
emissions standards, 1:255
fish farming, 1:306, 307, 308–309
forests, 1:325
fossil fuel combustion impacts, 1:327, *327*
global warming, 1:*368*
green movement, 1:376, 378, 381–382
groundwater, 1:*398*
hazardous waste, 1:414
hunting practices, 1:428
industrial water use, 1:*460*
industrialization in emerging economies, 1:462–466, *463, 464*
inland fisheries, 1:467
IPCC 2007 Report, 1:485, 488
irrigation, 1:494
Kyoto Protocol, 1:500, 501, 502
land use, 1:*521*
maps and atlases, 2:539
natural resource management, 2:*599*
nuclear test ban treaties, 2:*617*
overgrazing, 2:657
paper and wood pulp, 2:664
recycling, 2:697
toxic waste, 2:791
war and conflict-related environmental destruction, 2:825
waste transfer and dumping, 2:829, *830*
water pollution, 1:*398*
China Ready To Leap From Industrial To Information-Age Economy, 1:***464–466***
Chittenango snails, 1:*261*
Chlorine
chemical spills, 1:100
chlorofluorocarbons, 1:104
industrial pollution, 1:455
Montreal Protocol, 2:574, 575
wastewater treatment technologies, 2:833, 834
water pollution, 2:838
Chlorofluorocarbons (CFCs), 1:**102–104,** *103*
air pollution, 1:8, 9
atmosphere, 1:43
Clean Air Act of 1970, 1:109
climate change, 1:129
Earth Summit (1992), 1:212
geochemistry, 1:340
greenhouse effect, 1:384
greenhouse gases, 1:391
industrial pollution, 1:456
Montreal Protocol, 2:574, 575
ozone hole, 2:659, 660
ozone layer, 2:662–663
United Nations policy and activism, 2:810
Chlorophyll, 1:291, 490, 2:549, 551
Cholera, 1:*187,* 397, 2:837, 840
Cholinesterase, 1:470
Christy, John, 1:136
Chromatography, 1:510
Chrysotile, 1:39
CIA (Central Intelligence Agency), 1:342
Cichlid fish, 1:482
Ciguatea fish poisoning, 2:700
Ciguatoxin, 2:700
Cinergy, 1:3
Circulation, atmospheric. *See* Atmospheric circulation
Circulation, ocean. *See* Ocean circulation and currents
Cirrus clouds, 1:49, 50
CITES (Convention on International Trade in Endangered Species of Wild Fauna and Flora), 1:**105–107,** *106*
endangered species, 1:261
international environmental law, 1:477
reef ecosystems, 2:705
Cities. *See* Urban areas
Civilian Conservation Corps (CCC), 1:317–318
Clarren, Sterling, 2:782–783
Classification (species), 1:60, 244
Clay, 2:743
Clayoquot Sound, 1:308
Clean Air Act Amendments of 1977, 1:109
Clean Air Act Amendments of 1990, 1:109, 110, 2:740
Clean Air Act of 1963, 1:108
Clean Air Act of 1970, 1:**108–111,** ***110–111***
acid rain, 1:3–4
air pollution, 1:10
carbon dioxide emissions, 1:87, 88
Clean Air Mercury Rule, 1:113
Environmental Protection Agency, 1:286–287
industrial pollution, 1:457–458
passage, 1:209–210
Clean Air Mercury Rule of 2005, 1:**112–114**
Clean-Air Rule Targets Existing Diesel-Truck Fleet, 1:***255–256***
Clean Air Rules of 2004, 1:112
Clean Air Task Force, 1:92
Clean Development Mechanism (CDM), 1:**115–117,** *117,* 182, 183–184
Clean energy, 1:**118–121**
climate change, 1:133
coal resource use, 1:147–148, 149
coastal zones, 1:156–159
corporate green movement, 1:174–177
dams, 1:182
geothermal resources, 1:354, 355–356
green movement, 1:380

greenhouse effect, 1:388
greenhouse gases, 1:394
human impacts, 1:425
industrial pollution, 1:455
nuclear power, 2:611, 612
Our Common Future, 2:814
solar power, 2:750, 751–752
sustainable development, 2:771
tidal or wave power, 2:784–787
tides, 2:788
wind and wind power, 2:888–889

Clean Energy vs. Whales: How To Choose?, 1:***120–121***

Clean Water Act, 1:**122–124,** *123*
ecosystems, 1:246
industrial pollution, 1:457
industrial water use, 1:460
marine water quality, 2:551
nonpoint-source pollution, 2:602
runoff, 2:717
wastewater treatment technologies, 2:833
water pollution, 2:837
wetlands, 2:866

Clean Water Restoration Act, 1:123–124

Cleanup, ecological. *See specific types of remediation*

Clear-cutting, 1:**125–127**
agricultural practice impacts, 1:6
deforestation, 1:191
forest resources, 1:317
as logging, 1:536–537
natural resource management, 2:598, 599
rain forest destruction, 2:686
reforestation, 2:706
resource extraction, 2:712, 713

Clemson University, 1:*98*

Climate. *See* Weather and climate

Climate, ancient. *See* Paleoclimate

Climate: Warming Skeptics: Is the Research Too Political?, 1:***135–137***

Climate change, 1:**128–138**
acid rain, 1:1–2
agricultural practice impacts, 1:5, 6
air pollution, 1:8, 10
alternative fuel impacts, 1:14, 16
Antarctica, 1:21, 23–24
Arctic, 1:36–38
atmosphere, 1:41–43
atmospheric circulation, 1:44, 45–46
aviation emissions, 1:49
biodiversity, 1:60, 61–65
biofuels, 1:67, 68, 72
carbon dioxide, 1:83–84
carbon dioxide emissions, 1:86
carbon sequestration, 1:90, 91
clean energy, 1:120–121
climate modeling, 1:139, 140
coal resource use, 1:147, 148
coastal ecosystems, 1:150, 152
coastal zones, 1:155
conservation, 1:168
coral reefs and corals, 1:171–172
corporate green movement, 1:175–176
dams, 1:181–183
deforestation, 1:190
dendrochronology, 1:193
desertification, 1:194, 196
dust storms, 1:207–208
ecodisasters, 1:232–234
ecosystem diversity, 1:241
ecosystems, 1:245
El Niño and La Niña, 1:247–251
endangered species, 1:259
Environmental Protection Agency, 1:286–287
environmental protests, 1:290
extinction and extirpation, 1:299–300
floods, 1:314, 315–316
fossil fuel combustion impacts, 1:328
Gaia hypothesis, 1:333–334
geologic history, 1:349
glacial retreat, 1:357, 359, 360
glaciation, 1:362
global warming, 1:128, 129–131, 366–370
greenhouse effect, 1:128, 129, 130, 385–386
greenhouse gases, 1:390–395
groundwater quality, 1:401
habitat alteration, 1:*405*, 406
habitat loss, 1:411–412
history, 1:128–129
horticulture, 1:420
human impacts, 1:424, 425–426
hurricanes, 1:430, 432
hydrologic cycle, 1:445–447
ice ages, 1:448–451
ice cores, 1:452–453
impacts, 1:132–133
Intergovernmental Panel on Climate Change, 1:130, 131, 135–137, 474–475
international environmental law, 1:479
IPCC 2007 Report, 1:485–488
iron fertilization, 1:490
lakes, 1:517
logging, 1:536
mathematical modeling and simulation, 2:557
media, 2:559, 560–561
migratory species, 2:568, 569
National Oceanic and Atmospheric Administration, 2:582
natural resource management, 2:598
North Atlantic Oscillation, 2:608–609
nuclear power, 2:610, 612
ocean circulation and currents, 2:620, 622
oceanography, 2:630
ozone layer, 2:663
photography, 2:669, 670
pollinators, 2:672, 673
predator-prey relations, 2:678
radiative forcing, 2:680
red tide, 2:700
resource extraction, 2:713
saltwater intrusion, 2:719
seasonal migration, 2:730–731
snow and ice cover, 2:741
sustainable development, 2:771
temperature records, 2:779–780
theories, 1:129–132
United Nations Conference on the Human Environment, 2:801
United Nations Convention on the Law of the Sea, 2:804–805
United Nations Framework Convention on Climate Change, 2:806–809
United Nations policy and activism, 2:810–812
United States, 1:133–135
volcanoes, 2:819
Walden, 2:822
war and conflict-related environmental destruction, 2:827–828
water resources, 2:844
weather and climate, 2:860
weather extremes, 2:861–862, 863
wildfire control, 2:875, 876
wildfires, 2:879

Climate Change Impacts on teh United States: The Potential Consequences Of Climate Variability and Change, 1:***134–135***

Climate-Change Paradox: Greenhouse Gas Is Big Oil Boon, 1:***394–395***

Climate Change Science Program, 2:780

Climate forcing. *See* Radiative forcing

Climate Group, 1:177, 443

Climate modeling, 1:**139–141**
atmospheric circulation, 1:46
climate change, 1:130, 131
closed ecology experiments, 1:142, 143
El Niño and La Niña, 1:250, 251
floods, 1:315–316
Gaia hypothesis, 1:334
global warming, 1:368–369
greenhouse effect, 1:386
greenhouse gases, 1:393
hydrologic cycle, 1:446, 447
Intergovernmental Panel on Climate Change, 1:475

IPCC 2007 Report, 1:488
as mathematical modeling and simulation, 2:557
North Atlantic Oscillation, 2:608
ocean circulation and currents, 2:622
radiative forcing, 2:681
sea level rise, 2:721
surveying, 2:768
tundra, 2:798, 799
weather and climate, 2:859, 860
weather extremes, 2:861–862
Climate models, IPCC 2007 Report, 1:485
Climax forests, 2:706
Clinton, Bill, 1:210, 2:887
hybrid vehicles, 1:441
Kyoto Protocol, 1:501
recreational use and environmental destruction, 2:691
Clinton Climate Initiative, 1:210
Clinton Global Initiative, 1:64, 177
Cloning, 1:*7*
Closed ecology experiments, 1:**142–144,** *143*
Clouded leopards, 1:*324*
Clouds
cirrus, 1:49, 50
hydrologic cycle, 1:445
ozone hole, 2:659
precipitation, 2:674
radiative forcing, 2:681
Cnidarians, 2:542
CO. *See* Carbon monoxide (CO)
Co-generation, 1:119
CO_2. *See* Carbon dioxide (CO_2)
Coachella Valley, California, 1:*399*
Coal resource use, 1:**145–149,** *147*
air pollution, 1:9
alternative fuel impacts, 1:14
Antarctica, 1:21
biofuels, 1:67–68, 69
biostratigraphy, 1:79
carbon sequestration, 1:92
Clean Air Mercury Rule, 1:112–113
clean energy, 1:119
emissions standards, 1:255
fossil fuel combustion impacts, 1:326, 327
industrial pollution, 1:455
industrial water use, 1:460
Coalition for Clean Air, 1:256
Coalition to Save the Preserves (CSP), 1:228
Coast Guard, 2:643
Coastal ecosystems, 1:**150–153**
Antarctica, 1:21
aquatic ecosystems, 1:30–31, 32
bays and estuaries, 1:53–54
coastal zones, 1:154–159
ecosystem diversity, 1:240, 241
ecosystems, 1:243, *244,* 246
endangered species, 1:260–261
estuaries, 1:294–297
ocean tides, 2:626–627
reef ecosystems, 2:702–705
sea level rise, 2:722
tsunami impacts, 2:795
United Nations Convention on the Law of the Sea, 2:804
Coastal plain estuaries, 1:294
Coastal Zone Act Reauthorization Amendments, 2:602–603, 717
Coastal Zone Renewable Energy Act of 2003, 1:***156–159***
Coastal zones, 1:**154–159**
aquifers, 1:34
atmospheric inversions, 1:47
climate change, 1:133, 134
coastal ecosystems, 1:150–152
Coastal Zone Renewable Energy Act, 1:156–159
commercial fisheries, 1:160
deserts, 1:197
divisions, 1:154–155
dredging, 1:202, 203–204
El Niño and La Niña, 1:249
estuaries, 1:294–297
migratory species, 2:569
National Oceanic and Atmospheric Administration, 2:582
nonpoint-source pollution, 2:602–603
ocean tides, 2:627
oceanography, 2:628–630
oceans and coastlines, 2:633
runoff, 2:717
saltwater intrusion, 2:719–720
sea level rise, 2:722
threats, 1:155–156
wetlands, 2:864, 865, 866
Coastlines. *See* Oceans and coastlines
Coasts. *See* Oceans and coastlines
CoastWatch, 1:*151*
Coca-Cola Enterprises, 1:442–444
Cod, 1:160–161, 306, 2:546, 599, 652, 653
Coffee Bean & Tea Leaf, 2:648
Cole, David, 1:88
Coliform bacteria, 2:833
Colombia, 2:794
Colonization (species), 1:496, 497
Colorado
dams, 1:184
dust storms, 1:208
earthquakes, 1:220
eco-terrorism, 1:*226*
natural reserves and parks, 2:592
Columbia River, 1:119–121, 221, 2:683–684
Commerce Department, 2:876
Commercial fisheries, 1:**160–161**
fish farming, 1:306–309
inland fisheries, 1:467–468
marine ecosystems, 2:544–545
overfishing, 2:652–653
predator-prey relations, 2:678
Commoner, Barry, 1:375
Communities (biology), 1:239, 242
Community Climate System Model, version 3 (CCSM3), 1:140–141
Competition, ecological. *See* Ecological competition
Competitive exclusion, 1:236–237, 238, 408
Compliance officers, environmental, 1:97
Composting, 1:*403*
biodegredation, 1:57
organic farming, 2:649
recycling, 2:695
solid waste treatment technologies, 2:753, 754
Compounds, organic. *See* Organic compounds
Comprehensive Environmental Response, Compensation, and Liability Act (CERCLA), 1:**162–165,** ***164–165***
Environmental Protection Agency, 1:284
groundwater quality, 1:402–403
hazardous waste, 1:414–417
spill remediation, 2:759
Superfund site, 2:763
toxic waste, 2:791
water pollution, 2:840
Comprehensive Test Ban Treaty (CTBT), 2:615, 616, 617
Computational biology, 1:511, 512
Computer cartography. *See* Geographic Information Systems (GIS)
Computers, 1:139–141, 252
Conceptual models, 2:556–557
Concordant age, 1:348
Condensation, 1:430, 445, 2:674
Condensation trails (aircraft). *See* Contrails
Condors, 1:262
Conference of Parties, 1:106
Conference on Environment and Development (UNCED). *See* Earth Summit (1992)
Conference on the Human Environment. *See* United Nations Conference on the Human Environment (1972)

Confined animal feeding operations. *See* Factory farms, adverse effects
Congressional Act of 1872, 2:596
Coniferous forests, 1:323, 324
Connecticut, 2:562
Conservation, 1:**166–169**
biodiversity, 1:64
biogeography, 1:75
Bureau of Land Management, 1:81
CITES, 1:105–107
ecosystems, 1:245–246
endangered species, 1:259, 261–262
environmental activism, 1:265–268
environmental protests, 1:292
environmental science careers, 1:95, 98
forest resources, 1:317–318
forests, 1:325
genetic diversity, 1:335, 336–337
global warming, 1:370
green movement, 1:375, 376, 381–382
habitat alteration, 1:406
habitat loss, 1:409, 412
human impacts, 1:423, 425, 426
International Convention for the Regulation of Whaling, 2:867–870
international environmental law, 1:477, 478
island ecosystems, 1:496, 498
migratory species, 2:568, 569
National Oceanic and Atmospheric Administration, 2:582
National Park Service Organic Act, 2:585–587
natural reserves and parks, 2:589–596
non-scientist contributions, 2:604–607
predator-prey relations, 2:678
recreational use and environmental destruction, 2:690–691
reef ecosystems, 2:705
reforestation, 2:707, 710
seasonal migration, 2:730, 731
soil resources, 2:748
species reintroduction programs, 2:755
sustainable development, 2:769–772
tsunami impacts, 2:795, 796
United Nations Convention on the Law of the Sea, 2:804
Walden, 2:822
watersheds, 2:858
wetlands, 2:864, 866
Wilderness Act of 1964, 2:872–873
wildlife population management, 2:880–881
wildlife protection policies and legislation, 2:883–885
wildlife refuge, 2:886–887
See also specific types of conservation
Conservation easements, 2:592
Conservation International, 1:262, 376
Conservation Law Foundation, 1:88
The Conservation of Natural Resources (Van Hise), 1:168
Consumerism, 1:379–381
Contamination. *See specific types of contamination and pollution*
Continental drift, 1:61, 338, 448, 449
Continental shelves, 2:628, 803, 804
Contiguous zones, 2:803
Contrails, 1:50, 51
Controlled burns, 2:874, 878
Controlled hunting. *See* Hunting practices
Convention Concerning the Protection of the World Cultural and Natural Heritage, 1:261
Convention for the Conservation of Antarctic Seals, 1:26
Convention for the Regulation of Antarctic Mineral Resource Activities, 1:26
Convention on Biodiversity. *See* Convention on Biological Diversity
Convention on Biological Diversity
Earth Summit (2002), 1:213, 214
endangered species, 1:261–262
island ecosystems, 1:498
natural reserves and parks, 2:590–591
pharmaceutical development resources, 2:667
wildlife protection policies and legislation, 2:883–885
Convention on International Trade in Endangered Species of Wild Fauna and Flora (CITES). *See* CITES (Convention on International Trade in Endangered Species of Wild Fauna and Flora)
Convention on Migratory Species, 1:501, 503, 2:568, 569, 731, 884
Convention on the Conservation of Antarctic Marine Living Resources, 1:26
Convention on Wetlands of International Importance, 1:261
Cook, James, 2:628
Cool cloud forests, 1:321
Cooling, climatic. *See* Climate change
Cooling water, 1:457, 461, *461*, 2:550, 849
Cooney, Philip, 2:578
Copepods, 2:542
Coquina Beach, Florida, 2:*701*
Coral atolls, 1:170, 2:703
Coral bleaching
coral reefs and corals, 1:171, *171*, 172
greenhouse effect, 1:388, 2:634
habitat loss, 1:411
marine ecosystems, 2:544
reef ecosystems, 2:704
Coral polyps, 1:170, 171–172, 2:543, 634
Coral reefs and corals, 1:**170–173**, *171*
acid rain, 1:2
aquatic ecosystems, 1:32
biodiversity, 1:62, 63
biostratigraphy, 1:80
carbon dioxide emissions, 1:86
climate change, 1:133
coastal ecosystems, 1:151–152
ecodisasters, 1:234
ecosystem diversity, 1:239, 240, 241
El Niño and La Niña, 1:248
endangered species, 1:262, 263
greenhouse effect, 1:387–388
habitat loss, 1:411
marine ecosystems, 2:543–544, *544*
marine fisheries, 2:547
natural reserves and parks, 2:593
oceanography, 2:628
oceans and coastlines, 2:632, 634
predator-prey relations, 2:678
reef ecosystems, 2:702–705
tsunami impacts, 2:795, 796
Cordero, José F., 2:783–784
Core (Earth), 1:338, 354
Coriolis force, 1:45, 431, 2:620
Corn
alternative fuel impacts, 1:16, 19
biofuels, 1:67, 68, 69, 70–71, 72
bioremediation, 1:78
Corn ethanol. *See* Ethanol
Coronado, Rodney Adam, 1:229
Corporate green movement, 1:**174–178**
acid rain, 1:2–4
emissions standards, 1:257
environmental protests, 1:290
green movement and, 1:377
hazardous waste, 1:414
hybrid vehicles, 1:442–444
organic and locally grown foods, 2:647–648

pharmaceutical development resources, 2:666, 667–668
Correa, Rafael, 2:593
Corzine, Jon, 1:257
Costa Rica, 1:62, 2:*600,* 667, *708*
Cotton, 1:78, *402*
Council for Environmental Quality, 2:577–578
Cousins, Gregory T., 2:643
Cousteau, Jacques-Yves, 2:*629,* 685
Cousteau, Jean-Michel, 2:543
Cranberries, 2:866
Cranes, whooping, 2:*592*
Cranfield, Mississippi, 1:93
Crayfish, 1:483
Creutzfeld-Jacob disease, 1:305
Crime, environmental. *See* Environmental crime
Croatia, 2:825
Crocidolite, 1:39
Crop rotation, 2:744
Crops. *See* Agricultural practice
Crude oil. *See* Oil
Crust (Earth), 1:338, 340, 2:818
Crustaceans, 1:30, 31, 55, 86, 2:542
Crutzen, Paul J., 1:*339*
CSP (Coalition to Save the Preserves), 1:228
CTBT (Comprehensive Test Ban Treaty), 2:615, 616, 617
Ctenophore, 1:237
Cuban Missile Crisis, 2:616, 617
Culls. *See* Hunting practices
Cultural practices and environmental destruction, 1:**179–180**
hunting practices, 1:427–429
industrialization in emerging economies, 1:464–466
natural reserves and parks, 2:589–590, 593
sustainable development, 2:772
Cumberland Island, 2:*591*
Cun Yanfang, 1:381–382
Curbside recycling. *See* Recycling
Curitiba, Brazil, 1:*63*
Currents, ocean. *See* Ocean circulation and currents
Curuá-Una Dam, 1:182
Cusurichi, Julio, 1:*275*
Cuvier, Georges, 1:345
Cuyahoga River, 1:*123,* 2:837
Cyanide, 1:*31*
Cyanobacteria, 1:11
Cystic fibrosis, 2:701
Czech Republic, 1:*531*

D

D. radiodurans. See Deinococcus radiodurans
da Vinci, Leonardo, 2:750
Dams, 1:**181–185**
aquatic ecosystems, 1:32
clean energy, 1:119–121
conservation, 1:167
deforestation, 1:190
environmental science careers, 1:96
floods, 1:314, 315
freshwater and freshwater ecosystems, 1:330–331
habitat loss, 1:409
hydrologic cycle, 1:446–447
industrial water use, 1:*460*
industrialization in emerging economies, 1:463
lakes, 1:515
natural resource management, 2:598
recreational use and environmental destruction, 2:*692*
rivers and waterways, 2:715
streamflow, 2:762
Tennessee Valley Authority v. Hill, 2:776–778
tidal or wave power, 2:785
water supply and demand, 2:852
Dareste, Camille, 2:782
Darwin, Charles
biodiversity, 1:60
biostratigraphy, 1:79
coral reefs and corals, 1:170
ecological competition, 1:237
ecosystems, 1:244
genetic diversity, 1:335
island ecosystems, 1:496
oceanography, 2:628
reef ecosystems, 2:703
Data layers, 1:351
Dating, tree. *See* Dendrochronology
Dauphin Island, Alabama, 1:*440*
Davis, Ken, 1:444
Davis, Marc Leslie, 1:229
The Day After Tomorrow (motion picture), 2:559
Daylight, 1:*248*
DDT (Dichloro-diphenyl-trichloroethane), 1:**186–188**
bioremediation, 1:76
environmental protests, 1:289
genetic diversity, 1:336–337
industrial water use, 1:460
insecticide use, 1:469, 470, 471
media, 2:559
Silent Spring, 2:734–735, 736, 737
soil contamination, 2:747
sustainable development, 2:769
water pollution, 2:839
De la Beche, Henry, 1:346
Dead Sea, 2:*723*
Dead zones
algal blooms, 1:13
commercial fisheries, 1:155
conservation, 1:168
ecodisasters, 1:231
marine ecosystems, 2:543, 544
organic farming, 2:650
photography, 2:670
Watershed Protection and Flood Prevention Act of 1954, 2:855
Death Valley, California, 1:*249*
Debris, marine. *See* Marine debris
Debt for nature, 1:325
Deca-BDE, 1:152
Decay, radioactive, 1:348
Deciduous trees, 1:323
Declaration of the Conference on the Human Environment
Earth Summit (1992), 1:212
Earth Summit (2002), 1:215
international environmental law, 1:478, 479
Our Common Future, 2:800
United Nations Conference on the Human Environment, 2:800
Decomposers, 1:11, 12, 32, 242, 245, 2:544
Deep-sea mining. *See* Mining and quarrying impacts
Deepak Spinners Limited, 1:116
Deforestation, 1:**189–191**
acid rain, 1:2
agricultural practice impacts, 1:5, 6–7
alternative fuel impacts, 1:16
biodiversity, 1:60, 63
biofuels, 1:71
carbon dioxide, 1:83
carbon dioxide emissions, 1:85, 86
carbon sequestration, 1:90–91
cultural practices and environment destruction, 1:179–180
desertification, 1:194, 195
ecosystems, 1:246
endangered species, 1:259
forest resources, 1:324, 325
forests and, 1:317
freshwater and freshwater ecosystems, 1:331
global warming, 1:367
green movement, 1:381–382
greenhouse effect, 1:383
greenhouse gases, 1:390
habitat alteration, 1:404, 405, 406
habitat loss, 1:409, 411

human impacts, 1:423
ice ages, 1:450
landslides, 1:527
logging, 1:536–537
photography, 2:670
vegetation cycles, 2:816, 817
vs. reforestation, 2:706, 707, 709
war and conflict-related environmental destruction, 2:827
See also Clear-cutting

Degraded water, 1:459, 460, 2:849

Degredation, biological. *See* Biodegredation

Deinococcus radiodurans, 1:58, 77

Delivery Companies Switch To Hybrids, 1:***442–444***

Deltas, 1:202, *330,* 331, 2:725–728, 851–852

Demilitarized zones, 1:25–26, 27

Demonstrations, environmental. *See* Environmental protests

Denan, Ethiopia, 1:*133*

Denbury Resources, 1:395

Dendroarchaelogy, 1:193

Dendrochronology, 1:**192–93,** *193*

Dendroclimatology, 1:193

Denmark, 2:756

Density, 1:236

Deoxyribonucleic acid (DNA). *See* DNA

Department of Agriculture (USDA)
biofuels, 1:69
DDT, 1:186
Massachusetts v. Environmental Protection Agency, 1:287
organic and locally grown foods, 2:646
Watershed Protection and Flood Prevention Act of 1954, 2:854–855
watersheds, 2:858
wildfire control, 2:876

Department of Commerce, 2:876

Department of Energy, 1:287, 395

Department of Homeland Security, 2:578

Department of the Interior, 1:81
National Park Service Organic Act, 2:586, 587
natural reserves and parks, 2:596
TVA v. Hill, 2:776–777
Watershed Protection and Flood Prevention Act of 1954, 2:854
wildfire control, 2:875, 876

Department of Transportation, 1:287

Depleted uranium (DU), 2:825

Depletion (Aquifers). *See* Recharging (Aquifers)

Deposit laws. *See* Bottle bills

Depth, ocean. *See* Ocean depth

Depth hoars, 2:741

Desert tortoise, 1:*82*

Desertification, 1:**194–196,** *195*
deserts, 1:198
drought, 1:206
dust storms, 1:207
ecodisasters, 1:232
ecosystems, 1:245–246
grasslands, 1:373
greenhouse effect, 1:387
habitat loss, 1:409
organic farming, 2:650
overgrazing, 2:656, 657
resource extraction, 2:712
shrublands, 2:732
soil chemistry, 2:744
soil resources, 2:748, 749

Deserts, 1:**197–199**
Antarctica, 1:21
biodiversity, 1:62
desertification, 1:194–196
dust storms, 1:207
ecosystems, 1:243, 244, 245–246
migratory species, 2:568
tundra, 2:798
war and conflict-related environmental destruction, 2:825

Destruction, rain forest. *See* Rain forest destruction

Detritus, 1:55

Developed nations. *See specific nations*

Developing countries
Clean Development Mechanism, 1:115–117
climate change, 1:133
deforestation, 1:189
Earth Summit (1992), 1:213
El Niño and La Niña, 1:251
electronics waste, 1:252, 253
emissions standards, 1:255
environmental activism, 1:268
environmental crime, 1:280
environmental protests, 1:290
forest resources, 1:325
forests, 1:319
habitat loss, 1:409
hazardous waste, 1:414
hunting practices, 1:427
industrial water use, 1:460
industrialization in emerging economies, 1:462–466
inland fisheries, 1:467
international environmental law, 1:478, 479–480
IPCC 2007 Report, 1:486
Kyoto Protocol, 1:500, 501
landslides, 1:528
marine fisheries, 2:547
mining and quarrying impacts, 2:573
Montreal Protocol, 2:575
natural reserves and parks, 2:593
Our Common Future, 2:813, 814
pharmaceutical development resources, 2:667
recycling, 2:697
saltwater intrusion, 2:720
soil contamination, 2:747
spill remediation, 2:757, 759
surface water, 2:766, 767
sustainable development, 2:770–771
toxic waste, 2:790, 791
United Nations Conference on the Human Environment, 2:800
United Nations Framework Convention on Climate Change, 2:807
waste transfer and dumping, 2:829–830, 831
wastewater treatment technologies, 2:834
water conservation, 2:835
water pollution, 2:840
water supply and demand, 2:849, 850
weather extremes, 2:862
See also specific nations

Development, sustainable. *See* Sustainable development

Development of land. *See* Land use

Diamond, Jared, 2:612

Diapirs, plutonic, 2:818

Diarrhetic shellfish poisoning, 2:700

Diatoms, 1:11, 32, 151, 2:627

DiCaprio, Leonardo, 1:*377*

Dichloro-diphenyl-trichloroethane (DDT). *See* DDT (Dichloro-diphenyl-trichloroethane)

Dickens, Charles, 2:739

Diesel fuel, alternative. *See* Biodiesel

Dikes (embankments). *See* Levees

DIME (Dual Independent Map Encoding), 1:341–342

Dinoflagellates, 1:11, 12–13, 2:699, 700

Dinosaurs, 1:298, *347,* 2:819

Dioxin
air pollution, 1:10
bioremediation, 1:76
chemical spills, 1:*101*
herbicides, 1:418
industrial pollution, 1:455
solid waste treatment technologies, 2:753, 754
Superfund site, 2:764

Direct action, 1:289

Disasters, environmental. *See* Ecodisasters

Discordant age, 1:348
Dispersal (species), 1:496, 497
Dispersants, 2:637, 642–643, 757–758
Distillation (biofuels), 1:70
Disturbance severity, 2:710
Diversity, biological. *See* Biodiversity
Diversity, ecosystem. *See* Ecosystem diversity
Diversity, genetic. *See* Genetic diversity
DNA
  biogeography, 1:74
  chlorofluorocarbons, 1:102–103
  genetic diversity, 1:335, 337
  laboratory methods in environmental science, 1:510, 511–512
  ozone hole, 2:658, 659–660
  ozone layer, 2:661–662
Dobson Spectrophotometers, 2:*662*
Dodo birds, 1:61
Dolphins, 1:61, 2:*653*
Domoic acid, 1:13, 2:699
Donora, Pennsylvania, 1:108
Don't Make a Wave Committee. *See* Greenpeace
Dorsey, Michael, 1:380
Dow Chemical Company, 2:736
Drainage basins, 1:**200–201,** *201*
  earthquakes, 1:219
  landfills, 1:524
  rivers and waterways, 2:714
Drainage control. *See* Flood control and floodplains
Dredging, 1:**202–204,** *203*
  estuaries, 1:296
  habitat alteration, 1:405
  habitat loss, 1:409
  reef ecosystems, 2:704
  sea level rise, 2:727
Drift, continental. *See* Continental drift
Driftnets, 1:224
Drilling, oil. *See* Oil
Drinking alcohol. *See* Ethanol
Drinking water. *See* Water resources
Drizzle, 2:674
Drought, 1:**205–206,** *206*
  biodiversity, 1:64
  climate change, 1:133, *133*
  coastal zones, 1:155
  deforestation, 1:190
  desertification, 1:194, *195*
  deserts, 1:197
  dust storms, 1:208
  ecodisasters, 1:232
  El Niño and La Niña, 1:250, 251
  greenhouse effect, 1:383, 387
  groundwater, 1:396
  groundwater quality, 1:401
  horticulture, 1:420
  irrigation, 1:493
  precipitation, 2:674
  soot, 1:148
  vegetation cycles, 2:817
  water conservation, 2:835
  water resources, 2:844
Drugs. *See* Pharmaceuticals and personal care products (PPCPs)
Drumlin, 1:*450*
Dry spells, 1:205
Drylands, 1:194–196
DU (Depleted uranium), 2:825
Dual Independent Map Encoding (DIME), 1:341–342
Dubos, Rene J., 1:*167*
Ducks. *See* Waterfowl
Ducks Unlimited, 1:167, 2:605, 881
Duckweed, 2:834
Duke Energy, 1:3
Dumping, waste. *See* Waste transfer and dumping
Duncan, Ian, 1:93
Dunes, sand. *See* Sand dunes
Dung. *See* Animal waste
Dust. *See* Particulates
Dust Bowl
  cultural practices and environment destruction, 1:179
  desertification, 1:194
  drought, 1:205
  ecodisasters, 1:232
  photography, 2:669
Dust storms, 1:**207–208**
  agricultural practice impacts, 1:5
  ecodisasters, 1:232
Dzerzhinsk, Russia, 1:*101*

E

E. coli, 1:397, 2:765, 838
E-waste. *See* Electronics waste
Eagles, bald. *See* Bald eagles
Early Anthropocene Hypothesis, 1:450
Earth Day, 1:**209–211,** *210, 211,* 289, 2:*633*
Earth Day Network, 1:210
Earth First!, 1:*147,* 168, 224, 227, 2:*665*
Earth Liberation Front (ELF), 1:225, *226,* 227–229, 267, 378
Earth Summit (1992), 1:**212–214,** *214*
  Earth Summit (2002), 1:215, 216
  endangered species, 1:262
  international environmental law, 1:478–479
  Kyoto Protocol, 1:500
  Our Common Future, 2:814–815
  United Nations Conference on the Human Environment, 2:801
  United Nations Framework Convention on Climate Change, 2:806
  United Nations policy and activism, 2:810
Earth Summit (2002), 1:**215–217,** *217*
  international environmental law, 1:479
  Our Common Future, 2:814, 815
  United Nations Conference on the Human Environment, 2:801
Earth Too Warm? Bury the CO_2, 1:**92–93**
Earth Track, 1:72
Earthquakes, 1:**218–223,** *221*
  causes, 1:219–220
  ecodisasters, 1:233
  floods, 1:315
  Geographic Information Systems (GIS), 1:*342*
  hazards, 1:220–223
  history, 1:218–219
  landslides, 1:527
  overgrazing, 2:657
  tsunami impacts, 2:793, *794–795*
Earth's atmosphere. *See* Atmosphere
Earth's chemistry. *See* Geochemistry
Earth's core. *See* Core (Earth)
Earth's crust. *See* Crust (Earth)
Earth's mantle. *See* Mantle (Earth)
Earth's orbit. *See* Orbit (Earth)
Earth's rotation. *See* Rotation (Earth)
Easements, conservation, 2:592
Easterlies (winds), 1:44–45, 248, 249
Eastern Northern Pacific gray whales, 2:730
Eaton Corp., 1:444
EC. *See* European Commission (EC)
Echinoderms, 2:542
Echo sounding, 2:629
Eco-sabotage. *See* Eco-terrorism
Eco-terrorism, 1:**224–229**
  environmental activism, 1:267
  environmental protests, 1:289–291
  Fedeal Bureau of Investigation, 1:226–229
  green movement, 1:378–379
  history, 1:224–225
  legal issues, 1:225–226
  radioactive waste, 2:683

Eco-tourism. *See* Tourism
Ecodefense (Foreman), 1:224
Ecodisasters, 1:**230–235,** *233*
climate change, 1:232–234
endangered species, 1:259
environmental protests, 1:292
floods, 1:314–316
hazardous waste, 1:414–417
history, 1:230–233
Hurricane Katrina, 1:437–440
insecticide use, 1:*471*
medical waste, 2:*564*
nuclear power, 2:610, 611, 612–614
Oil Pollution Acts, 2:636, 637
oil spills, 2:643
radioactive waste, 2:683
reforestation, 1:234–235
sustainable development, 2:770
tsunami impacts, 2:793–796
United Nations Convention on the Law of the Sea, 2:804
See also Earthquakes; Floods; Hurricanes; Landslides; Wildfires
Ecological competition, 1:**236–238**
coastal ecosystems, 1:152
coastal zones, 1:155–156
coral reefs and corals, 1:171
ecosystems, 1:243
endangered species, 1:259
habitat alteration, 1:404, 406
habitat loss, 1:408, 410
invasive species, 1:481–483
island ecosystems, 1:497
predator-prey relations, 2:678
reef ecosystems, 2:703
wildfires, 2:877
Ecological indicators, 1:497
Ecologists Union. *See* Nature Conservancy
Ecology
beginnings, 1:244
environmental activism, 1:*266*
industrial, 2:772
invasive species, 1:482
vegetation cycles, 2:816
Ecology experiments, closed. *See* Closed ecology experiments
The Ecology of Invasions by Animals and Plants (Elton), 1:482
Ecosystem diversity, 1:**239–241**
biodiversity, 1:60–64
biogeography, 1:74–75
genetic diversity, 1:335–337
marine ecosystems, 2:541–542, 543
reef ecosystems, 2:702–703, 704
Ecosystems, 1:**242–246**
alternative fuel impacts, 1:14
Antarctica, 1:21, 26
Arctic, 1:38
biodiversity, 1:60, 62, 63
climate change, 1:134
closed ecology experiments, 1:142–143
earthquakes, 1:218
ecological competition, 1:236–238
ecosystem diversity, 1:239–241
extinction and extirpation, 1:298–299
floods, 1:314
forest resources, 1:319
global warming, 1:369
grasslands, 1:372, 373
greenhouse effect, 1:387
habitat alteration, 1:404–406, 406
habitat loss, 1:409
international environmental law, 1:477
invasive species, 1:481–483
logging, 1:537
mathematical modeling and simulation, 2:556–557
National Park Service Organic Act, 2:586
natural resource management, 2:599
overgrazing, 2:655–657
pollinators, 2:673
predator-prey relations, 2:676–679
shrublands, 2:732
snow and ice cover, 2:741, 742
soil contamination, 2:746
species reintroduction programs, 2:756
spill remediation, 2:757, 758
tundra, 2:798–799
vegetation cycles, 2:816–817
war and conflict-related environmental destruction, 2:825, 827
wetlands, 2:864, 865
wildfires, 2:877–878
wildlife population management, 2:880–881
See also specific types of ecosystems
Ecotage. *See* Eco-terrorism
Ecotones, 2:864
Ecozones, 1:74
Ecuador, 1:*491,* 2:593
Edward I, King of England, 1:279
EEZs (Exclusive economic zones), 2:804
Effluent, 1:54, *398,* 2:832–833
Egypt, 1:48, 397, 494
El Niño and La Niña, 1:**247–251,** *248, 249, 250*
atmospheric circulation, 1:44, 45
coastal ecosystems, 1:152
coral reefs and corals, 1:171
greenhouse effect, 1:388
hurricanes, 1:432
reef ecosystems, 2:704
weather extremes, 2:862
El Salvador, 1:*34*
Electricity
clean energy, 1:118, 119
coal resource use, 1:145, 146, 149
nuclear power, 2:610, 611
solar power, 2:750, 751–752
wind and wind power, 2:888–889
Electronic Numerical Integrator and Computer (ENIAC), 1:139
Electronics waste, 1:**252–253,** *253*
hazardous waste, 1:414
health, 1:152
landfills, 1:525
recycling, 2:696–697
solid waste treatment technologies, 2:754
waste transfer and dumping, 2:829
Electrophoresis, 1:510
Elements (chemistry), 1:338, 2:624
ELF. *See* Earth Liberation Front (ELF)
Ellerman, Douglas Joshua, 1:228–229
Elton, Charles S., 1:482
Emanuel, Kerry, 1:433, 434
Embers, 2:878
Emergency Management, Office of, 1:162
Emergency Planning and Community Right-to-Know Act of 1986, 1:100
Emergency Wetlands Resource Act, 1:246
Emerging economies, industrialization. *See* Industrialization in emerging economies
Emerson, Ralph Waldo, 2:821
EMETIC (Evan Mecham Eco-Terrorist International Conspiracy), 1:229
Emissions. *See specific types of emissions*
Emissions standards, 1:**254–258**
carbon dioxide emissions, 1:87–89
Clean Air Act of 1970, 1:109, *110*
Clean Air Mercury Rule, 1:112–113
Clean Water Act, 1:123
corporate green movement, 1:175
Environmental Protection Agency, 1:283, 286–287
global warming, 1:370
greenhouse gases, 1:393–394

hybrid vehicles, 1:442
industrial pollution, 1:457
industrialization in emerging economies, 1:462, 464
international environmental law, 1:479
IPCC 2007 Report, 1:486–487
Kyoto Protocol, 1:500–508
laboratory methods in environmental science, 1:513
light pollution, 1:528, 529
liquefied natural gas resource use, 1:534
resource extraction, 2:713
smog, 2:740
United Nations Conference on the Human Environment, 2:801
United Nations Convention on the Law of the Sea, 2:804
United Nations Framework Convention on Climate Change, 2:806–809
Emissions trading. *See* Pollution credits
Encroachment, saltwater. *See* Saltwater intrusion
Endangered species, 1:**259–264**
biodiversity, 1:60–61, 62, 63, 64
Bureau of Land Management, 1:*82*
CITES, 1:105–107
clean energy, 1:120–121
cultural practices and environment destruction, 1:*180*
Endangered Species Act of 1973, 1:262–263
extinction and extirpation, 1:298–300
forests, 1:325
genetic diversity, 1:336–337
green movement, 1:381
habitat alteration, 1:*405*, 406, *406*
human impacts, 1:259–262, 426
inland fisheries, 1:467, *468*
International Convention for the Regulation of Whaling, 2:867–870
international environmental law, 1:477
invasive species, 1:483
island ecosystems, 1:496
marine fisheries, 2:546, 547
reforestation, 2:706, 708, 710
species reintroduction programs, 2:755
Tennessee Valley Authority v. Hill, 2:776–778
water supply and demand, 2:851–852
wildlife population management, 2:880, 881
wildlife protection policies and legislation, 2:883, 885
Endangered Species Act of 1973 (ESA)
endangered species and, 1:262–263
genetic diversity, 1:336–337
Tennessee Valley Authority v. Hill, 2:776–778
Endemic species, 1:497, 498
Endocrine disrupters, 1:456, 460, 2:790
Endres, Ed, 1:*511*
Energy. *See specific types of energy*
Energy Department, 1:287, 395
Energy Policy Act of 2005, 1:***16–20***
Energy recovery, 2:753–754
Energy Star program, 1:283
Energy Watch Group, 1:146
Engineering. *See specific types of engineering*
Engineers. *See specific types of engineers*
Engines
alternative fuel impacts, 1:14, 15, 16
aviation emissions, 1:49, 51
coal resource use, 1:145
hybrid vehicles, 1:441
England
acid rain, 1:1
air pollution, 1:10
biostratigraphy, 1:79
ecodisasters, 1:231
environmental crime, 1:279
environmental protests, 1:*290*
groundwater, 1:397
North Atlantic Oscillation, 2:*609*
nuclear power, 2:*613*
oil spills, 2:641
See also United Kingdom
Enhanced oil recovery (EOR), 1:394–395
ENIAC (Electronic Numerical Integrator and Computer), 1:139
Entertainment, environmental. *See* Media: Environmentally based news and entertainment
Environment Canada, 1:526
Environment News Service, 1:148
Environmental activism, 1:**265–268**
carbon dioxide emissions, 1:87, 88
coastal ecosystems, 1:*151*
Earth Day, 1:209–210
Earth Summit (1992), 1:212–214
Earth Summit (2002), 1:215–217
eco-terrorism, 1:224–229
environmental protests, 1:289–291, 292
forest resources, 1:317–319
global warming, 1:370
green movement, 1:375–382
International Convention for the Regulation of Whaling, 2:867, 868
media, 2:559
non-scientist contributions, 2:604–607
organic and locally grown foods, 2:645–648
organic farming, 2:649–651
Our Common Future, 2:813–815
recreational use and environmental destruction, 2:690–691
recycling, 2:693–697
seasonal migration, 2:730
Silent Spring, 2:734
sustainable development, 2:769–772
United Nations Conference on the Human Environment, 2:800–801
United Nations policy and activism, 2:810–812
Walden, 2:821–823
Environmental assessments, 1:**269–271**
corporate green movement, 1:175
environmental science careers, 1:96
National Environmental Policy Act, 2:577–581
wildlife protection policies and legislation, 2:883
Environmental benefits and liabilities of petroleum resource use, 1:**272–278**
drilling, 1:273–274
exploration, 1:273
history, 1:272
production, 1:274–275
refining, 1:275
transportation, 1:275
United States v. Standard Oil (1966), 1:276–278
Environmental Capital Partners, 1:177
Environmental careers. *See* Careers in environmental science
Environmental compliance officers, 1:97
Environmental crime, 1:224–229, **279–281**
Environmental Defense Fund, 1:88, 255, 395, 2:548
Environmental destruction, culture and. *See* Cultural practices and environmental destruction
Environmental engineering, 2:772
Environmental engineers, 1:96
Environmental estrogens. *See* Endocrine disrupters

Environmental health scientists, 1:96–97
Environmental impact assessments. *See* Environmental assessments
Environmental Integrity Project, 1:148
Environmental Investigation Agency, 1:376
Environmental law, international. *See* International environmental law
Environmental news and entertainment. *See* Media: Environmentally based news and entertainment
Environmental photography, 2:**669–671,** 822
Environmental protection. *See* Conservation
Environmental Protection Agency (EPA), 1:**282–285**
- acid rain, 1:2–3
- agricultural practice impacts, 1:6
- air pollution, 1:10
- algal blooms, 1:*12*
- biodegradation, 1:57
- bioremediation, 1:76, 78
- carbon dioxide emissions, 1:87, 88
- chemical spills, 1:100
- Clean Air Act of 1970, 1:108, 109
- Clean Air Mercury Rule, 1:112–113
- Clean Water Act, 1:123
- Comprehensive Environmental Response, Compensation, and Liability Act, 1:162
- electronics waste, 1:252
- emissions standards, 1:254, 255, 256, 257
- environmental crime, 1:279, 280
- environmental science careers, 1:95–96
- formation, 1:209
- groundwater quality, 1:402
- hazardous waste, 1:413
- Hurricane Katrina, 1:437, 438, 439, 440
- industrial water use, 1:460
- insecticide use, 1:472
- landfills, 1:522–525, 523, 1:525
- *Massachusetts v. Environmental Protection Agency* (2007), 1:286–287
- media, 2:559
- medical waste, 2:563–564
- mining and quarrying impacts, 2:571
- National Environmental Policy Act, 1:577
- National Oceanic and Atmospheric Administration, 2:582
- nonpoint-source pollution, 2:602
- Oil Pollution Acts, 2:636
- oil spills, 2:643
- smog, 2:740
- Superfund site, 2:763–764
- teratogens, 2:784
- toxic waste, 2:790
- water pollution, 2:838, 840
- weather extremes, 2:861
- wetlands, 2:866

Environmental Protection Agency (EPA) regulation: *Massachusetts v. Environmental Protection Agency* (2007), 1:109, **286–288**
Environmental protests, 1:**289–293,** *290, 291*
- aviation emissions, 1:*51*
- biodiversity, 1:*63*
- biofuels, 1:*69*
- carbon dioxide, 1:*84*
- coal resource use, 1:*147*
- conservation, 1:168
- dust storms, 1:*208*
- Earth Day, 1:209, 210
- eco-terrorism, 1:224–229
- electronics waste, 1:*253*
- environmental activism, 1:265–268
- environmental crime, 1:*280*
- genetic diversity, 1:*270, 337*
- green movement, 1:375–382
- greenhouse gases, 1:*393*
- habitat loss, 1:*410*
- International Convention for the Regulation of Whaling, 2:868
- lakes, 1:516
- land use, 1:*521*
- logging, 1:*537, 538*
- nuclear power, 2:610, 611–612, 612, 613
- nuclear test ban treaties, 2:*617, 618*
- oil spills, 2:*633*
- organic and locally grown foods, 2:*651*
- overfishing, 2:653
- paper and wood pulp, 2:*665*
- radioactive waste, 2:*684*
- rain forest destruction, 2:*687*
- recycling, 2:*695, 696*
- rivers and waterways, 2:*715*
- soil contamination, 2:*747*
- sustainable development, 2:*771*
- toxic waste, 2:*791, 792*
- war and conflict-related environmental destruction, 2:*827*
- wildlife protection policies and legislation, 2:*884, 885*

Environmental science
- careers, 1:95–99
- laboratory methods, 1:509–513
- media, 2:559–560
- non-scientist contributions, 2:604–607

Environmental scientists. *See* Careers in environmental science
Environmental stewardship. *See* Environmental activism
Environmental Systems Research Institute (ESRI), 1:342
Environmentalism. *See* Environmental activism
Enzymes, 1:57, 76
Eons, geologic. *See* Geologic history
EOR (Enhanced oil recovery), 1:394–395
EPA. *See* Environmental Protection Agency (EPA)
EPA Establishes Hazardous Waste Enforcement and Emergency Response System; Names 60 New Sites, 1:*283–284*
EPA's Record Settlement with Utility Could Lead to Other Deals, 1:*2–4*
Epicenters, 1:218, 220
Epifauna, 2:633
Epilimnion, 1:329
Epochs, geologic. *See* Geologic history
Equator, 1:44, 61, 2:674
Equator Principles, 1:175
Equinox, spring, 1:209
Eras, geologic. *See* Geologic history
Erika (ship), 2:642
Eritrea, 2:825
Erosion
- agricultural practice impacts, 1:5, 6
- carbon sequestration, 1:91
- clear-cutting, 1:125, 126
- coastal ecosystems, 1:150–151
- dredging, 1:202
- drought, 1:205, 206
- dust storms, 1:208
- ecodisasters, 1:232
- ecosystems, 1:245, 246
- forests, 1:324, 325
- grasslands, 1:373
- habitat loss, 1:409
- Hurricane Katrina, 1:439
- Lake Tahoe, 1:291–293
- landslides, 1:527
- logging, 1:536
- marine water quality, 2:549
- organic farming, 2:650
- overgrazing, 2:656
- precipitation, 2:675
- rain forest destruction, 2:686
- recreational use and environmental destruction, 2:691
- reef ecosystems, 2:704

resource extraction, 2:712
runoff, 2:716, 717
sea level rise, 2:726
soil chemistry, 2:744
soil resources, 2:748–749
tsunami impacts, 2:795
vegetation cycles, 2:817

Erosion From Tahoe Fire May Hurt Lake's Health, 1:***291–293***

Ertl, Gerhard, 1:513

Eruptions, volcanic. *See* Volcanoes

ESA. *See* Endangered Species Act of 1973 (ESA)

Escherichia coli. See E. coli

Eskers, 1:362

ESRI (Environmental Systems Research Institute), 1:342

Estrogens, environmental. *See* Endocrine disrupters

Estuaries, 1:**294–297**
algal blooms, 1:11
aquatic ecosystems, 1:30–31, 32
coastal ecosystems, 1:150
marine ecosystems, 2:543, 544
ocean salinity, 2:624, 625
water supply and demand, 2:851–852
See also Bays and estuaries; *specific types of estuaries*

Ethanol
alternative fuel impacts, 1:14, 15, 16, 19
biofuels, 1:67, 68, 69–72

Ethiopia, 1:*133,* 2:827

Ethyl alcohol. *See* Ethanol

Ethylene, 2:878

Etsy, Dan, 1:177

EU. *See* European Union (EU)

Euphotic zone, 1:31–32, 294, 2:542, 632

Euphrates River, 1:331

Europe
acid rain, 1:1–2
air pollution, 1:8
aviation emissions, 1:51
biofuels, 1:68, 70
clear-cutting, 1:125
climate change, 1:133
coal resource use, 1:145
dams, 1:183–184
ecodisasters, 1:232
El Niño and La Niña, 1:250
electronics waste, 1:253
emissions standards, 1:255, 257, 258
environmental crime, 1:279
environmental protests, 1:292
glacial retreat, 1:360
green movement, 1:375
hunting practices, 1:427
hydrologic cycle, 1:447
ice ages, 1:451
irrigation, 1:494
Kyoto Protocol, 1:500
maps and atlases, 2:539
natural resource management, 2:598
North Atlantic Oscillation, 2:608, 609
nuclear power, 2:610
organic and locally grown foods, 2:645, 646
overfishing, 2:653
precipitation, 2:675
recycling, 2:694
wetlands, 2:866
wind and wind power, 2:888

European Bank for Reconstruction and Development, 1:183

European Commission (EC)
acid rain, 1:1
aviation emissions, 1:51
glacial retreat, 1:360
Kyoto Protocol, 1:502

European Investment Bank, 1:183

European Space Agency, 1:10, 142, 462

European Union (EU), 1:51, 183–184, 491

Eustatic sea level rise. *See* Sea level rise

Eutrophication
algal blooms, 1:11–12
aquatic ecosystems, 1:32
genetic diversity, 1:336
habitat alteration, 1:406
habitat loss, 1:410
irrigation, 1:495
lakes, 1:514, 515
marine ecosystems, 2:544
marine water quality, 2:550, 551
nonpoint-source pollution, 2:602
oceans and coastlines, 2:634
runoff, 2:716, 717
tsunami impacts, 2:795
wastewater treatment technologies, 2:833
water pollution, 2:837, 840
watersheds, 2:856
wetlands, 2:865

Evan Mecham Eco-Terrorist International Conspiracy (EMETIC), 1:229

Evaporation
bays and estuaries, 1:53
greenhouse effect, 1:385, 387
greenhouse gases, 1:391
hurricanes, 1:430
hydrologic cycle, 1:445, 447
ice ages, 1:448
irrigation, 1:494
ocean salinity, 2:624–625
precipitation, 2:674
radiative forcing, 2:681
surface water, 2:764
water resources, 2:844

Evaporites, 2:570

Even As Economy Lags, Corporate 'Green' Push May Advance, 1:***177***

Everglades National Park, 2:*603, 717,* 720, 766, 865, 866

Evergreens, 1:323

Evolution
benthic ecosystems, 1:55, 56
biodiversity, 1:60
biogeography, 1:74–75
ecological competition, 1:237
ecosystems, 1:244
extinction and extirpation, 1:298, 299
Gaia hypothesis, 1:333
genetic diversity, 1:335
island ecosystems, 1:496
predator-prey relations, 2:678
wildfires, 2:877

Exclusive economic zones (EEZs), 2:804

Exhaust. *See* Air pollution

Exotic species. *See* Invasive species

Exploration
Antarctica, 1:21–23, 26
benthic ecosystems, 1:55, 56
oceanography, 2:629

Explosions, volcanic. *See* Volcanoes

Exports of waste. *See* Waste transfer and dumping

Extinction and extirpation, 1:**298–301**
biodiversity, 1:60–63, 64
carbon dioxide emissions, 1:86
CITES, 1:105, 106
climate change, 1:133
coral reefs and corals, 1:170–172
cultural practices and environment destruction, 1:179
ecological competition, 1:237
ecosystem diversity, 1:240–241
ecosystems, 1:245, 246
endangered species, 1:259
genetic diversity, 1:336
global warming, 1:369
greenhouse effect, 1:383
habitat alteration, 1:404
habitat loss, 1:412
human impacts, 1:426
hunting practices, 1:428
inland fisheries, 1:467
invasive species, 1:481
island ecosystems, 1:497–498
predator-prey relations, 2:678
Silent Spring, 2:734
species reintroduction programs, 2:755
volcanoes, 2:819
wildlife population management, 2:881

wildlife protection policies and legislation, 2:883
wildlife refuge, 2:886
See also Endangered species
Extirpation. *See* Extinction and extirpation
Extraction, resource. *See* Resource extraction
Extreme Ice Survey, 2:670
Extreme weather events. *See* Weather extremes
Exxon Valdez (ship)
biodegradation, 1:57
bioremediation, 1:76, *77*
ecodisasters, 1:230
Oil Pollution Acts, 2:636, 637
oil spills, 2:*642,* 643
spill remediation, 2:757, 758, 759–760
Eye (hurricanes), 1:431

F

FAA (Federal Aviation Administration), 2:583
FACE (Forests Absorbing Carbon-dioxide Emissions), 1:323
Factory farms, adverse effects of, 1:**302–305,** *303, 304*
agricultural practice impacts, 1:5–7
fish farming, 1:306–307
groundwater quality, 1:*403*
organic and locally grown foods, 2:645
organic farming, 2:649
Silent Spring, 2:734
water pollution, 2:838
watersheds, 2:856–857
Fairview Gardens, 2:648
Fallout, radioactive, 2:615–616, 617
Famine. *See* Food
FAO. *See* United Nations Food and Agriculture Organization (FAO)
Farming. *See* Agricultural practice impacts; *specific types of farming*
Farms, factory. *See* Factory farms, adverse effects of
Farms, operation of. *See* Agricultural practice impacts
FAS (Fetal alcohol syndrome), 2:782–783
Fault movements. *See* Tectonic plates
Faults (geology), 1:218, 219, 220, 221
Fauna. *See* Animals
FBI. *See* Federal Bureau of Investigation
FDA (Food and Drug Administration), 2:783
Fearnside, Philip, 1:182
Feces. *See* Animal waste
Federal Aviation Administration (FAA), 2:583
Federal Bureau of Investigation, 1:224, 225, 226–229, 378
Federal Emergency Management Agency, 1:437, 438
Federal Facilities Response and Reuse Office, 1:162
Federal Insecticide, Fungicide, and Rodenticide Act (FIFRA), 1:***471–473***
Federal Land Policy and Management Act of 1976, 2:873
Federal Water Pollution Control Act. *See* Clean Water Act
Federation of American Scientists, 2:616–617
FedEx, 1:443
Feedback, 1:234, 235, 385, 391, 2:681
FEMA. *See* Federal Emergency Management Agency
Fences, 2:578
Fens, 1:330
Feral cats, 1:481
Fermentation, 1:70
See also Cellulosic fermentation
Ferrel cell, 1:45
Fertilization (pollination). *See* Pollinators
Fertilization, iron. *See* Iron fertilization
Fertilizers
agricultural practice impacts, 1:5, 7
biodegradation, 1:57
biodiversity, 1:62, 63
biofuels, 1:69
bioremediation, 1:76, *77*
coastal ecosystems, 1:152
coastal zones, 1:155
groundwater quality, 1:*402*
irrigation, 1:493, 494
organic and locally grown foods, 2:645
organic farming, 2:649, 650, 651
Silent Spring, 2:734
soil chemistry, 2:743, 744
soil resources, 2:748
toxic waste, 2:791
wastewater treatment technologies, 2:832
Fetal abnormalities. *See* Birth defects
Fetal alcohol syndrome (FAS), 2:782–783
Fiber optics, 2:688, 689, 892
FIFRA (Federal Insecticide, Fungicide, and Rodenticide Act), 1:***471–473***
Filtration, water, 2:865, 866
Financing, 1:175
Finches, 1:237, 335, 481, 496
Finland, 2:684
Fire retardants, 2:878–879
Firebreaks, 2:874–875
Fires, 1:*123*
earthquakes, 1:223
eco-terrorism, 1:225, 227–228
ecodisasters, 1:231, 233
mining and quarrying impacts, 2:572
war and conflict-related environmental destruction, 2:825, *826*
wildfires, 2:877–879
Firewood. *See* Wood
Fish
agricultural practice impacts, 1:5, 6
algal blooms, 1:11, 12, 13
aquatic ecosystems, 1:31
benthic ecosystems, 1:55
biodiversity, 1:61, 63
coastal ecosystems, 1:150, 152
coastal zones, 1:154, 156
commercial fisheries, 1:160–161
conservation, 1:166
coral reefs and corals, 1:170–171
ecological competition, 1:237
estuaries, 1:294, 296, 297
fish farming, 1:306–309
freshwater and freshwater ecosystems, 1:329, 331
habitat alteration, 1:404–405
human impacts, 1:423–424
industrial pollution, 1:457
industrial water use, 1:460
inland fisheries, 1:467–468
invasive species, 1:482, 483
iron fertilization, 1:491
irrigation, 1:494
lakes, 1:516
marine ecosystems, 2:541, 542, 543, 544–545
marine fisheries, 2:546–548
marine water quality, 2:552
natural resource management, 2:599
North Atlantic Oscillation, 2:608, 609
ocean tides, 2:627
oceans and coastlines, 2:632, 633, 634
overfishing, 2:652–653
predator-prey relations, 2:678
reef ecosystems, 2:703, 704
Silent Spring, 2:734
teratogens, 2:784

United Nations Convention on the Law of the Sea, 2:803, 804
water pollution, 2:840
water supply and demand, 2:851–852
See also specific types of fish
Fish and Wildlife Coordination Act, 2:886
Fish and Wildlife Service (FWS)
coastal ecosystems, 1:150
endangered species, 1:262
species reintroduction programs, 2:755
Tennessee Valley Authority v. Hill, 2:776, 778
Wilderness Act of 1964, 2:873
wildlife refuge, 2:886
Fish farming, 1:**306–309,** *307, 308*
commercial fisheries, 1:*161*
industrialization in emerging economies, 1:463
inland fisheries, 1:467
marine fisheries, 2:547
overfishing, 2:653
Fish hatcheries. *See* Fish farming
Fisher, Howard T., 1:341
Fisher, Richard T., 2:707
Fisheries. *See specific types of fisheries*
Fishing, excessive. *See* Overfishing
Fishing stock. *See* Fish
Fission-track dating, 1:348
Fitness, species. *See* Evolution
Fixed air. *See* Carbon dioxide
Fjords, 1:53, 294
Flash floods, 1:315
Fleming, Alexander, 1:336, 2:666
Flood control and floodplains, 1:**310–313**
irrigation, 1:493
rivers and waterways, 2:714
sea level rise, 2:725–728
streamflow, 2:761, 762
Watershed Protection and Flood Prevention Act of 1954, 2:854–855
wetlands, 2:864–865
Floods, 1:*311,* **314–316,** *519*
coastal ecosystems, 1:150
coastal zones, 1:155
deforestation, 1:190
drainage basins, 1:200
dredging, 1:203
earthquakes, 1:222
flood control and floodplains, 1:310–313
freshwater and freshwater ecosystems, 1:330, 331
glacial retreat, 1:359
greenhouse effect, 1:383
habitat alteration, 1:405
habitat loss, 1:409
Hurricane Katrina, 1:439
precipitation, 2:674–675
runoff, 2:717
water resources, 2:844
weather extremes, 2:862
Floodwalls. *See* Levees and sea walls
Floor, ocean. *See* Benthic ecosystems
Flora. *See* Plants (organisms)
Floriculture, 1:420
Florida
aquifers, 1:34
atmospheric inversions, 1:48
benthic ecosystems, 1:*56*
carbon dioxide emissions, 1:88
coal resource use, 1:148
corporate green movement, 1:177
dams, 1:184
ecological competition, 1:237
endangered species, 1:263
groundwater quality, 1:*402*
hurricanes, 1:*432*
insecticide use, 1:*470*
land use, 1:*520*
natural reserves and parks, 2:590
saltwater intrusion, 2:720
Superfund site, 2:764
surface water, 2:766
wetlands, 2:866
wildlife refuge, 2:886
Florida Power & Light, 1:177
Flow, stream. *See* Streamflow
Flowers. *See* Plants (organisms)
Food
agricultural practice impacts, 1:5
alternative fuel impacts, 1:16, 19
biodiversity, 1:61
biofuels, 1:67, 70, 71
groundwater quality, 1:403
industrial pollution, 1:457
organic and locally grown foods, 2:645–648
teratogens, 2:784
war and conflict-related environmental destruction, 2:827–828
Food and Agriculture Organization (FAO). *See* United Nations Food and Agriculture Organization (FAO)
Food and Drug Administration (FDA), 2:783
Food chains and webs
Antarctica, 1:21, 26
biodiversity, 1:61
carbon dioxide emissions, 1:86
coastal ecosystems, 1:150, 152
coastal zones, 1:156
DDT, 1:187
ecosystems, 1:242, 246
estuaries, 1:294
habitat alteration, 1:404–405
industrial pollution, 1:457
insecticide use, 1:469, 471
invasive species, 1:482
iron fertilization, 1:490
marine ecosystems, 2:541, 544
marine fisheries, 2:546–547
marine water quality, 2:549
oceans and coastlines, 2:632
predator-prey relations, 2:676, 677, 678
runoff, 2:716, 717
Silent Spring, 2:734
snow and ice cover, 2:741, 742
soil contamination, 2:746
species reintroduction programs, 2:755
wetlands, 2:864
Food shortages. *See* Food
Food webs. *See* Food chains and webs
Foraminiferans, 1:55, 80, 86, 722–723
Forcing, radiative. *See* Radiative forcing
Ford, Henry, 1:14, 441
Forecasting
El Niño and La Niña, 1:251
landslides, 1:528
National Oceanic and Atmospheric Administration, 2:582–583
weather and climate, 2:859–860
Foreman, David, 1:224
Forest clearing. *See* Clear-cutting
Forest fires. *See* Wildfires
Forest reduction. *See* Deforestation
Forest resources, 1:**317–320**
deforestation, 1:189, 190–191
ecodisasters, 1:234–235
forests and, 1:321–325
logging, 1:536–537
paper and wood pulp, 2:664–665
reforestation, 2:706–710
wildfire control, 2:874–876
Forest restoration. *See* Reforestation
Forest Service. *See* National Forest Service
Forestland. *See* Forests
Forestry. *See* Forests
Forests, 1:**321–325**
acid rain, 1:1
agricultural practice impacts, 1:5, 6–7
biodiversity, 1:62, 63
carbon dioxide, 1:83
carbon dioxide emissions, 1:85
carbon sequestration, 1:90–91
clear-cutting, 1:125–127
climate change, 1:134
conservation, 1:166, 167
deforestation, 1:189–191

dendrochronology, 1:192–193
ecodisasters, 1:234–235
ecosystem diversity, 1:240
ecosystems, 1:243, 244–245, 246
endangered species, 1:260
forest resources and, 1:317–319
habitat alteration, 1:406
logging, 1:536–537
natural reserves and parks, 2:590
natural resource management, 2:598
precipitation, 2:675
reforestation, 2:706–710
vegetation cycles, 2:816
volcanoes, 2:820
war and conflict-related environmental destruction, 2:827
wildfire control, 2:874–876
wildfires, 2:877–879
See also specific types of forests

Forests Absorbing Carbon-dioxide Emissions (FACE), 1:323

Formal biostratigraphy. *See* Biostratigraphy

Fossil fuel combustion impacts, 1:**326–328**
acid rain, 1:1, 2
alternative fuel impacts, 1:14, 15, 16
atmosphere, 1:42
aviation emissions, 1:49
biofuels, 1:67, 68, 69, 70–71
carbon dioxide, 1:83
carbon dioxide emissions, 1:85, 86
carbon sequestration, 1:90, 91
clean energy, 1:118–119
climate change, 1:130
coal resource use, 1:147
dams, 1:182
dust storms, 1:207
ecosystems, 1:243
floods, 1:315
glacial retreat, 1:359
global warming, 1:366, 367
greenhouse effect, 1:383, 388
greenhouse gases, 1:390, 393, 395
human impacts, 1:423, 425–426
ice cores, 1:452
liquefied natural gas resource use, 1:533
organic farming, 2:650
radiative forcing, 2:681
sustainable development, 2:771

Fossils
biogeography, 1:75
biostratigraphy, 1:79–80
carbon dioxide emissions, 1:85
coastal zones, 1:*156*
geologic history, 1:345, 346, 347
glaciation, 1:362
predator-prey relations, 2:676, 678
sea level rise, 2:722–723

Fourier, Joseph, 1:129

Fourth Assessment Report (2007). *See* IPCC 2007 Report

Fragmentation, habitat. *See* Habitat alteration

Framework Convention On Climate Change, 2:***807–809***

France
air pollution, *1:9*
Antarctic Treaty, 1:25, 27
ecodisasters, 1:230
environmental protests, 1:289, 290
green movement, 1:376, 378, 379
Kyoto Protocol, 1:503
nuclear power, 2:612
nuclear test ban treaties, 2:616, *618*
oil spills, 2:641, 642
tidal or wave power, 2:785, 788

Fraser, Neal, 2:647, 648

Freedom-of-the-seas doctrine, 2:803

Freon. *See* Chlorofluorocarbons (CFCs)

Fresh Kills landfill, New York, 1:522, 562

Freshwater and freshwater ecosystems, 1:**329–331**
acid rain, 1:1
algal blooms, 1:12–13
as aquatic ecosystems, 1:30, 31
aquifers, 1:33–35
bays and estuaries, 1:53–54
benthic ecosystems, 1:55, *56*
biodiversity, 1:61, 62
climate modeling, 1:*140*
coastal ecosystems, 1:150
coastal zones, 1:154, 155–156
conservation, 1:168
dredging, 1:202, 203
estuaries, 1:294–297
geologic history, 1:349
glaciation, 1:363
groundwater quality, 1:401, 402
hydrologic cycle, 1:447
industrial water use, 1:459
inland fisheries, 1:467-468
invasive species, 1:482
irrigation, 1:493, 494
Lake Tahoe, 1:291–293
lakes, 1:514–517
nonpoint-source pollution, 2:602
ocean circulation and currents, 2:620, 622
ocean salinity, 2:624
runoff, 2:716
saltwater intrusion, 2:719–720
water pollution, 2:838
water resources, 2:843
water supply and demand, 2:848
wetlands, 2:864, 866

Fresno, California, 1:522

Friends of the Earth, 1:266, 376

Fringing reefs, 2:703

Frogs, 2:*783*
Monteverde harlequin tree, 1:62, 300
yellow-legged, 1:*406*

Fuel cells, 1:16–17, *17*, 513

Fuel loads, 2:878

Fuels, alternative. *See* Alternative fuel impacts

Fuels, fossil. *See* Fossil fuel combustion impacts

Fuelwood. *See* Wood

Fumigants, 1:469

Fungi
biodegredation, 1:57–58
biodiversity, 1:62
bioremediation, 1:76
ecosystems, 1:242
pharmaceutical development resources, 2:666

Furans, 1:10, 76, 455

G

Gabriel, Sigmar, 2:612

Gaia: A New Look at Life on Earth (Lovelock), 1:332–333

Gaia hypothesis, 1:**332–334,** *334*

Galactopyranose, 1:509

Galápagos Islands, 1:481, 483, 491, 496, *498*, 2:593

Gamma rays, 2:683

Gandhi, Mohandas, 2:822, 823

Garbage, municipal. *See* Municipal waste

Gardening, 1:420, *495*

Gas, natural. *See* Natural gas

Gases, greenhouse. *See* Greenhouse gases

Gases, trace, 1:9, 41

Gasification, 2:753, 754

Gasoline. *See* Fossil fuel combustion impacts

Gause, Georgi, 1:236–237, 408

Gause's law. *See* Competitive exclusion

GCM. *See* General Circulation Model (GCM)

GE (General Electric), 1:174–175, 380

Geese. *See* Waterfowl

General Circulation Model (GCM), 1:139, 368–369, 386, 388, 393

General Electric (GE), 1:174–175, 380
General Land Office, 1:81
General Motors, 1:175
Generators, hydrofoil, 2:786
Genetic diversity, 1:**335–337**
  biogeography, 1:74–75
  ecosystem diversity, 1:239, 240
  organic farming, 2:650
  pharmaceutical development resources, 2:666
  pollinators, 2:673
Genetic engineering
  biodegredation, 1:57, 58
  bioremediation, 1:76–78
  Bureau of Land Management, 1:81–82
  herbicides, 1:418, 419
  horticulture, 1:420
  insecticide use, 1:470
  laboratory methods in environmental science, 1:512
  Oil Pollution Acts, 2:637
  oil spills, 2:641
  organic and locally grown foods, 2:645, *651*
  organic farming, 2:649
  spill remediation, 2:758
Genoa, Italy, 2:*651*
Genomes, 1:335, 337
Genotoxic waste. *See* Toxic waste
Geochemistry, 1:**338–340,** 2:743–744
Geocoding, 1:341
Geoffroy Saint-Hilaire, Étienne, 2:782
Geographic Information Retrieval (GIR), 1:342–343
Geographic Information Systems (GIS), 1:**341–344**
  geospatial analysis, 1:351
  maps and atlases, 2:539
  mathematical modeling and simulation, 2:557
  surveying, 2:767
Geography, biological. *See* Biogeography
Geologic carbon emission storage. *See* Carbon sequestration
Geologic history, 1:**345–350,** *347*
  air pollution, 1:8
  Antarctica, 1:22
  bays and estuaries, 1:53
  biodiversity, 1:60
  biogeography, 1:74
  biostratigraphy, 1:79–80
  carbon dioxide emissions, 1:85
  dendrochronology, 1:192–193
  development, 1:345–346
  dust storms, 1:207
  ecosystem diversity, 1:240
  endangered species, 1:259–260
  extinction and extirpation, 1:298, 299
  forests, 1:321
  Gaia hypothesis, 1:333
  glacial retreat, 1:357–358
  glaciation, 1:362–363
  global warming, 1:366
  hierarchy, 1:346–347
  ice ages, 1:448–450
  ice cores, 1:452–453, 454
  IPCC 2007 Report, 1:486
  ocean salinity, 2:624
  predator-prey relations, 2:676, 678
  radiometric age, 1:347–349
  sea level rise, 2:721–722
  United Nations policy and activism, 2:811
  volcanoes, 2:819
Geologic record. *See* Geologic history
Geologic Society of America (GSA), 1:347
Geologic time. *See* Geologic history
Geologic timescale. *See* Geologic history
Geological engineers, 1:96
Geological Survey (U.S.)
  Arctic, 1:38
  earthquakes, 1:218
  groundwater, 1:398, *399*
  landslides, 1:527
  resource extraction, 2:712
  water pollution, 2:839
Geophysical surveying. *See* Surveying
Georeferencing, 1:351–352
Georgia, 1:148, 2:*591*
Geospatial analysis, 1:**351–353**
  Geographic Information Systems (GIS), 1:341–344
  maps and atlases, 2:540
  photography, 2:669, 670
Geosphere, 1:83
Geothermal resources, 1:**354–356**
  clean energy, 1:119
  greenhouse gases, 1:391
Gerbera daisies, 1:*422*
Germany
  acid rain, 1:1
  Clean Development Mechanism, 1:116
  coal resource use, 1:146
  green movement, 1:379
  nuclear power, 2:612
  radioactive waste, 2:*684*
  recycling, 2:694
The Geysers, 1:354, 355
Ghana, 1:324
GIR (Geographic Information Retrieval), 1:342–343
Girard, Pierre, 2:784
GIS. *See* Geographic Information Systems (GIS)
Gishwati Forest, 2:708
Glacial ablation. *See* Glacial retreat
Glacial accumulation. *See* Glaciation
Glacial melting. *See* Glacial retreat
Glacial periods, 1:448, 449, 450, 451, 452, 490
Glacial retreat, 1:**357–361,** *358, 360*
  biodiversity, 1:63
  biofuels, 1:69
  carbon dioxide, 1:84
  climate change, 1:134
  ecosystem diversity, 1:240
  endangered species, 1:261
  fossil fuel combustion impacts, 1:328
  glaciation and, 1:362–365
  greenhouse effect, 1:383, 385, 388
  hydrologic cycle, 1:445
  ice ages, 1:448–451
  ice cores, 1:453
  Intergovernmental Panel on Climate Change, 1:475
  IPCC 2007 Report, 1:486, 488
  lakes, 1:514
  photography, 2:670
  reef ecosystems, 2:703
  sea level rise, 2:721, 723–724
  snow and ice cover, 2:742
  surface water, 2:765
  water resources, 2:844
Glaciation, 1:**362–365,** *449*
  atmospheric circulation, 1:45
  biodiversity, 1:63
  glacial retreat and, 1:357, 358
  greenhouse effect, 1:387
  ice ages, 1:448–451
Glacier ablation. *See* Glacial retreat
Glacier accumulation. *See* Glaciation
Glacier melting. *See* Glacial retreat
Glaciers
  Antarctica, 1:21, 24
  bays and estuaries, 1:53
  glacial retreat, 1:357–360
  glaciation, 1:362–365
  ice ages, 1:448, 451
  ice cores, 1:452, 453
  photography, 2:670
  water resources, 2:843
Glass, Jeff, 1:292
Glass recycling. *See* Recycling
Gleason, Herbert, 2:822
Gleick, Peter, 2:852
Glen Canyon Dam, 1:167
Global Biodiversity Strategy, 1:262
Global climate change. *See* Climate change
Global Island Partnership, 1:498

Global Positioning Systems (GPS), 1:*342, 343,* 351
Global Standard Stratigraphy Ages (GSSA), 1:346
Global Stratotype and Point (GSSP), 1:346, 3467
Global warming, 1:**366–371**
air pollution, 1:8
alternative fuel impacts, 1:16
Antarctica, 1:23–24, 26
aquatic ecosystems, 1:32
Arctic, 1:36, 37
atmosphere, 1:41–43
atmospheric circulation, 1:44, *45,* 45–46
aviation emissions, 1:50, 51
biodegredation, 1:57
biodiversity, 1:61, 62, 63, 64–65
biofuels, 1:69
biogeography, 1:74, 75
carbon dioxide, 1:83–84
carbon dioxide emissions, 1:85, 86
carbon sequestration, 1:90, 92
Clean Development Mechanism, 1:115–117
clean energy, 1:118–119
clear-cutting, 1:127
climate change, 1:128, 129–131
climate modeling, 1:140–141
coastal ecosystems, 1:152
coral reefs and corals, 1:170, *171,* 171–172
corporate green movement, 1:175–176
dams, 1:181–183
desertification, 1:194
drought, 1:205, 206
ecodisasters, 1:233–234
ecosystems, 1:245
El Niño and La Niña, 1:250, 251
emissions standards, 1:257–258
endangered species, 1:261, 263
environmental science careers, 1:99
extinction and extirpation, 1:300
flood control and floodplains, 1:310
floods, 1:315
fossil fuel combustion impacts, 1:328
Gaia hypothesis, 1:332, 334
geochemistry, 1:340
geologic history, 1:349
geothermal resources, 1:354
glacial retreat, 1:358, 360
glaciation, 1:363
greenhouse effect, 1:383–388
greenhouse gases, 1:366–368, 390–395
habitat alteration, 1:406
horticulture, 1:421
human impacts, 1:423, 424, *425,* 425–426
hurricanes, 1:430, 432-435
hydrologic cycle, 1:445
ice cores, 1:452–453, 454
Intergovernmental Panel on Climate Change, 1:474–475
international environmental law, 1:479
IPCC 2007 Report, 1:486–488
lakes, 1:517
landfills, 1:526
landslides, 1:528
logging, 1:537
marine ecosystems, 2:544
media, 2:559, 560–561
migratory species, 2:569
National Oceanic and Atmospheric Administration, 2:582, 583–584
non-scientist contributions, 2:605
North Atlantic Oscillation, 2:609
nuclear power, 2:611, 613
ozone hole, 2:660
photography, 2:670
pollinators, 2:673
predicted consequences, 1:368–370
radiative forcing, 2:681
rain forest destruction, 2:687
recycling, 2:693
red tide, 2:699
reforestation, 2:706
resource extraction, 2:713
sea level rise, 2:721–722, 726
shrublands, 2:732, 733
snow and ice cover, 2:741
surface water, 2:764
sustainable development, 2:771
temperature records, 2:779–780
tundra, 2:798, 799
United Nations Framework Convention on Climate Change, 2:806–809
Walden, 2:822
water pollution, 2:840
weather and climate, 2:860
weather extremes, 2:861–862
wildfires, 2:877, 879
wind and wind power, 2:889
Globalization, 1:481
Glucose, 1:90, 2:817
Glyphosphate, 1:418, 419
Goats, 1:481, 483, *499*
Golden monkeys, 1:381
Goldschmidt, Victor, 1:338
Gore, Al
biodiversity, 1:65
environmental activism, 1:268
environmental protests, 1:290
green movement, 1:376–377
greenhouse effect, 1:*386*
Intergovernmental Panel on Climate Change, 1:475
IPCC 2007 Report, 1:488
Kyoto Protocol, 1:501
non-scientist contributions, 2:605
Government Accounting Office (GAO), 2:876
Government regulation
acid rain, 1:2, 3
air pollution, 1:*9,* 10
aviation emissions, 1:51
carbon dioxide emissions, 1:87–88
Clean Air Act of 1970, 1:108–111
Clean Water Act, 1:122–124
coastal ecosystems, 1:152
coastal zones, 1:156
DDT, 1:186, 187, 188
emissions standards, 1:254–258
Environmental Protection Agency, 1:282–284, 286–287
environmental protests, 1:290
habitat loss, 1:412
hazardous waste, 1:413, 414
industrial pollution, 1:457–458
industrialization in emerging economies, 1:462, 463, 464
insecticide use, 1:469–470, 471–473
International Convention for the Regulation of Whaling, 2:867–870
international environmental law, 1:479
IPCC 2007 Report, 1:486–487
laboratory methods in environmental science, 1:509, 513
landfills, 1:523–526
light pollution, 1:528
marine fisheries, 2:548
marine water quality, 2:552–555
medical waste, 2:552, 562, 563–566
Montreal Protocol, 2:575
National Environmental Policy Act, 2:577–581
National Oceanic and Atmospheric Administration, 2:582–584
National Park Service Organic Act, 2:585–587
natural reserves and parks, 2:589–596
nonpoint-source pollution, 2:601, 602–603
Oil Pollution Acts, 2:636–639
oil spills, 2:643
organic and locally grown foods, 2:645
overfishing, 2:653
overgrazing, 2:656
ozone layer, 2:662
pharmaceutical development resources, 2:667
predator-prey relations, 2:678

recreational use and environmental destruction, 2:690–691
recycling, 2:695, 697
reef ecosystems, 2:705
reforestation, 2:707, 709, 710
resource extraction, 2:712
runoff, 2:717–718
Silent Spring, 2:735
smog, 2:739, 740
soil contamination, 2:745
spill remediation, 2:757, 758, 759
sustainable development, 2:772
Tennessee Valley Authority v. Hill, 2:776–778
teratogens, 2:783–784
toxic waste, 2:790, 791
tsunami impacts, 2:796
United Nations Conference on the Human Environment, 2:800–801
United Nations Convention on the Law of the Sea, 2:803–805
waste transfer and dumping, 2:829–830
wastewater treatment technologies, 2:833, 834
water conservation, 2:835
water pollution, 2:839, 840–842
Watershed Protection and Flood Prevention Act of 1954, 2:854–855
watersheds, 2:858
wetlands, 2:866
Wilderness Act of 1964, 2:872–873
wildfire control, 2:874–876
wildlife population management, 2:880
wildlife protection policies and legislation, 2:883–885

Government standards. *See* Government regulation

Grace Restaurant, 2:647

Grafting, 1:420

Grasses. *See* Grasslands

Grasslands, 1:**372–374,** *373, 374*
aviation emissions, 1:49
biodiversity, 1:60, 62
biofuels, 1:71
conservation, 1:167
deserts, 1:198
dust storms, 1:208
ecosystems, 1:243, 245
flood control and floodplains, 1:*312*
forests, 1:321
greenhouse effect, 1:387
invasive species, 1:483
overgrazing, 2:655–657
wildfires, 2:878

Grassroots environmental activism. *See* Environmental activism

Gravel, 2:743

Gravity, 2:626, 788–789

Grazers, 1:154

Grazing, excessive. *See* Overgrazing

Grazing capacity. *See* Carrying capacity

Grazing Service (U.S.), 1:81

Great Barrier Reef, 1:152, 171, *172,* 411, 2:543, 704, 705

Great Britain. *See* United Kingdom

Great Lakes
coastal zones, 1:155–156
ecological competition, 1:237
glacial retreat, 1:357
invasive species, 1:482, *483*
as lakes, 1:514, 516, 517

Great Lakes Water Quality Agreement, 1:517

Great Plains Energy, 1:177

Great Sand Dunes National Park, 2:592

Great skuas, 2:547

Greece, 1:*187, 308,* 522, *525*

Green Belt Movement, 2:773–774

Green chemistry, 1:414, 455, 2:666, 667–668

Green movement, 1:**375–382,** *378, 379*
biodiversity, 1:64
China, 1:376, 378, 381–382
corporate green movement, 1:174–177
environmental activism, 1:265–268
hazardous waste, 1:414
history, 1:375—377
hybrid vehicles, 1:441–442
industrialization in emerging economies, 1:*463*
organic and locally grown foods, 2:645–648
organic farming, 2:649–651
political influence, 1:377–379
popularity, 1:379–381
recycling, 2:693–697
Silent Spring, 2:734
sustainable development, 2:769–772
Walden, 2:821–823

Green parties (Politics), 1:266, 289, 377–379

Green power. *See* Clean energy

Green products. *See* Green movement

Green Restaurant Association, 2:648

Green Revolution, 1:418, 2:650

Greene, Nathanael, 1:72

Greenhouse effect, 1:**383–389**
acid rain, 1:2
atmosphere, 1:42
carbon dioxide, 1:83–84
carbon dioxide emissions, 1:85
Clean Development Mechanism, 1:115–117
climate change, 1:128, 129, 130, 132, 385–386
explained, 1:383–385
global warming, 1:366–368
greenhouse gases, 1:390–391
impacts, 1:386–388
Intergovernmental Panel on Climate Change, 1:475
mining and quarrying impacts, 2:573
radiative forcing, 2:680

Greenhouse gases, 1:**390–395,** *392*
acid rain, 1:2
agricultural practice impacts, 1:6
air pollution, 1:9
alternative fuel impacts, 1:14
Arctic, 1:37
atmosphere, 1:41–43
atmospheric circulation, 1:46
aviation emissions, 1:49, 51
biodegredation, 1:57
biodiversity, 1:64, 65
biofuels, 1:67, 69, 71, 72
carbon dioxide, 1:83
carbon dioxide emissions, 1:85, 87–88
carbon sequestration, 1:90, 91–92, 394–395
Clean Air Act of 1970, 1:108, *110*
Clean Development Mechanism, 1:115–117
clean energy, 1:118–119, 121
climate change, 1:128, 129, 131, 133, 135
coal resource use, 1:147
dams, 1:181, 182
Earth Summit (1992), 1:213–214
ecodisasters, 1:234–235
emissions standards, 1:254, 255, 257, 258
Environmental Protection Agency, 1:286–287
fossil fuel combustion impacts, 1:327
global warming, 1:366–370
greenhouse effect, 1:383–386, 388, 390–391
hurricanes, 1:430
hybrid vehicles, 1:442–443
ice ages, 1:449, 450
ice cores, 1:448, 453
impacts, 1:393–394
industrialization in emerging economies, 1:462–463
insolation, 1:390, 391–392
Intergovernmental Panel on Climate Change, 1:474, 475

international environmental law, 1:479
IPCC 2007 Report, 1:486, 488
Kyoto Protocol, 1:500–508
landfills, 1:526
logging, 1:537
Montreal Protocol, 2:574
North Atlantic Oscillation, 2:609
nuclear power, 2:612
radiative forcing, 2:680
recycling, 2:693
resource extraction, 2:712, 713
scientific disputes, 1:392–393
solar power, 2:750
temperature records, 2:780
tundra, 2:798, 799
United Nations Framework Convention on Climate Change, 2:806–809
United Nations policy and activism, 2:811
wind and wind power, 2:889
See also specific greenhouse gases

Greenland
air pollution, 1:8
Arctic, 1:36
carbon dioxide emissions, 1:85
climate change, 1:*131*
deserts, 1:197
ecosystem diversity, 1:240
geologic history, 1:349
glacial retreat, 1:357, 358, 359, 360
glaciation, 1:362, 363
global warming, 1:368
greenhouse effect, 1:385, 386, 388
greenhouse gases, 1:392
hydrologic cycle, 1:445, 447
ice cores, 1:452
IPCC 2007 Report, 1:486, 488
ocean circulation and currents, 2:622
sea level rise, 2:721–722, 723–724, 725

Greenpeace
environmental activism, 1:265–266, 267, *267*
environmental assessments, 1:*270*
environmental protests, 1:289–290, *290, 337*
green movement, 1:375–376, 378
Kyoto Protocol, 1:503
nuclear test ban treaties, 2:*617*
reforestation, 2:709
toxic waste, 2:*791*
waste transfer and dumping, 2:*831*
wildlife protection policies and legislation, 2:*885*

Greenwashing. *See* Corporate green movement

Gregg, Norman, 2:783

Gregoire, Christine, 1:121

Grinnell, George Bird, 2:606

Groundwater, 1:**396–400**
drainage basins, 1:200
groundwater quality and, 1:401–403
hydrologic cycle, 1:445
industrial water use, 1:460
irrigation, 1:493
landfills, 1:523, 524, 525
water pollution, 2:840
water resources, 2:843, 844
water supply and demand, 2:848, 849
wetlands, 2:864

Groundwater quality, 1:**401–403**
agricultural practice impacts, 1:5–6, 7
aquifers, 1:33–35
carbon sequestration, 1:92, 93
factory farms, 1:302, 303–304
groundwater and, 1:396
mining and quarrying impacts, 2:571
nuclear power, 2:612
radioactive waste, 2:684
runoff, 2:716
saltwater intrusion, 2:719–720
soil contamination, 2:745
toxic waste, 2:790
water resources, 2:844
watersheds, 2:857

Growth rings, 1:192

Grunion, 2:627

GSA (Geologic Society of America), 1:347

GSSA (Global Standard Stratigraphy Ages), 1:346

GSSP (Global Stratotype and Point), 1:346, 347

Gulf of Mexico
algal blooms, 1:13
dredging, 1:203–204
hurricanes, 1:433
marine ecosystems, 2:543
red tide, 2:699
Watershed Protection and Flood Prevention Act of 1954, 2:854, 855
watersheds, 2:856
weather extremes, 2:862

Gulf Stream, 2:620, 621, *621*

Gulf War, 1:231, 2:825

Gulfport, Mississippi, 1:314, 437, *437*

Gymnosperms. *See* Trees

Gyres, 2:620–622

H

H_2. *See* Water vapor (H_2)

Haber-Bosh process, 1:513

Habitat alteration, 1:**404–407**
biogeography, 1:75
clean energy, 1:119–120
coastal ecosystems, 1:152
coastal zones, 1:154, 156
deforestation, 1:190
endangered species, 1:261
genetic diversity, 1:336
global warming, 1:369
greenhouse effect, 1:387
habitat loss, 1:410
human impacts, 1:423, 425
industrial pollution, 1:457
migratory species, 2:568–569
wildlife refuge, 2:886

Habitat destruction. *See* Habitat loss

Habitat loss, 1:**408–412**
biodiversity, 1:60, 61
CITES, 1:106
desertification, 1:195
endangered species, 1:259, 260, 263
estuaries, 1:296
extinction and extirpation, 1:298–299, 299
flood control and floodplains, 1:313
forests, 1:321
genetic diversity, 1:336, 337
habitat alteration, 1:404, 406
human impacts, 1:423, 425, 426
marine fisheries, 2:547
marine water quality, 2:550
mining and quarrying impacts, 2:571
pollinators, 2:672, 673
reforestation, 2:708, 709
Tennessee Valley Authority v. Hill, 2:776–778

Habitats
coastal ecosystems, 1:151
estuaries, 1:294–295
forests, 1:323
genetic diversity, 1:335–336
habitat alteration, 1:404–407
habitat loss, 1:408–412
seasonal migration, 2:730–733
wildlife population management, 2:880, 881
wildlife protection policies and legislation, 2:884

Hadal zone, 2:542–542, 632

Hadley cell, 1:45

Haeckel, Ernst, 1:244

Haiti, 1:70, 179–180, *195,* 324

Half-lives, 1:348, 2:683

Halley, Linda, 2:648

Haloalkanes, 2:574

Halocarbons, 1:366

Halogenated pyrroles, 1:470

Halpern, David, 1:251

Hanford Nuclear Reservation, 2:683–684
Hannon, Allison, 1:177, 443
Harbors. *See* Estuaries
Harmful algal blooms (HAB). *See* Algal blooms
Harmonic tremors, 1:220
Harvard Forest, 2:707
Harvard Laboratory for Computer Graphics and Spatial Analysis, 1:341, 342
Hatcheries, fish. *See* Fish farming
Haughley Experiment, 2:649
Hawaii
air pollution, 1:8
climate change, 1:130
dams, 1:184
emissions standards, 1:257
endangered species, 1:260
fish farming, 1:306
island ecosystems, 1:*499*
landfills, 1:*524*
tsunami impacts, 2:793
volcanoes, 2:819
Hawaiian monk seals, 1:*484*
Hazard creep, 1:184
Hazardous waste, 1:**413–417**
asbestos contamination, 1:39–40
chemical spills, 1:100–101
Comprehensive Environmental Response, Compensation, and Liability Act, 1:162–165
Environmental Protection Agency, 1:283–284
groundwater, 1:398
Hurricane Katrina, 1:*437*
industrial water use, 1:459
industrialization in emerging economies, 1:464
marine water quality, 2:549–550, 551–552
medical waste, 2:562–566, *563*
nonpoint-source pollution, 2:601, 603
pharmaceutical development resources, 2:667
solid waste treatment technologies, 2:753, 754
Superfund site, 2:762–763
toxic waste *vs.,* 2:790
war and conflict-related environmental destruction, 2:825
waste transfer and dumping, 2:829
wastewater treatment technologies, 2:832–834
watersheds, 2:857
Haze, atmospheric. *See* Smog
Hazelwood, Joseph J., 2:643
HCFCs. *See* Hydrochlorofluorocarbons
Heal the Bay, 2:552
Health
air pollution, 1:8, 9–10
algal blooms, 1:12, 13
asbestos contamination, 1:39–40
atmospheric inversions, 1:48
biofuels, 1:67, 69
chemical spills, 1:100
chlorofluorocarbons, 1:102–103
Clean Air Act of 1970, 1:108, 109
Clean Air Mercury Rule, 1:112
coal resource use, 1:147
Comprehensive Environmental Response, Compensation, and Liability Act, 1:164
DDT, 1:186, 187, 188
dust storms, 1:207
earthquakes, 1:223
ecodisasters, 1:231, 232
ecological competition, 1:238
electronics waste, 1:152
Environmental Protection Agency, 1:287–287
environmental science careers, 1:96–97
factory farms, 1:302, 304
fish farming, 1:307–309
fossil fuel combustion impacts, 1:327
geospatial analysis, 1:352
groundwater, 1:396, 397–398
groundwater quality, 1:401, 402, 403
hazardous waste, 1:413–417
herbicides, 1:418
Hurricane Katrina, 1:437, 440
industrial pollution, 1:455, 456, 457
industrial water use, 1:460
industrialization in emerging economies, 1:462
insecticide use, 1:471
marine water quality, 2:550, 551, 552
medical waste, 2:563–564
nuclear power, 2:611
organic and locally grown foods, 2:645, 646–647
ozone hole, 2:658, 659–660
ozone layer, 2:661–662, 662
radioactive waste, 2:683, 684
red tide, 2:699
smog, 2:739
soil contamination, 2:746
surface water, 2:765
teratogens, 2:782–784
toxic waste, 2:790
tsunami impacts, 2:796
waste transfer and dumping, 2:829–830
wastewater treatment technologies, 2:832
water pollution, 2:837, 838, 839, 840
Health-care waste. *See* Medical waste
Health scientists, environmental, 1:96–97
Heat (Earth's temperature). *See* Weather and climate
Heat, solar. *See* Solar power
Heat resources of Earth. *See* Geothermal resources
Heat transport, ocean. *See* Ocean heat transport
Heat waves, 2:862
Heathrow Airport, 1:*51*
Heavy metals
biodegredation, 1:57
bioremediation, 1:*77*
fossil fuel combustion impacts, 1:327
hazardous waste, 1:414
industrial water use, 1:460
laboratory methods in environmental science, 1:512–513
landfills, 1:525
marine water quality, 2:550
medical waste, 2:563, 564
solid waste treatment technologies, 2:754
waste transfer and dumping, 2:829
watersheds, 2:857
Hebgen Lake, 1:221
Helium, 1:41
Helsinki Declaration on the Protection of the Ozone Layer, 2:810
Hematite, 1:491
Henderson Mine, 2:*572*
Henke, Christopher, 1:380
Herbicides, 1:**418–419**
invasive species, 1:483
organic and locally grown foods, 2:645, 646
organic farming, 2:649
shrublands, 2:733
Silent Spring, 2:734
Herbivores
ecosystems, 1:242, 245
grasslands, 1:372
marine ecosystems, 2:541, 544
HFCs. *See* Hydrofluorocarbons (HFCs)
High tides. *See* Tides
Hill, Hank, 2:776
Hill, Julia Butterfly, 1:226, *538*
Hill, Tennessee Valley Authority v. (1977). *See* Tellico Dam Project (Snail Darters) and Supreme Court Case (*TVA v. Hill,* 1977)

Hillaire-Marcel, Claude, 1:451
Himalayas, 1:358, 359, 360, *373*
Hinchman, Steve, 1:88–89
Hiroshima, Japan, 1:231, 2:827
History, geologic. *See* Geologic history
HIV (Human immunodeficiency virus), 2:563
Hockey stick graphs, 2:780, 811
Hog farm impacts. *See* Factory farms, adverse effects of
Hog farms. *See* Factory farms, adverse effects of
Hogs. *See* Pigs
Holdfasts, 1:151
Holding ponds. *See* Ponds
Hollan, Ian, 1:*529*
Holocene extinction event, 1:60, 299
Homeland Security, Department of, 2:578
Homeostasis, 1:332, 333
Honeybees. *See* Bees
Honeycreepers, 1:260
Hong Kong, 1:48, 2:*830*
Hooker Chemical and Plastics Company, 1:415, 2:746, 763
Hoover Dam, 1:81
Horticulture, 1:**420–422**
Hot springs, 1:354
Hotspot volcanoes, 2:539, 819
Houghton Mifflin Company, 2:736
Housing, 1:519–520
Hovorka, Sue, 1:92–93
How To Fight A Rising Sea, 2:*725–728*
Howard, Albert, 2:649, 650
Hudson Institute, 1:136, 137
Hughes, Malcolm K., 2:780
Human immunodeficiency virus (HIV), 2:563
Human impacts, 1:**423–426**
- air pollution, 1:8–9, 10
- Antarctica, 1:21–24, 26
- aquatic ecosystems, 1:30, 32
- aquifers, 1:33–35
- atmospheric circulation, 1:44, 46
- aviation emissions, 1:49, 50–51
- bays and estuaries, 1:54
- biodiversity, 1:60–61, 62, 63, 64
- biogeography, 1:75
- carbon dioxide, 1:83, 84
- carbon dioxide emissions, 1:85, 86, 88
- carbon sequestration, 1:90–91, 93
- clear-cutting, 1:125–127
- climate change, 1:128, 129, 130–131, 132
- coastal ecosystems, 1:152
- conservation, 1:168
- coral reefs and corals, 1:170, 171–172
- cultural practices and environment destruction, 1:179–180
- desertification, 1:194
- drought, 1:205
- dust storms, 1:207
- earthquakes, 1:220
- ecodisasters, 1:230–233, 233–234
- ecological competition, 1:237–238
- ecosystem diversity, 1:239, 240–241
- ecosystems, 1:245–246
- endangered species, 1:259–262
- environmental assessments, 1:269
- environmental science careers, 1:95, 96, 98
- estuaries, 1:294, 296–297
- extinction and extirpation, 1:299–300
- factory farms, 1:302–305
- floods, 1:314
- forests, 1:321, 324–325
- freshwater and freshwater ecosystems, 1:330
- Gaia hypothesis, 1:332, 333
- genetic diversity, 1:335, 336–337
- geospatial analysis, 1:352
- global warming, 1:366–370
- grasslands, 1:373
- green movement, 1:375
- greenhouse effect, 1:383, 385, 388
- greenhouse gases, 1:390, 392, 393–394
- groundwater, 1:*399*
- habitat alteration, 1:404–406
- habitat loss, 1:408–412
- hazardous waste, 1:413–417
- hunting practices, 1:427–429
- hurricanes, 1:430
- hydrologic cycle, 1:445–447
- ice ages, 1:450
- ice cores, 1:452–453
- industrial pollution, 1:455–458
- industrial water use, 1:459–461
- industrialization in emerging economies, 1:462–466
- Intergovernmental Panel on Climate Change, 1:474–475
- international environmental law, 1:478
- invasive species, 1:481–482, 483
- IPCC 2007 Report, 1:486, 487, 488
- lakes, 1:516–517
- land use, 1:518
- light pollution, 1:530
- logging, 1:536–537
- marine ecosystems, 2:543, 544–545
- marine water quality, 2:549–552
- media, 2:560
- migratory species, 2:569
- North Atlantic Oscillation, 2:609
- ocean salinity, 2:625
- oceanography, 2:630
- oceans and coastlines, 2:632, 634
- oil spills, 2:641–643
- overfishing, 2:652–653
- overgrazing, 2:655–657
- ozone hole, 2:658–660
- ozone layer, 2:661–663
- paper and wood pulp, 2:664–665
- pollinators, 2:672
- predator-prey relations, 2:678
- radiative forcing, 2:680, 681
- rain forest destruction, 2:686–687
- red tide, 2:700
- reef ecosystems, 2:704–705
- rivers and waterways, 2:714–715
- saltwater intrusion, 2:719
- sea level rise, 2:721
- snow and ice cover, 2:741
- soil contamination, 2:745–747
- solid waste treatment technologies, 2:753
- species reintroduction programs, 2:755–756
- surface water, 2:764
- sustainable development, 2:770, 772
- toxic waste, 2:790
- tsunami impacts, 2:795
- United Nations Conference on the Human Environment, 2:800–801
- United Nations policy and activism, 2:811
- vegetation cycles, 2:816
- war and conflict-related environmental destruction, 2:825–828
- water pollution, 2:837
- water resources, 2:843, 844
- water supply and demand, 2:848–850
- weather and climate, 2:860
- wetlands, 2:865
- Woods Hole Oceanographic Institution, 2:892

Human rights, 1:479
Humboldt Current, 1:249
Humidity, 2:674
Humus, 2:651, 743
Hunting practices, 1:**427–429,** *428, 429*
- biodiversity, 1:60, 61, 62
- cultural practices and environment destruction, 1:179

ecological competition, 1:237, 238
endangered species, 1:259, 260
extinction and extirpation, 1:299–300
genetic diversity, 1:336
International Convention for the Regulation of Whaling, 2:867–870
invasive species, 1:481
island ecosystems, 1:*498*
natural reserves and parks, 2:593
predator-prey relations, 2:678
wildlife population management, 2:880
wildlife protection policies and legislation, 2:883

Huntingdon Life Sciences, 1:225

Hurricane Ivan, 1:203–204

Hurricanes, 1:**430–436**
agricultural practice impacts, 1:6
algal blooms, 1:13
asbestos contamination, 1:*40*
climate change, 1:133
climate modeling, 1:139, 140
dust storms, 1:208
ecodisasters, 1:233
El Niño and La Niña, 1:250
factory farms, 1:303
floods, 1:315
formation, 1:430–432
global warming, 1:430, 432–435
National Hurricane Center, 1:247
National Oceanic and Atmospheric Administration, 2:584
watersheds, 2:857
weather extremes, 2:860–861

Hurricanes: Katrina environmental impacts, 1:*435,* **437–440**
atmospheric circulation, 1:44
dredging, 1:204
ecodisasters, 1:233, 234–235
ecosystems, 1:*244*
flood control and floodplains, 1:311, 313, *313*
floods, 1:314, *316*
weather extremes, 2:862

Hurtt, George, 1:234, 235

Hutchinson, Evelyn, 1:408

Hutton, James, 1:345

Hwang, Roland, 1:257

Hybrid vehicles, 1:**441–444,** *443*
air pollution, 1:10
alternative fuel impacts, 1:18–19
emissions standards, 1:254

Hydrocarbons
alternative fuel impacts, 1:15
atmospheric inversions, 1:47
biodegredation, 1:57
bioremediation, 1:76
chlorofluorocarbons, 1:100
emissions standards, 1:254
Oil Pollution Acts, 2:637
oil spills, 2:642
ozone hole, 2:659
ozone layer, 2:662

Hydrochlorofluorocarbons (HCFCs), 1:64, 2:574, 575, 659, 662

Hydroelectric power. *See* Hydropower

Hydrofluorocarbons (HFCs), 1:104, 286, 500, 2:574, 575

Hydrofoil generators, 2:786

Hydrogen, 1:16, 41, 452

Hydrographs, 2:761

Hydrologic cycle, 1:**445–447**
drainage basins, 1:200
drought, 1:205
ecosystems, 1:243
groundwater quality, 1:401
precipitation, 2:674
streamflow, 2:762
surface water, 2:764
vegetation cycles, 2:816
water resources, 2:843
water supply and demand, 2:848–849

Hydrologists, 1:97

Hydrophates, 2:864

Hydropower
aquatic ecosystems, 1:32
clean energy, 1:118, 119
dams, 1:181
flood control and floodplains, 1:310–311
hydrologic cycle, 1:446–447
natural resource management, 2:598
tidal or wave power, 2:784–787
See also Dams

Hydrosphere, 1:83, 340

Hydrothermal vents, 1:55, 56, 129, 2:543, 629, 891–892

Hypertrophic lakes, 1:516

Hypolimnion, 1:330

Hypothermia, 2:637, 642

Hypoxia (aquatic). *See* Dead zones

I

Ibero, Carlos, 2:*875*

Ice ages, 1:**448–451,** *449*
extinction and extirpation, 1:299
Gaia hypothesis, 1:333
geologic history, 1:349
glacial retreat, 1:357
glaciation, 1:362
ice cores, 1:452, 453
sea level rise, 2:721

Ice caps. *See* Snow and ice cover

Ice cores, 1:**452–454,** *453, 454*
acid rain, 1:8
Antarctica, 1:22, *349*
Arctic, 1:36
carbon dioxide emissions, 1:85
climate change, 1:130, 131
dendrochronology, 1:193
dust storms, 1:207
glaciation, 1:362
greenhouse effect, 1:386
greenhouse gases, 1:392
ice ages, 1:450
Intergovernmental Panel on Climate Change, 1:475

Ice cover. *See* Snow and ice cover

Ice crystals, 2:741

Ice sheets
glacial retreat, 1:357, 358, 359
glaciation, 1:362, 363
hydrologic cycle, 1:445, 447
ice ages, 1:448, 449
ice cores, 1:452, 453
IPCC 2007 Report, 1:488
sea level rise, 2:721, *723–724,* 725
See also Snow and ice cover

Ice shelves, 1:21, *1:23,* 1:23–24

Icebergs, 1:*131,* 359

Iceland
commercial fisheries, 1:160
ecodisasters, 1:232
geothermal resources, 1:354
International Convention for the Regulation of Whaling, 2:867–868
Kyoto Protocol, 1:500
North Atlantic Oscillation, 2:608

ICES (International Council for the Exploration of the Sea), 1:160

Idaho, 1:*468*

Igneous rocks, 1:338

Iguanas, 1:481

IGY. *See* International Geophysical Year (IGY)

Illinois, 1:88

Immelt, Jeffrey, 1:174

Incineration, 2:*563,* 564, 694, 753–754, 830

An Inconvenient Truth (motion picture), 1:65, 290, 377, 2:605, 780

India
air pollution, 1:3, 10
atmospheric inversions, 1:48
biodiversity, 1:64, 65
biofuels, 1:68
chemical spills, 1:100
chlorofluorocarbons, 1:*103*
Clean Development Mechanism, 1:115, 116
coal resource use, 1:146
dams, 1:*183*

drainage basins, 1:201
earthquakes, 1:*221*
ecodisasters, 1:231
electronics waste, 1:253
flood control and floodplains, 1:313–314
floods, 1:315
glacial retreat, 1:359, 360
global warming, 1:*425*
hunting practices, 1:428
inland fisheries, 1:467
IPCC 2007 Report, 1:485
irrigation, 1:494
Kyoto Protocol, 1:500, 502
organic farming, 2:649
smog, 2:739
tsunami impacts, 2:794, *796*
wastewater treatment technologies, 2:834
water supply and demand, 2:*850*

Indian Ocean
biofuels, 1:69
coal resource use, 1:148
commecial fisheries, 1:160
earthquakes, 1:222
ocean circulation and currents, 2:622
reef ecosystems, 2:704
tsunami impacts, 2:793, 794, 795

Indiana, 1:148, 2:*865*

Indians (Native Americans). *See* Native Americans

Indicators, ecological, 1:497

Indonesia, 1:*92, 480*
chlorofluorocarbons, 1:103
ecodisasters, 1:233
environmental crime, 1:*280*
factory farms, 1:*304*
flood control and floodplains, 1:313
forests, 1:324
industrialization in emerging economies, 1:462
inland fisheries, 1:467
medical waste, 2:*564*
tsunami impacts, 2:793, 794
wildfires, 2:879

Industrial ecology, 2:772

Industrial pollution, 1:**455–458**
acid rain, 1:1, 2–4
agricultural practice impacts, 1:5–6
air pollution, 1:8–9, 10
Antarctica, 1:21, 24
asbestos contamination, 1:39–40
carbon dioxide emissions, 1:86
Clean Water Act, 1:122, 123
Comprehensive Environmental Response, Compensation, and Liability Act, 1:162–165
cultural practices and environment destruction, 1:179, 180
dust storms, 1:207
ecodisasters, 1:231
emissions standards, 1:254–255
global warming, 1:367
groundwater, 1:397–398
hazardous waste, 1:413
human impacts, 1:424, 425
industrial water use, 1:459–461, 460
industrialization in emerging economies, 1:462–466
international environmental law, 1:477
mining and quarrying impacts, 2:570–573
nonpoint-source pollution, 2:601
rivers and waterways, 2:715
smog, 2:739
solid waste treatment technologies, 2:753
toxic waste, 2:790
tundra, 2:798
United Nations Conference on the Human Environment, 2:800
water pollution, 2:837, 838–839
watersheds, 2:858

Industrial Revolution, 1:83, 86, 462

Industrial water use, 1:**459–461**
aquifers, 1:34–35
industrial pollution, 1:457
liquefied natural gas resource use, 1:534–535
marine water quality, 2:550
mining and quarrying impacts, 2:570, 572
pharmaceutical development resources, 2:666, 667
wastewater treatment technologies, 2:832
water conservation, 2:835
water supply and demand, 2:849

Industrialization in emerging economies, 1:**462–466**
air pollution, 1:10
biodiversity, 1:64, 65
carbon dioxide emissions, 1:86–87
cultural practices and environment destruction, 1:180

Industrialized nations. *See specific nations*

Inertia, 2:626, 788

Infauna, 2:633

Infiltration (hydrologic cycle), 1:445

Influenza, avian, 1:302, 304

Informal biostratigraphy. *See* Biostratigraphy

Infrared light, 1:129, 366, 383, 384–385, 390–391

Inland fisheries, 1:**467–468**

Innocent passage, 2:804, 805

Insecticide use, 1:**469–473**
agricultural practice impacts, 1:5
bioremediation, 1:76
chemical spills, 1:100, 101
DDT, 1:186–188
ecosystem diversity, 1:241
environmental protests, 1:289
genetic diversity, 1:336–337
habitat loss, 1:410
natural reserves and parks, 2:593
natural resource management, 2:598
organic and locally grown foods, 2:645, 646
organic farming, 2:649
Silent Spring, 2:734–738
tundra, 2:798

Insects
biodiversity, 1:60, 61
DDT, 1:186–188
ecosystem diversity, 1:241
insecticide use, 1:469–473
invasive species, 1:482–483
predator-prey relations, 2:677, 678–679
See also specific insects

Insitute for Environmental Conflict Resolution, 2:577

Insolation, 1:390, 391–392
See also Solar power

Instruments, monitoring. *See* Real-time monitoring and reporting

Interglacial periods
ice ages, 1:448, 449, 450
ice cores, 1:452–453
sea level rise, 2:721, 722

Intergovernmental Panel on Climate Change (IPCC), 1:**474–476**
air pollution, 1:8
atmospheric circulation, 1:46
aviation emissions, 1:50, 51
biodiversity, 1:65
carbon dioxide, 1:83, 84
carbon sequestration, 1:91
climate change, 1:130, 131, 132, 135–137
climate modeling, 1:140–141
ecodisasters, 1:234
ecosystems, 1:245
extinction and extirpation, 1:300
global warming, 1:369
green movement, 1:376–377
greenhouse effect, 1:388
greenhouse gases, 1:*392,* 393
groundwater quality, 1:402
hurricanes, 1:435
hydrologic cycle, 1:445, 446, 447
IPCC 2007 Report, 1:485–488
mathematical modeling and simulation, 2:557
oceanography, 2:630
radiative forcing, 2:681
reef ecosystems, 2:704

sea level rise, 2:722, 724, 725, 726
snow and ice cover, 2:742
soot, 1:148
temperature records, 2:780
United Nations policy and activism, 2:810, 811–812
war and conflict-related environmental destruction, 2:827–828

Interior Department. *See* Department of the Interior

Internal waters, 2:805

International Atomic Energy Agency, 2:611

International Carbon Action Partnership, 1:257

International Center for Technology Assessment, 1:286

International Convention for the Regulation of Whaling, 2:**867–871**

International Cooperative Programme on Assessment and Monitoring of Air Pollution Effects on Forests, 1:1

International Council for the Exploration of the Sea (ICES), 1:160

International Energy Agency, 1:68, 133

International environmental law, 1:**477–480**
Antarctic Treaty, 1:25–29
Antarctica, 1:21–22, 24
atmosphere, 1:42–43
commercial fisheries, 1:160
industrialization in emerging economies, 1:462
International Convention for the Regulation of Whaling, 2:867–870
marine fisheries, 2:548
Montreal Protocol, 2:574–575
overfishing, 2:653
seasonal migration, 2:730, 731
United Nations Convention on the Law of the Sea, 2:803–805
United Nations Framework Convention on Climate Change, 2:806–809
United Nations policy and activism, 2:810–812
wildlife protection policies and legislation, 2:883–885

International Geophysical Year (IGY), 1:25, 26, 2:658

International relations
air pollution, 1:10
Antarctic, 1:21–22, 24, 25–29
climate change, 1:135
Earth Summit (1992), 1:212–214
Earth Summit (2002), 1:215–217
Geographic Information Systems (GIS), 1:343
Intergovernmental Panel on Climate Change, 1:474, 475
International Convention for the Regulation of Whaling, 2:867–870
international environmental law, 1:477–480
IPCC 2007 Report, 1:485
Kyoto Protocol, 1:500–508
migratory species, 2:568, 569
Montreal Protocol, 2:574–575
nuclear test ban treaties, 2:615–619
Our Common Future, 2:813–815
pharmaceutical development resources, 2:667
seasonal migration, 2:730, 731
sustainable development, 2:769–770
United Nations Conference on the Human Environment, 2:800–801
United Nations Convention on the Law of the Sea, 2:803–805
United Nations Framework Convention on Climate Change, 2:806–809
water supply and demand, 2:850
watersheds, 2:858
wildlife protection policies and legislation, 2:883–885

International Seabed Authority (ISA), 2:804

International trade, 1:105–107, 260

International Transaction Log, 1:116, 117

International Union for Conservation of Nature and Natural Resources (IUCN). *See* World Conservation Union (IUCN)

International Union of Geological Sciences (IUGS), 1:346

International waters, 2:803–804

International Whaling Commission (IWC), 2:867–868

Internet
Geographic Information Systems (GIS), 1:342–343
media, 2:560
photography, 2:669–670
real-time monitoring and reporting, 2:688, 689

Interspecific competition, 1:236

Interstate Air Quality Rule, 1:112

Intertidal zones, 1:151, 154–155, 295, 2:789, 864, 865

Intertropical convergence zone, 1:44–45

Intraspecific competition, 1:236

Introduced species. *See* Invasive species

Intrusion, saltwater. *See* Saltwater intrusion

Invasive species, 1:**481–484**
Bureau of Land Management, 1:81–82
coastal ecosystems, 1:152
coastal zones, 1:155–156
conservation, 1:168
ecodisasters, 1:234, 235
ecological competition, 1:237–238
endangered species, 1:260
fish farming, 1:306, 308
genetic diversity, 1:336
habitat loss, 1:410, 412
island ecosystems, 1:496, 498
predator-prey relations, 2:678
reforestation, 2:710

Inversions, atmospheric, 1:**47–48**, *48,* 2:739

Invertebrates, 1:30, 31, 55, 2:633

Investments, 1:174, 175–176, 177

IPCC. *See* Intergovernmental Panel on Climate Change (IPCC)

IPCC 2007 Report, 1:**485–489**, *487*
carbon dioxide, 1:83
climate change, 1:137
climate modeling, 1:140–141
ecodisasters, 1:234
Gaia hypothesis, 1:332, 334
global warming, 1:366, 369
greenhouse gases, 1:393
hurricanes, 1:435
hydrologic cycle, 1:445, 446, 447
Intergovernmental Panel on Climate Change, 1:474, 475
media, 2:560
radiative forcing, 2:681
rain forest destruction, 2:687
sea level rise, 2:724, 725, 726
soot, 1:148
surface water, 2:766
United Nations policy and activism, 2:812

Iran, *1:2,* 1:*483,* 2:611

Iraq, 2:825, *826*

Ireland, 2:784

Iron, 1:91

Iron fertilization, 1:**490–492**
oceanography, 2:630
red tide, 2:700

Iron hypothesis, 1:490

Irrigation, 1:**493–495**, *494, 495*
bioremediation, 1:77
water conservation, 2:835
water supply and demand, 2:851–852

Irwin, Steve, 2:*560*

Irwin, Terri, 2:*560*
ISA (International Seabed Authority), 2:804
Isham, Jon, 1:380–381
Island ecosystems, 1:**496–499**
  biogeography, 1:74
  coral reefs and corals, 1:170
  genetic diversity, 1:335
  invasive species, 1:481–482
  marine water quality, 2:*551*
  saltwater intrusion, 2:720
Isoseismic maps, 1:219
Israel, 2:825, *827*
Italy, 1:116, 290, 2:*651*, 830
IUCN. *See* World Conservation Union (IUCN)
The IUCN Red List of Threatened Animals (World Conservation Union), 1:261
The IUCN Red List of Threatened Plants (World Conservation Union), 1:261
IUGS (International Union of Geological Sciences), 1:346
Ivory-billed woodpeckers, 1:260
Ivory Coast, 1:414, 2:*656*, 791, 830
IWC (International Whaling Commission), 2:867–868

J

Jackson, William Henry, 2:669
Jamaica, 1:222
James River, 1:101
Jamieson, Dale, 1:380
Japan
  Antarctic Treaty, 1:25, 27
  closed ecology experiments, 1:143, *143*
  earthquakes, 1:222, 223
  ecodisasters, 1:231, 232
  environmental protests, 1:*290*
  hunting practices, 1:*428*
  industrial pollution, 1:457
  International Convention for the Regulation of Whaling, 2:867–868
  Kyoto Protocol, 1:500
  laboratory methods in environmental science, 1:510
  nuclear power, 2:612
  overfishing, 2:653
  solid waste treatment technologies, 2:753, 754
  tsunami impacts, 2:793
  war and conflict-related environmental destruction, 2:827
Jarboe, James, 1:225, 226
Jet aircraft. *See* Aircraft
Jet streams, 1:251, 432
Johannesburg Declaration on Sustainable Development, 1:215, 216, 479, 2:801, 815
Johnny Appleseed. *See* Chapman, John
Johnson, Lyndon, 2:872
Johnson, Phillip, 1:*151*

K

Kaloko Dam, 1:184
Kansas, 1:208, 240
Karenia brevis, 2:699, 700–701
Karner Blue butterfly, 2:755
Katrina, Hurricane. *See* Hurricanes: Katrina environmental impacts
Katrina Rated Largest U.S. Ecodisaster, 1:***234–235***
Keeling, Charles David, 1:85–86, 130
Keeling curve, 1:86, 130, 131
Kelp, 1:151, 155
Kelvin, Lord, 1:345
Kennedy, John F., 1:209, 2:616, 735, 769
Kentucky, 1:148, 2:763, 764
Kenya, 1:429, 2:593, 772
Kepone, 1:101
Keystone species, 1:336
Khrushchev, Nikita, 2:616
Kibira National Park, 2:708
Kightlinger, Jeff, 2:852
Kilamanjaro. *See* Mount Kilamanjaro
Killer whales. *See* Orca whales
Kinder Morgan, 1:394
King, Martin Luther Jr., 2:822, 823
King Coal's Crown Is Losing Some Luster, 1:***148–149***
Kings Canyon National Park, 2:670
Kinley, Richard, 1:64–65
Klaus, Vaclav, 1:65
$KMnO_4$ (Potassium permanganate), 1:398–399
Knossos, 1:397
Koch, Robert, 1:510
Koehler, Tom, 1:72
Korin, Anne, 1:72
Krakatoa, 2:793
Kranz, Dave, 2:852
Krill, 1:21, 26, 238, 2:544, 632
Krupp, Fred, 2:612
Kulongoski, Ted, 1:121
Kyoto Protocol, 1:**500–508**, ***503–508***
  atmosphere, 1:42–43
  biodiversity, 1:65
  carbon dioxide emissions, 1:86–87
  Clean Development Mechanism, 1:115, 117
  climate change, 1:131–132, 133
  Earth Summit (1992), 1:213–214
  emissions standards, 1:254
  global warming, 1:370
  greenhouse effect, 1:388
  greenhouse gases, 1:393
  history, 1:500–501
  impacts, 1:501–503
  international environmental law, 1:479
  Montreal Protocol, 2:575
  text, 1:503–508
  United Nations Conference on the Human Environment, 2:801
  United Nations Framework Convention on Climate Change, 2:806, 807
  United Nations policy and activism, 2:810–811

L

La Niña. *See* El Niño and La Niña
Laboratory methods in environmental science, 1:**509–513**, *511*
  geochemistry, 1:340
  marine water quality, 2:549, 551
  mathematical modeling and simulation, 2:*558*
Lagoons (artificial). *See* Ponds
Lagoons (natural). *See* Estuaries
Lake Baikal, 1:*516*, 517
Lake Beloye, 1:516
Lake Chad, 1:232, 2:844
Lake Erie, 1:*48*, 514, 2:840
Lake Kivu, 2:820
Lake Mead, 2:752
Lake Okeechobee, 1:*56*
Lake St. Clair, 1:*330*
Lake Superior, 1:*140*
Lake Tahoe, 1:291–293
Lake Tanganyika, 1:61
Lake Victoria, 1:482, 483
Lake Vostok, 1:22–23
Lakes, 1:**514–517**
  acid rain, 1:1
  Antarctica, 1:21, 22–23
  aquatic ecosystems, 1:30, 32
  bays and estuaries, 1:53
  benthic ecosystems, 1:55, *56*
  biodiversity, 1:60, 61
  climate modeling, 1:*140*
  coastal ecosystems, 1:150
  coastal zones, 1:154, 155–156
  dams, 1:181, 182

erosion, 1:291–293
freshwater and freshwater ecosystems, 1:329–330
glacial retreat, 1:357
glaciation, 1:362
hydrologic cycle, 1:447
inland fisheries, 1:467–468
invasive species, 1:482
IPCC 2007 Report, 1:486
water pollution, 2:839–840
water resources, 2:843–844
See also Freshwater and freshwater ecosystems; *specific lakes*
Lan Xue, 1:464
Land subsidence, 1:53, 219, 222, 439
Land use, 1:**518–521**
aquifers, 1:34
biodiversity, 1:63
biogeography, 1:75
Bureau of Land Management, 1:81
deforestation, 1:191
desertification, 1:194, 195–196
dust storms, 1:207, 208
endangered species, 1:259
flood control and floodplains, 1:310, 311, 312
forests, 1:324–325
genetic diversity, 1:336
habitat alteration, 1:404–406
human impacts, 1:423, 426
landfills, 1:522–526
migratory species, 2:568
natural reserves and parks, 2:592
nonpoint-source pollution, 2:602–603
overgrazing, 2:655–657
shrublands, 2:732, *733*
streamflow, 2:762
tsunami impacts, 2:796
tundra, 2:798–799
vegetation cycles, 2:816
wetlands, 2:865, 866
wildlife population management, 2:880
wildlife protection policies and legislation, 2:883
wildlife refuge, 2:886–887
Landfills, 1:**522–526**
biodegredation, 1:57–58
biofuels, 1:69
electronics waste, 1:252, 253
groundwater, 1:398
hazardous waste, 1:415
human impacts, 1:425
industrial pollution, 1:456
laboratory methods in environmental science, 1:513
medical waste, 2:562, 564
recycling, 2:693, 694, 697
solid waste treatment technologies, 2:753
Superfund site, 2:763–764
waste transfer and dumping, 2:829, 830
Landless Peoples Movement, 2:*771*
Landmines, 2:825
Landslides, 1:**527–529**
earthquakes, 1:221
tsunami impacts, 1:527, 2:793, 794, 795
Laplace, Pierre-Simon, 2:626
Largemouth bass, 1:336
Larossi, Brad, 1:184
Larsen B ice shelf, *1:23*
Larson, Ray R., 1:342
Las Vegas, Nevada, 1:396
Latin America, 1:115
Lava, 2:818, 819, 820
Laws. *See specific acts or laws*
Leachate, 2:829
Leaching, water. *See* Water pollution
Lead
air pollution, 1:10
Clean Air Act of 1970, 1:109
electronics waste, 1:252, 253
Superfund site, 1:*163*
toxic waste, 2:790, *791*
Lean burn technology, 1:17–18
Lebanon, 2:825
Lee, John M., 2:735
Lee, Ronnie, 1:224
Legislation. *See specific acts or laws*
Lentic water, 1:30, 329–330
Leopards, clouded, 1:*324*
Leopold, Also, 1:*266*
Levees and sea walls
flood control and floodplains, 1:310, 311–313, *313*
floods, 1:314
Hurricane Katrina, 1:439, 440
sea level rise, 2:725, 726, 727, 728
tsunami impacts, 2:796
water resources, 2:844
Liberia, 1:324
Licensing. *See* Government regulation
Light
aquatic ecosystems, 1:31, 32
Arctic, 1:36, 37
atmosphere, 1:42
light pollution, 1:530–532
marine water quality, 2:551
See also Infrared light; Ultraviolet light
Light pollution, 1:**530–532,** *532,* 2:731
Light trespass, 1:528
Lightbulbs, 1:174–175
LightHawk, 2:*540*
Lignin, 2:664
Lignite, 1:145
Lima, Ivan, 1:183
Limbaugh, Rush, 1:488
Lime (chemical), 1:*464*
Limestone. *See* Calcium carbonate ($CaCO_3$)
Limited Test Ban Treaty (LTBT). *See Treaty Banning Nuclear Weapon Tests in the Atmosphere, in Outer Space, and Under Water (Limited Test Ban Treaty)*
Limiting factors, 1:243
Limpets, 1:151, 155
Lindane, 1:76
Lingle, Linda, 1:257
Linnaeus, Carl, 1:60, 244
Lions, 2:677, 678
Liquefaction, 1:221–222, 533
Liquefied natural gas resource use, 1:**533–535,** *534*
Lithosphere, 1:219
Little Ice Age, 1:453
Little Tennessee River, 2:776–778
Little Wood Satyr butterflies, 2:*569*
Littoral regions, 1:30, 154, 330, 2:543, 633
Live Earth, 1:289, 290
Livestock, 1:302–305, 2:755–756
Loans, 1:175
Local extinction. *See* Extinction and extirpation
Local food. *See* Organic and locally grown foods
Locally grown foods. *See* Organic and locally grown foods
Log jams, 2:715
Logging, 1:**536–538**
clear-cutting, 1:125, 126, 127
conservation, 1:166
deforestation, 1:189, *190,* 191
endangered species, 1:260
forest resources, 1:318, *318*
forests and, 1:325
green movement, 1:381–382
natural resource management, 2:598, 599
paper and wood pulp, 2:664–665
pollinators, 2:672, 673
rain forest destruction, 2:686
reforestation, 2:706, 708
resource extraction, 2:712, 713
Logon, Ralph, 1:360
London, England, 1:10, 48, 108, 231, 397, 2:*609*
London Conference on Overfishing, 1:160
London Convention, 1:491

Los Angeles, California, 1:9, 47, 48, 255–256, 2:739
Lotic water, 1:30, 329
Lotka, Alfred, 2:678
Lotka-Volterra model, 2:678
Louisiana, 1:310, 313, *313,* 314, 437–440
Love, William, 2:745–746
Love Canal
  Comprehensive Environmental Response, Compensation, and Liability Act, 1:162, 164
  hazardous waste, 1:414–417
  soil contamination, 2:745–746
  Superfund site, 2:763, 764
  toxic waste, 2:790–791
The Love Canal Tragedy, 1:***415–417***
Lovelock, James, 1:332–333
Low tides. *See* Tides
LTBT. *See Treaty Banning Nuclear Weapon Tests in the Atmosphere, in Outer Space, and Under Water (Limited Test Ban Treaty)*
Lumbering. *See* Logging
Lung cancer, 1:36, 40
Lyell, Charles, 1:80
Lynx, Canadian, 2:*579*

M

Maathai, Wangari Muta, 2:772–774
MacArthur, Robert, 1:74, 496–497
MacTaggart, David, 1:266
Mad cow disease. *See* Bovine spongiform encephalopathy (BSE)
Madagascar, 1:*297,* 324, *502*
Magma, 1:220, 354–355, 2:818
Magnetometers, 2:767
Magnitude (earthquakes), 1:218–219
Magnuson-Stevens Fishery Conservation and Management Reauthorization Act of 2006, 2:548
Maine, 1:*113,* 126, 2:592
Make a Wave Committee, 1:265
Malaria, 1:186, 188, 469, 470, 2:679, 735
Malaysia, 1:324
Malmö Declaration, 2:811
Mammals, 1:30, 63
  See also specific mammals
Management of natural resources. *See* Natural resource management
Management of wildlife populations. *See* Wildlife population management
Manatees, 2:701
Mangroves
  habitat loss, 1:411
  marine ecosystems, 2:543, 544
  marine fisheries, 2:547
  oceans and coastlines, 2:632, 634
  tsunami impacts, 2:795, 796
  wetlands, 2:864
Manila, Philippines, 2:829–830
Manila Bay, 1:*54*
Manmade impacts. *See* Human impacts
Mann, Michael E., 2:780
Manning, Martin, 1:136–137
Mantell, Gideon, 1:80
Mantle (Earth), 1:338, 2:818
Manure. *See* Animal waste
Map in-map out model (MIMO), 1:341
Maps and atlases, 2:**539–540,** *540*
  Geographic Information Systems (GIS), 1:341
  geologic history, 1:346–347
  isoseismic, 1:219
  surveying, 2:767
Marble, 1:3
Margulis, Lynn, 1:332, 333
Marianas Trench, 2:892
Marine debris, 2:*543,* 552–555, 795
Marine Debris Research Prevention and Reduction Act of 2006, 2:***552–555***
Marine ecosystems, 2:**541–545,** *703*
  acid rain, 1:2
  algal blooms, 1:12, 13
  Antarctica, 1:21
  aquatic ecosystems, 1:30, 31–32
  bays and estuaries, 1:53–54
  benthic ecosystems, 1:55, 56
  biodiversity, 1:62, 63
  coastal ecosystems, 1:150–153
  coastal zones, 1:154–155
  conservation, 1:168
  coral reefs and corals, 1:170–172
  endangered species, 1:260–261
  fish farming, 1:306, 308
  greenhouse effect, 1:387–388
  habitat alteration, 1:404–405
  habitat loss, 1:411
  international environmental law, 1:478
  iron fertilization, 1:490–491
  liquefied natural gas resource use, 1:534–535
  marine fisheries, 2:546
  marine water quality, 2:549–552
  ocean salinity, 2:624–625
  oceanography, 2:628–630
  oceans and coastlines, 2:632–634
  Oil Pollution Acts, 2:636–639
  ozone hole, 2:660
  pharmaceutical development resources, 2:667
  predator-prey relations, 2:677
  reef ecosystems, 2:702–705
  runoff, 2:716
  tides, 2:789
  United Nations Convention on the Law of the Sea, 2:803
  Woods Hole Oceanographic Institution, 2:892
Marine fisheries, 2:**546–548**
  climate change, 1:133
  commercial fisheries, 1:160–161
  El Niño and La Niña, 1:249, 250
  inland fisheries, 1:467
  International Convention for the Regulation of Whaling, 2:867
  marine ecosystems, 2:544–545
  National Oceanic and Atmospheric Administration, 2:583, 584
  natural resource management, 2:599
  oceans and coastlines, 2:632, 634
  overfishing, 2:652–653
  predator-prey relations, 2:678
  reef ecosystems, 2:704
Marine science. *See* Oceanography
Marine snow, 1:55, 2:544
Marine Stewardship Council (MSC), 2:653
Marine water. *See* Oceans and coastlines
Marine water quality, 2:**549–555**
  coal resource use, 1:148
  Marine Debris Research Prevention and Reduction Act of 2006, 1:552–555
  nonpoint-source pollution, 2:602–603
  ocean salinity, 2:624–625
  oceans and coastlines, 2:634
  reef ecosystems, 2:704
  water pollution, 2:549–552
Marine worms. *See* Worms
Mars, 1:208, 332, 2:819
Marsh Arabs, 1:331
Marsh Inc., 1:225
Marshes, 1:295, 330, 331, 440
Martin, John, 1:490
Martinique, 2:819
Maryland, 1:4, 342, 2:887
Mashkow, George, 1:228
Mass extinction, 1:298
Massachusetts, 1:287, 398, 522, 2:*740*
Massachusetts Forests and Parks Association, 2:707
Massachusetts Institute of Technology (MIT), 1:51, 148, 149, 2:891

Massachusetts v. Environmental Protection Agency (2007). *See* Environmental Protection Agency (EPA) regulation: *Massachusetts v. Environmental Protection Agency* (2007)
Mathematical modeling and simulation, 2:**556–558**
climate modeling, 1:139–141
global warming, 1:368–369
greenhouse effect, 1:386
predator-prey relations, 2:678
Mature forests. *See* Climax forests
McConnell, John, 1:209
McCurdy, Dave, 1:88
McFarland, J. Horace, 2:585
McGuire Nuclear Station, 1:*461*
McIntyre, Jared, 1:228
McIntyre, S., 2:780
McKitrick, R., 2:780
McTaggart, David, 1:376
Measurement, real-time. *See* Real-time monitoring and reporting
Mechanization, 1:5
Media: Environmentally based news and entertainment, 2:**559–561**
corporate green movement, 1:174, 175
green movement, 1:375, 376, 377
IPCC 2007 Report, 1:487, 488
photography, 2:669
radioactive waste, 2:683
Silent Spring, 2:735
war and conflict-related environmental destruction, 2:825
Medical waste, 2:**562–567**
as hazardous waste, 2:562–566, *563*
Medical Waste Tracking Act of 1988, 2:562, 565–566
Medical Waste Tracking Act of 1988, 2:562, ***565–566***
Medicines. *See* Pharmaceuticals and personal care products (PPCPs)
Mediterranean Sea, 2:685, 793, 794, 825
MELISSA (Micro-Ecological Life Support System Alternative), 1:142, 143
Melting, glacial. *See* Glacial retreat
Melting, snow and ice. *See* Snow and ice cover
Meltwaters. *See* Glacial retreat
Mercator, Gerard, 2:539
Merck Inc., 2:667
Mercury
biodegradation, 1:58
bioremediation, 1:77
chemical spills, 1:100
Clean Air Mercury Rule, 1:112–113, *113*
electronics waste, 1:253
industrial pollution, 1:457
industrial water use, 1:460
medical waste, 2:564
teratogens, 2:783, 784
Meridional overturning circulation (MOC), 1:447, 2:620, 622
Merkel, Angela, 2:612
Mesopause, 1:42
Mesosphere, 1:41, 42
Mesothelioma, 1:40
Mesotrophic lakes, 1:516
Metagenomics, 1:337
Metal recycling. *See* Recycling
Metals, heavy. *See* Heavy metals
Metamorphic rocks, 1:339
Meteorites, 1:53, 348–349, 2:793, 795
Meters, monitoring. *See* Real-time monitoring and reporting
Methane (CH_4)
biodegredation, 1:57
biofuels, 1:67, 68, 69
carbon sequestration, 1:91
climate change, 1:129
coal resource use, 1:147
dams, 1:181, 182–183
Environmental Protection Agency, 1:286
glaciation, 1:363
global warming, 1:366, 367
greenhouse effect, 1:383, 384, 385, 388
greenhouse gases, 1:391, 393
ice ages, 1:449, 450
ice cores, 1:452
Kyoto Protocol, 1:500, 502–503
landfills, 1:522, 523, 525, 526
mining and quarrying impacts, 2:573
solid waste treatment technologies, 2:754
tundra, 2:798, 799
waste transfer and dumping, 2:830
Methanol, *1:17*, 1:70
Methyl isocyanate (MIC), 1:100, 231
Methylmercury, 2:784
Metropolitan areas. *See* Urban areas
Metropulos, Jim, 1:257–258
Mexican gray wolves, 2:755, *756*
Mexico
air pollution, 1:9
aquifers, 1:33, 34
atmospheric inversions, 1:48
migratory species, 2:569
Montreal Protocol, 2:*576*
pollinators, 2:672, 673
reforestation, 2:709
species reintroduction programs, 2:755
waste transfer and dumping, 2:829–830
water pollution, 2:*841*
Mexico City, Mexico, 1:48, 221
Miami, Florida, 1:*520*
MIC (Methyl isocyanate), 1:100, 231
Michigan, 1:*403*, 526
Michoacán jungle, 2:709
Micro-Ecological Life Support System Alternative (MELISSA), 1:142, 143
Microbes. *See* Microorganisms
Microbiologists, 1:97
Microfossils. *See* Fossils
Microgametophytes, 2:673
Microorganisms
Gaia hypothesis, 1:333
global warming, 1:369
groundwater quality, 1:401
invasive species, 1:481
lakes, 1:514, 516
marine ecosystems, 2:543
natural resource management, 2:599
oil spills, 2:641
surface water, 2:765
wastewater treatment technologies, 2:832–833
water pollution, 2:837, 838
wetlands, 2:865
See also specific types of microorganisms
Midway Atoll, 1:*484*
MIFI (Mississippi Insitute for Forest Inventory), 1:235
Migration, seasonal, 2:568–569, **730–731**
Migratory Bird Act, 1:123, 2:886
Migratory species, 2:**568–569**
aquatic ecosystems, 1:32
commercial fisheries, 1:160
conservation, 1:167
ecosystems, 1:246
flood control and floodplains, 1:313
freshwater and freshwater ecosystems, 1:331
ice ages, 1:448
island ecosystems, 1:496, 497
light pollution, 1:528, 529
pollinators, 2:672
reforestation, 2:709
seasonal migration, 2:730–733
tundra, 2:798
wetlands, 2:864, 865

wildlife protection policies and legislation, 2:884
wildlife refuge, 2:886, 887
Milankovitch, Milutin, 1:362, 449
Milankovitch cycles, 1:449, 450
Military-free zones. *See* Demilitarized zones
Milk, 1:186, 2:646
Millet, Margaret Katherine, 1:229
Milwaukee, Wisconsin, 2:838
Mimicry, 2:677
MIMO model, 1:341
Mineral Policy Center, 2:571–572
Minerals
Antarctica, 1:21, 22, 26
biostratigraphy, 1:79
Bureau of Land Management, 1:81
conservation, 1:166
geochemistry, 1:338, 340
groundwater quality, 1:401–402
mining and quarrying impacts, 2:570
soil chemistry, 2:743
United Nations Convention on the Law of the Sea, 2:804
Mini Earth, 1:*143*
Mining and quarrying impacts, **2:570–573**
air pollution, 1:8
Antarctica, 1:25, 26
coal resource use, 1:145, 146–147
deforestation, 1:190
geochemistry, 1:340
habitat alteration, 1:404, 405
habitat loss, 1:409
human impacts, 1:423, 424
industrial water use, 1:460
rain forest destruction, 2:686
resource extraction, 2:712
solid waste treatment technologies, 2:753
toxic waste, 2:790
tundra, 2:798, 799
United Nations Convention on the Law of the Sea, 2:804
Minnesota, 1:88
Minnows, 1:336
Mississippi, 1:93, 234–235, 314, 437, *437,* 438
Mississippi Insitute for Forest Inventory (MIFI), 1:235
Mississippi River, 1:13, 310, 314, 2:854, 855, 856
Missouri Prairie State Park, 1:*374*
MIT. *See* Massachusetts Institute of Technology (MIT)
Mitigation, climate change. *See* Climate change
Mixing, water. *See* Saltwater intrusion
Mobbing, 2:677–678
Mobro (barge), 2:694
MOC (Meridional overturning circulation), 1:447, 2:620, 622
Modeling, climate. *See* Climate modeling
Modeling, mathematical. *See* Mathematical modeling and simulation
Modified Mercalli scale, 1:219
Mold, 1:437, 440, 2:666
Molecular biology, 1:509–511
Mollisols, 2:748
Mollusks, 1:31, 62, 86, 154–155, 2:542
Molten rock. *See* Magma
Monarch butterflies, 2:568, *571,* 672, 673, 709, 730
Monitoring, real-time. *See* Real-time monitoring and reporting
Moniz, Ernest, 1:93
Monkeys, golden, 1:381
Monkeywrenching, 1:168, 227
Monoculture forests, 1:125, 318, 325, 537, 2:598, 706
Monsanto Chemical Company, 2:736
Monsoons, 1:140, 148, *183,* 232
Montana, 1:148, 221
Monterey Bay, 1:13
Monteverde harlequin tree frogs, 1:62, 300
Montreal Protocol on Substances that Deplete the Ozone Layer, **2:574–576**
atmosphere, 1:43
chlorofluorocarbons, 1:100, 103
geochemistry, 1:340
industrial pollution, 1:456
ozone hole, 2:658, 660
ozone layer, 2:662
United Nations policy and activism, 2:810
Montrose Chemical Corporation, 2:736
Moon, 1:154, 2:626, 788–789
Moors, 2:732
Moosewood Restaurant, 2:648
More Restaurants Are Going Green By Going Local, 2:**647–648**
Morgan, Al, 1:*58*
Mosquitoes, 1:186, 188, 469, 483, 2:679
Motor vehicles. *See* Vehicles
Mount Kenya, 2:820
Mount Kilamanjaro, 1:*359,* 2:820
Mount Pelee, 2:819
Mount Pinatubo, 2:819
Mount St. Helens, 1:*342, 364,* 2:819
Mountains
atmospheric inversions, 1:47
biodiversity, 1:61, 62, 63
coal resource use, 1:145
earthquakes, 1:219
glaciation, 1:362, 363
landslides, 1:527
precipitation, 2:674
snow and ice cover, 2:741, 742
MSC (Marine Stewardship Council), 2:653
Mt. Kilamanjaro. *See* Mount Kilamanjaro
Mt. Pelee. *See* Mount Pelee
Mt. Pinatubo. *See* Mount Pinatubo
Mt. St. Helens. *See* Mount St. Helens
Mudflats, 1:30–31, 295
Mudslides, 1:527
Muir, John
conservation, 1:166, 167
glaciation, 1:363–365
green movement, 1:375
natural reserves and parks, 2:590, 593–596
Mulch, 1:*58,* 292–293
Müller, Paul, 1:186, 469
Multilateral Fund for the Implementation of the Montreal Protocol, 2:575
Multiple Use, Sustained Yield Act (MUSY), 2:872
Mumbai, India, 1:48
Municipal waste, 2:693, 753, 829
Munk, Walter, 2:724
Murchison, Roderick, 1:346
MUSY (Multiple Use, Sustained Yield Act), 2:872
Myanmar, 1:315
Mystic Power Plant, 2:*740*

N

N_2. *See* Nitrogen (N_2)
N_2O (Nitrous oxide). *See* Nitrogen oxides
Nagasaki, Japan, 1:231, 2:827
Namibia, 1:*106,* 412
NAO (North Atlantic Oscillation), **2:608–609**
Naples, Italy, 2:830
NASA. *See* National Aeronautics and Space Administration (NASA)
National Academy of Sciences, 2:617, 681, 780

National Aeronautics and Space Administration (NASA)
- climate modeling, 1:141
- dust storms, 1:207
- ecosystem diversity, 1:*241*
- El Niño and La Niña, 1:251
- Gaia hypothesis, 1:332
- glacial retreat, 1:359
- greenhouse gases, 1:392
- ice cores, 1:454
- landslides, 1:528

National Assessment (U.S.), 1:133–135

National Audubon Society, 2:604, 606

National Biological Service, 1:246

National Center for Atmospheric Research, 1:37, 140, 206, 432, 2:879

National Center for Policy Analysis, 1:258

National Environmental Policy Act (NEPA), 1:265, 269, 2:**577–581,** 759

National Environmental Policy Act of 1969, 2:*578–581*

National Environmental Satellite, Data, and Information Service, 2:583

National Estuary Program, 1:296–297

National Forest Service, 1:166, 2:690, 872, 873, 875, 876

National government organizations (NGOs), 1:212, 213

National Ground Water Association (NGWA), 1:402

National Hurricane Center, 1:247

National Institute of Environmental Health Sciences, 1:438

National Institute of Polar Research, 1:*454*

National Institutes of Health, 2:696

National Marine Fisheries Service, 1:262–263, 2:583, 776

National Marine Park (Brazil), 1:*497*

National Ocean Service, 2:583

National Oceanic and Atmospheric Administration (NOAA), 2:**582–584,** *583*
- Arctic, 1:37–38
- dust storms, 1:208
- El Niño and La Niña, 1:251
- endangered species, 2:776
- hurricanes, 1:434
- ozone hole, 2:658
- red tide, 2:699
- watersheds, 2:858
- weather extremes, 2:861

National Oil and Hazardous Substances Pollution Contingency Plan, 1:162

National Organic Standards Board, 2:646

National Park Service
- green movement, 1:375
- National Park Service Organic Act, 2:585–587
- natural reserves and parks, 2:590, 593
- recreational use and environmental destruction, 2:691
- Wilderness Act of 1964, 2:872, 873

National Park Service Organic Act, 2:**585–588,** *586–588*

National parks. *See* Natural reserves and parks

National Pollutant Discharge Elimination System Permit (NPDES), 2:601

National Priorities List (U.S.), 1:162, 2:764

National Research Council, 1:72

National reserves. *See* Natural reserves and parks

National Response Center, 1:100

National Weather Service, 2:582–583

National Wildlife Refuge System Administration Act, 2:886

Native Americans, 1:113, 354, 2:589, 657, 874

Natural disasters, ecological. *See* Ecodisasters

Natural gas
- air pollution, 1:9
- alternative fuel impacts, 1:14
- Antarctica, 1:21, 22
- biofuels, 1:67, 68, 69
- Bureau of Land Management, 1:81
- fossil fuel combustion impacts, 1:326
- liquefied natural gas resource use, 1:533–535
- marine ecosystems, 2:545

Natural reserves and parks, 2:**589–597**
- Antarctic Treaty, 1:25, 26
- Antarctica, 2:591
- conflicts, 2:593
- conservation, 1:166, 167–169
- flood control and floodplains, 1:310
- forest resources, 1:317–318, 319
- habitat alteration, 1:406
- habitat loss, 1:412
- history, 2:589–591
- Muir, John, 2:590, 593–596
- National Park Service Organic Act, 2:585–587
- photography, 2:669, 670
- private preservation, 2:591–592
- recreational use and environmental destruction, 2:690–691
- reef ecosystems, 2:705
- reforestation, 2:706, 707, 709–710
- sea level rise, 2:728
- tourism, 2:589–590, 593
- *Walden,* 2:822
- Wilderness Act of 1964, 2:872–873
- wildlife refuge, 2:886–887
- Yellowstone National Park, 2:590, 593–596

Natural resource management, **2:598–600**
- Bureau of Land Management, 1:82
- conservation, 1:166–169
- Earth Summit (1992), 1:213
- Earth Summit (2002), 1:216
- ecosystems, 1:245–246
- environmental assessments, 1:269–271
- Environmental Protection Agency, 1:282–284
- forest resources, 1:317–319
- forests, 1:325
- international environmental law, 1:478, 479
- mining and quarrying impacts, 2:570
- reforestation, 2:707, 710
- soil resources, 2:748–749
- sustainable development, 2:769–772
- tsunami impacts, 2:795, 796
- United Nations Conference on the Human Environment, 2:800
- United Nations Convention on the Law of the Sea, 2:804
- *Walden,* 2:822
- water resources, 2:843–847
- water supply and demand, 2:848–852
- Watershed Protection and Flood Prevention Act of 1954, 2:854–855
- watersheds, 2:858
- wildfire control, 2:874–876

Natural Resources Conservation Service, 2:874

Natural Resources Defense Council (NRDC), 1:3, 72, 88, 256, 257, 2:852

Natural selection, 1:335

Nature Conservancy
- conservation, 1:167–168
- green movement, 1:376
- greenhouse effect, 1:381, 382

island ecosystems, 1:498
marine water quality, 2:*551*
natural reserves and parks, 2:591–592
wildlife population management, 2:881

Natuurmonumenten, 2:728

NAVSTAR. *See* Global Positioning Systems (GPS)

Navy (U.S.), 2:578

NEFF (New England Forestry Foundation), 2:707

The Negative (Adams), 2:670

Nelson, Gaylord, 1:209, 289

Nematodes, 2:542

NEPA. *See* National Environmental Policy Act (NEPA)

Netherlands
flood control and floodplains, 1:310–311
pollinators, 2:673
sea level rise, 2:725–728
wind and wind power, 2:888

Neurotoxin shellfish poisoning. *See* Shellfish poisoning

Neurotoxins, 2:699, 700

Neuse River, 1:13

Nevada, 1:148, 149, 396, 2:685

New England, 2:706–708
See also specific New England states

New England Forestry Foundation (NEFF), 2:707

New Jersey
aquifers, 1:34
Clean Air Mercury Rule, 1:113
dams, 1:184
ecodisasters, 1:231
emissions standards, 1:257
land use, 1:519–520
landfills, 1:523
medical waste, 2:562
reforestation, 2:709–710

New Jersey Coastal Plain Aquifer, 1:34

New Jersey Pinelands. *See* Pinelands National Reserve

New Mexico
carbon dioxide emissions, 1:88
coal resource use, 1:148
dendrochronology, 1:*193*
dust storms, 1:208
earthquakes, 1:220
emissions standards, 1:257
species reintroduction programs, 2:755–756

New Orleans, Louisiana
dredging, 1:203–204
flood control and floodplains, 1:310, 311, 313, *313*
floods, 1:314, *316*
Hurricane Katrina, 1:*40*, 437–440, *438*

New River, 1:303

New Source Review, 1:3–4

New Tool to Fight Global Warming: Endangered Species Act?, 1:***262–263***

New York
atmospheric inversions, 1:*48*
Clean Air Mercury Rule, 1:113
Comprehensive Environmental Response, Compensation, and Liability Act, 1:162, 164
emissions standards, 1:257
hazardous waste, 1:414–417
landfills, 1:523
medical waste, 2:562
natural reserves and parks, 2:590
recreational use and environmental destruction, 2:*692*
soil contamination, 2:745–746
Superfund site, 2:763, 764

New York, New York, 1:442–444, *443*, 2:562

The New Yorker (Periodical), 2:736

New Zealand
Antarctic Treaty, 1:25, 27
biodiversity, 1:60
emissions standards, 1:257
geothermal resources, 1:354
green movement, 1:378
habitat loss, 1:412
invasive species, 1:481
overfishing, 2:652, 653

Newcomen, Thomas, 1:68, 145

Newman, Randy, 2:837

News, environmental. *See* Media: Environmentally based news and entertainment

Newsom, Gavin, 2:695

Newton, Isaac, 2:626

NGOs (Non-governmental organizations), 1:212, 213, 216

NGWA (National Ground Water Association), 1:402

Niches
coastal zones, 1:155
ecological competition, 1:236
ecosystem diversity, 1:239, 240
ecosystems, 1:243
habitat alteration, 1:404
habitat loss, 1:408
marine ecosystems, 2:543
reef ecosystems, 2:702

Nigeria, 1:230–231

Nile perch, 1:410, 482, 483

Nile River, 1:314, 315

La Niña. *See* El Niño and La Niña

El Niño and La Niña. *See* El Niño and La Niña

Nitrates, 2:816

Nitrogen (N_2)
agricultural practice impacts, 1:5
algal blooms, 1:11
atmosphere, 1:41
aviation emissions, 1:49
biodiversity, 1:63
emissions standards, 1:254
greenhouse effect, 1:384
vegetation cycles, 2:816
wastewater treatment technologies, 2:833

Nitrogen cycle, 2:816

Nitrogen dioxide (NO_2), 1:9, 10, 109, 462

Nitrogen fixation, 2:816

Nitrogen oxides
acid rain, 1:1–2, 3, 4
air pollution, 1:9, 10
alternative fuel impacts, 1:15
atmosphere, 1:41
aviation emissions, 1:49, 50, 51
carbon sequestration, 1:91
climate change, 1:129
emissions standards, 1:254
Environmental Protection Agency, 1:286
fossil fuel combustion impacts, 1:327
global warming, 1:366, 367
greenhouse effect, 1:384, 385
greenhouse gases, 1:391, 393
Kyoto Protocol, 1:500
ozone hole, 2:659
ozone layer, 2:662

Nitrous oxide (N_2O). *See* Nitrogen oxides

Nixon, Richard M., 1:166, 265, 2:582, 776

No-till farming. *See* Tilling

NO_2. *See* Nitrogen dioxide (NO_2)

NOAA. *See* National Oceanic and Atmospheric Administration (NOAA)

Nobel Prize, 1:475, 488, 513, 2:605, 772–773

Noise pollution, 1:49, 51, 2:690, 691, 889

Non-governmental organizations (NGOs), 1:212, 213, 216

Non-native species. *See* Invasive species

Non-scientist contributions to nature and environment studies, **2:604–607**

Nonpoint-source pollution, **2:601–603**
aquifers, 1:33, 35
runoff, 2:717
water pollution, 2:837, 838–839
watersheds, 2:857

Nordmann's greenshanks, 1:313
North America
biodiversity, 1:60
clear-cutting, 1:125
deforestation, 1:189
El Niño and La Niña, 1:250
forests, 1:324
grasslands, 1:373
migratory species, 2:569
natural resource management, 2:598
precipitation, 2:675
seasonal migration, 2:731
wetlands, 2:865
North Atlantic Ocean. *See* Atlantic Ocean
North Atlantic Oscillation (NAO), 2:**608–609**
North Carolina
algal blooms, 1:12–13
coal resource use, 1:148
factory farms, 1:302, 303
fossil fuel combustion impacts, 1:*328*
water pollution, 1:6
watersheds, 2:857
North Dakota, 1:395
North Pole. *See* Arctic
North Sea, 2:726, 728
Northeast Passage, 2:805
Northern forests. *See* Boreal forests
Northern Hemisphere
atmospheric circulation, 1:44, 45
biodiversity, 1:63
glacial retreat, 1:357
ice ages, 1:448
lakes, 1:514
North Atlantic Oscillation, 2:608
ocean circulation and currents, 2:620
Northern lights, 1:42
Northern Ocean, 2:624
Northern right whales, 2:731
Northern spotted owls, 1:260, 406
Northwest Passage, 1:38, 2:804–805
Northwest Power and Conservation Council, 1:121
Norway
Antarctic Treaty, 1:25, 27
emissions standards, 1:257
International Convention for the Regulation of Whaling, 2:867–868
tidal or wave power, 2:784
Nova Scotia, 1:126–127
NPDES (National Pollutant Discharge Elimination System Permit), 2:601
NRDC. *See* Natural Resources Defense Council (NRDC)
NTBT. *See Treaty Banning Nuclear Weapon Tests in the Atmosphere, in Outer Space, and Under Water (Limited Test Ban Treaty)*
Nuclear power, 2:**610–614, 683–685**
Nuclear power plant accidents. *See* Ecodisasters
Nuclear reactors. *See* Nuclear power
Nuclear test ban treaties, 2:**615–619**
Antarctic Treaty, 1:25, 26
earthquakes, 1:220
Nuclear Test Ban Treaty (NTBT). *See Treaty Banning Nuclear Weapon Tests in the Atmosphere, in Outer Space, and Under Water (Limited Test Ban Treaty)*
Nuclear warfare. *See* War and conflict-related environmental destruction
Nuclear waste. *See* Radioactive waste
Nuclear weapons
environmental activism, 1:265–266, 267
environmental protests, 1:289
green movement, 1:375, 376
nuclear power, 2:610, 611, 612
nuclear test ban treaties, 2:615–619
radioactive waste, 2:683
war and conflict-related environmental destruction, 2:827
Nuclides, 2:683
Nutrient cycling, 1:261
Nyungwe National Park, 2:708

O

Oaks, 1:*322*
Ocean chemistry. *See* Water chemistry
Ocean circulation and currents, 2:**620–623,** *621*
Antarctica, 1:21
atmospheric circulation, 1:44–46
bays and estuaries, 1:54
biodiversity, 1:63
climate modeling, 1:140
El Niño and La Niña, 1:247–251
geologic history, 1:349
glaciation, 1:362
global warming, 1:369
greenhouse effect, 1:385
greenhouse gases, 1:391
hydrologic cycle, 1:445, 447
ice ages, 1:451
ice cores, 1:452
marine ecosystems, 2:541
North Atlantic Oscillation, 2:608
oceanography, 2:628, 630
oceans and coastlines, 2:632
tidal or wave power, 2:784
Ocean conveyor belt. *See* Meridional overturning circulation (MOC)
Ocean currents. *See* Ocean circulation and currents
Ocean depth, 1:31–32, 2:628, 795
Ocean deserts, 1:490
Ocean Dumping Ban Act, 2:551
Ocean floor. *See* Benthic ecosystems
Ocean heat transport, 2:620
Ocean hydropower. *See* Tidal or wave power
Ocean salinity, 2:**624–625**
aquatic ecosystems, 1:30, 31
coastal zones, 1:154
ecodisasters, 1:232
Gaia hypothesis, 1:333
groundwater quality, 1:402
marine ecosystems, 2:541
ocean circulation and currents, 2:620
oceans and coastlines, 2:632
reef ecosystems, 2:702
Ocean tides, 2:**626–627,** *627*
coastal ecosystems, 1:150–151
coastal zones, 1:154–155
natural resource management, 2:598
sea level rise, 2:723, 724
tidal or wave power, 2:784–787
Ocean waves
bays and estuaries, 1:53
coastal ecosystems, 1:150–151
estuaries, 1:294, 295
flood control and floodplains, 1:312
floods, 1:314
ocean tides, 2:626
tidal or wave power, 2:784–787
tsunami impacts, 2:793–796
See also Tsunami impacts
Oceanic and Atmospheric Research, Office of, 2:583
Oceanographers, 1:97–98, 2:628, 685
Oceanography, 2:**628–631**
El Niño and La Niña, 1:251
environmental science careers, 1:97–98
marine ecosystems, 2:541
Woods Hole Oceanographic Institution, 2:891–893
Oceanology. *See* Oceanography
Oceans and coastlines, 2:**632–635,** *703*
acid rain, 1:2
algal blooms, 1:11, 12
Arctic, 1:36–38
atmospheric circulation, 1:44, 45
atmospheric inversions, 1:47

bays and estuaries, 1:53
biodegredation, 1:57
biodiversity, 1:60, 62, 63
bioremediation, 1:76
carbon dioxide, 1:83
carbon dioxide emissions, 1:85, 86
carbon sequestration, 1:90, 91
climate change, 1:128, 129, 132, 133, 134
coal resource use, 1:148
coastal ecosystems, 1:150–152, *151*
coastal zones, 1:154–159
commercial fisheries, 1:160–161
conservation, 1:168
coral reefs and corals, 1:170–172
dredging, 1:202
ecosystems, 1:243, 246
El Niño and La Niña, 1:247–251
environmental science careers, 1:97–98
estuaries, 1:294–297
flood control and floodplains, 1:310, 311–312
geospatial analysis, 1:352
glacial retreat, 1:357, 359
greenhouse effect, 1:383, 385, 387–388
greenhouse gases, 1:391
habitat alteration, 1:404–406
habitat loss, 1:411
hurricanes, 1:430
ice ages, 1:448
international environmental law, 1:478
iron fertilization, 1:490–491
marine ecosystems, 2:541–545
marine fisheries, 2:546–548
marine water quality, 2:549–555
National Oceanic and Atmospheric Administration, 2:582–584
ocean circulation and currents, 2:620–623
ocean salinity, 2:624–625
ocean tides, 2:626–627
oceanography, 2:628–630
oil spills, 2:641–643
overfishing, 2:652–653
ozone hole, 2:660
precipitation, 2:674
red tide, 2:699–701
reef ecosystems, 2:702–705
runoff, 2:716, 717
saltwater intrusion, 2:719–720
sea level rise, 2:721–728
surface water, 2:766
surveying, 2:768
tidal or wave power, 2:784–787
tides, 2:788–789
tsunami impacts, 2:793–796
United Nations Convention on the Law of the Sea, 2:803–805
water resources, 2:843
water supply and demand, 2:848
weather extremes, 2:861–862
wetlands, 2:864
Woods Hole Oceanographic Institution, 2:891–893
See also specific oceans

ODS (Ozone-depleting substances), 2:574, 575

ODYSSEY GIS, 1:342

OECD (Organization for Economic Cooperation and Development), 1:464–465

Oeschger, Hans, 1:*424*

Office of Emergency Management, 1:162

Office of Oceanic and Atmospheric Research, 2:583

Office of Solid Waste and Emergency Response, 1:162

Office of Superfund Remediation and Technology Innovation, 1:162

Ogallala Aquifer, 1:34–35, 403

Ogden, Mark, 1:184

Ohio, 1:*48,* 148, 184, 2:837

Oil
alternative fuel impacts, 1:14–15
Antarctica, 1:21, 22
biodegredation, 1:57
biofuels, 1:67
Bureau of Land Management, 1:81
carbon sequestration, 1:93
Comprehensive Environmental Response, Compensation, and Liability Act, 1:162, 164
conservation, 1:166
corporate green movement, 1:*176,* 177
environmental benefits and liabilities of petroleum resource use, 1:272–278
fossil fuel combustion impacts, 1:326
greenhouse gases, 1:394–395
industrial pollution, 1:455
laboratory methods in environmental science, 1:511
marine ecosystems, 2:545
marine water quality, 2:549, 552
Oil Pollution Acts, 2:636–639
oil spills, 2:641–643
resource extraction, 2:712
tundra, 2:798, 799
United Nations Convention on the Law of the Sea, 2:803, 804
wildlife refuge, 2:887
See also Vegetable oil

Oil Pollution Act of 1990, 2:*637–639*

Oil Pollution Acts, 2:**636–640,** 643

Oil recovery, enhanced, 1:394–395

Oil sands project, 2:712, 713

Oil slicks, 1:230, 231, 2:642

Oil Spill Recovery Institute (OSRI), 2:637

Oil spill remediation. *See* Spill remediation

Oil spills, 2:**641–644,** *642*
biodegredation, 1:57
bioremediation, 1:76, *77*
ecodisasters, 1:230–231
groundwater, 1:398
Hurricane Katrina, 1:437
laboratory methods in environmental science, 1:511
Oil Pollution Acts, 2:636–639
spill remediation, 2:757–758, 759–780
tides, 2:789
war and conflict-related environmental destruction, 2:825
wildlife protection policies and legislation, 2:*884*

Okadaic acid, 2:700

Oklahoma, 1:*163*

Old-growth forests. *See* Primeval forests

Oligotrophic lakes, 1:515–516

Olympus Mons, 2:819

On Civil Disobedience (Thoreau), 2:822, 823

On the Origin of Species (Darwin), 1:244

OPEC (Organization of the Petroleum Exporting Countries), 1:15, 68

Open Rack Vaporization (ORV), 1:534–535

Operation Deep Freeze, 1:*370*

Orange roughy, 2:652, 653

Orangutans, 1:*299*

Orbit (Earth), 1:61, 333, 362, 448, 449, 452

Orca whales, 1:119–121

Orchards, 2:709

Oregon, 1:120–121, *151,* 257

Ores, 2:570, 571, 572

Oreskes, Naomi, 1:137

Organic and locally grown foods, 2:**645–648,** *647*
environmental protests, 2:*651*
organic farming, 2:649–651

Organic compounds, 1:57–58, 455–456, 460
See also Volatile organic compounds (VOCs)

Organic Consumers Association, 2:646

Organic farming: Environmental science and philosophy, 2:**649–651,** *650*

agricultural practice impacts, 1:6
organic and locally grown foods, 2:645–648
Organic Foods Production Act, 2:646
Organic Remains of the County of Wiltshire (Benett), 1:80
Organic Trade Association, 2:646
Organization for Economic Cooperation and Development (OECD), 1:464–465
Organization of the Petroleum Exporting Countries (OPEC), 1:15, 68
Organochlorines, 1:469–470, 471
Organophosphates, 1:470
Origin of Species (Darwin), 1:79
Orinoco turtles, 1:*262*
Ortelius, Abraham, 2:539
ORV (Open Rack Vaporization), 1:534–535
Osborne, Rich, 1:120
Oscillation, North Atlantic. *See* North Atlantic Oscillation (NAO)
Oshman, Michael, 2:648
OSRI (Oil Spill Recovery Institute), 2:637
Oued Tansift, 1:*198*
Our Common Future (United Nations World Commission on Environment and Development). *See* United Nations World Commission on Environment and Development: *Our Common Future* (1987)
Our National Parks (Muir), 2:593, ***594–596***
Outreach programs, environmental. *See* Environmental activism
Overburden, 1:147, 2:570
Overfishing, 2:**652–654**
aquatic ecosystems, 1:32
biodiversity, 1:61, 62, 63
coastal ecosystems, 1:152
coastal zones, 1:156
commercial fisheries, 1:160–161
coral reefs and corals, 1:171
ecological competition, 1:237
ecosystems, 1:246
endangered species, 1:259
fish farming, 1:306
genetic diversity, 1:336
habitat alteration, 1:404
human impacts, 1:423–424
inland fisheries, 1:467–468
marine ecosystems, 2:544–545
marine fisheries, 2:546, 548
natural resource management, 2:599
oceans and coastlines, 2:632, 634
predator-prey relations, 2:678
reef ecosystems, 2:704
United Nations Convention on the Law of the Sea, 2:803
Overgrazing, 2:**655–657**
agricultural practice impacts, 1:5, 6
desertification, 1:194, 195
deserts, 1:198
dust storms, 1:208
ecosystems, 1:245, 246
grasslands, 1:373
habitat alteration, 1:404
Overhunting. *See* Hunting practices
Overkill hypothesis, 1:299
Overpeck, Jonathan, 2:722
Overseas Development Institute, 1:467
Ovules, 2:672
Owls, 1:260, 406, 2:*882*
Oxbow lakes, 1:515
Oxisols, 2:748
Oxygen
agricultural practice impacts, 1:5
algal blooms, 1:11, 12
atmosphere, 1:41
aviation emissions, 1:49
benthic ecosystems, 1:55
biodegredation, 1:57
biodiversity, 1:61
bioremediation, 1:77
carbon dioxide, 1:83
clear-cutting, 1:126
ecosystems, 1:245
fish farming, 1:307
greenhouse effect, 1:384
ice cores, 1:452
industrial water use, 1:461
lakes, 1:516
landfills, 1:523, 524
marine ecosystems, 2:541
marine water quality, 2:550, 551
water pollution, 2:840
wetlands, 2:864, 865
Ozone (O_3). *See* Ozone layer
Ozone-depleting substances (ODS), 2:574, 575, 658, 659, 660
Ozone hole, 2:**658–660**
Antarctica, 1:21, 24
chlorofluorocarbons, 1:102, 103, 104
Clean Air Act of 1970, 1:110
geochemistry, 1:340
industrial pollution, 1:456
Montreal Protocol, 2:574, 575
ozone layer, 2:661–663
Ozone layer, 2:**661–663**
air pollution, 1:8, 9–10
atmosphere, 1:41–42, 43
atmospheric inversions, 1:47
aviation emissions, 1:49, 50
Clean Air Act of 1970, 1:109
fossil fuel combustion impacts, 1:327
geochemistry, 1:340
greenhouse effect, 1:384, 385
greenhouse gases, 1:391
Intergovernmental Panel on Climate Change, 1:474
laboratory methods in environmental science, 1:513
Montreal Protocol, 2:574–575
ozone hole, 2:658–660
radiative forcing, 2:681
smog, 2:*739*
United Nations policy and activism, 2:810

P

Pacific Institute, 2:852
Pacific islands, 1:481
Pacific Ocean, 2:*584*
atmospheric circulation, 1:44, 45, 46
coastal ecosystems, 1:152
commercial fisheries, 1:160
ecosystems, 1:246
El Niño and La Niña, 1:247–251
marine ecosystems, 2:541
marine water quality, 2:*551*
mathematical modeling and simulation, 2:556–557
ocean circulation and currents, 2:620, 621–622
real-time monitoring and reporting, 2:688–689
red tide, 2:699
tsunami impacts, 2:793, 794
volcanoes, 2:819
Woods Hole Oceanographic Institution, 2:891, 892
Pacific yews, 1:325
Pack-ice melting, Arctic. *See* Arctic darkening and pack-ice melting
Packaging recycling. *See* Recycling
Pagara, India, 1:116
PAHs. *See* Polyaromatic hydrocarbons (PAHs)
PAHs (Polyaromatic hydrocarbons), 1:57, *416*
Paine, R. T., 1:241
Pakistan, 1:*208*, 494, 2:657, 850
Paleobiogeography, 1:75
Paleoclimate
geologic history, 1:349
glaciation, 1:362
greenhouse gases, 1:392
hydrologic cycle, 1:447
Intergovernmental Panel on Climate Change, 1:475
iron fertilization, 1:490
sea level rise, 2:721–723

temperature records, 2:*779*
Paleoecology, 1:79
Paleontological stratigraphy. *See* Biostratigraphy
Palomar Observatory, 1:529
Pamlico Sound, 1:12–13
Pampas, 1:372
Paper and wood pulp, 2:**664–665**
industrial pollution, 1:455–456
recycling, 2:695–696
reforestation, 2:706
Papua New Guinea, 1:527
Paralytic shellfish poisoning. *See* Shellfish poisoning
Parasites, 1:308
Pardo, Arvid, 2:803
Paris, France, *1:9*
Park Service (U.S.). *See* National Park Service
Parks. *See* Natural reserves and parks
Partial Test Ban Treaty (PTBT). *See Treaty Banning Nuclear Weapon Tests in the Atmosphere, in Outer Space, and Under Water (Limited Test Ban Treaty)*
Particulates
air pollution, 1:9, 10
alternative fuel impacts, 1:15
Arctic, 1:36
Clean Air Act of 1970, 1:109
dust storms, 1:207
ecodisasters, 1:232
emissions standards, 1:254
industrial pollution, 1:457
iron fertilization, 1:490
solid waste treatment technologies, 2:754
Partnership for a New Generation of Vehicles (PNGV), 1:441
Passaic River, 2:*715*
Passenger pigeons, 1:61
Passerine birds, 1:75
Pastures. *See* Grasslands
Patents, 1:*337*
Patricio, Stephen, 2:852
Pavley, Fran, 1:257
Pawa, Matt, 1:88
PBDEs (Polybrominated diphenyl ethers), 1:152
PCBs. *See* Polychlorinated biphenyls (PCBs)
Peace Parks Foundation, 1:412
Peat, 1:330, 2:743, 866
Pecans, 1:*402*
Peer review, 1:474, 475, 485, 488
Pelagic zone, 2:542, 632
Pelican Island National Wildlife Refuge, 2:590, 886
Pelosi, Nancy, 1:72
Penalties. *See* Settlements
Penguins, 1:21, 22, 23, 2:593, *622*
Penicillin, 1:336, 2:666
Pennsylvania, 1:108, 148, *156*, 184, 225, 226
PepsiCo, 1:444
Perata, Don, 2:852
Perching birds. *See* Passerine birds
Perennials, 1:420
Perfluorocarbons, 1:500
Periods, geologic. *See* Geologic history
Permafrost, 1:132, 388, 486, 2:742, 798, 799
Permeability zones, 1:33–34
Permits. *See* Government regulation
Peroxyacetyl nitrate, 1:9
Persepolis, 1:*2*
Persia. *See* Iran
Persian Gulf, 2:641, 642, 825
Persistant organic pollutants, 1:10
Persistent contrails. *See* Contrails
Personal care products. *See* Pharmaceuticals and personal care products (PPCPs)
Peru, 1:160, *275*, 359
Pesticide use against insects. *See* Insecticide use
Pesticides for weeds. *See* Herbicides
Pests (Insects). *See* Insects
Petroleum. *See* Oil
Petroleum-based fuels. *See* Fossil fuel combustion impacts
Petroleum resource use, environmental benefits and liabilities of. *See* Environmental benefits and liabilities of petroleum resource use
Pets, 1:*210*
Petten, Netherlands, 2:728
Pfiesteria piscicida, 1:12–13
pH. *See* Acidity
Phaeopigment, 2:549, 551
Pharmaceutical development resources, 2:**666–668**
Pharmaceuticals and personal care products (PPCPs)
biodegredation, 1:57
forests, 1:325
groundwater, 1:397–398
medical waste, 2:563, 564
pharmaceutical development resources, 2:666–668
teratogens, 2:783–784
water pollution, 2:837, 839
See also specific pharmaceuticals
Philippines, 1:*180*
bays and estuaries, 1:*54*
toxic waste, 2:*792*
tsunami impacts, 2:793
volcanoes, 2:819, *820*
waste transfer and dumping, 2:829–830
Phillips, Kathryn, 1:255
Phosphates, 2:840
Phosphorus, 1:5, 11, 2:833
Photochemical smog. *See* Smog
Photography, environmental, 2:**669–671**, 822
Photosynthesis
algal blooms, 1:11, 12
aquatic ecosystems, 1:31, 32
atmosphere, 1:41
carbon dioxide, 1:83
carbon sequestration, 1:90
coastal ecosystems, 1:151
coral reefs and corals, 1:170
deforestation, 1:190
ecosystem diversity, 1:240
ecosystems, 1:242, 245
estuaries, 1:294
forests, 1:321–322
greenhouse gases, 1:392
marine ecosystems, 2:541, 542, 544
mathematical modeling and simulation, 2:556–557
oceans and coastlines, 2:632
ozone layer, 2:663
reef ecosystems, 2:702, 704
runoff, 2:716
vegetation cycles, 2:817
Phototrophs, 1:142
Photovoltaic cells, 2:750–751
Physical stratigraphy, 1:79
Physicians for Social Responsibility, 1:113
Phytoplankton
algal blooms, 1:11
aquatic ecosystems, 1:30, 32
carbon dioxide emissions, 1:86
carbon sequestration, 1:90, 91
coral reefs and corals, 1:171
ecosystem diversity, 1:240
ecosystems, 1:245
iron fertilization, 1:490–491
marine ecosystems, 2:541, 542, 544
marine water quality, 2:550, 551
mathematical modeling and simulation, 2:556–557
nonpoint-source pollution, 2:602
oceans and coastlines, 2:632
ozone hole, 2:660
red tide, 2:674, 700
runoff, 2:717
Phytoremediation, 1:77, 512–513, 2:754
Picher, Oklahoma, 1:*163*

Picking Up Where 'Silent Spring' Left Off, 2:*737–738*
Pig farm impacts. *See* Factory farms, adverse effects of
Pig farms. *See* Factory farms, adverse effects of
Pigs, 1:302, *303,* 481, 483
Pinatubo. *See* Mount Pinatubo
Pine forests. *See* Boreal forests
Pinelands National Reserve, 2:709–710
Pipelines, 1:230–231, 394–395, *516,* 2:798
Planes (aircraft). *See* Aircraft
Plankton
  biodiversity, 1:61, 62, 63
  biostratigraphy, 1:80
  coastal zones, 1:154
  coral reefs and corals, 1:170
  El Niño and La Niña, 1:249
  liquefied natural gas resource use, 1:534–535
  See also Phytoplankton; Zooplankton
Planktos Inc., 1:491, 2:630
Plant-based fuel impacts. *See* Alternative fuel impacts
Plants (organisms)
  biodiversity, 1:60, 61, 62, 63
  biofuels, 1:67, 69
  bioremediation, 1:76, *77–78*
  carbon dioxide, 1:83
  carbon dioxide emissions, 1:85, 86
  carbon sequestration, 1:90
  CITES, 1:105–107
  climate change, 1:133
  closed ecology experiments, 1:142
  coal resource use, 1:145
  coastal ecosystems, 1:150, 151–152
  coastal zones, 1:154, 155
  coral reefs and corals, 1:170
  desertification, 1:194, 196
  deserts, 1:197–198
  ecological competition, 1:237
  ecosystem diversity, 1:240
  ecosystems, 1:242, 243, 244–245
  endangered species, 1:261
  estuaries, 1:294
  forests, 1:321, 322, 323, 325
  freshwater and freshwater ecosystems, 1:330, 331
  global warming, 1:369
  greenhouse effect, 1:383, 386–387
  greenhouse gases, 1:390, 391, 392
  habitat alteration, 1:404, 405, 406
  habitat loss, 1:410–411
  herbicides, 1:418–419
  horticulture, 1:420–422
  hydrologic cycle, 1:445
  invasive species, 1:481, 483
  IPCC 2007 Report, 1:486
  irrigation, 1:493–494, 495
  laboratory methods in environmental science, 1:511
  marine ecosystems, 2:541, 542, 543–544
  mining and quarrying impacts, 2:571
  non-scientist contributions, 2:604
  North Atlantic Oscillation, 2:608
  oil spills, 2:641
  overgrazing, 2:655–657
  ozone layer, 2:663
  pharmaceutical development resources, 2:666, 667
  pollinators, 2:672–673
  predator-prey relations, 2:678
  radiative forcing, 2:680
  shrublands, 2:732–733
  snow and ice cover, 2:741
  soil chemistry, 2:742–743
  vegetation cycles, 2:816–817
  wetlands, 2:864
  wildfires, 2:877–878
Plasmids, 1:76
Plass, Gilbert, 1:474
Plastic, 1:455–456
Pleistocene Epoch, 1:362
Plutonic diapirs, 2:818
Plutonium, 2:610, 611, 683, 684
PNGV (Partnership for a New Generation of Vehicles), 1:441
Poaching. *See* Hunting practices
Point source pollution, 1:123, 457, 2:601, 716, 838
Poisoning, shellfish, 2:700
Poisons. *See* Toxins
Polar bears, 1:*37,* 38, 2:677, 885
Polar cells, 1:45
Polar ice caps. *See* Snow and ice cover
Polar ice sheets. *See* Ice sheets
Polar regions. *See* Antarctic issues and challenges; Arctic
Polar stratospheric clouds, 2:659
Polar wandering. *See* Rotation (Earth)
Polar winds, 1:44, 45
Polders, 1:310–311, 2:726–727, 728, 888
Poles (regions). *See* Antarctic issues and challenges; Arctic
Policy, United Nations. *See* United Nations policy and activism
Political parties, Green. *See* Green parties
The Politics of Ethanol Outshine Its Costs, 1:*71–72*
Pollen, 2:672, 673
Pollinators, 2:**672–673**
Pollutants
  inorganic, 2:839–840
  primary, 1:9, 10
  See also specific pollutants
Polluter pays principle, 1:477
Pollution. *See specific types of pollution*
Pollution credits
  Clean Air Mercury Rule, 1:112
  Clean Development Mechanism, 1:115–116
  dams, 1:183–184
  Kyoto Protocol, 1:501, 502
  United Nations Framework Convention on Climate Change, 2:807
Polyaromatic hydrocarbons (PAHs), 1:57, *416*
Polybrominated diphenyl ethers (PBDEs), 1:152
Polychlorinated biphenyls (PCBs), 1:57, 76, 456, 2:758, 790, 839
Polyps, coral. *See* Coral polyps
Pombo, Richard, 2:586
Pomology, 1:420
Ponds
  agricultural practice impacts, 1:6
  aquatic ecosystems, 1:30
  factory farms, 1:302, 303
  fish farming, 1:308–309
  flood control and floodplains, 1:312
  freshwater and freshwater ecosystems, 1:329–330
  groundwater, 1:398
  water resources, 2:843
  watersheds, 2:857
Poppies, 1:*419*
Population dynamics, 2:678
Population management, wildlife. *See* Wildlife population management
Population models, 2:557
Porsche, Ferdinand, 1:441
Port Arthur, Texas, 2:*833*
Port Royal, Jamaica, 1:222
Portugal, 2:793
Post-glacial rebound, 2:724–725
Posthoorn, Roel, 2:728
Potassium permanganate ($KMnO_4$), 1:398–399
Poultry farm impacts. *See* Factory farms, adverse effects of
Pounds, J. Alan, 1:62
Power. *See specific types of power*

Power plants
acid rain, 1:2–4
Clean Air Mercury Rule, 1:112–113
clean energy, 1:119
coal resource use, 1:148–149
corporate green movement, 1:177
environmental protests, 1:292
geothermal resources, 1:354, 355
industrial water use, 1:459, 461, *461*
nuclear power, 2:610–614
ocean tides, 2:627
radioactive waste, 2:683, 684
PPCPs. *See* Pharmaceuticals and personal care products (PPCPs)
Prairies, 1:372, *374,* 2:864, 865
Precession, 1:362
Precipitation, 2:**674–675**
acid rain, 1:1–4
Antarctica, 1:21
atmosphere, 1:41
atmospheric circulation, 1:45
clear-cutting, 1:126
climate change, 1:133
deforestation, 1:190
deserts, 1:197
drainage basins, 1:200
drought, 1:205–206
ecosystems, 1:243, 244
El Niño and La Niña, 1:249, 250
floods, 1:314, 315
forests, 1:323
global warming, 1:369
grasslands, 1:372
greenhouse effect, 1:383, 386–387
greenhouse gases, 1:390
groundwater, 1:396
habitat alteration, 1:406
hurricanes, 1:430
hydrologic cycle, 1:445
ice ages, 1:448
IPCC 2007 Report, 1:486
landslides, 1:528
nonpoint-source pollution, 2:601
runoff, 2:716, 717
snow and ice cover, 2:741
surface water, 2:764
vegetation cycles, 2:817
water resources, 2:843
weather extremes, 2:862
Predator-prey relations, 2:**676–679**
aquatic ecosystems, 1:32
biodiversity, 1:63
ecological competion, 1:238
ecosystem diversity, 1:240–241
ecosystems, 1:242, 245
endangered species, 1:260
grasslands, 1:372
invasive species, 1:481, 482
marine ecosystems, 2:541, 544
marine fisheries, 2:548
mathematical modeling and simulation, 2:557
National Park Service Organic Act, 2:586
oceans and coastlines, 2:632
reef ecosystems, 2:703
species reintroduction programs, 2:755–756
wildlife population management, 2:880
Predicting. *See* Forecasting
Preservation, private, 2:591–592
Preservation areas. *See* Natural reserves and parks; Wildlife refuge
Preservation efforts. *See* Conservation
Pressure, atmospheric. *See* Atmospheric pressure
Pressure, water, 2:541
Prevailing westerlies, 1:44, 45
Prey-predator relations. *See* Predator-prey relations
Primary pollutants, 1:9, 10
Primeval forests, 1:321, 325, 406, 2:706
Prince Edward Island, 1:13, 2:720
Prince William Sound, 1:57, 76, 230, 2:643
The Print (Adams), 2:670
Privacy rights, 1:343
Private preservation, 2:591–592
Probes, monitoring. *See* Real-time monitoring and reporting
Problem Dams On the Rise in US, 1:***184***
Processed oil. *See* Oil
Products, green. *See* Green movement
Progress in California On Curbing Emissions, 1:***257–258***
Pronatura, 1:360
Property rights, 1:190, 399, 2:592, 691
Protected areas. *See* Natural reserves and parks
Protection, environmental. *See* Conservation
Protection, wildlife. *See* Wildlife protection
Proteomics, 1:512
Protests, environmental. *See* Environmental protests
Protists, 1:11
Protocol on Environmental Protection to the Antarctic Treaty, 1:26
Proxy records, 2:779
Pseudo-nitzschia, 1:13
Pseudotachylyte, 1:219
PTBT. *See Treaty Banning Nuclear Weapon Tests in the Atmosphere, in Outer Space, and Under Water (Limited Test Ban Treaty)*
Ptolemy, Claudius, 2:539
Public health. *See* Health
Public Health Service (U.S.), 1:108
Public Law 109-58. *See* Energy Policy Act of 2005
Public utilities. *See* Power plants
Puerto Rico, 1:*171,* 527, 2:562
Puget Sound, 1:120
Pulp, wood. *See* Paper and wood pulp
Pulse generators. *See* Hydrofoil generators
Pumps, 1:397, 2:843
Pyroclastic flow, 2:819
Pytheas, 2:626

Q

Qatar, 1:*534*
Qualitative models, 2:556–557
Quality, groundwater. *See* Groundwater quality
Quality, water (marine). *See* Marine water quality
Quarrying impacts. *See* Mining and quarrying impacts
Quinn, Tim, 2:851

R

Rabbits, 2:678
Radiation. *See specific types of radiation*
Radiative forcing, 2:**680–682**
climate change, 1:129, 130, 132
fossil fuel combustion impacts, 1:327
global warming, 1:367
soot, 1:148
Radioactive clocks, 1:348
Radioactive decay, 1:348
Radioactive fallout, 2:615–616, 617
Radioactive waste, 1:*415,* 2:**683–685**
Antarctica, 1:28
biodegredation, 1:58
bioremediation, 1:77
ecodisasters, 1:231–232
environmental protests, 1:292
marine ecosystems, 2:544
medical waste, 2:563, 564
nuclear power, 2:610, 611, 612

solid waste treatment technologies, 2:754
war and conflict-related environmental destruction, 2:825, 827
water pollution, 2:839
Radioisotopes, 1:57
Radiolaria, 1:55
Radiometric dating, 1:192–193, 347–349
Radionuclides, 1:348–349
Rahmstorf, Stefan, 2:725
Rain, acid. *See* Acid rain
Rain forest destruction, 2:**686–687**
agricultural practice impacts, 1:5, 6
alternative fuel impacts, 1:16
biodiversity, 1:60, 61, 63
biofuels, 1:71
biogeography, 1:75
clear-cutting, 1:125, 126, 127
ecosystems, 1:246
endangered species, 1:260
extinction and extirpation, 1:299
human impacts, 1:423
logging, 1:536, 537
mining and quarrying impacts, 2:*572*
natural resource management, 2:598, 599
photography, 2:669
resource extraction, 2:713
Rain forests, 1:*324,* 2:*600*
ecosystems, 1:243–244
as forests, 1:321–322, 324
genetic diversity, 1:335
greenhouse effect, 1:387
Kyoto Protocol, 1:*502*
natural reserves and parks, 2:592
pharmaceutical development resources, 2:667
precipitation, 2:674
See also Rain forest destruction
Rainfall. *See* Precipitation
Rainwater collection. *See* Recycling
Rammelkamp, Matthew, 1:228
Ramsar Convention on Wetlands, 2:569, 866, 884
Rangelands, 1:246, 372
Rapanos v. United States (2006), 1:123
Rapeseed oil. *See* Canola oil
RARE, 1:381
Rats, 1:481
Rayleigh effect, 1:528–529
RCRA. *See* Resource Conservation and Recovery Act (RCRA)
Reactors, nuclear. *See* Nuclear power
Reagan, Ronald, 1:81
Real-time monitoring and reporting, 2:**688–689**
climate modeling, 1:140, 141
El Niño and La Niña, 1:249, 251
geochemistry, 1:340
Geographic Information Systems (GIS), 1:*342, 343*
geospatial analysis, 1:351, 352
glacial retreat, 1:357–358
Hurricane Katrina, 1:438
hurricanes, 1:434–435
industrial pollution, 1:455
industrialization in emerging economies, 1:463
laboratory methods in environmental science, 1:509, 513
landfills, 1:525
landslides, 1:528
marine water quality, 2:549, 551, 552
medical waste, 2:562, 565
nonpoint-source pollution, 2:602
nuclear test ban treaties, 2:617
oceanography, 2:629–630
ozone hole, 2:658
pharmaceutical development resources, 2:667
photography, 2:669, 670
red tide, 2:700
sea level rise, 2:723
spill remediation, 2:759
streamflow, 2:761, 762
surveying, 2:767–768
temperature records, 2:779, 780
tsunami impacts, 2:794, 796
United Nations Conference on the Human Environment, 2:801
weather and climate, 2:859
weather extremes, 2:861–862
wildlife population management, 2:881
Woods Hole Oceanographic Institution, 2:891, 892, *892*
Reasonable use (water rights), 1:399
Recharging (aquifers), 1:33–35, *34,* 396, 403, 2:843, 844
Recreational use and environmental destruction, 2:**690–692**
Antarctica, 1:21, 22, 23
bays and estuaries, 1:54
habitat alteration, 1:405–406
habitat loss, 1:412
lakes, 1:515
National Park Service Organic Act, 2:585, 586
natural reserves and parks, 2:593
oceans and coastlines, 2:632
reef ecosystems, 2:704–705
Recycling, 2:**693–698**
beverage containers, 2:696
electronics, 2:696–697
electronics waste, 1:252, 253
history, 2:693–694
human impacts, 1:426
impacts, 2:697
industrialization in emerging economies, 1:*463*
paper, 2:695–696
paper and wood pulp, 2:664, 665
precipitation, 2:*675*
San Francisco program, 2:694–695
solid waste treatment technologies, 2:753
waste transfer and dumping, 2:829, 830
water conservation, 2:835
water resources, 2:844
Red-cockaded woodpecker, 1:260
Red squirrels, 1:262
Red tide, 1:11, 12, 2:**699–701,** *701*
Red water fern, 1:483, *483*
Redwoods, 2:877
Reef ecosystems, 2:**702–705,** *704*
acid rain, 1:2
aquatic ecosystem, 1:32
biodiversity, 1:62
coral reefs and corals, 1:170–172
ecosystem diversity, 1:241
habitat loss, 1:411
marine ecosystems, 2:543–544, *544*
oceans and coastlines, 2:632, 634
Reefs, coral. *See* Coral reefs and corals
Refined oil. *See* Oil
Reforestation, 2:**706–711**
clear-cutting, 1:126
deforestation, 1:189
ecodisasters, 1:234–235
forest resources, 1:317–319
forests and, 1:325
habitat alteration, 1:406
habitat loss, 1:409
Mexico, 2:709
New England, 2:706–708
New Jersey, 2:709–710
Rwanda, 2:708
sustainable development, 2:773–774
vegetation cycles, 2:816, 817
wildfires, 2:877–878
Refuge, wildlife, 2:**886–887**
Refugees, 1:133
Refuse Act of 1899, 1:122, 279
Refuse-derived fuel, 2:753–754
Regan, Julie, 1:292
Regulation, government. *See* Government regulation
Reid, Harry, 1:148, 149
Reintroduction of species. *See* Species reintroduction programs

Relentless Advocate 'Greens' Rural China, Village By Village, 1:*381–382*
Remediation. *See specific types of remediation*
Remote sensing, 1:189
REMUS, 2:*892*
Rendell, Edward, 1:225
Renewable energy. *See* Clean energy
Renewable Fuels Standard (U.S.), 1:71–72
Replanting, forest. *See* Reforestation
Report of the Secretary-General's High-Level Panel on Threats, Challenges, and Change (United Nations), 1:225
Reporting, environmental. *See* Media: Environmentally based news and entertainment
Reporting, real-time. *See* Real-time monitoring and reporting
Reptiles, 1:497
Research. *See specific research topics*
Reserves, coal. *See* Coal resource use
Reserves, mineral. *See* Minerals
Reserves, oil. *See* Oil
Reservoirs (groundwater). *See* Aquifers
Reservoirs (surface water), 2:844
Residence time, 1:367
Resource Conservation and Recovery Act (RCRA), 1:402, 414, 2:840
Resource extraction, 2:**712–713**
  Antarctica, 1:22, 26
  geochemistry, 1:340
  mining and quarrying impacts, 2:570–573
  recycling, 2:693
  United Nations Convention on the Law of the Sea, 2:803
Respiration, 1:83, 90, 243, 2:817
Restaurants, 2:647–648
Return intervals, 2:710
Reuter, John, 1:292–293
Reynolds, Harry, 2:707
Rhine River, 1:*471*
Rhinoceros, 1:428
Rhode Island, 1:88, 2:562
Ribbon seals, 1:*405*
Rice, 1:*143*, 2:865–866
Richter, Charles, 1:218
Richter scale, 1:218–219
  See also Earthquakes
Rifts, 2:820
Rights, property. *See* Property rights
Ring of Fire, 2:819, 820
Rings, tree, 1:192
Rio Declaration on Environment and Development, 1:212, 213, 215, 216, 478–479, 2:814
Riparian zones, 1:329
Rising sea level. *See* Sea level rise
Rising Tide North America, 1:*147*
River Ganges, 1:201
River Paraguay, 1:*203*
Rivers and Harbors Act, 1:276
Rivers and waterways, 2:**714–715**
  algal blooms, 1:11, 13
  aquatic ecosystems, 1:30
  bays and estuaries, 1:53
  benthic ecosystems, 1:55
  biodiversity, 1:60
  dams, 1:181
  deserts, 1*198*
  dredging, 1:202, 203, *203*
  estuaries, 1:294–297
  flood control and floodplains, 1:310, 311
  freshwater and freshwater ecosystems, 1:329
  glacial retreat, 1:359
  habitat alteration, 1:405
  hydrologic cycle, 1:447
  inland fisheries, 1:467–468
  lakes, 1:514, 515
  nonpoint-source pollution, 2:601–603
  ocean salinity, 2:624
  runoff, 2:716
  sea level rise, 2:725–728
  streamflow, 2:761–762
  water pollution, 2:837
  water resources, 2:843, 844–847
  watersheds, 2:856–857
  See also specific rivers
Roads, 1:108–409, 405
Roberts, Gareth, 1:395
Robinson, Michelle, 1:88
Rocks
  coastal ecosystems, 1:151
  geochemistry, 1:338–340
  geologic history, 1:345, 346, 347–349
  molten, 1:220, 354–355, 2:818
  soil chemistry, 2:743–744
  See also specific types of rocks
Rocky Mountains, 1:134, 2:604, 673
Rodale, J. J., 2:649
Rodale Institute, 2:649
Roh Moo-hyun, 1:*84*
Romania, 1:*31*
Rome (ancient), 1:202, 396–397
Romm, Joseph, 1:395
Roosevelt, Franklin D., 2:585
Roosevelt, Theodore, 1:166, 168, 2:590, 886
Roseate terns, 1:*433*
Ross Sea, 1:22
Rotation (Earth)
  atmospheric circulation, 1:45
  glaciation, 1:362
  ice ages, 1:449
  ocean circulation and currents, 2:620
  ocean tides, 2:626
  sea level rise, 1:362, 2:724
  tides, 2:788
Rothbert, P., 2:736
Rothrock, Dorothy, 1:257
Rubber recycling. *See* Recycling
Ruddiman, William, 1:450
Runoff, 2:**716–718**
  agricultural practice impacts, 1:5–6, *7*
  algal blooms, 1:11, 13
  carbon sequestration, 1:91
  clear-cutting, 1:125–126
  drainage basins, 1:200, 201
  estuaries, 1:294
  factory farms, 1:303–304
  groundwater, 1:397–398
  human impacts, 1:423, 425
  hydrologic cycle, 1:445, 447
  industrialization in emerging economies, 1:463
  irrigation, 1:494, 495
  laboratory methods in environmental science, 1:513
  Lake Tahoe, 1:291–292
  lakes, 1:514, 516
  logging, 1:536
  marine water quality, 2:549
  natural resource management, 2:598
  nonpoint-source pollution, 2:601, 602, 603
  ocean salinity, 2:624, 625
  organic farming, 2:650
  rain forest destruction, 2:686
  red tide, 2:700
  reef ecosystems, 2:704
  resource extraction, 2:712
  sea level rise, 2:723
  soil chemistry, 2:744
  soil resources, 2:748
  streamflow, 2:761–762
  tsunami impacts, 2:795
  wastewater treatment technologies, 2:832
  water pollution, 2:837, 838
  water resources, 2:843
  water supply and demand, 2:848
  watersheds, 2:856, 857
  wetlands, 2:865
Russia
  Antarctic Treaty, 1:25, 27
  Antarctica, 1:21, 22–23

Arctic, 1:26
chemical spills, 1:*101*
closed ecology experiments, 1:142–143
desertification, 1:195
International Convention for the Regulation of Whaling, 2:867–868
Kyoto Protocol, 1:500, 501, 502
lakes, 1:515
nuclear power, 2:610, 613
nuclear test ban treaties, 2:615, 616, 617
overfishing, 2:653
radioactive waste, 2:683
tundra, 2:798
United Nations Convention on the Law of the Sea, 2:805
wetlands, 2:866
Rwanda, 2:708

S

SACROC (Scurry Area Canyon Reef Operators Committee), 1:92, 93, 394
Safety, nuclear. *See* Nuclear power
Saffir-Simpson Scale, 1:431
Sahara Desert, 1:207, 208
Salicylic acid. *See* Aspirin
Salinity, 1:30, 31, 53, 514, 515, 2:650
See also Ocean salinity
Salmon
acid rain, 1:1
aquatic ecosystems, 1:32
clean energy, 1:119–121
fish farming, 1:306, 308
inland fisheries, 1:*468*
Salmonella, 1:238, 304
Salstein, David, 1:*248*
Salt concentation of water. *See* Salinity
Salt Lake City, Utah, 1:48
Salt marshes. *See* Marshes
Saltation, 1:207
Saltiness, ocean. *See* Ocean salinity
Salts, ocean. *See* Ocean salinity
Saltwater ecosystems. *See* Marine ecosystems
Saltwater intrusion, 2:**719–720**
aquifers, 1:34
bays and estuaries, 1:53
biodiversity, 1:61
coastal ecosystems, 1:150
estuaries, 1:294–297, 295
groundwater quality, 1:402
marine ecosystems, 2:541, 543
ocean circulation and currents, 2:620, 622
ocean salinity, 2:624
oceans and coastlines, 2:632
tsunami impacts, 2:795
wetlands, 2:864
Samuel, Justin, 1:228
San Francisco, California
Earth Day, 1:209
earthquakes, 1:218, 222, 223
estuaries, 1:295
recycling, 2:694–695
San Joaquin Valley Air Pollution Control District, 1:256
San Salvador, El Salvador, 1:*34*
Sand
coastal ecosystems, 1:150–151
dust storms, 1:207–208
earthquakes, 1:221
estuaries, 1:295
ocean salinity, 2:625
oceans and coastlines, 2:633
resource extraction, 2:712, 713
soil chemistry, 2:743
"A Sand County Almanac, and Sketches Here and There" (Leopold), 1:266
Sand dunes, 2:*634, 727, 795, 796*
Sand storms, 1:207–208
Sandalow, David, 1:65
Sandflats, 1:295
Sandia National Laboratory, 1:533–534
Sandstone, 1:92, 93
Santa Ana winds, 2:877, 878
Santa Catalina Island, 1:498
São Paulo, Brazil, 1:48
SARA. *See* Superfund Amendments and Reauthorization Act (SARA)
Sarcopterygiian, 1:*156*
Sardar Sarovar Dam, 1:*183*
Satellites
dust storms, 1:208
geospatial analysis, 1:351
National Oceanic and Atmospheric Administration, 2:583, 584, *584*
photography, 2:669, 670
real-time monitoring and reporting, 2:688
sea level rise, 2:723
temperature records, 2:780
weather and climate, 2:859
Saturation point, 2:674
Saudi Arabia, 1:488
Savannas, 1:372
Save the Whales, 1:376
SAX-40, 1:51
Saxitoxin, 2:699, 700
Scandinavia, 1:1, 189, 515
Scavenging, 2:829–830
Schistosomiasis. *See* Bilharzia
Schmidt, Gavin, 1:136
Schwarzenegger, Arnold, 1:65, *110,* 255, 257–258, 2:852
Science and Survival (Commoner), 1:375
Scorched earth tactics, 2:825
Scotland, 1:1, *307,* 2:732
Screw, Archimedes'. *See* Archimedes' screw
Scripps Institution of Oceanography, 2:628, 678, 741, 751–752, 891
Scurry Area Canyon Reef Operators Committee (SACROC), 1:92, 93, 394
SCV (Submerged Combustion Vaporization), 1:534–535
Sea anemones, 1:55
Sea cucumbers, 1:55
Sea floor. *See* Benthic ecosystems
Sea level rise, 2:**721–729**
Antarctica, 1:21, 23–24
Arctic, 1:38
bays and estuaries, 1:53
carbon dioxide emissions, 1:86
causes, 2:723–724
climate change, 1:130, 132, 134
coastal zones, 1:155
effects, 2:724–725
enigma, 2:724
Environmental Protection Agency, 1:286
fossil fuel combustion impacts, 1:328
glacial retreat, 1:357, 359, 360
global warming, 1:368, *369*
greenhouse effect, 1:383, 388
greenhouse gases, 1:390
groundwater quality, 1:402
habitat alteration, 1:406
history, 2:721–722
hydrologic cycle, 1:445
ice ages, 1:448
IPCC 2007 Report, 1:486
measurement, 2:722–723
migratory species, 2:569
Netherlands, 2:725–728
ocean circulation and currents, 2:622
projections, 2:725
reef ecosystems, 2:703
saltwater intrusion, 2:719–720
surface water, 2:766
tides, 2:788
Sea lice, 1:306, 308
Sea life. *See* Marine ecosystems
Sea otters, 2:643
Sea power. *See* Tidal or wave power
Sea sediment cores, 1:453
Sea Shepherd Conservation Society, 1:224

Sea surface temperature (SST), 1:432, 433
Sea turtles, 1:*180*
Sea-viewing Wide Field-of-view Sensor (SeaWiFS), 1:*241*
Sea walls. *See* Levees and sea walls
Seals (animals), 1:26, *429, 484,* 2:643, 677
Seas. *See* Oceans and coastlines
Seasonal migration, 2:568–569, **730–731**
Seawalls. *See* Levees and sea walls
Seaweed, 1:*92,* 171, 509
SeaWiFS (Sea-viewing Wide Field-of-view Sensor), 1:*241*
Secondary pollutants, 1:9, 10
Secretary of Agriculture. *See* Department of Agriculture (USDA)
Secretary of the Interior. *See* Department of the Interior
Sedgwick, Adam, 1:346
Sediment
  benthic ecosystems, 1:55
  drainage basins, 1:200
  dredging, 1:202–204
  earthquakes, 1:219
  estuaries, 1:295
  geochemistry, 1:338–339
  glaciation, 1:362
  marine water quality, 2:549, 551
  nonpoint-source pollution, 2:601
  reef ecosystems, 2:702
  resource extraction, 2:712, 713
  runoff, 2:716–717
  soil chemistry, 2:744
  tsunami impacts, 2:795
Sediment cores, sea. *See* Sea sediment cores
Sediment loading, 1:171–172
Sedimentary rocks, 1:79–80, 193, 338–339, 2:624
Seechi disk depth, 2:551
Seeding, 1:90, 91, 2:630, 799
Seismic sea waves. *See* Tsunami impacts
Seismic waves, 1:220
Seismographs, 1:218–219
Semi-arid regions, 1:205, 208, 245, 261
Semidiurnal tides, 2:788
Sensors, monitoring. *See* Real-time monitoring and reporting
September 11, 2001 attack, 1:*233,* 283
Septic tanks, 2:833
Sequestration, carbon. *See* Carbon sequestration
Serbia, 1:*31*
Sessile species, 2:549, 702
Sessions, William, 1:88
Set America Free Coalition, 1:72
Settlements, 1:2–3, 4, 503
1750 value, 1:393
Sewage. *See* Hazardous waste
Sewage contamination. *See* Water pollution
Sewer systems. *See* Wastewater treatment technologies
Sharks, 2:541, 542, 544, 678
Sharpe, Sara, 1:256
Sharps (medical waste), 2:563
Sheep, 2:655
Shellenberger, Michael, 1:380
Shellfish, 1:12, 457, 2:700
Shellfish poisoning, 2:700
Shelter wood harvesting, 1:191
Shenandoah National Park, 1:4
Shephard, Jim, 1:235
Shield volcanoes, 2:819
Shishmaref, Alaska, 2:741
Shoaling effect, 2:795
Shopping bags, 1:*378,* 379
Shorelines, marine. *See* Oceans and coastlines
Shrublands, 2:**732–733**
  tundra, 2:798
  wildfires, 2:878
Siberia, 1:*516*
Siegel, David, 2:647
Sierra Club
  carbon dioxide emissions, 1:88
  conservation, 1:166, 167
  emissions standards, 1:257–258
  green movement, 1:380
  light pollution, 1:529
  non-scientist contributions, 2:605
  wildlife population management, 2:881
Sierra Nevada: The John Muir Trail (Adams), 2:670
Silence of the Songbirds, 2:737–738
Silent Aircraft Initiative, 1:51
Silent Spring (Carson), 2:**734–738**
  agricultural practice impacts, 1:5
  DDT, 1:186, 187
  environmental activism, 1:265
  environmental assessments, 1:269–271
  environmental protests, 1:289
  environmental science careers, 1:95
  green movement, 1:375
  insecticide use, 1:471
  media, 2:559
  organic and locally grown foods, 2:645
  soil contamination, 2:745
  sustainable development, 2:769
  water pollution, 2:838
  wildlife protection policies and legislation, 2:883
'Silent Spring' Is Now Noisy Summer, 2:*736–737*
Silicon, 2:751
Silt, 2:743
Silviculture, 1:125, 537
Simon, Fred, 1:465–466
Simonson, Kenneth, 1:177
Simulation, mathematical. *See* Mathematical modeling and simulation
Singer, S. Fred, 1:136
Sinkholes, 1:34
Sinks, carbon. *See* Carbon sequestration
Sleet, 2:674
Sludge, 2:832
Smelting, 1:8, 2:571, 572
Smith, Robert Angus, 2:675
Smith, Tom K., 2:736
Smith, William, 1:80, 345
Smithsonian Institution, 1:168
Smog, 2:**739–740,** *740*
  acid rain, 1:2, 3, 4
  air pollution, 1:9, 10
  alternative fuel impacts, 1:15
  atmospheric inversions, 1:47, 48, *48*
  Clean Air Act of 1970, 1:108, 109
  ecodisasters, 1:231
  emissions standards, 1:256, 258
  environmental crime, 1:279
  Environmental Protection Agency, 1:*284*
  fossil fuel combustion impacts, 1:327, *327*
  industrial pollution, 1:455, 457, 458
  light pollution, 1:529
  mining and quarrying impacts, 2:572
  volcanoes, 2:819
Smoke. *See* Particulates
Smokestack pollution. *See* Air pollution
Smoking, 1:40
Smoldering, 2:878
Smyth, Rebecca, 1:92, 93
Snail darters, 2:776–778
Snails
  benthic ecosystems, 1:55
  biodiversity, 1:61
  coastal ecosystems, 1:151
  coastal zones, 1:154
  endangered species, 1:*261*

Snake River, 1:120, 121
Snakes, 1:*210,* 336
Snetsinger, Phoebe, 2:*605*
Snow, John, 1:397, 2:837
Snow and ice cover, 2:**741–742**
Antarctica, 1:21, 22–24, *23*
Arctic, 1:36–38, *37*
atmospheric circulation, 1:45
biodiversity, 1:64
carbon dioxide, 1:84
climate change, 1:*131,* 132
geologic history, 1:349
glacial retreat, 1:357
glaciation, 1:362, 363, *364*
global warming, 1:368
greenhouse effect, 1:385, 388
hydrologic cycle, 1:445, 447
ice ages, 1:448–451
ice cores, 1:452
IPCC 2007 Report, 1:486, 488
laboratory methods in environmental science, 1:*511*
ocean circulation and currents, 2:620, 622
precipitation, 2:674
radiative forcing, 2:680, 681
sea level rise, 2:721
water resources, 2:843
See also Marine snow
Snowfall. *See* Precipitation
Snowmobiles, 2:690, 691
Snowpack. *See* Snow and ice cover
Snowy plovers, 1:*238*
SO_2. *See* Smog
Sodium content of water. *See* Salinity
Software, monitoring. *See* Real-time monitoring and reporting
Soil chemistry, 2:**743–744**
acid rain, 1:1
clear-cutting, 1:125, 126
coastal zones, 1:154
dust storms, 1:208
geochemistry, 1:340
greenhouse effect, 1:387, 388
soil resources, 2:748
volcanoes, 2:820
wetlands, 2:864
Soil contamination, 2:**745–747**
biodegredation, 1:57
bioremediation, 1:76, 77, 78
Comprehensive Environmental Response, Compensation, and Liability Act, 1:163
groundwater, 1:397
groundwater quality, 1:401
laboratory methods in environmental science, 1:511, 512–513
soil chemistry, 2:743
spill remediation, 2:757–758, 759
war and conflict-related environmental destruction, 2:825
Soil resources, 2:**748–750**
agricultural practice impacts, 1:5, 6, 7
alternative fuel impacts, 1:19
coastal ecosystems, 1:150–151
deforestation, 1:189, 190
desertification, 1:194
deserts, 1:197, 198
earthquakes, 1:221–222
ecosystems, 1:245, 246
endangered species, 1:259
environmental science careers, 1:95
flood control and floodplains, 1:313
floods, 1:314–315
forests, 1:322, 323, 324, 325
grasslands, 1:372–373
habitat alteration, 1:404
irrigation, 1:493–494
logging, 1:536
organic and locally grown foods, 2:645, 646
organic farming, 2:649, 650, 651
overgrazing, 2:656
rain forest destruction, 2:686
recreational use and environmental destruction, 2:691
shrublands, 2:732, 733
soil chemistry, 2:743–744
vegetation cycles, 2:816, 817
See also Erosion
Solar panels, 2:750–751
Solar power, 2:**750–752**
air pollution, 1:8, 10
atmosphere, 1:41, 42, *42*
atmospheric circulation, 1:45
aviation emissions, 1:49, 50
biofuels, 1:67
chlorofluorocarbons, 1:100
clean energy, 1:119
climate change, 1:129
closed ecology experiments, 1:142
coral reefs and corals, 1:170
deforestation, 1:190
dust storms, 1:207
Gaia hypothesis, 1:333
geochemistry, 1:340
glaciation, 1:362
global warming, 1:366
greenhouse effect, 1:383–385
greenhouse gases, 1:390–393
hurricanes, 1:430
hydrologic cycle, 1:445
ice ages, 1:449
ozone hole, 2:658–660
radiative forcing, 2:680–681
solar power, 2:750–752
tidal or wave power, 2:784
tundra, 2:799
weather and climate, 2:859
Solid Waste Agency of Northern Cook County v. U.S. Army Corps of Engineers (2001), 1:123
Solid Waste and Emergency Response, Office of, 1:162
Solid Waste Disposal Act of 1965, 1:523, 2:565
Solid waste treatment technologies, 1:522, 2:**753–754**
Somerville, Richard, 1:136
Sonar, 2:578, 677
Songbirds. *See* Birds
Soon, W., 2:780
Soot
Arctic, 1:36, 37
biofuels, 1:68–69
coal resource use, 1:148
glacial retreat, 1:359
radiative forcing, 2:681
wildfires, 2:879
Sounds (estuaries). *See* Estuaries
South Africa
Antarctic Treaty, 1:25, 27
Earth Summit (2002), 1:*217*
environmental protests, 2:*771*
habitat loss, 1:412
oil spills, 2:641–642
South African lovegrass, 1:483
South America, 1:249, 250, 528
South Coast Air Quality Management District (California), 1:255, 256
South Korea, 1:*84*
electronics waste, 1:*253*
flood control and floodplains, 1:313
greenhouse gases, 1:*393*
nuclear test ban treaties, 2:*618*
recycling, 2:*695*
South Korean Ecological Youth, 1:*393*
South Pole. *See* Antarctic issues and challenges
Southern Company, 1:3
Southern Hemisphere
atmospheric circulation, 1:44, 45
El Niño and La Niña, 1:247–248
ice ages, 1:449
ocean circulation and currents, 2:620
Southern lights, 1:42
Southern Ocean, 1:160, 490
Southern Oscillation. *See* El Niño and La Niña
Sovereignty, 1:477, 2:803
Soviet Union. *See* Russia
SO_x. *See* Sulfur oxides (SO_x)

Soybeans, 1:78
Space flight, 1:142, 143
Spain, 2:*875*
Speciation, 1:60, 335, 408, 496
Species
  bays and estuaries, 1:53
  benthic ecosystems, 1:55, 56
  biodiversity, 1:60–61, 62, 63, 64
  biogeography, 1:74–75
  biostratigraphy, 1:79
  DDT, 1:187
  deserts, 1:197–198
  ecological competition, 1:236–238
  ecosystem diversity, 1:239–241
  ecosystems, 1:242–246
  extinction and extirpation, 1:298–300
  genetic diversity, 1:335–337
  greenhouse effect, 1:387
  habitat alteration, 1:404–406
  habitat loss, 1:408
  marine ecosystems, 2:542, 543
  oceans and coastlines, 2:633
  reef ecosystems, 2:702–703
  See also Endangered species; Endemic species; Invasive species; Keystone species; Sessile species
Species-area relationship, 1:298–299
Species at Risk Act (Canada), 2:883
Species diversity. *See* Biodiversity
Species reintroduction programs, 2:586, 593, 678–679, **755–756**
Spill remediation, 2:**757–760**
  biodegredation, 1:57, 58
  bioremediation, 1:76, 77
  chemical spills, 1:100–101
  ecodisasters, 1:230
  industrialization in emerging economies, 1:*464*
  laboratory methods in environmental science, 1:511
  Oil Pollution Acts, 2:636–639
  oil spills, 2:641
  soil contamination, 2:745, 746
Splash zones, 1:154, 155
Spodosols, 2:748
Sponges (organisms), 1:55, 2:542
Spoon-billed sandpipers, 1:313
Sprawl, urban and suburban. *See* Urban and suburban sprawl
Spring equinox, 1:209
Springs, underground. *See* Artesian springs and wells
Sri Lanka, 1:*222*
  flood control and floodplains, 1:313
  floods, 1:315
  insecticide use, 1:470
  landslides, 1:*528*
  saltwater intrusion, 2:720
  sustainable development, 2:771
  tsunami impacts, 2:794, 795
SST (Sea surface temperature), 1:432, 433
St. Clair Flats, Michigan, 1:*330*
St. Joseph Peninsula State Park, 2:*634*
Standard Oil, United States v. (1966), 1:***276–278***
Standard Oil Co., 1:276–278
Standards, government. *See* Government regulation
Stang, Dorothy, 2:*687*
Starfish, 1:55, 241
Stars, 1:349, 530–531
Statement of Forest Principles, 1:212, 213
States Are Closer To Trimming Autos' CO_2 Emissions, 1:***87–89***
Statistics. *See specific research topics*
Staudt, Chris, 1:177
Steam fields, 1:355
Steam plants, 1:355
Steam vents, 1:354
Steffen, Alex, 1:380
Stein, Winnie, 2:648
Steiner, Rudolph, 2:649
Steppes, 1:372
Stewardship, environmental. *See* Environmental activism
Stockard, Charles R., 2:782
Stockholm Conference. *See* United Nations Conference on the Human Environment (1972)
Stomata, 1:323
Stony corals, 1:32, 2:543, 634
Stop Huntingdon Animal Cruelty, 1:225
Storage, carbon. *See* Carbon sequestration
Storage, radioactive waste. *See* Radioactive waste
Storm surges, 1:313, 314, 437, 2:726, 727–728, 865
Storm water, 2:832
Storms
  atmospheric circulation, 1:44
  biodiversity, 1:64
  climate modeling, 1:140, 141
  coastal ecosystems, 1:152
  coastal zones, 1:155
  dust storms, 1:207–208
  El Niño and La Niña, 1:247, 250
  greenhouse effect, 1:383
  hurricanes, 1:430, 431
  National Oceanic and Atmospheric Administration, 2:582–584
  North Atlantic Oscillation, 2:608
Strata, 1:79, 80
Stratigraphy, 1:345, 347–349
  See also Biostratigraphy
Stratosphere, 1:41–42, 47, 100, 104, 2:658–660, 661
Stratospheric ozone layer. *See* Ozone layer
Streamflow, 2:**761–762**
Streams
  aquatic ecosystems, 1:30
  benthic ecosystems, 1:55
  freshwater and freshwater ecosystems, 1:329
  rivers and waterways, 2:714
  streamflow, 2:761–762
  water pollution, 2:839
  water resources, 2:843
Strip mining, 1:145, 146–147
Strip-tilling. *See* Tilling
Strong, Maurice, 2:806
Stronkhoorst, Joost, 2:727
Stutchbury, Bridget, 2:737–738
Subduction, 1:219, 2:818–819
Submarine landslides. *See* Landslides
Submerged Combustion Vaporization (SCV), 1:534–535
Submerged shorelines, 1:53
Subsidence, land. *See* Land subsidence
Subsoil, 2:743, 748
Substrates, 1:151, 2:702
Succession (forestry), 1:189, 2:732, 816
Suckling, Kieran, 1:263
Sudan, 1:206, *206,* 315, 494, 2:825, 828
Sugar, 1:15
Sugarcane, 1:68, *69,* 70, 71
Sulfur
  acid rain, 1:1
  atmosphere, 1:41
  aviation emissions, 1:49
  coal resource use, 1:146–147
  ecodisasters, 1:231
  industrial pollution, 1:455, 457
  precipitation, 2:674, 675
  wetlands, 2:864
Sulfur dioxide (SO_2). *See* Smog
Sulfur hexafluoride, 1:500
Sulfur oxides (SO_x), 1:9, 49, 254
Sumatra, 2:794
Summers, Larry, 2:831
Sun
  coastal zones, 1:154
  ocean tides, 2:626

solar power, 2:750–752
tides, 2:788–789
Superfund Act. *See* Comprehensive Environmental Response, Compensation, and Liability Act (CERCLA)
Superfund Amendments and Reauthorization Act (SARA), 2:763
Superfund Remediation and Technology Innovation, Office of, 1:162
Superfund site, 1:*163*, 2:**763–764**
biodegredation, 1:57–58
bioremediation, 1:78
Comprehensive Environmental Response, Compensation, and Liability Act, 1:162–165
Environmental Protection Agency, 1:283–284
hazardous waste, 1:414–417
Hurricane Katrina, 1:437
laboratory methods in environmental science, 1:512
soil contamination, 2:745, 747
spill remediation, 2:759
toxic waste, 2:791
Superior Anhausner Foods, 2:648
Superposition, 1:348
Supervolcanoes, 2:818
Supreme Court (India), 1:3
Supreme Court (U.S.)
carbon dioxide emissions, 1:87
Clean Air Act of 1970, 1:109
Clean Water Act, 1:123
Environmental Protection Agency, 1:287
National Environmental Policy Act, 2:578
pollution control, 1:4
Tennessee Valley Authority v. Hill, 2:776, 777–778
United States v. Standard Oil, 1:276–278
Surface chemistry, 1:513
Surface water, 2:**765–766**
agricultural practice impacts, 1:5
Antarctica, 1:23
aquifers, 1:33, 35
drought, 1:205
factory farms, 1:302, 303–304
groundwater quality, 1:401
groundwater *vs.,* 1:396, 397, 398
hydrologic cycle, 1:445
industrial water use, 1:460
irrigation, 1:493
landfills, 1:523
soil chemistry, 2:744
tundra, 2:798
wastewater treatment technologies, 2:834
water resources, 2:843, 844
water supply and demand, 2:848, 849
wetlands, 2:865
Surface waves, 1:220
Surfrider Foundation, 2:552
Surveying, 2:**767–768**
photography, 2:669, 670
real-time monitoring and reporting, 2:688–689
Sustainable development, 2:**769–775**
alternative fuel impacts, 1:14, 16, 19
biodiversity, 1:65
commercial fisheries, 1:160
corporate green movement, 1:174–177
cultural practices and environment destruction, 1:180
deforestation, 1:189, 191
Earth Summit (1992), 1:212, 213
Earth Summit (2002), 1:215–217
fish farming, 1:307
forest resources, 1:318, 319
forests, 1:325
green movement, 1:380
groundwater, 1:399–400
habitat alteration, 1:404
habitat loss, 1:412
history, 2:769–770
inland fisheries, 1:467–468
International Convention for the Regulation of Whaling, 2:867–870
international environmental law, 1:478, 479
land use, 1:520–521
logging, 1:537
Maathai, Wangari Muta, 2:772–774
marine fisheries, 2:546, 547–548
National Oceanic and Atmospheric Administration, 2:582
natural resource management, 2:598–599
organic farming, 2:649, 651
Our Common Future, 2:813–815
overfishing, 2:653
overgrazing, 2:657
paper and wood pulp, 2:664, 665
reforestation, 2:707–708, 709
sea level rise, 2:725–728
soil chemistry, 2:743
soil resources, 2:748–749
types, 2:770–772
United Nations Conference on the Human Environment, 2:800–801
water resources, 2:844
wildlife population management, 2:880–881
wildlife protection policies and legislation, 2:883–885
Swamps, 1:330, 411
Sweden
precipitation, 2:675
Swiss Glacier Retreats At A Rapid Rate, 1:*360*
Switchgrass, 1:70, 71
Switzerland, 1:359–360, *450*
insecticide use, 1:*471*
toxic waste, 2:*792*
SYMAP (Synagraphic Mapping System), 1:341
Symbiosis, 1:151, 170, 411, 2:543, 544, 702
Synagraphic Mapping System (SYMAP), 1:341
Syringe Tide, 2:562

T

Tahoe Regional Planning Agency (TRPA), 1:292
Taiga. *See* Boreal forests
Taiwan, 1:462, 2:*696*
Taj Mahal, 1:3
Tamargo, José Luis Luege, 2:*576*
Tansley, Arthur, 1:237
Tar Creek, 1:*163*
Tar sands project. *See* Oil sands project
Tars, 1:272
Tax credits, 1:16–20
Taxa, 1:79
Taxol, 1:325
Technologies, wastewater treatment. *See* Wastewater treatment technologies
Technology, monitoring. *See* Real-time monitoring and reporting
Tectonic estuaries, 1:294
Tectonic plates, 1:53, 219–220, 338, 340, 362, 514
Television coverage, environmental. *See* Media: Environmentally based news and entertainment
Tellico Dam Project (Snail Darters) and Supreme Court Case (*TVA v. Hill,* 1977), 2:**776–778**
Temperate forests
ecosystems, 1:243, 244–245, 246
as forests, 1:321, 322–323, 325
Temperature, increased global. *See* Global warming
Temperature records (Information), 2:**779–781**
weather extremes, 2:861–862, 863

Tennessee, 1:100, 2:776–778
Tennessee River, 2:755
Tennessee Valley Authority (TVA), 2:776–778
Tennessee Valley Authority v. Hill (1977). *See* Tellico Dam Project (Snail Darters) and Supreme Court Case (*TVA v. Hill,* 1977)
Teratogens, 2:**782–784**
  DDT, 1:187, 188
  hazardous waste, 1:416
  insecticide use, 1:471
  toxic waste, 2:790
Terrestrial ecosystems. *See specific types of terrestrial ecosystems*
Terrestrial forests. *See* Forests
Territorial claims. *See* International relations
Territorial waters, 2:803–804
Terrorism, environmental. *See* Eco-terrorism
Texas, 1:92, 93, 148, 208, 394–395, 2:*833*
Thailand
  flood control and floodplains, 1:313
  floods, 1:315
  international environmental law, 1:*479*
  tsunami impacts, 2:794
  wetlands, 2:865–866
Thalidomide, 2:783
Thawing, Arctic. *See* Arctic darkening and pack-ice melting
The Theory of Island Biogeography (MacArthur and Wilson), 1:74
Thermal conversion process (waste treatment), 2:754
Thermal inversions. *See* Atmospheric inversions
Thermal plume, 1:461
Thermal resources (Earth). *See* Geothermal resources
Thermocline, 1:330
Thermohaline circulation, 1:452, 2:608, 620
Thermosphere, 1:42
Thistle, Utah, 1:527
Thomas, Chris, 1:64, 300
Thompson, John, 1:92
Thoreau, Henry David, 2:589, 821–823
The Threat of Eco-terrorism, 1:***226–229***
Threatened species. *See* Endangered species
Three Gorges Dam, 1:181, *460*
Threshold Test Ban Treaty (TTBT), 2:615, 616, 617
Throughflow, 1:200
Thunderstorms. *See* Storms
Tiber River, 1:396–397
Tibet, 1:*373*
Tidal or wave power, 2:**785–787**
  clean energy, 1:119
  coastal ecosystems, 1:150–151
  natural resource management, 2:598
  ocean tides, 2:626, 627
  tides, 2:788, 789
Tide pools, 1:155, 243
Tides, 2:626–627, **788–789**
Tierney, John, 2:697
Tigers, 1:106–107, *409,* 427–428
Tigris River, 1:331
Tilling, 1:6, *1:6*
Timber. *See* Logging
Time, 1:*248*
Time, geologic. *See* Geologic history
Timescale, geologic. *See* Geologic history
Tiphobia horei, 1:61
Tisa River, 1:*31*
Titan (moon), 1:208
Titanic, 2:891, 892
Tobler, Waldo R., 1:341
Toilets, 2:835, 849
Toluene, 1:58, 77
Topography, 1:200, 240, 2:878
Topsoil, 2:743, 748
Tornadoes, 1:249–250, 434
Torrey Canyon, 1:231, 2:641
Tortoises, 1:*82,* 481, *498*
Tourism
  Antarctica, 1:21, 22, 23
  aviation emissions, 1:49
  coral reefs and corals, 1:171
  habitat alteration, 1:405–406
  habitat loss, 1:412
  National Park Service Organic Act, 2:585
  natural reserves and parks, 2:589–590, 593
  reef ecosystems, 2:704
  reforestation, 2:708
  waste transfer and dumping, 2:830
Toxic waste, 2:**790–792**
  Antarctica, 1:26
  biodegredation, 1:57–58
  chemical spills, 1:100–101
  Comprehensive Environmental Response, Compensation, and Liability Act, 1:162–165
  dredging, 1:203
  electronics waste, 1:252–253
  industrialization in emerging economies, 1:462
  medical waste, 2:562–566
  soil contamination, 2:746
  solid waste treatment technologies, 2:753, 754
  Superfund site, 2:762–763
  See also Radioactive waste
Toxins
  algal blooms, 1:11, 12, 13
  aquifers, 1:35
  herbicides, 1:418
  Hurricane Katrina, 1:437, 440
  industrial pollution, 1:455–457
  industrial water use, 1:460
  insecticide use, 1:469, 470
  mathematical modeling and simulation, 2:557
  pharmaceutical development resources, 2:667
  red tide, 2:699
  toxic waste, 2:790
  See also specific toxins by name
Toyota Prius, 1:441, 442, *442*
Trace gases, 1:9, 41
Trade, international, 1:105–107, 260
Trade winds, 1:44–45, 248, 249
Transcendentalism, 2:821
Transfer, waste. *See* Waste transfer and dumping
Transgenic plants, 1:511
Transgenic soya, 1:*63*
Transit passage, 2:804, 805
Transpiration, 1:243, 2:816
Transportation Department, 1:287
Trash, municipal. *See* Municipal waste
Trawling, bottom, 1:404–405, 2:547, 670
Treaties. *See specific treaties*
Treaty Banning Nuclear Weapon Tests in the Atmosphere, in Outer Space, and Under Water (Limited Test Ban Treaty), 2:615, 616, 617, ***618–619***
Tree canopies. *See* Forests
Tree clearing. *See* Clear-cutting
Tree frogs. *See* Monteverde harlequin tree frogs
Tree rings, 1:192
Tree-sitters, 1:225–226
Tree-spiking, 1:224–225, 227, 228, 267
Trees
  carbon sequestration, 1:90–91
  clear-cutting, 1:125–127
  deforestation, 1:189–191
  dendrochronology, 1:192–193
  desertification, 1:196
  ecodisasters, 1:234–235
  ecosystems, 1:246

forest resources, 1:317–320
forests, 1:321–325, *322*
fossil fuel combustion impacts, 1:*328*
horticulture, 1:420–421
Hurricane Katrina, 1:*437*
hydrologic cycle, 1:*446*
logging, 1:536–537
paper and wood pulp, 2:664
reforestation, 2:706–710
sustainable development, 2:773–774
Trenberth, Kevin, 1:136, 432, 435
Triassic period, 1:*347*
Trihalomethanes, 2:839
Tropical cyclones, 1:430
Tropical depressions, 1:431
Tropical easterlies. *See* Trade winds
Tropical forests. *See* Rain forests
Tropical regions
atmospheric circulation, 1:44, 46
biodiversity, 1:63
biofuels, 1:71
clear-cutting, 1:126
climate change, 1:133
dams, 1:181, 182
deforestation, 1:189
ecosystem diversity, 1:240
El Niño and La Niña, 1:250
endangered species, 1:260, 262
grasslands, 1:373
habitat loss, 1:409
hurricanes, 1:430
weather extremes, 2:862
Tropical storms. *See* Storms
Tropopause, 1:41
Troposphere, 1:41–42, 47, 57, 104
TRPA (Tahoe Regional Planning Agency), 1:292
Trucking industry, 1:255–256
Trucks. *See* Vehicles
Truman, Harry S., 2:804, 866
Tsunami impacts, 2:**793–797**, *796*
earthquakes, 1:218, *222*
ecodisasters, 1:233
flood control and floodplains, 1:312–313
floods, 1:315
laboratory methods in environmental science, 1:*512*
landslides, 1:527
mathematical modeling and simulation, 2:*558*
medical waste, 2:*564*
saltwater intrusion, 2:720
Tsunami warning systems (TWS), 2:794, 796
TTBT (Threshold Test Ban Treaty), 2:615, 616, 617
Tube worms. *See* Worms
Tucker, Wayne, 1:235
Tuna, bluefin, 2:*547*
Tundra, 2:**798–799**
Arctic, 1:37
biodiversity, 1:63
Bureau of Land Management, 1:81
ecosystems, 1:243, 244
greenhouse effect, 1:388
Turbidity, 1:53, 2:716
Turbines, 1:119, 181, 355, 2:785–786, 888, 889
Turtles, 1:*180, 262,* 2:*887*
TVA (Tennessee Valley Authority), 2:776–778
TVA v. Hill (1977). *See* Tellico Dam Project (Snail Darters) and Supreme Court Case (*TVA v. Hill,* 1977)
Twenty in Ten plan, 1:287
2007 Fourth Assessment Report. *See* IPCC 2007 Report
2,4-dichlorophenoxyacetic acid, 1:418
2,4.5-trichlorophenoxyacetic acid (2,4.5-T), 1:418
TWS (Tsunami warning systems), 2:794, 796
TXU Corp., 1:149, 176
Tyndall, Mike, 1:3
Typhoons, 1:430
Typhus, 1:186

U

Uganda, 2:593
Ultisols, 2:748
Ultraviolet light, 1:100, 574, 2:658–660, 661–663, 838
UN. *See specific terms beginning with United Nations*
U.N. Revs Up Over Global Warming, 1:***64–65***
UNCED (United Nations Conference on Environment and Development). *See* Earth Summit (1992)
UNCHE (United Nations Conference on the Human Environment). *See* United Nations Conference on the Human Environment (1972)
UNCLOS (United Nations Convention on the Law of the Sea), 1:478, 2:**803–805**
Underdeveloped countries. *See* Developing countries
Underground carbon emission storage. *See* Carbon sequestration
Underground water resources. *See* Aquifers
Underwater landslides. *See* Landslides
Underwater vehicles, 2:629, 891, 892
UNEP. *See* United Nations Environment Programme (UNEP)
UNESCO (United Nations Educational, Scientific and Cultural Organization), 2:794
UNFCCC. *See* United Nations Framework Convention on Climate Change (UNFCCC)
Union Carbide, 1:100, 231
Union of Concerned Scientists, 1:88
Union of Soviet Socialist Republics (USSR). *See* Russia
United Kingdom
Antarctic Treaty, 1:25, 27
Antarctica, 1:22
atmospheric inversions, 1:48
aviation emissions, 1:51, *51*
commercial fisheries, 1:160
eco-terrorism, 1:224
ecodisasters, 1:231
environmental protests, 1:290
factory farms, 1:302–303
floods, 1:315
green movement, 1:376, 377, 379
hunting practices, 1:428
Kyoto Protocol, 1:502–503
media, 2:560–561
nuclear power, 2:*613*
ocean circulation and currents, 2:620
pollinators, 2:673
tidal or wave power, 2:785
wetlands, 2:866
wind and wind power, 2:889
United Nations Biosphere Conference (1968), 1:478
United Nations Commission on Sustainable Development, 1:213
United Nations Conference on Environment and Development (UNCED). *See* Earth Summit (1992)
United Nations Conference on the Human Environment (1972), 1:212, 215, 2:**800–802**
international environmental law, 1:478
Our Common Future, 2:813
sustainable development, 2:770
United Nations policy and activism, 2:810
United Nations Convention on Biodiversity. *See* Convention on Biological Diversity

United Nations Convention on Biological Diversity. *See* Convention on Biological Diversity

United Nations Convention on the Law of the Sea (UNCLOS), 1:478, **2:803–805**

United Nations Convention to Combat Desertification, 1:196

United Nations Educational, Scientific and Cultural Organization (UNESCO), 2:794

United Nations Environment Programme (UNEP)
- algal blooms, 1:13
- DDT, 1:188
- Earth Summit (2002), 1:215
- electronics waste, 1:253
- endangered species, 1:262
- glacial retreat, 1:360
- glaciation, 1:363
- Intergovernmental Panel on Climate Change, 1:474
- international environmental law, 1:478
- IPCC 2007 Report, 1:485
- ozone layer, 2:663
- saltwater intrusion, 2:720
- United Nations Conference on the Human Environment, 2:801
- United Nations policy and activism, 2:810, 811

United Nations Food and Agriculture Organization (FAO)
- commercial fisheries, 1:160
- ecosystems, 1:246
- forests, 1:321, 324
- habitat loss, 1:411
- human impacts, 1:424
- irrigation, 1:494
- marine fisheries, 2:547
- overfishing, 2:652–653

United Nations Framework Convention on Climate Change (UNFCCC), 1:131, **2:806–809**
- carbon dioxide emissions, 1:86
- Clean Development Mechanism, 1:115, 116, 117
- Earth Summit (1992), 1:213, 214
- global warming, 1:370
- greenhouse gases, 1:393
- international environmental law, 1:478, 479
- Kyoto Protocol, 1:500, 503
- Montreal Protocol, 2:575
- United Nations Conference on the Human Environment, 2:575
- United Nations policy and activism, 2:810–811

United Nations Intergovernmental Panel on Climate Change (IPCC). *See* Intergovernmental Panel on Climate Change (IPCC)

United Nations Plan of Action to Combat Desertification, 1:195, 196

United Nations policy and activism, **2:810–812**
- biodiversity, 1:64–65
- biofuels, 1:67, 69, 71
- carbon dioxide emissions, 1:86–87
- climate change, 1:137
- commercial fisheries, 1:160
- desertification, 1:194–195
- Earth Summit (1992), 1:212–214
- Earth Summit (2002), 1:215–217
- eco-terrorism, 1:225
- international environmental law, 1:477–480
- Kyoto Protocol, 1:500–508
- Montreal Protocol, 2:574–576
- sustainable development, 2:769–770
- water pollution, 2:840
- water resources, 2:844

United Nations University Institute for Environment and Human Security, 1:133

United Nations World Bank, 1:183, 534, 2:831

United Nations World Commission on Dams, 1:183–184

United Nations World Commission on Environment and Development (WCED)
- Earth Summit (1992), 1:212
- Earth Summit (2002), 1:215–216
- international environmental law, 1:478
- *Our Common Future,* 2:813–815
- United Nations Conference on the Human Environment, 2:801

United Nations World Commission on Environment and Development: *Our Common Future* (1987), **2:813–815**
- Earth Summit (1992), 1:212
- Earth Summit (2002), 1:216
- international environmental law, 1:478
- United Nations Conference on the Human Environment, 2:801

United Nations World Food Programme, 1:16

United Nations World Glacier Monitoring Service, 1:358

United Nations World Health Organization. *See* World Health Organization (WHO)

United Nations World Meteorological Organization. *See* World Meteorological Organization (WMO)

United Nations World Summit on Sustainable Development. *See* Earth Summit (2002)

United Nations World Water Development Report, 1:459, 460

United States (U.S.)
- acid rain, 1:1, 2–3
- agricultural practice impacts, 1:5, 6
- air pollution, 1:9, 10
- algal blooms, 1:*12,* 12–13
- alternative fuel impacts, 1:15, 16–20
- Antarctic Treaty, 1:25, 26, 27
- Antarctica, 1:22
- aquifers, 1:33, 34–35
- Arctic, 1:26
- asbestos contamination, 1:40
- aviation emissions, 1:49, 51
- biodegredation, 1:57–58
- biodiversity, 1:65
- biofuels, 1:67, 68, 69–70, 71–72
- bioremediation, 1:78
- Bureau of Land Management, 1:81–82
- carbon dioxide emissions, 1:86, 87–89
- carbon sequestration, 1:91, 92–93
- chemical spills, 1:100
- Clean Air Act of 1970, 1:108–111
- Clean Air Mercury Rule, 1:112–113
- Clean Development Mechanism, 1:117
- clean energy, 1:119
- Clean Water Act, 1:122–124
- climate change, 1:133–135
- climate modeling, 1:140, 141
- coal resource use, 1:145, 146, 147, 148–149
- coastal ecosystems, 1:151, 152
- coastal zones, 1:156–159
- Comprehensive Environmental Response, Compensation, and Liability Act, 1:162–165
- conservation, 1:166–169
- corporate green movement, 1:174–177
- cultural practices and environment destruction, 1:179, 180
- dams, 1:181, 184
- DDT, 1:186–187
- dredging, 1:202
- drought, 1:205
- dust storms, 1:208
- Earth Day, 1:209–210
- Earth Summit (2002), 1:216
- earthquakes, 1:218, 221, 222
- eco-terrorism, 1:224, 226–229
- ecodisasters, 1:231, 232, 233, 234–235

ecosystems, 1:246
El Niño and La Niña, 1:249, 251
electronics waste, 1:252, 253
emissions standards, 1:254–258
endangered species, 1:259–260, 262
environmental activism, 1:265
environmental crime, 1:279, 280
Environmental Protection Agency, 1:282–284, 286–287
environmental science careers, 1:95–96
factory farms, 1:302, 303
fish farming, 1:309
forest resources, 1:317–319
forests, 1:325
freshwater and freshwater ecosystems, 1:331
genetic diversity, 1:336–337
geothermal resources, 1:354, 356
global warming, 1:370
green movement, 1:375, 376, 377, 380
greenhouse effect, 1:388
greenhouse gases, 1:394–395
groundwater, 1:399
groundwater quality, 1:402–403
hazardous waste, 1:413, 414–417
horticulture, 1:420, *421*
Hurricane Katrina, 1:437
hybrid vehicles, 1:441–444
industrial pollution, 1:457–458
insecticide use, 1:471–473
International Convention for the Regulation of Whaling, 2:867
international environmental law, 1:477
invasive species, 1:483
IPCC 2007 Report, 1:485, 488
irrigation, 1:494
Kyoto Protocol, 1:500, 501–502
laboratory methods in environmental science, 1:512
lakes, 1:517
land use, 1:518–521
landfills, 1:522–525
landslides, 1:527, 528
light pollution, 1:528
logging, 1:536
marine fisheries, 2:548
media, 2:560–561
medical waste, 2:562, 563–566
migratory species, 2:569
National Environmental Policy Act, 2:577–581
National Oceanic and Atmospheric Administration, 2:582–584
National Park Service Organic Act, 2:585–587
natural reserves and parks, 2:589–596
nonpoint-source pollution, 2:601, 602–603
nuclear power, 2:610, 611–612, 613
nuclear test ban treaties, 2:615–617
ocean circulation and currents, 2:620
oceanography, 2:628, 629
Oil Pollution Acts, 2:636–639
oil spills, 2:641, 642, 643
organic and locally grown foods, 2:645, 646
organic farming, 2:649–650
overfishing, 2:653
overgrazing, 2:655, 656, 657
radioactive waste, 2:683–685
recycling, 2:693–697
red tide, 2:699
reforestation, 2:706–710
runoff, 2:716, 717–718
saltwater intrusion, 2:720
soil contamination, 2:745–747
soil resources, 2:748, 749
solar power, 2:752
solid waste treatment technologies, 2:753–754
species reintroduction programs, 2:755
spill remediation, 2:759
Superfund site, 2:762–763
sustainable development, 2:769, *772*
Tennessee Valley Authority v. Hill, 2:776–778
teratogens, 2:783–784
toxic waste, 2:790–791
United Nations Convention on the Law of the Sea, 2:803, 804, 805
United Nations Framework Convention on Climate Change, 2:807
waste transfer and dumping, 2:829
wastewater treatment technologies, 2:833, 834
water conservation, 2:835
water pollution, 2:837–838, 839, 840–842
water supply and demand, 2:850–851
Watershed Protection and Flood Prevention Act of 1954, 2:854–855
watersheds, 2:856, 858
weather extremes, 2:862
wetlands, 2:866
Wilderness Act of 1964, 2:872–873
wildfire control, 2:874–876
wildlife population management, 2:880
wildlife protection policies and legislation, 2:883, 885
wildlife refuge, 2:886–887
wind and wind power, 2:888
See also specific locations and sites
United States, Rapanos v (2006), 1:123
United States v. Standard Oil (1966), 1:*276–278*
University of Calgary, 1:161
University of Georgia, 1:*7*
University of Texas, 2:*675*
University of Twente, 2:835
University of Wisconsin, 1:168
Upconing, 2:719
Upper Fremont Glacier, 1:453
Upper Mississippi River Wild Life and Fish Refuge, 2:886
UPS, 1:443
Upwellings
El Niño and La Niña, 1:249
environmental science careers, 1:98
marine fisheries, 2:546
ocean circulation and currents, 2:620, 621–622
oceanography, 2:628
oceans and coastlines, 2:632
surveying, 2:768
Uranium, depleted, 2:825
Urban, Noel, 1:*140*
Urban and Regional Information Systems Association (URISA), 1:341
Urban and suburban sprawl, 1:520
Urban areas
aquifers, 1:34
atmospheric inversions, 1:47
biodiversity, 1:63
endangered species, 1:259
flood control and floodplains, 1:312
forest resources, 1:319
groundwater quality, 1:402
horticulture, 1:421
human impacts, 1:424–425
hybrid vehicles, 1:443
industrialization in emerging economies, 1:462
land use, 1:518
light pollution, 1:530–531
Our Common Future, 2:814
runoff, 2:716, 717
smog, 2:739–740
soil contamination, 2:745
watersheds, 2:856
wildfires, 2:877, 878
Urban heat island effect, 1:392, 421, *446*
URISA (Urban and Regional Information Systems Association), 1:341
U.S. *See* United States (U.S.); *specific terms beginning with U.S.*

U.S. Agricultural Research Service, 1:512
U.S. Army Corps of Engineers, 1:221, 313, 2.866
U.S. Army Corps of Engineers v.Solid Waste Agency of Northern Cook County (2001), 1:123
U.S. Bureau of Fisheries, 2:736
U.S. Bureau of the Census, 1:341–342
U.S. Central Intelligence Agency. *See* Central Intelligence Agency (CIA)
U.S. Department of Commerce, 2:876
U.S. Department of Energy, 1:287, 395
U.S. Department of Homeland Security, 2:578
U.S. Department of Transportation, 1:287
U.S. Estuary Program, 1:296–297
U.S. Federal Emergency Management Agency, 1:437, 438
U.S. Geological Survey. *See* Geological Survey (U.S.)
U.S. Insitute for Environmental Conflict Resolution, 2:577
U.S. National Institutes of Health, 2:696
U.S. National Park Service. *See* National Park Service
U.S. Navy, 2:578
U.S. Public Health Service, 1:108
U.S. Supreme Court. *See* Supreme Court (U.S.)
USDA. *See* Department of Agriculture (USDA)
USGS. *See* Geological Survey (U.S.)
USSR. *See* Russia
Utah, 1:88, *176,* 257, 527
Utility companies. *See* Power plants
UV light. *See* Ultraviolet light

V

Vaccines, 2:783–784
Vail, Colorado, 1:225, *226*
Valley of the Drums, 2:763, 764
Valleys, 1:53
Van Helmont, Johannes, 1:83
Van Hise, Charles Richard, 1:168
van Winden, Alphons, 2:725–726
Vapor, water. *See* Water vapor (H_2)
Varves, 1:193
Vegetable oil, 1:14, 15, 67
Vegetation. *See* Plants (organisms)
Vegetation cycles, 2:**816–817,** 877–878
Vehicles, 1:*288*
  air pollution, 1:8, 10
  alternative fuel impacts, 1:*15,* 16–20
  atmospheric inversions, 1:47
  biofuels, 1:68
  carbon dioxide emissions, 1:87–89
  Clean Air Act of 1970, 1:109
  corporate green movement, 1:175
  emissions standards, 1:254, 255, 258
  Environmental Protection Agency, 1:286–287
  fossil fuel combustion impacts, 1:327
  geospatial analysis, 1:352
  green movement, 1:380
  habitat loss, 1:410
  Hurricane Katrina, 1:437–438
  laboratory methods in environmental science, 1:513
  See also Hybrid vehicles; Underwater vehicles
Velds, 1:372
Vellinga, Pier, 2:727
Velsicol Chemical Company, 2:736
Venezuela, 1:*262,* 527, 2:*572*
Venter, Craig, 2:630
Vents, hydrothermal. *See* Hydrothermal vents
Venus, 1:208
VENUS (Victoria Experimental Network Under the Sea), 2:688–689
Veolia Environmental Services, 2:*833*
Vermont, 1:87–88, *287, 456*
Vertebrates, 1:30
  See also specific vertebrates
Victoria Experimental Network Under the Sea (VENUS), 2:688–689
Vienna Convention for the Protection of the Ozone Layer (1985), 2:574–575, 810
Vinyl chloride, 2:790
Violence, 1:224–229
Virgin Islands, 1:*171*
Virginia, 1:4, 101, *147*
Virunga Mountains, 2:820
Viruses, 1:304
Vitrification, 2:754
Vivisection, 1:224
VOCs. *See* Volatile organic compounds
Volatile organic compounds (VOCs), 1:9, 10, 399, 457
Volcanoes, 2:**818–820**
  air pollution, 1:8
  atmosphere, 1:41
  bays and estuaries, 1:53
  biodiversity, 1:61
  carbon dioxide emissions, 1:86
  earthquakes, 1:220
  ecodisasters, 1:232–233
  extinction and extirpation, 1:298
  Geographic Information Systems (GIS), 1:*342*
  geothermal resources, 1:354–355
  greenhouse effect, 1:385
  ice ages, 1:449
  lakes, 1:515
  ocean salinity, 2:624
  radiative forcing, 2:680
  tsunami impacts, 2:793, 794, 795
  See also specific types of volcanoes
Volterra, Vito, 2:678
von Storch, Hans, 2:780
Vortex, 1:430, 432
Vostok ice core, 1:452

W

Walden (Thoreau), 2:**821–824,** *823*
Walke, John, 1:3–4
Walker, Gilbert, 1:247
Walkerton, Ontario, Canada, 1:397, 2:838
Wallace, Alfred Russel, 1:74
Wan Gang, 1:465
Wangari Maathai Nobel Lecture, 2:*773–774*
War and conflict-related environmental destruction, 2:**825–828**
  drought, 1:206
  ecodisasters, 1:231, 234
  nuclear power, 2:610
  sustainable development, 2:771
Wardens, 1:98
Warming, global. *See* Global warming
Warming, greenhouse. *See* Greenhouse effect
Warner, John, 2:667
Washington, 1:120–121, 220, *256,* 257
Waste. *See specific types of waste*
Waste transfer and dumping, 2:**829–831**
  Comprehensive Environmental Response, Compensation, and Liability Act, 1:164
  electronics waste, 1:252, 253
  hazardous waste, 1:413–414, 415

industrialization in emerging economies, 1:464
landfills, 1:526
marine ecosystems, 2:544
marine water quality, 2:551–555
recycling, 2:694
soil contamination, 2:746
solid waste treatment technologies, 2:753
spill remediation, 2:757
Superfund site, 2:763
toxic waste, 2:790
United Nations Convention on the Law of the Sea, 2:804
Waste treatment. *See specific types of waste treatment*
Wastewater treatment technologies, 2:**832–834**
Clean Water Act, 1:122
industrial water use, 1:459
marine water quality, 2:549–550
water conservation, 2:835, 836
water pollution, 2:839
wetlands, 2:865
Water, ground. *See* Groundwater
Water, storm. *See* Storm water
Water, surface. *See* Surface water
Water chemistry
bays and estuaries, 1:53
biodiversity, 1:61, 62, 63
carbon dioxide, 1:83
coral reefs and corals, 1:172
environmental science careers, 1:98
marine ecosystems, 2:541
ocean salinity, 2:624–625
oceanography, 2:628, 630
Woods Hole Oceanographic Institution, 2:892
Water conservation, 2:**835–836**
Clean Water Act, 1:122–124
drought, 1:205, 206
El Niño and La Niña, 1:251
irrigation, 1:493, 494
United Nations Convention on the Law of the Sea, 2:804
water resources, 2:843–847
water supply and demand, 2:848–852
Watershed Protection and Flood Prevention Act of 1954, 2:854–855
Water consumption, domestic. *See* Water supply and demand
Water consumption, industrial. *See* Industrial water use
Water contamination. *See* Water pollution
Water cycle. *See* Hydrologic cycle
Water demand. *See* Water supply and demand
Water ecosystems. *See* Aquatic ecosystems
Water filtration, 2:865, 866
Water footprint, 2:835
Water for injection (WFI), 2:667
Water hyacinths, 1:483
Water mixing. *See* Saltwater intrusion
Water pollution, 2:**837–842**
agricultural practice impacts, 1:5–6, 7
algal blooms, 1:11–13
Antarctica, 1:21, 22–23
aquatic ecosystems, 1:30, 32
aquifers, 1:33
bays and estuaries, 1:*54*
biodegradation, 1:57
bioremediation, 1:76, 77
chemical spills, 1:100, 101
China, 1:*398*
Clean Water Act, 1:122–124
clear-cutting, 1:125–126
coal resource use, 1:146–147
disease prevention, 2:838
drainage basins, 1:200, 201
ecosystems, 1:246
endangered species, 1:259
environmental science careers, 1:97
estuaries, 1:294, 296–297
factory farms, 1:302, 303–304
groundwater, 1:396, 397–400
groundwater quality, 1:401–403
habitat loss, 1:410
hazardous waste, 1:413
history, 2:837–838
human impacts, 1:425
Hurricane Katrina, 1:438
impacts, 2:840
industrial pollution, 1:455, 456–457
industrial water use, 1:459, 460
industrialization in emerging economies, 1:462, 463
inorganic pollutants, 2:839–840
international environmental law, 1:478
irrigation, 1:493, 494–495
laboratory methods in environmental science, 1:512
Lake Tahoe, 1:291–293
lakes, 1:514, 516–517
landfills, 1:525
marine ecosystems, 2:543, *543*, 544
marine water quality, 2:549–552
mining and quarrying impacts, 2:570, 571–572
nonpoint-source pollution, 2:601–603, 838–839
oceans and coastlines, 2:632, 634
Oil Pollution Acts, 2:636–639
oil spills, 2:641–643
radioactive waste, 2:683–684
rain forest destruction, 2:686
red tide, 2:700
reef ecosystems, 2:704
rivers and waterways, 2:715
runoff, 2:716–718
saltwater intrusion, 2:719–720
seasonal migration, 2:731
spill remediation, 2:759
surface water, 2:765
toxic waste, 2:790
tsunami impacts, 2:795, 796
tundra, 2:798
United Nations Convention on the Law of the Sea, 2:803, 804
United States v. Standard Oil, 1:276–278
war and conflict-related environmental destruction, 2:825
wastewater treatment technologies, 2:832, 833
water conservation, 2:835–836
Water Quality Act of 1965, 2:840–842
water supply and demand, 2:850
watersheds, 2:856, 857
wetlands, 2:866
Water Pollution Control Act, 2:551
Water power. *See* Hydropower
Water pressure, 2:541
Water quality, groundwater. *See* Groundwater quality
Water quality, marine. *See* Marine water quality
Water Quality Act of 1965, 2:840, ***841–842***
Water Quality Act of 1987, 1:123, 2:551
Water Quality Improvement Act of 1970, 1:122
Water resources, 2:*836*, **843–847**
agricultural practice impacts, 1:5–6
aquifers, 1:33–35
biofuels, 1:72
Bureau of Land Management, 1:81
climate change, 1:135
desertification, 1:195
drainage basins, 1:200–201
drought, 1:205
groundwater, 1:396–400
groundwater quality, 1:401–403
human impacts, 1:423, 425
hydrologic cycle, 1:445–447
industrial water use, 1:460
industrialization in emerging economies, 1:463
irrigation, 1:493, 494
medical waste, 2:564
mining and quarrying impacts, 2:571

natural resource management, 2:598
non-scientist contributions, 2:605
precipitation, 2:*675*
saltwater intrusion, 2:719–720
surface water, 2:766
United Nations Convention on the Law of the Sea, 2:803–805
war and conflict-related environmental destruction, 2:827–828
wastewater treatment technologies, 2:834
water conservation, 2:835–836
water pollution, 2:837–840
water supply and demand, 2:848–852
Watershed Protection and Flood Prevention Act of 1954, 2:854–855
wetlands, 2:864–866

Water rights, 1:399

Water shortages. *See* Water supply and demand

Water Sourcebook Series, 2:850–851

Water supply and demand, 2:**848–853**
aquifers, 1:33–35
California, 2:851–852
Clean Water Act, 1:122–124
climate change, 1:133, 134
drought, 1:205–206
glacial retreat, 1:359, 360
groundwater, 1:396
human impacts, 1:425, 2:848–850
irrigation, 1:493, 494
water conservation, 2:835
water resources, 2:843–844
Water Sourcebook Series, 2:850–851

Water table, 1:33, 396, 2:844

Water treatment. *See* Wastewater treatment technologies

Water use, domestic. *See* Water supply and demand

Water use, industrial. *See* Industrial water use

Water vapor (H_2)
atmosphere, 1:41
aviation emissions, 1:49
biofuels, 1:70
clear-cutting, 1:126
climate change, 1:129
drought, 1:205
ecosystems, 1:243
geothermal resources, 1:355
global warming, 1:366, 367–368
greenhouse effect, 1:384, 385
greenhouse gases, 1:390, 391
hurricanes, 1:430–431
hydrologic cycle, 1:445
precipitation, 2:674
radiative forcing, 2:681
rain forest destruction, 2:687
snow and ice cover, 2:741
surface water, 2:765
vegetation cycles, 2:816

Water wheels. *See* Hydropower

Water withdrawal, domestic. *See* Water supply and demand

Water withdrawal, industrial. *See* Industrial water use

Watercourses. *See* Rivers and waterways

Waterfowl, 1:167, 295, 2:864, 886, 887

Watermills. *See* Hydropower

WaterSense program, 1:283

Watershed Protection and Flood Prevention Act of 1954 (WPFPA), **2:854–855,** 858

Watersheds, 2:**856–858**
aquifers, 1:33–35
conservation, 1:167
drainage basins, 1:200
marine ecosystems, 2:543
streamflow, 2:762
Watershed Protection and Flood Prevention Act of 1954, 2:854–855
wetlands, 2:864

Waterways. *See* Rivers and waterways

Waterwheels. *See* Hydropower

Wave power. *See* Tidal or wave power

Waves, ocean. *See* Ocean waves

Waves, seismic. *See* Seismic waves

WCED (World Commission on Environment and Development). *See* United Nations World Commission on Environment and Development (WCED)

Weather and climate, 2:**859–860**
Antarctica, 1:21–24
Arctic, 1:36–38
atmosphere, 1:41–42
atmospheric circulation, 1:44–46
atmospheric inversions, 1:47–48
aviation emissions, 1:50
biofuels, 1:69
carbon dioxide, 1:83–84
clear-cutting, 1:125
climate change, 1:128–137
climate modeling, 1:139–141
coastal zones, 1:155
deserts, 1:197
drought, 1:205–206
ecodisasters, 1:233, 234
ecosystem diversity, 1:240
ecosystems, 1:243
El Niño and La Niña, 1:247–251
floods, 1:315–316
fossil fuel combustion impacts, 1:328
Gaia hypothesis, 1:333
glaciation, 1:363
global warming, 1:368, 369
greenhouse effect, 1:383, 385–388
greenhouse gases, 1:390, 391, 392–394
hydrologic cycle, 1:445–446
ice ages, 1:448–451
ice cores, 1:452–453
Intergovernmental Panel on Climate Change, 1:474–475
IPCC 2007 Report, 1:485–488
National Oceanic and Atmospheric Administration, 2:582–584
North Atlantic Oscillation, 2:608–609
ocean circulation and currents, 2:622
radiative forcing, 2:680–681
real-time monitoring and reporting, 2:688
temperature records, 2:779–780
volcanoes, 2:819
wildfires, 2:877
Woods Hole Oceanographic Institution, 2:891, 893
See also Paleoclimate

Weather balloons, 1:*45*

Weather extremes, 2:**861–863**
deserts, 1:197
ecodisasters, 1:232
floods, 1:315–316
fossil fuel combustion impacts, 1:328
global warming, 1:370
greenhouse effect, 1:388
greenhouse gases, 1:390, 392
ice cores, 1:452–453
IPCC 2007 Report, 1:486
National Oceanic and Atmospheric Administration, 2:582–584
water resources, 2:844

Weathering, 1:339–340, 2:743

Weaver, Andrew, 1:451

Web. *See* Internet

Webs, food. *See* Food chains and webs

Webster, P. J., 1:433–434, 435

Webster, William, 1:225

Weedkillers. *See* Herbicides

Wegener, Alfred, 1:338

Weirs, 1:312

Wells, 1:33–34, 35, 401, 2:719
See also Artesian springs and wells

Wells Fargo, 1:175

Wen Jiabao, 1:465

Werner, Abraham, 1:345

West Antarctic Penisula. *See* Antarctic issues and challenges
West Virginia, 1:100, 148, 2:770
Westerlies (winds), 1:44
Western Boreal Forest, 1:167
Western Climate Initiative, 1:257
Wetlands, 2:**864–866,** *865*
aviation emissions, 1:49
biodiversity, 1:60
biofuels, 1:*69*
Clean Water Act, 1:123–124
climate change, 1:134
conservation, 1:167
dredging, 1:202–204
ecosystems, 1:*244,* 246
estuaries, 1:294–295
flood control and floodplains, 1:310, 313
freshwater and freshwater ecosystems, 1:330, 331
genetic diversity, 1:336
habitat alteration, 1:405, 406
Hurricane Katrina, 1:440
invasive species, 1:483, *483*
landfills, 1:522, 523
recreational use and environmental destruction, 2:691
saltwater intrusion, 2:719, 720
seasonal migration, 2:730
surface water, 2:766
tundra, 2:798
wastewater treatment technologies, 2:834
water conservation, 2:836
wildlife protection policies and legislation, 2:884
wildlife refuge, 2:886
See also specific types of wetlands
WFI (Water for injection), 2:667
Whale oil, 1:272
Whales
aquatic ecosystems, 1:30, 31
endangered species, 1:119–121, 259–260
environmental activism, 1:267
green movement, 1:376
hunting practices, 1:*428*
International Convention for the Regulation of Whaling, 2:867–870
marine ecosystems, 2:542
oil spills, 2:643
predator-prey relations, 2:677
seasonal migration, 2:730, 731
See also specific types of whales
Whaling, International Convention for the Regulation of, 2:**867–871**
White asbestos, 1:39
Whitman, Christine Todd, 1:283
WHO. *See* World Health Organization (WHO)
WHOI. *See* Woods Hole Oceanographic Institution (WHOI)
Whole Foods Market, 1:379, 380, 2:653
Whooping cranes, 2:*592*
Wicker, Bill, 1:71–72
Wild and Scenic Rivers Act, 2:844–847, ***845–847***
Wild boars, 2:756
Wilderness Act of 1964, 2:**872–873**
Wilderness reserves. *See* Natural reserves and parks
Wildfire control, 2:**874–876,** 877–879
Wildfires, 2:*875,* **877–879**
clear-cutting, 1:126
climate modeling, 1:141
deforestation, 1:189, 190
drought, 1:205, 206
ecodisasters, 1:235
forest resources, 1:317
grasslands, 1:372
greenhouse effect, 1:387
habitat loss, 1:410–411
Lake Tahoe, 1:291–293
landslides, 1:527
rain forest destruction, 2:686
reforestation, 2:706, 710
wildfire control, 2:874–876
Wildlands, 1:319
Wildlife population management, 2:**880–882**
environmental science careers, 1:98
overfishing, 2:652, 653
species reintroduction programs, 2:755–756
Wildlife protection policies and legislation, 2:**883–885**
Antarctic Treaty, 1:26
Antarctica, 1:26
CITES, 1:105–107
conservation, 1:166
endangered species, 1:262–263
environmental science careers, 1:62
genetic diversity, 1:336–337
habitat loss, 1:412
international environmental law, 1:478
migratory species, 2:568, 569
natural reserves and parks, 2:589, 593
Oil Pollution Acts, 2:636–637
recreational use and environmental destruction, 2:691
reforestation, 2:708
species reintroduction programs, 2:755–756
spill remediation, 2:759
wildlife refuge, 2:886–887
Wildlife refuge, 2:**886–887**
Wildlife reserves. *See* Natural reserves and parks
Wilkins ice shelf, 1:24
Williams, Tom, 1:3
Wilson, Edward, 1:74, 496–497
Wilson, Woodrow, 2:585
Wind and wind power, 2:**888–890**
acid rain, 1:1, *2*
air pollution, 1:10
Antarctica, 1:21
atmospheric circulation, 1:44–45
clean energy, 1:118, 119
climate change, 1:133
drought, 1:205
dust storms, 1:207
El Niño and La Niña, 1:247–251
flood control and floodplains, 1:310–311
hurricanes, 1:431, 432
tidal or wave power, 2:786
wildfires, 2:877, 878
See also Polar winds; Trade winds
Wind cells, 1:45
Wind shear, 1:432
Windmills. *See* Wind and wind power
Wisconsin, 1:168, 2:838
Wisconsin Paper Council, 2:664
WMO. *See* World Meteorological Organization (WMO)
Wolfort, Henk, 2:728
Wolves, 2:586, 593, 755–756, 880, *881*
Wood
air pollution, 1:9
biodiversity, 1:63
biofuels, 1:67, 68
clean energy, 1:118
deforestation, 1:189, 191
forests, 1:323–324
Wood pulp. *See* Paper and wood pulp
Woods Hole Oceanographic Institution (WHOI), 2:**891–893,** *892*
oceanography, 2:628, 629, 630
red tide, 2:699, 700
weather extremes, 2:861
Wöppelmann, G., 2:724
World Bank, 1:183, 534, 2:831
World Climate Conference (1979), 1:474
World Coal Institute, 1:145
World Commission on Dams, 1:183–184
World Commission on Environment and Development (WCED). *See* United Nations World Commission

on Environment and Development (WCED)
World Conservation Strategy, 2:770
World Conservation Union (IUCN)
CITES, 1:105
dams, 1:183
endangered species, 1:261, 262
extinction and extirpation, 1:299
international environmental law, 1:477
invasive species, 1:483
natural reserves and parks, 2:590, 591
World Energy Council, 1:146
World Food Programme, 1:16
World Glacier Monitoring Service, 1:358
World Health Organization (WHO)
air pollution, 1:10
aquifers, 1:35
DDT, 1:186
factory farms, 1:304
groundwater quality, 1:403
medical waste, 2:562–563, *564*
smog, 2:739
water pollution, 2:840
water resources, 2:844
World Land Trust, 2:592
World Meteorological Organization (WMO)
climate change, 1:130, 132
greenhouse effect, 1:388
Intergovernmental Panel on Climate Change, 1:474
IPCC 2007 Report, 1:485
temperature records, 2:780
weather extremes, 2:861, 863
World Nuclear Association, 2:611
World Resources Institute, 1:262, 462
World Trade Center bombings, 1:*233*, 283
World Watch Institute, 1:376
World Water Council, 1:315
World Wide Web. *See* Internet
World Wildlife Fund (WWF)
biodiversity, 1:62
conservation, 1:168
endangered species, 1:262
green movement, 1:380
non-scientist contributions, 2:605
wildlife protection policies and legislation, 2:*884*
Worms, 1:55, 2:542
WPFPA. *See* Watershed Protection and Flood Prevention Act of 1954 (WPFPA)
WWF. *See* World Wildlife Fund (WWF)
Wyoming, 1:148, 453

X

Xanthan, 2:*817*
Xenobiotics, 1:455–456, 460

Y

Yellow-legged frogs, 1:*406*
Yellowstone National Park
conservation, 1:166
National Park Service Organic Act, 2:585, 586
natural reserves and parks, 2:590, 593–596
photography, 2:669
recreational use and environmental destruction, 2:691
volcanoes, 2:818, 819
wildlife refuge, 2:886
Yellowstone National Park Act, 2:**596**
Yellowstone Protective Act, 2:886
Yosemite Glaciers, 1:**363–365**
"Yosemite Glaciers" (Muir), 1:363–365
Yosemite National Park, 2:*587*
glaciation, 1:363–365
natural reserves and parks, 2:590
photography, 2:670
Wilderness Act of 1964, 2:872
wildlife refuge, 2:886
Yosemite Reserves Act, 2:872
Young, Peter, 1:228
Yucca Mountain, 1:*415*, 2:685, 839
Yugoslavia, 2:825

Z

Zahniser, Hans, 2:873
Zaire, 2:820
Zebra mussels, 1:155–156, 237, 410, 482, *483*
Zebras, 2:*677*
Zhang Gang, 1:465
Zoning, 1:519–520
Zoological Society of London, 1:62
Zoologists, 1:98
Zooplankton, 1:30, 336, 2:608–609, *677*
Zooxanthellae
coastal ecosystems, 1:151
coral reefs and corals, 1:170, 171, 172
reef ecosystems, 2:702, 704
Zuiderzee Works, 1:311
Zuidplaspolder, Netherlands, 2:*726–727*
Zurita, Fernando, 1:*491*